LARGE STRAIN FINITE ELEMENT METHOD

LARGE STRAIN FINITE ELEMENT METHOD

A PRACTICAL COURSE

Antonio Munjiza
Queen Mary, University of London

Esteban Rougier
Los Alamos National Laboratory, US

Earl E. Knight
Los Alamos National Laboratory, US

WILEY

This edition first published 2015

Registered Office
John Wiley & Sons Ltd, The Atrium, Southern Gate, Chichester, West Sussex, PO19 8SQ, United Kingdom

For details of our global editorial offices, for customer services and for information about how to apply for permission to reuse the copyright material in this book please see our website at www.wiley.com.

Library of Congress Cataloging-in-Publication Data

Munjiza, Antonio A.
Large strain finite element method : a practical course / Antonio Munjiza, Esteban Rougier, Earl E. Knight.
pages cm
Includes bibliographical references and index.
ISBN 978-1-118-40530-7 (cloth)
1. Finite element method. 2. Stress-strain curves. 3. Deformations (Mechanics)–Mathematical models.
I. Rougier, Esteban. II. Knight, Earl E. III. Title.
QC20.7.F56M86 2015
620.1′1230151825–dc23

2014027705

A catalogue record for this book is available from the British Library.

ISBN: 9781118405307

Set in 10/12pt Times by SPi Publisher Services, Pondicherry, India

1 2015

Antonio Munjiza would like to dedicate this book to Jasna and Boney.

Esteban Rougier would like to dedicate this book to his wife Sole and to his sons Ignacio and Matias.

Earl E. Knight would like to dedicate this book to the love of his life, his best friend and confidante, Cheryl Marie.

Contents

Preface xiii

Acknowledgements xv

PART ONE FUNDAMENTALS 1

1 Introduction **3**
1.1 Assumption of Small Displacements 3
1.2 Assumption of Small Strains 6
1.3 Geometric Nonlinearity 6
1.4 Stretches 8
1.5 Some Examples of Large Displacement Large Strain Finite Element Formulation 8
1.6 The Scope and Layout of the Book 13
1.7 Summary 13

2 Matrices **15**
2.1 Matrices in General 15
2.2 Matrix Algebra 16
2.3 Special Types of Matrices 21
2.4 Determinant of a Square Matrix 22
2.5 Quadratic Form 24
2.6 Eigenvalues and Eigenvectors 24
2.7 Positive Definite Matrix 26
2.8 Gaussian Elimination 26
2.9 Inverse of a Square Matrix 28
2.10 Column Matrices 30
2.11 Summary 32

3 Some Explicit and Iterative Solvers **35**
3.1 The Central Difference Solver 35
3.2 Generalized Direction Methods 43
3.3 The Method of Conjugate Directions 50
3.4 Summary 63

4 Numerical Integration **65**
4.1 Newton-Cotes Numerical Integration 65
4.2 Gaussian Numerical Integration 67
4.3 Gaussian Integration in 2D 70
4.4 Gaussian Integration in 3D 71
4.5 Summary 72

5 Work of Internal Forces on Virtual Displacements **75**
5.1 The Principle of Virtual Work 75
5.2 Summary 78

PART TWO PHYSICAL QUANTITIES **79**

6 Scalars **81**
6.1 Scalars in General 81
6.2 Scalar Functions 81
6.3 Scalar Graphs 82
6.4 Empirical Formulas 82
6.5 Fonts 83
6.6 Units 83
6.7 Base and Derived Scalar Variables 85
6.8 Summary 85

7 Vectors in 2D **87**
7.1 Vectors in General 87
7.2 Vector Notation 91
7.3 Matrix Representation of Vectors 91
7.4 Scalar Product 92
7.5 General Vector Base in 2D 93
7.6 Dual Base 94
7.7 Changing Vector Base 95
7.8 Self-duality of the Orthonormal Base 97
7.9 Combining Bases 98
7.10 Examples 104
7.11 Summary 108

8 Vectors in 3D **109**
8.1 Vectors in 3D 109
8.2 Vector Bases 111
8.3 Summary 114

9 Vectors in n-Dimensional Space **117**
9.1 Extension from 3D to 4-Dimensional Space 117
9.2 The Dual Base in 4D 118

9.3 Changing the Base in 4D 120
9.4 Generalization to n-Dimensional Space 121
9.5 Changing the Base in n-Dimensional Space 124
9.6 Summary 127

10 First Order Tensors **129**
10.1 The Slope Tensor 129
10.2 First Order Tensors in 2D 131
10.3 Using First Order Tensors 132
10.4 Using Different Vector Bases in 2D 134
10.5 Differential of a 2D Scalar Field as the First Order Tensor 137
10.6 First Order Tensors in 3D 141
10.7 Changing the Vector Base in 3D 142
10.8 First Order Tensor in 4D 143
10.9 First Order Tensor in n-Dimensions 147
10.10 Differential of a 3D Scalar Field as the First Order Tensor 149
10.11 Scalar Field in n-Dimensional Space 152
10.12 Summary 153

11 Second Order Tensors in 2D **155**
11.1 Stress Tensor in 2D 155
11.2 Second Order Tensor in 2D 158
11.3 Physical Meaning of Tensor Matrix in 2D 159
11.4 Changing the Base 161
11.5 Using Two Different Bases in 2D 163
11.6 Some Special Cases of Stress Tensor Matrices in 2D 167
11.7 The First Piola-Kirchhoff Stress Tensor Matrix 168
11.8 The Second Piola-Kirchhoff Stress Tensor Matrix 169
11.9 Summary 174

12 Second Order Tensors in 3D **175**
12.1 Stress Tensor in 3D 175
12.2 General Base for Surfaces 179
12.3 General Base for Forces 182
12.4 General Base for Forces and Surfaces 184
12.5 The Cauchy Stress Tensor Matrix in 3D 186
12.6 The First Piola-Kirchhoff Stress Tensor Matrix in 3D 186
12.7 The Second Piola-Kirchhoff Stress Tensor Matrix in 3D 188
12.8 Summary 189

13 Second Order Tensors in nD **191**
13.1 Second Order Tensor in *n*-Dimensions 191
13.2 Summary 200

PART THREE DEFORMABILITY AND MATERIAL MODELING **201**

14 Kinematics of Deformation in 1D **203**
14.1 Geometric Nonlinearity in General 203
14.2 Stretch 205
14.3 Material Element and Continuum Assumption 208

14.4 Strain 209
14.5 Stress 213
14.6 Summary 214

15 Kinematics of Deformation in 2D **217**
15.1 Isotropic Solids 217
15.2 Homogeneous Solids 217
15.3 Homogeneous and Isotropic Solids 217
15.4 Nonhomogeneous and Anisotropic Solids 218
15.5 Material Element Deformation 221
15.6 Cauchy Stress Matrix for the Solid Element 225
15.7 Coordinate Systems in 2D 227
15.8 The Solid- and the Material-Embedded Vector Bases 228
15.9 Kinematics of 2D Deformation 229
15.10 2D Equilibrium Using the Virtual Work of Internal Forces 231
15.11 Examples 235
15.12 Summary 238

16 Kinematics of Deformation in 3D **241**
16.1 The Cartesian Coordinate System in 3D 241
16.2 The Solid-Embedded Coordinate System 241
16.3 The Global and the Solid-Embedded Vector Bases 243
16.4 Deformation of the Solid 244
16.5 Generalized Material Element 246
16.6 Kinematic of Deformation in 3D 247
16.7 The Virtual Work of Internal Forces 249
16.8 Summary 255

17 The Unified Constitutive Approach in 2D **257**
17.1 Introduction 257
17.2 Material Axes 259
17.3 Micromechanical Aspects and Homogenization 260
17.4 Generalized Homogenization 263
17.5 The Material Package 264
17.6 Hyper-Elastic Constitutive Law 265
17.7 Hypo-Elastic Constitutive Law 266
17.8 A Unified Framework for Developing Anisotropic Material Models in 2D 267
17.9 Generalized Hyper-Elastic Material 267
17.10 Converting the Munjiza Stress Matrix to the Cauchy Stress Matrix 274
17.11 Developing Constitutive Laws 279
17.12 Generalized Hypo-Elastic Material 288
17.13 Unified Constitutive Approach for Strain Rate and Viscosity 292
17.14 Summary 293

18 The Unified Constitutive Approach in 3D **295**
18.1 Material Package Framework 295
18.2 Generalized Hyper-Elastic Material 295
18.3 Generalized Hypo-Elastic Material 299

18.4 Developing Material Models 302
18.5 Calculation of the Cauchy Stress Tensor Matrix 302
18.6 Summary 312

PART FOUR THE FINITE ELEMENT METHOD IN 2D 315

19 2D Finite Element: Deformation Kinematics Using the Homogeneous Deformation Triangle 317
19.1 The Finite Element Mesh 317
19.2 The Homogeneous Deformation Finite Element 317
19.3 Summary 326

20 2D Finite Element: Deformation Kinematics Using Iso-Parametric Finite Elements 327
20.1 The Finite Element Library 327
20.2 The Shape Functions 327
20.3 Nodal Positions 330
20.4 Positions of Material Points inside a Single Finite Element 331
20.5 The Solid-Embedded Vector Base 332
20.6 The Material-Embedded Vector Base 334
20.7 Some Examples of 2D Finite Elements 337
20.8 Summary 340

21 Integration of Nodal Forces over Volume of 2D Finite Elements 343
21.1 The Principle of Virtual Work in the 2D Finite Element Method 343
21.2 Nodal Forces for the Homogeneous Deformation Triangle 348
21.3 Nodal Forces for the Six-Noded Triangle 352
21.4 Nodal Forces for the Four-Noded Quadrilateral 353
21.5 Summary 355

22 Reduced and Selective Integration of Nodal Forces over Volume of 2D Finite Elements 357
22.1 Volumetric Locking 357
22.2 Reduced Integration 358
22.3 Selective Integration 359
22.4 Shear Locking 362
22.5 Summary 364

PART FIVE THE FINITE ELEMENT METHOD IN 3D 365

23 3D Deformation Kinematics Using the Homogeneous Deformation Tetrahedron Finite Element 367
23.1 Introduction 367
23.2 The Homogeneous Deformation Four-Noded Tetrahedron Finite Element 368
23.3 Summary 377

24 3D Deformation Kinematics Using Iso-Parametric Finite Elements 379
24.1 The Finite Element Library 379
24.2 The Shape Functions 379

24.3 Nodal Positions 381
24.4 Positions of Material Points inside a Single Finite Element 382
24.5 The Solid-Embedded Infinitesimal Vector Base 383
24.6 The Material-Embedded Infinitesimal Vector Base 386
24.7 Examples of Deformation Kinematics 387
24.8 Summary 392

25 Integration of Nodal Forces over Volume of 3D Finite Elements 393
25.1 Nodal Forces Using Virtual Work 393
25.2 Four-Noded Tetrahedron Finite Element 396
25.3 Reduce Integration for Eight-Noded 3D Solid 399
25.4 Selective Stretch Sampling-Based Integration for the Eight-Noded Solid Finite Element 400
25.5 Summary 401

26 Integration of Nodal Forces over Boundaries of Finite Elements 403
26.1 Stress at Element Boundaries 403
26.2 Integration of the Equivalent Nodal Forces over the Triangle Finite Element 404
26.3 Integration over the Boundary of the Composite Triangle 407
26.4 Integration over the Boundary of the Six-Noded Triangle 408
26.5 Integration of the Equivalent Internal Nodal Forces over the Tetrahedron Boundaries 409
26.6 Summary 412

PART SIX THE FINITE ELEMENT METHOD IN 2.5D 415

27 Deformation in 2.5D Using Membrane Finite Elements 417
27.1 Solids in 2.5D 417
27.2 The Homogeneous Deformation Three-Noded Triangular Membrane Finite Element 419
27.3 Summary 438

28 Deformation in 2.5D Using Shell Finite Elements 439
28.1 Introduction 439
28.2 The Six-Noded Triangular Shell Finite Element 440
28.3 The Solid-Embedded Coordinate System 441
28.4 Nodal Coordinates 442
28.5 The Coordinates of the Finite Element's Material Points 443
28.6 The Solid-Embedded Infinitesimal Vector Base 444
28.7 The Solid-Embedded Vector Base versus the Material-Embedded Vector Base 447
28.8 The Constitutive Law 449
28.9 Selective Stretch Sampling Based Integration of the Equivalent Nodal Forces 449
28.10 Multi-Layered Shell as an Assembly of Single Layer Shells 455
28.11 Improving the CPU Performance of the Shell Element 456
28.12 Summary 462

Index 463

Preface

The conventional finite element method is based on the assumption that structural system displacements under load are small and that the structural material does not stretch much under that load. Arguably, the small strain, small displacements-based finite element method is not of much use in modern scientific, engineering and technological applications. Even in classic structural engineering applications, the conventional finite element method is hardly applicable. This shift has occurred because design codes and standards have changed in recent years to include the ultimate limit state, i.e., considerations of structural collapse. As a consequence, one now has to consider both large strains (plastic strains) and large displacements. In other state of the art applications of the finite element method, finite element simulations are increasingly becoming an integral part of the so-called virtual experimentation, examples of these are biological, medical science, material science, process engineering, military and many other applications of the finite element method. In these applications the finite element simulation has to reproduce reality (as opposed to approximating reality), together with possible emergent properties such as flow, damage, failure, collapse, yield, etc.

In this context, not even the higher order theories and their finite element realizations are suitable representations of the physical realities involved. The answer is an exact formulation that encompasses an exact representation of large displacements, large strains, and material properties including anisotropy. Such a theory, when implemented in a finite element software package, must cover 2D solids, 3D solids, and 2.5D shell and membrane static and dynamic simulations.

Theoretical aspects of these formulations were resolved in the 1960s and 1970s. The finite element adaptation of these theoretical formulations has mostly taken place during the 1990s and early years of the 21st century. This work has resulted in a large body of scientific papers that have described it as the next generation of finite element packages. Nevertheless, the subject has remained a mystery for undergraduate students, postgraduate students, practicing engineers and scientist and even for users and developers of finite element software.

This book is written with the key objective of "demystifying" the subject, making it easy for students, engineers and software developers to master the minute details of the finite element method that incorporates large strains, large displacements, and material nonlinearity.

The book is written in such a way that it provides a pathway to master all the method's related subjects starting with matrices, systems of equations, scalar and vectors and progressing onto tensors of the first order, and tensors of the second order. With this knowledge base in hand, the book provides an engineering-based approach to deformation kinematics that avoids the often confusing mathematical jargon yet concentrates on the physics and uses mathematics only when necessary. At this stage, the reader is made familiar with a generalized framework for developing large strains based nonlinear material laws. This is done without any reference to the finite element method, having in mind, for example, a material modeler whose job is to solely develop material laws.

Finally, the book presents the large strain large displacement based finite element method including 2D solid, 3D solid, 2.5D membrane, plate and shell problems. These are explained in such detail that they contain all the necessary mathematical equations, algorithms and formulae that can be readily implemented into the finite element method. As such, they should be of great value for developers of finite element packages. They will also provide users of finite element packages with an enhanced understanding of the algorithmic, theoretical and formulation aspects of the finite element software they are using.

The authors hope that the book will ultimately benefit practicing engineers, scientists, undergraduate students, master students and PhD students in diverse fields of related applied subjects.

Acknowledgements

The authors would like to express their gratitude to the publishers, Wiley, for their excellent support. We would also like to thank our numerous colleagues and research collaborators from all over the world; the USA, China, Japan, Germany, Italy, Canada, France, Korea, Australia, India and the UK. Our thanks also go to current and previous PhD students as well as Postdoctoral researchers. Special thanks go to Professor J.R. Williams from MIT, Professor Bibhu Mohanty from University of Toronto, Professor Graham Mustoe from Colorado School of Mines, Robert P. Swift, Theodore C. Carney, Jennifer Wilson, and Wendee M. Brunish from Los Alamos National Laboratory, Professor F. Aliabadi from Imperial College London, Dr. Ing Harald Kruggel-Emden from Ruhr-University, Dr. Paul Cleary from CSIRO, Australia, Roger Owen from University of Wales, Rene De Borst from the University of Glasgow, Peter Wriggers from the University of Hannover, Eugenio Oñate from the Technical University of Catalonia, Peter Cundall, Marcio Muniz de Farias from the University of Brasilia, Herbert Mang from the Austrian Academy of Science, Eduardo de Souza Neto and Y Feng from Swansea University, Bernardin Peros from the University of Split, Karin Hing from the University of London, Robert Zimmerman from Imperial College London, Collin Thornton from University of Birmingham, and many others who have helped make this book better.

Part One
Fundamentals

1

Introduction

1.1 Assumption of Small Displacements

University courses often introduce the small displacement assumption in an implicit way, without explaining to the students that it is applicable only in special cases. Take, for example, the structural system shown in Figure 1.1. Point P is attached to the ground using two straight rods that are pin-connected both to the ground and to each other. In a first-year stress analysis course this problem would be solved using the solution procedure shown in Figure 1.2, thus yielding

$$f_1 = f_2 = \frac{1}{2}\frac{f}{\sin\bar{\theta}} = \frac{1}{2}\frac{f}{\sin 45}. \tag{1.1}$$

Note that in Equation (1.1) force f_1 is a function of angle $\bar{\theta}$, where $\bar{\theta}$ is 45°. However, with this approach, one has ignored the fact that the rods can deform (shorten) under load, thus resulting in a downward displacement of point P. This deformation, in turn changes the initial angle $\bar{\theta}$ into $\tilde{\theta}$, Figure 1.3.

In real life, the actual equilibrium of forces occurs not on the initial geometry, but on the deformed geometry. The internal forces (forces between atoms) move with the geometry (changed position of atoms) – the internal forces $\mathbf{f}_1$ and $\mathbf{f}_2$ rotate with the corresponding rods and are always parallel to their corresponding rods. The load $\mathbf{f}$ also moves with point P. Finally the state of equilibrium shown in Figure 1.4 is reached, when

$$f_1 = f_2 = \frac{1}{2}\frac{f}{\sin\tilde{\theta}} > \frac{1}{2}\frac{f}{\sin 45}. \tag{1.2}$$

Large Strain Finite Element Method: A Practical Course, First Edition. Antonio Munjiza, Esteban Rougier and Earl E. Knight.

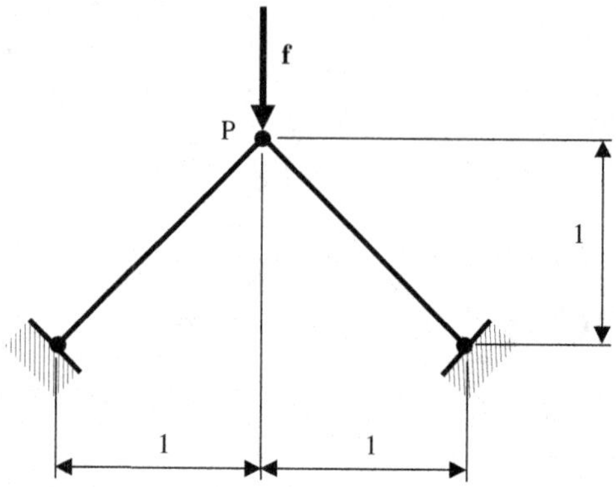

Figure 1.1 A two member truss

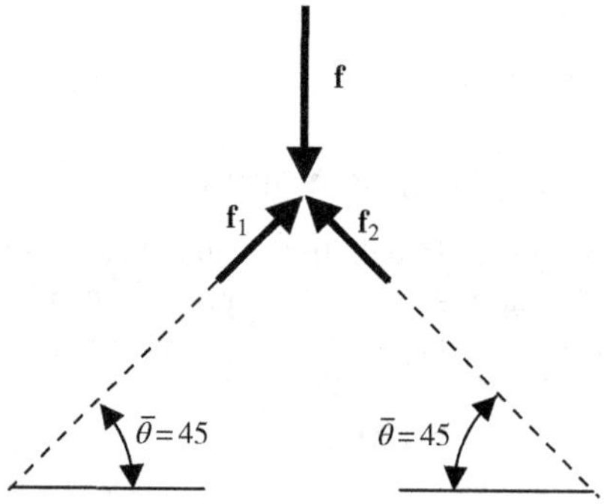

Figure 1.2 Equilibrium of forces at point P

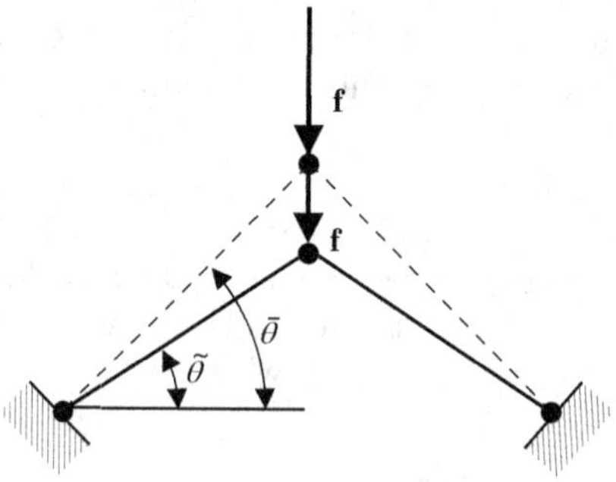

Figure 1.3 The initial (dashed lines) and the current (solid lines) geometries

In other words, the internal forces obtained using Equation (1.1) are wrong and the correct internal forces are obtained using Equation (1.2). The problem with Equation (1.2) is that the geometry of the system is a function of the internal forces, which in turn are a function of the geometry. This yields an implicit equilibrium formulation

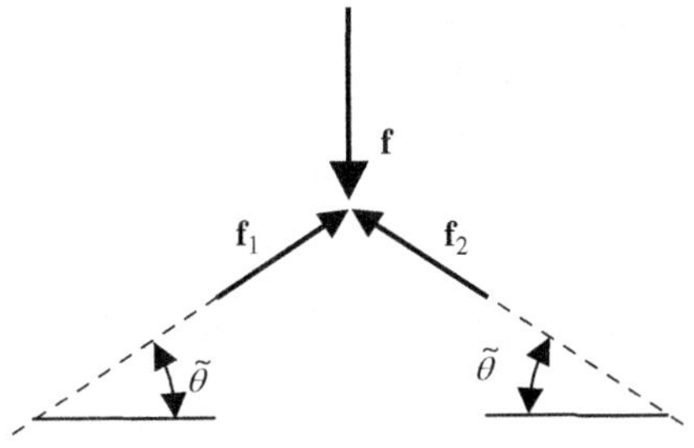

Figure 1.4 Equilibrium of the deformed system

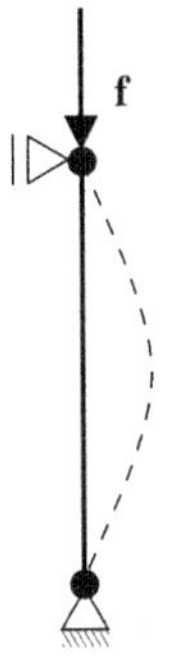

Figure 1.5 Instability of the equilibrium (buckling of struts)

$$f_1 = \frac{1}{2}\frac{f}{\sin\tilde{\theta}(f_1)}$$
$$f_2 = \frac{1}{2}\frac{f}{\sin\tilde{\theta}(f_2)}. \tag{1.3}$$

The formulation shown in Equation (1.3) is obviously nonlinear. As such, it can be difficult to resolve. For this reason, the formulation given by Equation (1.1) is often used instead, which is fine only when:

a. the initial and the deformed geometry are nearly identical – this case occurs only when the displacements are infinitesimally small or tolerably small in practical applications.
b. the displacements do not change progressively with the applied load – infinitesimally small perturbations (inaccuracies) in the geometry will not lead to disproportionally large changes in the internal forces, Figure 1.5.

The situation shown in Figure 1.5 is called instability of the equilibrium. Buckling of struts is only one example of an unstable equilibrium. There exists an entire field of applied science dedicated to the analysis of structural stability. It is often formulated in terms of modal analysis, which students usually find difficult to understand, although the concept is relatively simple:

For a given load, there may exist a particular deformed shape in which the structure has in a sense "escaped" from under the load: for example, in Figure 1.5 the load has stayed vertical, but the strut has moved (escaped) sideways and, as such, it does not support the load any longer.

In order to solve the above problems, the second order formulation was developed. It is not a general formulation, but rather a patchwork of application-specific formulations that address either the problem of large displacements or structural stability. Some classical examples are the deformation of a rope, the deformation of membranes, and the deformation of slender structures used in civil engineering, aerospace engineering and naval architecture.

As an alternative, the theoretically exact generalized large displacements approach was introduced in the late 1990s and early years of the 21st century and it has gained significant popularity. The idea is relatively simple:

Always consider equilibrium using the deformed geometry of the solid.

The resulting formulation is called the large displacement formulation, for it represents the exact equilibrium of internal and external forces regardless of the size of the displacements. As such, it captures: (a) equilibrium of systems with small displacements, (b) any instability of equilibrium, and (c) equilibrium of systems with large displacements. In contrast to the 2nd order theory, this is the exact theoretical formulation. It is, by default, nonlinear.

1.2 Assumption of Small Strains

In order to simplify how one solves solid deformation problems, the assumption of small strains through the engineering strain is often introduced. Many times this approach is utilized without any thought of explaining that it is only valid if the strains are infinitesimally small. For this purpose, the strain is often defined as engineering strain

$$\varepsilon = \frac{\Delta L}{\bar{L}} = \frac{\tilde{L} - \bar{L}}{\bar{L}}, \tag{1.4}$$

where ΔL is the elongation of a rod of initial length $\bar{L}$ and deformed length $\tilde{L}$. The assumption of small strains is only valid in exceptional circumstances such as deformation of glass at room temperature and similar materials.

When it comes to plastics, rubber, metals, clay, gels, granular materials, glass fibers, carbon fibers, biological tissues, mechanics of cells (such as red blood cells), bitumen, kerogen, and many other materials of modern technology, modern industry, modern science and modern engineering, the assumption of small strains is simply not valid.

In order to rectify the problem for specific applications, various second order formulations have been developed. These parallel the second order formulations for large displacements and are in general applicable only to a specific narrowly defined problem.

In contrast, in this book the theoretically exact generalized large strain formulation is explained.

1.3 Geometric Nonlinearity

The large strain formulation combines naturally with the large displacement formulation. The result is a formulation that reproduces a theoretically exact solution (as opposed to the second order formulation) for both large displacements and large strains.

As such, it addresses (in an exact manner) geometric nonlinearity. Geometric nonlinearity by definition includes nonlinear aspects of deformation that arise from large displacements and/or large strains.

The theoretically exact formulation is based on the multiplicative decomposition of deformation. The concept is relatively simple:

Write the current coordinates of the material points of the solid as

$$\tilde{x} = \tilde{x}(\xi, \eta, \zeta), \tag{1.5}$$

where (ξ, η, ζ) somehow uniquely define a given material point and as such do not change with deformation. Now, Equation (1.5) can be written as

$$\tilde{x} = \tilde{x}_S(\tilde{x}_R(\tilde{x}_T(\xi, \eta, \zeta))), \tag{1.6}$$

where $\tilde{x}_T$ represents the material points' translation. It is followed by the rotation $\tilde{x}_R$ and the stretch $\tilde{x}_S$. In other words, the function $\tilde{x}$ is a composition of three functions

$$\tilde{x} = \tilde{x}_S \circ \tilde{x}_R \circ \tilde{x}_T, \tag{1.7}$$

where $\tilde{x}_S$ stretches the solid, $\tilde{x}_R$ rotates the solid and $\tilde{x}_T$ translates the solid. It is like one person first comes and translates material point P. The second person comes and rotates the solid. Finally the third person stretches the solid. Only the third stage causes internal forces in the material and can, for example, break the material.

The function $\tilde{x}$ describes the deformation of the solid body and is therefore called the deformation function or simply, deformation. The deformation $\tilde{x}$ is made from the composition of translation, rotation and stretch in any order. Translation and rotation move the solid as though it was rigid. As such, they do not stretch the solid. In contrast, stretch changes the shape (the geometry) of the solid. In an infinitesimally close vicinity of a given material point P, all these functions are de-facto linear functions of coordinates x, y and z, as shown in Figure 1.6.

This leads to the multiplicative decomposition of rotation and stretch. First, the translation is removed from the deformation and what is left is decomposed into a product of stretch and rotation. In addition, stretch is expressed as a product of different types of stretches, such as volumetric stretch, shear stretch, elastic stretch, plastic stretch, etc.

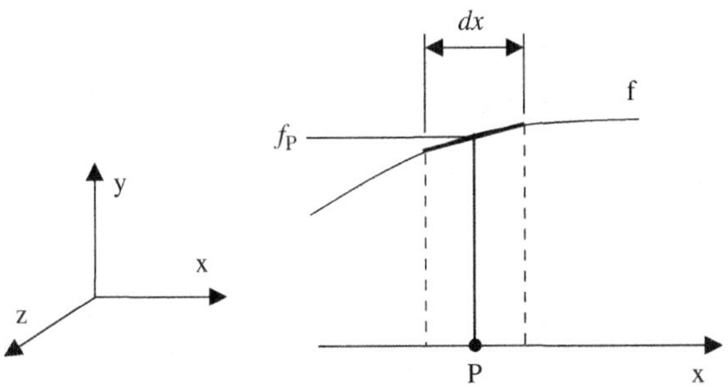

Figure 1.6 Linearity of deformation in the infinitesimal vicinity of a given point P: Note that for infinitesimally small dx any function f(x) reduces to f(x) = $f_P + \alpha x$

1.4 Stretches

In large strain large displacements deformation, it is convenient to formulate the problem not in terms of strains, but in terms of stretches. The reasons for this are as follows:

a. Stretch is well represented by a second order tensor.
b. Stretch is easily calculated from the deformation function.
c. Stretch is easily separated from rotation.
d. Stretch can be further decomposed into an elastic part, a plastic part, a volumetric part, etc.
e. Multiplicative decomposition of stretches comes naturally.
f. Any type of strains (strain measures) can be calculated from the stretches.
g. Stretches are applicable to nonlinear material formulations including nonlinear anisotropic materials.

1.5 Some Examples of Large Displacement Large Strain Finite Element Formulation

Biological Tissue. In Figure 1.7 a 3D finite element based simulation of blood plasma containing red blood cells is shown. It is evident that individual red blood cells stretch significantly and consequently their shape is changing.

Membrane Structures. In Figure 1.8 a 2.5D finite element simulation of a membrane structure subject to an initial velocity is shown (such as a flag on a mast). In this case, the material of the structure does not undergo large strains, i.e., it does not stretch a lot. However, despite this, the displacements are extremely large. This problem can be categorized as a large displacement, small strain type of problem. Nevertheless, the simulation results shown are obtained by using the large displacement, large strains finite element formulation.

2D Solid Structures. In Figure 1.9 a problem similar to a rubber cylinder hitting the ground is shown. This is a 2D solid problem consisting of large displacements and large strains. The results shown are obtained using the finite element formulation described in detail in this book.

(a)

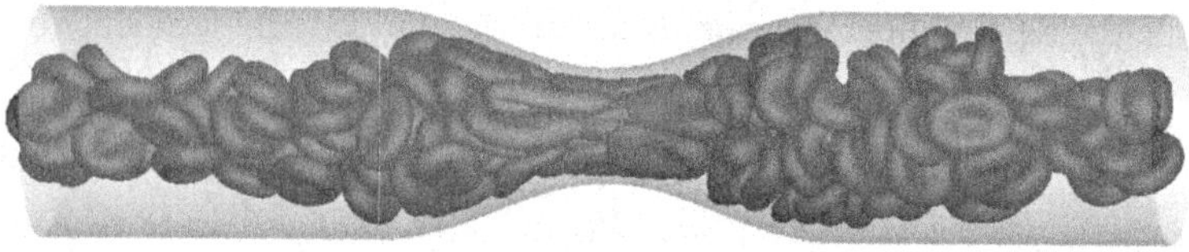

(b)

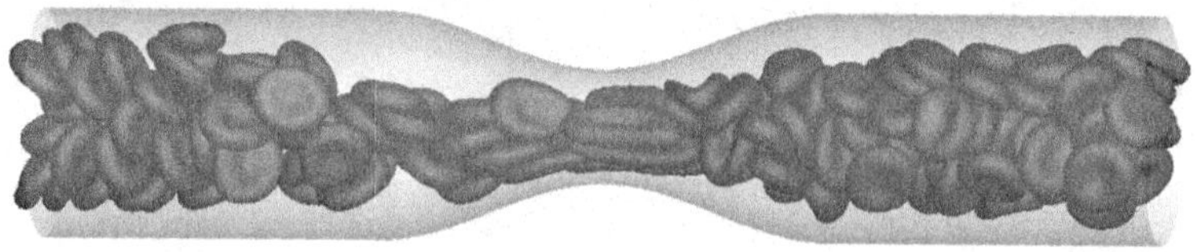

(c)

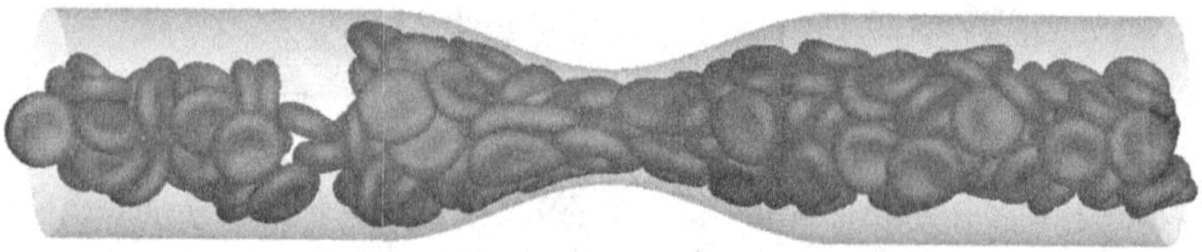

Figure 1.7 The flow of red blood cells accompanied by significant stretching, large translations and large rotations, i.e. complete geometric nonlinearity

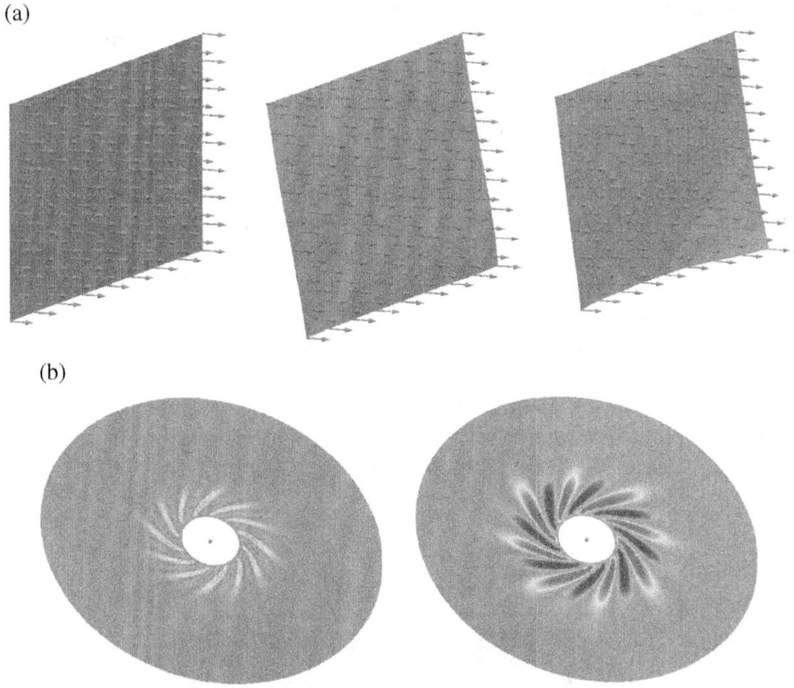

Figure 1.8 (a) A deformation sequence of a flag-like membrane subject to large displacements but small strains; (b) results of geometrically nonlinear analysis of a circular membrane

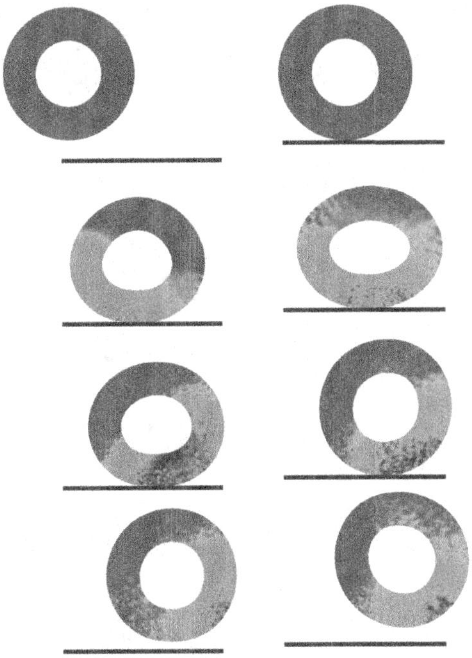

Figure 1.9 A sequence of deformation of a 2D rubber cylinder hitting the ground

Large Displacements Large Strains Shells. Shells are usually termed as 2.5D problems. The finite element formulation for shells, even for small strains and small displacements, is complex. In this book it is demonstrated that the large strain large displacement formulation can be relatively simple to understand, to implement and to use for shells. A typical dynamics problem using shells is shown in Figure 1.10.

Nonlinear Materials. This book covers geometric (large strains large displacements) nonlinearities in combination with arbitrary anisotropic material nonlinearity. As such, the solid can undergo plastic deformation or damage based nonlinearity leading to localized failure and fracture. For example, in Figure 1.11, the fracture of a glass panel is obtained using the 2.5D large displacement, large strain, nonlinear material based finite element formulation.

In a similar way, the damage based failure sequence of a 2.5D shell is shown in Figure 1.12.

3D Solids. In Figure 1.13 a problem similar to a tennis ball hitting a circular plate (shell) is shown. This is a 3D solid problem consisting of large displacements and large strains. The results shown are obtained using the finite element formulation described in detail in this book. A cross-sectional view of the same problem is shown in Figure 1.14.

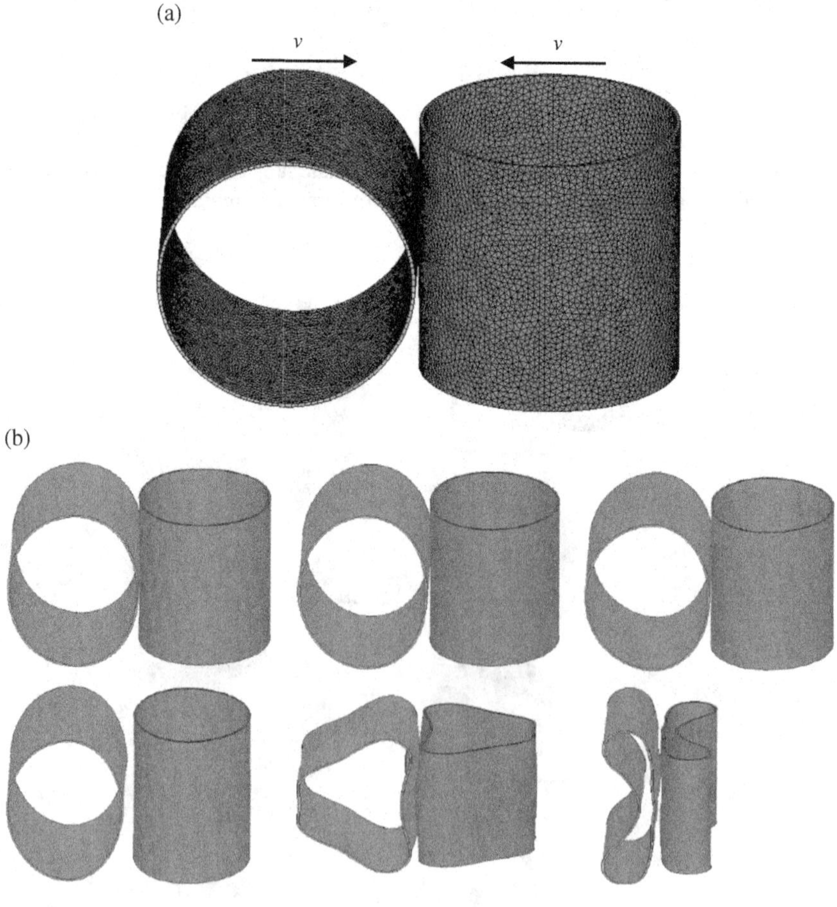

Figure 1.10 (a) Impact of two shells. (b) Deformation sequence obtained using the shell formulation described in detail in this book

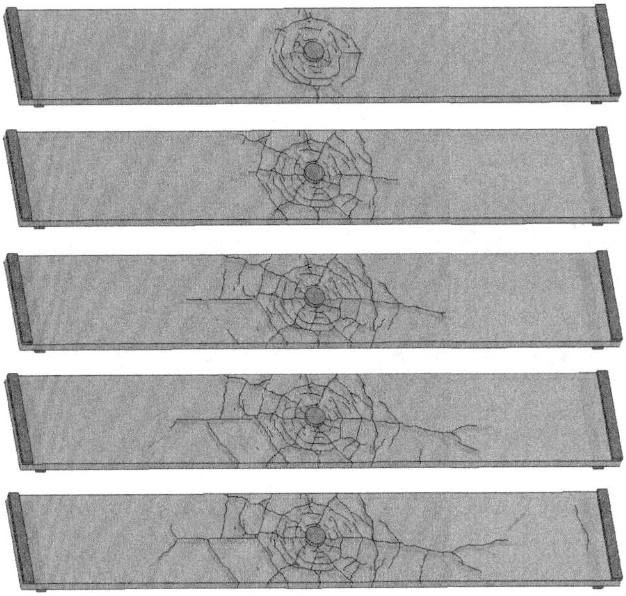

Figure 1.11 Damage based nonlinear material finite element simulation showing fracture of a window screen

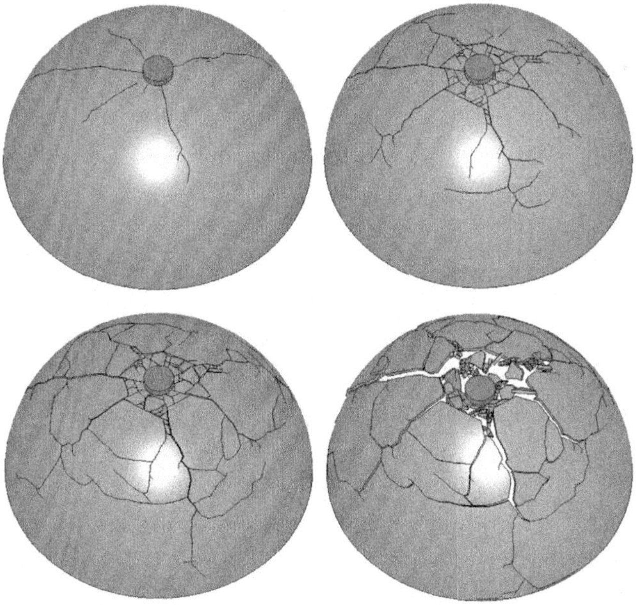

Figure 1.12 Damage based fracture of a spherical shell subject to a penny shaped impactor

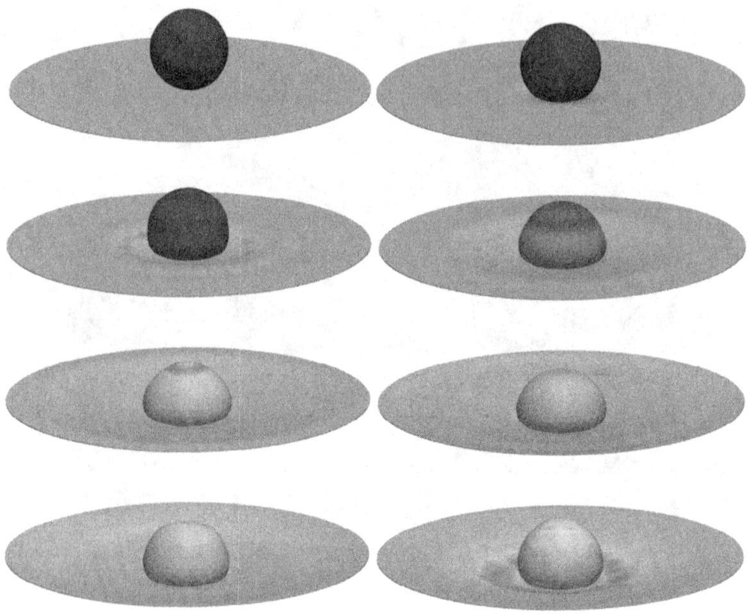

Figure 1.13 The deformation sequence for a tennis ball (3D solid) hitting a circular plate (2.5D shell). Both the shell and the 3D solid deform considerably (large displacements combined with large strains)

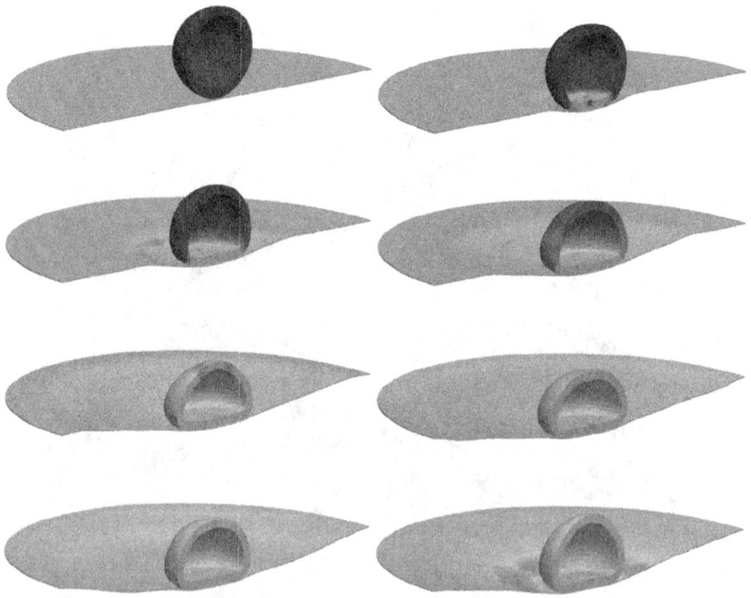

Figure 1.14 A sequence of deformation (a cross section view) showing a tennis ball hitting a circular plate (shell)

1.6 The Scope and Layout of the Book

In this book, the theoretically exact formulation for large strain large displacements based simulations using multiplicative decomposition is presented. This includes: 2D finite element method, 3D finite element method, 2.5D finite element method (plates, shells and membranes), static finite element analysis, transient dynamics finite element analysis, as well as a generalized nonlinear anisotropic material formulation including hyper-elastic, hypo-elastic and unified constitutive formulation. The book is written in such a manner that it is self-contained in the sense that a reader is not required to have any previous knowledge of the subject.

In the first part of the book some essential mathematical tools are covered including a hands-on approach to matrix algebra.

In the second part of the book some necessary physics concepts are introduced including vectors, first order tensors and second order tensors. These are deliberately separated from the first part of the book in order to emphasize that they are not mathematical constructs, but physical realities. As a consequence, relatively easy-to-master approaches for tensorial calculus in 2D, 3D, 4D and nD spaces are presented with an aim of familiarizing the reader with the modern notion of tensorial algebra, thus demystifying tensorial calculus itself.

In the third part of the book deformation in 1D is explained (mainly for didactical reasons) in order to help the reader to grasp the subject. Also in this part, 1D deformation kinematics is extended to 2D. This is naturally followed by an extension to 3D. At the end of the third part a unified approach to formulating constitutive laws for general anisotropic materials subject to large displacements is presented.

In the fourth, fifth and sixth parts of the book deformation kinematics is presented using the finite element method in 2D, 3D and 2.5D respectively. In this manner, the finite element formulation is completely separated from the stress calculation (constitutive law), thus enabling material modelers to work completely independent from any finite element developers. This is especially important in modern industry wherein finite element packages are often off-the-shelf commercial packages, while the material models employed may be proprietary and developed completely independently from the finite element package itself.

Internal forces over finite elements are represented using stress tensors (obtained from the material package that uses constitutive laws developed by the material modelers). From these, the equivalent nodal forces are calculated using stress integration over either volume or boundaries of finite elements. These procedures are explained in Chapters 21, 25 and 26.

In this process, a special role is also given to selective stretch sampling, which is in essence a generalization of both reduced and selective integration, Chapter 22.

The resulting nonlinear algebraic equations are solved using suitable algebraic equation solvers. This explicit iterative approach to solving the equations is generally preferred. Some of these are explained in Chapter 3, including the dynamic relaxation and conjugate directions methods.

1.7 Summary

In this chapter, an introduction to geometric and material nonlinearities in general has been provided. This was followed by the scope and layout of the book.

Finally, examples obtained using the finite element formulation described in this book have been shown. The main aim of these examples is to not only make the reader interested in the subject but also to illustrate the power of the large strain large displacement finite element method and its inherent advantages:

a. There are no restraints to the size of displacements.
b. There are no restraints to the size of the strains.

c. Finite element-independent material formulations consisting of anisotropy by default, plasticity, viscosity, etc. are implemented in a format friendly to a material model developer who has no experience with finite elements.

Further Reading

[1] Munjiza, A., K. R. F. Andrews, and J. K. White (1998) Combined single and smeared crack model in combined finite-discrete element method. *Int. J. Numer. Methods Eng.*, **44:** 41–57.

[2] Munjiza, A., J. P. Latham, and K. F. Andrews (2000) Detonation gas model for combined finite-discrete element simulation of fracture and fragmentation. *Int. J. Numer. Meth. Eng.*, **49:** 1495–1520.

[3] Munjiza, A. (2004) *The Combined Finite-Discrete Element Method.* Chichester: John Wiley & Sons, Ltd.

[4] Munjiza A., Lei Z., Divic V. and Peros, B. (2013) Fracture and fragmentation of thin shells using the combined finite-discrete element method. *International Journal for Numerical Methods in Engineering,* **95**(6): 478–98.

[5] Mustoe, G. G. W., J. R. Williams, G. Hocking, and K. Worgan (1988) *Penetration and Fracturing of Brittle Plates under Dynamic Impact.* INTERA Technologies, Inc.

[6] Owen, D. R. J., A. Munjiza and N. Bicanic (1992) A finite element–discrete element approach to the simulation of rode blasting problems. *Proceedings FEMSA-92, 11th Symposium on Finite Element Methods,* South Africa, Cape Town, 39–59.

[7] Owen, D. R. J. and Y. T. Feng (2001) Parallelised finite/discrete element simulation of multi-fracturing solids and discrete systems. *Engineering Computations,* **18**(3/4): 557–76.

[8] Owen, D. R. J., Y. T. Feng, J. Yu, and D. Peric (2001) Finite/discrete element analysis of multi-fracture and multi-contact phenomena. *Lecture Notes in Computer Science,* **1981:** 484–505.

[9] Rougier E., Knight E. E., Lei Z., Bartoli, G., Betti, M. and Munjiza, A. (2013) Preserving significant historical structures with the help of computational mechanics of discontinua. *Particle-Based Methods III: Fundamentals and Applications – Proceedings of the 3rd International Conference on Particle-based Methods Fundamentals and Applications, Particles.*

[10] Xu, D., Kaliviotis, E., Munjiza, A., Avital, E., Ji, C. and Williams, J. (2013) Large scale simulation of red blood cell aggregation in shear flows. *J Biomech.*, **46**(11): 1810–17.

2

Matrices

2.1 Matrices in General

In engineering and science one very often deals with large sets of numbers. It is convenient to arrange these numbers into rectangular forms called matrices. As an example, a person may have bought two bananas, three apples, and six pears. These three numbers can be conveniently put into a column matrix such as

$$\begin{bmatrix} 2 \\ 3 \\ 6 \end{bmatrix}. \tag{2.1}$$

In addition, the person may have paid 10 pence for each banana, 6 pence for each apple and 4 pence for each pear. The total amount paid is

$$2 \cdot 10 + 3 \cdot 6 + 6 \cdot 4 = 62\,\text{pence}. \tag{2.2}$$

This can be written using matrices as follows

$$\begin{bmatrix} 10 & 6 & 4 \end{bmatrix} \begin{bmatrix} 2 \\ 3 \\ 6 \end{bmatrix}, \tag{2.3}$$

where

$$\begin{bmatrix} 10 & 6 & 4 \end{bmatrix} \tag{2.4}$$

Large Strain Finite Element Method: A Practical Course, First Edition. Antonio Munjiza, Esteban Rougier and Earl E. Knight.

is a row matrix and

$$\begin{bmatrix} 2 \\ 3 \\ 6 \end{bmatrix} \tag{2.5}$$

is a column matrix and the multiplication of these two matrices is defined such that each quantity is multiplied by its corresponding unit price.

Column and row matrices are special kinds of matrices. Another special kind of matrix is the square matrix. A 2 by 2 square matrix is given by:

$$\begin{bmatrix} 3 & 6 \\ 7 & 5 \end{bmatrix}. \tag{2.6}$$

By analogy, a 3 by 3 square matric is given by:

$$\begin{bmatrix} 3 & 6 & 7 \\ 9 & 8 & 3 \\ 2 & 4 & 15 \end{bmatrix}. \tag{2.7}$$

In the above equations, the individual numbers are called matrix components or components of the matrix. A 2 by 2 square matrix has four components. A 3 by 3 square matrix has 9 components, a 10 by 10 square matrix has 100 components, a 1000 by 1000 square matrix has one million components, etc.

In Equation (2.7), components 3, 8, and 15 form the diagonal of the matrix. Very often, square matrices are symmetric relative to the diagonal;

$$\begin{bmatrix} 3 & 2 \\ 2 & 7 \end{bmatrix}; \quad \begin{bmatrix} 3 & 2 & 4 \\ 2 & 7 & 5 \\ 4 & 5 & 9 \end{bmatrix}. \tag{2.8}$$

The most general case of a matrix is a rectangular matrix made of m rows by n columns,

$$\mathbf{A} = \begin{bmatrix} a_{11} & a_{12} & a_{13} & \dots & a_{1n} \\ a_{21} & a_{22} & a_{23} & \dots & a_{2n} \\ a_{31} & a_{32} & a_{33} & \dots & a_{3n} \\ \vdots & \vdots & \vdots & \ddots & \vdots \\ a_{m1} & a_{m2} & a_{m3} & \dots & a_{mn} \end{bmatrix}. \tag{2.9}$$

2.2 Matrix Algebra

The key reason for introducing matrices is to simplify the handling of large sets of numbers; with this in mind, a number of operations are defined for matrices. These operations are within the field of matrix algebra.

Addition of Matrices. Two rectangular matrices **A** and **B** which have the same number of rows and columns can be added together. This is done by adding their corresponding components, thus

$$\begin{bmatrix} 1 & 2 \\ 3 & 4 \end{bmatrix} + \begin{bmatrix} 5 & 6 \\ 7 & 8 \end{bmatrix} = \begin{bmatrix} (1+5) & (2+6) \\ (3+7) & (4+8) \end{bmatrix} = \begin{bmatrix} 6 & 8 \\ 10 & 12 \end{bmatrix}. \quad (2.10)$$

This operation can also be written as

$$\mathbf{A} + \mathbf{B} = \mathbf{C}, \quad (2.11)$$

where

$$\mathbf{A} = \begin{bmatrix} 1 & 2 \\ 3 & 4 \end{bmatrix}; \mathbf{B} = \begin{bmatrix} 5 & 6 \\ 7 & 8 \end{bmatrix}; \mathbf{C} = \begin{bmatrix} 6 & 8 \\ 10 & 12 \end{bmatrix}. \quad (2.12)$$

The easiest way to understand this operation is to consider an example where John has bought 1 banana, 2 apples, and 3 pears, i.e.,

$$\mathbf{A} = \begin{bmatrix} 1 \\ 2 \\ 3 \end{bmatrix}. \quad (2.13)$$

John's partner has bought 4 bananas, 5 apples, and 6 pears, i.e.,

$$\mathbf{B} = \begin{bmatrix} 4 \\ 5 \\ 6 \end{bmatrix}. \quad (2.14)$$

John and his partner together have then bought:

$$\mathbf{C} = \mathbf{A} + \mathbf{B} = \begin{bmatrix} 1 \\ 2 \\ 3 \end{bmatrix} + \begin{bmatrix} 4 \\ 5 \\ 6 \end{bmatrix} = \begin{bmatrix} 5 \\ 7 \\ 9 \end{bmatrix}, \quad (2.15)$$

i.e. 5 bananas, 7 apples, and 9 pears.

Subtraction of Matrices. The easiest way to understand this operation is to consider an example where John's partner bought 5 bananas, 7 apples, and 9 pears,

$$\mathbf{C} = \begin{bmatrix} 5 \\ 7 \\ 9 \end{bmatrix}, \quad (2.16)$$

but John ate 1 banana, 2 apples, and 3 pears, i.e.,

$$\mathbf{A} = \begin{bmatrix} 1 \\ 2 \\ 3 \end{bmatrix}. \quad (2.17)$$

Thus, one is left with 4 bananas, 5 apples, and 6 pears, i.e.,

$$\mathbf{B}=\mathbf{C}-\mathbf{A}=\begin{bmatrix}5\\7\\9\end{bmatrix}-\begin{bmatrix}1\\2\\3\end{bmatrix}=\begin{bmatrix}4\\5\\6\end{bmatrix}. \tag{2.18}$$

Multiplication of Matrices by a Scalar. One can consider the following example: John went to shop on six consecutive days and each time he bought 1 banana, 2 apples, and 3 pears. How much did he buy altogether? Each day John bought

$$\mathbf{A}=\begin{bmatrix}1\\2\\3\end{bmatrix}. \tag{2.19}$$

He repeated this 6 times, thus John bought:

$$\mathbf{B}=6\mathbf{A}=\mathbf{A}6=6\begin{bmatrix}1\\2\\3\end{bmatrix}=\begin{bmatrix}6\cdot 1\\6\cdot 2\\6\cdot 3\end{bmatrix}=\begin{bmatrix}6\\12\\18\end{bmatrix}. \tag{2.20}$$

In a similar way, one can divide a matrix by a scalar:

$$\mathbf{C}=\frac{\mathbf{A}}{2}=\frac{\begin{bmatrix}1\\2\\3\end{bmatrix}}{2}=\begin{bmatrix}1/2\\2/2\\3/2\end{bmatrix}=\begin{bmatrix}0.5\\1.0\\1.5\end{bmatrix}. \tag{2.21}$$

Transposition of a Matrix. Very often it is convenient to turn rows of a matrix into columns and vice versa. The column matrix

$$\mathbf{A}=\begin{bmatrix}3\\7\\8\end{bmatrix} \tag{2.22}$$

can be turned into row matrix

$$\mathbf{B}=\begin{bmatrix}3 & 7 & 8\end{bmatrix}, \tag{2.23}$$

where

$$\mathbf{B}=\mathbf{A}^{\mathrm{T}}=\begin{bmatrix}3\\7\\8\end{bmatrix}^{\mathrm{T}}=\begin{bmatrix}3 & 7 & 8\end{bmatrix}. \tag{2.24}$$

In a similar manner

$$\mathbf{B}=\begin{bmatrix}3 & 7 & 8\end{bmatrix} \tag{2.25}$$

can be turned into

$$\mathbf{A} = \begin{bmatrix} 3 \\ 7 \\ 8 \end{bmatrix}, \tag{2.26}$$

where

$$\mathbf{A} = \mathbf{B}^{\mathrm{T}} = [3 \;\; 7 \;\; 8]^{\mathrm{T}} = \begin{bmatrix} 3 \\ 7 \\ 8 \end{bmatrix}. \tag{2.27}$$

In general,

$$\begin{bmatrix} 3 & 7 \\ 9 & 4 \\ 8 & 6 \end{bmatrix}^{\mathrm{T}} = \begin{bmatrix} 3 & 9 & 8 \\ 7 & 4 & 6 \end{bmatrix}, \tag{2.28}$$

where the rows have become columns and the columns have become rows. In other words, the transposition operation simply rearranges the components of a matrix.

Matrix Multiplication. It is often convenient to multiply two matrices. For example, John has bought 2 bananas and 3 apples:

$$\mathbf{A} = \begin{bmatrix} 2 \\ 3 \end{bmatrix}. \tag{2.29}$$

He paid 6 pence for each banana and 7 pence for each apple

$$\mathbf{B} = [6 \;\; 7]. \tag{2.30}$$

Thus, all together, John has paid:

$$\mathbf{C} = \mathbf{BA} = [6 \;\; 7]\begin{bmatrix} 2 \\ 3 \end{bmatrix} = 6 \cdot 2 + 7 \cdot 3 = 33. \tag{2.31}$$

In general matrices **A** and **B** are multiplied by multiplying each row of matrix **A** with each column of matrix **B**. Thus,

$$\mathbf{C} = \mathbf{AB}, \tag{2.32}$$

where

$$c_{ij} = \left(i^{\text{th}} \text{ row of matrix } \mathbf{A}\right)\left(j^{\text{th}} \text{ column of matrix } \mathbf{B}\right). \tag{2.33}$$

For this operation to work, the number of columns of matrix **A** has to be equal to the number of rows of matrix **B**. For instance,

$$\mathbf{A} = \begin{bmatrix} 1 & 4 \\ 2 & 5 \\ 3 & 6 \end{bmatrix}; \quad \mathbf{B} = \begin{bmatrix} 7 & 8 & 9 \\ 10 & 11 & 12 \end{bmatrix}, \tag{2.34}$$

$$\begin{aligned} \mathbf{C} = \mathbf{AB} &= \begin{bmatrix} 1 & 4 \\ 2 & 5 \\ 3 & 6 \end{bmatrix} \begin{bmatrix} 7 & 8 & 9 \\ 10 & 11 & 12 \end{bmatrix} \\ &= \begin{bmatrix} 1\cdot 7 + 4\cdot 10 & 1\cdot 8 + 4\cdot 11 & 1\cdot 9 + 4\cdot 12 \\ 2\cdot 7 + 5\cdot 10 & 2\cdot 8 + 5\cdot 11 & 2\cdot 9 + 5\cdot 12 \\ 3\cdot 7 + 6\cdot 10 & 3\cdot 8 + 6\cdot 11 & 3\cdot 9 + 6\cdot 12 \end{bmatrix} \\ &= \begin{bmatrix} 47 & 52 & 57 \\ 64 & 71 & 78 \\ 81 & 90 & 99 \end{bmatrix}. \end{aligned} \tag{2.35}$$

In a similar manner, for

$$\mathbf{A} = \begin{bmatrix} 1 & 2 & 3 \\ 4 & 5 & 6 \end{bmatrix}; \quad \mathbf{B} = \begin{bmatrix} 7 \\ 8 \\ 9 \end{bmatrix} \tag{2.36}$$

$$\begin{aligned} \mathbf{C} = \mathbf{AB} &= \begin{bmatrix} 1 & 2 & 3 \\ 4 & 5 & 6 \end{bmatrix} \begin{bmatrix} 7 \\ 8 \\ 9 \end{bmatrix} \\ &= \begin{bmatrix} 1\cdot 7 + 2\cdot 8 + 3\cdot 9 \\ 4\cdot 7 + 5\cdot 8 + 6\cdot 9 \end{bmatrix} \\ &= \begin{bmatrix} 50 \\ 122 \end{bmatrix}. \end{aligned} \tag{2.37}$$

This is often written as

$$\begin{aligned} \mathbf{C} = \mathbf{AB} &= \begin{bmatrix} 1 & 2 & 3 \\ 4 & 5 & 6 \end{bmatrix} \begin{bmatrix} 7 \\ 8 \\ 9 \end{bmatrix} \\ &= 7 \begin{bmatrix} 1 \\ 4 \end{bmatrix} + 8 \begin{bmatrix} 2 \\ 5 \end{bmatrix} + 9 \begin{bmatrix} 3 \\ 6 \end{bmatrix} \\ &= \begin{bmatrix} 50 \\ 122 \end{bmatrix}. \end{aligned} \tag{2.38}$$

It is worth noting that:

$$\mathbf{AB} \neq \mathbf{BA}. \tag{2.39}$$

2.3 Special Types of Matrices

Square Matrix. A square matrix has the same number of rows and columns:

$$\begin{bmatrix} 5 & 4 \\ 1 & 3 \end{bmatrix}; \quad \begin{bmatrix} 5 & 8 & 7 \\ 3 & 4 & 6 \\ 9 & 5 & 8 \end{bmatrix}. \tag{2.40}$$

Diagonal Matrix. A diagonal matrix has all components equal to zero, except those components that are on the diagonal:

$$\begin{bmatrix} 2 & 0 & 0 \\ 0 & 5 & 0 \\ 0 & 0 & -7 \end{bmatrix}. \tag{2.41}$$

Identity Matrix. An identity matrix is a square matrix with all components equal to zero, except those on the diagonal, which are equal to one:

$$\mathbf{I} = \begin{bmatrix} 1 & 0 & 0 \\ 0 & 1 & 0 \\ 0 & 0 & 1 \end{bmatrix}. \tag{2.42}$$

Zero Matrix. A zero matrix is usually a column matrix with all components equal to zero:

$$\mathbf{0} = \begin{bmatrix} 0 \\ 0 \\ \vdots \\ 0 \end{bmatrix}. \tag{2.43}$$

Symmetric Matrix. A symmetric matrix is a square matrix with non-diagonal components being equal to the corresponding components on the other side of the diagonal:

$$\begin{bmatrix} 4 & 13 & 20 \\ 13 & 8 & -12 \\ 20 & -12 & -2 \end{bmatrix}. \tag{2.44}$$

Upper Triangular Matrix. An upper triangular matrix is a square matrix with all components below the diagonal being equal to zero:

$$\begin{bmatrix} 19 & 7 & 4 \\ 0 & 6 & 16 \\ 0 & 0 & 73 \end{bmatrix}. \tag{2.45}$$

Lower Triangular Matrix. A lower triangular matrix is a square matrix with all components above the diagonal being equal to zero:

$$\begin{bmatrix} 19 & 0 & 0 \\ 7 & 6 & 0 \\ 4 & 16 & 73 \end{bmatrix}. \tag{2.46}$$

2.4 Determinant of a Square Matrix

The determinant of a 2 by 2 matrix is calculated as follows:

$$\det\begin{bmatrix} a_{11} & a_{12} \\ a_{21} & a_{22} \end{bmatrix} = \begin{vmatrix} a_{11} & a_{12} \\ a_{21} & a_{22} \end{vmatrix} = a_{11}a_{22} - a_{12}a_{21}. \tag{2.47}$$

The determinant of a 3 by 3 matrix is calculated using the following equation:

$$\begin{aligned} \det\begin{bmatrix} a_{11} & a_{12} & a_{13} \\ a_{21} & a_{22} & a_{23} \\ a_{31} & a_{32} & a_{33} \end{bmatrix} &= a_{11}\begin{vmatrix} a_{22} & a_{23} \\ a_{32} & a_{33} \end{vmatrix} - a_{12}\begin{vmatrix} a_{21} & a_{23} \\ a_{31} & a_{33} \end{vmatrix} + a_{13}\begin{vmatrix} a_{21} & a_{22} \\ a_{31} & a_{32} \end{vmatrix} \\ &= a_{11}(a_{22}a_{33} - a_{32}a_{23}) \\ &\quad - a_{12}(a_{21}a_{33} - a_{31}a_{23}) \\ &\quad + a_{13}(a_{21}a_{32} - a_{31}a_{22}) \end{aligned} \tag{2.48}$$

The determinant of a 4 by 4 matrix is calculated using the following equation:

$$\begin{aligned} \det\begin{bmatrix} a_{11} & a_{12} & a_{13} & a_{14} \\ a_{21} & a_{22} & a_{23} & a_{24} \\ a_{31} & a_{32} & a_{33} & a_{34} \\ a_{41} & a_{42} & a_{43} & a_{44} \end{bmatrix} &= a_{11}\begin{vmatrix} a_{22} & a_{23} & a_{24} \\ a_{32} & a_{33} & a_{34} \\ a_{42} & a_{43} & a_{44} \end{vmatrix} - a_{12}\begin{vmatrix} a_{21} & a_{23} & a_{24} \\ a_{31} & a_{33} & a_{34} \\ a_{41} & a_{43} & a_{44} \end{vmatrix} \\ &\quad + a_{13}\begin{vmatrix} a_{21} & a_{22} & a_{24} \\ a_{31} & a_{32} & a_{34} \\ a_{41} & a_{42} & a_{44} \end{vmatrix} - a_{14}\begin{vmatrix} a_{21} & a_{22} & a_{23} \\ a_{31} & a_{32} & a_{33} \\ a_{41} & a_{42} & a_{43} \end{vmatrix} \end{aligned} \tag{2.49}$$

The Determinant of a n by n Matrix. The minor m_{ij} of a square matrix **A** is defined as

$$m_{ij} = \det \mathbf{M}, \tag{2.50}$$

where **M** is a matrix obtained from eliminating the ith row and the jth column of matrix **A**. For example, if

$$\mathbf{A} = \begin{bmatrix} a_{11} & a_{12} & a_{13} & a_{14} & a_{15} \\ a_{21} & a_{22} & a_{23} & a_{24} & a_{25} \\ a_{31} & a_{23} & a_{33} & a_{34} & a_{35} \\ a_{41} & a_{24} & a_{34} & a_{44} & a_{45} \\ a_{51} & a_{25} & a_{35} & a_{54} & a_{55} \end{bmatrix}. \tag{2.51}$$

Then m_{42} is given by

$$m_{42} = \det \mathbf{M} = \det \begin{bmatrix} a_{11} & \blacksquare & a_{13} & a_{14} & a_{15} \\ a_{21} & \blacksquare & a_{23} & a_{24} & a_{25} \\ a_{31} & \blacksquare & a_{33} & a_{34} & a_{35} \\ \blacksquare & \blacksquare & \blacksquare & \blacksquare & \blacksquare \\ a_{51} & \blacksquare & a_{35} & a_{54} & a_{55} \end{bmatrix}$$
$$= \det \begin{bmatrix} a_{11} & a_{13} & a_{14} & a_{15} \\ a_{21} & a_{23} & a_{24} & a_{25} \\ a_{31} & a_{33} & a_{34} & a_{35} \\ a_{51} & a_{35} & a_{54} & a_{55} \end{bmatrix}. \tag{2.52}$$

The determinant of a square matrix can therefore be defined as:

$$\det \mathbf{A} = \sum_{j=1}^{n} m_{1j}(-1)^{1+j} a_{1j}. \tag{2.53}$$

In this manner, the determinant of a 1 by 1 matrix is the actual matrix component itself. The determinant of a 2 by 2 matrix becomes

$$\begin{aligned} \det \mathbf{A} &= (-1)^{1+1} m_{11} a_{11} + (-1)^{1+2} m_{12} a_{12} \\ &= m_{11} a_{11} - m_{12} a_{12} \\ &= a_{11} a_{22} - a_{12} a_{21}. \end{aligned} \tag{2.54}$$

The determinant of a 3 by 3 matrix is:

$$\begin{aligned} \det \mathbf{A} &= (-1)^{1+1} m_{11} a_{11} + (-1)^{1+2} m_{12} a_{12} + (-1)^{1+3} m_{13} a_{13} \\ &= m_{11} a_{11} - m_{12} a_{12} + m_{13} a_{13} \\ &= a_{11} \begin{vmatrix} a_{22} & a_{23} \\ a_{32} & a_{33} \end{vmatrix} - a_{12} \begin{vmatrix} a_{21} & a_{23} \\ a_{31} & a_{33} \end{vmatrix} + a_{13} \begin{vmatrix} a_{21} & a_{22} \\ a_{31} & a_{32} \end{vmatrix}. \end{aligned} \tag{2.55}$$

While the determinant of a 4 by 4 matrix is:

$$\begin{aligned} \det \mathbf{A} &= (-1)^{1+1} m_{11} a_{11} + (-1)^{1+2} m_{12} a_{12} \\ &\quad + (-1)^{1+3} m_{13} a_{13} + (-1)^{1+4} m_{14} a_{14} \\ &= m_{11} a_{11} - m_{12} a_{12} + m_{13} a_{13} - m_{14} a_{14} \\ &= a_{11} \begin{vmatrix} a_{22} & a_{23} & a_{24} \\ a_{32} & a_{33} & a_{34} \\ a_{42} & a_{43} & a_{44} \end{vmatrix} - a_{12} \begin{vmatrix} a_{21} & a_{23} & a_{24} \\ a_{31} & a_{33} & a_{34} \\ a_{41} & a_{43} & a_{44} \end{vmatrix} \\ &\quad + a_{13} \begin{vmatrix} a_{21} & a_{22} & a_{24} \\ a_{31} & a_{32} & a_{34} \\ a_{41} & a_{42} & a_{44} \end{vmatrix} - a_{14} \begin{vmatrix} a_{21} & a_{22} & a_{23} \\ a_{31} & a_{32} & a_{33} \\ a_{41} & a_{42} & a_{43} \end{vmatrix}. \end{aligned} \tag{2.56}$$

By continuing the process described above the determinant of a square matrix of any size can be calculated. Also, instead of using row 1, one can use any other row or column and its minors.

2.5 Quadratic Form

For a given square matrix **A** and column matrix **x** of the same size, the quadratic form is defined by:

$$\mathbf{x}^T\mathbf{A}\mathbf{x}. \tag{2.57}$$

For instance,

$$\begin{bmatrix} 1 & 2 \end{bmatrix} \begin{bmatrix} 2 & 3 \\ 3 & 7 \end{bmatrix} \begin{bmatrix} 1 \\ 2 \end{bmatrix} \tag{2.58}$$

is equal to:

$$\begin{bmatrix} 1 & 2 \end{bmatrix} \begin{bmatrix} 8 \\ 17 \end{bmatrix} = 42. \tag{2.59}$$

2.6 Eigenvalues and Eigenvectors

When multiplying a column matrix and a square matrix

$$\mathbf{b} = \mathbf{A}\mathbf{x} \tag{2.60}$$

it can be proven that there exists a column matrix **y** such that

$$\mathbf{A}\mathbf{y} = \lambda\mathbf{y}. \tag{2.61}$$

This can be written as

$$(\mathbf{A} - \lambda\mathbf{I})\mathbf{y} = 0. \tag{2.62}$$

One way the above equation can be satisfied is to have matrix **y** with all components equal to zero (the trivial solution). The alternative is to have the nontrivial solution, in which case

$$\det(\mathbf{A} - \lambda\mathbf{I}) = 0. \tag{2.63}$$

Equation (2.63) is the characteristic equation, which, for a n by n matrix yields n roots

$$\lambda_1, \lambda_2, \lambda_3, \ldots, \lambda_n. \tag{2.64}$$

These roots are called eigenvalues. For each eigenvalue λ_i there is an associated $\mathbf{y}_i$ such that

$$(\mathbf{A} - \lambda_i\mathbf{I})\mathbf{y}_i = 0, \tag{2.65}$$

where $\mathbf{y}_i$ is called eigenmatrix. The max absolute value of λ for a given matrix is called the spectral radius.

For example, if

$$\mathbf{A} = \begin{bmatrix} 4.0 & -2.236 \\ -2.236 & 8 \end{bmatrix} \tag{2.66}$$

the eigenvalues are obtained as follows:

$$\det\begin{bmatrix} 4.0-\lambda & -2.236 \\ -2.236 & 8.0-\lambda \end{bmatrix} = 0. \tag{2.67}$$

Equation (2.67) expands into

$$(4.0-\lambda)(8.0-\lambda)-5.0=0, \tag{2.68}$$

which, when solved gives:

$$\begin{aligned} \lambda_1 &= 3.0 \\ \lambda_2 &= 9.0. \end{aligned} \tag{2.69}$$

Substituting the value of λ_1 into equation (2.65) one obtains

$$\begin{bmatrix} 4.0-3.0 & -2.236 \\ -2.236 & 8.0-3.0 \end{bmatrix} \begin{bmatrix} y_{11} \\ y_{21} \end{bmatrix} = \begin{bmatrix} 0.0 \\ 0.0 \end{bmatrix}, \tag{2.70}$$

which is the same as

$$1.0 \cdot y_{11} - 2.236 \cdot y_{21} = 0 \tag{2.71}$$

$$-2.236 \cdot y_{11} + 5.0 \cdot y_{21} = 0. \tag{2.72}$$

From Equation (2.71) it follows that

$$y_{21} = \frac{1}{2.236} y_{11} = 0.447 y_{11}. \tag{2.73}$$

This means that matrix

$$\mathbf{y}_1 = \begin{bmatrix} y_{11} \\ y_{21} \end{bmatrix} = \begin{bmatrix} 1.0 \\ 0.447 \end{bmatrix} \tag{2.74}$$

is an eigenmatrix corresponding to the eigenvalue λ_1. Therefore, the matrices defined by

$$\mathbf{y}_1 = a \begin{bmatrix} 1.0 \\ 0.447 \end{bmatrix}, \tag{2.75}$$

where a is scaling factor, are also eigenmatrices of matrix **A**. In a similar manner, substituting the value of λ_2 into Equation (2.65) yields

$$\begin{bmatrix} 4.0-9.0 & -2.236 \\ -2.236 & 8.0-9.0 \end{bmatrix} \begin{bmatrix} y_{12} \\ y_{22} \end{bmatrix} = \begin{bmatrix} 0.0 \\ 0.0 \end{bmatrix}, \tag{2.76}$$

which is the same as

$$-5.0 \cdot y_{12} - 2.236 \cdot y_{22} = 0, \tag{2.77}$$

$$-2.236 \cdot y_{12} - 1.0 \cdot y_{22} = 0. \tag{2.78}$$

From Equation (2.77) it follows that

$$y_{22} = -\frac{5.0}{2.236} y_{12} = -2.236 y_{12}. \tag{2.79}$$

This means that matrix

$$\mathbf{y}_2 = \begin{bmatrix} y_{12} \\ y_{22} \end{bmatrix} = \begin{bmatrix} 1.0 \\ -2.236 \end{bmatrix} \tag{2.80}$$

is an eigenmatrix corresponding to the eigenvalue λ_2. Therefore, the family of matrices defined by

$$\mathbf{y}_2 = b \begin{bmatrix} 1.0 \\ -2.236 \end{bmatrix}, \tag{2.81}$$

where b is scaling factor, are also eigenmatrices of matrix **A**.

2.7 Positive Definite Matrix

A positive definite matrix is any symmetric matrix that has all its eigenvalues greater than zero, i.e.,

$$\lambda_i > 0 \ \forall \ i. \tag{2.82}$$

This means that the quadratic form is greater than zero, i.e.,

$$\mathbf{x}^T \mathbf{A} \mathbf{x} > 0 \tag{2.83}$$

for any nonzero column matrix **x**.

2.8 Gaussian Elimination

Gaussian elimination is used to solve a system of linear algebraic equations

$$\mathbf{A}\mathbf{x} = \mathbf{b}, \tag{2.84}$$

where **A** is a positive definite square matrix, **b** is a known column matrix and **x** is an unknown column matrix.

$$\begin{bmatrix} a_{11} & a_{12} & a_{13} & \dots & a_{1n} \\ a_{21} & a_{22} & a_{23} & \dots & a_{2n} \\ a_{31} & a_{32} & a_{33} & \dots & a_{3n} \\ \vdots & \vdots & \vdots & \ddots & \vdots \\ a_{n1} & a_{n2} & a_{n3} & \dots & a_{nn} \end{bmatrix} \begin{bmatrix} x_1 \\ x_2 \\ x_3 \\ \vdots \\ x_n \end{bmatrix} = \begin{bmatrix} b_1 \\ b_2 \\ b_3 \\ \vdots \\ b_n \end{bmatrix}. \tag{2.85}$$

The first row (equation) is divided by the pivot (a_{11}).

$$\begin{bmatrix} 1 & {}^1a_{12} & {}^1a_{13} & \dots & {}^1a_{1n} \\ a_{21} & a_{22} & a_{23} & \dots & a_{2n} \\ a_{31} & a_{32} & a_{33} & \dots & a_{3n} \\ \vdots & \vdots & \vdots & \ddots & \vdots \\ a_{n1} & a_{n2} & a_{n3} & \dots & a_{nn} \end{bmatrix} \begin{bmatrix} x_1 \\ x_2 \\ x_3 \\ \vdots \\ x_n \end{bmatrix} = \begin{bmatrix} {}^1b_1 \\ b_2 \\ b_3 \\ \vdots \\ b_n \end{bmatrix}, \tag{2.86}$$

where

$${}^1a_{12} = \frac{a_{12}}{a_{11}}; \dots ; \; {}^1a_{1n} = \frac{a_{1n}}{a_{11}}; \; {}^1b_1 = \frac{b_1}{a_{11}}. \tag{2.87}$$

Then, the components of the second row are modified as follows,

$$\begin{aligned} {}^1a_{2i} &= a_{2i} - {}^1a_{1i}a_{21} \\ {}^1b_2 &= b_2 - {}^1b_1 a_{21}. \end{aligned} \tag{2.88}$$

This yields

$$\begin{bmatrix} 1 & {}^1a_{12} & {}^1a_{13} & \dots & {}^1a_{1n} \\ 0 & {}^1a_{22} & {}^1a_{23} & \dots & {}^1a_{2n} \\ a_{31} & a_{32} & a_{33} & \dots & a_{3n} \\ \vdots & \vdots & \vdots & \ddots & \vdots \\ a_{n1} & a_{n2} & a_{n3} & \dots & a_{nn} \end{bmatrix} \begin{bmatrix} x_1 \\ x_2 \\ x_3 \\ \vdots \\ x_n \end{bmatrix} = \begin{bmatrix} {}^1b_1 \\ {}^1b_2 \\ b_3 \\ \vdots \\ b_n \end{bmatrix}. \tag{2.89}$$

The same process is repeated for the rest of the rows, yielding

$$\begin{bmatrix} 1 & {}^1a_{12} & {}^1a_{13} & \dots & {}^1a_{1n} \\ 0 & {}^1a_{22} & {}^1a_{23} & \dots & {}^1a_{2n} \\ 0 & {}^1a_{32} & {}^1a_{33} & \dots & {}^1a_{3n} \\ \vdots & \vdots & \vdots & \ddots & \vdots \\ 0 & {}^1a_{n2} & {}^1a_{n3} & \dots & {}^1a_{nn} \end{bmatrix} \begin{bmatrix} x_1 \\ x_2 \\ x_3 \\ \vdots \\ x_n \end{bmatrix} = \begin{bmatrix} {}^1b_1 \\ {}^1b_2 \\ {}^1b_3 \\ \vdots \\ {}^1b_n \end{bmatrix}. \tag{2.90}$$

Now the second equation is divided by the pivot (${}^1a_{22}$). It is then in turn multiplied by ${}^1a_{i2}$ and subtracted to the corresponding equations below it, thus producing:

$$\begin{bmatrix} 1 & {}^1a_{12} & {}^1a_{13} & \dots & {}^1a_{1n} \\ 0 & 1 & {}^2a_{23} & \dots & {}^2a_{2n} \\ 0 & 0 & {}^2a_{33} & \dots & {}^2a_{3n} \\ \vdots & \vdots & \vdots & \ddots & \vdots \\ 0 & 0 & {}^2a_{n3} & \dots & 1 \end{bmatrix} \begin{bmatrix} x_1 \\ x_2 \\ x_3 \\ \vdots \\ x_n \end{bmatrix} = \begin{bmatrix} {}^1b_1 \\ {}^2b_2 \\ {}^2b_3 \\ \vdots \\ {}^2b_n \end{bmatrix}. \tag{2.91}$$

By repeating this process for all pivot equations, an upper triangular matrix is obtained

$$\begin{bmatrix} 1 & {}^1a_{12} & {}^1a_{13} & \dots & {}^1a_{1n} \\ 0 & 1 & {}^2a_{23} & \dots & {}^2a_{2n} \\ 0 & 0 & 1 & \dots & {}^3a_{3n} \\ \vdots & \vdots & \vdots & \ddots & \vdots \\ 0 & 0 & 0 & \dots & 1 \end{bmatrix} \begin{bmatrix} x_1 \\ x_2 \\ x_3 \\ \vdots \\ x_n \end{bmatrix} = \begin{bmatrix} {}^1b_1 \\ {}^2b_2 \\ {}^3b_3 \\ \vdots \\ {}^nb_n \end{bmatrix}. \tag{2.92}$$

Using an analog procedure for the equations above the diagonal, the following equation is obtained,

$$\begin{bmatrix} 1 & 0 & 0 & \dots & 0 \\ 0 & 1 & 0 & \dots & 0 \\ 0 & 0 & 1 & \dots & 0 \\ \vdots & \vdots & \vdots & \ddots & \vdots \\ 0 & 0 & 0 & \dots & 1 \end{bmatrix} \begin{bmatrix} x_1 \\ x_2 \\ x_3 \\ \vdots \\ x_n \end{bmatrix} = \begin{bmatrix} {}^{n-1}b_1 \\ {}^{n-2}b_2 \\ {}^{n-3}b_3 \\ \vdots \\ {}^nb_n \end{bmatrix}, \tag{2.93}$$

thus yielding the unknowns.

2.9 Inverse of a Square Matrix

A square matrix with determinant equal to zero is called a singular square matrix. A square matrix with nonzero determinant is called a nonsingular square matrix. Any nonsingular square matrix $\mathbf{A}$ has a uniquely defined inverse matrix

$$\mathbf{B} = \mathbf{A}^{-1}, \tag{2.94}$$

such that

$$\mathbf{A}^{-1}\mathbf{A} = \mathbf{A}\mathbf{A}^{-1} = \mathbf{I}. \tag{2.95}$$

Gaussian elimination can be used to calculate the inverse matrix. This is done by starting from

$$\mathbf{Ax} = \mathbf{b} \tag{2.96}$$

and writing it as

$$\mathbf{Ax} = \mathbf{Ib}. \tag{2.97}$$

By performing Gaussian elimination on the above equation, one obtains

$$\mathbf{Ix} = \mathbf{A}^{-1}\mathbf{b}. \tag{2.98}$$

Example,

$$\begin{bmatrix} 3 & 2 \\ 2 & 7 \end{bmatrix} \begin{bmatrix} x_1 \\ x_2 \end{bmatrix} = \begin{bmatrix} 1 & 0 \\ 0 & 1 \end{bmatrix} \begin{bmatrix} 3 \\ 4 \end{bmatrix}. \tag{2.99}$$

Step 1:

$$\begin{bmatrix} 1 & 2/3 \\ 2 & 7 \end{bmatrix} \begin{bmatrix} x_1 \\ x_2 \end{bmatrix} = \begin{bmatrix} 1/3 & 0 \\ 0 & 1 \end{bmatrix} \begin{bmatrix} 3 \\ 4 \end{bmatrix}. \tag{2.100}$$

Step 2:

$$\begin{bmatrix} 1 & 2/3 \\ 0 & 17/3 \end{bmatrix} \begin{bmatrix} x_1 \\ x_2 \end{bmatrix} = \begin{bmatrix} 1/3 & 0 \\ -2/3 & 1 \end{bmatrix} \begin{bmatrix} 3 \\ 4 \end{bmatrix}. \tag{2.101}$$

Step 3:

$$\begin{bmatrix} 1 & 2/3 \\ 0 & 1 \end{bmatrix} \begin{bmatrix} x_1 \\ x_2 \end{bmatrix} = \begin{bmatrix} 1/3 & 0 \\ -2/17 & 3/17 \end{bmatrix} \begin{bmatrix} 3 \\ 4 \end{bmatrix}. \tag{2.102}$$

Step 4:

$$\begin{bmatrix} 1 & 0 \\ 0 & 1 \end{bmatrix} \begin{bmatrix} x_1 \\ x_2 \end{bmatrix} = \begin{bmatrix} 7/17 & -2/17 \\ -2/17 & 3/17 \end{bmatrix} \begin{bmatrix} 3 \\ 4 \end{bmatrix}, \tag{2.103}$$

which means that

$$\mathbf{A}^{-1} = \begin{bmatrix} 7/17 & -2/17 \\ -2/17 & 3/17 \end{bmatrix}. \tag{2.104}$$

Of course, no changes have been done on column matrices **x** and **b**, which means that they need not to be considered at all.

Very often one needs to calculate

$$\mathbf{B} = \mathbf{A}^{-1} \tag{2.105}$$

and

$$\mathbf{C} = \mathbf{B}^{\mathrm{T}} = \left(\mathbf{A}^{-1}\right)^{\mathrm{T}} = \left(\mathbf{A}^{\mathrm{T}}\right)^{-1}, \tag{2.106}$$

which is conveniently written as

$$\mathbf{C} = \mathbf{A}^{-\mathrm{T}}. \tag{2.107}$$

2.10 Column Matrices

Column matrices play a special role in many applications and together with square matrices they are the most often used matrices.

Multiplication by Square Matrix. Multiplication of a square matrix by a column matrix

$$\mathbf{b} = \mathbf{A}\mathbf{x} \tag{2.108}$$

is defined by

$$\begin{bmatrix} b_1 \\ b_2 \\ b_3 \\ \vdots \\ b_n \end{bmatrix} = \begin{bmatrix} a_{11} & a_{12} & a_{13} & \dots & a_{1n} \\ a_{21} & a_{22} & a_{23} & \dots & a_{2n} \\ a_{31} & a_{32} & a_{33} & \dots & a_{3n} \\ \vdots & \vdots & \vdots & \ddots & \vdots \\ a_{n1} & a_{n2} & a_{n3} & \dots & a_{nn} \end{bmatrix} \begin{bmatrix} x_1 \\ x_2 \\ x_3 \\ \vdots \\ x_n \end{bmatrix}$$
$$= x_1 \begin{bmatrix} a_{11} \\ a_{21} \\ a_{31} \\ \vdots \\ a_{n1} \end{bmatrix} + x_2 \begin{bmatrix} a_{12} \\ a_{22} \\ a_{32} \\ \vdots \\ a_{n2} \end{bmatrix} + x_3 \begin{bmatrix} a_{13} \\ a_{23} \\ a_{33} \\ \vdots \\ a_{n3} \end{bmatrix} + \dots + x_n \begin{bmatrix} a_{1n} \\ a_{2n} \\ a_{3n} \\ \vdots \\ a_{nn} \end{bmatrix} \tag{2.109}$$

Therefore, it produces a linear combination of the columns of the square matrix.

Inner Product. The inner product is also called the scalar product or the dot product. The inner product of two column matrices **a** and **b** is defined as

$$\mathbf{a} \bullet \mathbf{b} = \mathbf{b} \bullet \mathbf{a} = a_1 b_1 + a_2 b_2 + a_3 b_3 + \dots + a_n b_n = \mathbf{a}^{\mathrm{T}} \mathbf{b} = \mathbf{b}^{\mathrm{T}} \mathbf{a}. \tag{2.110}$$

Orthogonal Column Matrices. Column matrices **a** and **b** are said to be orthogonal if

$$\mathbf{a} \bullet \mathbf{b} = 0. \tag{2.111}$$

For instance,

$$\begin{bmatrix} 1 \\ 1 \end{bmatrix} \text{ and } \begin{bmatrix} -2 \\ 2 \end{bmatrix} \tag{2.112}$$

are orthogonal

$$\begin{bmatrix} 1 \\ 1 \end{bmatrix} \bullet \begin{bmatrix} -2 \\ 2 \end{bmatrix} = 1(-2) + 1(2) = 0. \tag{2.113}$$

It is also convenient to define the inner (dot) product between a row matrix

$$\mathbf{a} = [a_1 \;\; a_2 \;\; a_3 \;\; \dots \;\; a_n] \tag{2.114}$$

and a column matrix

$$\mathbf{b} = \begin{bmatrix} b_1 \\ b_2 \\ b_3 \\ \vdots \\ b_n \end{bmatrix}, \tag{2.115}$$

such that

$$\mathbf{a} \bullet \mathbf{b} = \mathbf{b} \bullet \mathbf{a} = a_1 b_1 + a_2 b_2 + a_3 b_3 + \cdots + a_n b_n. \tag{2.116}$$

A-Orthogonal Column Matrices. Consider a positive definite symmetric square matrix

$$\mathbf{A} = \begin{bmatrix} a_{11} & a_{12} & a_{13} & \ldots & a_{1n} \\ a_{21} & a_{22} & a_{23} & \ldots & a_{2n} \\ a_{31} & a_{32} & a_{33} & \ldots & a_{3n} \\ \vdots & \vdots & \vdots & \ddots & \vdots \\ a_{n1} & a_{n2} & a_{n3} & \ldots & a_{nn} \end{bmatrix} \tag{2.117}$$

and column matrices

$$\mathbf{b} = \begin{bmatrix} b_1 \\ b_2 \\ b_3 \\ \vdots \\ b_n \end{bmatrix} \quad \text{and} \quad \mathbf{c} = \begin{bmatrix} c_1 \\ c_2 \\ c_3 \\ \vdots \\ c_n \end{bmatrix}. \tag{2.118}$$

Matrices **b** and **c** are said to be **A**-orthogonal if column matrix

$$\mathbf{d} = \mathbf{A}\mathbf{b} \tag{2.119}$$

is orthogonal to the column matrix **c**, thus

$$\mathbf{c} \bullet \mathbf{d} = \mathbf{d} \bullet \mathbf{c} = 0, \tag{2.120}$$

which means that

$$\mathbf{c} \bullet (\mathbf{A}\mathbf{b}) = (\mathbf{A}\mathbf{b}) \bullet \mathbf{c} = 0. \tag{2.121}$$

By substituting from equation (2.110)

$$\mathbf{c} \bullet (\mathbf{A}\mathbf{b}) = \mathbf{c}^{\mathrm{T}}(\mathbf{A}\mathbf{b}) = \mathbf{c}^{\mathrm{T}}\mathbf{A}\mathbf{b} = (\mathbf{A}\mathbf{b})^{\mathrm{T}}\mathbf{c}. \tag{2.122}$$

Because **A** is symmetric,

$$(\mathbf{A}\mathbf{b})^{\mathrm{T}} = \mathbf{b}^{\mathrm{T}}\mathbf{A}^{\mathrm{T}} = \mathbf{b}^{\mathrm{T}}\mathbf{A}, \tag{2.123}$$

this yields

$$\mathbf{c} \bullet (\mathbf{A}\mathbf{b}) = \mathbf{b}^{\mathrm{T}}\mathbf{A}\mathbf{c} = \mathbf{c}^{\mathrm{T}}\mathbf{A}\mathbf{b}. \tag{2.124}$$

Unit Length Column Matrix. The length of a column matrix

$$\mathbf{a} = \begin{bmatrix} a_1 \\ a_2 \\ a_3 \\ \vdots \\ a_n \end{bmatrix} \tag{2.125}$$

is defined as

$$|\mathbf{a}| = \sqrt{\mathbf{a} \bullet \mathbf{a}} = \sqrt{a_1^2 + a_2^2 + a_3^2 + \cdots + a_n^2}. \tag{2.126}$$

A column matrix **a** can be normalized to achieve a unit length column matrix

$$\mathbf{a}^0 = \frac{1}{\sqrt{\mathbf{a} \bullet \mathbf{a}}}\mathbf{a}. \tag{2.127}$$

Orthonormal Column Matrices. Column matrices **a** and **b** (each of which is of unit length) that are orthogonal to each other are said to be orthonormal

$$\mathbf{a} \bullet \mathbf{a} = \mathbf{b} \bullet \mathbf{b} = 1; \quad \mathbf{a} \bullet \mathbf{b} = 0. \tag{2.128}$$

A-Orthonormal Column Matrices. Column matrices **b** and **c** are said to be **A**-orthonormal (where **A** is a positive definite symmetric square matrix) if

$$\begin{aligned} \mathbf{b} \bullet (\mathbf{A}\mathbf{c}) = \mathbf{c} \bullet (\mathbf{A}\mathbf{b}) = 0 \\ \mathbf{b} \bullet (\mathbf{A}\mathbf{b}) = \mathbf{c} \bullet (\mathbf{A}\mathbf{c}) = 1. \end{aligned} \tag{2.129}$$

2.11 Summary

In this chapter matrices of numbers have been introduced together with the basic algebraic operations on matrices. Matrices are heavily used throughout this book and a reader that is not familiar with matrices is advised to study this chapter. By mastering this chapter, the reader will have very little difficulty in reading matrix equations in this book.

Further Reading

[1] Abadir, K. M. and Magnus, J. R. (2005) *Matrix Algebra*. Cambridge, Cambridge University Press.
[2] Abraham, R., Marsden, J. E and Ratiu, T. S. (1988) *Manifolds, Tensor Analysis, and Applications*. New York, Springer-Verlag.
[3] Bellman, R. (1987) *Introduction to Matrix Analysis*. Society for Industrial and Applied Mathematics.
[4] Bronstein, I. N. and Semendyayev, S. (2007) *Handbook of Mathematics*. New York, Springer.

[5] Golub, G. H. and van Van Loan, C. F. (1996) *Matrix Computations*, 3rd edn. Baltimore, MD, Johns Hopkins University Press.
[6] Johnson, C. (2009) *Numerical Solutions of Partial Differential Equations by the Finite Element Method.* New York, Dover Publications.
[7] Matthews, P. C. (1998) *Vector Calculus.* London, Springer.
[8] Rudin, W. (1976) *The Principles of Mathematical Analysis.* New York, McGraw-Hill Science/Engineering/Math.
[9] Saad, Y. (2003) *Iterative Methods for Sparse Linear Systems*, 2nd edn, Society for Industrial and Applied Mathematics.

3

Some Explicit and Iterative Solvers

3.1 The Central Difference Solver

The central difference solver is an explicit conditionally stable solver used to solve a system of differential equations given by

$$\mathbf{Kx} + \mathbf{M\ddot{x}} = \mathbf{b}, \tag{3.1}$$

where **x** is the column matrix of unknown displacements, $\ddot{\mathbf{x}}$ is the column matrix of unknown accelerations, **b** is the column matrix of known loads and **K** is the stiffness matrix, such that **Kx** represents the equivalent internal forces. It is worth noting that dynamic finite element analysis often produces a system of this kind.

In the above system the displacements are unknown functions of time, i.e.,

$$\mathbf{x} = \begin{bmatrix} x_1 \\ x_2 \\ x_3 \\ \vdots \\ x_n \end{bmatrix} = \begin{bmatrix} \mathrm{x}_1(t) \\ \mathrm{x}_2(t) \\ \mathrm{x}_3(t) \\ \vdots \\ \mathrm{x}_n(t) \end{bmatrix}. \tag{3.2}$$

Large Strain Finite Element Method: A Practical Course, First Edition. Antonio Munjiza, Esteban Rougier and Earl E. Knight.

The column matrix of unknown accelerations is therefore

$$\mathbf{a} = \ddot{\mathbf{x}} = \begin{bmatrix} a_1 \\ a_2 \\ a_3 \\ \vdots \\ a_n \end{bmatrix} = \begin{bmatrix} \ddot{x}_1 \\ \ddot{x}_2 \\ \ddot{x}_3 \\ \vdots \\ \ddot{x}_n \end{bmatrix} = \begin{bmatrix} \frac{d^2 x_1}{dt^2} \\ \frac{d^2 x_2}{dt^2} \\ \frac{d^2 x_3}{dt^2} \\ \vdots \\ \frac{d^2 x_n}{dt^2} \end{bmatrix}. \tag{3.3}$$

The known external loads are given by the following column matrix

$$\mathbf{b} = \begin{bmatrix} b_1 \\ b_2 \\ b_3 \\ \vdots \\ b_n \end{bmatrix}. \tag{3.4}$$

The internal forces are calculated using the stiffness matrix

$$\mathbf{s} = \begin{bmatrix} s_1 \\ s_2 \\ s_3 \\ \vdots \\ s_n \end{bmatrix} = \begin{bmatrix} k_{11} & k_{12} & k_{13} & \dots & k_{1n} \\ k_{21} & k_{22} & k_{23} & \dots & k_{2n} \\ k_{31} & k_{32} & k_{33} & \dots & k_{3n} \\ \vdots & \vdots & \vdots & \ddots & \vdots \\ k_{n1} & k_{n2} & k_{n3} & \dots & k_{nn} \end{bmatrix} \begin{bmatrix} x_1 \\ x_2 \\ x_3 \\ \vdots \\ x_n \end{bmatrix}, \tag{3.5}$$

where the components of the stiffness matrix are in general a function of the displacements

$$k_{ij} = \mathrm{k}_{ij}(x_1, x_2, x_3, \dots x_n). \tag{3.6}$$

Equation (3.1) represents a dynamic equilibrium of internal forces **s**, inertial forces **g**, and external loads **b**, where the inertial forces are given by

$$\mathbf{g} = \begin{bmatrix} g_1 \\ g_2 \\ g_3 \\ \vdots \\ g_n \end{bmatrix} = \begin{bmatrix} m_1 & 0 & 0 & \dots & 0 \\ 0 & m_2 & 0 & \dots & 0 \\ 0 & 0 & m_3 & \dots & 0 \\ \vdots & \vdots & \vdots & \ddots & \vdots \\ 0 & 0 & 0 & \dots & m_n \end{bmatrix} \begin{bmatrix} \ddot{x}_1 \\ \ddot{x}_2 \\ \ddot{x}_3 \\ \vdots \\ \ddot{x}_n \end{bmatrix}, \tag{3.7}$$

with a diagonal mass matrix being employed, i.e.,

$$\mathbf{M}=\begin{bmatrix} m_1 & 0 & 0 & \dots & 0 \\ 0 & m_2 & 0 & \dots & 0 \\ 0 & 0 & m_3 & \dots & 0 \\ \vdots & \vdots & \vdots & \ddots & \vdots \\ 0 & 0 & 0 & \dots & m_n \end{bmatrix}. \tag{3.8}$$

The fact that the mass matrix is diagonal enables explicit integration. There are a whole range of integration schemes available in literature. The central difference scheme is the most often used scheme for finite element dynamic analysis. It is also used to obtain static solutions by employing dynamic relaxation techniques.

The scheme uses a discretization in the time domain, which is implemented through the time step, h, Figure 3.1.

The integration scheme is broken into the following steps:

Step 1: Calculate the current internal and external forces

$$\mathbf{s}_{cur}=\mathbf{K}_{cur}\mathbf{x}_{cur}=\begin{bmatrix} s_{1cur} \\ s_{2cur} \\ s_{3cur} \\ \vdots \\ s_{ncur} \end{bmatrix}=\begin{bmatrix} k_{11cur} & k_{12cur} & k_{13cur} & \dots & k_{1ncur} \\ k_{21cur} & k_{22cur} & k_{23cur} & \dots & k_{2ncur} \\ k_{31cur} & k_{32cur} & k_{33cur} & \dots & k_{3ncur} \\ \vdots & \vdots & \vdots & \ddots & \vdots \\ k_{n1cur} & k_{n2cur} & k_{n3cur} & \dots & k_{nncur} \end{bmatrix}\begin{bmatrix} x_{1cur} \\ x_{2cur} \\ x_{3cur} \\ \vdots \\ x_{ncur} \end{bmatrix} \tag{3.9}$$

i.e.,

$$\mathbf{f}_{cur}=\mathbf{b}_{cur}-\mathbf{s}_{cur}. \tag{3.10}$$

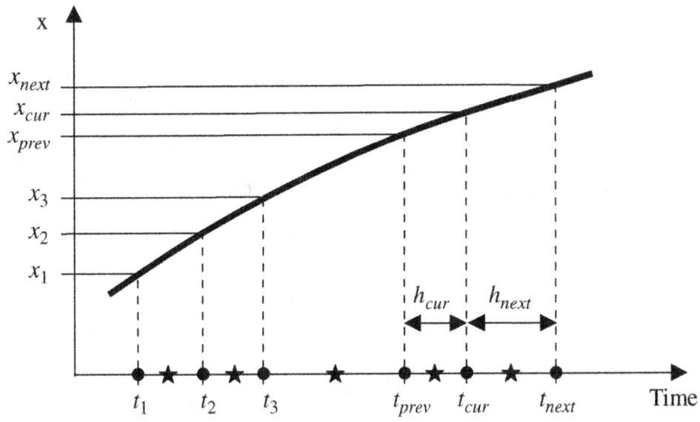

Figure 3.1 Discretization of displacements and velocities in time. Displacement and acceleration sampling points are shown using (•), while velocity sampling points are shown using (★)

Step 2: calculate the current acceleration

$$\mathbf{a}_{cur} = \begin{bmatrix} a_{1cur} \\ a_{2cur} \\ a_{3cur} \\ \vdots \\ a_{ncur} \end{bmatrix} = \begin{bmatrix} f_{1cur}/m_1 \\ f_{2cur}/m_2 \\ f_{3cur}/m_3 \\ \vdots \\ f_{ncur}/m_n \end{bmatrix}. \tag{3.11}$$

Step 3: calculate the next velocity

$$\begin{aligned} \mathbf{v}_{next} = \begin{bmatrix} v_{1next} \\ v_{2next} \\ v_{3next} \\ \vdots \\ v_{nnext} \end{bmatrix} &= \mathbf{v}_{curr} + \mathbf{a}_{cur} \left(\frac{h_{cur} + h_{next}}{2} \right) \\ &= \begin{bmatrix} v_{1curr} \\ v_{2curr} \\ v_{3curr} \\ \vdots \\ v_{ncurr} \end{bmatrix} + \left(\frac{h_{cur} + h_{next}}{2} \right) \begin{bmatrix} a_{1cur} \\ a_{2cur} \\ a_{3cur} \\ \vdots \\ a_{ncur} \end{bmatrix}. \end{aligned} \tag{3.12}$$

Step 4: Calculate the next displacement

$$\begin{aligned} \mathbf{x}_{next} = \begin{bmatrix} x_{1next} \\ x_{2next} \\ x_{3next} \\ \vdots \\ x_{nnext} \end{bmatrix} &= \mathbf{x}_{curr} + \mathbf{v}_{cur} h_{next} \\ &= \begin{bmatrix} x_{1curr} \\ x_{2curr} \\ x_{3curr} \\ \vdots \\ x_{ncurr} \end{bmatrix} + h_{next} \begin{bmatrix} v_{1cur} \\ v_{2cur} \\ v_{3cur} \\ \vdots \\ v_{ncur} \end{bmatrix}. \end{aligned} \tag{3.13}$$

Stability of the Central Difference Scheme. In the case of a linear 1D system with no external loads, the central difference time integration scheme reduces to

Step 1: Calculate the internal force

$$f_{cur} = -kx_{cur}. \tag{3.14}$$

Step 2: Calculate the current acceleration

$$a_{cur} = \frac{f_{cur}}{m} = \frac{-kx_{cur}}{m}. \tag{3.15}$$

Step 3: Calculate the next velocity

$$v_{next} = v_{cur} + a_{cur}\left(\frac{h_{cur} + h_{next}}{2}\right) = v_{cur} - \frac{kx_{cur}}{m}\left(\frac{h_{cur} + h_{next}}{2}\right), \tag{3.16}$$

or simply

$$\left(\frac{h_{cur} + h_{next}}{2}\right)v_{next} = \left(\frac{h_{cur} + h_{next}}{2}\right)v_{cur} - \frac{k}{m}\left(\frac{h_{cur} + h_{next}}{2}\right)^2 x_{cur}. \tag{3.17}$$

Step 4: Calculate the next displacement

$$\begin{aligned} x_{next} &= x_{cur} + v_{next}h \\ &= v_{cur}h + \left(x_{cur} - \frac{kx_{cur}}{m}h^2\right) \\ &= v_{cur}h + \left(1 - \frac{k}{m}h^2\right)x_{cur} \end{aligned} \tag{3.18}$$

or

$$x_{next} = v_{cur}h + \left(1 - \frac{k}{m}h^2\right)x_{cur}. \tag{3.19}$$

Equations (3.17) and (3.19) can be written in a matrix form

$$\begin{bmatrix} vh \\ x \end{bmatrix}_{next} = \begin{bmatrix} 1 & -h^2k/m \\ 1 & 1-h^2k/m \end{bmatrix}\begin{bmatrix} vh \\ x \end{bmatrix}_{current}. \tag{3.20}$$

This is a recursive formula, which can be written using eigenvalues λ. The eigenvalues are obtained from

$$\begin{vmatrix} 1-\lambda & -h^2k/m \\ 1 & 1-h^2k/m-\lambda \end{vmatrix} = 0, \tag{3.21}$$

which yields the following equation

$$(1-\lambda)(1-h^2k/m-\lambda) + h^2k/m = 0. \tag{3.22}$$

The roots of the above equation yield two eigenvalues λ

$$\lambda_{1,2} = \frac{-(h^2k/m-2) \pm \sqrt{(h^2k/m-2)^2-4}}{2}. \tag{3.23}$$

As it can be seen, eigenvalues $\lambda_{1,2}$ depend on the time step h. For the time step

$$h = \sqrt{\frac{4m}{k}}, \tag{3.24}$$

both eigenvalues are equal to

$$\lambda_{1,2} = -1. \tag{3.25}$$

The spectral radius is therefore

$$_{\max}|\lambda_{1,2}| = 1. \tag{3.26}$$

For any

$$h < \sqrt{\frac{4m}{k}}, \tag{3.27}$$

Equation (3.23) yields

$$\lambda_{1,2} = \frac{-(h^2k/m-2) \pm i\sqrt{4-(h^2k/m-2)^2}}{2} \tag{3.28}$$

and the spectral radius

$$_{\max}|\lambda_{1,2}| = \frac{1}{2}\sqrt{(h^2k/m-2)^2 + \left[4-(h^2k/m-2)^2\right]} = 1. \tag{3.29}$$

In the case that

$$h > \sqrt{\frac{4m}{k}}, \tag{3.30}$$

the eigenvalues are given by

$$\lambda_{1,2} = \frac{-(h^2k/m-2) \pm \sqrt{(h^2k/m-2)^2-4}}{2}. \tag{3.31}$$

This gives the spectral radius

$$_{\max}|\lambda_{1,2}| = \frac{(h^2k/m-2) + \sqrt{(h^2k/m-2)^2-4}}{2} > 1. \tag{3.32}$$

This spectral radius is greater than 1 for any

$$h > \sqrt{\frac{4m}{k}}. \tag{3.33}$$

The graph of the spectral radius as a function of h is shown in Figure 3.2.

Example. The system shown in Figure 3.3 is solved using the central difference time integration scheme. In Figure 3.4 the results obtained using different time step sizes are shown.

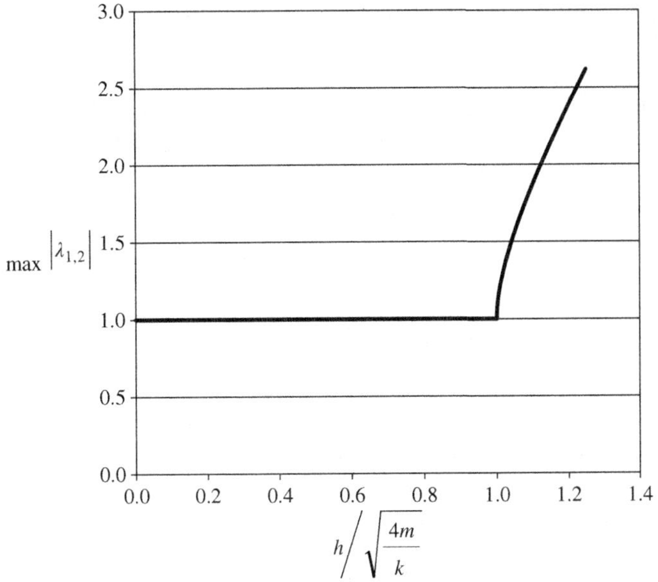

Figure 3.2 Conditional stability of the central difference explicit time integration scheme. For $h < \sqrt{4m/k}$ the scheme is stable with the spectral radius being exactly 1. For $h > \sqrt{4m/k}$ the scheme is unstable

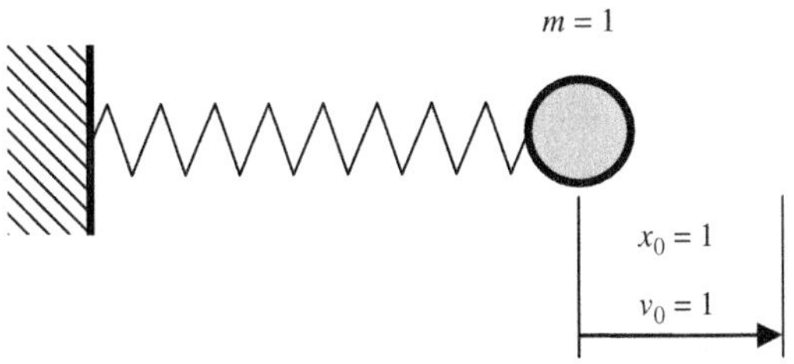

Figure 3.3 A 1D mass-spring system

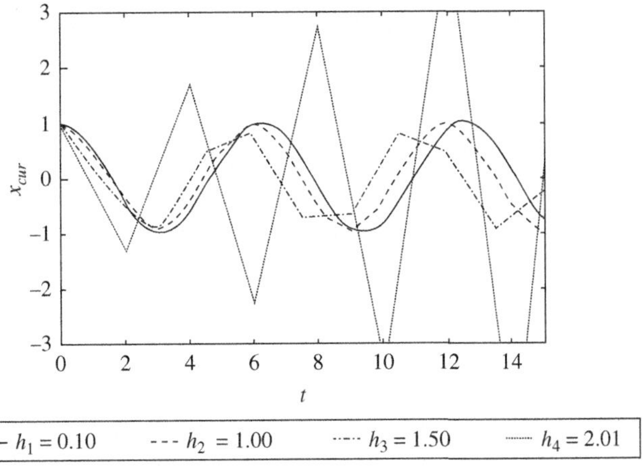

Figure 3.4 Demonstration of the stability of the central difference time integration scheme using a 1D system

Introducing Damping. Very often, dynamic systems are damped. In essence there are two types of damping:

a. External Damping Forces. This is a type of damping produced by external factors, such as air, water, etc. In general, this kind of damping is described by the damping forces given by

$$\mathbf{d} = \begin{bmatrix} d_1 \\ d_2 \\ d_3 \\ \vdots \\ d_n \end{bmatrix} = \begin{bmatrix} \mathrm{d}_1(v_1) \\ \mathrm{d}_2(v_2) \\ \mathrm{d}_3(v_3) \\ \vdots \\ \mathrm{d}_n(v_n) \end{bmatrix}. \tag{3.34}$$

b. Internal Damping Forces. This is a type of damping produced by internal factors, such as interaction between micro-structural elements of the solid medium. In general, this kind of damping is described by the damping forces given by

$$\mathbf{d} = \begin{bmatrix} d_1 \\ d_2 \\ d_3 \\ \vdots \\ d_n \end{bmatrix} = \begin{bmatrix} \mathrm{d}_1(v_1, v_2, v_3, \ldots, v_n) \\ \mathrm{d}_2(v_1, v_2, v_3, \ldots, v_n) \\ \mathrm{d}_3(v_1, v_2, v_3, \ldots, v_n) \\ \vdots \\ \mathrm{d}_n(v_1, v_2, v_3, \ldots, v_n) \end{bmatrix}. \tag{3.35}$$

The implementation of the central difference time integration scheme for the external damping is done using the implicit formulation

$$\mathbf{f}_{cur} = \mathbf{b}_{cur} - \mathbf{s}_{cur} + \mathbf{d}_{next}, \tag{3.36}$$

where $\mathbf{s}_{cur}$ is given by Equation (3.9) and $\mathbf{d}_{next}$ is given by

$$\mathbf{d}_{next} = \begin{bmatrix} d_{1next} \\ d_{2next} \\ d_{3next} \\ \vdots \\ d_{nnext} \end{bmatrix} = \begin{bmatrix} \mathrm{d}_{1next}(v_{1next}) \\ \mathrm{d}_{2next}(v_{2next}) \\ \mathrm{d}_{3next}(v_{3next}) \\ \vdots \\ \mathrm{d}_{nnext}(v_{nnext}) \end{bmatrix}. \tag{3.37}$$

From these forces, the current acceleration is calculated as follows

$$\ddot{\mathbf{x}}_{cur} = \begin{bmatrix} \ddot{x}_{1cur} \\ \ddot{x}_{2cur} \\ \ddot{x}_{3cur} \\ \vdots \\ \ddot{x}_{ncur} \end{bmatrix} = \begin{bmatrix} f_{1cur}/m_1 \\ f_{2cur}/m_2 \\ f_{3cur}/m_3 \\ \vdots \\ f_{ncur}/m_n \end{bmatrix}. \tag{3.38}$$

When this acceleration is substituted into step 3, it yields $\mathbf{v}_{next}$ with damping forces being calculated using $\mathbf{v}_{next}$, therefore it is an implicit formulation.

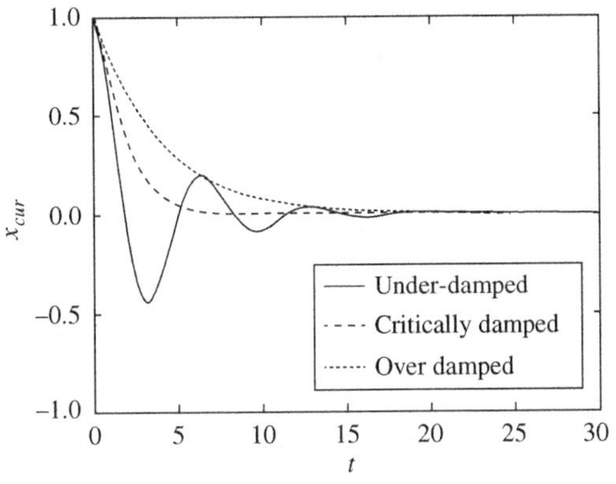

Figure 3.5 Dynamic relaxation of a static system

For internal damping, the damping forces are calculated using the explicit formulation,

$$\mathbf{d}_{cur} = \begin{bmatrix} d_{1cur} \\ d_{2cur} \\ d_{3cur} \\ \vdots \\ d_{ncur} \end{bmatrix} = \begin{bmatrix} \mathrm{d}_{1cur}(v_{1cur}, v_{2cur}, v_{3cur}, \ldots, v_{ncur}) \\ \mathrm{d}_{2cur}(v_{1cur}, v_{2cur}, v_{3cur}, \ldots, v_{ncur}) \\ \mathrm{d}_{3cur}(v_{1cur}, v_{2cur}, v_{3cur}, \ldots, v_{ncur}) \\ \vdots \\ \mathrm{d}_{ncur}(v_{1cur}, v_{2cur}, v_{3cur}, \ldots, v_{ncur}) \end{bmatrix}. \tag{3.39}$$

Dynamic Relaxation. Using a damped dynamic system, in conjunction with static loads, and subjecting it to transient dynamic motion eventually leads to a state of rest and by definition a state of static equilibrium wherein the inertia forces are equal to zero, Figure 3.5.

This is often conveniently used to obtain the equilibrium solution for static loads. Thus, instead of looking for static equilibrium, i.e., solving the static system (inertia free)

$$\mathbf{Kx} = \mathbf{b}, \tag{3.40}$$

one considers dynamic equilibrium by solving the dynamic system

$$\mathbf{Kx} + \mathbf{M}\ddot{\mathbf{x}} + \mathbf{C}\dot{\mathbf{x}} = \mathbf{b}, \tag{3.41}$$

where $\mathbf{M}$ is the lumped (diagonal mass) matrix, $\mathbf{b}$ is the external static load and $\mathbf{x}$ are the unknown static displacements while $\dot{\mathbf{x}}$ and $\ddot{\mathbf{x}}$ are the transient velocity and transient acceleration respectively. Both the transient velocity and the transient acceleration converge to zero as the system loses energy and reaches static equilibrium.

3.2 Generalized Direction Methods

In static problems the finite element method involves solving a system of algebraic equations. Consider for example the following system of linear algebraic equations

$$\mathbf{Ax} = \mathbf{b}. \tag{3.42}$$

The n-dimensional space is called a hyper-space. In two dimensions, this space is relatively easy to visualize and describe by a Cartesian coordinate system (x_1, x_2).

The column matrix

$$\begin{bmatrix} x_1 \\ x_2 \\ x_3 \\ \vdots \\ x_n \end{bmatrix} \tag{3.43}$$

represents a hyper-point in the n-dimensional space. For example, in 2D it represents point

$$\begin{bmatrix} x_1 \\ x_2 \end{bmatrix}, \tag{3.44}$$

which is the solution for the system of equations $\mathbf{Ax} = \mathbf{b}$.

Matrix $\mathbf{A}$ can be broken into rows such that

$$\mathbf{A} = \begin{bmatrix} a_{11} & a_{12} & a_{13} & \cdots & a_{1n} \\ a_{21} & a_{22} & a_{23} & \cdots & a_{2n} \\ a_{31} & a_{32} & a_{33} & \cdots & a_{3n} \\ \vdots & \vdots & \vdots & \ddots & \vdots \\ a_{n1} & a_{n2} & a_{n3} & \cdots & a_{nn} \end{bmatrix} = \begin{bmatrix} \mathbf{a}_1 \\ \mathbf{a}_2 \\ \mathbf{a}_3 \\ \vdots \\ \mathbf{a}_n \end{bmatrix}. \tag{3.45}$$

Following this, the product $\mathbf{Ax}$ can be written as follows

$$\mathbf{Ax} = \begin{bmatrix} \mathbf{a}_1 \bullet \mathbf{x} \\ \mathbf{a}_2 \bullet \mathbf{x} \\ \mathbf{a}_3 \bullet \mathbf{x} \\ \vdots \\ \mathbf{a}_n \bullet \mathbf{x} \end{bmatrix} = \begin{bmatrix} b_1 \\ b_2 \\ b_3 \\ \vdots \\ b_n \end{bmatrix}, \tag{3.46}$$

where

$$\mathbf{a}_1 \bullet \mathbf{x} \tag{3.47}$$

is the dot product between row matrix $\mathbf{a}_1$ and column matrix $\mathbf{x}$. Accordingly, each of the algebraic equations (3.45) becomes

$$\begin{aligned} \mathbf{a}_1 \bullet \mathbf{x} &= b_1 \\ \mathbf{a}_2 \bullet \mathbf{x} &= b_2 \\ \mathbf{a}_3 \bullet \mathbf{x} &= b_3 \\ &\vdots \\ \mathbf{a}_n \bullet \mathbf{x} &= b_n \end{aligned} \tag{3.48}$$

or simply

$$\begin{aligned} z_1 &= z_1(x_1,x_2,x_3,\ldots x_n) = \mathbf{a}_1 \bullet \mathbf{x} - b_1 = 0 \\ z_2 &= z_2(x_1,x_2,x_3,\ldots x_n) = \mathbf{a}_2 \bullet \mathbf{x} - b_2 = 0 \\ z_3 &= z_3(x_1,x_2,x_3,\ldots x_n) = \mathbf{a}_3 \bullet \mathbf{x} - b_3 = 0 \\ &\vdots \\ z_n &= z_n(x_1,x_2,x_3,\ldots x_n) = \mathbf{a}_n \bullet \mathbf{x} - b_n = 0. \end{aligned} \tag{3.49}$$

where functions z_1, z_2, z_3, …, z_n are in actual fact scalar fields which assign a real number (i.e., z_1, z_2, z_3, …, z_n) to each hyper-point in space. This is best demonstrated in 2D where

$$z_1 = z_1(x_1, x_2) \tag{3.50}$$

represents (for constant **A**) a plane. The contour lines of this plane are shown in Figure 3.6. In a similar manner, the scalar field $z_2(x_1, x_2)$ represents another plane shown in Figure 3.7.

Provided that one is sitting at point $\mathbf{x}_0$ (initial solution), there is an infinite number of ways to arrive at point $\mathbf{x}$ (the exact solution of the system $\mathbf{Ax} = \mathbf{b}$), Figure 3.8.

In order to "see" where these lines intersect, it is enough to "rise up" at point $\mathbf{x}_0$ and "look around" to "see" both solid lines and their intersection point $\mathbf{x}$. Following this space "scouting" step, one visually decides to travel along the direction

$$\mathbf{x} - \mathbf{x}_0 = \begin{bmatrix} x_1 \\ x_2 \end{bmatrix} - \begin{bmatrix} x_{1,0} \\ x_{2,0} \end{bmatrix}, \tag{3.51}$$

thus arriving at $\mathbf{x}$ in a single step.

As seen above, solving a 2D system of equations has involved "scouting" in 3D space and "observing" the intersections of 3D surfaces (planes).

By analogy, it follows that solving a 3D system of equations involves surfaces in 4D space and "scouting" in 4D space. In general, for nD space (a system of n by n equations), one deals with $(n+1)$D surfaces, called hyper-surfaces. By starting from a given point $\mathbf{x}_0$, one has to "navigate" their way to point $\mathbf{x}$.

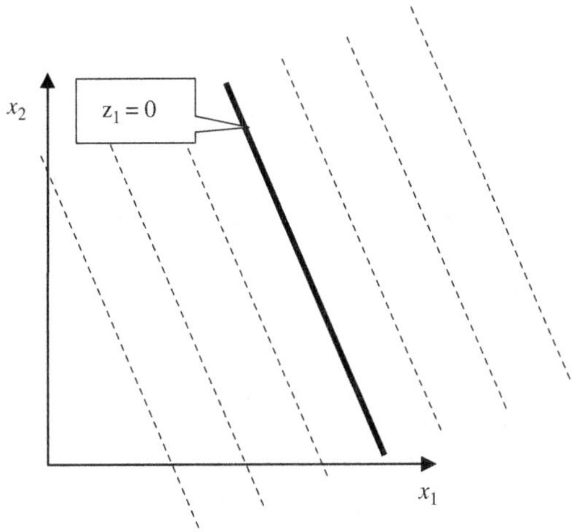

Figure 3.6 Contour lines of the surface (plane) $z_1 = z_1(x_1, x_2)$

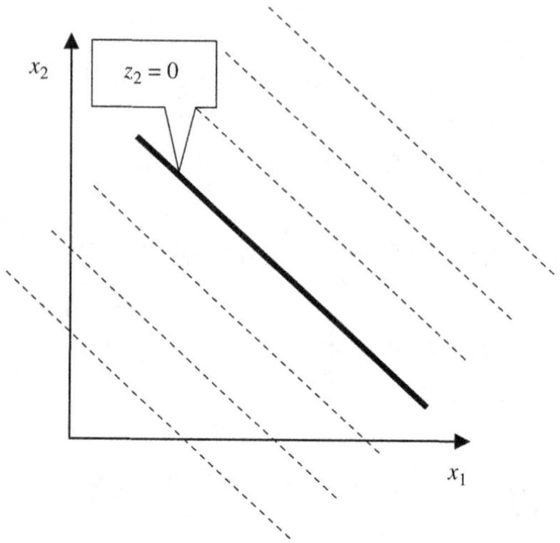

Figure 3.7 Contour lines of the surface (plane) $z_2 = z_2(x_1, x_2)$

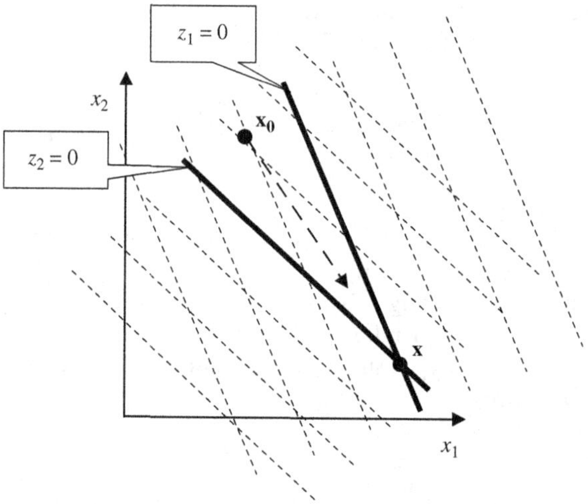

Figure 3.8 Navigating the 2D space

A number of methods can be devised to accomplish the nD "navigation" task. As a consequence, there is a whole family of iterative direction methods. Some of these methods are illustrated below, using a 2D (2 by 2) system of equations.

Gauss-Seidel Method. The Gauss-Seidel method considers one surface at the time. One first travels in the direction of axis x_1 towards the zero line of surface z_1, Figure 3.9.

This is followed by a move in the x_2 direction towards the line $z_2 = 0$, Figure 3.10. By repeating these steps one gets closer and closer to point $\mathbf{x}$, Figure 3.11.

Some of the advantages of the Gauss-Seidel method are:

1. At each iteration, only $\mathbf{a} \bullet \mathbf{x}$ is calculated, instead of $\mathbf{Ax}$ (which is n times more expensive to do).
2. There is no need to do anything special to calculate directions.

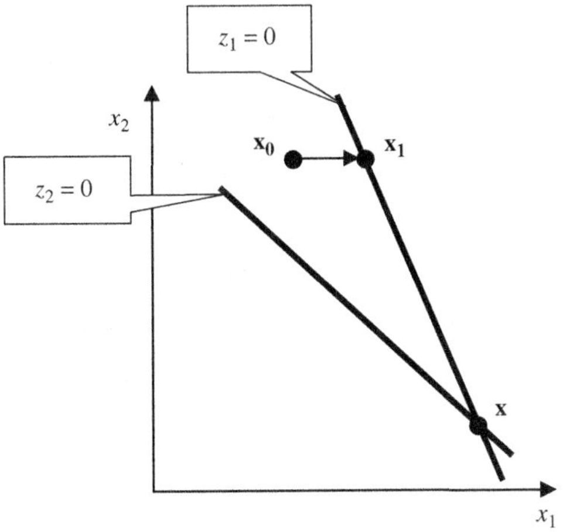

Figure 3.9 Traveling in x_1 direction towards line $z_1 = 0$

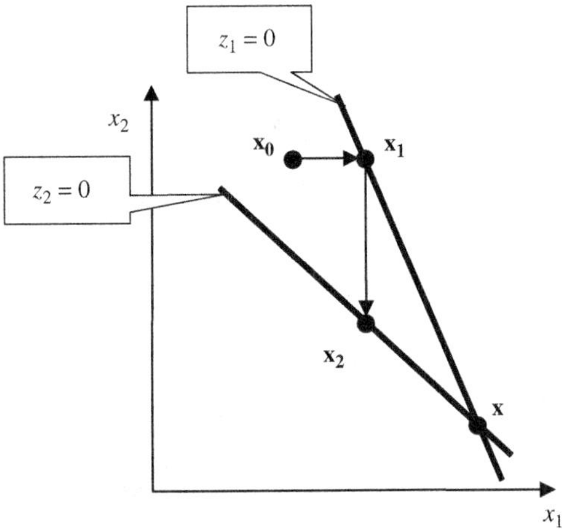

Figure 3.10 Traveling in x_2 direction towards line $z_2 = 0$

The main difficulty with the Gauss-Seidel method is that it may diverge, i.e., not converge to the exact solution. For example, if one was to swap the equations

$$\left.\begin{array}{l} z_1 = 0 \\ z_2 = 0 \end{array}\right\} \Rightarrow \left\{\begin{array}{l} z_2 = 0 \\ z_1 = 0, \end{array}\right. \tag{3.52}$$

i.e.,

$$\begin{array}{l} \bar{z}_1 = 0 \\ \bar{z}_2 = 0, \end{array} \tag{3.53}$$

where $\bar{z}_1 = z_2$ and $\bar{z}_2 = z_1$, and then perform the Gauss-Seidel iterations, the evolution of the iterative process as shown in Figure 3.12, clearly demonstrates the divergence from the exact solution $\mathbf{x}$.

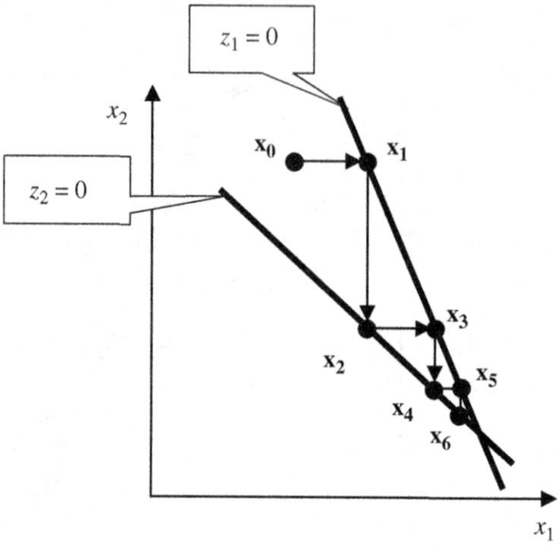

Figure 3.11 Gauss-Seidel iteration process in 2D

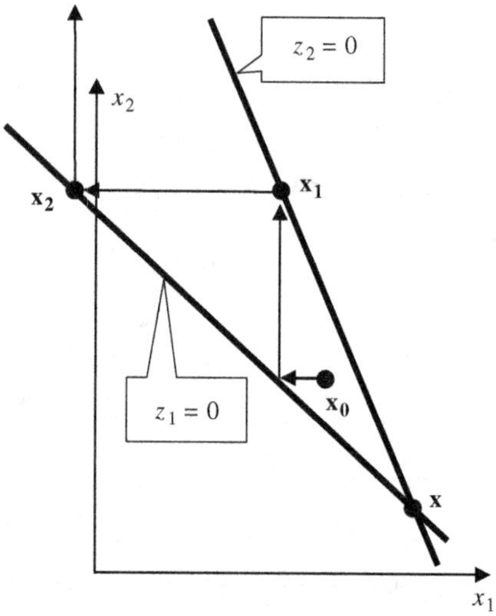

Figure 3.12 Diverging Gauss-Seidel iterations

Nevertheless, the method is used extensively in different applications under different names. A well known Gauss-Seidel application for the solution of 2D structural systems is in an iterative procedure called the Cross method.

An extension of the Gauss-Seidel method to 3D space is straightforward with 4D hyper-lines

$$\begin{aligned} z_1 &= 0 \\ z_2 &= 0 \\ z_3 &= 0 \end{aligned} \tag{3.54}$$

being considered for each direction x_1, x_2 and x_3 respectively.

An extension to nD space has to consider $(n+1)$D hyper-surfaces

$$\begin{aligned} z_1 &= \mathbf{a}_1 \bullet \mathbf{x} \\ z_2 &= \mathbf{a}_2 \bullet \mathbf{x} \\ z_3 &= \mathbf{a}_3 \bullet \mathbf{x} \\ &\vdots \\ z_n &= \mathbf{a}_n \bullet \mathbf{x} \end{aligned} \tag{3.55}$$

and $(n+1)$D hyper-lines

$$\begin{aligned} z_1 &= 0 \\ z_2 &= 0 \\ z_3 &= 0 \\ &\vdots \\ z_n &= 0. \end{aligned} \tag{3.56}$$

Method of a-Orthogonal Directions. It is relatively easy to devise a Gauss-Seidel like iteration method that follows directions orthogonal to the lines

$$\begin{aligned} z_1 &= 0 \\ z_2 &= 0 \end{aligned} \tag{3.57}$$

as shown in Figure 3.13.

Unlike the Gauss-Seidel method, this method never diverges. The line-orthogonal directions are obtained as

$$\begin{aligned} \mathbf{d}_1 &= \mathbf{a}_1 \\ \mathbf{d}_2 &= \mathbf{a}_2 \end{aligned}. \tag{3.58}$$

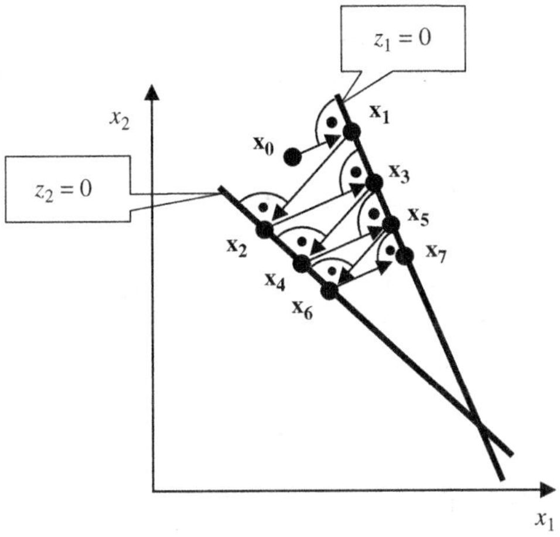

Figure 3.13 The method of directions orthogonal to hyper-lines $z_1, z_2, z_3, \ldots, z_n$

Extension to nD space is straightforward. The **a**-orthogonal directions are simply

$$\begin{aligned}
&\mathbf{d}_1 = \mathbf{a}_1\\
&\mathbf{d}_2 = \mathbf{a}_2\\
&\mathbf{d}_3 = \mathbf{a}_3\\
&\vdots\\
&\mathbf{d}_n = \mathbf{a}_n\\
&\mathbf{d}_{n+1} = \mathbf{a}_1\\
&\mathbf{d}_{n+2} = \mathbf{a}_2\\
&\mathbf{d}_{n+3} = \mathbf{a}_3\\
&\vdots\\
&\mathbf{d}_{n+n} = \mathbf{a}_n\\
&\mathbf{d}_{n+n+1} = \mathbf{a}_1\\
&\mathbf{d}_{n+n+2} = \mathbf{a}_2\\
&\text{etc.}
\end{aligned} \tag{3.59}$$

The final solution after many iterations is therefore given by

$$\begin{aligned}
\mathbf{x}_{n+n+1} = \mathbf{x}_0 &+ \alpha_1\mathbf{d}_1 + \alpha_2\mathbf{d}_2 + \alpha_3\mathbf{d}_3 + \cdots + \alpha_n\mathbf{d}_n\\
&+ \alpha_{n+1}\mathbf{d}_{n+1} + \alpha_{n+2}\mathbf{d}_{n+2} + \alpha_{n+3}\mathbf{d}_{n+3} + \cdots + \alpha_{n+n}\mathbf{d}_{n+n}\\
&+ \alpha_{n+n+1}\mathbf{d}_{n+n+1} + \alpha_{n+n+2}\mathbf{d}_{n+n+2} + \alpha_{n+n+3}\mathbf{d}_{n+n+3} + \cdots + \alpha_{n+n+n}\mathbf{d}_{n+n+n}\\
&+ \cdots
\end{aligned} \tag{3.60}$$

The scaling factor for each of these directions is obtained by considering that by traveling in a given **d** direction, one has to reach the corresponding line z.

Generalization of the Direction-Based Solvers. In theory, any number of direction-based solvers can be devised. The key difference between them is in the way one evaluates the traveling direction at each step. The Gauss-Seidel has the simplest procedure for evaluating the traveling directions. The method of **a**-orthogonal directions has directions obtained from row matrices $\mathbf{a}_1$, $\mathbf{a}_2$, $\mathbf{a}_3$, $\mathbf{a}_4$, …, $\mathbf{a}_n$.

Extension of the direction-based solvers to nonlinear systems is straight forward – the contour lines become curved, while the travelling directions stay as straight lines.

3.3 The Method of Conjugate Directions

In the previous section, two directions-based iterative methods for solving systems of linear algebraic equations were introduced. The problem with both methods is that a relatively slow convergence may occur. In order to improve the convergence speed, it is convenient to introduce directions that are **A**-orthogonal. In further text, a system of linear algebraic equations

$$\mathbf{A}\mathbf{x} = \mathbf{b} \tag{3.61}$$

(where **A** is a positive definite matrix called "stiffness matrix," **b** is a known column matrix called "load" and **x** is the unknown column matrix called "displacement") is considered.

The solution to this system can be written as a linear combination of column matrices (also called directions)

$$\mathbf{x} = \alpha_1 \mathbf{d}_1 + \alpha_2 \mathbf{d}_2 + \alpha_3 \mathbf{d}_3 + \cdots + \alpha_n \mathbf{d}_n, \tag{3.62}$$

where $\alpha_1, \alpha_2, \alpha_3, \ldots, \alpha_n$ are scaling factors and $\mathbf{d}_1, \mathbf{d}_2, \mathbf{d}_3, \ldots, \mathbf{d}_n$ are directions chosen in such a way that

$$\begin{aligned} \mathbf{d}_i \bullet (\mathbf{A}\mathbf{d}_j) &= \mathbf{d}_i^{\mathrm{T}} \mathbf{A}\mathbf{d}_j > 0 \quad \text{for } i = j \\ \mathbf{d}_i \bullet (\mathbf{A}\mathbf{d}_j) &= \mathbf{d}_i^{\mathrm{T}} \mathbf{A}\mathbf{d}_j = 0 \quad \text{for } i \neq j, \end{aligned} \tag{3.63}$$

i.e., these column matrices are **A**-orthogonal to each other (this is also termed as **A**-conjugate). Thus, the iterative solution procedure is termed the method of **A**-orthogonal directions. It is also sometimes called the method of conjugate directions. In its generalized form, it is called the method of conjugate gradients.

By substituting equation (3.62) into (3.61) one obtains

$$\alpha_1 \mathbf{A}\mathbf{d}_1 + \alpha_2 \mathbf{A}\mathbf{d}_2 + \alpha_3 \mathbf{A}\mathbf{d}_3 + \cdots + \alpha_n \mathbf{A}\mathbf{d}_n = \mathbf{b} \tag{3.64}$$

or

$$\mathbf{b} = \tilde{\mathbf{q}}_1 + \tilde{\mathbf{q}}_2 + \tilde{\mathbf{q}}_3 + \cdots + \tilde{\mathbf{q}}_n, \tag{3.65}$$

where

$$\begin{aligned} \tilde{\mathbf{q}}_1 &= \alpha_1 \mathbf{A}\mathbf{d}_1 \\ \tilde{\mathbf{q}}_2 &= \alpha_2 \mathbf{A}\mathbf{d}_2 \\ \tilde{\mathbf{q}}_3 &= \alpha_3 \mathbf{A}\mathbf{d}_3 \\ &\vdots \\ \tilde{\mathbf{q}}_n &= \alpha_n \mathbf{A}\mathbf{d}_n. \end{aligned} \tag{3.66}$$

This means that the load **b** is actually split into components that correspond to individual directions.

By applying a dot product with $\mathbf{d}_1$ to equation (3.64) one obtains

$$\alpha_1 \mathbf{d}_1 \bullet (\mathbf{A}\mathbf{d}_1) + \alpha_2 \mathbf{d}_1 \bullet (\mathbf{A}\mathbf{d}_2) + \alpha_3 \mathbf{d}_1 \bullet (\mathbf{A}\mathbf{d}_3) + \cdots + \alpha_n \mathbf{d}_1 \bullet (\mathbf{A}\mathbf{d}_n) = \mathbf{d}_1 \bullet \mathbf{b}. \tag{3.67}$$

After substitution of the **A**-orthogonality condition from equation (3.63) one obtains

$$\alpha_1 \mathbf{d}_1 \bullet (\mathbf{A}\mathbf{d}_1) = \mathbf{d}_1 \bullet \mathbf{b}, \tag{3.68}$$

which yields the scaling factor α_1

$$\alpha_1 = \frac{\mathbf{d}_1 \bullet \mathbf{b}}{\mathbf{d}_1 \bullet (\mathbf{A}\mathbf{d}_1)}. \tag{3.69}$$

By proceeding with the above dot products by $\mathbf{d}_2, \mathbf{d}_3, \mathbf{d}_4, \ldots, \mathbf{d}_n$, one obtains all the scaling coefficients

$$\begin{aligned}
\alpha_1 &= \frac{\mathbf{d}_1 \bullet \mathbf{b}}{\mathbf{d}_1 \bullet (\mathbf{A}\mathbf{d}_1)} \\
\alpha_2 &= \frac{\mathbf{d}_2 \bullet \mathbf{b}}{\mathbf{d}_2 \bullet (\mathbf{A}\mathbf{d}_2)} \\
\alpha_3 &= \frac{\mathbf{d}_3 \bullet \mathbf{b}}{\mathbf{d}_3 \bullet (\mathbf{A}\mathbf{d}_3)} \\
\alpha_4 &= \frac{\mathbf{d}_4 \bullet \mathbf{b}}{\mathbf{d}_4 \bullet (\mathbf{A}\mathbf{d}_4)} \\
&\vdots \\
\alpha_n &= \frac{\mathbf{d}_n \bullet \mathbf{b}}{\mathbf{d}_n \bullet (\mathbf{A}\mathbf{d}_n)}.
\end{aligned} \tag{3.70}$$

In other words, thanks to **A**-orthogonality, the scaling factors are obtained from the dot product between the load and the direction. In this manner, the solution shown in Equation (3.62) simply becomes

$$\mathbf{x} = \frac{\mathbf{d}_1 \bullet \mathbf{b}}{\mathbf{d}_1 \bullet (\mathbf{A}\mathbf{d}_1)}\mathbf{d}_1 + \frac{\mathbf{d}_2 \bullet \mathbf{b}}{\mathbf{d}_2 \bullet (\mathbf{A}\mathbf{d}_2)}\mathbf{d}_2 + \frac{\mathbf{d}_3 \bullet \mathbf{b}}{\mathbf{d}_3 \bullet (\mathbf{A}\mathbf{d}_3)}\mathbf{d}_3 + \cdots + \frac{\mathbf{d}_n \bullet \mathbf{b}}{\mathbf{d}_n \bullet (\mathbf{A}\mathbf{d}_n)}\mathbf{d}_n \tag{3.71}$$

Iteration Process. Instead of obtaining all the directions at once, the displacements $\mathbf{x}$ are obtained through an iterative process. One first starts with the solution $\mathbf{x}_0$ at iteration 0. From this, one obtains the solution $\mathbf{x}_1$ at iteration 1. From $\mathbf{x}_1$ one obtains $\mathbf{x}_2$ at iteration 2, etc. At the nth iteration one obtains $\mathbf{x}_n$, which according to Equation (3.62) should be the exact solution (exact displacements).

The iterative procedure is as follows:

$$\begin{aligned}
&\mathbf{x}_0 = 0 \ \text{(or some other starting point)} \\
&\mathbf{x}_1 = \alpha_1 \mathbf{d}_1 \\
&\mathbf{x}_2 = \mathbf{x}_1 + \alpha_2 \mathbf{d}_2 \\
&\mathbf{x}_3 = \mathbf{x}_2 + \alpha_3 \mathbf{d}_3 \\
&\mathbf{x}_4 = \mathbf{x}_3 + \alpha_4 \mathbf{d}_4 \\
&\vdots \\
&\mathbf{x}_n = \mathbf{x}_{n-1} + \alpha_n \mathbf{d}_n,
\end{aligned} \tag{3.72}$$

which means that:

$$\begin{aligned}
&\mathbf{x}_0 = 0 \\
&\mathbf{x}_1 = \alpha_1 \mathbf{d}_1 \\
&\mathbf{x}_2 = \alpha_1 \mathbf{d}_1 + \alpha_2 \mathbf{d}_2 \\
&\mathbf{x}_3 = \alpha_1 \mathbf{d}_1 + \alpha_2 \mathbf{d}_2 + \alpha_3 \mathbf{d}_3 \\
&\mathbf{x}_4 = \alpha_1 \mathbf{d}_1 + \alpha_2 \mathbf{d}_2 + \alpha_3 \mathbf{d}_3 + \alpha_4 \mathbf{d}_4 \\
&\vdots \\
&\mathbf{x}_n = \alpha_1 \mathbf{d}_1 + \alpha_2 \mathbf{d}_2 + \alpha_3 \mathbf{d}_3 + \alpha_4 \mathbf{d}_4 + \cdots + \alpha_n \mathbf{d}_n
\end{aligned} \tag{3.73}$$

The unbalanced part of the load **b** at each of the above iterations is given by

$$
\begin{aligned}
&\mathbf{b}_0 = \mathbf{A}\mathbf{x}_0 = \mathbf{b} \\
&\mathbf{b}_1 = \mathbf{b} - \mathbf{A}\mathbf{x}_1 = \mathbf{b} - \alpha_1 \mathbf{A}\mathbf{d}_1 \\
&\mathbf{b}_2 = \mathbf{b} - \mathbf{A}\mathbf{x}_2 = \mathbf{b} - \alpha_1 \mathbf{A}\mathbf{d}_1 - \alpha_2 \mathbf{A}\mathbf{d}_2 \\
&\mathbf{b}_3 = \mathbf{b} - \mathbf{A}\mathbf{x}_3 = \mathbf{b} - \alpha_1 \mathbf{A}\mathbf{d}_1 - \alpha_2 \mathbf{A}\mathbf{d}_2 - \alpha_3 \mathbf{A}\mathbf{d}_3 \\
&\mathbf{b}_4 = \mathbf{b} - \mathbf{A}\mathbf{x}_4 = \mathbf{b} - \alpha_1 \mathbf{A}\mathbf{d}_1 - \alpha_2 \mathbf{A}\mathbf{d}_2 - \alpha_3 \mathbf{A}\mathbf{d}_3 - \alpha_4 \mathbf{A}\mathbf{d}_4 \\
&\vdots \\
&\mathbf{b}_n = \mathbf{b} - \mathbf{A}\mathbf{x}_n = \mathbf{b} - \alpha_1 \mathbf{A}\mathbf{d}_1 - \alpha_2 \mathbf{A}\mathbf{d}_2 - \alpha_3 \mathbf{A}\mathbf{d}_3 - \alpha_4 \mathbf{A}\mathbf{d}_4 - \cdots - \alpha_n \mathbf{A}\mathbf{d}_n = 0
\end{aligned}
\tag{3.74}
$$

One can rewrite Equation (3.74) and obtain the unbalanced load at each iteration in the following format

$$
\begin{aligned}
&\mathbf{b}_0 = \alpha_1 \mathbf{A}\mathbf{d}_1 + \alpha_2 \mathbf{A}\mathbf{d}_2 + \alpha_3 \mathbf{A}\mathbf{d}_3 + \alpha_4 \mathbf{A}\mathbf{d}_4 + \cdots + \alpha_n \mathbf{A}\mathbf{d}_n \\
&\mathbf{b}_1 = \alpha_2 \mathbf{A}\mathbf{d}_2 + \alpha_3 \mathbf{A}\mathbf{d}_3 + \alpha_4 \mathbf{A}\mathbf{d}_4 + \cdots + \alpha_n \mathbf{A}\mathbf{d}_n \\
&\mathbf{b}_2 = \alpha_3 \mathbf{A}\mathbf{d}_3 + \alpha_4 \mathbf{A}\mathbf{d}_4 + \cdots + \alpha_n \mathbf{A}\mathbf{d}_n \\
&\mathbf{b}_3 = \alpha_4 \mathbf{A}\mathbf{d}_4 + \cdots + \alpha_n \mathbf{A}\mathbf{d}_n \\
&\vdots \\
&\mathbf{b}_n = 0
\end{aligned}
\tag{3.75}
$$

The unbalanced parts of the column vector **b** after each iteration are called residual loads or simply residuals. These residuals represent the portion of the column matrix **b** (load) that has not yet been balanced by the corresponding approximate solution **x**.

The iteration process described above is relatively simple provided that, at each iteration, one knows the direction in which to move in n-D space. With the Gauss-Seidel method, finding this direction was trivial. With the conjugate directions method, a more elaborate procedure is needed. In order to better understand this procedure, it is important to study some key properties of both directions and residuals.

The key logical steps in deriving (and understanding) the method of conjugate directions are as follows:

1. Start with the system of n linear algebraic equations with n unknowns

$$\mathbf{A}\mathbf{x} = \mathbf{b} \tag{3.76}$$

   (where **A** is a positive definite matrix, **x** is a column matrix called displacement and **b** is a column matrix called load), and assume the solution **x** in a form

$$\mathbf{x} = \alpha_1 \mathbf{d}_1 + \alpha_2 \mathbf{d}_2 + \alpha_3 \mathbf{d}_3 + \cdots + \alpha_n \mathbf{d}_n. \tag{3.77}$$

   When substituted into equation (3.76) this produces

$$\mathbf{b} = \alpha_1 \mathbf{A}\mathbf{d}_1 + \alpha_2 \mathbf{A}\mathbf{d}_2 + \alpha_3 \mathbf{A}\mathbf{d}_3 + \cdots + \alpha_n \mathbf{A}\mathbf{d}_n, \tag{3.78}$$

   which means that the column matrix **x** is represented as a linear combination of column matrices $\mathbf{d}_1$, $\mathbf{d}_2$, $\mathbf{d}_3$, …, $\mathbf{d}_n$. These matrices are called directions (at each step of the solution they guide the hyper-direction to the next step). Each of these directions is scaled by the scaling factor α and the scaled directions are simply added together to obtain the solution **x**.
2. Assume directions to be mutually **A**-orthogonal (**A**-conjugate) column matrices

$$\begin{aligned}&\mathbf{d}_i\mathbf{A}\mathbf{d}_i>0 \quad \forall i=1,2,3,\ldots,n\\&\mathbf{d}_i\mathbf{A}\mathbf{d}_j>0 \quad \forall i\neq j\end{aligned} \tag{3.79}$$

and use **A**-orthogonality properties to calculate the scaling factors:

$$\begin{aligned}&\mathbf{d}_1\bullet\mathbf{b}=\alpha_1\mathbf{d}_1\bullet(\mathbf{A}\mathbf{b}_1)\\&\mathbf{d}_2\bullet\mathbf{b}=\alpha_2\mathbf{d}_2\bullet(\mathbf{A}\mathbf{b}_2)\\&\mathbf{d}_3\bullet\mathbf{b}=\alpha_3\mathbf{d}_3\bullet(\mathbf{A}\mathbf{b}_3)\\&\vdots\\&\mathbf{d}_n\bullet\mathbf{b}=\alpha_n\mathbf{d}_n\bullet(\mathbf{A}\mathbf{b}_n),\end{aligned} \tag{3.80}$$

which yields the scaling factors

$$\begin{aligned}\alpha_1&=\frac{\mathbf{d}_1\bullet\mathbf{b}}{\mathbf{d}_1\bullet(\mathbf{A}\mathbf{b}_1)}\\\alpha_2&=\frac{\mathbf{d}_2\bullet\mathbf{b}}{\mathbf{d}_2\bullet(\mathbf{A}\mathbf{b}_2)}\\\alpha_3&=\frac{\mathbf{d}_3\bullet\mathbf{b}}{\mathbf{d}_3\bullet(\mathbf{A}\mathbf{b}_3)}\\&\vdots\\\alpha_n&=\frac{\mathbf{d}_n\bullet\mathbf{b}}{\mathbf{d}_n\bullet(\mathbf{A}\mathbf{b}_n)}.\end{aligned} \tag{3.81}$$

3. Apply the solver as an iterative process with residuals and displacements obtained after each iteration as shown in Table 3.1 and Table 3.2.
4. Calculate the directions as illustrated in Figure 3.14.

Table 3.1 Displacements and residuals at each iteration

Iteration	Displacement	Residual
0	$\mathbf{x}_0=0$	$\mathbf{b}_0=\mathbf{b}$
1	$\mathbf{x}_1=\alpha_1\mathbf{d}_1$	$\mathbf{b}_1=\mathbf{b}-\alpha_1\mathbf{A}\mathbf{d}_1$ $=\alpha_2\mathbf{A}\mathbf{d}_2+\alpha_3\mathbf{A}\mathbf{d}_3+\cdots+\alpha_n\mathbf{A}\mathbf{d}_n$
2	$\mathbf{x}_2=\alpha_1\mathbf{d}_1+\alpha_2\mathbf{d}_2$	$\mathbf{b}_2=\mathbf{b}-\alpha_1\mathbf{A}\mathbf{d}_1-\alpha_2\mathbf{A}\mathbf{d}_2$ $=\alpha_3\mathbf{A}\mathbf{d}_3+\alpha_4\mathbf{A}\mathbf{d}_4+\cdots+\alpha_n\mathbf{A}\mathbf{d}_n$
3	$\mathbf{x}_3=\alpha_1\mathbf{d}_1+\alpha_2\mathbf{d}_2+\alpha_3\mathbf{d}_3$	$\mathbf{b}_3=\mathbf{b}-\alpha_1\mathbf{A}\mathbf{d}_1-\alpha_2\mathbf{A}\mathbf{d}_2-\alpha_3\mathbf{A}\mathbf{d}_3$ $=\alpha_4\mathbf{A}\mathbf{d}_4+\alpha_5\mathbf{A}\mathbf{d}_5+\cdots+\alpha_n\mathbf{A}\mathbf{d}_n$
$\vdots$	$\vdots$	$\vdots$
k	$\mathbf{x}_k=\alpha_1\mathbf{d}_1+\alpha_2\mathbf{d}_2+\alpha_3\mathbf{d}_3+\cdots+\alpha_k\mathbf{d}_k$	$\mathbf{b}_k=\mathbf{b}-\alpha_1\mathbf{A}\mathbf{d}_1-\alpha_2\mathbf{A}\mathbf{d}_2-\alpha_3\mathbf{A}\mathbf{d}_3-\cdots-\alpha_k\mathbf{A}\mathbf{d}_k$ $=\alpha_{k+1}\mathbf{A}\mathbf{d}_{k+1}+\alpha_{k+2}\mathbf{A}\mathbf{d}_{k+2}+\cdots+\alpha_n\mathbf{A}\mathbf{d}_n$
$\vdots$	$\vdots$	$\vdots$
n	$\mathbf{x}_n=\alpha_1\mathbf{d}_1+\alpha_2\mathbf{d}_2+\alpha_3\mathbf{d}_3+\cdots+\alpha_k\mathbf{d}_k+\cdots\alpha_n\mathbf{d}_n$	$\mathbf{b}_n=\mathbf{b}-\alpha_1\mathbf{A}\mathbf{d}_1-\alpha_2\mathbf{A}\mathbf{d}_2-\alpha_3\mathbf{A}\mathbf{d}_3-\cdots-\alpha_n\mathbf{A}\mathbf{d}_n$ $=0$

Table 3.2 An alternative way of showing residuals at each iteration

Iteration	$\mathbf{x}_k$	$\mathbf{b}_k$
0		$\alpha_1\mathbf{Ad}_1 + \alpha_2\mathbf{Ad}_2 + \alpha_3\mathbf{Ad}_3 + \alpha_4\mathbf{Ad}_4 + \cdots + \alpha_k\mathbf{Ad}_k + \cdots + \alpha_n\mathbf{Ad}_n$
1	$\alpha_1\mathbf{d}_1$	$\alpha_2\mathbf{Ad}_2 + \alpha_3\mathbf{Ad}_3 + \alpha_4\mathbf{Ad}_4 + \cdots + \alpha_k\mathbf{Ad}_k + \cdots + \alpha_n\mathbf{Ad}_n$
2	$\alpha_1\mathbf{d}_1 + \alpha_2\mathbf{d}_2$	$\alpha_3\mathbf{Ad}_3 + \alpha_4\mathbf{Ad}_4 + \cdots + \alpha_k\mathbf{Ad}_k + \cdots + \alpha_n\mathbf{Ad}_n$
3	$\alpha_1\mathbf{d}_1 + \alpha_2\mathbf{d}_2 + \alpha_3\mathbf{d}_3$	$\alpha_4\mathbf{Ad}_4 + \cdots + \alpha_k\mathbf{Ad}_k + \cdots + \alpha_n\mathbf{Ad}_n$
$\vdots$	$\vdots$	$\vdots$
k	$\alpha_1\mathbf{d}_1 + \alpha_2\mathbf{d}_2 + \alpha_3\mathbf{d}_3 + \cdots + \alpha_k\mathbf{d}_k$	$\alpha_{k+1}\mathbf{Ad}_{k+1} + \cdots + \alpha_n\mathbf{Ad}_n$
$\vdots$	$\vdots$	$\vdots$
n	$\alpha_1\mathbf{d}_1 + \alpha_2\mathbf{d}_2 + \alpha_3\mathbf{d}_3 + \cdots + \alpha_k\mathbf{d}_k + \cdots + \alpha_n\mathbf{d}_n$	0

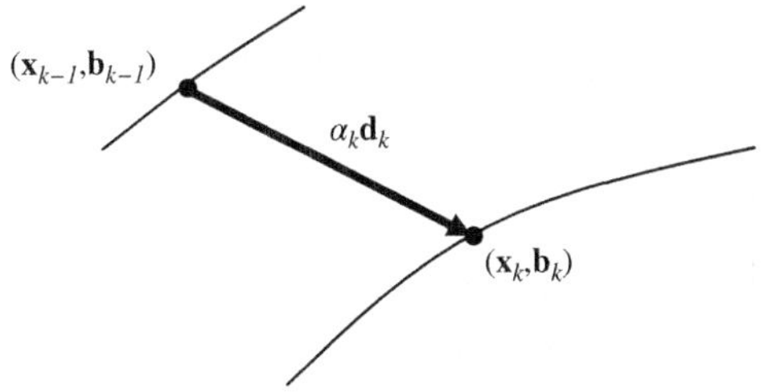

Figure 3.14 Illustration of the iteration process: One "arrives" at the solution $\mathbf{x}_k$ by "travelling" from the solution $\mathbf{x}_{k-1}$ in the direction $\mathbf{d}_k$. As one progresses on the "way" from $\mathbf{x}_{k-1}$ to $\mathbf{x}_k$, the residual continuously reduces. Thus, after arriving at $\mathbf{x}_k$ the residual reduces to $\mathbf{b}_k$

At this stage, it is worth observing (from Table 3.3 and the **A**-orthogonality condition) that the dot product between directions and residuals at each iteration are as follows:

$$
\begin{array}{lllll}
\mathbf{d}_1 \bullet \mathbf{b}_1 = 0 & & & & \\
\mathbf{d}_1 \bullet \mathbf{b}_2 = 0 & \mathbf{d}_2 \bullet \mathbf{b}_2 = 0 & & & \\
\mathbf{d}_1 \bullet \mathbf{b}_3 = 0 & \mathbf{d}_2 \bullet \mathbf{b}_3 = 0 & & & \\
\vdots & \vdots & \ddots & & \\
\mathbf{d}_1 \bullet \mathbf{b}_k = 0 & \mathbf{d}_2 \bullet \mathbf{b}_k = 0 & \ldots & \mathbf{d}_k \bullet \mathbf{b}_k = 0 & \\
\vdots & \vdots & \vdots & \vdots & \ddots \\
\mathbf{d}_1 \bullet \mathbf{b}_n = 0 & \mathbf{d}_2 \bullet \mathbf{b}_n = 0 & \ldots & \mathbf{d}_k \bullet \mathbf{b}_n = 0 & \ldots \quad \mathbf{d}_n \bullet \mathbf{b}_n = 0
\end{array}
\tag{3.82}
$$

In other words, a given direction $\mathbf{d}_k$ is orthogonal to all residuals $\mathbf{b}_j$ where $j \geq k$. From this it follows that a given residual $\mathbf{b}_j$ is orthogonal to all directions $\mathbf{b}_k$ where $k \leq j$.

One can say that the residual at the current iteration k is orthogonal to all directions $\mathbf{d}_j$ already used; all directions already used are orthogonal to all future residuals.

Although this sounds complicated, it is a direct result of using only the directions that are **A**-orthogonal. All the zero dot products are conveniently shown by the following matrix

Table 3.3 The directions at each step of the iteration process

Iteration	Direction $\mathbf{d}_k$	A-Orthogonality Condition $\mathbf{d}_k \bullet (\mathbf{A}\mathbf{d}_{k-1})$	β_k obtained from A-Orthogonality Condition
1	$\mathbf{d}_1 = \mathbf{b}_0 = \mathbf{b}$		
2	$\mathbf{d}_2 = \mathbf{b}_1 - \beta_2 \mathbf{d}_1$	$\mathbf{b}_1 \bullet (\mathbf{A}\mathbf{d}_1) - \beta_2 \mathbf{d}_1 \bullet (\mathbf{A}\mathbf{d}_1)$	$\dfrac{\mathbf{b}_1 \bullet (\mathbf{A}\mathbf{d}_1)}{\mathbf{d}_1 \bullet (\mathbf{A}\mathbf{d}_1)}$
3	$\mathbf{d}_3 = \mathbf{b}_2 - \beta_3 \mathbf{d}_2$	$\mathbf{b}_2 \bullet (\mathbf{A}\mathbf{d}_2) - \beta_3 \mathbf{d}_2 \bullet (\mathbf{A}\mathbf{d}_2)$	$\dfrac{\mathbf{b}_2 \bullet (\mathbf{A}\mathbf{d}_2)}{\mathbf{d}_2 \bullet (\mathbf{A}\mathbf{d}_2)}$
4	$\mathbf{d}_4 = \mathbf{b}_3 - \beta_4 \mathbf{d}_3$	$\mathbf{b}_3 \bullet (\mathbf{A}\mathbf{d}_3) - \beta_4 \mathbf{d}_3 \bullet (\mathbf{A}\mathbf{d}_3)$	$\dfrac{\mathbf{b}_3 \bullet (\mathbf{A}\mathbf{d}_3)}{\mathbf{d}_3 \bullet (\mathbf{A}\mathbf{d}_3)}$
$\vdots$	$\vdots$	$\vdots$	$\vdots$
k	$\mathbf{d}_k = \mathbf{b}_{k-1} - \beta_k \mathbf{d}_{k-1}$	$\mathbf{b}_{k-1} \bullet (\mathbf{A}\mathbf{d}_{k-1}) - \beta_k \mathbf{d}_{k-1} \bullet (\mathbf{A}\mathbf{d}_{k-1})$	$\dfrac{\mathbf{b}_{k-1} \bullet (\mathbf{A}\mathbf{d}_{k-1})}{\mathbf{d}_{k-1} \bullet (\mathbf{A}\mathbf{d}_{k-1})}$
$\vdots$	$\vdots$	$\vdots$	$\vdots$
n	$\mathbf{d}_n = \mathbf{b}_{n-1} - \beta_n \mathbf{d}_{n-1}$	$\mathbf{b}_{n-1} \bullet (\mathbf{A}\mathbf{d}_{n-1}) - \beta_n \mathbf{d}_{n-1} \bullet (\mathbf{A}\mathbf{d}_{n-1})$	$\dfrac{\mathbf{b}_{n-1} \bullet (\mathbf{A}\mathbf{d}_{n-1})}{\mathbf{d}_{n-1} \bullet (\mathbf{A}\mathbf{d}_{n-1})}$

$$\begin{bmatrix} \mathbf{d}_1 \bullet \mathbf{b}_1 & & & & & & \\ \mathbf{d}_1 \bullet \mathbf{b}_2 & \mathbf{d}_2 \bullet \mathbf{b}_2 & & & & & \\ \mathbf{d}_1 \bullet \mathbf{b}_3 & \mathbf{d}_2 \bullet \mathbf{b}_3 & \mathbf{d}_3 \bullet \mathbf{b}_3 & & & & \\ \vdots & \vdots & \vdots & \ddots & & & \\ \mathbf{d}_1 \bullet \mathbf{b}_k & \mathbf{d}_2 \bullet \mathbf{b}_k & \mathbf{d}_3 \bullet \mathbf{b}_k & \dots & \mathbf{d}_k \bullet \mathbf{b}_k & & \\ \vdots & \vdots & \vdots & \vdots & \vdots & \ddots & \\ \mathbf{d}_1 \bullet \mathbf{b}_n & \mathbf{d}_2 \bullet \mathbf{b}_n & \mathbf{d}_3 \bullet \mathbf{b}_n & \dots & \mathbf{d}_k \bullet \mathbf{b}_n & \dots & \mathbf{d}_n \bullet \mathbf{b}_n \end{bmatrix} \tag{3.83}$$

The columns of the above matrix signify that a given direction is orthogonal to all the residuals not reached yet. The rows of the above matrix signify that a given residual is orthogonal to all the directions already used.

At each iteration k, the direction is conveniently calculated using the following formula

$$\begin{aligned} \mathbf{d}_1 &= -\mathbf{b}_0 \\ \mathbf{d}_2 &= -\mathbf{b}_1 + \beta_{2,1}\mathbf{d}_1 \\ \mathbf{d}_3 &= -\mathbf{b}_2 + \beta_{3,2}\mathbf{d}_2 + \beta_{3,1}\mathbf{d}_1 \\ &\vdots \\ \mathbf{d}_k &= -\mathbf{b}_k + \beta_{k,k-1}\mathbf{d}_{k-1} + \beta_{k,k-2}\mathbf{d}_{k-2} + \cdots + \beta_{k,1}\mathbf{d}_1 \end{aligned} \tag{3.84}$$

By substituting these into Equation (3.83) one obtains

$$\begin{bmatrix} \mathbf{b}_0 \bullet \mathbf{b}_1 & & & & \\ \mathbf{b}_0 \bullet \mathbf{b}_2 & \mathbf{b}_1 \bullet \mathbf{b}_2 & & & \\ \mathbf{b}_0 \bullet \mathbf{b}_3 & \mathbf{b}_1 \bullet \mathbf{b}_3 & \mathbf{b}_2 \bullet \mathbf{b}_3 & & \\ \vdots & \vdots & \vdots & \ddots & \\ \mathbf{b}_0 \bullet \mathbf{b}_n & \mathbf{b}_1 \bullet \mathbf{b}_n & \mathbf{b}_2 \bullet \mathbf{b}_n & \dots & \mathbf{b}_n \bullet \mathbf{b}_n \end{bmatrix}. \tag{3.85}$$

Thus, thanks to using formula (3.84) for calculating directions, the residuals are orthogonal to each other, which simply means that

$$\mathbf{b}_i \bullet \mathbf{b}_j = 0 \text{ for } i \neq j. \tag{3.86}$$

In order to exploit the above orthogonality, the directions are calculated in the following sequence:

a. The direction $\mathbf{d}_1$

$$\mathbf{d}_1 = -\mathbf{b}_0 = -\mathbf{b}. \tag{3.87}$$

b. The direction $\mathbf{d}_2$

$$\mathbf{d}_2 = -\mathbf{b}_1 + \beta_{2,1}\mathbf{d}_1. \tag{3.88}$$

As the direction $\mathbf{d}_2$ has to be $\mathbf{A}$-orthogonal to the direction $\mathbf{d}_1$, i.e.,

$$(-\mathbf{b}_1 + \beta_{2,1}\mathbf{d}_1) \bullet (\mathbf{A}\mathbf{d}_1) = 0. \tag{3.89}$$

It follows that

$$\beta_{2,1} = \beta_2 = \frac{\mathbf{b}_1 \bullet (\mathbf{A}\mathbf{d}_1)}{\mathbf{d}_1 \bullet (\mathbf{A}\mathbf{d}_1)}. \tag{3.90}$$

c. The direction $\mathbf{d}_3$

$$\mathbf{d}_3 = -\mathbf{b}_2 + \beta_{3,2}\mathbf{d}_2 + \beta_{3,1}\mathbf{d}_1. \tag{3.91}$$

This direction has to be $\mathbf{A}$-orthogonal to both directions $\mathbf{d}_1$ and $\mathbf{d}_2$, i.e.,

$$(-\mathbf{b}_2 + \beta_{3,2}\mathbf{d}_2 + \beta_{3,1}\mathbf{d}_1) \bullet (\mathbf{A}\mathbf{d}_2) = 0 \tag{3.92}$$

$$(-\mathbf{b}_2 + \beta_{3,2}\mathbf{d}_2 + \beta_{3,1}\mathbf{d}_1) \bullet (\mathbf{A}\mathbf{d}_1) = 0. \tag{3.93}$$

Since

$$\mathbf{d}_2 \bullet (\mathbf{A}\mathbf{d}_1) = 0 \text{ and } \mathbf{d}_1 \bullet (\mathbf{A}\mathbf{d}_2) = 0, \tag{3.94}$$

it follows that

$$-\mathbf{b}_2 \bullet (\mathbf{A}\mathbf{d}_2) + \beta_{3,2}\mathbf{d}_2 \bullet (\mathbf{A}\mathbf{d}_2) = 0 \tag{3.95}$$

$$-\mathbf{b}_2 \bullet (\mathbf{A}\mathbf{d}_1) + \beta_{3,1}\mathbf{d}_1 \bullet (\mathbf{A}\mathbf{d}_1) = 0, \tag{3.96}$$

which yields

$$\beta_{3,2} = \beta_3 = \frac{\mathbf{b}_2 \bullet (\mathbf{A}\mathbf{d}_2)}{\mathbf{d}_2 \bullet (\mathbf{A}\mathbf{d}_2)} \quad \text{and} \quad \beta_{3,1} = \frac{\mathbf{b}_2 \bullet (\mathbf{A}\mathbf{d}_1)}{\mathbf{d}_1 \bullet (\mathbf{A}\mathbf{d}_1)} \tag{3.97}$$

and since

$$\mathbf{A}\mathbf{d}_1 = \frac{1}{\alpha_1}(\mathbf{b}_0 - \mathbf{b}_1), \tag{3.98}$$

one obtains

$$\beta_{3,1} = \frac{1}{\alpha_1}\frac{\mathbf{b}_2\bullet(\mathbf{b}_0-\mathbf{b}_1)}{\mathbf{d}_1\bullet(\mathbf{A}\mathbf{d}_1)}. \tag{3.99}$$

But because $\mathbf{b}_0$ and $\mathbf{b}_1$ are orthogonal to all $\mathbf{b}_2$ it follows that

$$\beta_{3,1} = 0. \tag{3.100}$$

By following the above procedure for $\mathbf{d}_4$, $\mathbf{d}_5$, $\mathbf{d}_6$, …, $\mathbf{d}_k$, …, $\mathbf{d}_n$ one obtains

$$\mathbf{d}_k = -\mathbf{b}_{k-1} + \beta_{k,k-1}\mathbf{d}_{k-1} = -\mathbf{b}_{k-1} + \beta_k\mathbf{d}_{k-1}, \tag{3.101}$$

where

$$\beta_k = \frac{\mathbf{b}_{k-1}\bullet(\mathbf{A}\mathbf{d}_{k-1})}{\mathbf{d}_{k-1}\bullet(\mathbf{A}\mathbf{d}_{k-1})}. \tag{3.102}$$

The directions obtained at each iteration using these formulas are shown in Table 3.3.
It is worth mentioning that this is not the only formula for β_k; due to all the orthogonalities explained before, β_k can be calculated as

$$\beta_k = -\frac{1}{\alpha_{k-1}}\frac{\mathbf{b}_{k-1}\bullet\mathbf{b}_{k-1}}{\mathbf{d}_{k-1}\bullet(\mathbf{A}\mathbf{d}_{k-1})}. \tag{3.103}$$

Also, since

$$\mathbf{A}\mathbf{d}_{k-1} = (\mathbf{b}_{k-1} - \mathbf{b}_k)\frac{1}{\alpha_{k-1}}, \tag{3.104}$$

it follows that

$$\beta_k = -\frac{\mathbf{b}_{k-1}\bullet\mathbf{b}_{k-1}}{\mathbf{d}_{k-1}\bullet(\mathbf{b}_{k-1}-\mathbf{b}_k)}. \tag{3.105}$$

In essence, there are many ways to calculate β_k; some alternative expressions are given in Table 3.4.

Solving a System of Nonlinear Algebraic Equations. A system of nonlinear algebraic equations

$$\mathbf{A}\mathbf{x} = \mathbf{b} \tag{3.106}$$

is characterized by the matrix **A** that is a function of **x**, i.e., each of the components of the matrix **A** are a function of **x**. In actual fact, in the nonlinear finite element method considered in this book, the matrix **A** is not even assembled and the following equation is arrived at

$$\mathbf{f}(\mathbf{x}) = \mathbf{b}, \tag{3.107}$$

Table 3.4 Alternative ways to calculate coefficient β_k

$\beta_k = \beta_{k,k-1} = \dfrac{\mathbf{b}_{k-1} \bullet (\mathbf{A}\mathbf{d}_{k-1})}{\mathbf{d}_{k-1} \bullet (\mathbf{A}\mathbf{d}_{k-1})}$	(3.108)
$\beta_k = \dfrac{\mathbf{b}_{k-1} \bullet \frac{1}{\alpha_{k-1}} (\mathbf{b}_{k-1} - \mathbf{b}_{k-2})}{\mathbf{d}_{k-1} \bullet (\mathbf{A}\mathbf{b}_{k-1})}$	(3.109)
$\beta_k = \dfrac{\mathbf{b}_{k-1} \bullet \mathbf{b}_{k-1}}{\mathbf{d}_{k-1} \bullet (\mathbf{A}\mathbf{b}_{k-1})} \dfrac{1}{\alpha_{k-1}}$	(3.110)
$\beta_k = \dfrac{\mathbf{b}_{k-1} \bullet \mathbf{b}_{k-1}}{\mathbf{d}_{k-1} \bullet (\mathbf{A}\mathbf{b}_{k-1})} \dfrac{1}{\dfrac{\mathbf{d}_{k-1} \bullet \mathbf{b}_{k-1}}{\mathbf{d}_{k-1} \bullet (\mathbf{A}\mathbf{d}_{k-1})}}$	(3.111)
$\beta_k = -\dfrac{\mathbf{b}_{k-1} \bullet \mathbf{b}_{k-1}}{\mathbf{d}_{k-1} \bullet \mathbf{b}_{k-1}}$	(3.112)
$\beta_k = -\dfrac{\mathbf{b}_{k-1} \bullet \mathbf{b}_{k-1}}{(-\mathbf{b}_{k-2} + \beta_{k-1}, (k-2) \mathbf{d}_{k-2}) \bullet \mathbf{b}_{k-1}}$	(3.113)
$\beta_k = \dfrac{\mathbf{b}_{k-1} \bullet \mathbf{b}_{k-1}}{\mathbf{b}_{k-2} \bullet \mathbf{b}_{k-2}}$	(3.114)
$\beta_k = \dfrac{\mathbf{b}_{k-1} \bullet (\mathbf{b}_{k-1} - \mathbf{b}_{k-2})}{\mathbf{d}_{k-1} \bullet (\mathbf{b}_{k-1} - \mathbf{b}_{k-2})}$	(3.115)

which can be written as

$$\begin{bmatrix} f_1(\mathbf{x}_1, \mathbf{x}_2, \mathbf{x}_3, \ldots, \mathbf{x}_n) \\ f_2(\mathbf{x}_1, \mathbf{x}_2, \mathbf{x}_3, \ldots, \mathbf{x}_n) \\ f_3(\mathbf{x}_1, \mathbf{x}_2, \mathbf{x}_3, \ldots, \mathbf{x}_n) \\ \vdots \\ f_n(\mathbf{x}_1, \mathbf{x}_2, \mathbf{x}_3, \ldots, \mathbf{x}_n) \end{bmatrix} = \begin{bmatrix} b_1 \\ b_2 \\ b_3 \\ \vdots \\ b_n \end{bmatrix}. \tag{3.116}$$

The column matrix **b** can be called "external loads" and the column matrix **f** can be called "internal forces," while the column matrix **x** can be called displacements. With this approach, the internal forces are obtained from displacements and should be equal to the external loads.

After arriving to $\mathbf{x}_{k-1}$ by following the direction $\mathbf{d}_{k-1}$, the solution procedure for moving from $\mathbf{x}_{k-1}$ to $\mathbf{x}_k$ is as follows:

1. Calculate the residual $\mathbf{b}_{k-1} = \mathbf{b} - f(\mathbf{x}_{k-1})$
2. Obtain the direction $\mathbf{d}_k = \mathbf{b}_{k-1} + \beta_k \mathbf{d}_{k-1}$
3. Calculate $\mathbf{x}_k = \alpha_k \mathbf{d}_k$

The first question that arises is how should β_k be calculated. There is a whole menu of formulas for this as given in Table 3.4. The second question that arises is how to calculate α_k. Unfortunately, there is no formulas or an easy way to calculate α_k.

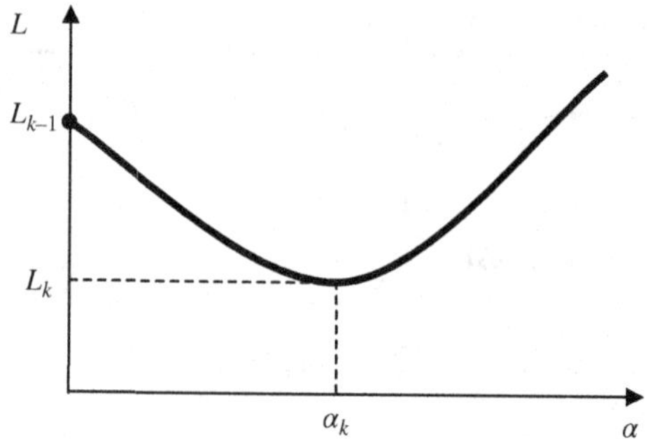

Figure 3.15 Obtaining L_k by making the residual as small as possible

One way to proceed, however, is to calculate the length of the residual $\mathbf{b}_k$

$$L_k = \mathbf{b}_k \bullet \mathbf{b}_k \tag{3.117}$$

and move in the $\mathbf{d}_k$ direction as long as L_k keeps decreasing, see Figure 3.15. This process is equivalent to finding a minimum of a function of one variable

$$L = \mathrm{L}(\alpha) \tag{3.118}$$

Instability and Preconditioning. The conjugate directions method will converge in n-steps for a linear system with a well conditioned matrix **A**. For matrices with a large condition number, the method does not converge in n-steps. This occurs because the process for obtaining conjugate directions is progressively sensitive to rounding errors such that small rounding errors will "derail" the process completely before $\mathbf{d}_n$ is reached. For small problems (small n) this does not happen. For large n, this always happens. One can compare this situation to the following problem:

Start with a real number s and do the loop:

```
s = 1.0
LOOP
   s_k = s_{k-1} s
END LOOP
For s = 1 after 100 steps one obtains s_100 = 1.
For s = 1.01 after 10 steps one obtains s_100 = (1.01)^100 = 2.70
```

In other words, the small initial error in s has resulted in a significant error in s_{100}. In a sense, this is similar to compound interest rates: If two people start with an initial sum of 1 dollar each, at interest rates 2% and 10% respectively, the second person will, after 300 years, be much richer than the first person. Now, compare this to the fact that finite element simulations routinely result in systems of one million or even one billion equations, which means one million or one billion increments (compare this to 300 interest increments – 300 years), which clearly illustrates the role of the rounding error.

Although the conjugate directions method should theoretically achieve the correct result after n steps, for the above reasons it does not. As such, it is considered an iterative method and is terminated only when a desired accuracy is achieved. In order to shorten the number of iterations, the matrix **A** is sometimes multiplied by a diagonal matrix **P** such that

$$\mathbf{P}^{-1}(\mathbf{A}\mathbf{x} - \mathbf{b}) = 0 \tag{3.119}$$

is solved. This results in

$$\mathbf{P}^{-1}\mathbf{A}\mathbf{x} = \mathbf{P}^{-1}\mathbf{b} \tag{3.120}$$

or

$$\bar{\mathbf{A}}\mathbf{x} = \bar{\mathbf{b}}, \tag{3.121}$$

where the condition number of $\bar{\mathbf{A}}$ is smaller than the condition number of $\mathbf{A}$. For nonlinear systems, the preconditioner $\mathbf{P}$ may change as the iteration proceeds. This is called flexible preconditioning.

In order to understand how pre-conditioning works, it is important to understand why it is needed in the first place. It is needed primarily because the process of obtaining $\mathbf{A}$-orthogonal directions is very sensitive to small perturbations. This is termed instability. At each step the residual is calculated as

$$\mathbf{b}_0 = \mathbf{b} - \mathbf{A}\mathbf{x} = \mathbf{b}. \tag{3.122}$$

Now the next direction $\mathbf{d}_1$ contains $\mathbf{b}_0$ as its component. The next displacement $\mathbf{x}_1$ in turn contains $\mathbf{d}_1$ as its component. This means that the next displacement $\mathbf{x}_1$ contains $\mathbf{b}_0$ as its component. This leads to the conclusion that the next residual $\mathbf{b}_1$ contains $\mathbf{A}\mathbf{b}_0$ as its component, and the direction $\mathbf{d}_2$ contains $\mathbf{b}_1$ and so does $\mathbf{x}_2$, which means that the residual $\mathbf{b}_2$ contains $\mathbf{A}(\mathbf{A}\mathbf{b}_0)$ as its component.

By similar reasoning the residual $\mathbf{b}_3$ contains $\mathbf{A}\mathbf{b}_2$ as its component, which means that it contains $\mathbf{A}(\mathbf{A}(\mathbf{A}\mathbf{b}_0))$. By the time k iterations are reached, one has $\mathbf{A}^k\mathbf{b}_0$ being an integral part of the direction $\mathbf{d}_n$. The best way to understand this multiplication is to use eigenvalues of the matrix $\mathbf{A}$, which means that

$$\mathbf{b}_0 = \mathbf{v}_1 + \mathbf{v}_2 + \mathbf{v}_3 + \cdots + \mathbf{v}_n, \tag{3.123}$$

where

$$\begin{aligned}
&\mathbf{A}\mathbf{v}_1 = \lambda_1\mathbf{v}_1 \quad \lambda_1 < \lambda_2\\
&\mathbf{A}\mathbf{v}_2 = \lambda_2\mathbf{v}_2 \quad \lambda_2 < \lambda_3\\
&\mathbf{A}\mathbf{v}_3 = \lambda_3\mathbf{v}_3 \quad \lambda_3 < \lambda_4\\
&\vdots\\
&\mathbf{A}\mathbf{v}_n = \lambda_n\mathbf{v}_n \quad \lambda_{\text{n-1}} < \lambda_n.
\end{aligned} \tag{3.124}$$

In a similar manner

$$\begin{aligned}
\mathbf{A}^k\mathbf{b}_0 &= \mathbf{A}^k(\mathbf{v}_1 + \mathbf{v}_2 + \mathbf{v}_3 + \cdots + \mathbf{v}_n)\\
&= \lambda_1^k\mathbf{v}_1 + \lambda_2^k\mathbf{v}_2 + \lambda_3^k\mathbf{v}_3 + \cdots + \lambda_n^k\mathbf{v}_n.
\end{aligned} \tag{3.125}$$

By using

$$\bar{\mathbf{A}} = \frac{1}{\lambda_n}\mathbf{A} \quad \text{instead of } \mathbf{A}, \tag{3.126}$$

one obtains the following equation instead of Equation (3.125)

$$\bar{\mathbf{A}}^k\mathbf{b}_0 = \bar{\mathbf{A}}^k(\mathbf{v}_1 + \mathbf{v}_2 + \mathbf{v}_3 + \cdots + \mathbf{v}_n)$$
$$= \left(\frac{\lambda_1}{\lambda_n}\right)^k \mathbf{v}_1 + \left(\frac{\lambda_2}{\lambda_n}\right)^k \mathbf{v}_2 + \left(\frac{\lambda_3}{\lambda_n}\right)^k \mathbf{v}_3 + \cdots + \left(\frac{\lambda_n}{\lambda_n}\right)^k \mathbf{v}_n. \quad (3.127)$$

From Equation (3.124) it follows that

$$\frac{\lambda_1}{\lambda_n} < \frac{\lambda_2}{\lambda_n} < \frac{\lambda_3}{\lambda_n} < \cdots < \frac{\lambda_n}{\lambda_n} \; ; \; \frac{\lambda_n}{\lambda_n} = 1. \quad (3.128)$$

This means that after k iterations (for sufficiently large k) all terms converge to zero, while the term

$$\left(\frac{\lambda_n}{\lambda_n}\right)^k \quad (3.129)$$

remains equal to 1. The result is that the direction $\mathbf{d}_k$ is dominated by $\mathbf{v}_n$ (is "derailed" towards $\mathbf{v}_n$).

This kind of instability is not different from the instability one encounters with the central difference time integration scheme: for large time steps the scheme becomes unstable resulting in high frequency components that grow exponentially (much like compound interests) from one step to the next.

It is worth mentioning that with extremely large systems of equations, not even Gaussian elimination is immune to the rounding errors and to the amplification of the higher frequency modes. As a consequence, iterative smoothing procedures have to be employed even after the Gaussian elimination process.

Conjugate Gradient Method. Consider the quadratic form

$$\mathrm{f}(\mathbf{x}) = \frac{1}{2}\mathbf{x}^{\mathrm{T}}\mathbf{A}\mathbf{x} - \mathbf{x}^{\mathrm{T}}\mathbf{b}, \quad (3.130)$$

where $\mathbf{A}$ is a positive definite symmetric matrix. The minimum for this function is found at the point where the slope of the surface $\mathrm{f}(\mathbf{x})$ is zero, which yields a system of equations

$$\mathbf{A}\mathbf{x} - \mathbf{b} = 0. \quad (3.131)$$

In other words, the minimum of function $\mathrm{f}(\mathbf{x})$ and the solution of the system of equations shown in Equation (3.131) coincide. Therefore, instead of proceeding by trying to solve the system of equations one tries to find the minimum of function f. This can be done by using conjugate column matrices that represent individual directions as shown in Figure 3.16.

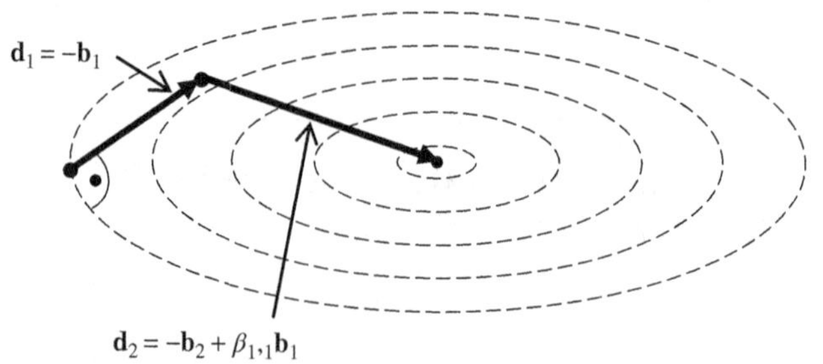

Figure 3.16 The minimum of a quadratic form using the conjugate gradient method

This method is called the conjugate gradient method. When solving a system of equations, the method is the same as the conjugate directions method. However, it is readily used in other applications as well. The advantage of this approach is that the iteration process is easily visualized as shown in Figure 3.16. Nevertheless, this visualization is only possible in 2D and can be achieved to some extent in 3D, which means that it is only helpful for a system of two linear algebraic equations with two unknowns. Jumping to a 4 by 4 system or to a 10 000 by 10 000 system is not straightforward.

3.4 Summary

In this chapter, it has been proven that for a system of linear algebraic equations

$$\mathbf{Ax} = \mathbf{b} \tag{3.132}$$

(where **A** is a positive definite matrix), a solver equivalent to the conjugate gradient method can be derived without ever having to consider the minimization problem. The obtained solver is called the conjugate directions method (or method of conjugate directions) and although its algorithmic procedure is nearly identical to the conjugate gradients method; its derivation and logical constructs are different.

Further Reading

[1] Bathe, K. J. (1982) *Finite Element Procedures Engineering Analysis.* New York, Prentice-Hall.

[2] Bathe, K. J. (1996) *Finite Element Procedures.* New York, Prentice-Hall.

[3] Belytschko, T. (1983) An overview of semidiscretization and time integration procedures. In *Computational Methods for Transient Analysis,* T. Belytschko and T. J. R. Hughes (eds), Amsterdam, Elsevier Science, pp. 1–65.

[4] Golub, G. H. and van Van Loan, C. F. (1996) *Matrix Computations*, 3rd edn. New York, Johns Hopkins University Press.

[5] Hestenes, M. R. (1969) Multipliers and gradient methods. *Journal of Optimization Theory and Applications*, **4**(5): 303–20.

[6] Johnson, C. (2009) *Numerical Solutions of Partial Differential Equations by the Finite Element Method.* New York, Dover Publications.

[7] Saad, Y. (2003) *Iterative Methods for Sparse Linear Systems*, 2nd edn. Society for Industrial and Applied Mathematics.

[8] Schwarz, H. R. (1989) *Numerical Analysis: A Comprehensive Introduction.* Chichester, John Wiley & Sons, Ltd.

[9] Zienkiewicz, O. C. (1971) *The Finite Element Method in Engineering Science.* New York, McGraw-Hill.

[10] Zienkiewicz, O. C. and Taylor, R. L. (2005) *The Finite Element Method Set.* Oxford, Elsevier Science.

4

Numerical Integration

Numerical integration is extensively used in the finite element method. Most often Gaussian integration is employed. An example of a typical 1D integration problem is given by

$$b = \int_{-1}^{1} \mathrm{f}(x)dx. \tag{4.1}$$

The numerical value of b represents the area underneath the curve (shaded portion) shown in Figure 4.1.

4.1 Newton-Cotes Numerical Integration

One of the simplest numerical integration approaches is the Newton-Cotes integration formula.

The main steps of the Newton-Cotes quadrature are as follows:

a. Select n points between the integration limits and calculate the function at those points, Figure 4.2.
b. Using these points, define a polynomial of degree $n-1$,

$$\mathrm{p}(x) = a_0 + a_1 x + a_2 x^2 + \cdots + a_{n-1} x^{n-1}. \tag{4.2}$$

c. Analytically integrate the above shown polynomial.

Large Strain Finite Element Method: A Practical Course, First Edition. Antonio Munjiza, Esteban Rougier and Earl E. Knight.

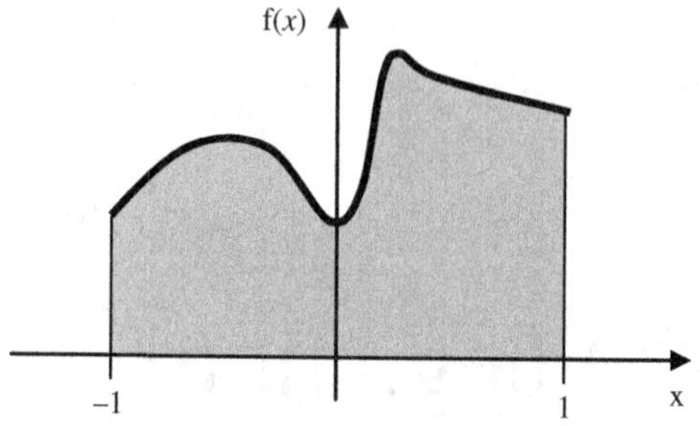

Figure 4.1 Graphical representation of a typical 1D integration problem

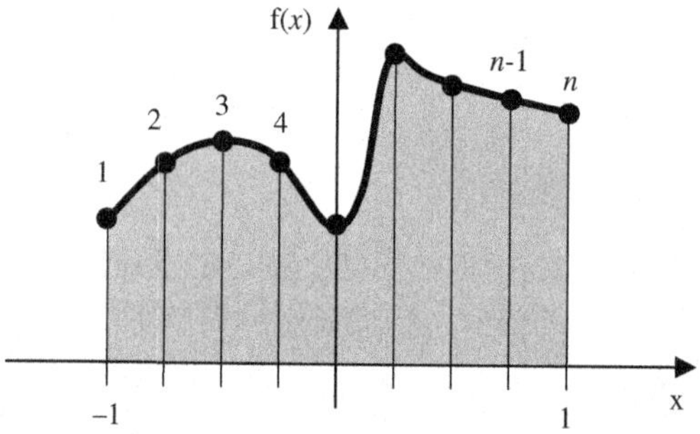

Figure 4.2 Newton-Cotes integration points

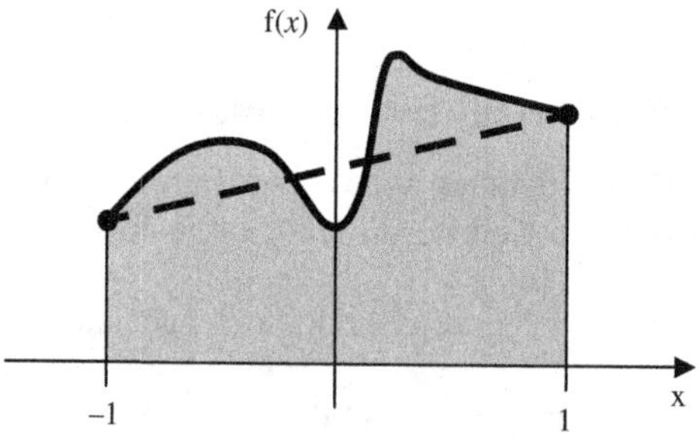

Figure 4.3 Newton-Cotes integration using two integration points

Two Points Newton-Cotes Integration Formula. In the case of $n = 2$ the polynomial is of degree $n - 1 = 2 - 1 = 1$, Figure 4.3.

This yields a polynomial defined by the integration points

$$p(x) = \frac{f_1 + f_2}{2} + \frac{f_1 - f_2}{2}x, \tag{4.3}$$

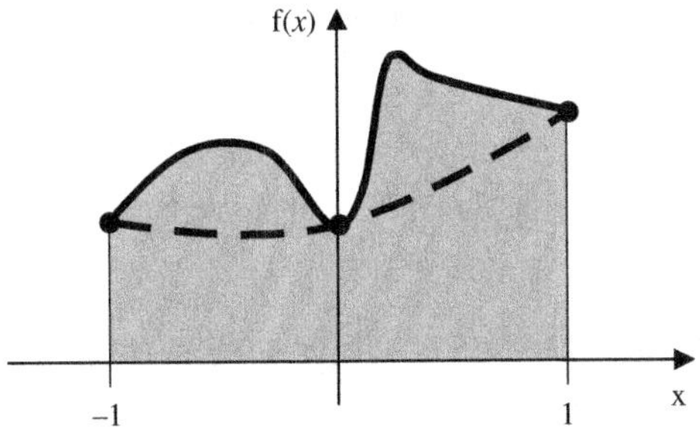

Figure 4.4 Newton-Cotes integration using three integration points

which after the integration produces

$$\int_{-1}^{1} \mathrm{p}(x) = \frac{f_1 + f_2}{2} x\Big|_{-1}^{1} + \frac{f_1 - f_2}{2}\frac{x^2}{2}\Bigg|_{-1}^{1} \\ = f_1 + f_2. \tag{4.4}$$

Three Points Newton-Cotes Integration Formula. In the case of three integration points the polynomial is of degree $n - 1 = 3 - 1 = 2$, Figure 4.4.

The polynomial that passes through all three integration points is given by

$$\mathrm{p}(x) = f_1 \frac{x}{2}(x-1) - f_2(x-1)(x+1) + f_1 \frac{x}{2}(x+1), \tag{4.5}$$

which after integration yields

$$\int_{-1}^{1} \mathrm{p}(x) = \frac{f_1 x^3}{2\,3}\Bigg|_{-1}^{1} - f_2\left(\frac{x^3}{3} - x\right)\Bigg|_{-1}^{1} + \frac{f_3 x^3}{2\,3}\Bigg|_{-1}^{1} \\ = \frac{1}{3}f_1 + \frac{4}{3}f_2 + \frac{1}{3}f_3. \tag{4.6}$$

4.2 Gaussian Numerical Integration

In comparison to the Newton-Cotes integration, Gaussian numerical integration achieves better accuracy with a smaller number of points, thus improving the speed of the numerical integration process.

The idea is to select n integration points x_i and weights h_i such that the exact answer for polynomials up to a degree $2n - 1$ is obtained, i.e.,

$$\int_{-1}^{1} \mathrm{f}(x)dx = \sum_{1}^{n} h_i \mathrm{f}(x_i) \\ = \sum_{1}^{n} h_i f_i. \tag{4.7}$$

One Gauss Point Numerical Integration. For one integration point (also called Gauss point) Equation (4.7) produces

$$\int_{-1}^{1} \mathrm{f}(x)dx = h_1 \mathrm{f}(x_1) = h_1 f_1. \tag{4.8}$$

This equation should produce the exact result for a polynomial of degree

$$2n-1 = 2\cdot 1 - 1 = 1. \tag{4.9}$$

Thus, for

$$\mathrm{f}(x) = a_0 + a_1 x, \tag{4.10}$$

one gets

$$\int_{-1}^{1} \mathrm{f}(x)dx = 2a_0. \tag{4.11}$$

This must equal the result obtained by equation (4.8), i.e.,

$$2a_0 = h_1(a_0 + a_1 x_1), \tag{4.12}$$

which yields

$$a_0(2-h_1) - h_1 a_1 x_1 = 0. \tag{4.13}$$

In order for Equation (4.13) to be valid, for any a_0 and any a_1, it is necessary that

$$\begin{aligned} (2-h_1) &= 0 \\ h_1 a_1 x_1 &= 0. \end{aligned} \tag{4.14}$$

The above equation yields

$$\begin{aligned} h_1 &= 2 \\ x_1 &= 0. \end{aligned} \tag{4.15}$$

After substituting into Equation (4.8) the one Gauss point numerical integration formula is obtained

$$\int_{-1}^{1} \mathrm{f}(x)dx = 2\mathrm{f}(0) = 2f_1. \tag{4.16}$$

Two Gauss Points Numerical Integration. For $n = 2$, the Gaussian integration produces the exact result for a polynomial of degree

$$2n-1 = 4\cdot 1 - 1 = 3. \tag{4.17}$$

This means that

$$\int_{-1}^{1} f(x)dx = h_1 f(x_1) + h_2 f(x_2) = h_1 f_1 + h_2 f_2 \tag{4.18}$$

should equal the exact value of

$$\int_{-1}^{1} \left(a_0 + a_1 x + a_2 x^2 + a_3 x^3\right)dx. \tag{4.19}$$

After integration, one obtains:

$$\begin{aligned} 2a_0 + a_2\frac{2}{3} &= h_1 f(x_1) + h_2 f(x_2) \\ &= h_1\left(a_0 + a_1 x_1 + a_2 x_1^2 + a_3 x_1^3\right) \\ &\quad + h_2\left(a_0 + a_1 x_2 + a_2 x_2^2 + a_3 x_2^3\right). \end{aligned} \tag{4.20}$$

The above equation must be satisfied for every a_0, which yields:

$$h_1 + h_2 = 2. \tag{4.21}$$

The same equation must be satisfied for every a_1, thus

$$h_1 x_1 + h_2 x_2 = 0. \tag{4.22}$$

In a similar manner, the same equation must be satisfied for every a_2, which yields

$$h_1 x_1^2 + h_2 x_2^2 = \frac{2}{3}. \tag{4.23}$$

Finally, the same equation must be satisfied for every a_3, i.e.,

$$h_1 x_1^3 + h_2 x_2^3 = 0. \tag{4.24}$$

After solving Equations (4.21), (4.22), (4.23) and (4.24), one obtains

$$h_1 = h_2 = 1 \tag{4.25}$$

and

$$-x_1 = x_2 = \frac{1}{\sqrt{3}}. \tag{4.26}$$

When substituted into Equation (4.18) these yield the two point Gauss integration formula

$$\int_{-1}^{1} f(x)dx = f\left(-\frac{1}{\sqrt{3}}\right) + f\left(\frac{1}{\sqrt{3}}\right) = f_1 + f_2. \tag{4.27}$$

4.3 Gaussian Integration in 2D

In a two dimensional space, one is often interested in obtaining the following integral, Figure 4.5.

$$\int_{-1}^{1}\left(\int_{-1}^{1} f(x,y)dx\right)dy. \tag{4.28}$$

The integration in 2D is done by using the formulas obtained for the 1D integration.

One Gauss Point Integration in 2D. The simplest Gaussian integration in 2D uses one integration point (Gauss point) in x direction and one integration point in y direction (see Figure 4.6), which reduces to a single integration point with

$$x_1 = y_1 = 0 \quad \text{and} \quad h_1 = 4. \tag{4.29}$$

This yields

$$\int_{-1}^{1}\left(\int_{-1}^{1} f(x,y)dx\right)dy = 4f(0,0) = 4f_1. \tag{4.30}$$

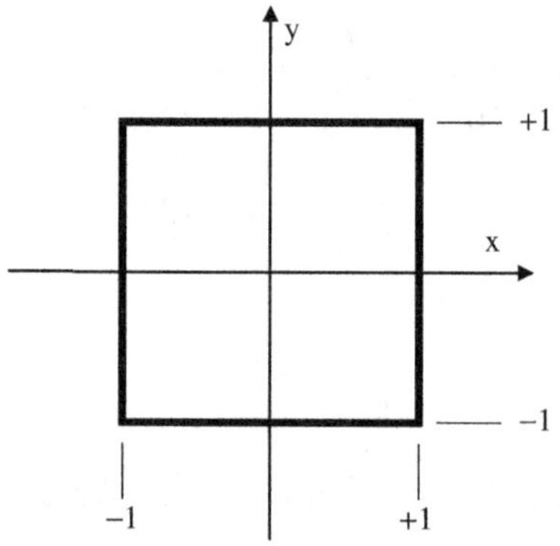

Figure 4.5 A 2D Gaussian integration domain

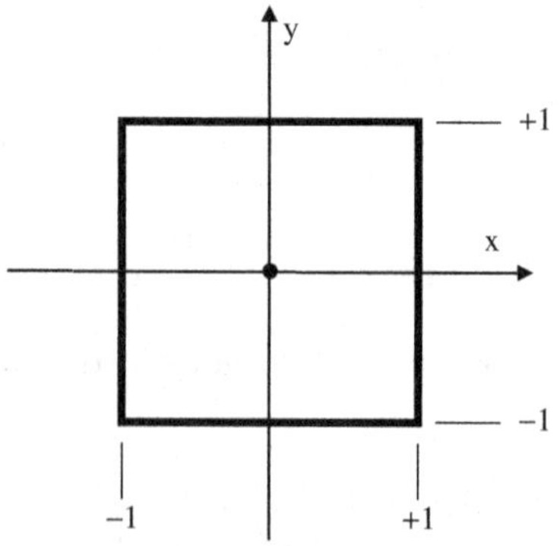

Figure 4.6 One Gauss point integration in 2D

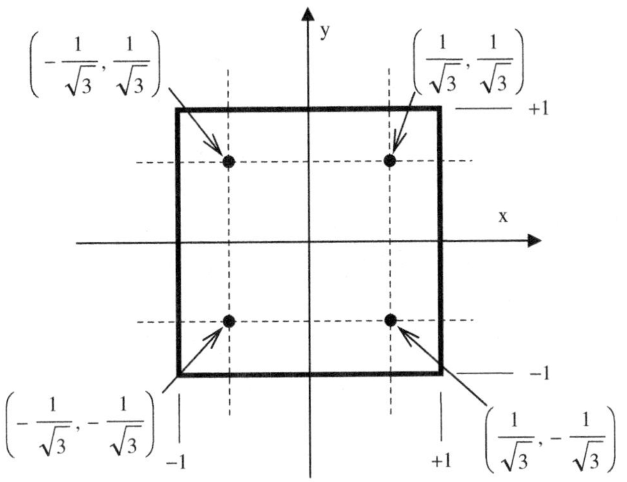

Figure 4.7 Four Gauss points integration in 2D

Four Gauss Points Integration in 2D. In 2D a better accuracy of integration is obtained by choosing two Gauss points in the x direction and two Gauss points in the y direction, Figure 4.7.

This is the case of a 2 by 2 Gauss point integration or four Gauss points integration. By using the 1D integration formula, Equation (4.27), one obtains

$$\begin{aligned}\int_{-1}^{1}\left(\int_{-1}^{1} \mathrm{f}(x,y)dx\right)dy &= \mathrm{f}\left(-\frac{1}{\sqrt{3}}, -\frac{1}{\sqrt{3}}\right) + \mathrm{f}\left(\frac{1}{\sqrt{3}}, -\frac{1}{\sqrt{3}}\right) \\ &\quad + \mathrm{f}\left(\frac{1}{\sqrt{3}}, \frac{1}{\sqrt{3}}\right) + \mathrm{f}\left(-\frac{1}{\sqrt{3}}, \frac{1}{\sqrt{3}}\right) \\ &= f_1 + f_2 + f_3 + f_4.\end{aligned} \tag{4.31}$$

4.4 Gaussian Integration in 3D

Gaussian integration in 3D is used to obtain the following integral

$$\int_{-1}^{1}\left(\int_{-1}^{1}\left(\int_{-1}^{1} \mathrm{f}(x,y,z)dx\right)dy\right)dz = \sum_{i=0}^{n} h_i \mathrm{f}(x_i, y_i, z_i), \tag{4.32}$$

where (x_i, y_i, z_i) are the Gauss points and h_i are the weights. The simplest formula is the one Gauss point integration

$$\int_{-1}^{1}\left(\int_{-1}^{1}\left(\int_{-1}^{1} \mathrm{f}(x,y,z)dx\right)dy\right)dz = 8\mathrm{f}(0,0,0) = 8f_1. \tag{4.33}$$

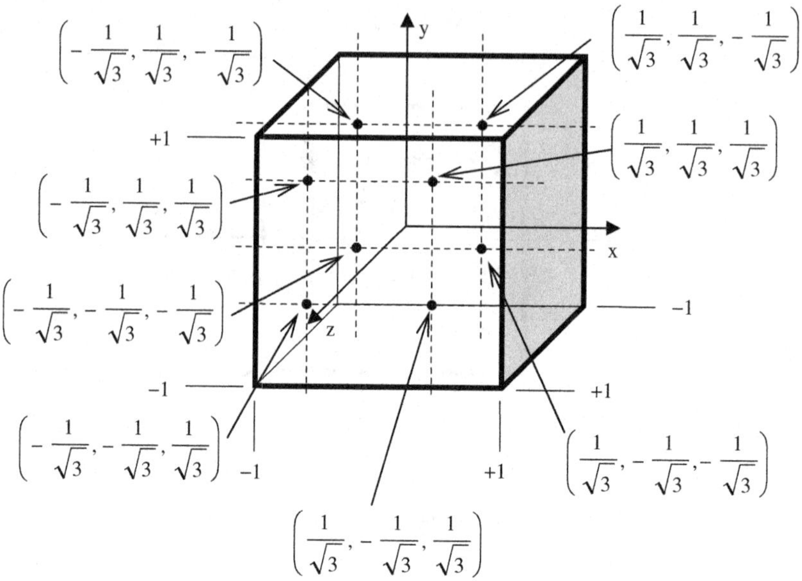

Figure 4.8 Eight Gauss points integration in 3D

A more accurate alternative to Equation (4.33) is the one given by the 2 by 2 by 2 integration formula, also called the eight point integration formula, Figure 4.8,

$$
\begin{aligned}
\int_{-1}^{1}\left(\int_{-1}^{1}\left(\int_{-1}^{1} f(x,y,z)dx\right)dy\right)dz = {} & f\left(-\frac{1}{\sqrt{3}}, -\frac{1}{\sqrt{3}}, -\frac{1}{\sqrt{3}}\right) + f\left(\frac{1}{\sqrt{3}}, -\frac{1}{\sqrt{3}}, -\frac{1}{\sqrt{3}}\right) \\
& + f\left(-\frac{1}{\sqrt{3}}, \frac{1}{\sqrt{3}}, -\frac{1}{\sqrt{3}}\right) + f\left(\frac{1}{\sqrt{3}}, \frac{1}{\sqrt{3}}, -\frac{1}{\sqrt{3}}\right) \\
& + f\left(-\frac{1}{\sqrt{3}}, -\frac{1}{\sqrt{3}}, \frac{1}{\sqrt{3}}\right) + f\left(\frac{1}{\sqrt{3}}, -\frac{1}{\sqrt{3}}, \frac{1}{\sqrt{3}}\right) \\
& + f\left(-\frac{1}{\sqrt{3}}, \frac{1}{\sqrt{3}}, \frac{1}{\sqrt{3}}\right) + f\left(\frac{1}{\sqrt{3}}, \frac{1}{\sqrt{3}}, \frac{1}{\sqrt{3}}\right) \\
= {} & f_1 + f_2 + f_3 + f_4 + f_5 + f_6 + f_7 + f_8
\end{aligned}
\tag{4.34}
$$

4.5 Summary

In this chapter numerical integration in 1D, 2D and 3D space has been introduced and special attention has been paid to Gaussian integration. Gaussian integration is a very CPU-efficient way of integrating functions over simplex domains. It is used extensively in finite element analysis and the readers are advised to familiarize themselves with it.

Further Reading

[1] Bathe, K. J. (1982) *Finite Element Procedures Engineering Analysis*. New York, Prentice-Hall.
[2] Bathe, K. J. (1996) *Finite Element Procedures*. New York, Prentice-Hall.
[3] Hamming, R. W. (1973) *Numerical Methods for Scientists and Engineers*. New York, Dover Publications.
[4] Hildebrand, F. B. (1987) *Introduction to Numerical Analysis*, 2nd edn. New York, Dover Publication.
[5] Johnson, C. (2009) *Numerical Solutions of Partial Differential Equations by the Finite Element Method*. New York, Dover Publications.
[6] Schwarz, H. R. (1989) *Numerical Analysis: A Comprehensive Introduction*. Chichester, John Wiley & Sons, Ltd.
[7] Zienkiewicz, O. C. (1971) *The Finite Element Method in Engineering Science*. New York, McGraw-Hill.
[8] Zienkiewicz, O. C. and Taylor, R. L. (2005) *The Finite Element Method Set*. Oxford, Elsevier Science.

5

Work of Internal Forces on Virtual Displacements

5.1 The Principle of Virtual Work

A material point P loaded with two opposing forces $\mathbf{f}_1$ and $\mathbf{f}_2$ is shown in Figure 5.1. The easiest way to check that the point is in a state of equilibrium is to determine if

$$\sum \mathbf{H} = 0, \tag{5.1}$$

$$\sum \mathbf{V} = 0, \tag{5.2}$$

where the first equation represents the sum of the "horizontal forces," while the second equation represents the sum of the "vertical forces." Thus,

$$\begin{aligned} &\sum \mathbf{H} = 0 \Rightarrow \mathbf{f}_1 - \mathbf{f}_2 = 0; \;\; \mathbf{f}_1 = \mathbf{f}_2 \\ &\sum \mathbf{V} = 0 \Rightarrow 0 = 0. \end{aligned} \tag{5.3}$$

This is the simplest possible way to check whether or not the static system is in equilibrium. However, it is not always convenient or even possible to determine if Equations (5.1) and (5.2) hold true and therefore one has to resort to other means for checking equilibrium, such as the potential energy method or the method of residuals.

In the context of the finite element method, the more often used approach is to check the work of real forces on virtual displacement. The idea is relatively simple:

a. Move the point P by an infinitesimally small displacement. An infinitesimally small displacement is, in engineering terms, a displacement that is as close to zero as possible, but never zero. This displacement is called virtual displacement.

Large Strain Finite Element Method: A Practical Course, First Edition. Antonio Munjiza, Esteban Rougier and Earl E. Knight.

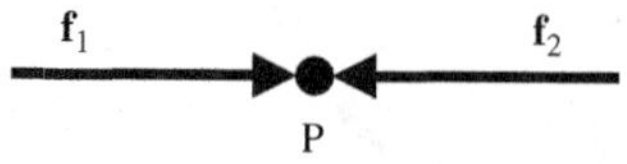

Figure 5.1 An example of equilibrium of a single material point

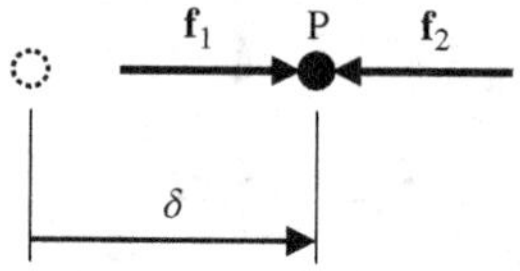

Figure 5.2 Virtual displacement

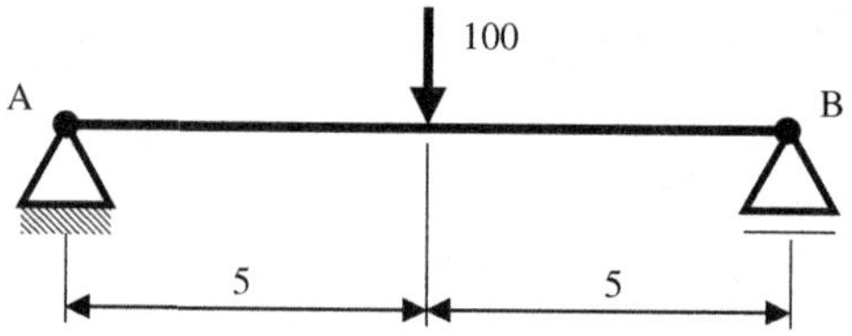

Figure 5.3 A simply supported beam

b. Calculate the work of all (real) forces on virtual displacements. This work should be zero if the system is in a state of equilibrium.

Virtual Work for a Single Material Point. For the example shown in Figure 5.1, one can give any virtual displacement; for example a horizontal virtual displacement equal to δ is given in Figure 5.2 where

$$\begin{aligned} &\delta > 0 \text{ and} \\ &\delta < \varepsilon \text{ for every } \varepsilon > 0. \end{aligned} \tag{5.4}$$

The virtual work by definition is given by

$$f_1\delta - f_2\delta = 0 \tag{5.5}$$

or

$$\delta(f_1 - f_2) = 0. \tag{5.6}$$

From Equation (5.4) it follows that δ is never zero, which implies that

$$f_1 - f_2 = 0. \tag{5.7}$$

Virtual Work for a Simply Supported Beam. An example of a simply supported beam is given in Figure 5.3. Calculate the reaction at support B.

The solution of the problem is as follows:

a. Release support B and replace it with a reaction force, $\mathbf{r}_B$, see Figure 5.4.
b. Introduce a virtual displacement of 1 im (where im is an infinitesimally small unit for length, i.e., "infimeter") at the place where support B was, see Figure 5.5.
c. Calculate the virtual work done by all forces on the given virtual displacement

$$100\ 5.0 - r_B\ 10.0 = 0, \tag{5.8}$$

which yields the reaction force at support B

$$r_B = 50.0. \tag{5.9}$$

Bending Moment. For the simply supported beam shown in Figure 5.6 calculate the bending moment at point C.
Solution:

a. Remove the bending connection at point C and replace it with a bending moment, i.e., introduce a hinge at point C, Figure 5.7.
b. Supply a virtual displacement at point C, Figure 5.8.
c. Calculate the virtual work (Note that the moment works on rotation.)

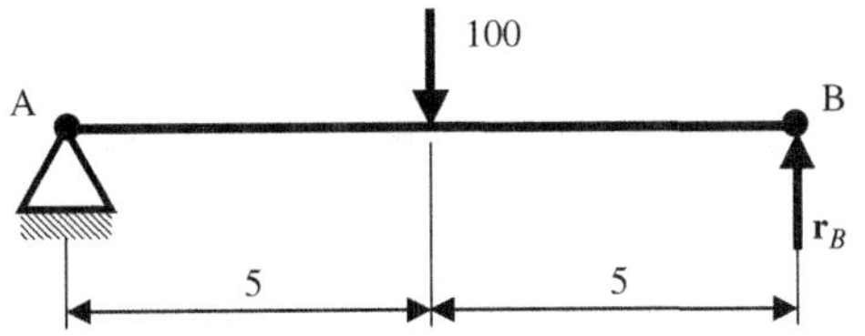

Figure 5.4 Support B is released and replaced by reaction force, $\mathbf{r}_B$

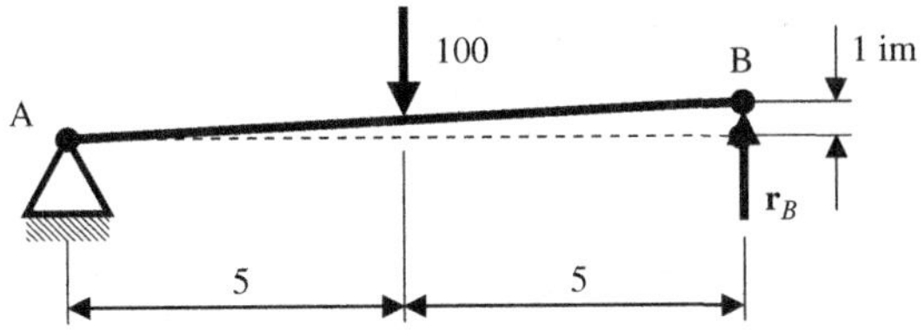

Figure 5.5 Virtual displacement

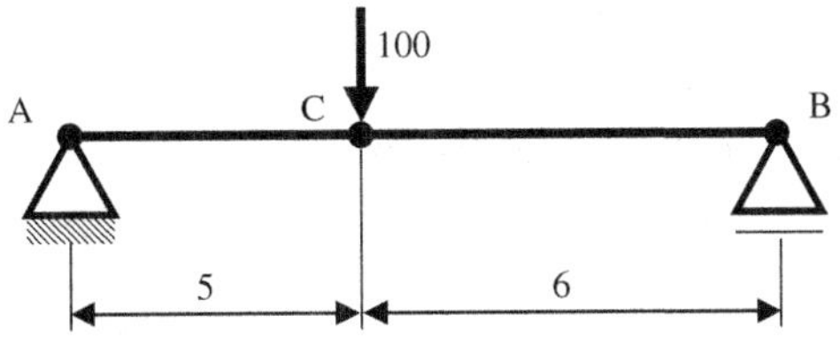

Figure 5.6 A simply supported beam

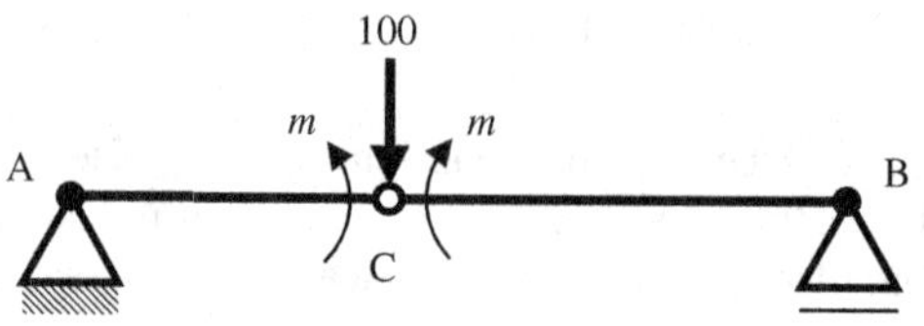

Figure 5.7 A hinge being introduced at point C

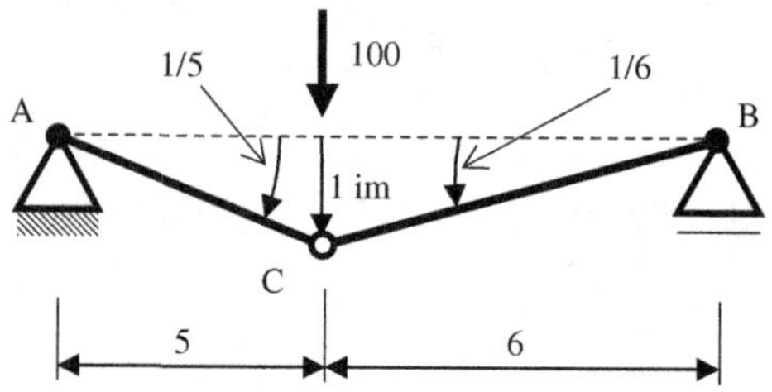

Figure 5.8 A virtual displacement

$$
\begin{aligned}
&100 \quad 1 - m\frac{1}{5} - m\frac{1}{6} = 0 \\
&m = 272.73.
\end{aligned}
\tag{5.10}
$$

It is evident that there is no need to calculate the reactions in order to calculate the bending moment.

5.2 Summary

In this chapter the work of real forces on virtual displacements has been introduced. It is then demonstrated that this virtual work can be used to check the equilibrium of a static system. First, an example of equilibrium of a single material point has been shown using the principle of virtual work. This is followed by some more complicated examples of a simply supported beam. The principle of virtual work is extensively used in the finite element method.

Further Reading

[1] Bathe, K. J. (1982) *Finite Element Procedures Engineering Analysis.* New York, Prentice-Hall.
[2] Bathe, K. J. (1996) *Finite Element Procedures.* New York, Prentice-Hall.
[3] Dvorkin, E. N. and Goldschmit, M. B. (2005) *Nonlinear Continua.* New York, Springer.
[4] Johnson, C. (2009) *Numerical Solutions of Partial Differential Equations by the Finite Element Method.* New York, Dover Publications.
[5] Zienkiewicz, O. C. (1971) *The Finite Element Method in Engineering Science.* New York, McGraw-Hill.
[6] Zienkiewicz, O. C. and Taylor, R. L. (2005) *The Finite Element Method Set.* Oxford, Elsevier Science.

Part Two

Physical Quantities

6

Scalars

6.1 Scalars in General

Physical quantities such as time, length, mass, temperature, etc., are usually quantified by a single real number. This number is usually multiplied by a unit, for example

$$t = 25.0 \text{ s}, \tag{6.1}$$

where t is a scalar variable equal to duration of one second multiplied by the real number 25.0.

In a similar manner, a given length a, is described by multiplication of a real number and a predefined length unit called the meter:

$$a = 36.3 \text{ m}. \tag{6.2}$$

A speed variable, v, can be defined as:

$$v = 65.8 \text{ m/s}. \tag{6.3}$$

A mass variable, m, can be defined as:

$$m = 22.5 \text{ kg}. \tag{6.4}$$

6.2 Scalar Functions

Very often one scalar variable is a function of another scalar variable. For example, the speed of a car is usually a function of time

$$v = v(t). \tag{6.5}$$

Large Strain Finite Element Method: A Practical Course, First Edition. Antonio Munjiza, Esteban Rougier and Earl E. Knight.

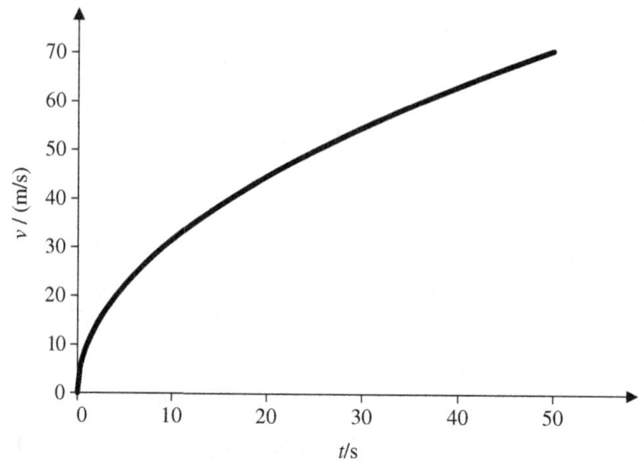

Figure 6.1 Plotting the graph of a scalar function

An extended formulation may look like

$$v = \left(28.0 + 3.0\frac{t}{s}\right)\frac{m}{s}. \tag{6.6}$$

Note that 28.0 is multiplied by m/s, while time t is divided into seconds and then multiplied by 3 m/s.

6.3 Scalar Graphs

Very often scalar functions, such as speed, are a result of lengthy computer simulations. Consequently, the obtained results are given as a table of numbers.

In a similar manner, modern digital data acquisition systems can provide a table of numbers via experimental observation of scalar variables. In order to present these to the human eye, function graphs are used. In Figure 6.1 a graph of rocket speed as a function of time is provided.

Note that on the horizontal axis the time t divided by seconds is shown; thus the values on the horizontal axis are nondimensional. On the vertical axis the rocket speed v divided by meters per second is shown making the numbers on the vertical axis nondimensional.

This is a standard convention (ISO standard) utilized for plotting scalar variables. Sometimes in literature, a number of nonstandard representations are erroneously used and they should be avoided.

6.4 Empirical Formulas

Very often empirical or similar functions for scalar variables are derived. For example, the air temperature for a certain geographical point may be given as

$$T = \left[300.0 + 20.0\sin\left(\frac{2\pi}{365}\frac{t}{\text{day}}\right)\right]\text{K}. \tag{6.7}$$

Every scalar variable in Equation (6.7) has been put into the nondimensional format by dividing them into units. A similar formula, like the one shown in Equation (6.8) would make no sense without the additional "manual" on how to use it. This is why scalar formulas must be written as demonstrated by Equation (6.7) and not as shown by Equation (6.8).

$$T = \left[300.0 + 20.0 \sin\left(\frac{2\pi t}{365}\right)\right]. \tag{6.8}$$

6.5 Fonts

When writing mathematical equations using scalar variables the following general guidelines apply:

1. Variables such as t, T, m, etc. are written with regular italic font (never bold, not even in a title – note that a bold symbol would refer to a completely different variable).
2. Real numbers are written with regular font (never bold, never slanted, not even in a table).
3. Units such as m, s, m/s are written with regular font (never bold and never slanted).
4. Functions are written with regular font.
5. Matrices are written with bold font (never slanted).

For instance **m**, m and m are three different symbols; the first is a matrix, the second is a mass variable and the third is the unit for length as shown in Equation (6.9), which simply states that mass m is proportional to the length a, where a is divided into meters and the constant of proportionality is 250.0, while the result is in kg:

$$m = \left(250.0 \frac{a}{\mathrm{m}}\right) \mathrm{kg}. \tag{6.9}$$

6. Never replace lower case letter in a unit with a capital, for instance Kg in place of kg would be a nonsense; actually it would mean Kelvin multiplied by gram.

6.6 Units

When writing units it is worth describing prefixes as shown in Tables 6.1, 6.2, 6.3 and 6.4.

Table 6.1 Unit prefixes greater than one

Designation	Description	Multiplier
Y	For yotta	10^{24}
Z	For zetta	10^{21}
E	For exa	10^{18}
P	For peta	10^{15}
T	For tera	10^{12}
G (never g)	For giga	10^{9}
M (never m)	For mega	10^{6}
k (never K)	For kilo	10^{3}
h	For hecta	10^{2}
da	For deca	10^{1}

Table 6.2 Unit prefixes less than one

Designation	Description	Multiplier
d (never D)	For deci	10^{-1}
c	For centi	10^{-2}
m	For mili	10^{-3}
μ	For micro	10^{-6}
n	For nano	10^{-9}
p	For pico	10^{-12}
f	For femto	10^{-15}
a	For atto	10^{-18}
z	For zepto	10^{-21}
y	For yocto	10^{-24}

Table 6.3 Examples of units with prefixes greater than one

Length	Mass	Time	Temperature	Luminous Intensity
Ym	Yg	Ys	YK	Ycd
Zm	Zg	Zs	ZK	Zcd
Em	Eg	Es	EK	Ecd
Pm	Pg	Ps	PK	Pcd
Tm	Tg	Ts	TK	Tcd
Gm	Gg	Gs	GK	Gcd
Mm	Mg	Ms	MK	Mcd
km	kg	ks	kK	kcd
hm	hg	hs	hK	hcd
dam	dag	das	daK	dacd

Table 6.4 Examples of units with prefixes less than one

Length	Mass	Time	Temperature	Luminous Intensity
Dm	dg	ds	dK	dcd
Cm	cg	cs	cK	ccd
Mm	mg	ms	mK	mcd
μm	μg	μs	μK	μcd
Nm	ng	ns	nK	ncd
Pm	pg	ps	pK	pcd
Fm	fg	fs	fK	fcd
Am	ag	as	aK	acd
Zm	zg	zs	zK	zcd
Ym	yg	ys	yK	ycd

Most often for mechanics problems the only unit sets needed are length, mass and time and all other units are derived from these three. In some problems one may also need temperature T, which is measured in Kelvin K (never k). In other problems units may be introduced as needed but the basic concept is as described above.

6.7 Base and Derived Scalar Variables

When running finite element simulations one has to supply some input data such as mass m, length a, speed v, time t, etc. For consistency, one chooses a unit set for the input parameters with base (fundamental) variables (SI, CGS, etc.) such that:

$$\text{time}/\text{s}\,;\ \text{length}/\text{m}\,;\ \text{mass}/\text{kg} \tag{6.10}$$

or

$$\text{time}/\text{ms}\,;\ \text{length}/\text{cm}\,;\ \text{mass}/\text{kg} \tag{6.11}$$

or any other combination of time, length and mass units.

All derived units, such as speed, are then obtained from these base units,

$$v = \frac{\text{length}}{\text{time}}. \tag{6.12}$$

If the base units shown in Equation (6.10) are used, then

$$v = \frac{\text{length} \cdot \text{m}}{\text{time} \cdot \text{s}} = \frac{\text{length}}{\text{time}} \frac{\text{m}}{\text{s}}. \tag{6.13}$$

In a similar manner, if the base units shown in equation (6.11) are adopted then

$$v = \frac{\text{length} \cdot \text{cm}}{\text{time} \cdot \text{ms}} = \frac{\text{length}}{\text{time}} \frac{\text{cm}}{\text{ms}}. \tag{6.14}$$

The unit for the magnitude of acceleration is given by

$$a = \frac{\text{length}}{\text{time}}/\text{time}. \tag{6.15}$$

If the base units shown in Equation (6.10) are used then the unit for acceleration is

$$\frac{\text{m}}{\text{s}}/\text{s} = \frac{\text{m}}{\text{s}^2}. \tag{6.16}$$

In a similar manner, if the base units shown in Equation (6.11) are adopted then the unit for acceleration is

$$\frac{\text{cm}}{\text{ms}}/\text{ms} = \frac{\text{cm}}{(\text{ms})^2}. \tag{6.17}$$

For any other combination of the base units, the corresponding unit for the derived scalar variable is obtained.

6.8 Summary

In this chapter some important fundamentals about how to use scalar quantities have been summarized. These include scalar variables, scalar functions, scalar graphs and scalar formulas.

The standard convention for writing scalar variables has also been explained together with a detailed description of how to use units with scalar variables.

Further Reading

[1] Abraham, R., Marsden, J. E and Ratiu, T. S. (1988) *Manifolds, Tensor Analysis, and Applications*. New York, Springer-Verlag.
[2] Bronstein, I. N. and Semendyayev, S. (2007) *Handbook of Mathematics*. New York, Springer.
[3] International Bureau of Weights and Measures (2006) *The International System of Units*, 8th edn. Bureau International des Poids et Mesures.
[4] ISO 80000-1:2009 (2009).
[5] Johnson, C. (2009) *Numerical Solutions of Partial Differential Equations by the Finite Element Method*. New York, Dover Publications.
[6] Kusse, B. R. and Westwig, E. A. (2006) *Mathematical Physics: Applied Mathematics for Scientists and Engineers*, 2nd edn. Wiley-VCH Verlag GmbH & Co.
[7] Matthews, P. C. (1998) *Vector Calculus*. London, Springer.
[8] Rudin, W. (1976) *The Principles of Mathematical Analysis*. New York, McGraw-Hill Science/Engineering/Math.
[9] Thompson, A. and Taylor, B. N. (2008) *Guide for the Use of the International System of Units (SI)* (Special publication 811). Gaithersburg, MD, National Institute of Standards and Technology.

7

Vectors in 2D

7.1 Vectors in General

Take the following example: a person has moved in space from point A to point B; without concern for how the person has moved, only the starting and the ending points are identified as shown by the arrow in Figure 7.1.

Subsequently, the same person moves from point B to point C, as shown in Figure 7.2. The final result is that the person has moved from point A to point C as shown in Figure 7.3 and the arrow indicates that the person started at point A and is now at point C. The person could keep moving and eventually end up at a final destination point F, Figure 7.4.

The total movement or "displacement" of the person is from point A to point F. However, this total displacement was made by a number of smaller moves (displacements) as shown in Figure 7.5.

In other words, the individual displacements are "added" or "summed" together as shown analytically in Equation (7.1) and graphically in Figure 7.6:

$$\mathbf{r} = \mathbf{r}_1 + \mathbf{r}_2 + \mathbf{r}_3 + \mathbf{r}_4 + \mathbf{r}_5 + \mathbf{r}_6 + \mathbf{r}_7 + \mathbf{r}_8 \tag{7.1}$$

It can be easily demonstrated that the associativity of the summation shown in Equation (7.1) is valid, i.e.,

$$\mathbf{r} = (\mathbf{r}_1 + \mathbf{r}_2 + \mathbf{r}_3) + (\mathbf{r}_4 + \mathbf{r}_5 + \mathbf{r}_6 + \mathbf{r}_7) + \mathbf{r}_8. \tag{7.2}$$

Large Strain Finite Element Method: A Practical Course, First Edition. Antonio Munjiza, Esteban Rougier and Earl E. Knight.

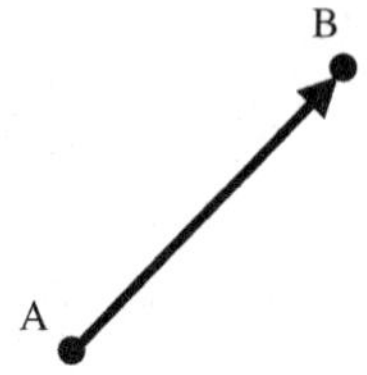

Figure 7.1 Graphical representation of a person moving from point A to point B

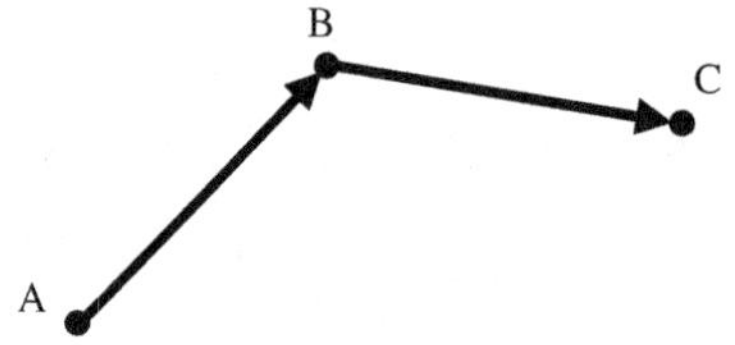

Figure 7.2 Graphical representation of a person moving from point A to point B and then to point C

Figure 7.3 Graphical representation of a person moving from point A to point C

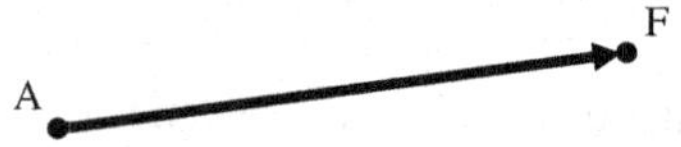

Figure 7.4 Graphical representation of a person moving from point A to point F

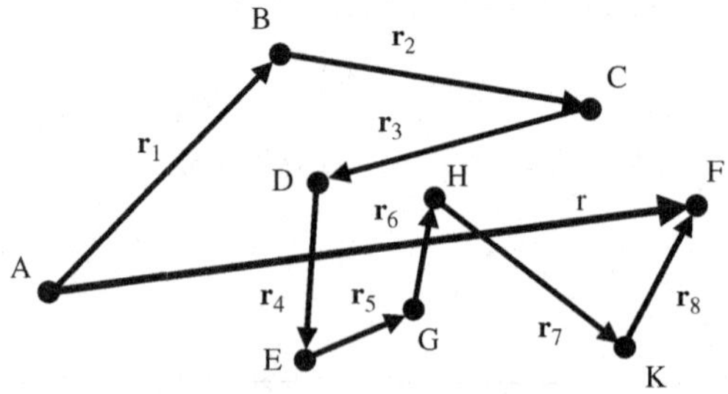

Figure 7.5 Graphical representation of a person moving from point A to point F with all the intermediate points

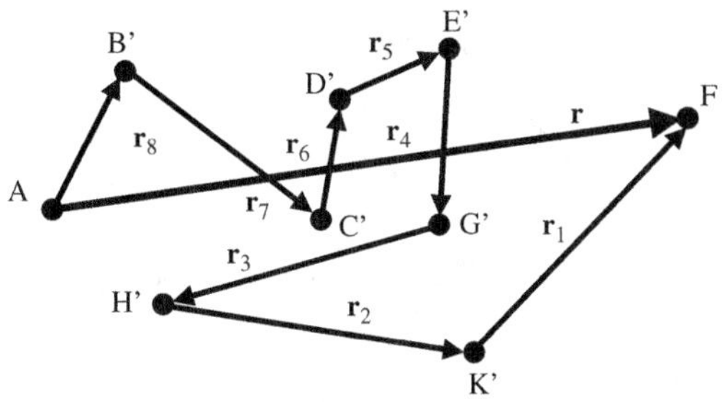

Figure 7.6 Graphical representation of the sum of the displacements and the commutations

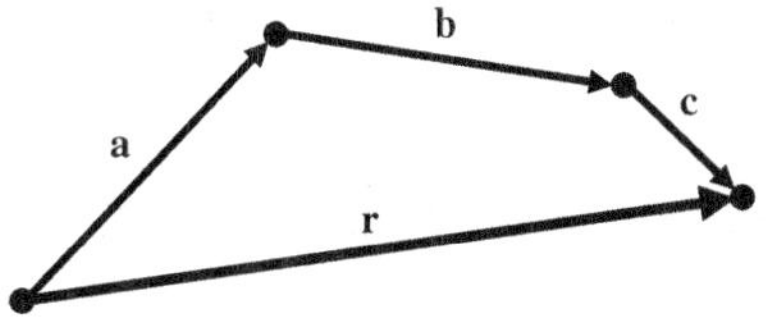

Figure 7.7 Adding vectors

Also, the commutativity of Equation (7.1) is valid, i.e.,

$$\begin{aligned}\mathbf{r} &= \mathbf{r}_1 + \mathbf{r}_2 + \mathbf{r}_3 + \mathbf{r}_4 + \mathbf{r}_5 + \mathbf{r}_6 + \mathbf{r}_7 + \mathbf{r}_8 \\ &= \mathbf{r}_8 + \mathbf{r}_7 + \mathbf{r}_6 + \mathbf{r}_5 + \mathbf{r}_4 + \mathbf{r}_3 + \mathbf{r}_2 + \mathbf{r}_1 \\ &= \mathbf{r}_1 + \mathbf{r}_6 + \mathbf{r}_7 + \mathbf{r}_4 + \mathbf{r}_5 + \mathbf{r}_2 + \mathbf{r}_3 + \mathbf{r}_8 \\ &= etc.\end{aligned} \tag{7.3}$$

In other words, by starting from point A and making eight displacements $\mathbf{r}_1$, …, $\mathbf{r}_8$; the person will arrive at their destination point F regardless of the order in which the person makes the moves, as long as all the moves are made.

It is self-evident that the displacement, as shown in Equation (7.3) and defined above, is a physical quantity. In order to represent the displacement a straight line with an arrow is really needed.

The displacement is therefore not a scalar; it is a new type of physical reality, which is called the vector.

In 2D, vectors are easily described using lines with arrows. Adding them together is done by simply "sticking" the end of one arrow to the beginning of the next one, thus forming a polygon of vectors as shown in Figure 7.7.

The result shown in Figure 7.7 can be written by:

$$\mathbf{r} = \mathbf{a} + \mathbf{b} + \mathbf{c}, \tag{7.4}$$

where $\mathbf{r}$ is the resultant displacement achieved after making three individual displacements respectively (**a**, **b** and **c**). Of course,

$$\mathbf{r} = \mathbf{a} + \mathbf{c} + \mathbf{b}, \tag{7.5}$$

as shown in Figure 7.8.

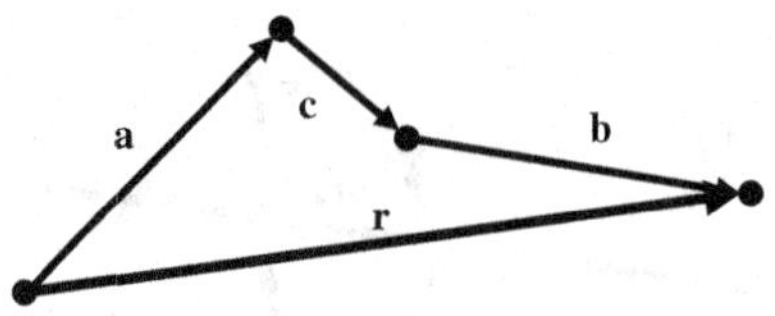

Figure 7.8 Polygon of vectors

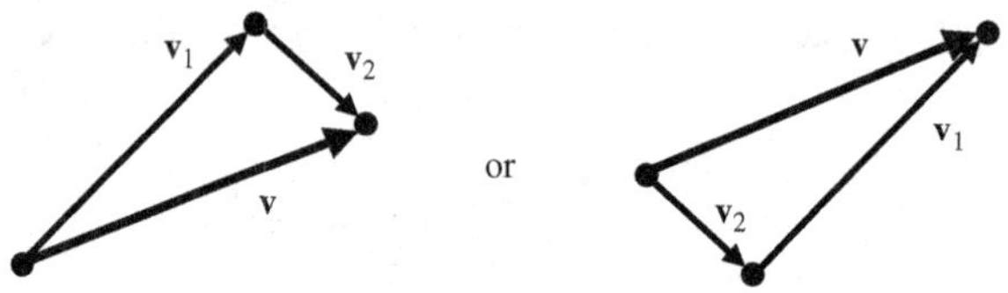

Figure 7.9 Sum of velocities for a point

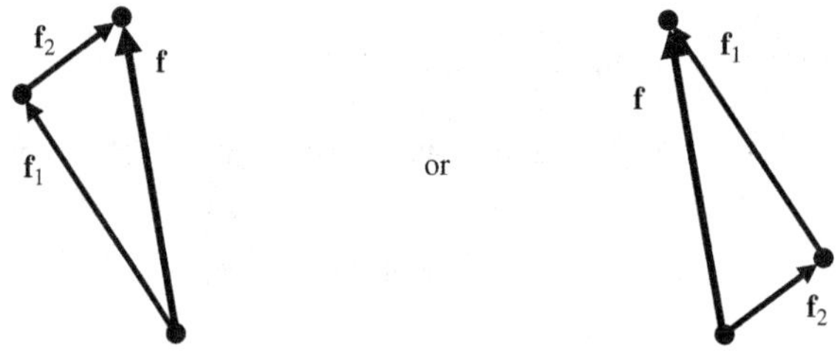

Figure 7.10 Sum of two forces on a point

By thinking of vectors as displacements, one can get a clearer idea of their associativity and commutativity properties. That stated, apart from displacements, velocity, acceleration, and force also behave as vectors. Thus, if a point has a velocity $\mathbf{v}_1$ and velocity $\mathbf{v}_2$ is then added, the total velocity of the point is as seen in Figure 7.9,

$$\mathbf{v} = \mathbf{v}_1 + \mathbf{v}_2 = \mathbf{v}_2 + \mathbf{v}_1. \tag{7.6}$$

If a force $\mathbf{f}_1$ is applied to a point and a force $\mathbf{f}_2$ is added to it, the total force $\mathbf{f}$ is as shown in Figure 7.10,

$$\mathbf{f} = \mathbf{f}_1 + \mathbf{f}_2 = \mathbf{f}_2 + \mathbf{f}_1. \tag{7.7}$$

In the case that a number of forces $\mathbf{f}_1$, $\mathbf{f}_2$, $\mathbf{f}_3$, …, $\mathbf{f}_n$ has been added to a specific point, the resultant force is

$$\mathbf{f} = \mathbf{f}_1 + \mathbf{f}_2 + \mathbf{f}_3 + \cdots + \mathbf{f}_n, \tag{7.8}$$

where the polygon of forces is shown in Figure 7.11.

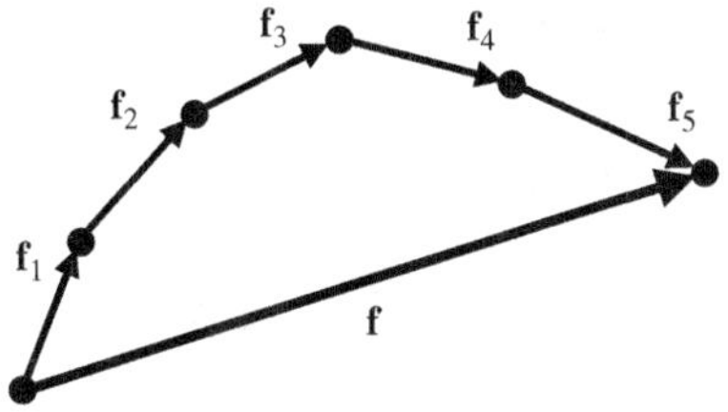

Figure 7.11 Polygon of forces for a point

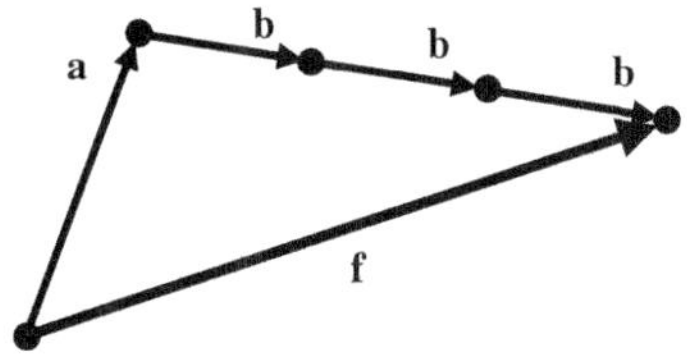

Figure 7.12 Sum of vector **a** plus three times vector **b**

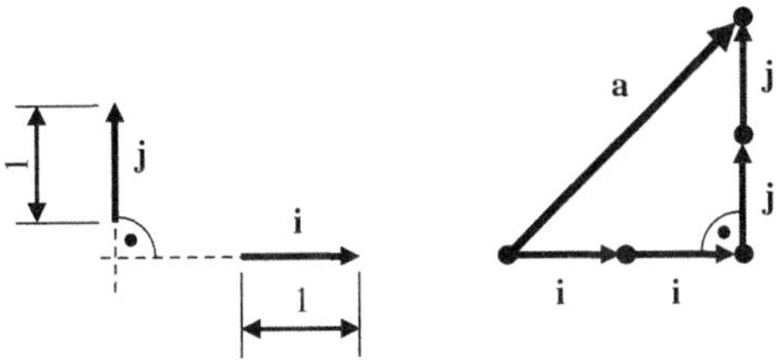

Figure 7.13 Base vectors forming an orthonormal base

7.2 Vector Notation

When writing vector variable equations one should use a non-slanted bold font (preferably lower case letters) typeset convention

$$\mathbf{f} = \mathbf{a} + 3\mathbf{b}. \tag{7.9}$$

In the above equation vector **b** has been multiplied by three and added to vector **a**, as shown in Figure 7.12.

Multiplication of a vector by a real number simply preserves the direction of the vector and accordingly changes its length. For points, non-slanted non-bold (preferably upper case) font is used.

7.3 Matrix Representation of Vectors

With an understanding of the physics of vector summation and scalar multiplication, one can now see how two base vectors in 2D can represent any other vector. In Figure 7.13, two orthogonal base vectors with a length (magnitude) equal to one are shown.

Vectors **i** and **j** form an orthonormal base $[\mathbf{i}\ \mathbf{j}]$ and any other vector **r** can be represented by

$$\mathbf{r} = r_i\mathbf{i} + r_j\mathbf{j}, \tag{7.10}$$

where r_i and r_j are scalars. In Figure 7.13 the vector

$$\mathbf{a} = 2\mathbf{i} + 2\mathbf{j} \tag{7.11}$$

is shown.

It is very often convenient to put real numbers r_i and r_j into a column matrix

$$\begin{bmatrix} r_i \\ r_j \end{bmatrix} \tag{7.12}$$

and base vectors **i** and **j** into a row matrix

$$[\mathbf{i}\ \mathbf{j}]. \tag{7.13}$$

Vector **r** is then represented as follows:

$$\mathbf{r} = [\mathbf{i}\ \mathbf{j}]\begin{bmatrix} r_i \\ r_j \end{bmatrix} = r_i\mathbf{i} + r_j\mathbf{j}. \tag{7.14}$$

This is conveniently written as

$$\mathbf{r} = \begin{bmatrix} r_i \\ r_j \end{bmatrix}, \tag{7.15}$$

where the base $[\mathbf{i}\ \mathbf{j}]$ is implied from the context.

The column matrix

$$\begin{bmatrix} r_i \\ r_j \end{bmatrix} \tag{7.16}$$

is the matrix of real numbers called vector components r_i, r_j in the base $[\mathbf{i}\ \mathbf{j}]$. This matrix is not the vector itself, it is only a convenient representation of the vector

$$\mathbf{r} = r_i\mathbf{i} + r_j\mathbf{j}. \tag{7.17}$$

One can think of the matrix as a tidy way of storing the real numbers (vector components).

7.4 Scalar Product

The scalar product (also called dot product) of two vectors in 2D is defined as

$$s = \mathbf{a} \bullet \mathbf{b} = \mathbf{b} \bullet \mathbf{a} = ab\cos\alpha, \tag{7.18}$$

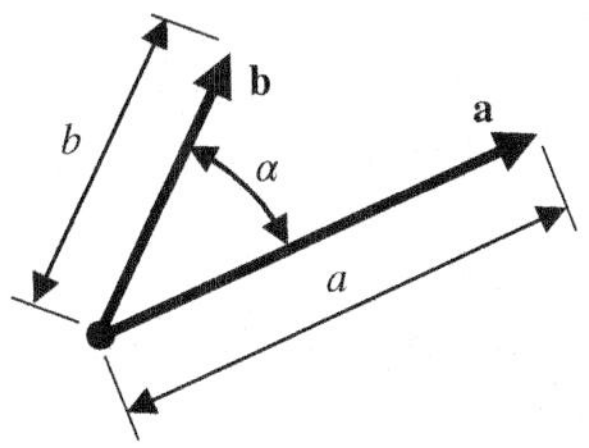

Figure 7.14 The scalar (inner or dot) product

where a and b are the magnitudes of vectors **a** and **b** respectively and α is the angle between the vectors, as shown in Figure 7.14.

In the case that

$$\begin{aligned}\mathbf{a} &= a_i\mathbf{i} + a_j\mathbf{j} \\ \mathbf{b} &= b_i\mathbf{i} + b_j\mathbf{j}\end{aligned}, \tag{7.19}$$

the dot product is

$$\begin{aligned} s &= \left(a_i\mathbf{i} + a_j\mathbf{j}\right)\bullet\left(b_i\mathbf{i} + b_j\mathbf{j}\right) \\ &= \left(a_ib_i\right)\left(\mathbf{i}\bullet\mathbf{i}\right) + \left(a_jb_i\right)\left(\mathbf{i}\bullet\mathbf{j}\right) + \left(a_ib_j\right)\left(\mathbf{i}\bullet\mathbf{j}\right) + \left(a_jb_j\right)\left(\mathbf{j}\bullet\mathbf{j}\right). \end{aligned} \tag{7.20}$$

For the orthonormal base, by definition (Equation (7.18))

$$\begin{aligned}\mathbf{i}\bullet\mathbf{i} &= \cos 0 = 1 \\ \mathbf{i}\bullet\mathbf{j} &= \cos\frac{\pi}{2} = 0\end{aligned}, \tag{7.21}$$

which means that

$$\mathbf{a}\bullet\mathbf{b} = a_ib_i + a_jb_j. \tag{7.22}$$

Once again, this is valid only for the orthonormal base $[\mathbf{i} \ \ \mathbf{j}]$.

7.5 General Vector Base in 2D

In the most general case, base vectors are neither unit vectors nor orthogonal to each other, Figure 7.15.

A particular vector **a** is uniquely defined as

$$\mathbf{a} = a_\xi\boldsymbol{\xi} + a_\eta\boldsymbol{\eta}, \tag{7.23}$$

where $[\boldsymbol{\xi} \ \ \boldsymbol{\eta}]$ is a base of nonorthogonal nonunit vectors. Equation (7.23) can be written in a matrix form as:

$$\mathbf{a} = [\boldsymbol{\xi} \ \ \boldsymbol{\eta}]\begin{bmatrix} a_\xi \\ a_\eta \end{bmatrix} = a_\xi\boldsymbol{\xi} + a_\eta\boldsymbol{\eta}. \tag{7.24}$$

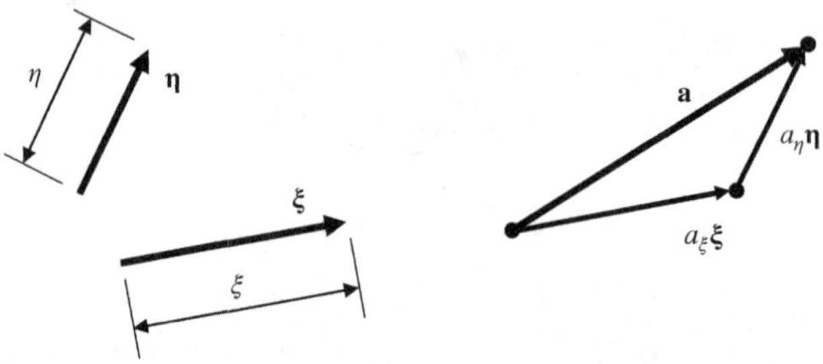

Figure 7.15 A general vector base $[\boldsymbol{\xi} \;\; \boldsymbol{\eta}]$

This is often shortened to

$$\mathbf{a} = \begin{bmatrix} a_\xi \\ a_\eta \end{bmatrix}, \tag{7.25}$$

where the base $[\boldsymbol{\xi} \;\; \boldsymbol{\eta}]$ should somehow be known from the context – one first introduces the base and then describes the vector using the matrix of its components.

Note that Equation (7.25) formally equates vector **a** and matrix

$$\begin{bmatrix} a_\xi \\ a_\eta \end{bmatrix}. \tag{7.26}$$

However, vector **a** and the matrix shown in Equation (7.26) should never be confused; the former is physically a vector representing displacement, or force, or velocity, or acceleration, or some other physical quantity; the latter is only a table of numbers with no physical meaning – it is conveniently used to store the components of the vector **a** in base $[\boldsymbol{\xi} \;\; \boldsymbol{\eta}]$.

It is worth noting that for a given vector **f** representing the force on point A is always the same, irrespective of the vector base used. Nevertheless, it can be represented using an infinite number of different vector bases. For each of these bases a matrix of components corresponding to a particular base can be derived. This means that any particular vector can be described using an infinite number of different matrices that each store the vector components for a particular base.

7.6 Dual Base

The two bases, $[\boldsymbol{\xi} \;\; \boldsymbol{\eta}]$ and $[\underline{\boldsymbol{\xi}} \;\; \underline{\boldsymbol{\eta}}]$, shown in Figure 7.16 are dual to each other if:

$$\begin{aligned} \boldsymbol{\xi} \bullet \underline{\boldsymbol{\xi}} &= \underline{\boldsymbol{\xi}} \bullet \boldsymbol{\xi} = 1 \\ \boldsymbol{\eta} \bullet \underline{\boldsymbol{\eta}} &= \underline{\boldsymbol{\eta}} \bullet \boldsymbol{\eta} = 1 \\ \underline{\boldsymbol{\xi}} \bullet \boldsymbol{\eta} &= \boldsymbol{\eta} \bullet \underline{\boldsymbol{\xi}} = 0 \\ \underline{\boldsymbol{\eta}} \bullet \boldsymbol{\xi} &= \boldsymbol{\xi} \bullet \underline{\boldsymbol{\eta}} = 0. \end{aligned} \tag{7.27}$$

In other words vectors $\underline{\boldsymbol{\eta}}$ and $\boldsymbol{\xi}$ are orthogonal to each other and so are vectors $\underline{\boldsymbol{\xi}}$ and $\boldsymbol{\eta}$.

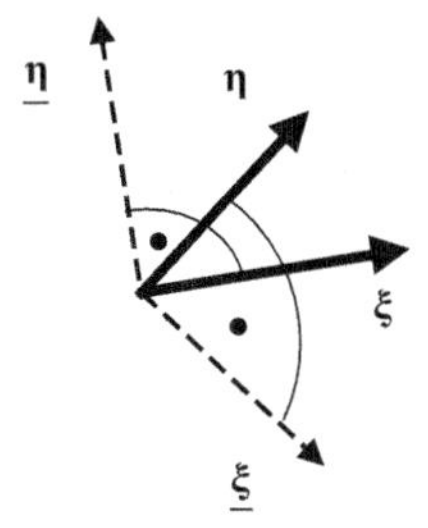

Figure 7.16 The dual bases

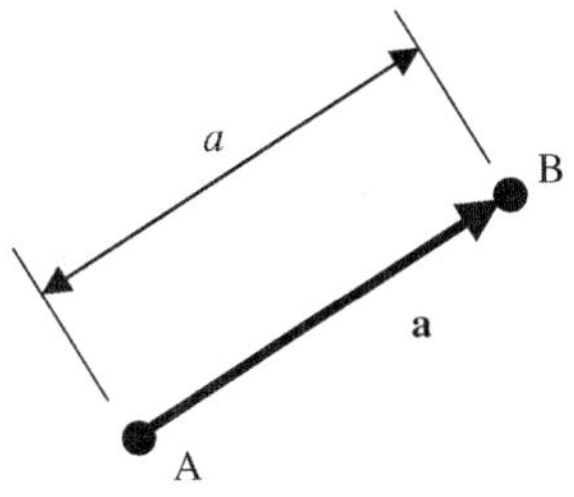

Figure 7.17 The vector representing the displacement of a person from point A to point B

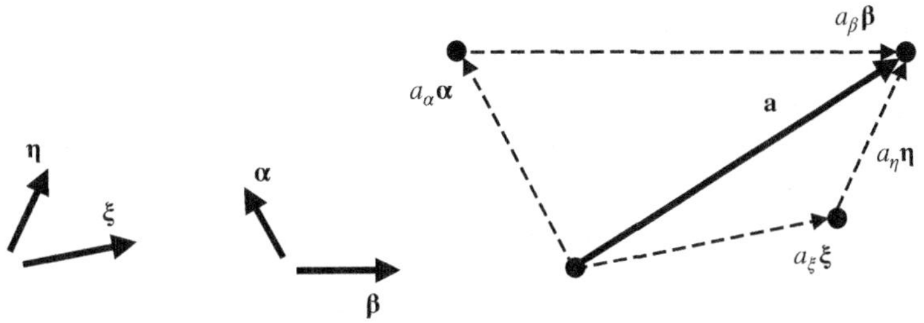

Figure 7.18 Components of vector **a** in bases $[\boldsymbol{\xi} \;\; \boldsymbol{\eta}]$ and $[\boldsymbol{\alpha} \;\; \boldsymbol{\beta}]$

7.7 Changing Vector Base

Very often there is a need to represent the same vector using two or more different bases. In some old books this is usually called "vector transformation." This is a somewhat misleading terminology for, despite the change of the base, the vector stays the same and does not go through any "transformation".

In this book, in order to make the subject matter easier to understand, the term transformation is deliberately avoided and the change of the vector base is introduced instead.

Let us assume that displacement of a person from point A to another point B is described by the vector **a**, as shown in Figure 7.17.

Both the length (magnitude of the displacement) and the direction of the displacement are well defined by the line and arrow. In theory, one would need no other description of the vector **a**. However, it is sometimes convenient to represent vector **a** as a linear combination of base vectors (sum of scaled base vectors).

For example, in Figure 7.18 vector **a** is described using two bases $[\boldsymbol{\xi} \;\; \boldsymbol{\eta}]$ and $[\boldsymbol{\alpha} \;\; \boldsymbol{\beta}]$.

Thus,

$$\mathbf{a} = a_\xi \boldsymbol{\xi} + a_\eta \boldsymbol{\eta} = a_\alpha \boldsymbol{\alpha} + a_\beta \boldsymbol{\beta}. \tag{7.28}$$

Using the matrix notation and base $[\boldsymbol{\xi} \;\; \boldsymbol{\eta}]$, the vector $\mathbf{a}$ can be represented as

$$\mathbf{a} = [\boldsymbol{\xi} \;\; \boldsymbol{\eta}] \begin{bmatrix} a_\xi \\ a_\eta \end{bmatrix} = \begin{bmatrix} a_\xi \\ a_\eta \end{bmatrix}. \tag{7.29}$$

At the same time, using the shortened matrix notation and base $[\boldsymbol{\alpha} \;\; \boldsymbol{\beta}]$, the same vector $\mathbf{a}$ can be represented as

$$\mathbf{a} = [\boldsymbol{\alpha} \;\; \boldsymbol{\beta}] \begin{bmatrix} a_\alpha \\ a_\beta \end{bmatrix} = \begin{bmatrix} a_\alpha \\ a_\beta \end{bmatrix}. \tag{7.30}$$

Of course the equal sign in Equations (7.29) and (7.30) means: vector $\mathbf{a}$ is made of a_α vectors $\boldsymbol{\alpha}$ and a_β vectors $\boldsymbol{\beta}$, the same vector is made of a_ξ vectors $\boldsymbol{\xi}$ and a_η vectors $\boldsymbol{\eta}$. In other words, although the equation reads

$$\mathbf{a} = \begin{bmatrix} a_\alpha \\ a_\beta \end{bmatrix}, \tag{7.31}$$

it never means that the vector $\mathbf{a}$ is equal to a matrix. It would make no sense in any case. It just states that vector $\mathbf{a}$ is represented by the matrix shown in Equation (7.31) in the base $[\boldsymbol{\alpha} \;\; \boldsymbol{\beta}]$, where the base employed must be clear from its given context.

By using the scalar product of vector $\mathbf{a}$ and the base vectors of the dual base $[\underline{\boldsymbol{\xi}} \;\; \underline{\boldsymbol{\eta}}]$, one obtains

$$\begin{aligned} \mathbf{a} \bullet \underline{\boldsymbol{\xi}} &= (a_\xi \boldsymbol{\xi} + a_\eta \boldsymbol{\eta}) \bullet \underline{\boldsymbol{\xi}} \\ &= a_\xi \left(\boldsymbol{\xi} \bullet \underline{\boldsymbol{\xi}} \right) + a_\eta \left(\boldsymbol{\eta} \bullet \underline{\boldsymbol{\xi}} \right) \\ &= a_\xi, \end{aligned} \tag{7.32}$$

$$\begin{aligned} \mathbf{a} \bullet \underline{\boldsymbol{\eta}} &= (a_\xi \boldsymbol{\xi} + a_\eta \boldsymbol{\eta}) \bullet \underline{\boldsymbol{\eta}} \\ &= a_\xi \left(\boldsymbol{\xi} \bullet \underline{\boldsymbol{\eta}} \right) + a_\eta \left(\boldsymbol{\eta} \bullet \underline{\boldsymbol{\eta}} \right) \\ &= a_\eta. \end{aligned} \tag{7.33}$$

In other words when one changes the vector base, the components of the corresponding matrix are obtained by employing the scalar product with the vectors of the dual base.

$$\mathbf{a} = \begin{bmatrix} a_\xi \\ a_\eta \end{bmatrix} = \begin{bmatrix} \mathbf{a} \bullet \underline{\boldsymbol{\xi}} \\ \mathbf{a} \bullet \underline{\boldsymbol{\eta}} \end{bmatrix} \tag{7.34}$$

and

$$\mathbf{a} = \begin{bmatrix} a_\alpha \\ a_\beta \end{bmatrix} = \begin{bmatrix} \mathbf{a} \bullet \underline{\boldsymbol{\alpha}} \\ \mathbf{a} \bullet \underline{\boldsymbol{\beta}} \end{bmatrix}. \tag{7.35}$$

In a similar manner

$$\mathbf{a} = \begin{bmatrix} a_{\underline{\xi}} \\ a_{\underline{\eta}} \end{bmatrix} = \begin{bmatrix} \mathbf{a} \bullet \boldsymbol{\xi} \\ \mathbf{a} \bullet \boldsymbol{\eta} \end{bmatrix} \tag{7.36}$$

and

$$\mathbf{a} = \begin{bmatrix} a_{\underline{\alpha}} \\ a_{\underline{\beta}} \end{bmatrix} = \begin{bmatrix} \mathbf{a} \bullet \boldsymbol{\alpha} \\ \mathbf{a} \bullet \boldsymbol{\beta} \end{bmatrix}. \tag{7.37}$$

This explains why the dual vector base was introduced and why it is conveniently used in obtaining vector components.

7.8 Self-duality of the Orthonormal Base

The dual base

$$\begin{bmatrix} \underline{\mathbf{i}} & \underline{\mathbf{j}} \end{bmatrix} \tag{7.38}$$

to the orthonormal base

$$\begin{bmatrix} \mathbf{i} & \mathbf{j} \end{bmatrix} \tag{7.39}$$

is the base $\begin{bmatrix} \mathbf{i} & \mathbf{j} \end{bmatrix}$ itself, see Figure 7.19. This is because

$$\begin{aligned} \mathbf{i} \bullet \mathbf{i} &= 1 \\ \mathbf{j} \bullet \mathbf{j} &= 1 \\ \mathbf{i} \bullet \mathbf{j} &= 0 \\ \mathbf{j} \bullet \mathbf{i} &= 0 \end{aligned} \tag{7.40}$$

Using the orthonormal base $\begin{bmatrix} \mathbf{i} & \mathbf{j} \end{bmatrix}$ any vector $\mathbf{a}$ can be shown as

$$\mathbf{a} = \begin{bmatrix} \mathbf{i} & \mathbf{j} \end{bmatrix} \begin{bmatrix} a_i \\ a_j \end{bmatrix} \tag{7.41}$$

or simply,

$$\mathbf{a} = \begin{bmatrix} a_i \\ a_j \end{bmatrix}. \tag{7.42}$$

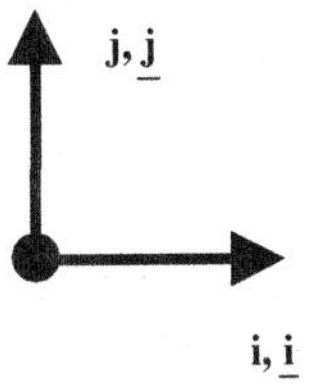

Figure 7.19 Dual orthonormal bases

By using the scalar product of the vector **a** and dual base vector $\underline{\mathbf{i}}$:

$$\begin{aligned}\mathbf{a}\bullet\underline{\mathbf{i}} &= (a_i\mathbf{i}+a_j\mathbf{j})\bullet\underline{\mathbf{i}}\\ &= (a_i\mathbf{i}+a_j\mathbf{j})\bullet\mathbf{i}\\ &= a_i.\end{aligned} \tag{7.43}$$

In a similar manner

$$\begin{aligned}\mathbf{a}\bullet\underline{\mathbf{j}} &= (a_i\mathbf{i}+a_j\mathbf{j})\bullet\underline{\mathbf{j}}\\ &= (a_i\mathbf{i}+a_j\mathbf{j})\bullet\mathbf{j}\\ &= a_j,\end{aligned} \tag{7.44}$$

thus,

$$\mathbf{a} = \begin{bmatrix}\mathbf{a}\bullet\mathbf{i}\\ \mathbf{a}\bullet\mathbf{j}\end{bmatrix}. \tag{7.45}$$

In short, the orthonormal base is dual to itself, which makes the task of calculating vector components even easier.

7.9 Combining Bases

Instead of using a graphical representation of a vector, a matrix of vector components using a predefined global orthonormal base $[\mathbf{i}\ \ \mathbf{j}]$ is often used

$$\mathbf{a} = \begin{bmatrix}a_i\\ a_j\end{bmatrix}. \tag{7.46}$$

Any other general base $[\boldsymbol{\xi}\ \ \boldsymbol{\eta}]$ is then described using this global base

$$\boldsymbol{\xi} = \xi_i\mathbf{i}+\xi_j\mathbf{j} = \begin{bmatrix}\xi_i\\ \xi_j\end{bmatrix}, \tag{7.47}$$

$$\boldsymbol{\eta} = \eta_i\mathbf{i}+\eta_j\mathbf{j} = \begin{bmatrix}\eta_i\\ \eta_j\end{bmatrix}. \tag{7.48}$$

In a similar manner another general base $[\boldsymbol{\alpha}\ \ \boldsymbol{\beta}]$ can be introduced as needed and its base vectors $\boldsymbol{\alpha}$ and $\boldsymbol{\beta}$ can be described using the orthonormal base $[\mathbf{i}\ \ \mathbf{j}]$ as follows:

$$\boldsymbol{\alpha} = \alpha_i\mathbf{i}+\alpha_j\mathbf{j} = \begin{bmatrix}\alpha_i\\ \alpha_j\end{bmatrix}, \tag{7.49}$$

$$\boldsymbol{\beta} = \beta_i\mathbf{i}+\beta_j\mathbf{j} = \begin{bmatrix}\beta_i\\ \beta_j\end{bmatrix}. \tag{7.50}$$

In principle, any number of generally nonorthonormal bases (all expressed using the same orthonormal base $[\mathbf{i}\ \ \mathbf{j}]$) can be introduced as conveniently required.

For clarity reasons the components of the base vectors are stored using a square matrix. The base $[\boldsymbol{\xi}\ \ \boldsymbol{\eta}]$ simply becomes

$$[\boldsymbol{\xi} \;\; \boldsymbol{\eta}] = [\mathbf{i} \;\; \mathbf{j}] \begin{bmatrix} \xi_i & \eta_i \\ \xi_j & \eta_j \end{bmatrix} \tag{7.51}$$

and the base $[\boldsymbol{\alpha} \;\; \boldsymbol{\beta}]$ becomes

$$[\boldsymbol{\alpha} \;\; \boldsymbol{\beta}] = [\mathbf{i} \;\; \mathbf{j}] \begin{bmatrix} \alpha_i & \beta_i \\ \alpha_j & \beta_j \end{bmatrix}. \tag{7.52}$$

By further shortening the notation one obtains

$$[\boldsymbol{\xi} \;\; \boldsymbol{\eta}] = \begin{bmatrix} \xi_i & \eta_i \\ \xi_j & \eta_j \end{bmatrix} \tag{7.53}$$

and

$$[\boldsymbol{\alpha} \;\; \boldsymbol{\beta}] = \begin{bmatrix} \alpha_i & \beta_i \\ \alpha_j & \beta_j \end{bmatrix}, \tag{7.54}$$

where columns of the square matrix describe the individual base vectors in the global base $[\mathbf{i} \;\; \mathbf{j}]$. The dual bases are given by

$$[\underline{\boldsymbol{\xi}} \;\; \underline{\boldsymbol{\eta}}] = [\mathbf{i} \;\; \mathbf{j}] \begin{bmatrix} \underline{\xi}_i & \underline{\eta}_i \\ \underline{\xi}_j & \underline{\eta}_j \end{bmatrix} = \begin{bmatrix} \underline{\xi}_i & \underline{\eta}_i \\ \underline{\xi}_j & \underline{\eta}_j \end{bmatrix}, \tag{7.55}$$

$$[\underline{\boldsymbol{\alpha}} \;\; \underline{\boldsymbol{\beta}}] = [\mathbf{i} \;\; \mathbf{j}] \begin{bmatrix} \underline{\alpha}_i & \underline{\beta}_i \\ \underline{\alpha}_j & \underline{\beta}_j \end{bmatrix} = \begin{bmatrix} \underline{\alpha}_i & \underline{\beta}_i \\ \underline{\alpha}_j & \underline{\beta}_j \end{bmatrix}. \tag{7.56}$$

The duality conditions (see Equation (7.27)) dictate that

$$\begin{aligned} \underline{\boldsymbol{\alpha}} \bullet \boldsymbol{\alpha} &= \alpha_i \underline{\alpha}_i + \alpha_j \underline{\alpha}_j = 1 \\ \underline{\boldsymbol{\beta}} \bullet \boldsymbol{\beta} &= \beta_i \underline{\beta}_i + \beta_j \underline{\beta}_j = 1 \\ \underline{\boldsymbol{\alpha}} \bullet \boldsymbol{\beta} &= \underline{\alpha}_i \beta_i + \underline{\alpha}_j \beta_j = 0 \\ \underline{\boldsymbol{\beta}} \bullet \boldsymbol{\alpha} &= \underline{\beta}_i \alpha_i + \underline{\beta}_j \alpha_j = 0 \end{aligned} \tag{7.57}$$

and also

$$\begin{aligned} \underline{\boldsymbol{\xi}} \bullet \boldsymbol{\xi} &= \xi_i \underline{\xi}_i + \xi_j \underline{\xi}_j = 1 \\ \underline{\boldsymbol{\eta}} \bullet \boldsymbol{\eta} &= \eta_i \underline{\eta}_i + \eta_j \underline{\eta}_j = 1 \\ \underline{\boldsymbol{\xi}} \bullet \boldsymbol{\eta} &= \underline{\xi}_i \eta_i + \underline{\xi}_j \eta_j = 0 \\ \underline{\boldsymbol{\eta}} \bullet \boldsymbol{\xi} &= \underline{\eta}_i \xi_i + \underline{\eta}_j \xi_j = 0. \end{aligned} \tag{7.58}$$

The duality condition (7.57) can also be written in a matrix form

$$\begin{bmatrix} \alpha_i & \alpha_j \\ \beta_i & \beta_j \end{bmatrix} \begin{bmatrix} \underline{\alpha}_i & \underline{\beta}_i \\ \underline{\alpha}_j & \underline{\beta}_j \end{bmatrix} = \mathbf{I} = \begin{bmatrix} 1 & 0 \\ 0 & 1 \end{bmatrix} \tag{7.59}$$

and

$$\begin{bmatrix} \xi_i & \xi_j \\ \eta_i & \eta_j \end{bmatrix} \begin{bmatrix} \underline{\xi}_i & \underline{\eta}_i \\ \underline{\xi}_j & \underline{\eta}_j \end{bmatrix} = \mathbf{I} = \begin{bmatrix} 1 & 0 \\ 0 & 1 \end{bmatrix}. \tag{7.60}$$

By multiplying Equations (7.59) and (7.60) from the left by

$$\begin{bmatrix} \alpha_i & \alpha_j \\ \beta_i & \beta_j \end{bmatrix}^{-1} \quad \text{and} \quad \begin{bmatrix} \xi_i & \xi_j \\ \eta_i & \eta_j \end{bmatrix}^{-1}, \tag{7.61}$$

respectively, one obtains

$$\begin{bmatrix} \underline{\alpha}_i & \underline{\beta}_i \\ \underline{\alpha}_j & \underline{\beta}_j \end{bmatrix} = \begin{bmatrix} \alpha_i & \alpha_j \\ \beta_i & \beta_j \end{bmatrix}^{-1} \quad \text{and} \quad \begin{bmatrix} \underline{\xi}_i & \underline{\eta}_i \\ \underline{\xi}_j & \underline{\eta}_j \end{bmatrix} = \begin{bmatrix} \xi_i & \xi_j \\ \eta_i & \eta_j \end{bmatrix}^{-1} \tag{7.62}$$

or

$$\begin{bmatrix} \underline{\alpha}_i & \underline{\beta}_i \\ \underline{\alpha}_j & \underline{\beta}_j \end{bmatrix} = \left(\begin{bmatrix} \alpha_i & \beta_i \\ \alpha_j & \beta_j \end{bmatrix}^{-1} \right)^{\mathrm{T}} = \begin{bmatrix} \alpha_i & \beta_i \\ \alpha_j & \beta_j \end{bmatrix}^{-\mathrm{T}} \tag{7.63}$$

and

$$\begin{bmatrix} \underline{\xi}_i & \underline{\eta}_i \\ \underline{\xi}_j & \underline{\eta}_j \end{bmatrix} = \left(\begin{bmatrix} \xi_i & \eta_i \\ \xi_j & \eta_j \end{bmatrix}^{-1} \right)^{\mathrm{T}} = \begin{bmatrix} \xi_i & \eta_i \\ \xi_j & \eta_j \end{bmatrix}^{-\mathrm{T}}, \tag{7.64}$$

where –T stands for inverse transpose matrix. Also,

$$\begin{bmatrix} \underline{\xi}_i & \underline{\xi}_j \\ \underline{\eta}_i & \underline{\eta}_j \end{bmatrix} = \begin{bmatrix} \xi_i & \eta_i \\ \xi_j & \eta_j \end{bmatrix}^{-1}. \tag{7.65}$$

As any vector $\mathbf{a}$ is initially identified using the global base $[\mathbf{i} \;\; \mathbf{j}]$, Equation (7.46), the same vector in the nonorthonormal general base $[\boldsymbol{\alpha} \;\; \boldsymbol{\beta}]$ is given by

$$\begin{aligned} \mathbf{a} &= a_\alpha \boldsymbol{\alpha} + a_\beta \boldsymbol{\beta} \\ &= (\mathbf{a} \cdot \underline{\boldsymbol{\alpha}}) \boldsymbol{\alpha} + (\mathbf{a} \cdot \underline{\boldsymbol{\beta}}) \boldsymbol{\beta} \\ &= \left(\begin{bmatrix} \underline{\alpha}_i & \underline{\alpha}_j \end{bmatrix} \begin{bmatrix} a_i \\ a_j \end{bmatrix} \right) \boldsymbol{\alpha} + \left(\begin{bmatrix} \underline{\beta}_i & \underline{\beta}_j \end{bmatrix} \begin{bmatrix} a_i \\ a_j \end{bmatrix} \right) \boldsymbol{\beta} \\ &= [\boldsymbol{\alpha} \;\; \boldsymbol{\beta}] \left(\begin{bmatrix} \underline{\alpha}_i & \underline{\alpha}_j \\ \underline{\beta}_i & \underline{\beta}_j \end{bmatrix} \begin{bmatrix} a_i \\ a_j \end{bmatrix} \right) \\ &= \begin{bmatrix} \underline{\alpha}_i & \underline{\alpha}_j \\ \underline{\beta}_i & \underline{\beta}_j \end{bmatrix} \begin{bmatrix} a_i \\ a_j \end{bmatrix}. \end{aligned} \tag{7.66}$$

By substituting Equation (7.62) into equation (7.66), one obtains

$$\mathbf{a}=\begin{bmatrix} a_\alpha \\ a_\beta \end{bmatrix}=\begin{bmatrix} \alpha_i & \beta_i \\ \alpha_j & \beta_j \end{bmatrix}^{-1}\begin{bmatrix} a_i \\ a_j \end{bmatrix}. \tag{7.67}$$

It is worth noting that the equal sign does not mean that the vector **a** is equal to the matrix, but that it is represented by the matrix of its components in the base $[\boldsymbol{\alpha} \;\; \boldsymbol{\beta}]$.

In short, any given vector **a** described by a global orthonormal base

$$\mathbf{a}=\begin{bmatrix} a_i \\ a_j \end{bmatrix} \tag{7.68}$$

can also be described using any other generally non-orthonormal base

$$[\boldsymbol{\alpha} \;\; \boldsymbol{\beta}]=\begin{bmatrix} \alpha_i & \beta_i \\ \alpha_j & \beta_j \end{bmatrix}, \tag{7.69}$$

where the new matrix of vector components is simply

$$\mathbf{a}=\begin{bmatrix} a_\alpha \\ a_\beta \end{bmatrix}=\begin{bmatrix} \alpha_i & \beta_i \\ \alpha_j & \beta_j \end{bmatrix}^{-1}\begin{bmatrix} a_i \\ a_j \end{bmatrix}. \tag{7.70}$$

For the base $[\boldsymbol{\xi} \;\; \boldsymbol{\eta}]$ the matrix

$$\mathbf{a}=\begin{bmatrix} a_\xi \\ a_\eta \end{bmatrix} \tag{7.71}$$

is obtained in a similar manner, thus:

$$\begin{aligned}\mathbf{a}&=[\boldsymbol{\xi} \;\; \boldsymbol{\eta}]\begin{bmatrix} a_\xi \\ a_\eta \end{bmatrix}\\ &=\begin{bmatrix} a_\xi \\ a_\eta \end{bmatrix}\\ &=\begin{bmatrix} \underline{\xi}_i & \underline{\xi}_j \\ \underline{\eta}_i & \underline{\eta}_j \end{bmatrix}\begin{bmatrix} a_i \\ a_j \end{bmatrix}\\ &=\begin{bmatrix} \xi_i & \eta_i \\ \xi_j & \eta_j \end{bmatrix}^{-1}\begin{bmatrix} a_i \\ a_j \end{bmatrix}.\end{aligned} \tag{7.72}$$

It can be proven that

$$\begin{bmatrix} \underline{\xi}_i & \underline{\xi}_j \\ \underline{\eta}_i & \underline{\eta}_j \end{bmatrix}=\begin{bmatrix} i_\xi & j_\xi \\ i_\eta & j_\eta \end{bmatrix}=\begin{bmatrix} \xi_i & \eta_i \\ \xi_j & \eta_j \end{bmatrix}^{-1}. \tag{7.73}$$

This means that

$$\mathbf{i}=[\boldsymbol{\xi}\ \ \boldsymbol{\eta}]\begin{bmatrix} i_\xi \\ i_\eta \end{bmatrix}=\begin{bmatrix} i_\xi \\ i_\eta \end{bmatrix} \quad \text{and} \quad \mathbf{j}=[\boldsymbol{\xi}\ \ \boldsymbol{\eta}]\begin{bmatrix} j_\xi \\ j_\eta \end{bmatrix}=\begin{bmatrix} j_\xi \\ j_\eta \end{bmatrix}. \tag{7.74}$$

When substituted into Equation (7.72) this yields

$$\begin{aligned}\mathbf{a}=\begin{bmatrix} a_\xi \\ a_\eta \end{bmatrix}&=\begin{bmatrix} i_\xi & j_\xi \\ i_\eta & j_\eta \end{bmatrix}\begin{bmatrix} a_i \\ a_j \end{bmatrix}\\ &=a_i\begin{bmatrix} i_\xi \\ i_\eta \end{bmatrix}+a_j\begin{bmatrix} j_\xi \\ j_\eta \end{bmatrix}.\end{aligned} \tag{7.75}$$

In this manner, Equation (7.75) acquires physical meaning for vector **a** is made of a_i vectors **i** plus a_j vectors **j**, where **i** is in turn made of i_ξ vectors $\boldsymbol{\xi}$ plus i_η vectors $\boldsymbol{\eta}$, while vector **j** is made of j_ξ vectors $\boldsymbol{\xi}$ and j_η vectors $\boldsymbol{\eta}$. This is illustrated in Figure 7.20.

Very often the vector **a** is initially given in the general base $[\boldsymbol{\alpha}\ \ \boldsymbol{\beta}]$ only and not in the orthonormal base $[\mathbf{i}\ \ \mathbf{j}]$, thus

$$\mathbf{a}=[\boldsymbol{\alpha}\ \ \boldsymbol{\beta}]\begin{bmatrix} a_\alpha \\ a_\beta \end{bmatrix}=\begin{bmatrix} a_\alpha \\ a_\beta \end{bmatrix} \tag{7.76}$$

and one needs to represent the same vector in the general base $[\boldsymbol{\xi}\ \ \boldsymbol{\eta}]$

$$\mathbf{a}=[\boldsymbol{\xi}\ \ \boldsymbol{\eta}]\begin{bmatrix} a_\xi \\ a_\eta \end{bmatrix}=\begin{bmatrix} a_\xi \\ a_\eta \end{bmatrix}. \tag{7.77}$$

By multiplying by the dual vectors $[\underline{\boldsymbol{\xi}}\ \ \underline{\boldsymbol{\eta}}]$

$$\begin{aligned}\mathbf{a}\bullet\underline{\boldsymbol{\xi}}&=(a_\xi\boldsymbol{\xi}+a_\eta\boldsymbol{\eta})\bullet\underline{\boldsymbol{\xi}}\\ \mathbf{a}\bullet\underline{\boldsymbol{\eta}}&=(a_\xi\boldsymbol{\xi}+a_\eta\boldsymbol{\eta})\bullet\underline{\boldsymbol{\eta}},\end{aligned} \tag{7.78}$$

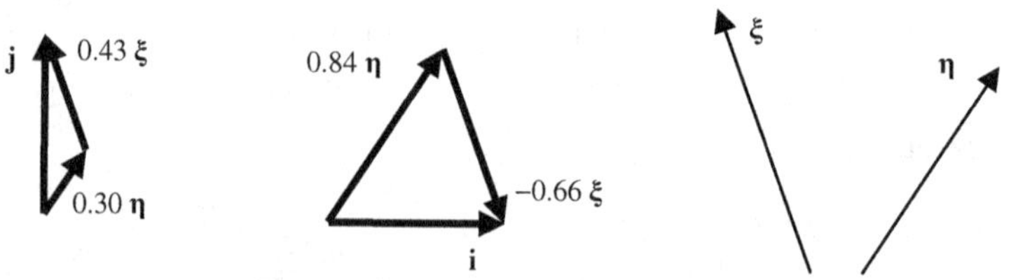

Figure 7.20 Base vectors **i** and **j** given in terms of base vectors ξ and η

i.e.,

$$\mathbf{a} \bullet \underline{\boldsymbol{\xi}} = a_\xi \\ \mathbf{a} \bullet \underline{\boldsymbol{\eta}} = a_\eta \tag{7.79}$$

or

$$\mathbf{a} = \begin{bmatrix} a_\xi \\ a_\eta \end{bmatrix} = \begin{bmatrix} \mathbf{a} \bullet \underline{\boldsymbol{\xi}} \\ \mathbf{a} \bullet \underline{\boldsymbol{\eta}} \end{bmatrix}, \tag{7.80}$$

where

$$\begin{aligned} a_\xi &= \mathbf{a} \bullet \underline{\boldsymbol{\xi}} \\ &= (a_\alpha \boldsymbol{\alpha} + a_\beta \boldsymbol{\beta}) \bullet \underline{\boldsymbol{\xi}} \\ &= a_\alpha \left(\boldsymbol{\alpha} \bullet \underline{\boldsymbol{\xi}}\right) + a_\beta \left(\boldsymbol{\beta} \bullet \underline{\boldsymbol{\xi}}\right) \end{aligned} \tag{7.81}$$

$$\begin{aligned} a_\eta &= \mathbf{a} \bullet \underline{\boldsymbol{\eta}} \\ &= (a_\alpha \boldsymbol{\alpha} + a_\beta \boldsymbol{\beta}) \bullet \underline{\boldsymbol{\eta}} \\ &= a_\alpha \left(\boldsymbol{\alpha} \bullet \underline{\boldsymbol{\eta}}\right) + a_\beta \left(\boldsymbol{\beta} \bullet \underline{\boldsymbol{\eta}}\right). \end{aligned} \tag{7.82}$$

This can be written in the matrix form:

$$\begin{bmatrix} a_\xi \\ a_\eta \end{bmatrix} = \begin{bmatrix} \boldsymbol{\alpha} \bullet \underline{\boldsymbol{\xi}} & \boldsymbol{\beta} \bullet \underline{\boldsymbol{\xi}} \\ \boldsymbol{\alpha} \bullet \underline{\boldsymbol{\eta}} & \boldsymbol{\beta} \bullet \underline{\boldsymbol{\eta}} \end{bmatrix} \begin{bmatrix} a_\alpha \\ a_\beta \end{bmatrix}. \tag{7.83}$$

Since

$$\begin{bmatrix} \boldsymbol{\alpha} \bullet \underline{\boldsymbol{\xi}} & \boldsymbol{\beta} \bullet \underline{\boldsymbol{\xi}} \\ \boldsymbol{\alpha} \bullet \underline{\boldsymbol{\eta}} & \boldsymbol{\beta} \bullet \underline{\boldsymbol{\eta}} \end{bmatrix} = \begin{bmatrix} \underline{\xi}_i & \underline{\xi}_j \\ \underline{\eta}_i & \underline{\eta}_j \end{bmatrix} \begin{bmatrix} \alpha_i & \beta_i \\ \alpha_j & \beta_j \end{bmatrix}, \tag{7.84}$$

it follows that

$$\begin{bmatrix} a_\xi \\ a_\eta \end{bmatrix} = \left(\begin{bmatrix} \underline{\xi}_i & \underline{\xi}_j \\ \underline{\eta}_i & \underline{\eta}_j \end{bmatrix} \begin{bmatrix} \alpha_i & \beta_i \\ \alpha_j & \beta_j \end{bmatrix} \right) \begin{bmatrix} a_\alpha \\ a_\beta \end{bmatrix}. \tag{7.85}$$

Therefore, if the components of vector a in the general base $[\boldsymbol{\alpha} \;\; \boldsymbol{\beta}]$ are known, the components of the same vector in the general base $[\boldsymbol{\xi} \;\; \boldsymbol{\eta}]$ are given by:

$$\begin{bmatrix} a_\xi \\ a_\eta \end{bmatrix} = \left(\begin{bmatrix} \xi_i & \eta_i \\ \xi_j & \eta_j \end{bmatrix}^{-1} \begin{bmatrix} \alpha_i & \beta_i \\ \alpha_j & \beta_j \end{bmatrix} \right) \begin{bmatrix} a_\alpha \\ a_\beta \end{bmatrix}. \tag{7.86}$$

By analogy,

$$\begin{bmatrix} a_\alpha \\ a_\beta \end{bmatrix} = \left(\begin{bmatrix} \alpha_i & \beta_i \\ \alpha_j & \beta_j \end{bmatrix}^{-1} \begin{bmatrix} \xi_i & \eta_i \\ \xi_j & \eta_j \end{bmatrix} \right) \begin{bmatrix} a_\xi \\ a_\eta \end{bmatrix}. \tag{7.87}$$

7.10 Examples

Example 1. An ant has moved from its nest located at point A to a source of food located at point B as shown in Figure 7.21.

a. Describe the displacement vector **r** using the orthonormal base $[\mathbf{i}\ \ \mathbf{j}]$.

Solution:

$$\mathbf{r} = [\mathbf{i}\ \ \mathbf{j}]\begin{bmatrix} r_i \\ r_j \end{bmatrix}$$

$$= [\mathbf{i}\ \ \mathbf{j}]\begin{bmatrix} \mathbf{r}\bullet\underline{\mathbf{i}} \\ \mathbf{r}\bullet\underline{\mathbf{j}} \end{bmatrix} = \begin{bmatrix} \mathbf{r}\bullet\mathbf{i} \\ \mathbf{r}\bullet\mathbf{j} \end{bmatrix} = [\mathbf{i}\ \ \mathbf{j}]\begin{bmatrix} 10\cdot\cos 30 \\ 10\cdot\sin 30 \end{bmatrix} = [\mathbf{i}\ \ \mathbf{j}]\begin{bmatrix} 8.66 \\ 5.00 \end{bmatrix} \tag{7.88}$$

$$= \begin{bmatrix} 8.66 \\ 5.00 \end{bmatrix}.$$

b. Describe vector **r** using the general base

$$[\boldsymbol{\alpha}\ \ \boldsymbol{\beta}] = [\mathbf{i}\ \ \mathbf{j}]\begin{bmatrix} 2 & 2 \\ 3 & 8 \end{bmatrix}. \tag{7.89}$$

Solution: The base $[\boldsymbol{\alpha}\ \ \boldsymbol{\beta}]$ is shown in Figure 7.21.

$$[\boldsymbol{\alpha}\ \ \boldsymbol{\beta}] = \begin{bmatrix} \alpha_i & \beta_i \\ \alpha_j & \beta_j \end{bmatrix} = \begin{bmatrix} 2 & 2 \\ 3 & 8 \end{bmatrix}. \tag{7.90}$$

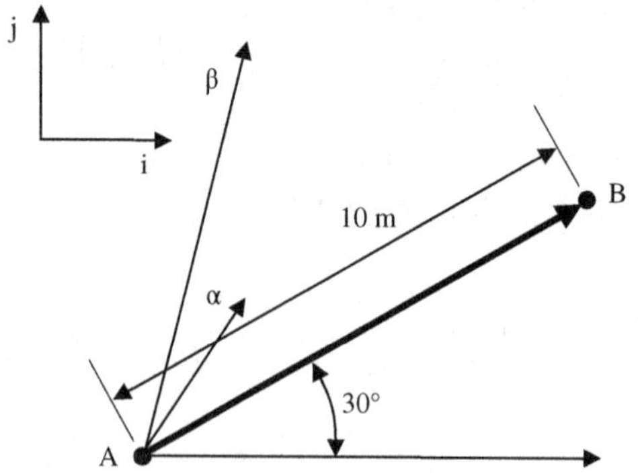

Figure 7.21 Displacement of the ant from A to B

The dual base is

$$\begin{aligned}\begin{bmatrix}\underline{\boldsymbol{\alpha}}\\ \underline{\boldsymbol{\beta}}\end{bmatrix} &= \begin{bmatrix}\underline{\alpha}_i & \underline{\alpha}_j\\ \underline{\beta}_i & \underline{\beta}_j\end{bmatrix}\begin{bmatrix}\mathbf{i}\\ \mathbf{j}\end{bmatrix} = \begin{bmatrix}\underline{\alpha}_i & \underline{\alpha}_j\\ \underline{\beta}_i & \underline{\beta}_j\end{bmatrix}\\ &= \begin{bmatrix}\alpha_i & \beta_i\\ \alpha_j & \beta_j\end{bmatrix}^{-1}\\ &= \begin{bmatrix}2 & 2\\ 3 & 8\end{bmatrix}^{-1}\\ &= \begin{bmatrix}0.8 & -0.2\\ -0.3 & 0.2\end{bmatrix}.\end{aligned} \tag{7.91}$$

This yields

$$\begin{aligned}\mathbf{r} = [\boldsymbol{\alpha}\ \ \boldsymbol{\beta}]\begin{bmatrix}r_\alpha\\ r_\beta\end{bmatrix} &= \begin{bmatrix}\underline{\alpha}_i & \underline{\alpha}_j\\ \underline{\beta}_i & \underline{\beta}_j\end{bmatrix}\begin{bmatrix}r_i\\ r_j\end{bmatrix}\\ &= \begin{bmatrix}0.8 & -0.2\\ -0.3 & 0.2\end{bmatrix}\begin{bmatrix}8.66\\ 5.00\end{bmatrix} = 8.66\begin{bmatrix}0.8\\ -0.3\end{bmatrix} + 5.00\begin{bmatrix}-0.2\\ 0.2\end{bmatrix}\\ &= \begin{bmatrix}5.92\\ -1.60\end{bmatrix}.\end{aligned} \tag{7.92}$$

Alternatively:

$$[\mathbf{i}\ \ \mathbf{j}] = [\boldsymbol{\alpha}\ \ \boldsymbol{\beta}]\begin{bmatrix}i_\alpha & j_\alpha\\ i_\beta & j_\beta\end{bmatrix}, \tag{7.93}$$

where

$$\begin{aligned}\begin{bmatrix}i_\alpha & j_\alpha\\ i_\beta & j_\beta\end{bmatrix} &= \begin{bmatrix}\alpha_i & \beta_i\\ \alpha_j & \beta_j\end{bmatrix}^{-1}\\ &= \begin{bmatrix}2 & 2\\ 3 & 8\end{bmatrix}^{-1} = \begin{bmatrix}0.8 & -0.2\\ -0.3 & 0.2\end{bmatrix}.\end{aligned} \tag{7.94}$$

In other words,

$$\mathbf{i} = [\boldsymbol{\alpha}\ \ \boldsymbol{\beta}]\begin{bmatrix}0.8\\ -0.3\end{bmatrix};\ \mathbf{j} = [\boldsymbol{\alpha}\ \ \boldsymbol{\beta}]\begin{bmatrix}-0.2\\ 0.2\end{bmatrix}. \tag{7.95}$$

Thus,

$$\begin{aligned}\mathbf{r} &= r_i\mathbf{i} + r_j\mathbf{j}\\ &= \begin{bmatrix}0.8 & -0.2\\ -0.3 & 0.2\end{bmatrix}\begin{bmatrix}r_i\\ r_j\end{bmatrix}\\ &= \begin{bmatrix}0.8 & -0.2\\ -0.3 & 0.2\end{bmatrix}\begin{bmatrix}8.66\\ 5.00\end{bmatrix} = \begin{bmatrix}5.92\\ -1.60\end{bmatrix}.\end{aligned} \tag{7.96}$$

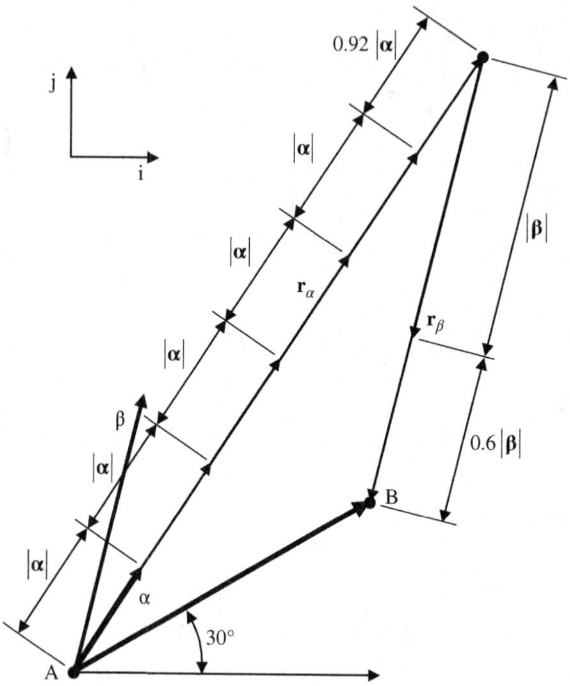

Figure 7.22 Representation of vector **r** in the general base $[\boldsymbol{\alpha}\ \ \boldsymbol{\beta}]$

Example 2. A force vector **f** is given by

$$\mathbf{f} = [\boldsymbol{\alpha}\ \ \boldsymbol{\beta}]\begin{bmatrix} 3 \\ 15 \end{bmatrix}, \tag{7.97}$$

where

$$[\boldsymbol{\alpha}\ \ \boldsymbol{\beta}] = [\mathbf{i}\ \ \mathbf{j}]\begin{bmatrix} 2 & 2 \\ 3 & 8 \end{bmatrix} \tag{7.98}$$

and $[\mathbf{i}\ \ \mathbf{j}]$ is the orthonormal base. Calculate the force components in the new general base $[\boldsymbol{\xi}\ \ \boldsymbol{\eta}]$, where

$$[\boldsymbol{\xi}\ \ \boldsymbol{\eta}] = [\mathbf{i}\ \ \mathbf{j}]\begin{bmatrix} 1 & 0.7 \\ -1 & 1 \end{bmatrix}. \tag{7.99}$$

Solution:

$$\begin{aligned}\mathbf{f} &= [\boldsymbol{\alpha}\ \ \boldsymbol{\beta}]\begin{bmatrix}3\\15\end{bmatrix}\\ &= \left([\mathbf{i}\ \ \mathbf{j}]\begin{bmatrix}2 & 2\\3 & 8\end{bmatrix}\right)\begin{bmatrix}3\\15\end{bmatrix}\\ &= [\mathbf{i}\ \ \mathbf{j}]\left(\begin{bmatrix}2 & 2\\3 & 8\end{bmatrix}\begin{bmatrix}3\\15\end{bmatrix}\right) = [\mathbf{i}\ \ \mathbf{j}]\left(3\begin{bmatrix}2\\3\end{bmatrix}+15\begin{bmatrix}2\\8\end{bmatrix}\right)\\ &= [\mathbf{i}\ \ \mathbf{j}]\begin{bmatrix}36\\129\end{bmatrix}.\end{aligned} \tag{7.100}$$

Also,

$$\begin{aligned}\mathbf{f} &= [\boldsymbol{\xi}\ \ \boldsymbol{\eta}]\begin{bmatrix}f_\xi\\f_\eta\end{bmatrix} = [\boldsymbol{\xi}\ \ \boldsymbol{\eta}]\begin{bmatrix}\underline{\boldsymbol{\xi}}\bullet\mathbf{f}\\\underline{\boldsymbol{\eta}}\bullet\mathbf{f}\end{bmatrix}\\ &= [\boldsymbol{\xi}\ \ \boldsymbol{\eta}]\left(\begin{bmatrix}\underline{\xi}_i & \underline{\xi}_j\\\underline{\eta}_i & \underline{\eta}_j\end{bmatrix}\begin{bmatrix}f_i\\f_j\end{bmatrix}\right) = [\boldsymbol{\xi}\ \ \boldsymbol{\eta}]\left(\begin{bmatrix}\xi_i & \eta_i\\\xi_j & \eta_j\end{bmatrix}^{-1}\begin{bmatrix}f_i\\f_j\end{bmatrix}\right),\end{aligned} \tag{7.101}$$

which after substitution of the numerical values yields

$$\begin{aligned}\mathbf{f} &= [\boldsymbol{\xi}\ \ \boldsymbol{\eta}]\left(\begin{bmatrix}1 & 0.7\\-1 & 1\end{bmatrix}^{-1}\begin{bmatrix}36\\129\end{bmatrix}\right)\\ &= [\boldsymbol{\xi}\ \ \boldsymbol{\eta}]\begin{bmatrix}-31.9\\97.1\end{bmatrix}\\ &= \begin{bmatrix}-31.9\\97.1\end{bmatrix}.\end{aligned} \tag{7.102}$$

Alternatively,

$$\begin{aligned}\mathbf{f} &= [\boldsymbol{\xi}\ \ \boldsymbol{\eta}]\begin{bmatrix}f_\xi\\f_\eta\end{bmatrix} = [\boldsymbol{\xi}\ \ \boldsymbol{\eta}]\begin{bmatrix}\underline{\boldsymbol{\xi}}\bullet\mathbf{f}\\\underline{\boldsymbol{\eta}}\bullet\mathbf{f}\end{bmatrix}\\ &= [\boldsymbol{\xi}\ \ \boldsymbol{\eta}]\left(\begin{bmatrix}\underline{\xi}_i & \underline{\xi}_j\\\underline{\eta}_i & \underline{\eta}_j\end{bmatrix}\begin{bmatrix}\alpha_i & \beta_i\\\alpha_j & \beta_j\end{bmatrix}\right)\begin{bmatrix}f_\alpha\\f_\beta\end{bmatrix}\\ &= [\boldsymbol{\xi}\ \ \boldsymbol{\eta}]\left(\begin{bmatrix}\xi_i & \eta_i\\\xi_j & \eta_j\end{bmatrix}^{-1}\begin{bmatrix}\alpha_i & \beta_i\\\alpha_j & \beta_j\end{bmatrix}\right)\begin{bmatrix}f_\alpha\\f_\beta\end{bmatrix}.\end{aligned} \tag{7.103}$$

Substituting the numerical values

$$\begin{aligned}
\mathbf{f} &= [\xi \ \ \eta]\left(\begin{bmatrix} 1 & 0.7 \\ -1 & 1 \end{bmatrix}^{-1} \begin{bmatrix} 2 & 2 \\ 3 & 8 \end{bmatrix}\right)\begin{bmatrix} 3 \\ 15 \end{bmatrix} \\
&= [\xi \ \ \eta]\left(\frac{1}{1.7}\begin{bmatrix} 1 & -0.7 \\ 1 & 1 \end{bmatrix} \begin{bmatrix} 2 & 2 \\ 3 & 8 \end{bmatrix}\right)\begin{bmatrix} 3 \\ 15 \end{bmatrix} \\
&= [\xi \ \ \eta]\frac{1}{1.7}\begin{bmatrix} 1 & -0.7 \\ 1 & 1 \end{bmatrix} \begin{bmatrix} 36 \\ 129 \end{bmatrix} \\
&= [\xi \ \ \eta]\begin{bmatrix} -31.9 \\ 97.1 \end{bmatrix} \\
&= \begin{bmatrix} -31.9 \\ 97.1 \end{bmatrix}.
\end{aligned} \tag{7.104}$$

7.11 Summary

In this chapter, physical quantities known as vectors have been introduced. Both the graphical representation and matrix representation of vectors have been explained using either orthonormal or general vector bases.

Changing vector bases is demonstrated to be a relatively simple operation as well as switching from orthonormal to general bases, which basically reduces to a matrix multiplication operation. General bases were discussed in detail, for they usually simplify the mathematical representation of the physics and mechanics involved.

The reader should become familiar with these concepts, before moving on to the tensor chapters.

Further Reading

[1] Abadir, K. M. and Magnus, J. R. (2005) *Matrix Algebra*. Cambridge, Cambridge University Press.

[2] Abraham, R., Marsden, J. E. and Ratiu, T. S. (1988) *Manifolds, Tensor Analysis, and Applications*. New York, Springer-Verlag.

[3] Bronstein, I. N. and Semendyayev, S. (2007) *Handbook of Mathematics*. New York.

[4] Fleisch, D. (2012) *A Student's Guide to Vectors and Tensors*. New York, Cambridge University Press.

[5] Golub, G. H. and van Van Loan, C. F. (1996) *Matrix Computations*, 3rd edn. Baltimore, MD, Johns Hopkins University Press.

[6] Kusse, B. R. and Westwig, E. A. (2006) *Mathematical Physics: Applied Mathematics for Scientists and Engineers*, 2nd edn. Wiley-VCH Verlag GmbH & Co.

[7] Levi-Civita, T. (2013) *The Absolute Differential Calculus (Calculus of Tensors)*. New York, Dover Publications.

[8] Matthews, P. C. (1998) *Vector Calculus*. London, Springer.

[9] Munjiza, A. (2004) *The Combined Finite-Discrete Element Method*. Hoboken, NJ, John Wiley & Sons, Inc.

[10] Rudin, W. (1976) *The Principles of Mathematical Analysis*. New York, McGraw-Hill Science/Engineering/Math.

8

Vectors in 3D

8.1 Vectors in 3D

The easiest way to represent a vector in 2D is by a line with arrow. In 3D space this is an impractical way of describing displacements, velocities, forces, accelerations, etc. Instead, a 3D orthonormal base is introduced

$$[\mathbf{i} \ \ \mathbf{j} \ \ \mathbf{k}]. \tag{8.1}$$

For the sake of clarity in further text this base is referred to as the global base. A given vector **a** is described by a linear combination of three orthonormal vectors.

$$\mathbf{a} = a_i\mathbf{i} + a_j\mathbf{j} + a_k\mathbf{k}. \tag{8.2}$$

In a similar manner, a given vector **b** is given by:

$$\mathbf{b} = b_i\mathbf{i} + b_j\mathbf{j} + b_k\mathbf{k}. \tag{8.3}$$

A sum of vectors **a** and **b** becomes

$$\mathbf{a} + \mathbf{b} = \mathbf{b} + \mathbf{a} = (a_i + b_i)\mathbf{i} + (a_j + b_j)\mathbf{j} + (a_k + b_k)\mathbf{k}. \tag{8.4}$$

Vector **a** multiplied by a scalar s becomes:

$$s\mathbf{a} = \mathbf{a}s = sa_i\mathbf{i} + sa_j\mathbf{j} + sa_k\mathbf{k}. \tag{8.5}$$

Large Strain Finite Element Method: A Practical Course, First Edition. Antonio Munjiza, Esteban Rougier and Earl E. Knight.

The scalar product of vectors **a** and **b** (also called the dot or inner product) is given by:

$$\begin{aligned}\mathbf{a}\bullet\mathbf{b} &= \mathbf{b}\bullet\mathbf{a}\\ &= (a_i\mathbf{i}+a_j\mathbf{j}+a_k\mathbf{k})\cdot(b_i\mathbf{i}+b_j\mathbf{j}+b_k\mathbf{k})\\ &= (a_ib_i)(\mathbf{i}\bullet\mathbf{i})+(a_ib_j)(\mathbf{i}\bullet\mathbf{j})+(a_ib_k)(\mathbf{i}\bullet\mathbf{k})\\ &\quad+(a_jb_i)(\mathbf{j}\bullet\mathbf{i})+(a_jb_j)(\mathbf{j}\bullet\mathbf{j})+(a_jb_k)(\mathbf{j}\bullet\mathbf{k})\\ &\quad+(a_kb_i)(\mathbf{k}\bullet\mathbf{i})+(a_kb_j)(\mathbf{k}\bullet\mathbf{j})+(a_kb_k)(\mathbf{k}\bullet\mathbf{k})\\ &= a_ib_i+a_jb_j+a_kb_k.\end{aligned} \tag{8.6}$$

Equation (8.2) can also be written using matrices

$$\mathbf{a}=\begin{bmatrix}\mathbf{i} & \mathbf{j} & \mathbf{k}\end{bmatrix}\begin{bmatrix}a_i\\ a_j\\ a_k\end{bmatrix}=a_i\mathbf{i}+a_j\mathbf{j}+a_k\mathbf{k} \tag{8.7}$$

or simply

$$\mathbf{a}=\begin{bmatrix}a_i\\ a_j\\ a_k\end{bmatrix} \text{ and } \mathbf{b}=\begin{bmatrix}b_i\\ b_j\\ b_k\end{bmatrix}. \tag{8.8}$$

where the equal sign means that a vector is only represented by a matrix.

The sum of the two vectors becomes:

$$\mathbf{a}+\mathbf{b}=\begin{bmatrix}a_i\\ a_j\\ a_k\end{bmatrix}+\begin{bmatrix}b_i\\ b_j\\ b_k\end{bmatrix}=\begin{bmatrix}a_i+b_i\\ a_j+b_j\\ a_k+b_k\end{bmatrix}. \tag{8.9}$$

A vector multiplied by a scalar becomes

$$s\mathbf{a}=s\begin{bmatrix}a_i\\ a_j\\ a_k\end{bmatrix}=\begin{bmatrix}sa_i\\ sa_j\\ sa_k\end{bmatrix} \tag{8.10}$$

and the scalar product becomes:

$$\begin{aligned}\mathbf{a}\bullet\mathbf{b} &= \begin{bmatrix}a_i & a_j & a_k\end{bmatrix}\begin{bmatrix}b_i\\ b_j\\ b_k\end{bmatrix}\\ &= \begin{bmatrix}b_i & b_j & b_k\end{bmatrix}\begin{bmatrix}a_i\\ a_j\\ a_k\end{bmatrix}=a_ib_i+a_jb_j+a_kb_k.\end{aligned} \tag{8.11}$$

8.2 Vector Bases

Nonorthonormal Base. Apart from the global orthonormal base introduced in Equation (8.1) it is very often convenient and sometimes necessary to use other bases of generally nonunit vectors that are generally not orthogonal to each other. Such bases are called nonorthonormal bases or general bases.

Individual vectors of any nonorthonormal base are best represented as a linear combination of the global base vectors, such that

$$\begin{aligned}\boldsymbol{\alpha} &= \alpha_i\mathbf{i} + \alpha_j\mathbf{j} + \alpha_k\mathbf{k}\\ \boldsymbol{\beta} &= \beta_i\mathbf{i} + \beta_j\mathbf{j} + \beta_k\mathbf{k}\\ \boldsymbol{\gamma} &= \gamma_i\mathbf{i} + \gamma_j\mathbf{j} + \gamma_k\mathbf{k}\end{aligned} \tag{8.12}$$

or, in matrix form:

$$\begin{bmatrix}\boldsymbol{\alpha} & \boldsymbol{\beta} & \boldsymbol{\gamma}\end{bmatrix} = \begin{bmatrix}\mathbf{i} & \mathbf{j} & \mathbf{k}\end{bmatrix}\begin{bmatrix}\alpha_i & \beta_i & \gamma_i\\ \alpha_j & \beta_j & \gamma_j\\ \alpha_k & \beta_k & \gamma_k\end{bmatrix} = \begin{bmatrix}\alpha_i & \beta_i & \gamma_i\\ \alpha_j & \beta_j & \gamma_j\\ \alpha_k & \beta_k & \gamma_k\end{bmatrix}. \tag{8.13}$$

By using the base

$$\begin{bmatrix}\boldsymbol{\alpha} & \boldsymbol{\beta} & \boldsymbol{\gamma}\end{bmatrix} \tag{8.14}$$

any vector **a** can be represented as

$$\mathbf{a} = a_\alpha\boldsymbol{\alpha} + a_\beta\boldsymbol{\beta} + a_\gamma\boldsymbol{\gamma} = \begin{bmatrix}a_\alpha\\ a_\beta\\ a_\gamma\end{bmatrix}. \tag{8.15}$$

The Dual Base. At this point, it is convenient to introduce the dual base

$$\begin{bmatrix}\underline{\boldsymbol{\alpha}} & \underline{\boldsymbol{\beta}} & \underline{\boldsymbol{\gamma}}\end{bmatrix}, \tag{8.16}$$

such that

$$\begin{aligned}&\boldsymbol{\alpha}\bullet\underline{\boldsymbol{\alpha}} = \boldsymbol{\beta}\bullet\underline{\boldsymbol{\beta}} = \boldsymbol{\gamma}\bullet\underline{\boldsymbol{\gamma}} = 1\\ &\underline{\boldsymbol{\alpha}}\bullet\boldsymbol{\beta} = \underline{\boldsymbol{\alpha}}\bullet\boldsymbol{\gamma} = \underline{\boldsymbol{\beta}}\bullet\boldsymbol{\alpha} = \underline{\boldsymbol{\beta}}\bullet\boldsymbol{\gamma} = \underline{\boldsymbol{\gamma}}\bullet\boldsymbol{\alpha} = \underline{\boldsymbol{\gamma}}\bullet\boldsymbol{\beta} = 0,\end{aligned} \tag{8.17}$$

i.e.,

$$\begin{bmatrix}\boldsymbol{\alpha} & \boldsymbol{\beta} & \boldsymbol{\gamma}\end{bmatrix}\bullet\begin{bmatrix}\underline{\boldsymbol{\alpha}}\\ \underline{\boldsymbol{\beta}}\\ \underline{\boldsymbol{\gamma}}\end{bmatrix} = \begin{bmatrix}1 & 0 & 0\\ 0 & 1 & 0\\ 0 & 0 & 1\end{bmatrix} = \mathbf{I}, \tag{8.18}$$

where

$$\begin{bmatrix}\underline{\boldsymbol{\alpha}}\\ \underline{\boldsymbol{\beta}}\\ \underline{\boldsymbol{\gamma}}\end{bmatrix} = \begin{bmatrix}\underline{\boldsymbol{\alpha}} & \underline{\boldsymbol{\beta}} & \underline{\boldsymbol{\gamma}}\end{bmatrix}^{\mathrm{T}} = \begin{bmatrix}\underline{\alpha}_i & \underline{\alpha}_j & \underline{\alpha}_k\\ \underline{\beta}_i & \underline{\beta}_j & \underline{\beta}_k\\ \underline{\gamma}_i & \underline{\gamma}_j & \underline{\gamma}_k\end{bmatrix} \tag{8.19}$$

$$[\boldsymbol{\alpha}\ \boldsymbol{\beta}\ \boldsymbol{\gamma}] = \begin{bmatrix} \alpha_i & \beta_i & \gamma_i \\ \alpha_j & \beta_j & \gamma_j \\ \alpha_k & \beta_k & \gamma_k \end{bmatrix}. \tag{8.20}$$

It is worth noting that in Equation (8.19) each row represents one base vector of the dual base, while in Equation (8.20) each column of the matrix represents one base vector of the general base. Thus the orthogonality condition (8.17) becomes

$$\begin{bmatrix} \alpha_i & \beta_i & \gamma_i \\ \alpha_j & \beta_j & \gamma_j \\ \alpha_k & \beta_k & \gamma_k \end{bmatrix} \begin{bmatrix} \underline{\alpha}_i & \underline{\alpha}_j & \underline{\alpha}_k \\ \underline{\beta}_i & \underline{\beta}_j & \underline{\beta}_k \\ \underline{\gamma}_i & \underline{\gamma}_j & \underline{\gamma}_k \end{bmatrix} = \mathbf{I}, \tag{8.21}$$

which yields

$$\begin{bmatrix} \underline{\alpha}_i & \underline{\alpha}_j & \underline{\alpha}_k \\ \underline{\beta}_i & \underline{\beta}_j & \underline{\beta}_k \\ \underline{\gamma}_i & \underline{\gamma}_j & \underline{\gamma}_k \end{bmatrix} = \begin{bmatrix} \alpha_i & \beta_i & \gamma_i \\ \alpha_j & \beta_j & \gamma_j \\ \alpha_k & \beta_k & \gamma_k \end{bmatrix}^{-1}. \tag{8.22}$$

It is also worth noting that vectors **i**, **j** and **k** can be expressed using the base $[\boldsymbol{\alpha}\ \boldsymbol{\beta}\ \boldsymbol{\gamma}]$

$$[\mathbf{i}\ \mathbf{j}\ \mathbf{k}] = [\boldsymbol{\alpha}\ \boldsymbol{\beta}\ \boldsymbol{\gamma}] \begin{bmatrix} i_\alpha & j_\alpha & k_\alpha \\ i_\beta & j_\beta & k_\beta \\ i_\gamma & j_\gamma & k_\gamma \end{bmatrix}. \tag{8.23}$$

By substituting from Equation (8.13) one obtains

$$[\mathbf{i}\ \mathbf{j}\ \mathbf{k}] = [\mathbf{i}\ \mathbf{j}\ \mathbf{k}] \begin{bmatrix} \alpha_i & \beta_i & \gamma_i \\ \alpha_j & \beta_j & \gamma_j \\ \alpha_k & \beta_k & \gamma_k \end{bmatrix} \begin{bmatrix} i_\alpha & j_\alpha & k_\alpha \\ i_\beta & j_\beta & k_\beta \\ i_\gamma & j_\gamma & k_\gamma \end{bmatrix}. \tag{8.24}$$

This means that

$$\begin{bmatrix} \alpha_i & \beta_i & \gamma_i \\ \alpha_j & \beta_j & \gamma_j \\ \alpha_k & \beta_k & \gamma_k \end{bmatrix} \begin{bmatrix} i_\alpha & j_\alpha & k_\alpha \\ i_\beta & j_\beta & k_\beta \\ i_\gamma & j_\gamma & k_\gamma \end{bmatrix} = \mathbf{I}, \tag{8.25}$$

which yields

$$\begin{bmatrix} i_\alpha & j_\alpha & k_\alpha \\ i_\beta & j_\beta & k_\beta \\ i_\gamma & j_\gamma & k_\gamma \end{bmatrix} = \begin{bmatrix} \alpha_i & \beta_i & \gamma_i \\ \alpha_j & \beta_j & \gamma_j \\ \alpha_k & \beta_k & \gamma_k \end{bmatrix}^{-1}, \tag{8.26}$$

or simply

$$\begin{bmatrix} \underline{\alpha}_i & \underline{\alpha}_j & \underline{\alpha}_k \\ \underline{\beta}_i & \underline{\beta}_j & \underline{\beta}_k \\ \underline{\gamma}_i & \underline{\gamma}_j & \underline{\gamma}_k \end{bmatrix} = \begin{bmatrix} i_\alpha & j_\alpha & k_\alpha \\ i_\beta & j_\beta & k_\beta \\ i_\gamma & j_\gamma & k_\gamma \end{bmatrix} = \begin{bmatrix} \alpha_i & \beta_i & \gamma_i \\ \alpha_j & \beta_j & \gamma_j \\ \alpha_k & \beta_k & \gamma_k \end{bmatrix}^{-1}. \tag{8.27}$$

Changing Bases in 3D. A vector **a** in the base $[\boldsymbol{\alpha}\ \ \boldsymbol{\beta}\ \ \boldsymbol{\gamma}]$ is given as a linear combination of the base vectors $\boldsymbol{\alpha}$, $\boldsymbol{\beta}$ and $\boldsymbol{\gamma}$

$$\begin{aligned}\mathbf{a} &= a_\alpha \boldsymbol{\alpha} + a_\beta \boldsymbol{\beta} + a_\gamma \boldsymbol{\gamma} \\ &= [\boldsymbol{\alpha}\ \ \boldsymbol{\beta}\ \ \boldsymbol{\gamma}] \begin{bmatrix} a_\alpha \\ a_\beta \\ a_\gamma \end{bmatrix}.\end{aligned} \tag{8.28}$$

The base vectors $\boldsymbol{\alpha}$, $\boldsymbol{\beta}$ and $\boldsymbol{\gamma}$ are given by

$$\begin{aligned}\boldsymbol{\alpha} &= \alpha_i \mathbf{i} + \alpha_j \mathbf{j} + \alpha_k \mathbf{k} \\ \boldsymbol{\beta} &= \beta_i \mathbf{i} + \beta_j \mathbf{j} + \beta_k \mathbf{k} \\ \boldsymbol{\gamma} &= \gamma_i \mathbf{i} + \gamma_j \mathbf{j} + \gamma_k \mathbf{k}.\end{aligned} \tag{8.29}$$

When substituted into Equation (8.28), one obtains

$$\begin{aligned}\mathbf{a} &= a_\alpha [\alpha_i \mathbf{i} + \alpha_j \mathbf{j} + \alpha_k \mathbf{k}] \\ &\quad + a_\beta [\beta_i \mathbf{i} + \beta_j \mathbf{j} + \beta_k \mathbf{k}] \\ &\quad + a_\gamma [\gamma_i \mathbf{i} + \gamma_j \mathbf{j} + \gamma_k \mathbf{k}] \\ &= (a_\alpha \alpha_i + a_\beta \beta_i + a_\gamma \gamma_i)\mathbf{i} \\ &\quad + (a_\alpha \alpha_j + a_\beta \beta_j + a_\gamma \gamma_j)\mathbf{j} \\ &\quad + (a_\alpha \alpha_k + a_\beta \beta_k + a_\gamma \gamma_k)\mathbf{k}\end{aligned} \tag{8.30}$$

or in a matrix form:

$$\begin{aligned}\mathbf{a} &= [\mathbf{i}\ \ \mathbf{j}\ \ \mathbf{k}] \begin{bmatrix} a_i \\ a_j \\ a_k \end{bmatrix} \\ &= [\mathbf{i}\ \ \mathbf{j}\ \ \mathbf{k}] \left(\begin{bmatrix} \alpha_i & \beta_i & \gamma_i \\ \alpha_j & \beta_j & \gamma_j \\ \alpha_k & \beta_k & \gamma_k \end{bmatrix} \begin{bmatrix} a_\alpha \\ a_\beta \\ a_\gamma \end{bmatrix} \right) = \begin{bmatrix} \alpha_i & \beta_i & \gamma_i \\ \alpha_j & \beta_j & \gamma_j \\ \alpha_k & \beta_k & \gamma_k \end{bmatrix} \begin{bmatrix} a_\alpha \\ a_\beta \\ a_\gamma \end{bmatrix}.\end{aligned} \tag{8.31}$$

In a similar manner,

$$\begin{aligned}\begin{bmatrix} a_\alpha \\ a_\beta \\ a_\gamma \end{bmatrix} &= \begin{bmatrix} \alpha_i & \beta_i & \gamma_i \\ \alpha_j & \beta_j & \gamma_j \\ \alpha_k & \beta_k & \gamma_k \end{bmatrix}^{-1} \begin{bmatrix} a_i \\ a_j \\ a_k \end{bmatrix} = \begin{bmatrix} \underline{\alpha}_i & \underline{\alpha}_j & \underline{\alpha}_k \\ \underline{\beta}_i & \underline{\beta}_j & \underline{\beta}_k \\ \underline{\gamma}_i & \underline{\gamma}_j & \underline{\gamma}_k \end{bmatrix} \begin{bmatrix} a_i \\ a_j \\ a_k \end{bmatrix} \\ &= \begin{bmatrix} i_\alpha & j_\alpha & k_\alpha \\ i_\beta & j_\beta & k_\beta \\ i_\gamma & j_\gamma & k_\gamma \end{bmatrix} \begin{bmatrix} a_i \\ a_j \\ a_k \end{bmatrix} = a_i \begin{bmatrix} i_\alpha \\ i_\beta \\ i_\gamma \end{bmatrix} + a_j \begin{bmatrix} j_\alpha \\ j_\beta \\ j_\gamma \end{bmatrix} + a_k \begin{bmatrix} k_\alpha \\ k_\beta \\ k_\gamma \end{bmatrix}.\end{aligned} \tag{8.32}$$

The above are general formulas for expressing the same vector in any vector base by starting from the global base.

Example. A 3D vector **r** is given by

$$\mathbf{r}=[\mathbf{i}\ \ \mathbf{j}\ \ \mathbf{k}]\begin{bmatrix}1\\7\\4\end{bmatrix}. \tag{8.33}$$

Express this vector using the base

$$[\boldsymbol{\alpha}\ \ \boldsymbol{\beta}\ \ \boldsymbol{\gamma}]=\begin{bmatrix}2&1&0\\2&3&2\\1&1&4\end{bmatrix}. \tag{8.34}$$

Using Equation (8.32)

$$\begin{aligned}
\mathbf{r}&=\begin{bmatrix}r_\alpha\\r_\beta\\r_\gamma\end{bmatrix}\\
&=\begin{bmatrix}i_\alpha & j_\alpha & k_\alpha\\ i_\beta & j_\beta & k_\beta\\ i_\gamma & j_\gamma & k_\gamma\end{bmatrix}\begin{bmatrix}r_i\\r_j\\r_k\end{bmatrix}\\
&=\begin{bmatrix}\alpha_i & \beta_i & \gamma_i\\ \alpha_j & \beta_j & \gamma_j\\ \alpha_k & \beta_k & \gamma_k\end{bmatrix}^{-1}\begin{bmatrix}r_i\\r_j\\r_k\end{bmatrix}\\
&=\begin{bmatrix}2&1&0\\2&3&2\\1&1&4\end{bmatrix}^{-1}\begin{bmatrix}1\\7\\4\end{bmatrix}\\
&=\frac{1}{14}\begin{bmatrix}10&-4&2\\-6&8&-4\\-1&-1&4\end{bmatrix}\begin{bmatrix}1\\7\\4\end{bmatrix}\\
&=1\left(\frac{1}{14}\begin{bmatrix}10\\-6\\-1\end{bmatrix}\right)+7\left(\frac{1}{14}\begin{bmatrix}-4\\8\\-1\end{bmatrix}\right)+4\left(\frac{1}{14}\begin{bmatrix}2\\-4\\4\end{bmatrix}\right)\\
&=\begin{bmatrix}-5/7\\17/7\\4/7\end{bmatrix}.
\end{aligned} \tag{8.35}$$

8.3 Summary

In this chapter an extension of vectors to 3D space is presented. Both orthonormal and general vector bases in conjunction with matrices are used to represent vectors in 3D.

Obtaining the matrix of a given 3D vector in any general base turns out to be a simple algebraic operation on the matrices representing both the vector and the base vectors.

The above material proves the convenience of using matrices to represent vector components in any generalized or orthonormal base. A reader not familiar with matrix algebra should consider reading Chapter 2, Matrices.

Further Reading

[1] Abadir, K. M. and Magnus, J. R. (2005) *Matrix Algebra*. Cambridge, Cambridge University Press.

[2] Abraham, R., Marsden, J. E and Ratiu, T. S. (1988) *Manifolds, Tensor Analysis, and Applications*. New York, Springer-Verlag.

[3] Bronstein, I. N. and Semendyayev, S. (2007) *Handbook of Mathematics*. New York, Springer.

[4] Fleisch, D. (2012) *A Student's Guide to Vectors and Tensors*. New York, Cambridge University Press.

[5] Golub, G. H. and van Van Loan, C. F. (1996) *Matrix Computations*, 3rd edn. Baltimore, MD, Johns Hopkins University Press.

[6] Kusse, B. R. and Westwig, E. A. (2006) *Mathematical Physics: Applied Mathematics for Scientists and Engineers*, 2nd edn, Wiley-VCH Verlag GmbH & Co.

[7] Levi-Civita, T. (2013) *The Absolute Differential Calculus (Calculus of Tensors)*. New York, Dover Publications.

[8] Matthews, P. C. (1998) *Vector Calculus*. London, Springer.

[9] Munjiza, A. (2004) *The Combined Finite-Discrete Element Method*. Hoboken, NJ, John Wiley & Sons, Inc.

[10] Rudin, W. (1976) *The Principles of Mathematical Analysis*. New York, McGraw-Hill Science/Engineering/Math.

9

Vectors in n-Dimensional Space

9.1 Extension from 3D to 4-Dimensional Space

So far vectors in 2D and 3D space have been introduced. First 2D vectors were described in detail in 2D and for 3D space the third dimension was added. To ensure completeness, discussions on a fourth dimension are necessary.

Any vector **a** in 4D space is simply

$$\mathbf{a} = a_i\mathbf{i} + a_j\mathbf{j} + a_k\mathbf{k} + a_t\mathbf{t}, \tag{9.1}$$

where the base

$$\begin{bmatrix} \mathbf{i} & \mathbf{j} & \mathbf{k} & \mathbf{t} \end{bmatrix} \tag{9.2}$$

is the orthonormal base consisting of four unit vectors that are orthogonal to each other. In 2D the same vector **a** was described as a line on a sheet of paper, Figure 9.1. In 3D the vector **a** was described as a 3D line of certain length facing in some direction in 3D space, Figure 9.2.

In both 2D and 3D, by using the concept of displacements (for example the trajectory of an insect moving from point A to point B), the vectors can be added by using a polygon of vectors; which simply means one adds the lines together,

However, the human brain and eye cannot visualize a 4D space environment and therefore the only way to represent vectors in 4D is by using the linear combination of the base vectors as shown in Equation (9.1).

Large Strain Finite Element Method: A Practical Course, First Edition. Antonio Munjiza, Esteban Rougier and Earl E. Knight.

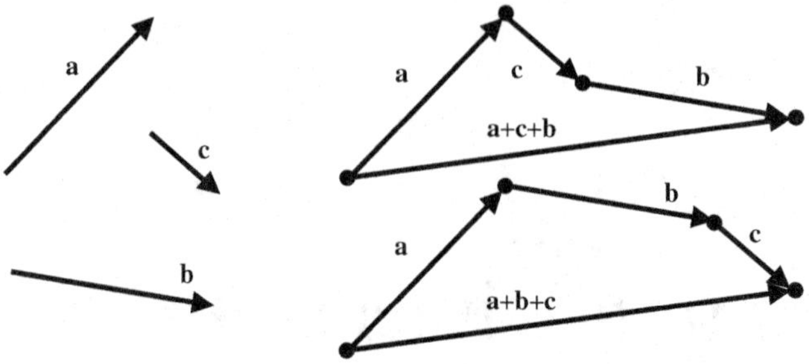

Figure 9.1 Adding vectors in 2D

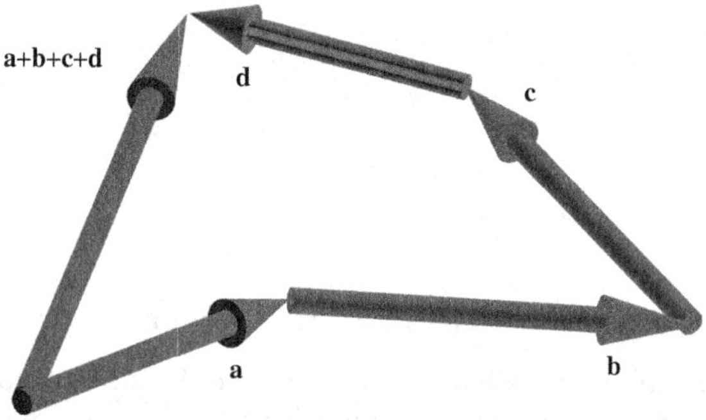

Figure 9.2 Adding vectors in 3D; a polygon of vectors is formed by arrow-ending rods attached to each other

Of course, different bases can be used. For instance, a general base of four nonunit nonorthogonal vectors would look as follows:

$$
\begin{aligned}
[\boldsymbol{\alpha}\ \boldsymbol{\beta}\ \boldsymbol{\gamma}\ \boldsymbol{\delta}] &= \begin{bmatrix} \alpha_i\mathbf{i}+\alpha_j\mathbf{j}+\alpha_k\mathbf{k}+\alpha_t\mathbf{t} \\ \beta_i\mathbf{i}+\beta_j\mathbf{j}+\beta_k\mathbf{k}+\beta_t\mathbf{t} \\ \gamma_i\mathbf{i}+\gamma_j\mathbf{j}+\gamma_k\mathbf{k}+\gamma_t\mathbf{t} \\ \delta_i\mathbf{i}+\delta_j\mathbf{j}+\delta_k\mathbf{k}+\delta_t\mathbf{t} \end{bmatrix}^{\mathrm{T}} \\
&= [\mathbf{i}\ \mathbf{j}\ \mathbf{k}\ \mathbf{t}] \begin{bmatrix} \alpha_i & \beta_i & \gamma_i & \delta_i \\ \alpha_j & \beta_j & \gamma_j & \delta_j \\ \alpha_k & \beta_k & \gamma_k & \delta_k \\ \alpha_t & \beta_t & \gamma_t & \delta_t \end{bmatrix}.
\end{aligned}
\tag{9.3}
$$

9.2 The Dual Base in 4D

Following the analogy with 3D, the dual base in 4D is given by

$$
\begin{bmatrix} \underline{\boldsymbol{\alpha}} \\ \underline{\boldsymbol{\beta}} \\ \underline{\boldsymbol{\gamma}} \\ \underline{\boldsymbol{\delta}} \end{bmatrix} = \begin{bmatrix} \underline{\alpha}_i & \underline{\alpha}_j & \underline{\alpha}_k & \underline{\alpha}_t \\ \underline{\beta}_i & \underline{\beta}_j & \underline{\beta}_k & \underline{\beta}_t \\ \underline{\gamma}_i & \underline{\gamma}_j & \underline{\gamma}_k & \underline{\gamma}_t \\ \underline{\delta}_i & \underline{\delta}_j & \underline{\delta}_k & \underline{\delta}_t \end{bmatrix} \begin{bmatrix} \mathbf{i} \\ \mathbf{j} \\ \mathbf{k} \\ \mathbf{t} \end{bmatrix} = \begin{bmatrix} \underline{\alpha}_i & \underline{\alpha}_j & \underline{\alpha}_k & \underline{\alpha}_t \\ \underline{\beta}_i & \underline{\beta}_j & \underline{\beta}_k & \underline{\beta}_t \\ \underline{\gamma}_i & \underline{\gamma}_j & \underline{\gamma}_k & \underline{\gamma}_t \\ \underline{\delta}_i & \underline{\delta}_j & \underline{\delta}_k & \underline{\delta}_t \end{bmatrix}.
\tag{9.4}
$$

This base must satisfy the duality condition

$$[\boldsymbol{\alpha}\ \boldsymbol{\beta}\ \boldsymbol{\gamma}\ \boldsymbol{\delta}]\bullet\begin{bmatrix}\underline{\boldsymbol{\alpha}}\\ \underline{\boldsymbol{\beta}}\\ \underline{\boldsymbol{\gamma}}\\ \underline{\boldsymbol{\delta}}\end{bmatrix}=\begin{bmatrix}1&0&0&0\\0&1&0&0\\0&0&1&0\\0&0&0&1\end{bmatrix}=\mathbf{I} \tag{9.5}$$

or

$$\begin{bmatrix}\alpha_i&\beta_i&\gamma_i&\delta_i\\ \alpha_j&\beta_j&\gamma_j&\delta_j\\ \alpha_k&\beta_k&\gamma_k&\delta_k\\ \alpha_t&\beta_t&\gamma_t&\delta_t\end{bmatrix}\begin{bmatrix}\underline{\alpha}_i&\underline{\alpha}_j&\underline{\alpha}_k&\underline{\alpha}_t\\ \underline{\beta}_i&\underline{\beta}_j&\underline{\beta}_k&\underline{\beta}_t\\ \underline{\gamma}_i&\underline{\gamma}_j&\underline{\gamma}_k&\underline{\gamma}_t\\ \underline{\delta}_i&\underline{\delta}_j&\underline{\delta}_k&\underline{\delta}_t\end{bmatrix}=\begin{bmatrix}1&0&0&0\\0&1&0&0\\0&0&1&0\\0&0&0&1\end{bmatrix}, \tag{9.6}$$

which yields

$$\begin{bmatrix}\underline{\boldsymbol{\alpha}}\\ \underline{\boldsymbol{\beta}}\\ \underline{\boldsymbol{\gamma}}\\ \underline{\boldsymbol{\delta}}\end{bmatrix}=\begin{bmatrix}\underline{\alpha}_i&\underline{\alpha}_j&\underline{\alpha}_k&\underline{\alpha}_t\\ \underline{\beta}_i&\underline{\beta}_j&\underline{\beta}_k&\underline{\beta}_t\\ \underline{\gamma}_i&\underline{\gamma}_j&\underline{\gamma}_k&\underline{\gamma}_t\\ \underline{\delta}_i&\underline{\delta}_j&\underline{\delta}_k&\underline{\delta}_t\end{bmatrix}=\begin{bmatrix}\alpha_i&\beta_i&\gamma_i&\delta_i\\ \alpha_j&\beta_j&\gamma_j&\delta_j\\ \alpha_k&\beta_k&\gamma_k&\delta_k\\ \alpha_t&\beta_t&\gamma_t&\delta_t\end{bmatrix}^{-1}. \tag{9.7}$$

As was done for the 3D case, one can write

$$[\mathbf{i}\ \mathbf{j}\ \mathbf{k}\ \mathbf{t}]=[\boldsymbol{\alpha}\ \boldsymbol{\beta}\ \boldsymbol{\gamma}\ \boldsymbol{\delta}]\begin{bmatrix}i_\alpha&j_\alpha&k_\alpha&t_\alpha\\ i_\beta&j_\beta&k_\beta&t_\beta\\ i_\gamma&j_\gamma&k_\gamma&t_\gamma\\ i_\delta&j_\delta&k_\delta&t_\delta\end{bmatrix}. \tag{9.8}$$

After substitution from Equation (9.3) this yields

$$\begin{bmatrix}i_\alpha&j_\alpha&k_\alpha&t_\alpha\\ i_\beta&j_\beta&k_\beta&t_\beta\\ i_\gamma&j_\gamma&k_\gamma&t_\gamma\\ i_\delta&j_\delta&k_\delta&t_\delta\end{bmatrix}=\begin{bmatrix}\alpha_i&\beta_i&\gamma_i&\delta_i\\ \alpha_j&\beta_j&\gamma_j&\delta_j\\ \alpha_k&\beta_k&\gamma_k&\delta_k\\ \alpha_t&\beta_t&\gamma_t&\delta_t\end{bmatrix}^{-1}, \tag{9.9}$$

which means that

$$\begin{bmatrix}\underline{\alpha}_i&\underline{\alpha}_j&\underline{\alpha}_k&\underline{\alpha}_t\\ \underline{\beta}_i&\underline{\beta}_j&\underline{\beta}_k&\underline{\beta}_t\\ \underline{\gamma}_i&\underline{\gamma}_j&\underline{\gamma}_k&\underline{\gamma}_t\\ \underline{\delta}_i&\underline{\delta}_j&\underline{\delta}_k&\underline{\delta}_t\end{bmatrix}=\begin{bmatrix}i_\alpha&j_\alpha&k_\alpha&t_\alpha\\ i_\beta&j_\beta&k_\beta&t_\beta\\ i_\gamma&j_\gamma&k_\gamma&t_\gamma\\ i_\delta&j_\delta&k_\delta&t_\delta\end{bmatrix}. \tag{9.10}$$

9.3 Changing the Base in 4D

A given 4D vector $\mathbf{a}$ can be described in the orthonormal base $[\mathbf{i}\ \ \mathbf{j}\ \ \mathbf{k}\ \ \mathbf{t}]$ as:

$$\mathbf{a} = a_i\mathbf{i} + a_j\mathbf{j} + a_k\mathbf{k} + a_t\mathbf{t} = [\mathbf{i}\ \ \mathbf{j}\ \ \mathbf{k}\ \ \mathbf{t}]\begin{bmatrix} a_i \\ a_j \\ a_k \\ a_t \end{bmatrix}. \tag{9.11}$$

The same vector in the general base $[\boldsymbol{\alpha}\ \ \boldsymbol{\beta}\ \ \boldsymbol{\gamma}\ \ \boldsymbol{\delta}]$ is given by

$$\mathbf{a} = a_\alpha\boldsymbol{\alpha} + a_\beta\boldsymbol{\beta} + a_\gamma\boldsymbol{\gamma} + a_\delta\boldsymbol{\delta} = [\boldsymbol{\alpha}\ \ \boldsymbol{\beta}\ \ \boldsymbol{\gamma}\ \ \boldsymbol{\delta}]\begin{bmatrix} a_\alpha \\ a_\beta \\ a_\gamma \\ a_\delta \end{bmatrix}. \tag{9.12}$$

By substituting

$$[\boldsymbol{\alpha}\ \ \boldsymbol{\beta}\ \ \boldsymbol{\gamma}\ \ \boldsymbol{\delta}] = [\mathbf{i}\ \ \mathbf{j}\ \ \mathbf{k}\ \ \mathbf{t}]\begin{bmatrix} \alpha_i & \beta_i & \gamma_i & \delta_i \\ \alpha_j & \beta_j & \gamma_j & \delta_j \\ \alpha_k & \beta_k & \gamma_k & \delta_k \\ \alpha_t & \beta_t & \gamma_t & \delta_t \end{bmatrix} \tag{9.13}$$

into Equation (9.12) the following is obtained

$$\mathbf{a} = [\mathbf{i}\ \ \mathbf{j}\ \ \mathbf{k}\ \ \mathbf{t}]\left(\begin{bmatrix} \alpha_i & \beta_i & \gamma_i & \delta_i \\ \alpha_j & \beta_j & \gamma_j & \delta_j \\ \alpha_k & \beta_k & \gamma_k & \delta_k \\ \alpha_t & \beta_t & \gamma_t & \delta_t \end{bmatrix}\begin{bmatrix} a_\alpha \\ a_\beta \\ a_\gamma \\ a_\delta \end{bmatrix}\right) \tag{9.14}$$

which must be the same as

$$\mathbf{a} = [\mathbf{i}\ \ \mathbf{j}\ \ \mathbf{k}\ \ \mathbf{t}]\begin{bmatrix} a_i \\ a_j \\ a_k \\ a_t \end{bmatrix}. \tag{9.15}$$

Thus,

$$\begin{aligned} \begin{bmatrix} a_i \\ a_j \\ a_k \\ a_t \end{bmatrix} &= \begin{bmatrix} \alpha_i & \beta_i & \gamma_i & \delta_i \\ \alpha_j & \beta_j & \gamma_j & \delta_j \\ \alpha_k & \beta_k & \gamma_k & \delta_k \\ \alpha_t & \beta_t & \gamma_t & \delta_t \end{bmatrix}\begin{bmatrix} a_\alpha \\ a_\beta \\ a_\gamma \\ a_\delta \end{bmatrix} \\ &= a_\alpha\begin{bmatrix} \alpha_i \\ \alpha_j \\ \alpha_k \\ \alpha_t \end{bmatrix} + a_\beta\begin{bmatrix} \beta_i \\ \beta_j \\ \beta_k \\ \beta_t \end{bmatrix} + a_\gamma\begin{bmatrix} \gamma_i \\ \gamma_j \\ \gamma_k \\ \gamma_t \end{bmatrix} + a_\delta\begin{bmatrix} \delta_i \\ \delta_j \\ \delta_k \\ \delta_t \end{bmatrix}. \end{aligned} \tag{9.16}$$

This notation enhances the process of writing equations directly from the first principles; for example, one reads Equation (9.16) as: vector **a** is made of a_α vectors $\boldsymbol{\alpha}$ plus a_β vectors $\boldsymbol{\beta}$ plus a_γ vectors $\boldsymbol{\gamma}$ plus a_δ vectors $\boldsymbol{\delta}$. Vectors $\boldsymbol{\alpha}$, $\boldsymbol{\beta}$, $\boldsymbol{\gamma}$, and $\boldsymbol{\delta}$ are in turn made of α_i vectors **i** plus α_j vectors **j** plus α_k vectors **k** plus α_t vectors **t**, etc.

In a similar manner

$$\begin{bmatrix} a_\alpha \\ a_\beta \\ a_\gamma \\ a_\delta \end{bmatrix} = \begin{bmatrix} \alpha_i & \beta_i & \gamma_i & \delta_i \\ \alpha_j & \beta_j & \gamma_j & \delta_j \\ \alpha_k & \beta_k & \gamma_k & \delta_k \\ \alpha_t & \beta_t & \gamma_t & \delta_t \end{bmatrix}^{-1} \begin{bmatrix} a_i \\ a_j \\ a_k \\ a_t \end{bmatrix}, \tag{9.17}$$

where

$$\begin{bmatrix} \alpha_i & \beta_i & \gamma_i & \delta_i \\ \alpha_j & \beta_j & \gamma_j & \delta_j \\ \alpha_k & \beta_k & \gamma_k & \delta_k \\ \alpha_t & \beta_t & \gamma_t & \delta_t \end{bmatrix}^{-1} = \begin{bmatrix} \underline{\alpha}_i & \underline{\alpha}_j & \underline{\alpha}_k & \underline{\alpha}_t \\ \underline{\beta}_i & \underline{\beta}_j & \underline{\beta}_k & \underline{\beta}_t \\ \underline{\gamma}_i & \underline{\gamma}_j & \underline{\gamma}_k & \underline{\gamma}_t \\ \underline{\delta}_i & \underline{\delta}_j & \underline{\delta}_k & \underline{\delta}_t \end{bmatrix} = \begin{bmatrix} i_\alpha & j_\alpha & k_\alpha & t_\alpha \\ i_\beta & j_\beta & k_\beta & t_\beta \\ i_\gamma & j_\gamma & k_\gamma & t_\gamma \\ i_\delta & j_\delta & k_\delta & t_\delta \end{bmatrix}. \tag{9.18}$$

Thus,

$$\begin{aligned} \begin{bmatrix} a_\alpha \\ a_\beta \\ a_\gamma \\ a_\delta \end{bmatrix} &= \begin{bmatrix} \underline{\alpha}_i & \underline{\alpha}_j & \underline{\alpha}_k & \underline{\alpha}_t \\ \underline{\beta}_i & \underline{\beta}_j & \underline{\beta}_k & \underline{\beta}_t \\ \underline{\gamma}_i & \underline{\gamma}_j & \underline{\gamma}_k & \underline{\gamma}_t \\ \underline{\delta}_i & \underline{\delta}_j & \underline{\delta}_k & \underline{\delta}_t \end{bmatrix} \begin{bmatrix} a_i \\ a_j \\ a_k \\ a_t \end{bmatrix} = \begin{bmatrix} i_\alpha & j_\alpha & k_\alpha & t_\alpha \\ i_\beta & j_\beta & k_\beta & t_\beta \\ i_\gamma & j_\gamma & k_\gamma & t_\gamma \\ i_\delta & j_\delta & k_\delta & t_\delta \end{bmatrix} \begin{bmatrix} a_i \\ a_j \\ a_k \\ a_t \end{bmatrix} \\ &= a_i \begin{bmatrix} i_\alpha \\ i_\beta \\ i_\gamma \\ i_\delta \end{bmatrix} + a_j \begin{bmatrix} j_\alpha \\ j_\beta \\ j_\gamma \\ j_\delta \end{bmatrix} + a_k \begin{bmatrix} k_\alpha \\ k_\beta \\ k_\gamma \\ k_\delta \end{bmatrix} + a_t \begin{bmatrix} t_\alpha \\ t_\beta \\ t_\gamma \\ t_\delta \end{bmatrix}. \end{aligned} \tag{9.19}$$

This reads directly as vector **a** is made of a_i vectors **i** plus a_j vectors **j** plus a_k vectors **k** plus $a_t$vectors **t**, where each of these vectors **i**, **j**, **k** and **t** is in turn made of i_α vectors $\boldsymbol{\alpha}$ plus i_β vectors $\boldsymbol{\beta}$ plus i_γ vectors $\boldsymbol{\gamma}$ plus i_δ vectors $\boldsymbol{\delta}$, etc.

9.4 Generalization to n-Dimensional Space

Generalization to more dimensions is quite straightforward. However, one can expect to have large matrices. In a general n-dimensional space the orthonormal base is given by

$$[\mathbf{i}_1 \;\; \mathbf{i}_2 \;\; \mathbf{i}_3 \;\; \cdots \;\; \mathbf{i}_n]. \tag{9.20}$$

A nonorthonormal base is given by

$$[\boldsymbol{\alpha}_1 \;\; \boldsymbol{\alpha}_2 \;\; \boldsymbol{\alpha}_3 \;\; \cdots \;\; \boldsymbol{\alpha}_n], \tag{9.21}$$

where

$$\begin{bmatrix}\boldsymbol{\alpha}_1 & \boldsymbol{\alpha}_2 & \boldsymbol{\alpha}_3 & \cdots & \boldsymbol{\alpha}_n\end{bmatrix}=\begin{bmatrix}\mathbf{i}_1 & \mathbf{i}_2 & \mathbf{i}_3 & \cdots & \mathbf{i}_n\end{bmatrix}\begin{bmatrix}\alpha_{i_1 1} & \alpha_{i_1 2} & \alpha_{i_1 3} & \cdots & \alpha_{i_1 n}\\ \alpha_{i_2 1} & \alpha_{i_2 2} & \alpha_{i_2 3} & \cdots & \alpha_{i_2 n}\\ \alpha_{i_3 1} & \alpha_{i_3 2} & \alpha_{i_3 3} & \cdots & \alpha_{i_3 n}\\ \vdots & \vdots & \vdots & \ddots & \vdots\\ \alpha_{i_n 1} & \alpha_{i_n 2} & \alpha_{i_n 3} & \cdots & \alpha_{i_n n}\end{bmatrix}. \tag{9.22}$$

It is worth noting that in the above matrix $\alpha_{i_k j}$ means the component of the base vector $\boldsymbol{\alpha}_j$ in the direction of the base vector $\mathbf{i}_k$. Equation (9.22) can be shortened to

$$\begin{bmatrix}\boldsymbol{\alpha}_1 & \boldsymbol{\alpha}_2 & \boldsymbol{\alpha}_3 & \cdots & \boldsymbol{\alpha}_n\end{bmatrix}=\begin{bmatrix}\alpha_{i_1 1} & \alpha_{i_1 2} & \alpha_{i_1 3} & \cdots & \alpha_{i_1 n}\\ \alpha_{i_2 1} & \alpha_{i_2 2} & \alpha_{i_2 3} & \cdots & \alpha_{i_2 n}\\ \alpha_{i_3 1} & \alpha_{i_3 2} & \alpha_{i_3 3} & \cdots & \alpha_{i_3 n}\\ \vdots & \vdots & \vdots & \ddots & \vdots\\ \alpha_{i_n 1} & \alpha_{i_n 2} & \alpha_{i_n 3} & \cdots & \alpha_{i_n n}\end{bmatrix}, \tag{9.23}$$

where the orthonormal base is implicitly assumed and each column of the matrix represents one base vector. The dual base

$$\begin{bmatrix}\underline{\boldsymbol{\alpha}}_1\\ \underline{\boldsymbol{\alpha}}_2\\ \underline{\boldsymbol{\alpha}}_3\\ \vdots\\ \underline{\boldsymbol{\alpha}}_n\end{bmatrix} \tag{9.24}$$

is given by

$$\begin{bmatrix}\underline{\boldsymbol{\alpha}}_1\\ \underline{\boldsymbol{\alpha}}_2\\ \underline{\boldsymbol{\alpha}}_3\\ \vdots\\ \underline{\boldsymbol{\alpha}}_n\end{bmatrix}=\begin{bmatrix}\underline{\alpha}_{i_1 1} & \underline{\alpha}_{i_2 1} & \underline{\alpha}_{i_3 1} & \cdots & \underline{\alpha}_{i_n 1}\\ \underline{\alpha}_{i_1 2} & \underline{\alpha}_{i_2 2} & \underline{\alpha}_{i_3 2} & \cdots & \underline{\alpha}_{i_n 2}\\ \underline{\alpha}_{i_1 3} & \underline{\alpha}_{i_2 3} & \underline{\alpha}_{i_3 3} & \cdots & \underline{\alpha}_{i_n 3}\\ \vdots & \vdots & \vdots & \ddots & \vdots\\ \underline{\alpha}_{i_1 n} & \underline{\alpha}_{i_2 n} & \underline{\alpha}_{i_3 n} & \cdots & \underline{\alpha}_{i_n n}\end{bmatrix}\begin{bmatrix}\mathbf{i}_1\\ \mathbf{i}_2\\ \mathbf{i}_3\\ \vdots\\ \mathbf{i}_n\end{bmatrix} \tag{9.25}$$

or simply

$$\begin{bmatrix}\underline{\boldsymbol{\alpha}}_1\\ \underline{\boldsymbol{\alpha}}_2\\ \underline{\boldsymbol{\alpha}}_3\\ \vdots\\ \underline{\boldsymbol{\alpha}}_n\end{bmatrix}=\begin{bmatrix}\underline{\alpha}_{i_1 1} & \underline{\alpha}_{i_2 1} & \underline{\alpha}_{i_3 1} & \cdots & \underline{\alpha}_{i_n 1}\\ \underline{\alpha}_{i_1 2} & \underline{\alpha}_{i_2 2} & \underline{\alpha}_{i_3 2} & \cdots & \underline{\alpha}_{i_n 2}\\ \underline{\alpha}_{i_1 3} & \underline{\alpha}_{i_2 3} & \underline{\alpha}_{i_3 3} & \cdots & \underline{\alpha}_{i_n 3}\\ \vdots & \vdots & \vdots & \ddots & \vdots\\ \underline{\alpha}_{i_1 n} & \underline{\alpha}_{i_2 n} & \underline{\alpha}_{i_3 n} & \cdots & \underline{\alpha}_{i_n n}\end{bmatrix}, \tag{9.26}$$

where each row of the matrix represents one vector of the dual base. The duality condition is by definition given by:

$$\begin{bmatrix} \underline{\alpha}_{i_1 1} & \underline{\alpha}_{i_2 1} & \underline{\alpha}_{i_3 1} & \cdots & \underline{\alpha}_{i_n 1} \\ \underline{\alpha}_{i_1 2} & \underline{\alpha}_{i_2 2} & \underline{\alpha}_{i_3 2} & \cdots & \underline{\alpha}_{i_n 2} \\ \underline{\alpha}_{i_1 3} & \underline{\alpha}_{i_2 3} & \underline{\alpha}_{i_3 3} & \cdots & \underline{\alpha}_{i_n 3} \\ \vdots & \vdots & \vdots & \ddots & \vdots \\ \underline{\alpha}_{i_1 n} & \underline{\alpha}_{i_2 n} & \underline{\alpha}_{i_3 n} & \cdots & \underline{\alpha}_{i_n n} \end{bmatrix} \begin{bmatrix} \alpha_{i_1 1} & \alpha_{i_1 2} & \alpha_{i_1 3} & \cdots & \alpha_{i_1 n} \\ \alpha_{i_2 1} & \alpha_{i_2 2} & \alpha_{i_2 3} & \cdots & \alpha_{i_2 n} \\ \alpha_{i_3 1} & \alpha_{i_3 2} & \alpha_{i_3 3} & \cdots & \alpha_{i_3 n} \\ \vdots & \vdots & \vdots & \ddots & \vdots \\ \alpha_{i_n 1} & \alpha_{i_n 2} & \alpha_{i_n 3} & \cdots & \alpha_{i_n n} \end{bmatrix} = \mathbf{I}. \tag{9.27}$$

From the duality condition, it follows that

$$\begin{bmatrix} \underline{\alpha}_{i_1 1} & \underline{\alpha}_{i_2 1} & \underline{\alpha}_{i_3 1} & \cdots & \underline{\alpha}_{i_n 1} \\ \underline{\alpha}_{i_1 2} & \underline{\alpha}_{i_2 2} & \underline{\alpha}_{i_3 2} & \cdots & \underline{\alpha}_{i_n 2} \\ \underline{\alpha}_{i_1 3} & \underline{\alpha}_{i_2 3} & \underline{\alpha}_{i_3 3} & \cdots & \underline{\alpha}_{i_n 3} \\ \vdots & \vdots & \vdots & \ddots & \vdots \\ \underline{\alpha}_{i_1 n} & \underline{\alpha}_{i_2 n} & \underline{\alpha}_{i_3 n} & \cdots & \underline{\alpha}_{i_n n} \end{bmatrix} = \begin{bmatrix} \alpha_{i_1 1} & \alpha_{i_1 2} & \alpha_{i_1 3} & \cdots & \alpha_{i_1 n} \\ \alpha_{i_2 1} & \alpha_{i_2 2} & \alpha_{i_2 3} & \cdots & \alpha_{i_2 n} \\ \alpha_{i_3 1} & \alpha_{i_3 2} & \alpha_{i_3 3} & \cdots & \alpha_{i_3 n} \\ \vdots & \vdots & \vdots & \ddots & \vdots \\ \alpha_{i_n 1} & \alpha_{i_n 2} & \alpha_{i_n 3} & \cdots & \alpha_{i_n n} \end{bmatrix}^{-1}. \tag{9.28}$$

One can also write

$$\mathbf{i} = \begin{bmatrix} \boldsymbol{\alpha}_1 & \boldsymbol{\alpha}_2 & \boldsymbol{\alpha}_3 & \cdots & \boldsymbol{\alpha}_n \end{bmatrix} \begin{bmatrix} i_{\alpha_1 1} \\ i_{\alpha_2 1} \\ i_{\alpha_3 1} \\ \vdots \\ i_{\alpha_n 1} \end{bmatrix}, \tag{9.29}$$

where

$$\begin{aligned} i_{\alpha_1 1} &= \underline{\boldsymbol{\alpha}}_1 \bullet \mathbf{i}_1 = \underline{\boldsymbol{\alpha}}_{i_1 1} \\ i_{\alpha_2 1} &= \underline{\boldsymbol{\alpha}}_2 \bullet \mathbf{i}_1 = \underline{\boldsymbol{\alpha}}_{i_1 2} \\ i_{\alpha_3 1} &= \underline{\boldsymbol{\alpha}}_3 \bullet \mathbf{i}_1 = \underline{\boldsymbol{\alpha}}_{i_1 3} \\ &\vdots \\ i_{\alpha_n 1} &= \underline{\boldsymbol{\alpha}}_n \bullet \mathbf{i}_1 = \underline{\boldsymbol{\alpha}}_{i_1 n}. \end{aligned} \tag{9.30}$$

By continuing this process for the rest of the base vectors $\mathbf{i}_1, \mathbf{i}_2, \mathbf{i}_3, \ldots, \mathbf{i}_n$, one obtains

$$\begin{bmatrix} \mathbf{i}_1 & \mathbf{i}_2 & \mathbf{i}_3 & \cdots & \mathbf{i}_n \end{bmatrix} = \begin{bmatrix} \boldsymbol{\alpha}_1 & \boldsymbol{\alpha}_2 & \boldsymbol{\alpha}_3 & \cdots & \boldsymbol{\alpha}_n \end{bmatrix} \begin{bmatrix} i_{\alpha_1 1} & i_{\alpha_1 2} & i_{\alpha_1 3} & \cdots & i_{\alpha_1 n} \\ i_{\alpha_2 1} & i_{\alpha_2 2} & i_{\alpha_2 3} & \cdots & i_{\alpha_2 n} \\ i_{\alpha_3 1} & i_{\alpha_3 2} & i_{\alpha_3 3} & \cdots & i_{\alpha_3 n} \\ \vdots & \vdots & \vdots & \ddots & \vdots \\ i_{\alpha_n 1} & i_{\alpha_n 2} & i_{\alpha_n 3} & \cdots & i_{\alpha_n n}, \end{bmatrix} \tag{9.31}$$

where

$$\begin{bmatrix} i_{\alpha_1 1} & i_{\alpha_1 2} & i_{\alpha_1 3} & \dots & i_{\alpha_1 n} \\ i_{\alpha_2 1} & i_{\alpha_2 2} & i_{\alpha_2 3} & \dots & i_{\alpha_2 n} \\ i_{\alpha_3 1} & i_{\alpha_3 2} & i_{\alpha_3 3} & \dots & i_{\alpha_3 n} \\ \vdots & \vdots & \vdots & \ddots & \vdots \\ i_{\alpha_n 1} & i_{\alpha_n 2} & i_{\alpha_n 3} & \dots & i_{\alpha_n n} \end{bmatrix} = \begin{bmatrix} \alpha_{i_1 1} & \alpha_{i_2 1} & \alpha_{i_3 1} & \cdots & \alpha_{i_n 1} \\ \alpha_{i_1 2} & \alpha_{i_2 2} & \alpha_{i_3 2} & \cdots & \alpha_{i_n 2} \\ \alpha_{i_1 3} & \alpha_{i_2 3} & \alpha_{i_3 3} & \cdots & \alpha_{i_n 3} \\ \vdots & \vdots & \vdots & \ddots & \vdots \\ \alpha_{i_1 n} & \alpha_{i_2 n} & \alpha_{i_3 n} & \cdots & \alpha_{i_n n} \end{bmatrix} = \begin{bmatrix} \alpha_{i_1 1} & \alpha_{i_1 2} & \alpha_{i_1 3} & \cdots & \alpha_{i_1 n} \\ \alpha_{i_2 1} & \alpha_{i_2 2} & \alpha_{i_2 3} & \cdots & \alpha_{i_2 n} \\ \alpha_{i_3 1} & \alpha_{i_3 2} & \alpha_{i_3 3} & \cdots & \alpha_{i_3 n} \\ \vdots & \vdots & \vdots & \ddots & \vdots \\ \alpha_{i_n 1} & \alpha_{i_n 2} & \alpha_{i_n 3} & \cdots & \alpha_{i_n n} \end{bmatrix}^{-1}. \tag{9.32}$$

9.5 Changing the Base in n-Dimensional Space

In a n-dimensional space, a given vector **a** is given by

$$\mathbf{a} = \begin{bmatrix} \mathbf{i}_1 & \mathbf{i}_2 & \mathbf{i}_3 & \cdots & \mathbf{i}_n \end{bmatrix} \begin{bmatrix} a_{i_1} \\ a_{i_2} \\ a_{i_3} \\ \vdots \\ a_{i_n} \end{bmatrix}, \tag{9.33}$$

where a_{i_j} represents the component of vector **a** in the direction of the base vector $\mathbf{i}_j$. The same vector in the general base $[\boldsymbol{\alpha}_1 \;\; \boldsymbol{\alpha}_2 \;\; \boldsymbol{\alpha}_3 \;\; \cdots \;\; \boldsymbol{\alpha}_n]$ is given by

$$\mathbf{a} = \begin{bmatrix} \boldsymbol{\alpha}_1 & \boldsymbol{\alpha}_2 & \boldsymbol{\alpha}_3 & \cdots & \boldsymbol{\alpha}_n \end{bmatrix} \begin{bmatrix} a_{\alpha_1} \\ a_{\alpha_2} \\ a_{\alpha_3} \\ \vdots \\ a_{\alpha_n} \end{bmatrix}. \tag{9.34}$$

By substituting Equation (9.22) into (9.34) one obtains

$$\mathbf{a} = \left(\begin{bmatrix} \mathbf{i}_1 & \mathbf{i}_2 & \mathbf{i}_3 & \cdots & \mathbf{i}_n \end{bmatrix} \begin{bmatrix} \alpha_{i_1 1} & \alpha_{i_1 2} & \alpha_{i_1 3} & \cdots & \alpha_{i_1 n} \\ \alpha_{i_2 1} & \alpha_{i_2 2} & \alpha_{i_2 3} & \cdots & \alpha_{i_2 n} \\ \alpha_{i_3 1} & \alpha_{i_3 2} & \alpha_{i_3 3} & \cdots & \alpha_{i_3 n} \\ \vdots & \vdots & \vdots & \ddots & \vdots \\ \alpha_{i_n 1} & \alpha_{i_n 2} & \alpha_{i_n 3} & \cdots & \alpha_{i_n n} \end{bmatrix} \right) \begin{bmatrix} a_{\alpha_1} \\ a_{\alpha_2} \\ a_{\alpha_3} \\ \vdots \\ a_{\alpha_n} \end{bmatrix}, \tag{9.35}$$

which yields

$$\mathbf{a} = \begin{bmatrix} \mathbf{i}_1 & \mathbf{i}_2 & \mathbf{i}_3 & \cdots & \mathbf{i}_n \end{bmatrix} \left(\begin{bmatrix} \alpha_{i_1 1} & \alpha_{i_1 2} & \alpha_{i_1 3} & \cdots & \alpha_{i_1 n} \\ \alpha_{i_2 1} & \alpha_{i_2 2} & \alpha_{i_2 3} & \cdots & \alpha_{i_2 n} \\ \alpha_{i_3 1} & \alpha_{i_3 2} & \alpha_{i_3 3} & \cdots & \alpha_{i_3 n} \\ \vdots & \vdots & \vdots & \ddots & \vdots \\ \alpha_{i_n 1} & \alpha_{i_n 2} & \alpha_{i_n 3} & \cdots & \alpha_{i_n n} \end{bmatrix} \begin{bmatrix} a_{\alpha_1} \\ a_{\alpha_2} \\ a_{\alpha_3} \\ \vdots \\ a_{\alpha_n} \end{bmatrix} \right) = \begin{bmatrix} \mathbf{i}_1 & \mathbf{i}_2 & \mathbf{i}_3 & \cdots & \mathbf{i}_n \end{bmatrix} \begin{bmatrix} a_{i_1} \\ a_{i_2} \\ a_{i_3} \\ \vdots \\ a_{i_n} \end{bmatrix}, \tag{9.36}$$

or simply

$$\begin{bmatrix} a_{i_1} \\ a_{i_2} \\ a_{i_3} \\ \vdots \\ a_{i_n} \end{bmatrix} = \begin{bmatrix} \alpha_{i_1 1} & \alpha_{i_1 2} & \alpha_{i_1 3} & \cdots & \alpha_{i_1 n} \\ \alpha_{i_2 1} & \alpha_{i_2 2} & \alpha_{i_2 3} & \cdots & \alpha_{i_2 n} \\ \alpha_{i_3 1} & \alpha_{i_3 2} & \alpha_{i_3 3} & \cdots & \alpha_{i_3 n} \\ \vdots & \vdots & \vdots & \ddots & \vdots \\ \alpha_{i_n 1} & \alpha_{i_n 2} & \alpha_{i_n 3} & \cdots & \alpha_{i_n n} \end{bmatrix} \begin{bmatrix} a_{\alpha_1} \\ a_{\alpha_2} \\ a_{\alpha_3} \\ \vdots \\ a_{\alpha_n} \end{bmatrix} = a_{\alpha_1} \begin{bmatrix} \alpha_{i_1 1} \\ \alpha_{i_2 1} \\ \alpha_{i_3 1} \\ \vdots \\ \alpha_{i_n 1} \end{bmatrix} + a_{\alpha_2} \begin{bmatrix} \alpha_{i_1 2} \\ \alpha_{i_2 2} \\ \alpha_{i_3 2} \\ \vdots \\ \alpha_{i_n 2} \end{bmatrix} + a_{\alpha_3} \begin{bmatrix} \alpha_{i_1 3} \\ \alpha_{i_2 3} \\ \alpha_{i_3 3} \\ \vdots \\ \alpha_{i_n 3} \end{bmatrix} + \cdots + a_{\alpha_n} \begin{bmatrix} \alpha_{i_1 n} \\ \alpha_{i_2 n} \\ \alpha_{i_3 n} \\ \vdots \\ \alpha_{i_n n} \end{bmatrix}. \tag{9.37}$$

Also,

$$\begin{bmatrix} a_{\alpha_1} \\ a_{\alpha_2} \\ a_{\alpha_3} \\ \vdots \\ a_{\alpha_n} \end{bmatrix} = \begin{bmatrix} \alpha_{i_1 1} & \alpha_{i_1 2} & \alpha_{i_1 3} & \cdots & \alpha_{i_1 n} \\ \alpha_{i_2 1} & \alpha_{i_2 2} & \alpha_{i_2 3} & \cdots & \alpha_{i_2 n} \\ \alpha_{i_3 1} & \alpha_{i_3 2} & \alpha_{i_3 3} & \cdots & \alpha_{i_3 n} \\ \vdots & \vdots & \vdots & \ddots & \vdots \\ \alpha_{i_n 1} & \alpha_{i_n 2} & \alpha_{i_n 3} & \cdots & \alpha_{i_n n} \end{bmatrix}^{-1} \begin{bmatrix} a_{i_1} \\ a_{i_2} \\ a_{i_3} \\ \vdots \\ a_{i_n} \end{bmatrix} = \begin{bmatrix} \underline{\alpha}_{i_1 1} & \underline{\alpha}_{i_2 1} & \underline{\alpha}_{i_3 1} & \cdots & \underline{\alpha}_{i_n 1} \\ \underline{\alpha}_{i_1 2} & \underline{\alpha}_{i_2 2} & \underline{\alpha}_{i_3 2} & \cdots & \underline{\alpha}_{i_n 2} \\ \underline{\alpha}_{i_1 3} & \underline{\alpha}_{i_2 3} & \underline{\alpha}_{i_3 3} & \cdots & \underline{\alpha}_{i_n 3} \\ \vdots & \vdots & \vdots & \ddots & \vdots \\ \underline{\alpha}_{i_1 n} & \underline{\alpha}_{i_2 n} & \underline{\alpha}_{i_3 n} & \cdots & \underline{\alpha}_{i_n n} \end{bmatrix} \begin{bmatrix} a_{i_1} \\ a_{i_2} \\ a_{i_3} \\ \vdots \\ a_{i_n} \end{bmatrix}$$

$$= a_{i_1} \begin{bmatrix} \underline{\alpha}_{i_1 1} \\ \underline{\alpha}_{i_1 2} \\ \underline{\alpha}_{i_1 3} \\ \vdots \\ \underline{\alpha}_{i_1 n} \end{bmatrix} + a_{i_2} \begin{bmatrix} \underline{\alpha}_{i_2 1} \\ \underline{\alpha}_{i_2 2} \\ \underline{\alpha}_{i_2 3} \\ \vdots \\ \underline{\alpha}_{i_2 n} \end{bmatrix} + a_{i_3} \begin{bmatrix} \underline{\alpha}_{i_3 1} \\ \underline{\alpha}_{i_3 2} \\ \underline{\alpha}_{i_3 3} \\ \vdots \\ \underline{\alpha}_{i_3 n} \end{bmatrix} + \cdots + a_{i_n} \begin{bmatrix} \underline{\alpha}_{i_n 1} \\ \underline{\alpha}_{i_n 2} \\ \underline{\alpha}_{i_n 3} \\ \vdots \\ \underline{\alpha}_{i_n n} \end{bmatrix} \tag{9.38}$$

$$= a_{i_1} \begin{bmatrix} i_{\alpha_1 1} \\ i_{\alpha_2 1} \\ i_{\alpha_3 1} \\ \vdots \\ i_{\alpha_n 1} \end{bmatrix} + a_{i_2} \begin{bmatrix} i_{\alpha_1 2} \\ i_{\alpha_2 2} \\ i_{\alpha_3 2} \\ \vdots \\ i_{\alpha_n 2} \end{bmatrix} + a_{i_3} \begin{bmatrix} i_{\alpha_1 3} \\ i_{\alpha_2 3} \\ i_{\alpha_3 3} \\ \vdots \\ i_{\alpha_n 3} \end{bmatrix} + \cdots + a_{i_n} \begin{bmatrix} i_{\alpha_1 n} \\ i_{\alpha_2 n} \\ i_{\alpha_3 n} \\ \vdots \\ i_{\alpha_n n} \end{bmatrix}.$$

Example. A 4D vector **r** is given by

$$\mathbf{r}=[\mathbf{i}\ \ \mathbf{j}\ \ \mathbf{k}\ \ \mathbf{t}]\begin{bmatrix}1\\2\\3\\4\end{bmatrix}. \tag{9.39}$$

Express the vector using the base

$$[\boldsymbol{\alpha}\ \ \boldsymbol{\beta}\ \ \boldsymbol{\gamma}\ \ \boldsymbol{\delta}]=\begin{bmatrix}1&0&0&1\\1&1&0&0\\0&1&1&1\\0&0&1&1\end{bmatrix}. \tag{9.40}$$

Solution

$$\begin{aligned}
\begin{bmatrix}r_\alpha\\r_\beta\\r_\gamma\\r_\delta\end{bmatrix}&=\begin{bmatrix}i_\alpha&j_\alpha&k_\alpha&t_\alpha\\i_\beta&j_\beta&k_\beta&t_\beta\\i_\gamma&j_\gamma&k_\gamma&t_\gamma\\i_\delta&j_\delta&k_\delta&t_\delta\end{bmatrix}\begin{bmatrix}r_i\\r_j\\r_k\\r_t\end{bmatrix}=\begin{bmatrix}\alpha_i&\beta_i&\gamma_i&\delta_i\\\alpha_j&\beta_j&\gamma_j&\delta_j\\\alpha_k&\beta_k&\gamma_k&\delta_k\\\alpha_t&\beta_t&\gamma_t&\delta_t\end{bmatrix}^{-1}\begin{bmatrix}r_i\\r_j\\r_k\\r_t\end{bmatrix}\\
&=\begin{bmatrix}1&0&0&1\\1&1&0&0\\0&1&1&1\\0&0&1&1\end{bmatrix}^{-1}\begin{bmatrix}1\\2\\3\\4\end{bmatrix}=\begin{bmatrix}0&1&-1&1\\0&0&1&-1\\-1&1&-1&2\\1&-1&1&1\end{bmatrix}\begin{bmatrix}1\\2\\3\\4\end{bmatrix}\\
&=1\begin{bmatrix}0\\0\\-1\\1\end{bmatrix}+2\begin{bmatrix}1\\0\\1\\-1\end{bmatrix}+3\begin{bmatrix}-1\\1\\-1\\1\end{bmatrix}+4\begin{bmatrix}1\\-1\\2\\1\end{bmatrix}\\
&=\begin{bmatrix}3\\-1\\6\\6\end{bmatrix},
\end{aligned} \tag{9.41}$$

where

$$\mathbf{i}=[\boldsymbol{\alpha}\ \ \boldsymbol{\beta}\ \ \boldsymbol{\gamma}\ \ \boldsymbol{\delta}]\begin{bmatrix}0\\0\\-1\\1\end{bmatrix}, \tag{9.42}$$

$$\mathbf{j}=[\boldsymbol{\alpha}\ \ \boldsymbol{\beta}\ \ \boldsymbol{\gamma}\ \ \boldsymbol{\delta}]\begin{bmatrix}1\\0\\1\\-1\end{bmatrix}, \tag{9.43}$$

$$\mathbf{k}=[\boldsymbol{\alpha}\ \boldsymbol{\beta}\ \boldsymbol{\gamma}\ \boldsymbol{\delta}]\begin{bmatrix}-1\\1\\-1\\1\end{bmatrix}, \tag{9.44}$$

$$\mathbf{t}=[\boldsymbol{\alpha}\ \boldsymbol{\beta}\ \boldsymbol{\gamma}\ \boldsymbol{\delta}]\begin{bmatrix}1\\-1\\2\\1\end{bmatrix}. \tag{9.45}$$

9.6 Summary

In this chapter the concept of a vector has been extended first to 4D space, followed by an extension to a general nD space. In both cases, the complete analogy with the 2D and 3D spaces has been preserved: this includes the orthonormal vector base, the general vector bases and the dual bases together with the formulas for calculations of the vector components.

Much like the 2D and 3D spaces, vectors have been represented by a given vector base. This has resulted in a convenient use of matrix algebra to represent vector operations. In today's computer era this is the most often used approach to represent vectors in practical applications.

Further Reading

[1] Abadir, K. M. and Magnus, J. R. (2005) *Matrix Algebra*. Cambridge, Cambridge University Press.
[2] Abraham, R., Marsden, J. E and Ratiu, T. S. (1988) *Manifolds, Tensor Analysis, and Applications*. New York, Springer-Verlag.
[3] Bronstein, I. N. and Semendyayev, S. (2007) *Handbook of Mathematics*. New York, Springer.
[4] Fleisch, D. (2012) *A Student's Guide to Vectors and Tensors*. New York, Cambridge University Press.
[5] Golub, G. H. and van Van Loan, C. F. (1996) *Matrix Computations*, 3rd edn. Baltimore, MD, Johns Hopkins University Press.
[6] Kusse, B. R. and Westwig, E. A. (2006) *Mathematical Physics: Applied Mathematics for Scientists and Engineers*, 2nd edn. Wiley-VCH Verlag GmbH & Co.
[7] Levi-Civita, T. (2013) *The Absolute Differential Calculus (Calculus of Tensors)*. New York, Dover Publications.
[8] Matthews, P. C. (1998) *Vector Calculus*. London, Springer.
[9] Munjiza, A. (2004) *The Combined Finite-Discrete Element Method*. Hoboken, NJ, Wiley.
[10] Rudin, W. (1976) *The Principles of Mathematical Analysis*. New York, McGraw-Hill Science/Engineering/Math.

10

First Order Tensors

10.1 The Slope Tensor

A flat, rooflike surface is given as shown in Figure 10.1. The surface is inclined relative to the x-y (horizontal) plane. A very practical question often arises: how much does the height h (distance from the plane x-y in z direction) of the surface increase when one moves from point O horizontally by vector

$$a = a_x\mathbf{i} + a_y\mathbf{j} = \begin{bmatrix} a_x \\ a_y \end{bmatrix}. \tag{10.1}$$

For the vector **a** shown in Figure 10.2 the height is h_a. If one moves from point O by say 2**a** the height is $2h_a$. In general, since the roof surface is flat, if one moves by $s\mathbf{a}$ from point O the height is sh_a.

In a similar manner, if one moves from point O by **b**, the height is h_b. Also if one moves by

$$\mathbf{c} = \mathbf{a} + \mathbf{b}, \tag{10.2}$$

the height is

$$h_c = h_a + h_b. \tag{10.3}$$

It follows that for any linear combination of vectors

$$\mathbf{c} = s_a\mathbf{a} + s_b\mathbf{b}, \tag{10.4}$$

Large Strain Finite Element Method: A Practical Course, First Edition. Antonio Munjiza, Esteban Rougier and Earl E. Knight.

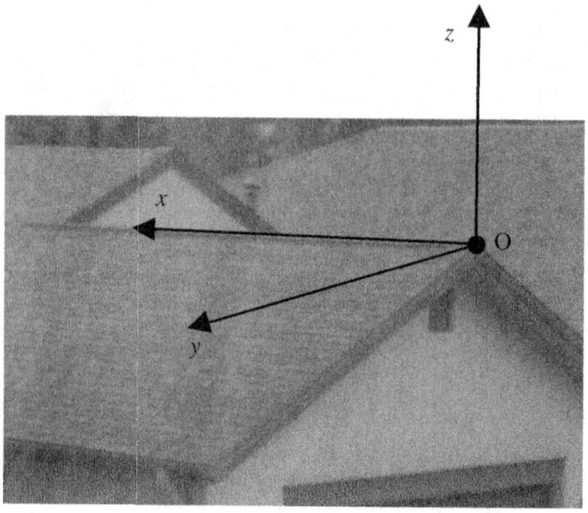

Figure 10.1 Flat roof-like surface

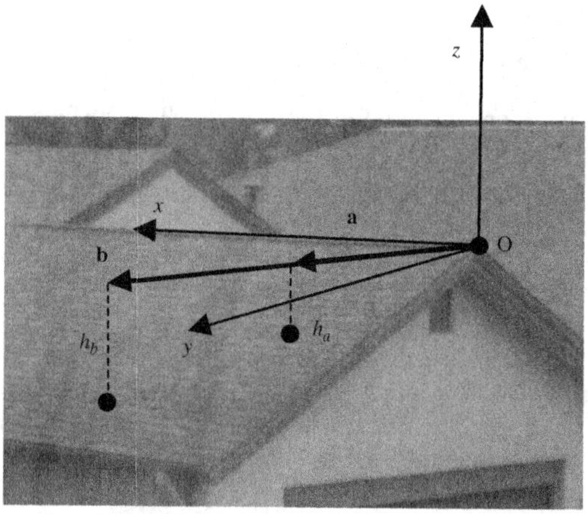

Figure 10.2 Distance to a flat surface

the height is simply

$$h_c = s_a h_a + s_b h_b. \tag{10.5}$$

It is evident that for every horizontal move (displacement) **a** from point O the height of the surface changes by h_a. There is therefore a mapping wherein for every single vector **a**, **b**, **c**, etc., there is a corresponding scalar h_a, h_b, h_c, etc., see Figure 10.3.

It is worth noting that the change in height of the roof is a physical quantity. One should therefore be able to provide an answer to the question: what is the change in the height of the roof at point O? This quantity is conveniently referred to as the slope.

Since a given displacement vector is given as a linear combination of the base vectors, i.e.,

$$\mathbf{a} = a_i\mathbf{i} + a_j\mathbf{j} \tag{10.6}$$

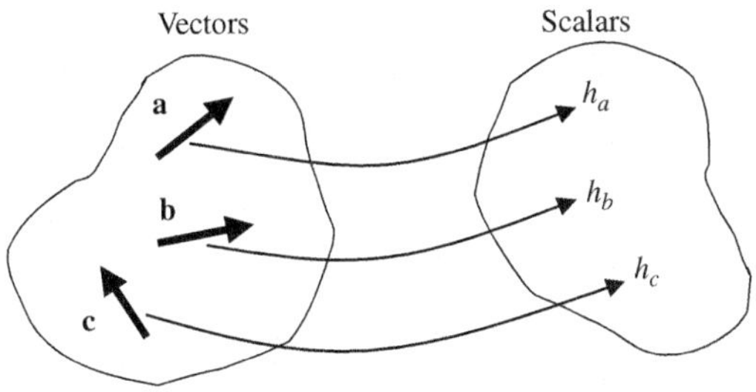

Figure 10.3 Mapping between vectors and scalars

and since the slope h is a linear mapping, then by definition

$$\begin{aligned} \mathrm{h}(\mathbf{a}) &= h_a \\ &= a_i \mathrm{h}(\mathbf{i}) + a_j \mathrm{h}(\mathbf{j}) \\ &= a_i h_i + a_j h_j. \end{aligned} \tag{10.7}$$

From the above discussion it is apparent that the slope is a linear mapping that, for a given horizontal displacement vector **a**, returns height h_a

$$h_a = \mathrm{h}(\mathbf{a}), \tag{10.8}$$

such that

$$\mathrm{h}(\mathbf{a} + s\mathbf{b}) = \mathrm{h}(\mathbf{a}) + s\mathrm{h}(\mathbf{b}). \tag{10.9}$$

Slope as defined above defines a physical reality; i.e., it quantifies how steep the hill or the road or the roof is. Slope as a physical quantity is definitely not a scalar. It is not a vector either. It is a new type of physical reality called a first order tensor. A first order tensor is defined as a linear mapping of vectors into scalars.

10.2 First Order Tensors in 2D

By definition, first order tensors are linear mappings of vectors into scalars

$$\begin{aligned} h &= \mathrm{h}(\mathbf{a}) \\ &= a_i \mathrm{h}(\mathbf{i}) + a_j \mathrm{h}(\mathbf{j}), \end{aligned} \tag{10.10}$$

where

$$\mathbf{a} = a_i \mathbf{i} + a_j \mathbf{j}. \tag{10.11}$$

As vector **a** can be expressed using any vector base such as

$$\mathbf{a} = a_\alpha \boldsymbol{\alpha} + a_\beta \boldsymbol{\beta} \quad \text{or} \quad \mathbf{a} = a_\xi \boldsymbol{\xi} + a_\eta \boldsymbol{\eta}. \tag{10.12}$$

The same tensor can be written as

$$\begin{aligned} h &= \mathrm{h}(\mathbf{a}) \\ &= a_\alpha \mathrm{h}(\boldsymbol{\alpha}) + a_\beta \mathrm{h}(\boldsymbol{\beta}) \end{aligned} \quad \text{or} \quad \begin{aligned} h &= \mathrm{h}(\mathbf{a}) \\ &= a_\xi \mathrm{h}(\boldsymbol{\xi}) + a_\eta \mathrm{h}(\boldsymbol{\eta}). \end{aligned} \tag{10.13}$$

Equations (10.10) and (10.13) can be conveniently written using matrices

$$\begin{aligned} h &= \begin{bmatrix} h_i & h_j \end{bmatrix} \begin{bmatrix} a_i \\ a_j \end{bmatrix} \\ &= \begin{bmatrix} h_\alpha & h_\beta \end{bmatrix} \begin{bmatrix} a_\alpha \\ a_\beta \end{bmatrix} \\ &= \begin{bmatrix} h_\xi & h_\eta \end{bmatrix} \begin{bmatrix} a_\xi \\ a_\eta \end{bmatrix}. \end{aligned} \tag{10.14}$$

Matrices

$$\begin{aligned} \begin{bmatrix} \mathrm{h}(\mathbf{i}) & \mathrm{h}(\mathbf{j}) \end{bmatrix} &= \begin{bmatrix} h_i & h_j \end{bmatrix} \\ \begin{bmatrix} \mathrm{h}(\boldsymbol{\alpha}) & \mathrm{h}(\boldsymbol{\beta}) \end{bmatrix} &= \begin{bmatrix} h_\alpha & h_\beta \end{bmatrix} \\ \begin{bmatrix} \mathrm{h}(\boldsymbol{\xi}) & \mathrm{h}(\boldsymbol{\eta}) \end{bmatrix} &= \begin{bmatrix} h_\xi & h_\eta \end{bmatrix} \end{aligned} \tag{10.15}$$

respectively represent the same tensor in vector bases

$$\begin{aligned} &\begin{bmatrix} \mathbf{i} & \mathbf{j} \end{bmatrix} \\ &\begin{bmatrix} \boldsymbol{\alpha} & \boldsymbol{\beta} \end{bmatrix} \\ &\begin{bmatrix} \boldsymbol{\xi} & \boldsymbol{\eta} \end{bmatrix}. \end{aligned} \tag{10.16}$$

It is important to emphasize that the slope tensor does not change with the change of the vector base. The slope tensor is a physical reality and the different matrices listed above are only convenient ways of describing the linear mapping that the slope tensor represents.

10.3 Using First Order Tensors

The linear mapping (tensor)

$$\begin{aligned} h &= \begin{bmatrix} \mathrm{h}(\mathbf{i}) & \mathrm{h}(\mathbf{j}) \end{bmatrix} \begin{bmatrix} a_i \\ a_j \end{bmatrix} = \begin{bmatrix} h_i & h_j \end{bmatrix} \begin{bmatrix} a_i \\ a_j \end{bmatrix} \\ &= \begin{bmatrix} \mathrm{h}(\boldsymbol{\alpha}) & \mathrm{h}(\boldsymbol{\beta}) \end{bmatrix} \begin{bmatrix} a_\alpha \\ a_\beta \end{bmatrix} = \begin{bmatrix} h_\alpha & h_\beta \end{bmatrix} \begin{bmatrix} a_\alpha \\ a_\beta \end{bmatrix} \end{aligned} \tag{10.17}$$

can also be written as

$$h = \mathbf{Ha}, \tag{10.18}$$

where the "product" does not mean multiplication but a linear mapping, i.e.,

$$h = \mathbf{Ha} = \mathrm{h}(\mathbf{a}), \tag{10.19}$$

while **H** is a tensor (linear mapping) and **a** is a vector; one could read Equation (10.19) as follows: Height h is a function **H** of **a**. In other words, this is just a convenient symbolic notation that makes equation writing tidier. It also makes it possible to say that the slope at point P is equal to the tensor **H**.

As tensor **H** represents a linear mapping, for a given base, it can be conveniently expressed using matrices.

$$\begin{aligned} h &= [\mathrm{h}(\mathbf{i}) \;\; \mathrm{h}(\mathbf{j})] \begin{bmatrix} a_i \\ a_j \end{bmatrix} \\ &= [h_i \;\; h_j] \begin{bmatrix} a_i \\ a_j \end{bmatrix} \\ &= \mathbf{Ha}. \end{aligned} \tag{10.20}$$

Very often, one writes

$$\mathbf{H} = [h_i \;\; h_j], \tag{10.21}$$

which means that tensor **H** is represented by (not equal to) matrix

$$[h_i \;\; h_j] \tag{10.22}$$

in the base

$$[\mathbf{i} \;\; \mathbf{j}]. \tag{10.23}$$

In a similar manner,

$$\begin{bmatrix} a_i \\ a_j \end{bmatrix} \tag{10.24}$$

is the matrix of the vector **a** obtained using the base $[\mathbf{i} \;\; \mathbf{j}]$. For different bases, different matrices are obtained. Thus,

$$\begin{aligned} \mathbf{H} &= [h_i \;\; h_j] \\ \mathbf{H} &= [h_\alpha \;\; h_\beta] \\ \mathbf{H} &= [h_\xi \;\; h_\eta] \end{aligned} \tag{10.25}$$

are three different matrices for the same tensor **H**.

It is always important to distinguish between the tensor and its matrices. The tensor describes the physical reality, such as slope, while any one of its matrices is just a convenient way for describing the given linear mapping using a given vector base.

Very often, especially in older books, there is very little distinction made between the two and the so-called "tensor transformation" is emphasized. This term can be quite confusing. A tensor is a physical reality. For example when one says that a particular mass is 10 kg, it cannot be changed to 5 kg using "transformation." When a force is 10 N in a certain direction, it cannot be changed by changing the vector base; the force stays the same. When a roof is built, its slope is well defined by the slope tensor **H**, which for every horizontal movement **a** returns the change in height. In a similar way, for a given point on a hill, the slope is defined by the unique slope tensor **H**. Of course, neither the roof, nor the hill, nor the slope **H** changes with a change in the vector base used. Only the tensor matrices change.

10.4 Using Different Vector Bases in 2D

Usually a global orthonormal base

$$[\mathbf{i}\ \ \mathbf{j}] \tag{10.26}$$

is used to represent vectors in 2D. Any vector **a** is thus given by

$$\begin{aligned} \mathbf{a} &= a_i\mathbf{i} + a_j\mathbf{j} \\ &= [\mathbf{i}\ \ \mathbf{j}]\begin{bmatrix} a_i \\ a_j \end{bmatrix} \\ &= \begin{bmatrix} a_i \\ a_j \end{bmatrix}, \end{aligned} \tag{10.27}$$

Where, in the last line of Equation (10.27), the base is implicitly assumed from the context.

A first order tensor **H** is a linear mapping of vectors into scalars.

$$\begin{aligned} h &= \mathrm{h}(\mathbf{a}) = \mathbf{Ha} \\ &= \mathrm{h}(\mathbf{i})a_i + \mathrm{h}(\mathbf{j})a_j \\ &= h_i a_i + h_j a_j \\ &= [h_i\ \ h_j]\begin{bmatrix} a_i \\ a_j \end{bmatrix} \\ &= [h_i\ \ h_j], \end{aligned} \tag{10.28}$$

where the matrix of tensor **H** is obtained using the vector base

$$[\mathbf{i}\ \ \mathbf{j}]. \tag{10.29}$$

Very often there is a practical need to use some other vector base such as

$$[\boldsymbol{\alpha}\ \ \boldsymbol{\beta}], \tag{10.30}$$

where $\boldsymbol{\alpha}$ and $\boldsymbol{\beta}$ are neither unit vectors nor are they orthogonal to each other.

$$\begin{aligned} [\boldsymbol{\alpha}\ \ \boldsymbol{\beta}] &= [\mathbf{i}\ \ \mathbf{j}]\begin{bmatrix} \alpha_i & \beta_i \\ \alpha_j & \beta_j \end{bmatrix} \\ &= \begin{bmatrix} \alpha_i & \beta_i \\ \alpha_j & \beta_j \end{bmatrix}. \end{aligned} \tag{10.31}$$

The vector **a** in the base

$$[\boldsymbol{\alpha}\ \ \boldsymbol{\beta}] \tag{10.32}$$

is given by

$$\begin{aligned}\mathbf{a} &= [\boldsymbol{\alpha}\ \ \boldsymbol{\beta}]\begin{bmatrix} a_\alpha \\ a_\beta \end{bmatrix} \\ &= a_\alpha \boldsymbol{\alpha} + a_\beta \boldsymbol{\beta} \\ &= a_\alpha(\alpha_i \mathbf{i} + \alpha_j \mathbf{j}) + a_\beta(\beta_i \mathbf{i} + \beta_j \mathbf{j}) \\ &= (a_\alpha \alpha_i + a_\beta \beta_i)\mathbf{i} + (a_\alpha \alpha_j + a_\beta \beta_j)\mathbf{j} \\ &= a_i \mathbf{i} + a_j \mathbf{j},\end{aligned} \tag{10.33}$$

where

$$\begin{aligned} a_i &= a_\alpha \alpha_i + a_\beta \beta_i \\ a_j &= a_\alpha \alpha_j + a_\beta \beta_j,\end{aligned} \tag{10.34}$$

or, in a matrix form:

$$\begin{bmatrix} a_i \\ a_j \end{bmatrix} = \begin{bmatrix} \alpha_i & \beta_i \\ \alpha_j & \beta_j \end{bmatrix}\begin{bmatrix} a_\alpha \\ a_\beta \end{bmatrix}. \tag{10.35}$$

From Equation (10.35) it follows that

$$\begin{bmatrix} a_\alpha \\ a_\beta \end{bmatrix} = \begin{bmatrix} \alpha_i & \beta_i \\ \alpha_j & \beta_j \end{bmatrix}^{-1}\begin{bmatrix} a_i \\ a_j \end{bmatrix}. \tag{10.36}$$

Tensor **H** is given by

$$h = \mathbf{Ha} = [h_i\ \ h_j]\begin{bmatrix} a_i \\ a_j \end{bmatrix}, \tag{10.37}$$

where **H** is the matrix of tensor **H** obtained using base $[\mathbf{i}\ \ \mathbf{j}]$. By substituting from Equation (10.35)

$$\begin{aligned} h &= [h_i\ \ h_j]\left(\begin{bmatrix} \alpha_i & \beta_i \\ \alpha_j & \beta_j \end{bmatrix}\begin{bmatrix} a_\alpha \\ a_\beta \end{bmatrix}\right) \\ &= \left([h_i\ \ h_j]\begin{bmatrix} \alpha_i & \beta_i \\ \alpha_j & \beta_j \end{bmatrix}\right)\begin{bmatrix} a_\alpha \\ a_\beta \end{bmatrix} \\ &= [h_\alpha\ \ h_\beta]\begin{bmatrix} a_\alpha \\ a_\beta \end{bmatrix} = \mathbf{Ha},\end{aligned} \tag{10.38}$$

where

$$[h_\alpha\ \ h_\beta] = [h_i\ \ h_j]\begin{bmatrix} \alpha_i & \beta_i \\ \alpha_j & \beta_j \end{bmatrix} \tag{10.39}$$

is the matrix of tensor **H** obtained using base $[\boldsymbol{\alpha}\ \ \boldsymbol{\beta}]$. From Equation (10.39) it follow that

$$[h_i \;\; h_j] = [h_\alpha \;\; h_\beta] \begin{bmatrix} \alpha_i & \beta_i \\ \alpha_j & \beta_j \end{bmatrix}^{-1}, \tag{10.40}$$

where the rows of the matrix

$$\begin{bmatrix} \alpha_i & \beta_i \\ \alpha_j & \beta_j \end{bmatrix}^{-1} \tag{10.41}$$

represent the vectors of the dual base (see Section 7.9)

$$\begin{bmatrix} \underline{\boldsymbol{\alpha}} \\ \underline{\boldsymbol{\beta}} \end{bmatrix} = \begin{bmatrix} \alpha_i & \beta_i \\ \alpha_j & \beta_j \end{bmatrix}^{-1}. \tag{10.42}$$

The columns of the same matrix in actual fact are

$$[\mathbf{i} \;\; \mathbf{j}] = [\boldsymbol{\alpha} \;\; \boldsymbol{\beta}] \begin{bmatrix} i_\alpha & j_\alpha \\ i_\beta & j_\beta \end{bmatrix}. \tag{10.43}$$

This means that

$$\begin{aligned} [h_i \;\; h_j] &= [h_\alpha \;\; h_\beta] \begin{bmatrix} i_\alpha & j_\alpha \\ i_\beta & j_\beta \end{bmatrix} \\ &= h_\alpha [i_\alpha \;\; j_\alpha] + h_\beta [i_\beta \;\; j_\beta], \end{aligned} \tag{10.44}$$

i.e., the base vectors **i** and **j** expressed using the base $[\boldsymbol{\alpha} \;\; \boldsymbol{\beta}]$.

In the case that two general bases are used such as

$$[\boldsymbol{\alpha} \;\; \boldsymbol{\beta}] \quad \text{and} \quad [\boldsymbol{\xi} \;\; \boldsymbol{\eta}], \tag{10.45}$$

such that

$$[\boldsymbol{\xi} \;\; \boldsymbol{\eta}] = [\boldsymbol{\alpha} \;\; \boldsymbol{\beta}] \begin{bmatrix} \xi_\alpha & \eta_\alpha \\ \xi_\beta & \eta_\beta \end{bmatrix} = \begin{bmatrix} \xi_\alpha & \eta_\alpha \\ \xi_\beta & \eta_\beta \end{bmatrix}, \tag{10.46}$$

then

$$h = \mathbf{Ha} = [h_\alpha \;\; h_\beta] \begin{bmatrix} a_\alpha \\ a_\beta \end{bmatrix}. \tag{10.47}$$

This should be the same as

$$h = \mathbf{Ha} = [h_\xi \;\; h_\eta] \begin{bmatrix} a_\xi \\ a_\eta \end{bmatrix}. \tag{10.48}$$

By combining Equations (10.47) and (10.48)

$$[h_\alpha \;\; h_\beta] \begin{bmatrix} a_\alpha \\ a_\beta \end{bmatrix} = [h_\xi \;\; h_\eta] \begin{bmatrix} a_\xi \\ a_\eta \end{bmatrix}. \tag{10.49}$$

By substituting from Equation (10.46)

$$
\begin{aligned}
\begin{bmatrix} h_\alpha & h_\beta \end{bmatrix}\begin{bmatrix} a_\alpha \\ a_\beta \end{bmatrix} &= \begin{bmatrix} h_\alpha & h_\beta \end{bmatrix}\left(\begin{bmatrix} \xi_\alpha & \eta_\alpha \\ \xi_\beta & \eta_\beta \end{bmatrix}\begin{bmatrix} a_\xi \\ a_\eta \end{bmatrix}\right) \\
&= \left(\begin{bmatrix} h_\alpha & h_\beta \end{bmatrix}\begin{bmatrix} \xi_\alpha & \eta_\alpha \\ \xi_\beta & \eta_\beta \end{bmatrix}\right)\begin{bmatrix} a_\xi \\ a_\eta \end{bmatrix} \\
&= \begin{bmatrix} h_\xi & h_\eta \end{bmatrix}\begin{bmatrix} a_\xi \\ a_\eta \end{bmatrix} = \mathbf{Ha},
\end{aligned}
\tag{10.50}
$$

where

$$
\begin{aligned}
\begin{bmatrix} h_\xi & h_\eta \end{bmatrix} &= \begin{bmatrix} h_\alpha & h_\beta \end{bmatrix}\begin{bmatrix} \xi_\alpha & \eta_\alpha \\ \xi_\beta & \eta_\beta \end{bmatrix} \\
&= h_\alpha\begin{bmatrix} \xi_\alpha & \eta_\alpha \end{bmatrix} + h_\beta\begin{bmatrix} \xi_\beta & \eta_\beta \end{bmatrix}.
\end{aligned}
\tag{10.51}
$$

In a similar manner,

$$
\begin{bmatrix} h_\alpha & h_\beta \end{bmatrix} = \begin{bmatrix} h_\xi & h_\eta \end{bmatrix}\begin{bmatrix} \xi_\alpha & \eta_\alpha \\ \xi_\beta & \eta_\beta \end{bmatrix}^{-1},
\tag{10.52}
$$

where

$$
\begin{bmatrix} \xi_\alpha & \eta_\alpha \\ \xi_\beta & \eta_\beta \end{bmatrix}^{-1} = \begin{bmatrix} \alpha_\xi & \beta_\xi \\ \alpha_\eta & \beta_\eta \end{bmatrix}.
\tag{10.53}
$$

Thus,

$$
\begin{aligned}
\begin{bmatrix} h_\alpha & h_\beta \end{bmatrix} &= \begin{bmatrix} h_\xi & h_\eta \end{bmatrix}\begin{bmatrix} \alpha_\xi & \beta_\xi \\ \alpha_\eta & \beta_\eta \end{bmatrix} \\
&= h_\xi\begin{bmatrix} \alpha_\xi & \beta_\xi \end{bmatrix} + h_\eta\begin{bmatrix} \alpha_\eta & \beta_\eta \end{bmatrix}.
\end{aligned}
\tag{10.54}
$$

10.5 Differential of a 2D Scalar Field as the First Order Tensor

A scalar field is a function which assigns a scalar value to each point $P = (x, y)$ in a 2D space. For example, a temperature field, $t(x, y)$, or the height of a hill, $h(x, y)$. In Figure 10.4 a typical scalar field is shown using shades of gray to represent different values in the field.

A scalar field is a logical extension of a function $f(x)$, Figure 10.5. By differentiating function $f(x)$ one obtains

$$
df = \left(\frac{df}{dx}\right) dx,
\tag{10.55}
$$

where df/dx represents the slope of the tangent to the graph of function f at point P.

In 2D the scalar $h(x, y)$ looks like a hill, Figure 10.6.

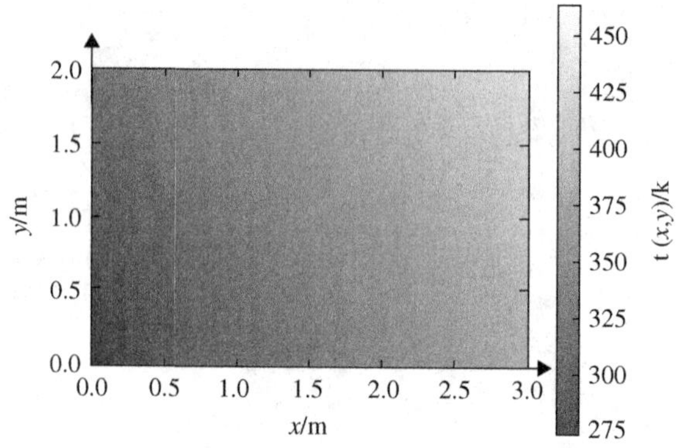

Figure 10.4 Example of a scalar field. Temperature distribution on a rectangular plate: $t = 273(1 + 0.1x + 0.2y)$

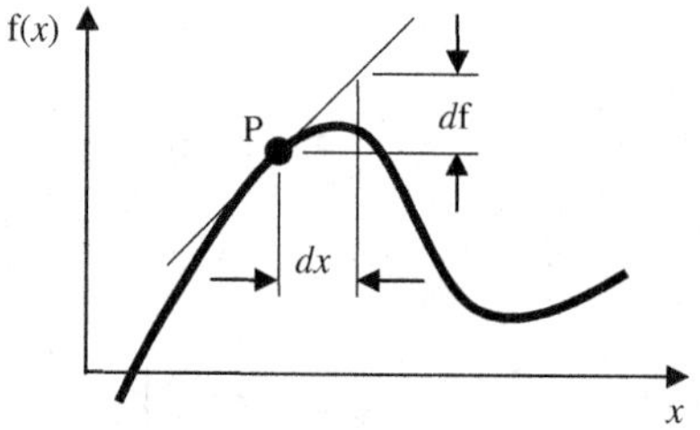

Figure 10.5 An example of a 1D scalar field

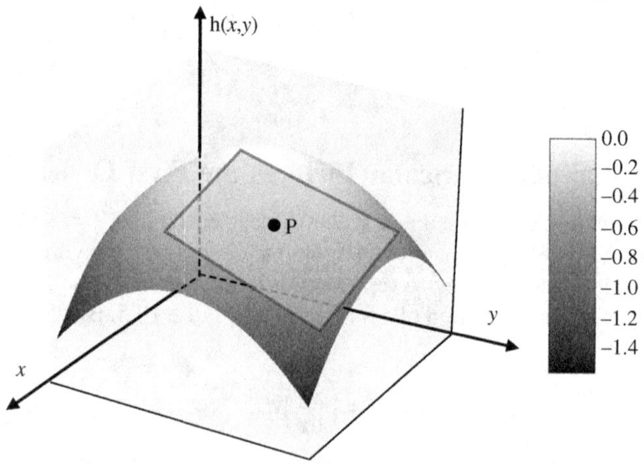

Figure 10.6 Example of a scalar field in 2D

At any point P a tangential plane to the surface can be produced. The slope of this plane is best described by using the orthonormal base $[\mathbf{i}\ \mathbf{j}]$ with the directions of vectors **i** and **j** coinciding with the coordinate axes x and y respectively.

The slope of the tangential plane at point P is given by tensor **H** such that in the base $[\mathbf{i}\ \mathbf{j}]$ it is defined by matrix

$$\mathbf{H} = \begin{bmatrix} \frac{\partial \mathrm{f}}{\partial x} & \frac{\partial \mathrm{f}}{\partial y} \end{bmatrix}. \tag{10.56}$$

For any vector

$$\mathbf{d} = [\mathbf{i}\ \mathbf{j}] \begin{bmatrix} dx \\ dy \end{bmatrix}, \tag{10.57}$$

the change in function f is given by

$$\begin{aligned} d\mathrm{f} = \mathbf{Hd} &= \begin{bmatrix} \frac{\partial \mathrm{f}}{\partial x} & \frac{\partial \mathrm{f}}{\partial y} \end{bmatrix} \begin{bmatrix} dx \\ dy \end{bmatrix} \\ &= \frac{\partial \mathrm{f}}{\partial x} dx + \frac{\partial \mathrm{f}}{\partial y} dy. \end{aligned} \tag{10.58}$$

Of course, $d\mathrm{f}$ is the differential of function f. From Equation (10.51), it is evident that the differential of a scalar field in 2D is a first order tensor.

For any infinitesimally short vector

$$\mathbf{da} = [\mathbf{i}\ \mathbf{j}] \begin{bmatrix} a_i ds \\ a_j ds \end{bmatrix} = ds[\mathbf{i}\ \mathbf{j}] \begin{bmatrix} a_i \\ a_j \end{bmatrix} = ds \begin{bmatrix} a_i \\ a_j \end{bmatrix} \tag{10.59}$$

the value ds can be conveniently chosen as 1 im (one infimeter, i.e., 1 infinitesimally small unit) which yields an infinitesimally short vector

$$\mathbf{da} = \begin{bmatrix} a_i \\ a_j \end{bmatrix} [\mathrm{im}]. \tag{10.60}$$

The differential of function f corresponding to the infinitesimally short vector **da** could be formally written as

$$\frac{\mathrm{df}}{\mathbf{da}}. \tag{10.61}$$

This is not a division by a vector but should be read as: the change of function f corresponding to an infinitesimally small displacement **da** from a given point P.

$$\frac{\mathrm{df}}{\mathbf{da}} = \begin{bmatrix} \frac{\partial \mathrm{f}}{\partial x} & \frac{\partial \mathrm{f}}{\partial y} \end{bmatrix} \begin{bmatrix} a_i \\ a_j \end{bmatrix}. \tag{10.62}$$

Now, one can formally write

$$\frac{\mathrm{df}}{\mathbf{di}}=\frac{\partial \mathrm{f}}{\partial x}$$
$$\frac{\mathrm{df}}{\mathbf{dj}}=\frac{\partial \mathrm{f}}{\partial y}. \tag{10.63}$$

By combining Equations (10.62) and (10.63) one can write

$$\begin{aligned}\frac{\mathrm{df}}{\mathbf{da}}&=\begin{bmatrix}\frac{\partial \mathrm{f}}{\partial x} & \frac{\partial \mathrm{f}}{\partial y}\end{bmatrix}\begin{bmatrix}a_i\\ a_j\end{bmatrix}\\ &=\mathbf{Da},\end{aligned} \tag{10.64}$$

where **D** is the first order tensor represented by the matrix

$$\mathbf{D}=\begin{bmatrix}\frac{\mathrm{df}}{\mathbf{di}} & \frac{\mathrm{df}}{\mathbf{dj}}\end{bmatrix}=\begin{bmatrix}\frac{\partial \mathrm{f}}{\partial x} & \frac{\partial \mathrm{f}}{\partial y}\end{bmatrix}. \tag{10.65}$$

Physically, tensor **D** is the slope of the tangential plane at any point $\mathrm{P}=(x,\ y)$. The slope of the tangential line to the function graph (scalar field in 1D) represents the first derivative of the function:

$$\mathrm{f}'(x)=\frac{\mathrm{df}}{\mathrm{d}x}. \tag{10.66}$$

In a similar manner, the slope of the tangential plane represents the first derivative of a 2D scalar field:

$$\mathrm{f}'(x,y)=\mathbf{D}=\begin{bmatrix}\frac{\mathrm{df}}{\mathbf{di}} & \frac{\mathrm{df}}{\mathbf{dj}}\end{bmatrix}=\begin{bmatrix}d_i & d_j\end{bmatrix}=\begin{bmatrix}\frac{\partial \mathrm{f}}{\partial x} & \frac{\partial \mathrm{f}}{\partial y}\end{bmatrix}. \tag{10.67}$$

From Equation (10.67) it follows that the first derivative of a 2D scalar field is a first order tensor. Provided that the matrix of tensor **D** is known in the base $[\mathbf{i}\ \ \mathbf{j}]$,

$$\mathbf{D}=\begin{bmatrix}\frac{\mathrm{df}}{\mathbf{di}} & \frac{\mathrm{df}}{\mathbf{dj}}\end{bmatrix}=\begin{bmatrix}d_i & d_j\end{bmatrix}, \tag{10.68}$$

the matrix of the tensor **D** in any other general base

$$\begin{aligned}\mathbf{D}&=\begin{bmatrix}\frac{\mathrm{df}}{\mathbf{d\boldsymbol{\alpha}}} & \frac{\mathrm{df}}{\mathbf{d\boldsymbol{\beta}}}\end{bmatrix}\\ &=\begin{bmatrix}d_\alpha & d_\beta\end{bmatrix}\end{aligned} \tag{10.69}$$

can be calculated using the expressions derived in Section 10.3

$$\begin{bmatrix}d_\alpha & d_\beta\end{bmatrix}=\begin{bmatrix}d_i & d_j\end{bmatrix}\begin{bmatrix}\alpha_i & \beta_i\\ \alpha_j & \beta_j\end{bmatrix}. \tag{10.70}$$

10.6 First Order Tensors in 3D

Using the same construct in extending vectors from 2D into 3D space, first order tensors are easily extended into 3D space. A first order tensor in 3D space is a linear mapping **H** which to any given vector

$$\begin{aligned}\mathbf{a} &= [\mathbf{i}\ \ \mathbf{j}\ \ \mathbf{k}]\begin{bmatrix} a_i \\ a_j \\ a_k \end{bmatrix} \\ &= \begin{bmatrix} a_i \\ a_j \\ a_k \end{bmatrix}\end{aligned} \tag{10.71}$$

assigns a scalar

$$h = \mathrm{h}(\mathbf{a}) = \mathbf{Ha}, \tag{10.72}$$

such that

$$\mathrm{h}(s\mathbf{a}) = s\mathrm{h}(\mathbf{a}) = s\mathbf{Ha} \tag{10.73}$$

and

$$\mathrm{h}(\mathbf{a}+\mathbf{b}) = \mathrm{h}(\mathbf{a}) + \mathrm{h}(\mathbf{b}) = \mathbf{Ha} + \mathbf{Hb}. \tag{10.74}$$

The easiest way to represent this linear mapping (tensor) is by using matrices. For a given global orthonormal base

$$[\mathbf{i}\ \ \mathbf{j}\ \ \mathbf{k}], \tag{10.75}$$

tensor **H** maps each of the base vectors into corresponding scalars as follows:

$$\begin{aligned} h_i &= \mathrm{h}(\mathbf{i}) = \mathbf{Hi} \\ h_j &= \mathrm{h}(\mathbf{j}) = \mathbf{Hj} \\ h_k &= \mathrm{h}(\mathbf{k}) = \mathbf{Hk}. \end{aligned} \tag{10.76}$$

A general vector

$$\mathbf{a} = a_i\mathbf{i} + a_j\mathbf{j} + a_k\mathbf{k} \tag{10.77}$$

is by definition mapped into

$$\begin{aligned} h_a &= \mathbf{Ha} = \mathrm{h}(\mathbf{a}) = a_i h_i + a_j h_j + a_k h_k \\ &= [h_i\ \ h_j\ \ h_k]\begin{bmatrix} a_i \\ a_j \\ a_k \end{bmatrix}, \end{aligned} \tag{10.78}$$

where

$$[h_i\ \ h_j\ \ h_k] \tag{10.79}$$

is the matrix of tensor **H** obtained using the base $[\mathbf{i}\ \ \mathbf{j}\ \ \mathbf{k}]$ and

$$\begin{bmatrix} a_i \\ a_j \\ a_k \end{bmatrix} \tag{10.80}$$

is the matrix of vector **a** obtained using the base $[\mathbf{i}\ \ \mathbf{j}\ \ \mathbf{k}]$.

Very often one simply writes

$$\mathbf{a} = \begin{bmatrix} a_i \\ a_j \\ a_k \end{bmatrix} \quad \text{and} \quad \mathbf{H} = [h_i\ \ h_j\ \ h_k], \tag{10.81}$$

where the equal sign does not mean that vector **a** or tensor **H** are equal to the corresponding matrices. Instead, it is inferred that vector **a** is represented by matrix

$$\begin{bmatrix} a_i \\ a_j \\ a_k \end{bmatrix} \tag{10.82}$$

and tensor **H** is represented by matrix

$$[h_i\ \ h_j\ \ h_k] \tag{10.83}$$

in the base $[\mathbf{i}\ \ \mathbf{j}\ \ \mathbf{k}]$.

10.7 Changing the Vector Base in 3D

As was done for the 2D case, one can also choose any vector base in 3D, for instance, the general base

$$\begin{aligned} [\boldsymbol{\alpha}\ \ \boldsymbol{\beta}\ \ \boldsymbol{\gamma}] &= [\mathbf{i}\ \ \mathbf{j}\ \ \mathbf{k}] \begin{bmatrix} \alpha_i & \beta_i & \gamma_i \\ \alpha_j & \beta_j & \gamma_j \\ \alpha_k & \beta_k & \gamma_k \end{bmatrix} \\ &= \begin{bmatrix} \alpha_i & \beta_i & \gamma_i \\ \alpha_j & \beta_j & \gamma_j \\ \alpha_k & \beta_k & \gamma_k \end{bmatrix}. \end{aligned} \tag{10.84}$$

By definition, for any vector **a**

$$h_a = [h_\alpha\ \ h_\beta\ \ h_\gamma] \begin{bmatrix} a_\alpha \\ a_\beta \\ a_\gamma \end{bmatrix} = [h_i\ \ h_j\ \ h_k] \begin{bmatrix} a_i \\ a_j \\ a_k \end{bmatrix}, \tag{10.85}$$

where

$$\begin{bmatrix} a_i \\ a_j \\ a_k \end{bmatrix} = \begin{bmatrix} \alpha_i & \beta_i & \gamma_i \\ \alpha_j & \beta_j & \gamma_j \\ \alpha_k & \beta_k & \gamma_k \end{bmatrix} \begin{bmatrix} a_\alpha \\ a_\beta \\ a_\gamma \end{bmatrix}, \tag{10.86}$$

by substituting Equation (10.86) into Equation (10.85), one obtains

$$\begin{aligned} h &= [h_i \;\; h_j \;\; h_k] \left(\begin{bmatrix} \alpha_i & \beta_i & \gamma_i \\ \alpha_j & \beta_j & \gamma_j \\ \alpha_k & \beta_k & \gamma_k \end{bmatrix} \begin{bmatrix} a_\alpha \\ a_\beta \\ a_\gamma \end{bmatrix} \right) \\ &= \left([h_i \;\; h_j \;\; h_k] \begin{bmatrix} \alpha_i & \beta_i & \gamma_i \\ \alpha_j & \beta_j & \gamma_j \\ \alpha_k & \beta_k & \gamma_k \end{bmatrix} \right) \begin{bmatrix} a_\alpha \\ a_\beta \\ a_\gamma \end{bmatrix} \\ &= [h_\alpha \;\; h_\beta \;\; h_\gamma] \begin{bmatrix} a_\alpha \\ a_\beta \\ a_\gamma \end{bmatrix}, \end{aligned} \tag{10.87}$$

where

$$\mathbf{H} = [h_\alpha \;\; h_\beta \;\; h_\gamma] = [h_i \;\; h_j \;\; h_k] \begin{bmatrix} \alpha_i & \beta_i & \gamma_i \\ \alpha_j & \beta_j & \gamma_j \\ \alpha_k & \beta_k & \gamma_k \end{bmatrix}. \tag{10.88}$$

From Equation (10.88) it also follows that

$$\mathbf{H} = [h_i \;\; h_j \;\; h_k] = [h_\alpha \;\; h_\beta \;\; h_\gamma] \begin{bmatrix} i_\alpha & j_\alpha & k_\alpha \\ i_\beta & j_\beta & k_\beta \\ i_\gamma & j_\gamma & k_\gamma \end{bmatrix}, \tag{10.89}$$

where

$$\begin{bmatrix} i_\alpha & j_\alpha & k_\alpha \\ i_\beta & j_\beta & k_\beta \\ i_\gamma & j_\gamma & k_\gamma \end{bmatrix} = \begin{bmatrix} \alpha_i & \beta_i & \gamma_i \\ \alpha_j & \beta_j & \gamma_j \\ \alpha_k & \beta_k & \gamma_k \end{bmatrix}^{-1}. \tag{10.90}$$

10.8 First Order Tensor in 4D

Very often in engineering and scientific practices a first order tensor in 4D is required in order to define certain properties for the physical realities of a specific problem. For example, one may need to associate time dependency within their problem realm.

A first order tensor in 4D is defined as a linear mapping that maps to each vector **a** in a 4D space a scalar h such that:

$$\mathrm{h}(s\mathbf{a}) = s\mathrm{h}(\mathbf{a}) = s\mathbf{Ha}, \tag{10.91}$$

$$\mathrm{h}(\mathbf{a}+\mathbf{b}) = \mathrm{h}(\mathbf{a}) + \mathrm{h}(\mathbf{b}) = \mathbf{Ha} + \mathbf{Hb}. \tag{10.92}$$

A vector in 4D space is best defined using a global orthonormal vector base

$$[\mathbf{i}\ \ \mathbf{j}\ \ \mathbf{k}\ \ \mathbf{t}], \tag{10.93}$$

so that

$$\begin{aligned}\mathbf{a} &= a_i\mathbf{i} + a_j\mathbf{j} + a_k\mathbf{k} + a_t\mathbf{t} \\ &= [\mathbf{i}\ \ \mathbf{j}\ \ \mathbf{k}\ \ \mathbf{t}]\begin{bmatrix} a_i \\ a_j \\ a_k \\ a_t \end{bmatrix} = \begin{bmatrix} a_i \\ a_j \\ a_k \\ a_t \end{bmatrix},\end{aligned} \tag{10.94}$$

where

$$\begin{bmatrix} a_i \\ a_j \\ a_k \\ a_t \end{bmatrix} \tag{10.95}$$

is the matrix of the vector **a** obtained using the base $[\mathbf{i}\ \ \mathbf{j}\ \ \mathbf{k}\ \ \mathbf{t}]$.

In order to define a linear mapping (tensor) as defined above, one must first obtain the mapped values for the base vectors,

$$\begin{aligned} h_i &= \mathrm{h}(\mathbf{i}) = \mathbf{Hi} \\ h_j &= \mathrm{h}(\mathbf{j}) = \mathbf{Hj} \\ h_k &= \mathrm{h}(\mathbf{k}) = \mathbf{Hk} \\ h_t &= \mathrm{h}(\mathbf{t}) = \mathbf{Ht}. \end{aligned} \tag{10.96}$$

By definition a general vector

$$\mathbf{a} = \begin{bmatrix} a_i \\ a_j \\ a_k \\ a_t \end{bmatrix} \tag{10.97}$$

is mapped into

$$\begin{aligned} h_a &= \mathrm{h}(\mathbf{a}) = \mathbf{Ha} = a_ih_i + a_jh_j + a_kh_k + a_th_t \\ &= [h_i\ \ h_j\ \ h_k\ \ h_t]\begin{bmatrix} a_i \\ a_j \\ a_k \\ a_t \end{bmatrix}, \end{aligned} \tag{10.98}$$

where

$$[h_i\ \ h_j\ \ h_k\ \ h_t] \tag{10.99}$$

is the matrix of the tensor **H** obtained using the base $[\mathbf{i}\ \ \mathbf{j}\ \ \mathbf{k}\ \ \mathbf{t}]$. Very often one simply writes:

$$\mathbf{H} = [h_i \ \ h_j \ \ h_k \ \ h_t], \tag{10.100}$$

which means that tensor **H** is represented by matrix $[h_i \ \ h_j \ \ h_k \ \ h_t]$ in base $[\mathbf{i}\ \ \mathbf{j}\ \ \mathbf{k}\ \ \mathbf{t}]$.

There may exist a need to introduce a new base

$$[\boldsymbol{\alpha}\ \ \boldsymbol{\beta}\ \ \boldsymbol{\gamma}\ \ \boldsymbol{\delta}], \tag{10.101}$$

such that

$$\begin{aligned} \boldsymbol{\alpha} &= \alpha_i\mathbf{i} + \alpha_j\mathbf{j} + \alpha_k\mathbf{k} + \alpha_t\mathbf{t} \\ \boldsymbol{\beta} &= \beta_i\mathbf{i} + \beta_j\mathbf{j} + \beta_k\mathbf{k} + \beta_t\mathbf{t} \\ \boldsymbol{\gamma} &= \gamma_i\mathbf{i} + \gamma_j\mathbf{j} + \gamma_k\mathbf{k} + \gamma_t\mathbf{t} \\ \boldsymbol{\delta} &= \delta_i\mathbf{i} + \delta_j\mathbf{j} + \delta_k\mathbf{k} + \delta_t\mathbf{t}, \end{aligned} \tag{10.102}$$

or

$$\begin{aligned} [\boldsymbol{\alpha}\ \ \boldsymbol{\beta}\ \ \boldsymbol{\gamma}\ \ \boldsymbol{\delta}] &= [\mathbf{i}\ \ \mathbf{j}\ \ \mathbf{k}\ \ \mathbf{t}] \begin{bmatrix} \alpha_i & \beta_i & \gamma_i & \delta_i \\ \alpha_j & \beta_j & \gamma_j & \delta_j \\ \alpha_k & \beta_k & \gamma_k & \delta_k \\ \alpha_t & \beta_t & \gamma_t & \delta_t \end{bmatrix} \\ &= \begin{bmatrix} \alpha_i & \beta_i & \gamma_i & \delta_i \\ \alpha_j & \beta_j & \gamma_j & \delta_j \\ \alpha_k & \beta_k & \gamma_k & \delta_k \\ \alpha_t & \beta_t & \gamma_t & \delta_t \end{bmatrix}. \end{aligned} \tag{10.103}$$

In such a case, since

$$\begin{aligned} \begin{bmatrix} a_i \\ a_j \\ a_k \\ a_t \end{bmatrix} &= \begin{bmatrix} \alpha_i & \beta_i & \gamma_i & \delta_i \\ \alpha_j & \beta_j & \gamma_j & \delta_j \\ \alpha_k & \beta_k & \gamma_k & \delta_k \\ \alpha_t & \beta_t & \gamma_t & \delta_t \end{bmatrix} \begin{bmatrix} a_\alpha \\ a_\beta \\ a_\gamma \\ a_\delta \end{bmatrix} \\ &= a_\alpha \begin{bmatrix} \alpha_i \\ \alpha_j \\ \alpha_k \\ \alpha_t \end{bmatrix} + a_\beta \begin{bmatrix} \beta_i \\ \beta_j \\ \beta_k \\ \beta_t \end{bmatrix} + a_\gamma \begin{bmatrix} \gamma_i \\ \gamma_j \\ \gamma_k \\ \gamma_t \end{bmatrix} + a_\delta \begin{bmatrix} \delta_i \\ \delta_j \\ \delta_k \\ \delta_t \end{bmatrix} \end{aligned} \tag{10.104}$$

and

$$h = \mathbf{Ha} = \mathrm{h}(\mathbf{a}) = [h_i \ \ h_j \ \ h_k \ \ h_t] \begin{bmatrix} a_i \\ a_j \\ a_k \\ a_t \end{bmatrix}, \tag{10.105}$$

it follows that

$$\begin{aligned}\mathbf{Ha} &= [h_i \;\; h_j \;\; h_k \;\; h_t]\left(\begin{bmatrix}\alpha_i & \beta_i & \gamma_i & \delta_i\\ \alpha_j & \beta_j & \gamma_j & \delta_j\\ \alpha_k & \beta_k & \gamma_k & \delta_k\\ \alpha_t & \beta_t & \gamma_t & \delta_t\end{bmatrix}\begin{bmatrix}a_\alpha\\ a_\beta\\ a_\gamma\\ a_\delta\end{bmatrix}\right)\\ &= \left([h_i \;\; h_j \;\; h_k \;\; h_t]\begin{bmatrix}\alpha_i & \beta_i & \gamma_i & \delta_i\\ \alpha_j & \beta_j & \gamma_j & \delta_j\\ \alpha_k & \beta_k & \gamma_k & \delta_k\\ \alpha_t & \beta_t & \gamma_t & \delta_t\end{bmatrix}\right)\begin{bmatrix}a_\alpha\\ a_\beta\\ a_\gamma\\ a_\delta\end{bmatrix} \qquad (10.106)\\ &= [h_\alpha \;\; h_\beta \;\; h_\gamma \;\; h_\delta]\begin{bmatrix}a_\alpha\\ a_\beta\\ a_\gamma\\ a_\delta\end{bmatrix},\end{aligned}$$

where

$$\begin{aligned}\mathbf{H} &= [h_\alpha \;\; h_\beta \;\; h_\gamma \;\; h_\delta]\\ &= [h_i \;\; h_j \;\; h_k \;\; h_t]\begin{bmatrix}\alpha_i & \beta_i & \gamma_i & \delta_i\\ \alpha_j & \beta_j & \gamma_j & \delta_j\\ \alpha_k & \beta_k & \gamma_k & \delta_k\\ \alpha_t & \beta_t & \gamma_t & \delta_t\end{bmatrix} = \begin{bmatrix}\mathbf{H}_\alpha\\ \mathbf{H}_\beta\\ \mathbf{H}_\gamma\\ \mathbf{H}_\delta\end{bmatrix}\end{aligned} \qquad (10.107)$$

is the new matrix of the tensor **H**, this time defined by using the base

$$[\boldsymbol{\alpha} \;\; \boldsymbol{\beta} \;\; \boldsymbol{\gamma} \;\; \boldsymbol{\delta}]. \qquad (10.108)$$

From Equation (10.107) it follows that

$$\begin{aligned}\mathbf{H} &= [h_i \;\; h_j \;\; h_k \;\; h_t]\\ &= [h_\alpha \;\; h_\beta \;\; h_\gamma \;\; h_\delta]\begin{bmatrix}i_\alpha & j_\alpha & k_\alpha & t_\alpha\\ i_\beta & j_\beta & k_\beta & t_\beta\\ i_\gamma & j_\gamma & k_\gamma & t_\gamma\\ i_\delta & j_\delta & k_\delta & t_\delta\end{bmatrix},\end{aligned} \qquad (10.109)$$

where

$$\begin{bmatrix}i_\alpha & j_\alpha & k_\alpha & t_\alpha\\ i_\beta & j_\beta & k_\beta & t_\beta\\ i_\gamma & j_\gamma & k_\gamma & t_\gamma\\ i_\delta & j_\delta & k_\delta & t_\delta\end{bmatrix} = \begin{bmatrix}\alpha_i & \beta_i & \gamma_i & \delta_i\\ \alpha_j & \beta_j & \gamma_j & \delta_j\\ \alpha_k & \beta_k & \gamma_k & \delta_k\\ \alpha_t & \beta_t & \gamma_t & \delta_t\end{bmatrix}^{-1}. \qquad (10.110)$$

10.9 First Order Tensor in n-Dimensions

Generalization of the first order tensor to an n-dimensional space is straight forward. Any vector in the n-D space is defined through vectors of the vector base

$$\begin{bmatrix}\mathbf{i}_1 & \mathbf{i}_2 & \mathbf{i}_3 & \cdots & \mathbf{i}_n\end{bmatrix}, \tag{10.111}$$

$$\mathbf{a} = a_{i_1}\mathbf{i}_1 + a_{i_2}\mathbf{i}_2 + a_{i_3}\mathbf{i}_3 + \cdots + a_{i_n}\mathbf{i}_n. \tag{10.112}$$

Each of the base vectors is then mapped into a scalar using tensor (linear mapping) **H** such that

$$\begin{aligned}
h_{i_1} &= \mathrm{h}(\mathbf{i}_1) = \mathbf{H}\mathbf{i}_1\\
h_{i_2} &= \mathrm{h}(\mathbf{i}_2) = \mathbf{H}\mathbf{i}_2\\
h_{i_3} &= \mathrm{h}(\mathbf{i}_3) = \mathbf{H}\mathbf{i}_3\\
&\vdots\\
h_{i_n} &= \mathrm{h}(\mathbf{i}_n) = \mathbf{H}\mathbf{i}_n.
\end{aligned} \tag{10.113}$$

A general vector **a** is by definition mapped into a scalar given by

$$\begin{aligned}
h = \mathrm{h}(\mathbf{a}) = \mathbf{H}\mathbf{a} &= a_{i_1}h_{i_1} + a_{i_2}h_{i_2} + a_{i_3}h_{i_3} + \cdots + a_{i_n}h_{i_n}\\
&= \begin{bmatrix}h_{i_1} & h_{i_2} & h_{i_3} & \cdots & h_{i_n}\end{bmatrix}\begin{bmatrix}a_{i_1}\\ a_{i_2}\\ a_{i_3}\\ \vdots\\ a_{i_n}\end{bmatrix}.
\end{aligned} \tag{10.114}$$

When changing the vector base to say,

$$\begin{bmatrix}\boldsymbol{\alpha}_1 & \boldsymbol{\alpha}_2 & \boldsymbol{\alpha}_3 & \cdots & \boldsymbol{\alpha}_n\end{bmatrix}, \tag{10.115}$$

each of the vectors of the new base can be expressed using the base

$$\begin{bmatrix}\mathbf{i}_1 & \mathbf{i}_2 & \mathbf{i}_3 & \cdots & \mathbf{i}_n\end{bmatrix}. \tag{10.116}$$

Thus,

$$\begin{aligned}
\begin{bmatrix}\boldsymbol{\alpha}_1 & \boldsymbol{\alpha}_2 & \boldsymbol{\alpha}_3 & \cdots & \boldsymbol{\alpha}_n\end{bmatrix} &= \begin{bmatrix}\mathbf{i}_1 & \mathbf{i}_2 & \mathbf{i}_3 & \cdots & \mathbf{i}_n\end{bmatrix}\begin{bmatrix}\alpha_{i_1 1} & \alpha_{i_1 2} & \alpha_{i_1 3} & \cdots & \alpha_{i_1 n}\\ \alpha_{i_2 1} & \alpha_{i_2 2} & \alpha_{i_2 3} & \cdots & \alpha_{i_2 n}\\ \alpha_{i_3 1} & \alpha_{i_3 2} & \alpha_{i_3 3} & \cdots & \alpha_{i_3 n}\\ \vdots & \vdots & \vdots & \ddots & \vdots\\ \alpha_{i_n 1} & \alpha_{i_n 2} & \alpha_{i_n 3} & \cdots & \alpha_{i_n n}\end{bmatrix}\\
&= \begin{bmatrix}\alpha_{i_1 1} & \alpha_{i_1 2} & \alpha_{i_1 3} & \cdots & \alpha_{i_1 n}\\ \alpha_{i_2 1} & \alpha_{i_2 2} & \alpha_{i_2 3} & \cdots & \alpha_{i_2 n}\\ \alpha_{i_3 1} & \alpha_{i_3 2} & \alpha_{i_3 3} & \cdots & \alpha_{i_3 n}\\ \vdots & \vdots & \vdots & \ddots & \vdots\\ \alpha_{i_n 1} & \alpha_{i_n 2} & \alpha_{i_n 3} & \cdots & \alpha_{i_n n}\end{bmatrix}.
\end{aligned} \tag{10.117}$$

Vector **a** in the new base is given by

$$\mathbf{a} = a_{\alpha_1}\boldsymbol{\alpha}_1 + a_{\alpha_2}\boldsymbol{\alpha}_2 + a_{\alpha_3}\boldsymbol{\alpha}_3 + \cdots + a_{\alpha_n}\boldsymbol{\alpha}_n = \begin{bmatrix} \boldsymbol{\alpha}_1 & \boldsymbol{\alpha}_2 & \boldsymbol{\alpha}_3 & \cdots & \boldsymbol{\alpha}_n \end{bmatrix} \begin{bmatrix} a_{\alpha_1} \\ a_{\alpha_2} \\ a_{\alpha_3} \\ \vdots \\ a_{\alpha_n} \end{bmatrix}. \tag{10.118}$$

This vector is by definition mapped into

$$\begin{aligned}
\mathbf{Ha} &= \begin{bmatrix} h_{i_1} & h_{i_2} & h_{i_3} & \cdots & h_{i_n} \end{bmatrix} \begin{bmatrix} a_{i_1} \\ a_{i_2} \\ a_{i_3} \\ \vdots \\ a_{i_n} \end{bmatrix} \\
&= \begin{bmatrix} h_{i_1} & h_{i_2} & h_{i_3} & \cdots & h_{i_n} \end{bmatrix} \left(\begin{bmatrix} \alpha_{i_1 1} & \alpha_{i_1 2} & \alpha_{i_1 3} & \cdots & \alpha_{i_1 n} \\ \alpha_{i_2 1} & \alpha_{i_2 2} & \alpha_{i_2 3} & \cdots & \alpha_{i_2 n} \\ \alpha_{i_3 1} & \alpha_{i_3 2} & \alpha_{i_3 3} & \cdots & \alpha_{i_3 n} \\ \vdots & \vdots & \vdots & \ddots & \vdots \\ \alpha_{i_n 1} & \alpha_{i_n 2} & \alpha_{i_n 3} & \cdots & \alpha_{i_n n} \end{bmatrix} \begin{bmatrix} a_{\alpha_1} \\ a_{\alpha_2} \\ a_{\alpha_3} \\ \vdots \\ a_{\alpha_n} \end{bmatrix} \right) \\
&= \left(\begin{bmatrix} h_{i_1} & h_{i_2} & h_{i_3} & \cdots & h_{i_n} \end{bmatrix} \begin{bmatrix} \alpha_{i_1 1} & \alpha_{i_1 2} & \alpha_{i_1 3} & \cdots & \alpha_{i_1 n} \\ \alpha_{i_2 1} & \alpha_{i_2 2} & \alpha_{i_2 3} & \cdots & \alpha_{i_2 n} \\ \alpha_{i_3 1} & \alpha_{i_3 2} & \alpha_{i_3 3} & \cdots & \alpha_{i_3 n} \\ \vdots & \vdots & \vdots & \ddots & \vdots \\ \alpha_{i_n 1} & \alpha_{i_n 2} & \alpha_{i_n 3} & \cdots & \alpha_{i_n n} \end{bmatrix} \right) \begin{bmatrix} a_{\alpha_1} \\ a_{\alpha_2} \\ a_{\alpha_3} \\ \vdots \\ a_{\alpha_n} \end{bmatrix} \\
&= \begin{bmatrix} h_{\alpha_1} & h_{\alpha_2} & h_{\alpha_3} & \cdots & h_{\alpha_n} \end{bmatrix} \begin{bmatrix} a_{\alpha_1} \\ a_{\alpha_2} \\ a_{\alpha_3} \\ \vdots \\ a_{\alpha_n} \end{bmatrix},
\end{aligned} \tag{10.119}$$

where

$$\begin{aligned}
\mathbf{H} &= \begin{bmatrix} h_{\alpha_1} & h_{\alpha_2} & h_{\alpha_3} & \cdots & h_{\alpha_n} \end{bmatrix} \\
&= \begin{bmatrix} h_{i_1} & h_{i_2} & h_{i_3} & \cdots & h_{i_n} \end{bmatrix} \begin{bmatrix} \alpha_{i_1 1} & \alpha_{i_1 2} & \alpha_{i_1 3} & \cdots & \alpha_{i_1 n} \\ \alpha_{i_2 1} & \alpha_{i_2 2} & \alpha_{i_2 3} & \cdots & \alpha_{i_2 n} \\ \alpha_{i_3 1} & \alpha_{i_3 2} & \alpha_{i_3 3} & \cdots & \alpha_{i_3 n} \\ \vdots & \vdots & \vdots & \ddots & \vdots \\ \alpha_{i_n 1} & \alpha_{i_n 2} & \alpha_{i_n 3} & \cdots & \alpha_{i_n n} \end{bmatrix}
\end{aligned} \tag{10.120}$$

is the new matrix of tensor **H** obtained using the base

$$[\boldsymbol{\alpha}_1 \ \ \boldsymbol{\alpha}_2 \ \ \boldsymbol{\alpha}_3 \ \ \cdots \ \ \boldsymbol{\alpha}_n]. \tag{10.121}$$

From Equation (10.120) it also follows that

$$\begin{aligned}\mathbf{H} &= [h_{i_1} \ \ h_{i_2} \ \ h_{i_3} \ \ \cdots \ \ h_{i_n}] \\ &= [h_{\alpha_1} \ \ h_{\alpha_2} \ \ h_{\alpha_3} \ \ \cdots \ \ h_{\alpha_n}] \begin{bmatrix} i_{\alpha_1 1} & i_{\alpha_1 2} & i_{\alpha_1 3} & \cdots & i_{\alpha_1 n} \\ i_{\alpha_2 1} & i_{\alpha_2 2} & i_{\alpha_2 3} & \cdots & i_{\alpha_2 n} \\ i_{\alpha_3 1} & i_{\alpha_3 2} & i_{\alpha_3 3} & \cdots & i_{\alpha_3 n} \\ \vdots & \vdots & \vdots & \ddots & \vdots \\ i_{\alpha_n 1} & i_{\alpha_n 2} & i_{\alpha_n 3} & \cdots & i_{\alpha_n n} \end{bmatrix},\end{aligned} \tag{10.122}$$

where

$$\begin{bmatrix} i_{\alpha_1 1} & i_{\alpha_1 2} & i_{\alpha_1 3} & \cdots & i_{\alpha_1 n} \\ i_{\alpha_2 1} & i_{\alpha_2 2} & i_{\alpha_2 3} & \cdots & i_{\alpha_2 n} \\ i_{\alpha_3 1} & i_{\alpha_3 2} & i_{\alpha_3 3} & \cdots & i_{\alpha_3 n} \\ \vdots & \vdots & \vdots & \ddots & \vdots \\ i_{\alpha_n 1} & i_{\alpha_n 2} & i_{\alpha_n 3} & \cdots & i_{\alpha_n n} \end{bmatrix} = \begin{bmatrix} \alpha_{i_1 1} & \alpha_{i_1 2} & \alpha_{i_1 3} & \cdots & \alpha_{i_1 n} \\ \alpha_{i_2 1} & \alpha_{i_2 2} & \alpha_{i_2 3} & \cdots & \alpha_{i_2 n} \\ \alpha_{i_3 1} & \alpha_{i_3 2} & \alpha_{i_3 3} & \cdots & \alpha_{i_3 n} \\ \vdots & \vdots & \vdots & \ddots & \vdots \\ \alpha_{i_n 1} & \alpha_{i_n 2} & \alpha_{i_n 3} & \cdots & \alpha_{i_n n} \end{bmatrix}^{-1}. \tag{10.123}$$

10.10 Differential of a 3D Scalar Field as the First Order Tensor

A scalar field is a function which assigns a scalar value to each point $P=(x, y, z)$ of the 3D space. For example, a temperature field, $t(x, y, z)$, or the height of a hill, $h(x, y, z)$.

The slope of the tangential plane at point P is given by tensor **H** such that in the base $[\mathbf{i} \ \ \mathbf{j} \ \ \mathbf{k}]$ it is defined by matrix

$$\mathbf{D} = \begin{bmatrix} \frac{\partial \mathrm{f}}{\partial x} & \frac{\partial \mathrm{f}}{\partial y} & \frac{\partial \mathrm{f}}{\partial z} \end{bmatrix} = [\mathrm{df}_i \ \ \mathrm{df}_j \ \ \mathrm{df}_k]. \tag{10.124}$$

For any vector

$$\mathbf{d} = [\mathbf{i} \ \ \mathbf{j} \ \ \mathbf{k}] \begin{bmatrix} dx \\ dy \\ dz \end{bmatrix}, \tag{10.125}$$

the change in function f is given by

$$\mathrm{df} = \mathbf{Dd} = [\mathrm{df}_i \ \ \mathrm{df}_j \ \ \mathrm{df}_k] \begin{bmatrix} dx \\ dy \\ dz \end{bmatrix} = \begin{bmatrix} \frac{\partial \mathrm{f}}{\partial x} & \frac{\partial \mathrm{f}}{\partial y} & \frac{\partial \mathrm{f}}{\partial z} \end{bmatrix} \begin{bmatrix} dx \\ dy \\ dz \end{bmatrix} = \frac{\partial \mathrm{f}}{\partial x}dx + \frac{\partial \mathrm{f}}{\partial y}dy + \frac{\partial \mathrm{f}}{\partial z}dz. \tag{10.126}$$

Infinitesimally Small Vectors. From Equation (10.126), it is evident that the differential of a scalar field in 3D is a first order tensor. An infinitesimally small vector **a** can be obtained as follows

$$\mathbf{da} = [\mathbf{i} \ \ \mathbf{j} \ \ \mathbf{k}] \begin{bmatrix} a_i ds \\ a_j ds \\ a_k ds \end{bmatrix} = ds \begin{bmatrix} a_i \\ a_j \\ a_k \end{bmatrix}, \tag{10.127}$$

where ds is infinitesimally small and conveniently chosen as 1 im (one infi-meter, i.e. 1 infinitesimally small unit). This yields

$$\mathbf{da} = \begin{bmatrix} a_i \\ a_j \\ a_k \end{bmatrix} [\text{im}]. \tag{10.128}$$

The differential of function f for the infinitesimally small vector d**a** is formally written as

$$\frac{\text{df}}{\mathbf{da}} \tag{10.129}$$

(note that this is not a division by a vector, but a convenient notation, which should be read as "increment in function f if one moves by infinitesimal vector d**a** given by Equation (10.128)) and is given by

$$\frac{\text{df}}{\mathbf{da}} = \begin{bmatrix} \frac{\text{df}}{\mathbf{di}} & \frac{\text{df}}{\mathbf{dj}} & \frac{\text{df}}{\mathbf{dk}} \end{bmatrix} \begin{bmatrix} a_i \\ a_j \\ a_k \end{bmatrix}, \tag{10.130}$$

where

$$\begin{aligned} \frac{\text{df}}{\mathbf{di}} &= \frac{\partial \text{f}}{\partial x} \\ \frac{\text{df}}{\mathbf{dj}} &= \frac{\partial \text{f}}{\partial y} \\ \frac{\text{df}}{\mathbf{dk}} &= \frac{\partial \text{f}}{\partial z} \end{aligned} \tag{10.131}$$

are increments in function f when one moves by infinitesimal vectors

$$\begin{aligned} \mathbf{di} &= \mathbf{i} \ (\text{im}) \\ \mathbf{dj} &= \mathbf{j} \ (\text{im}) \\ \mathbf{dk} &= \mathbf{k} \ (\text{im}) \end{aligned} \tag{10.132}$$

respectively. By definition it follows that

$$\frac{\text{df}}{\mathbf{da}} = \begin{bmatrix} \frac{\text{df}}{\mathbf{di}} & \frac{\text{df}}{\mathbf{dj}} & \frac{\text{df}}{\mathbf{dk}} \end{bmatrix} \begin{bmatrix} a_i \\ a_j \\ a_k \end{bmatrix} = \begin{bmatrix} \frac{\partial \text{f}}{\partial x} & \frac{\partial \text{f}}{\partial y} & \frac{\partial \text{f}}{\partial z} \end{bmatrix} \begin{bmatrix} a_i \\ a_j \\ a_k \end{bmatrix} = \mathbf{Da} = [d_i \ \ d_j \ \ d_k] \begin{bmatrix} a_i \\ a_j \\ a_k \end{bmatrix}. \tag{10.133}$$

The first order tensor **D** is represented by matrix

$$\mathbf{D}=\begin{bmatrix} d_i & d_j & d_k \end{bmatrix}=\begin{bmatrix} \frac{\mathrm{df}}{\mathbf{di}} & \frac{\mathrm{df}}{\mathbf{dj}} & \frac{\mathrm{df}}{\mathbf{dk}} \end{bmatrix}=\begin{bmatrix} \frac{\partial \mathrm{f}}{\partial x} & \frac{\partial \mathrm{f}}{\partial y} & \frac{\partial \mathrm{f}}{\partial z} \end{bmatrix}. \tag{10.134}$$

Tensor **D** is the slope of the 3D "tangential plane" at any point P = (x, y, z). As the slope of the tangential line to the function graph (scalar field in 1D) represents the first derivative of the function

$$\mathrm{f}'(x)=\frac{\partial \mathrm{f}}{\partial x}, \tag{10.135}$$

in a similar way, the slope of the 3D tangential plane represents the first derivative of a 3D scalar field

$$\begin{aligned} \mathrm{f}'(x,y,z)=\mathbf{D}&=\begin{bmatrix} \frac{\mathrm{df}}{\mathbf{di}} & \frac{\mathrm{df}}{\mathbf{dj}} & \frac{\mathrm{df}}{\mathbf{dk}} \end{bmatrix}\\ &=\begin{bmatrix} \frac{\partial \mathrm{f}}{\partial x} & \frac{\partial \mathrm{f}}{\partial y} & \frac{\partial \mathrm{f}}{\partial z} \end{bmatrix}\\ &=\begin{bmatrix} d_i & d_j & d_k \end{bmatrix}. \end{aligned} \tag{10.136}$$

The above matrix components can be interpreted as: d_i is the increment of function f when one moves in **i** direction by 1 im; d_j is the increment of function f when one moves in **j** direction by 1 im; d_k is the increment of function f when one moves in **k** direction by 1 im.

The first derivative of a 3D scalar field is a first order tensor. Provided that the matrix of tensor **D** is known in the base $[\mathbf{i}\ \ \mathbf{j}\ \ \mathbf{k}]$, the matrix of the tensor **D** in any other general base can be calculated. Formally, one can write

$$\begin{aligned} \mathbf{D}&=\begin{bmatrix} \frac{\mathrm{df}}{\mathbf{d}\boldsymbol{\alpha}} & \frac{\mathrm{df}}{\mathbf{d}\boldsymbol{\beta}} & \frac{\mathrm{df}}{\mathbf{d}\boldsymbol{\gamma}} \end{bmatrix}\\ &=\begin{bmatrix} d_\alpha & d_\beta & d_\gamma \end{bmatrix}\\ &=\begin{bmatrix} d_i & d_j & d_k \end{bmatrix}\begin{bmatrix} \alpha_i & \beta_i & \gamma_i \\ \alpha_j & \beta_j & \gamma_j \\ \alpha_k & \beta_k & \gamma_k \end{bmatrix}. \end{aligned} \tag{10.137}$$

In a similar manner,

$$\mathbf{D}=\begin{bmatrix} d_i & d_j & d_k \end{bmatrix}=\begin{bmatrix} d_\alpha & d_\beta & d_\gamma \end{bmatrix}\begin{bmatrix} i_\alpha & j_\alpha & k_\alpha \\ i_\beta & j_\beta & k_\beta \\ i_\gamma & j_\gamma & k_\gamma \end{bmatrix}, \tag{10.138}$$

where

$$\begin{bmatrix} i_\alpha & j_\alpha & k_\alpha \\ i_\beta & j_\beta & k_\beta \\ i_\gamma & j_\gamma & k_\gamma \end{bmatrix}=\begin{bmatrix} \alpha_i & \beta_i & \gamma_i \\ \alpha_j & \beta_j & \gamma_j \\ \alpha_k & \beta_k & \gamma_k \end{bmatrix}^{-1}. \tag{10.139}$$

10.11 Scalar Field in n-Dimensional Space

By analogy to the 3D space, a scalar field in a nD space is represented by a function

$$f(x_1, x_2, x_3, \ldots, x_n). \tag{10.140}$$

The differential of this function associated with an infinitesimal displacement $\mathbf{i}_1$ im (infi-meters) is given by

$$\frac{\mathrm{df}}{\mathrm{d}\mathbf{i}_1} = \frac{\partial \mathrm{f}}{\partial x}. \tag{10.141}$$

In a similar manner, for other base vectors

$$[\mathbf{i}_1 \;\; \mathbf{i}_2 \;\; \mathbf{i}_3 \;\; \cdots \;\; \mathbf{i}_n], \tag{10.142}$$

the differential of function f is defined by the following tensor

$$\mathbf{D} = [d_{i_1} \;\; d_{i_2} \;\; d_{i_3} \;\; \cdots \;\; d_{i_n}] = \begin{bmatrix} \frac{\mathrm{df}}{\mathrm{d}\mathbf{i}_1} & \frac{\mathrm{df}}{\mathrm{d}\mathbf{i}_2} & \frac{\mathrm{df}}{\mathrm{d}\mathbf{i}_3} & \cdots & \frac{\mathrm{df}}{\mathrm{d}\mathbf{i}_n} \end{bmatrix}. \tag{10.143}$$

For any other vector

$$\mathbf{da} = \begin{bmatrix} a_{i_1} \\ a_{i_2} \\ a_{i_3} \\ \vdots \\ a_{i_n} \end{bmatrix}, \tag{10.144}$$

it follows by definition that

$$\mathbf{D} = \frac{\mathrm{df}}{\mathbf{da}}\mathbf{da} = \begin{bmatrix} \frac{\mathrm{df}}{\mathrm{d}\mathbf{i}_1} & \frac{\mathrm{df}}{\mathrm{d}\mathbf{i}_2} & \frac{\mathrm{df}}{\mathrm{d}\mathbf{i}_3} & \cdots & \frac{\mathrm{df}}{\mathrm{d}\mathbf{i}_n} \end{bmatrix} \begin{bmatrix} a_{i_1} \\ a_{i_2} \\ a_{i_3} \\ \vdots \\ a_{i_n} \end{bmatrix} = \begin{bmatrix} \frac{\partial \mathrm{f}}{\partial x_1} & \frac{\partial \mathrm{f}}{\partial x_2} & \frac{\partial \mathrm{f}}{\partial x_3} & \cdots & \frac{\partial \mathrm{f}}{\partial x_n} \end{bmatrix} \begin{bmatrix} a_{i_1} \\ a_{i_2} \\ a_{i_3} \\ \vdots \\ a_{i_n} \end{bmatrix}, \tag{10.145}$$

which maps vector **a** into **da**.

The same differential in the base

$$[\boldsymbol{\alpha}_1 \;\; \boldsymbol{\alpha}_2 \;\; \boldsymbol{\alpha}_3 \;\; \cdots \;\; \boldsymbol{\alpha}_n] \tag{10.146}$$

is given by

$$\mathbf{D} = \frac{\mathrm{d}f}{\mathbf{da}}\mathbf{da} = \begin{bmatrix} \dfrac{\mathrm{d}f}{\mathrm{d}\boldsymbol{\alpha}_1} & \dfrac{\mathrm{d}f}{\mathrm{d}\boldsymbol{\alpha}_2} & \dfrac{\mathrm{d}f}{\mathrm{d}\boldsymbol{\alpha}_3} & \cdots & \dfrac{\mathrm{d}f}{\mathrm{d}\boldsymbol{\alpha}_n} \end{bmatrix} \begin{bmatrix} a_{\alpha_1} \\ a_{\alpha_2} \\ a_{\alpha_3} \\ \vdots \\ a_{\alpha_n} \end{bmatrix}, \tag{10.147}$$

where

$$\begin{bmatrix} \dfrac{\mathrm{d}f}{\mathrm{d}\boldsymbol{\alpha}_1} & \dfrac{\mathrm{d}f}{\mathrm{d}\boldsymbol{\alpha}_2} & \dfrac{\mathrm{d}f}{\mathrm{d}\boldsymbol{\alpha}_3} & \cdots & \dfrac{\mathrm{d}f}{\mathrm{d}\boldsymbol{\alpha}_n} \end{bmatrix} = \begin{bmatrix} \dfrac{\mathrm{d}f}{\mathrm{d}\mathbf{i}_1} & \dfrac{\mathrm{d}f}{\mathrm{d}\mathbf{i}_2} & \dfrac{\mathrm{d}f}{\mathrm{d}\mathbf{i}_3} & \cdots & \dfrac{\mathrm{d}f}{\mathrm{d}\mathbf{i}_n} \end{bmatrix} \begin{bmatrix} \alpha_{i_1 1} & \alpha_{i_1 2} & \alpha_{i_1 3} & \cdots & \alpha_{i_1 n} \\ \alpha_{i_2 1} & \alpha_{i_2 2} & \alpha_{i_2 3} & \cdots & \alpha_{i_2 n} \\ \alpha_{i_3 1} & \alpha_{i_3 2} & \alpha_{i_3 3} & \cdots & \alpha_{i_3 n} \\ \vdots & \vdots & \vdots & \ddots & \vdots \\ \alpha_{i_n 1} & \alpha_{i_n 2} & \alpha_{i_n 3} & \cdots & \alpha_{i_n n} \end{bmatrix} \tag{10.148}$$

and also

$$\begin{bmatrix} \dfrac{\mathrm{d}f}{\mathrm{d}\mathbf{i}_1} & \dfrac{\mathrm{d}f}{\mathrm{d}\mathbf{i}_2} & \dfrac{\mathrm{d}f}{\mathrm{d}\mathbf{i}_3} & \cdots & \dfrac{\mathrm{d}f}{\mathrm{d}\mathbf{i}_n} \end{bmatrix} = \begin{bmatrix} \dfrac{\mathrm{d}f}{\mathrm{d}\boldsymbol{\alpha}_1} & \dfrac{\mathrm{d}f}{\mathrm{d}\boldsymbol{\alpha}_2} & \dfrac{\mathrm{d}f}{\mathrm{d}\boldsymbol{\alpha}_3} & \cdots & \dfrac{\mathrm{d}f}{\mathrm{d}\boldsymbol{\alpha}_n} \end{bmatrix} \begin{bmatrix} i_{\alpha_1 1} & i_{\alpha_1 2} & i_{\alpha_1 3} & \cdots & i_{\alpha_1 n} \\ i_{\alpha_2 1} & i_{\alpha_2 2} & i_{\alpha_2 3} & \cdots & i_{\alpha_2 n} \\ i_{\alpha_3 1} & i_{\alpha_3 2} & i_{\alpha_3 3} & \cdots & i_{\alpha_3 n} \\ \vdots & \vdots & \vdots & \ddots & \vdots \\ i_{\alpha_n 1} & i_{\alpha_n 2} & i_{\alpha_n 3} & \cdots & i_{\alpha_n n} \end{bmatrix}, \tag{10.149}$$

where

$$\begin{bmatrix} i_{\alpha_1 1} & i_{\alpha_1 2} & i_{\alpha_1 3} & \cdots & i_{\alpha_1 n} \\ i_{\alpha_2 1} & i_{\alpha_2 2} & i_{\alpha_2 3} & \cdots & i_{\alpha_2 n} \\ i_{\alpha_3 1} & i_{\alpha_3 2} & i_{\alpha_3 3} & \cdots & i_{\alpha_3 n} \\ \vdots & \vdots & \vdots & \ddots & \vdots \\ i_{\alpha_n 1} & i_{\alpha_n 2} & i_{\alpha_n 3} & \cdots & i_{\alpha_n n} \end{bmatrix} = \begin{bmatrix} \alpha_{i_1 1} & \alpha_{i_1 2} & \alpha_{i_1 3} & \cdots & \alpha_{i_1 n} \\ \alpha_{i_2 1} & \alpha_{i_2 2} & \alpha_{i_2 3} & \cdots & \alpha_{i_2 n} \\ \alpha_{i_3 1} & \alpha_{i_3 2} & \alpha_{i_3 3} & \cdots & \alpha_{i_3 n} \\ \vdots & \vdots & \vdots & \ddots & \vdots \\ \alpha_{i_n 1} & \alpha_{i_n 2} & \alpha_{i_n 3} & \cdots & \alpha_{i_n n} \end{bmatrix}^{-1}. \tag{10.150}$$

10.12 Summary

In this chapter a clear need for physical quantities that are neither scalars nor vectors has been demonstrated. These quantities are in actual fact a linear mapping of vectors into scalars. As such, they are called first order tensors.

As scalars quantities are best expressed using real numbers, first order tensors are best expressed using matrices. The same tensor can be expressed using different matrices depending on the vector base used to describe the vectors.

Vectors are expressed using column matrices that change by changing the vector base. In a similar manner, first order tensors are expressed using row matrices, which also change by changing the vector base.

The first order tensor has been illustrated using the slope tensor. In addition, the total differential of a scalar field has been defined using first order tensors.

All of the above have been explained using 2D, 3D, 4D and finally nD space.

Further Reading

[1] Abraham, R., Marsden, J. E. and Ratiu, T. S. (1988) *Manifolds, Tensor Analysis, and Applications*. New York, Springer-Verlag.

[2] Aris, R. (1962) *Vectors, Tensors and the Basic Equations of Fluid Mechanics*. New Jersey, Prentice Hall.

[3] Bronstein, I. N. and Semendyayev, S. (2007) *Handbook of Mathematics*. New York, Springer.

[4] de Souza Neto, E. A. (2004) On general isotropic tensor functions of one tensor. *International Journal for Numerical Methods in Engineering*, **61**(6), 880–95.

[5] Dvorkin, E. N. and Goldschmit, M. B. (2005) *Nonlinear Continua*. New York, Springer.

[6] Fleisch, D. (2012) *A Student's Guide to Vectors and Tensors*. New York, Cambridge University Press.

[7] Kusse, B. R. and Westwig, E. A. (2006) *Mathematical Physics: Applied Mathematics for Scientists and Engineers*, 2nd edn. Wiley-VCH Verlag GmbH & Co.

[8] Levi-Civita, T. (2013) *The Absolute Differential Calculus (Calculus of Tensors)*. New York, Dover Publications.

[9] Matthews, P. C. (1998) *Vector Calculus*. London, Springer.

11

Second Order Tensors in 2D

11.1 Stress Tensor in 2D

In classical engineering theory stress is defined as an internal force over a unit surface (it is worth noting that the surface is a vector). These internal forces can be, for example the forces by which individual atoms from different sides of the surface **a** pull each other, as shown schematically in Figure 11.1.

The resultant of these forces is the force **f**, which can be broken into two forces: orthogonal to surface vector **a** (also known as normal stress, $\boldsymbol{\sigma}$) and parallel to surface vector **a** (also known as shear stress, $\boldsymbol{\tau}$), Figure 11.2.

An engineer often encounters problems similar to the one shown in Figure 11.3. A flat sheet made of separate pieces is glued together and is loaded by a uniformly distributed load as shown. The question is: will this sheet break at point P under this load, see Figure 11.4?

In order for the sheet not to break along any of the glued surfaces, the glue strength keeping the surfaces together must be greater than the force trying to separate them.

The practical question therefore arises: for a given surface **a**, what is the internal force **f** on that surface? As surface **a** is defined by a vector, it has its orientation and its magnitude a, as shown in Figure 11.5.

Also, force **f** is a vector with its own direction and magnitude. As previously discussed in this book, the best way to express vectors is to introduce the orthonormal vector base $[\mathbf{i}\ \ \mathbf{j}]$. Thus, surface **a** is given by

$$\begin{aligned}\mathbf{a} &= a_i\mathbf{i} + a_j\mathbf{j} \\ &= [\mathbf{i}\ \ \mathbf{j}]\begin{bmatrix} a_i \\ a_j \end{bmatrix} \\ &= \begin{bmatrix} a_i \\ a_j \end{bmatrix},\end{aligned} \tag{11.1}$$

Large Strain Finite Element Method: A Practical Course, First Edition. Antonio Munjiza, Esteban Rougier and Earl E. Knight.

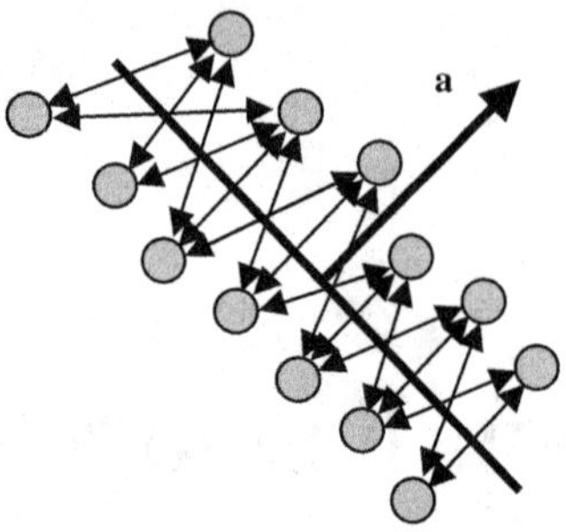

Figure 11.1 Internal forces over a unit surface **a**. Note that the surface is a vector. The circles indicate atoms and the arrows indicate internal forces between atoms

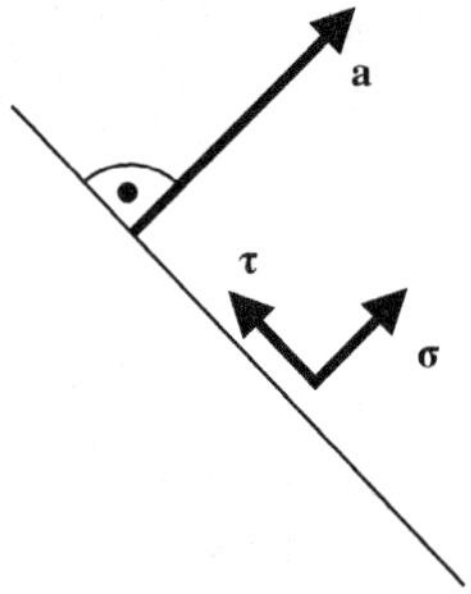

Figure 11.2 Normal (**σ**) and shear (**τ**) stresses on surface **a**

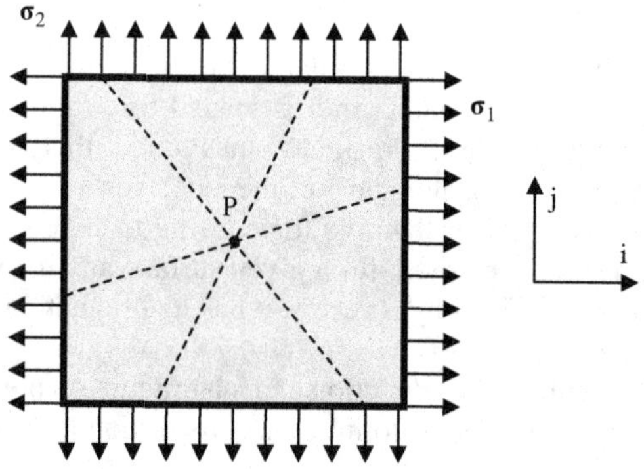

Figure 11.3 Flat thin sheet under load. The sheet is composed of six pieces that are glued together. The dashed lines represent the glued surfaces

while the force **f** is given by

$$\begin{aligned}\mathbf{f} &= f_i\mathbf{i} + f_j\mathbf{j} \\ &= [\mathbf{i}\ \ \mathbf{j}]\begin{bmatrix} f_i \\ f_j \end{bmatrix} \\ &= \begin{bmatrix} f_i \\ f_j \end{bmatrix}.\end{aligned} \tag{11.2}$$

Figure 11.4 Some possible cracks through the glue

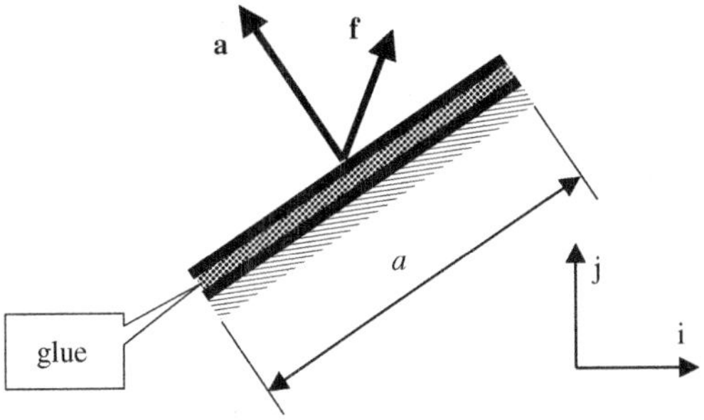

Figure 11.5 Surface **a** and its relationship with the internal force vector **f**

For the example shown in Figure 11.5 the internal force on surface **i** is

$$\mathbf{f_i} = \begin{bmatrix} \sigma_1 \\ 0 \end{bmatrix} \tag{11.3}$$

and the internal force on surface **j** is

$$\mathbf{f_j} = \begin{bmatrix} 0 \\ \sigma_2 \end{bmatrix}. \tag{11.4}$$

One should note that (since the force per unit area is constant in Figure 11.3) should one increase the surface size by a factor s, the corresponding internal force also increases by the same factor, i.e.,

$$\mathbf{f}(s\mathbf{a}) = s\mathbf{f}(\mathbf{a}). \tag{11.5}$$

In addition, the internal force on the sum of the surfaces **a** and **b** is equal to the sum of internal forces $\mathbf{f_a}$ and $\mathbf{f_b}$, as shown in Figure 11.6. Expressing this mathematically

$$\mathbf{f}(\mathbf{a}+\mathbf{b}) = \mathbf{f}(\mathbf{a}) + \mathbf{f}(\mathbf{b}). \tag{11.6}$$

From Equations (11.5) and (11.6) it follows that

$$\begin{aligned} \mathbf{f}(\mathbf{a}) &= a_i\mathbf{f}(\mathbf{i}) + a_j\mathbf{f}(\mathbf{j}) \\ &= a_i\mathbf{f_i} + a_j\mathbf{f_j}. \end{aligned} \tag{11.7}$$

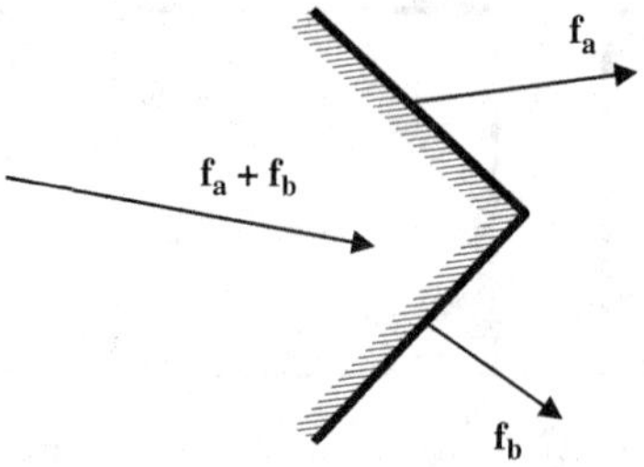

Figure 11.6 Force on the sum of two surfaces **a** and **b**

In other words, if the internal forces on surfaces **i** and **j** are known, the internal forces on any other surface

$$\mathbf{a} = \begin{bmatrix} a_i \\ a_j \end{bmatrix} \tag{11.8}$$

can be calculated using Equation (11.7).

Equation (11.7) represents a linear mapping of surfaces to forces; for a given surface **a**, it returns the force **f** associated with that surface.

Now stress at point P of the sheet shown in Figure 11.3 can be defined as a mapping such that every surface that passes through point P has a corresponding internal force **f** associated with that surface.

In order to check whether a particular glued surface that passes through point P would break, one needs to calculate the internal force on that surface

$$\mathbf{f} = \begin{bmatrix} \mathbf{f_i} & \mathbf{f_j} \end{bmatrix} \begin{bmatrix} a_i \\ a_j \end{bmatrix} \tag{11.9}$$

and compare it to the maximum force the glue can carry.

From Equation (11.9) it is evident that stress is a linear mapping of surface vectors into force vectors. As such, it is called tensor. Stress is therefore a second order tensor for it maps a vector to a vector, unlike a first order tensor which maps a vector to a scalar.

11.2 Second Order Tensor in 2D

The second order tensor is defined as a linear mapping of one space (set) of vectors to another space (set) of vectors. For example, the stress tensor is a linear mapping of surface vectors into force vectors; since for a given surface **a** it returns force **f**.

Vectors in 2D space are best expressed using an orthonormal base $[\mathbf{i} \;\; \mathbf{j}]$ also called the global base. Any surface **a** is a linear combination of base surfaces **i** and **j**

$$\begin{aligned} \mathbf{a} &= a_i\mathbf{i} + a_j\mathbf{j} \\ &= \begin{bmatrix} \mathbf{i} & \mathbf{j} \end{bmatrix} \begin{bmatrix} a_i \\ a_j \end{bmatrix} = \begin{bmatrix} a_i \\ a_j \end{bmatrix}. \end{aligned} \tag{11.10}$$

Any force **f** is a linear combination of base forces **i** and **j**

$$\begin{aligned}\mathbf{f} &= f_i\mathbf{i} + f_j\mathbf{j} \\ &= [\mathbf{i}\ \ \mathbf{j}]\begin{bmatrix} f_i \\ f_j \end{bmatrix} = \begin{bmatrix} f_i \\ f_j \end{bmatrix}.\end{aligned} \tag{11.11}$$

For a specific point P, the force on the base surfaces is either measured or calculated so that one has

$$\begin{aligned}\mathbf{f}(\mathbf{i}) &= \mathbf{f_i} \\ \mathbf{f}(\mathbf{j}) &= \mathbf{f_j}.\end{aligned} \tag{11.12}$$

Force on any other surface

$$\mathbf{a} = a_i\mathbf{i} + a_j\mathbf{j} \tag{11.13}$$

is by definition

$$\begin{aligned}\mathbf{f}(\mathbf{a}) &= a_i\mathbf{f}(\mathbf{i}) + a_j\mathbf{f}(\mathbf{j}) \\ &= a_i\mathbf{f_i} + a_j\mathbf{f_j} \\ &= [\mathbf{f_i}\ \ \mathbf{f_j}]\begin{bmatrix} a_i \\ a_j \end{bmatrix}.\end{aligned} \tag{11.14}$$

Equation (11.14) is in essence the same equation as the equivalent equation for the first order tensors. The only difference is that $\mathbf{f_i}$ and $\mathbf{f_j}$ are vectors and

$$[\mathbf{f_i}\ \ \mathbf{f_j}] \tag{11.15}$$

is the stress tensor matrix obtained using base $[\mathbf{i}\ \ \mathbf{j}]$ for both surfaces and forces. The matrix shown in Equation (11.15) is a matrix of vectors

$$\begin{aligned}[\mathbf{f_i}\ \ \mathbf{f_j}] &= \begin{bmatrix} \mathbf{i}f_{ii} & \mathbf{i}f_{ij} \\ \mathbf{j}f_{ji} & \mathbf{j}f_{jj} \end{bmatrix} \\ &= [\mathbf{i}\ \ \mathbf{j}]\begin{bmatrix} f_{ii} & f_{ij} \\ f_{ji} & f_{jj} \end{bmatrix} \\ &= \begin{bmatrix} f_{ii} & f_{ij} \\ f_{ji} & f_{jj} \end{bmatrix},\end{aligned} \tag{11.16}$$

where, in the last line of the equation, the base $[\mathbf{i}\ \ \mathbf{j}]$ is implicitly assumed.

Very often one conveniently writes

$$\mathbf{F} = \begin{bmatrix} f_{ii} & f_{ij} \\ f_{ji} & f_{jj} \end{bmatrix}, \tag{11.17}$$

where the equal sign does not mean that tensor **F** is equal to the matrix, but that tensor **F** is represented by the matrix in base $[\mathbf{i}\ \ \mathbf{j}]$.

11.3 Physical Meaning of Tensor Matrix in 2D

When studying vectors, it has been explained that for a given base

$$[\mathbf{i}\ \ \mathbf{j}], \tag{11.18}$$

any vector **a** is best represented by its matrix

$$\begin{bmatrix} a_i \\ a_j \end{bmatrix}, \tag{11.19}$$

where a_i is the force in the **i** direction and a_j is the force in the **j** direction. In other words, the components of a vector matrix have physical meaning.

In a similar manner, matrix components of a second order tensor in 2D have physical meaning. The first column of the matrix contains the components

$$\begin{bmatrix} f_{ii} \\ f_{ji} \end{bmatrix}, \tag{11.20}$$

which in actual fact represent force $\mathbf{f_i}$, i.e., the force on surface **i**. The second column of the matrix contains the components

$$\begin{bmatrix} f_{ij} \\ f_{jj} \end{bmatrix}, \tag{11.21}$$

which represent the force $\mathbf{f_j}$, which is the force on surface **j**.

For force component f_{ij}, the first index indicates the force direction and the second index indicates surface direction on which the force is acting. In other words, the force f_{ij} is the force in direction **i** on the surface **j**. Graphical representation of these is given in Figure 11.7.

By definition, it follows that any force applied to a given surface

$$\mathbf{a} = \begin{bmatrix} a_i \\ a_j \end{bmatrix} \tag{11.22}$$

is simply

$$\mathbf{f} = \begin{bmatrix} f_{ii} & f_{ij} \\ f_{ji} & f_{jj} \end{bmatrix} \begin{bmatrix} a_i \\ a_j \end{bmatrix}. \tag{11.23}$$

Equation (11.23) gives the internal force on any surface through a given point P. As such, it defines the state of internal forces at point P and is therefore called the stress tensor **F**; it is used to evaluate how "stressed" the material at point P is. The stress tensor **F** is therefore used to

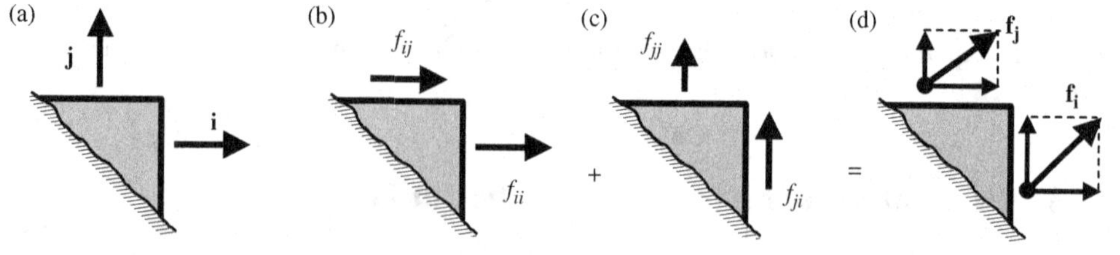

Figure 11.7 Graphical representation of: the surfaces **i** and **j** (a) and the stress tensor matrix components (b, c and d) using the base $[\mathbf{i}\ \ \mathbf{j}]$

determine how much the material deforms under a particular loading condition and also whether the material would yield or break, and how it would break.

The stress tensor **F** is represented by the matrix

$$\mathbf{F} = \begin{bmatrix} f_{ii} & f_{ij} \\ f_{ji} & f_{jj} \end{bmatrix}, \tag{11.24}$$

when base $[\mathbf{i} \ \ \mathbf{j}]$ is used for both surfaces and forces.

11.4 Changing the Base

Using the global base is not always sufficient and very often additional vector bases may be needed. In general, these bases are made of nonunit vectors that are nonorthogonal to each other, such as

$$[\boldsymbol{\alpha} \ \ \boldsymbol{\beta}], \tag{11.25}$$

where

$$\boldsymbol{\alpha} = \begin{bmatrix} \alpha_i \\ \alpha_j \end{bmatrix}; \ \boldsymbol{\beta} = \begin{bmatrix} \beta_i \\ \beta_j \end{bmatrix}; \ [\boldsymbol{\alpha} \ \ \boldsymbol{\beta}] = \begin{bmatrix} \alpha_i & \beta_i \\ \alpha_j & \beta_j \end{bmatrix}. \tag{11.26}$$

The matrix of surface **a** expressed using this base is

$$\mathbf{a} = \begin{bmatrix} a_\alpha \\ a_\beta \end{bmatrix}, \tag{11.27}$$

where the surface itself is

$$\begin{aligned} \mathbf{a} &= [\boldsymbol{\alpha} \ \ \boldsymbol{\beta}] \begin{bmatrix} a_\alpha \\ a_\beta \end{bmatrix} \\ &= [\mathbf{i} \ \ \mathbf{j}] \left(\begin{bmatrix} \alpha_i & \beta_i \\ \alpha_j & \beta_j \end{bmatrix} \begin{bmatrix} a_\alpha \\ a_\beta \end{bmatrix} \right). \end{aligned} \tag{11.28}$$

The matrix of surface **a** expressed using base $[\mathbf{i} \ \ \mathbf{j}]$ is therefore

$$\begin{bmatrix} a_i \\ a_j \end{bmatrix} = \begin{bmatrix} \alpha_i & \beta_i \\ \alpha_j & \beta_j \end{bmatrix} \begin{bmatrix} a_\alpha \\ a_\beta \end{bmatrix}. \tag{11.29}$$

The stress on surface **a** is by definition

$$\begin{aligned} \mathbf{f} = \begin{bmatrix} f_i \\ f_j \end{bmatrix} &= \mathbf{f_i} a_i + \mathbf{f_j} a_j = a_i \begin{bmatrix} f_{ii} \\ f_{ji} \end{bmatrix} + a_j \begin{bmatrix} f_{ij} \\ f_{jj} \end{bmatrix} \\ &= \begin{bmatrix} f_{ii} & f_{ij} \\ f_{ji} & f_{jj} \end{bmatrix} \begin{bmatrix} a_i \\ a_j \end{bmatrix}. \end{aligned} \tag{11.30}$$

After substituting Equation (11.29) into (11.30)

$$\begin{bmatrix} f_i \\ f_j \end{bmatrix} = \begin{bmatrix} f_{ii} & f_{ij} \\ f_{ji} & f_{jj} \end{bmatrix} \left(\begin{bmatrix} \alpha_i & \beta_i \\ \alpha_j & \beta_j \end{bmatrix} \begin{bmatrix} a_\alpha \\ a_\beta \end{bmatrix} \right) = \left(\begin{bmatrix} f_{ii} & f_{ij} \\ f_{ji} & f_{jj} \end{bmatrix} \begin{bmatrix} \alpha_i & \beta_i \\ \alpha_j & \beta_j \end{bmatrix} \right) \begin{bmatrix} a_\alpha \\ a_\beta \end{bmatrix}. \tag{11.31}$$

This can be written as

$$\mathbf{f} = [\mathbf{f}_\alpha \ \ \mathbf{f}_\beta] \begin{bmatrix} a_\alpha \\ a_\beta \end{bmatrix} = a_\alpha \begin{bmatrix} f_{i\alpha} \\ f_{j\alpha} \end{bmatrix} + a_\beta \begin{bmatrix} f_{i\beta} \\ f_{j\beta} \end{bmatrix} = \begin{bmatrix} f_{i\alpha} & f_{i\beta} \\ f_{j\alpha} & f_{j\beta} \end{bmatrix} \begin{bmatrix} a_\alpha \\ a_\beta \end{bmatrix}, \tag{11.32}$$

where

$$\begin{bmatrix} f_{i\alpha} & f_{i\beta} \\ f_{j\alpha} & f_{j\beta} \end{bmatrix} = \begin{bmatrix} f_{ii} & f_{ij} \\ f_{ji} & f_{jj} \end{bmatrix} \begin{bmatrix} \alpha_i & \beta_i \\ \alpha_j & \beta_j \end{bmatrix}. \tag{11.33}$$

It is worth noting that the obtained internal force **f** is still expressed using the base

$$[\mathbf{i} \ \ \mathbf{j}], \tag{11.34}$$

where f_i is the force in the **i** direction on surface **a** and f_j is the force in the **j** direction on surface **a**, see Figure 11.8.

In order to obtain the force components for the base $[\boldsymbol{\alpha} \ \ \boldsymbol{\beta}]$, vector **f** has to be expressed using base

$$[\boldsymbol{\alpha} \ \ \boldsymbol{\beta}]. \tag{11.35}$$

As force **f** is given by

$$\mathbf{f} = f_i\mathbf{i} + f_j\mathbf{j} = f_\alpha\boldsymbol{\alpha} + f_\beta\boldsymbol{\beta}. \tag{11.36}$$

By substituting

$$\boldsymbol{\alpha} = \alpha_i\mathbf{i} + \alpha_j\mathbf{j} = \begin{bmatrix} \alpha_i \\ \alpha_j \end{bmatrix}; \ \ \boldsymbol{\beta} = \beta_i\mathbf{i} + \beta_j\mathbf{j} = \begin{bmatrix} \beta_i \\ \beta_j \end{bmatrix} \tag{11.37}$$

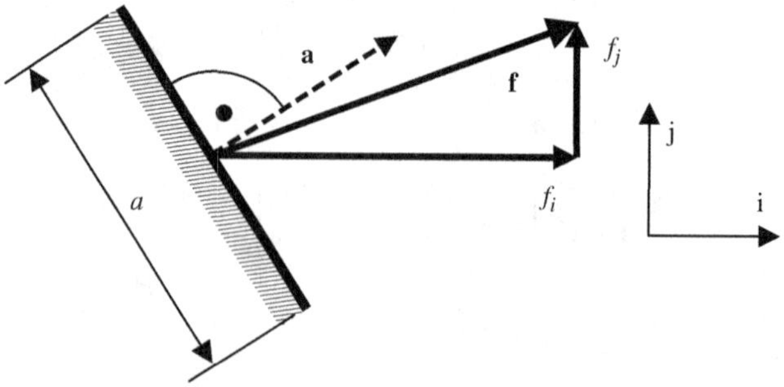

Figure 11.8 Internal force on surface **a**

into Equation (11.36) one obtains

$$\begin{aligned}\mathbf{f} &= f_\alpha(\alpha_i\mathbf{i}+\alpha_j\mathbf{j}) + f_\beta(\beta_i\mathbf{i}+\beta_j\mathbf{j}) \\ &= (f_\alpha\alpha_i + f_\beta\beta_i)\mathbf{i} + (f_\alpha\alpha_j + f_\beta\beta_j)\mathbf{j},\end{aligned} \tag{11.38}$$

which yields

$$\begin{bmatrix} f_i \\ f_j \end{bmatrix} = \begin{bmatrix} \alpha_i & \beta_i \\ \alpha_j & \beta_j \end{bmatrix} \begin{bmatrix} f_\alpha \\ f_\beta \end{bmatrix}. \tag{11.39}$$

After multiplication from the left by

$$\begin{bmatrix} \alpha_i & \beta_i \\ \alpha_j & \beta_j \end{bmatrix}^{-1}, \tag{11.40}$$

one obtains

$$\begin{aligned}\begin{bmatrix} f_\alpha \\ f_\beta \end{bmatrix} &= \begin{bmatrix} \alpha_i & \beta_i \\ \alpha_j & \beta_j \end{bmatrix}^{-1} \begin{bmatrix} f_i \\ f_j \end{bmatrix} = \begin{bmatrix} i_\alpha & j_\alpha \\ i_\beta & j_\beta \end{bmatrix} \begin{bmatrix} f_i \\ f_j \end{bmatrix} \\ &= f_i \begin{bmatrix} i_\alpha \\ i_\beta \end{bmatrix} + f_j \begin{bmatrix} j_\alpha \\ j_\beta \end{bmatrix}.\end{aligned} \tag{11.41}$$

By substituting Equation (11.31) into Equation (11.41)

$$\begin{aligned}\begin{bmatrix} f_\alpha \\ f_\beta \end{bmatrix} &= \begin{bmatrix} \alpha_i & \beta_i \\ \alpha_j & \beta_j \end{bmatrix}^{-1} \left(\begin{bmatrix} f_{ii} & f_{ij} \\ f_{ji} & f_{jj} \end{bmatrix} \begin{bmatrix} \alpha_i & \beta_i \\ \alpha_j & \beta_j \end{bmatrix} \right) \begin{bmatrix} a_\alpha \\ a_\beta \end{bmatrix} \\ &= \left(\begin{bmatrix} \alpha_i & \beta_i \\ \alpha_j & \beta_j \end{bmatrix}^{-1} \begin{bmatrix} f_{ii} & f_{ij} \\ f_{ji} & f_{jj} \end{bmatrix} \begin{bmatrix} \alpha_i & \beta_i \\ \alpha_j & \beta_j \end{bmatrix} \right) \begin{bmatrix} a_\alpha \\ a_\beta \end{bmatrix} \\ &= \begin{bmatrix} f_{\alpha\alpha} & f_{\alpha\beta} \\ f_{\beta\alpha} & f_{\beta\beta} \end{bmatrix} \begin{bmatrix} a_\alpha \\ a_\beta \end{bmatrix}\end{aligned} \tag{11.42}$$

where the tensor matrix

$$\begin{bmatrix} f_{\alpha\alpha} & f_{\alpha\beta} \\ f_{\beta\alpha} & f_{\beta\beta} \end{bmatrix} = \begin{bmatrix} \alpha_i & \beta_i \\ \alpha_j & \beta_j \end{bmatrix}^{-1} \begin{bmatrix} f_{ii} & f_{ij} \\ f_{ji} & f_{jj} \end{bmatrix} \begin{bmatrix} \alpha_i & \beta_i \\ \alpha_j & \beta_j \end{bmatrix} \tag{11.43}$$

describes the linear mapping of surfaces into forces (i.e., the tensor) using base $[\boldsymbol{\alpha} \ \ \boldsymbol{\beta}]$ for both surfaces and forces.

The physical meaning of these matrix components is shown in Figure 11.9.

11.5 Using Two Different Bases in 2D

As has been already explained, a linear mapping of one space of vectors to another space of vectors is best understood and demonstrated using the stress tensor. The stress tensor is a linear mapping of surfaces into internal forces. For a given surface, this mapping (i.e., stress tensor)

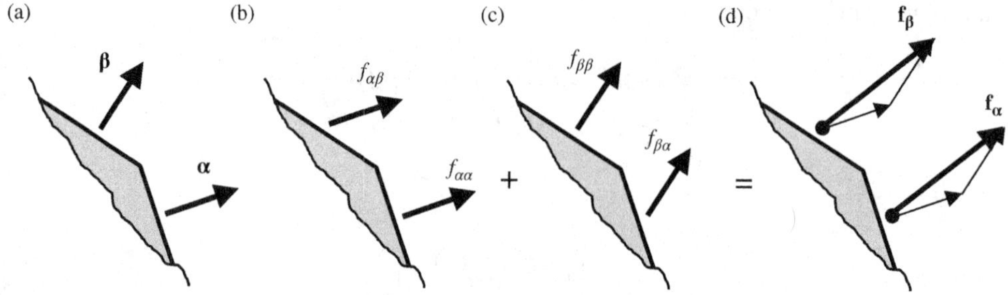

Figure 11.9 Graphical representation of: the surfaces **α** and **β** (a) and the stress tensor matrix components (b, c and d) using the base [**α** **β**]

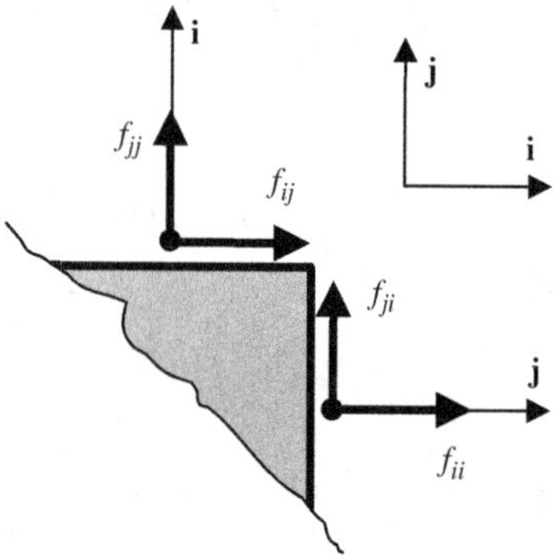

Figure 11.10 Forces on surfaces for base [**i** **j**]

produces the internal force acting on that surface, Figure 11.10. This force is of great importance in engineering practice; for example, if one wants to know whether a material would break along a particular surface.

Surfaces as vectors are best represented using a vector base. In a similar manner, forces as vectors are best represented using a vector base. So far, in this chapter the same vector base has been used for both surfaces and forces. However, it is not uncommon that one vector base is used to represent surfaces and a completely different vector base is used to represent forces, Figure 11.11.

In Figure 11.11, the surfaces are represented using the base [**α** **β**] while the forces are represented using the base [**ξ** **η**]. Both of these bases can be represented using the global base [**i** **j**] such that the base [**α** **β**] is represented by matrix

$$[\boldsymbol{\alpha}\ \ \boldsymbol{\beta}] = \begin{bmatrix} \alpha_i & \beta_i \\ \alpha_j & \beta_j \end{bmatrix}. \tag{11.44}$$

while base [**ξ** **η**] is represented by matrix

$$[\boldsymbol{\xi}\ \ \boldsymbol{\eta}] = \begin{bmatrix} \xi_i & \eta_i \\ \xi_j & \eta_j \end{bmatrix}, \tag{11.45}$$

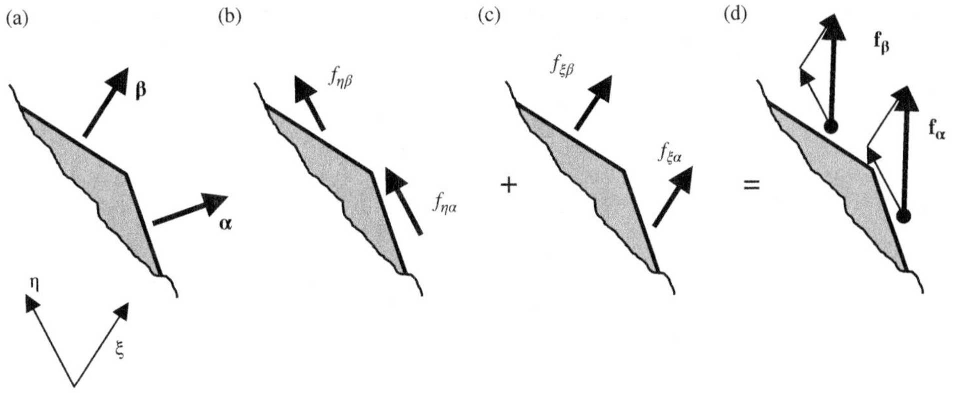

Figure 11.11 Forces and surfaces represented using different bases

where the equal sign does not mean that the vectors on the left hand side are equal to the matrix (this would make no sense in any case); one reads the equal sign as the vectors on the left hand side being represented by the matrix on the right hand side using the implicitly assumed (from the context) global vector base

$$[\mathbf{i}\ \ \mathbf{j}]. \tag{11.46}$$

Now, by starting with

$$\mathbf{f}(\mathbf{a}) = \mathbf{F}\mathbf{a} = \begin{bmatrix} f_i \\ f_j \end{bmatrix} = [\mathbf{f}_i\ \ \mathbf{f}_j] \begin{bmatrix} a_i \\ a_j \end{bmatrix} = \begin{bmatrix} f_{ii} & f_{ij} \\ f_{ji} & f_{jj} \end{bmatrix} \begin{bmatrix} a_i \\ a_j \end{bmatrix}, \tag{11.47}$$

where both forces and surfaces are represented using the global vector base (Figure 11.10), one arrives at

$$\begin{aligned} \mathbf{f}(\mathbf{a}) = \mathbf{F}\mathbf{a} &= \begin{bmatrix} f_\xi \\ f_\eta \end{bmatrix} = [\mathbf{f}_\alpha\ \ \mathbf{f}_\beta] \begin{bmatrix} a_\alpha \\ a_\beta \end{bmatrix} = a_\alpha \begin{bmatrix} f_{\xi\alpha} \\ f_{\eta\alpha} \end{bmatrix} + a_\beta \begin{bmatrix} f_{\xi\beta} \\ f_{\eta\beta} \end{bmatrix} \\ &= \begin{bmatrix} f_{\xi\alpha} & f_{\xi\beta} \\ f_{\eta\alpha} & f_{\eta\beta} \end{bmatrix} \begin{bmatrix} a_\alpha \\ a_\beta \end{bmatrix}, \end{aligned} \tag{11.48}$$

where surfaces are represented using the base $[\boldsymbol{\alpha}\ \ \boldsymbol{\beta}]$ and forces are represented using the base $[\boldsymbol{\xi}\ \ \boldsymbol{\eta}]$.

As base $[\boldsymbol{\alpha}\ \ \boldsymbol{\beta}]$ is represented by matrix

$$[\boldsymbol{\alpha}\ \ \boldsymbol{\beta}] = \begin{bmatrix} \alpha_i & \beta_i \\ \alpha_j & \beta_j \end{bmatrix}, \tag{11.49}$$

surface $\mathbf{a}$ is then given by

$$\mathbf{a} = \begin{bmatrix} a_i \\ a_j \end{bmatrix} = \begin{bmatrix} \alpha_i & \beta_i \\ \alpha_j & \beta_j \end{bmatrix} \begin{bmatrix} a_\alpha \\ a_\beta \end{bmatrix}. \tag{11.50}$$

By similar reasoning, as base $[\boldsymbol{\xi}\ \ \boldsymbol{\eta}]$ in the global base is represented (using the global base) by matrix

$$[\boldsymbol{\xi}\ \ \boldsymbol{\eta}] = \begin{bmatrix} \xi_i & \eta_i \\ \xi_j & \eta_j \end{bmatrix}, \tag{11.51}$$

force $\mathbf{f}$ is then given by

$$\mathbf{f} = \begin{bmatrix} f_i \\ f_j \end{bmatrix} = \begin{bmatrix} \xi_i & \eta_i \\ \xi_j & \eta_j \end{bmatrix} \begin{bmatrix} f_\xi \\ f_\eta \end{bmatrix}, \tag{11.52}$$

which after left hand multiplication by

$$\begin{bmatrix} \xi_i & \eta_i \\ \xi_j & \eta_j \end{bmatrix}^{-1} \tag{11.53}$$

yields

$$\begin{bmatrix} f_\xi \\ f_\eta \end{bmatrix} = \begin{bmatrix} \xi_i & \eta_i \\ \xi_j & \eta_j \end{bmatrix}^{-1} \begin{bmatrix} f_i \\ f_j \end{bmatrix} = \begin{bmatrix} i_\xi & j_\xi \\ i_\eta & j_\eta \end{bmatrix} = f_i \begin{bmatrix} i_\xi \\ i_\eta \end{bmatrix} + f_j \begin{bmatrix} j_\xi \\ j_\eta \end{bmatrix}, \tag{11.54}$$

where

$$\begin{bmatrix} i_\xi & j_\xi \\ i_\eta & j_\eta \end{bmatrix} = \begin{bmatrix} \xi_i & \eta_i \\ \xi_j & \eta_j \end{bmatrix}^{-1}. \tag{11.55}$$

By substituting Equation (11.47) into (11.54) one obtains

$$\begin{aligned} \mathbf{f}(\mathbf{a}) = \begin{bmatrix} f_\xi \\ f_\eta \end{bmatrix} &= \begin{bmatrix} i_\xi & j_\xi \\ i_\eta & j_\eta \end{bmatrix} \left(\begin{bmatrix} f_{ii} & f_{ij} \\ f_{ji} & f_{jj} \end{bmatrix} \begin{bmatrix} a_i \\ a_j \end{bmatrix} \right) \\ &= \left(\begin{bmatrix} \xi_i & \eta_i \\ \xi_j & \eta_j \end{bmatrix}^{-1} \begin{bmatrix} f_{ii} & f_{ij} \\ f_{ji} & f_{jj} \end{bmatrix} \right) \begin{bmatrix} a_i \\ a_j \end{bmatrix}. \end{aligned} \tag{11.56}$$

After substituting Equation (11.50) into (11.56)

$$\begin{aligned} \mathbf{f}(\mathbf{a}) = \mathbf{F}\mathbf{a} = \begin{bmatrix} f_\xi \\ f_\eta \end{bmatrix} &= \\ &= \left(\begin{bmatrix} \xi_i & \eta_i \\ \xi_j & \eta_j \end{bmatrix}^{-1} \begin{bmatrix} f_{ii} & f_{ij} \\ f_{ji} & f_{jj} \end{bmatrix} \right) \left(\begin{bmatrix} \alpha_i & \beta_i \\ \alpha_j & \beta_j \end{bmatrix} \begin{bmatrix} a_\alpha \\ a_\beta \end{bmatrix} \right) \\ &= \left(\begin{bmatrix} \xi_i & \eta_i \\ \xi_j & \eta_j \end{bmatrix}^{-1} \begin{bmatrix} f_{ii} & f_{ij} \\ f_{ji} & f_{jj} \end{bmatrix} \begin{bmatrix} \alpha_i & \beta_i \\ \alpha_j & \beta_j \end{bmatrix} \right) \begin{bmatrix} a_\alpha \\ a_\beta \end{bmatrix}, \end{aligned} \tag{11.57}$$

where the tensor **F** is represented by the tensor matrix

$$\mathbf{F}=\begin{bmatrix}\xi_i & \eta_i\\ \xi_j & \eta_j\end{bmatrix}^{-1}\begin{bmatrix}f_{ii} & f_{ij}\\ f_{ji} & f_{jj}\end{bmatrix}\begin{bmatrix}\alpha_i & \beta_i\\ \alpha_j & \beta_j\end{bmatrix}. \tag{11.58}$$

Components of this matrix are obtained using vector base $[\boldsymbol{\alpha} \ \ \boldsymbol{\beta}]$ for the surfaces and vector base $[\boldsymbol{\xi} \ \ \boldsymbol{\eta}]$ for the forces. Also, the surface vector matrix

$$\mathbf{a}=\begin{bmatrix}a_\alpha\\ a_\beta\end{bmatrix} \tag{11.59}$$

is obtained using base $[\boldsymbol{\alpha} \ \ \boldsymbol{\beta}]$ while the force vector matrix

$$\mathbf{f}=\begin{bmatrix}f_\xi\\ f_\eta\end{bmatrix} \tag{11.60}$$

is obtained using base $[\boldsymbol{\xi} \ \ \boldsymbol{\eta}]$. One should note that none of the bases employed has to be orthonormal. In other words, all the bases employed are general bases with nonunit vectors that are at arbitrary angles to each other.

11.6 Some Special Cases of Stress Tensor Matrices in 2D

In the previous section, it has been demonstrated that any general base can be used to represent surfaces and another general base can be used to represent forces. By changing the vector base for surfaces and the vector base for forces a whole range of stress tensor matrices can be obtained. The most straightforward one is the global base of orthonormal vectors $[\mathbf{i} \ \ \mathbf{j}]$.

The obtained stress tensor matrix is simply

$$\mathbf{C}=\begin{bmatrix}c_{ii} & c_{ij}\\ c_{ji} & c_{jj}\end{bmatrix}, \tag{11.61}$$

where for each of the components c_{ij} the first index indicates the force's direction and the second index indicates the surface's direction, on which the force is acting. The first column is therefore the force on surface **i**. The second column is the force on surface **j**. The first row represents all the forces in **i** direction. The second row represents all the forces in **j** direction. Matrix **C** is called the Cauchy stress tensor matrix.

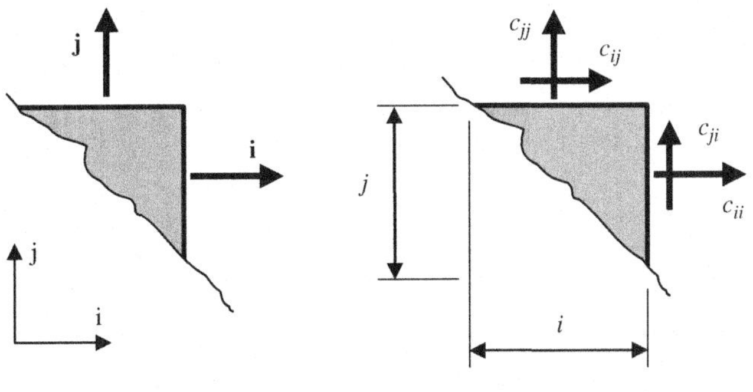

Figure 11.12 Global base surfaces (left) and stress components obtained using the global base (right)

As the vectors **i** and **j** of the vector base are orthonormal to each other, it happens that the above matrix's diagonal represents forces in the direction of the two surfaces. These are called normal forces (for they are in the direction of the surface vector). For the same reason, the off diagonal terms of the matrix represent forces that are orthogonal to their surfaces (orthogonal to the surface vector). These are called tangential or shear forces.

As vectors **i** and **j** are unit vectors, surfaces **i** and **j** are also unit surfaces. Thus, the terms of the stress tensor matrix also represent internal force per unit surface. This is a historical engineering definition of stress often taught to first year engineering students. Of course, it is a much narrower definition of stress than the stress definition presented in this chapter. As such, the old definition is applicable to a limited set of engineering problems and it often hinders efforts to explain theoretical concepts such as rubber elasticity or many other problems faced by modern engineers or computational mechanics experts or material scientists. For these reasons, the old definition of stress is not used in this book.

The above defined stress tensor matrix is usually called the *Cauchy Stress Tensor Matrix*, or simply *Cauchy Stress Matrix*. In old books the above matrix is sometimes called Cauchy stress. This is wrong; it is always important to keep in mind that there is only one stress (stress tensor) representing the internal forces at a given material point of a solid body (i.e., representing a physical reality), and there are many different stress matrices. A matrix is a mathematical construct, while a tensor is a physical quantity. A mass can be equal to 5 kg while a displacement can be equal to vector **b**; in a similar manner, the stress at point P is equal to tensor **C**.

11.7 The First Piola-Kirchhoff Stress Tensor Matrix

One special case of vector bases often used is mentioned in literature under the name: *the First Piola-Kirchhoff Stress Matrix*. A global orthonormal base $[\mathbf{i}\ \ \mathbf{j}]$ is used to represent forces and a non-orthonormal base is used to represent surfaces, Figure 11.13.

By starting from the Cauchy stress matrix

$$\mathbf{f} = \begin{bmatrix} f_i \\ f_j \end{bmatrix} = [\mathbf{c}_i\ \ \mathbf{c}_j] \begin{bmatrix} a_i \\ a_j \end{bmatrix} = a_i \begin{bmatrix} c_{ii} \\ c_{ji} \end{bmatrix} + a_j \begin{bmatrix} c_{ij} \\ c_{jj} \end{bmatrix} = \begin{bmatrix} c_{ii} & c_{ij} \\ c_{ji} & c_{jj} \end{bmatrix} \begin{bmatrix} a_i \\ a_j \end{bmatrix} \tag{11.62}$$

and substituting

$$\begin{bmatrix} a_i \\ a_j \end{bmatrix} = \begin{bmatrix} \alpha_i & \beta_i \\ \alpha_j & \beta_j \end{bmatrix} \begin{bmatrix} a_\alpha \\ a_\beta \end{bmatrix}, \tag{11.63}$$

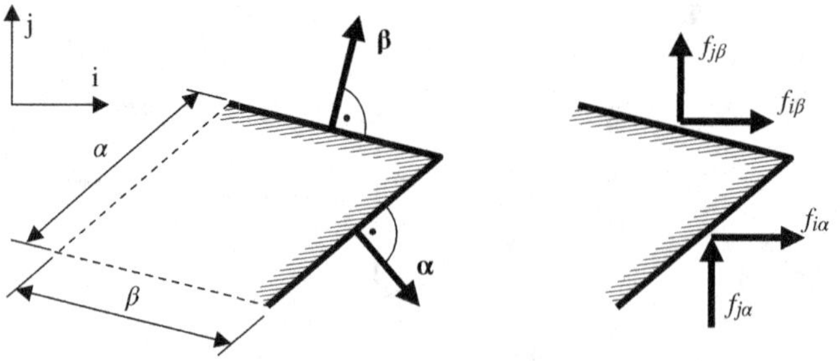

Figure 11.13 Graphical representation of the first Piola-Kirchhoff stress tensor matrix: surfaces (left) and forces (right)

one obtains

$$\begin{aligned}\mathbf{f}=\begin{bmatrix}f_i\\f_j\end{bmatrix}&=\begin{bmatrix}c_{ii}&c_{ij}\\c_{ji}&c_{jj}\end{bmatrix}\left(\begin{bmatrix}\alpha_i&\beta_i\\\alpha_j&\beta_j\end{bmatrix}\begin{bmatrix}a_\alpha\\a_\beta\end{bmatrix}\right)\\&=\left(\begin{bmatrix}c_{ii}&c_{ij}\\c_{ji}&c_{jj}\end{bmatrix}\begin{bmatrix}\alpha_i&\beta_i\\\alpha_j&\beta_j\end{bmatrix}\right)\begin{bmatrix}a_\alpha\\a_\beta\end{bmatrix}\\&=\begin{bmatrix}p_{i\alpha}&p_{i\beta}\\p_{j\alpha}&p_{j\beta}\end{bmatrix}\begin{bmatrix}a_\alpha\\a_\beta\end{bmatrix}.\end{aligned}\tag{11.64}$$

The obtained stress matrix **P** is given by

$$\mathbf{P}=\begin{bmatrix}p_{i\alpha}&p_{i\beta}\\p_{j\alpha}&p_{j\beta}\end{bmatrix}=\begin{bmatrix}c_{ii}&c_{ij}\\c_{ji}&c_{jj}\end{bmatrix}\begin{bmatrix}\alpha_i&\beta_i\\\alpha_j&\beta_j\end{bmatrix}\tag{11.65}$$

and is called the first Piola-Kirchhoff stress tensor matrix. From Equation (11.65) one also obtains

$$\begin{aligned}\mathbf{C}=\begin{bmatrix}c_{ii}&c_{ij}\\c_{ji}&c_{jj}\end{bmatrix}&=\begin{bmatrix}p_{i\alpha}&p_{i\beta}\\p_{j\alpha}&p_{j\beta}\end{bmatrix}\begin{bmatrix}\alpha_i&\beta_i\\\alpha_j&\beta_j\end{bmatrix}^{-1}\\&=\begin{bmatrix}p_{i\alpha}&p_{i\beta}\\p_{j\alpha}&p_{j\beta}\end{bmatrix}\begin{bmatrix}i_\alpha&j_\alpha\\i_\beta&j_\beta\end{bmatrix},\end{aligned}\tag{11.66}$$

where

$$\begin{bmatrix}i_\alpha&j_\alpha\\i_\beta&j_\beta\end{bmatrix}=\begin{bmatrix}\alpha_i&\beta_i\\\alpha_j&\beta_j\end{bmatrix}^{-1}.\tag{11.67}$$

In order to better "understand" Equation (11.66), it is helpful to consider that

$$\mathbf{f}_\mathbf{i}=\begin{bmatrix}c_{ii}\\c_{ji}\end{bmatrix}=\begin{bmatrix}p_{i\alpha}&p_{i\beta}\\p_{j\alpha}&p_{j\beta}\end{bmatrix}\begin{bmatrix}i_\alpha\\i_\beta\end{bmatrix}=i_\alpha\begin{bmatrix}p_{i\alpha}\\p_{j\alpha}\end{bmatrix}+i_\beta\begin{bmatrix}p_{i\beta}\\p_{j\beta}\end{bmatrix}\tag{11.68}$$

and

$$\mathbf{f}_\mathbf{j}=\begin{bmatrix}c_{ij}\\c_{jj}\end{bmatrix}=\begin{bmatrix}p_{i\alpha}&p_{i\beta}\\p_{j\alpha}&p_{j\beta}\end{bmatrix}\begin{bmatrix}j_\alpha\\j_\beta\end{bmatrix}=j_\alpha\begin{bmatrix}p_{i\alpha}\\p_{j\alpha}\end{bmatrix}+j_\beta\begin{bmatrix}p_{i\beta}\\p_{j\beta}\end{bmatrix},\tag{11.69}$$

which is nothing but the definition of the stress tensor itself. In other words, these equations are very intuitive and full of meaning. It is therefore well worthwhile to spend the time to understand them.

11.8 The Second Piola-Kirchhoff Stress Tensor Matrix

In some applications of solid mechanics it is convenient to represent surfaces using the base

$$[\boldsymbol{\alpha}\ \ \boldsymbol{\beta}],\tag{11.70}$$

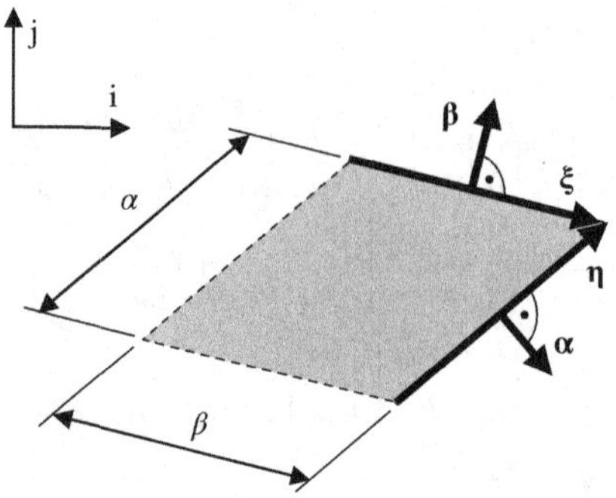

Figure 11.14 Graphical representation of the bases used to represent surfaces and forces

while forces are represented using the base

$$[\boldsymbol{\xi}\ \ \boldsymbol{\eta}], \tag{11.71}$$

as shown in Figure 11.14.

In the global orthonormal base $[\mathbf{i}\ \ \mathbf{j}]$ the matrices of both bases are given as follows

$$[\boldsymbol{\xi}\ \ \boldsymbol{\eta}] = \begin{bmatrix} \xi_i & \eta_i \\ \xi_j & \eta_j \end{bmatrix};\ [\boldsymbol{\alpha}\ \ \boldsymbol{\beta}] = \begin{bmatrix} \alpha_i & \beta_i \\ \alpha_j & \beta_j \end{bmatrix}, \tag{11.72}$$

where vectors $\boldsymbol{\alpha}$ and $\boldsymbol{\beta}$ are parallel to the dual base vectors

$$[\underline{\boldsymbol{\xi}}\ \ \underline{\boldsymbol{\eta}}], \tag{11.73}$$

in such a way that

$$\begin{aligned}[\boldsymbol{\alpha}\ \ \boldsymbol{\beta}] &= g[\underline{\boldsymbol{\xi}}\ \ \underline{\boldsymbol{\eta}}] \\ &= g\left(\begin{bmatrix} \xi_i & \eta_i \\ \xi_j & \eta_j \end{bmatrix}^{\mathrm{T}}\right)^{-1} \\ &= g\begin{bmatrix} \xi_i & \eta_i \\ \xi_j & \eta_j \end{bmatrix}^{-\mathrm{T}}. \end{aligned} \tag{11.74}$$

From geometric considerations the shaded area shown in Figure 11.14 is given by

$$g = \det\begin{bmatrix} \xi_i & \eta_i \\ \xi_j & \eta_j \end{bmatrix}. \tag{11.75}$$

The same area is given as a scalar product

$$g = \boldsymbol{\xi}\cdot\boldsymbol{\alpha} \tag{11.76}$$

as

$$\boldsymbol{\xi}\cdot\underline{\boldsymbol{\xi}} = 1 \tag{11.77}$$

and $\underline{\boldsymbol{\xi}}$ and $\boldsymbol{\alpha}$ are parallel, by multiplying both sides of Equation (11.77) by g, one obtains

$$\boldsymbol{\xi}\cdot\underline{\boldsymbol{\xi}}g = g. \tag{11.78}$$

By comparing Equation (11.78) with Equation (11.76) it can be concluded that

$$\boldsymbol{\xi}\cdot\boldsymbol{\alpha} = \boldsymbol{\xi}\cdot\underline{\boldsymbol{\xi}}g. \tag{11.79}$$

As $\boldsymbol{\alpha}$ and $\underline{\boldsymbol{\xi}}$ are parallel, this means that

$$\boldsymbol{\alpha} = \underline{\boldsymbol{\xi}}g. \tag{11.80}$$

By a similar reasoning

$$\boldsymbol{\beta} = \underline{\boldsymbol{\eta}}g. \tag{11.81}$$

Thus,

$$\begin{aligned}[\boldsymbol{\alpha}\ \ \boldsymbol{\beta}] &= [\underline{\boldsymbol{\xi}}\ \ \underline{\boldsymbol{\eta}}]g \\ &= \left(\begin{bmatrix}\xi_i & \eta_i \\ \xi_j & \eta_j\end{bmatrix}^{\mathrm{T}}\right)^{-1} g\end{aligned} \tag{11.82}$$

or

$$\begin{bmatrix}\alpha_i & \beta_i \\ \alpha_j & \beta_j\end{bmatrix} = \begin{bmatrix}\xi_i & \eta_i \\ \xi_j & \eta_j\end{bmatrix}^{-\mathrm{T}} g. \tag{11.83}$$

By substituting Equation (11.83) into Equation (11.65)

$$\begin{aligned}\mathbf{P} = \begin{bmatrix}p_{i\alpha} & p_{i\beta} \\ p_{j\alpha} & p_{j\beta}\end{bmatrix} &= \begin{bmatrix}c_{ii} & c_{ij} \\ c_{ji} & c_{jj}\end{bmatrix}\left(\begin{bmatrix}\xi_i & \eta_i \\ \xi_j & \eta_j\end{bmatrix}^{-\mathrm{T}} g\right) \\ &= \begin{bmatrix}c_{ii} & c_{ij} \\ c_{ji} & c_{jj}\end{bmatrix}\begin{bmatrix}\xi_i & \eta_i \\ \xi_j & \eta_j\end{bmatrix}^{-\mathrm{T}}\left(\det\begin{bmatrix}\xi_i & \eta_i \\ \xi_j & \eta_j\end{bmatrix}\right).\end{aligned} \tag{11.84}$$

From Equation (11.84), by multiplying from the right hand side by

$$\begin{bmatrix}\xi_i & \eta_i \\ \xi_j & \eta_j\end{bmatrix}^{\mathrm{T}} \tag{11.85}$$

one obtains

$$\begin{bmatrix}c_{ii} & c_{ij} \\ c_{ji} & c_{jj}\end{bmatrix} = \frac{1}{g}\begin{bmatrix}p_{i\alpha} & p_{i\beta} \\ p_{j\alpha} & p_{j\beta}\end{bmatrix}\begin{bmatrix}\xi_i & \eta_i \\ \xi_j & \eta_j\end{bmatrix}^{\mathrm{T}}. \tag{11.86}$$

The force (expressed using vector base $[\mathbf{i} \ \ \mathbf{j}]$) on a given surface $\mathbf{a}$ (expressed using vector base $[\boldsymbol{\alpha} \ \ \boldsymbol{\beta}]$) is given by

$$\mathbf{f} = [\mathbf{i} \ \ \mathbf{j}] \begin{bmatrix} f_i \\ f_j \end{bmatrix} = \begin{bmatrix} f_i \\ f_j \end{bmatrix} = \mathbf{Pa} = \begin{bmatrix} p_{i\alpha} & p_{i\beta} \\ p_{j\alpha} & p_{j\beta} \end{bmatrix} \begin{bmatrix} a_\alpha \\ a_\beta \end{bmatrix}, \tag{11.87}$$

where the surface normals are already expressed using the base $[\boldsymbol{\alpha} \ \ \boldsymbol{\beta}]$. However, the forces are expressed using the base $[\mathbf{i} \ \ \mathbf{j}]$, thus

$$\begin{bmatrix} f_i \\ f_j \end{bmatrix} = \begin{bmatrix} \xi_i & \eta_i \\ \xi_j & \eta_j \end{bmatrix} \begin{bmatrix} f_\xi \\ f_\eta \end{bmatrix}. \tag{11.88}$$

By multiplying Equation (11.88) from the left by

$$\begin{bmatrix} \xi_i & \eta_i \\ \xi_j & \eta_j \end{bmatrix}^{-1}, \tag{11.89}$$

one obtains

$$\begin{bmatrix} f_\xi \\ f_\eta \end{bmatrix} = \begin{bmatrix} \xi_i & \eta_i \\ \xi_j & \eta_j \end{bmatrix}^{-1} \begin{bmatrix} f_i \\ f_j \end{bmatrix}. \tag{11.90}$$

After substitution from Equation (11.87), this yields

$$\begin{aligned} \begin{bmatrix} f_\xi \\ f_\eta \end{bmatrix} &= \begin{bmatrix} \xi_i & \eta_i \\ \xi_j & \eta_j \end{bmatrix}^{-1} \left(\begin{bmatrix} p_{i\alpha} & p_{i\beta} \\ p_{j\alpha} & p_{j\beta} \end{bmatrix} \begin{bmatrix} a_\alpha \\ a_\beta \end{bmatrix} \right) \\ &= \left(\begin{bmatrix} \xi_i & \eta_i \\ \xi_j & \eta_j \end{bmatrix}^{-1} \begin{bmatrix} p_{i\alpha} & p_{i\beta} \\ p_{j\alpha} & p_{j\beta} \end{bmatrix} \right) \begin{bmatrix} a_\alpha \\ a_\beta \end{bmatrix} \\ &= \begin{bmatrix} k_{\xi\alpha} & k_{\xi\beta} \\ k_{\eta\alpha} & k_{\eta\beta} \end{bmatrix} \begin{bmatrix} a_\alpha \\ a_\beta \end{bmatrix} = \mathbf{Ka}, \end{aligned} \tag{11.91}$$

where

$$\mathbf{K} = \begin{bmatrix} k_{\xi\alpha} & k_{\xi\beta} \\ k_{\eta\alpha} & k_{\eta\beta} \end{bmatrix}, \tag{11.92}$$

which is called the second Piola-Kirchhoff stress tensor matrix. By substituting Equation (11.84) for the first Piola-Kirchhoff stress matrix one obtains

$$\begin{aligned} \mathbf{K} = \begin{bmatrix} k_{\xi\alpha} & k_{\xi\beta} \\ k_{\eta\alpha} & k_{\eta\beta} \end{bmatrix} &= \begin{bmatrix} \xi_i & \eta_i \\ \xi_j & \eta_j \end{bmatrix}^{-1} \begin{bmatrix} p_{i\alpha} & p_{i\beta} \\ p_{j\alpha} & p_{j\beta} \end{bmatrix} \\ &= g \begin{bmatrix} \xi_i & \eta_i \\ \xi_j & \eta_j \end{bmatrix}^{-1} \begin{bmatrix} c_{ii} & c_{ij} \\ c_{ji} & c_{jj} \end{bmatrix} \begin{bmatrix} \xi_i & \eta_i \\ \xi_j & \eta_j \end{bmatrix}^{-\mathrm{T}}. \end{aligned} \tag{11.93}$$

In a similar manner

$$\mathbf{C}=\begin{bmatrix} c_{ii} & c_{ij} \\ c_{ji} & c_{jj} \end{bmatrix}=\frac{1}{g}\begin{bmatrix} \xi_i & \eta_i \\ \xi_j & \eta_j \end{bmatrix}^{-1}\begin{bmatrix} k_{\xi\alpha} & k_{\xi\beta} \\ k_{\eta\alpha} & k_{\eta\beta} \end{bmatrix}\begin{bmatrix} \xi_i & \eta_i \\ \xi_j & \eta_j \end{bmatrix}^{\mathrm{T}}. \tag{11.94}$$

Example. The Cauchy stress tensor matrix is given by

$$\mathbf{C}=\begin{bmatrix} 3 & 7 \\ 7 & 9 \end{bmatrix}. \tag{11.95}$$

a. Explain the physical meaning of the components of matrix **C**.
b. Calculate the internal force on surface **a**

$$\mathbf{a}=\begin{bmatrix} 0.1 \\ 0.1 \end{bmatrix}. \tag{11.96}$$

Solution:

a. The physical meaning of the components of matrix **C** are demonstrated graphically in Figure 11.15 and Figure 11.16.
b. The internal force on surface a is given by

$$\begin{aligned} \mathbf{f}_a=\mathbf{Ca}&=\begin{bmatrix} 3 & 7 \\ 7 & 9 \end{bmatrix}\begin{bmatrix} 0.1 \\ 0.1 \end{bmatrix} \\ &=\begin{bmatrix} 1.0 \\ 1.6 \end{bmatrix}. \end{aligned} \tag{11.97}$$

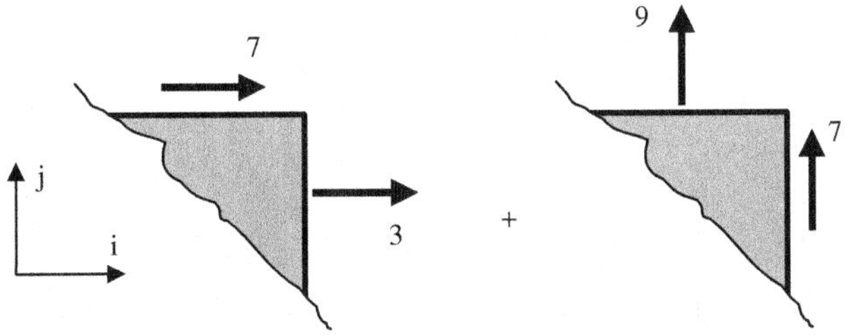

Figure 11.15 Physical meaning of matrix **C** components

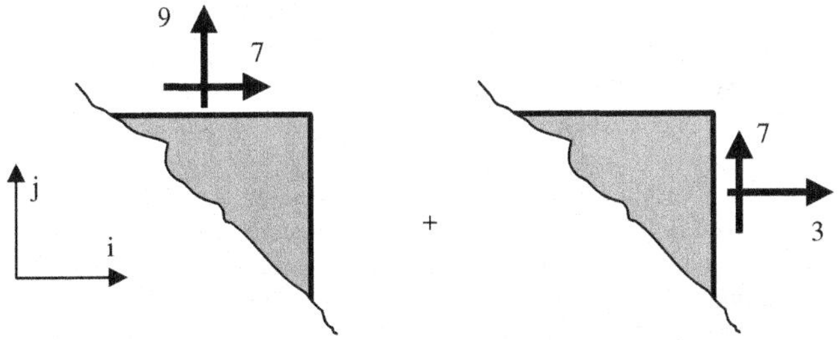

Figure 11.16 Physical meaning of matrix **C** components

11.9 Summary

The second order tensor has been introduced through the example of the stress tensor. The role of vector bases has been explained. Also the general procedure for different vector bases for both surfaces and forces has been presented in detail. No constraints on choosing general vector bases composed of nonorthogonal nonunit vectors were imposed.

The historical engineering definition of stress has also been explained, together with the most often employed considerations for stress tensors.

A clear difference has been introduced between the tensor itself and the representation of the tensor using a matrix of its components that correspond to specific vector bases for surfaces and forces.

Often some derivations have been deliberately repeated in order to emphasize that the final equations are not to be remembered; the derivation and the thought process behind the derivation should be understood instead. The idea is that the reader should be able to quickly arrive at the same equations on the back of an envelope. For these reasons, there may be plenty of repetitions; this is for the sake of always going back to the first principles, which are mostly contained in the definition of the second order tensor and stress tensor in particular.

The example of the stress tensor has been deliberately kept instead of going to abstractions. This is to emphasize that a tensor is not a mathematical abstract, but a physical quantity used by engineers on a daily basis.

The traditional notation of the stress tensor uses symbol $\boldsymbol{\sigma}$. Here the symbol **F** has been deliberately used to emphasize that stress is simply a linear mapping of surfaces to forces – thus **F**. The reader is encouraged to try not to think in terms of the traditional concept of $\boldsymbol{\sigma}$, but rather change to this new concept that was first introduced by Gurtin and Thrusdell, who are two of the more important names in twentieth-century Continuum Mechanics.

In this chapter, the spirit of Gurtin's concepts has been combined with a transparent engineering approach that provides detailed mathematical descriptions that are ready for computer implementations. By utilizing such an approach, the gap between continuum mechanics and computational mechanics is being bridged in a way that is understandable to the average engineer.

Extension to other tensors is straightforward. For instance, by analogy to Section 10.5, the differential of a 2D vector field has the following meaning: How much does vector **f** change if one moves from point P by **da**? This can be formally written as $\mathbf{D} = \mathbf{df}/\mathbf{da}$.

Further Reading

[1] Abraham, R., Marsden, J. E and Ratiu, T. S. (1988) *Manifolds, Tensor Analysis, and Applications*. New York, Springer-Verlag.
[2] Aris, R. (1962) *Vectors, Tensors and the Basic Equations of Fluid Mechanics*. New Jersey, Prentice Hall.
[3] Bronstein, I. N. and Semendyayev, S. (2007) *Handbook of Mathematics*. New York, Springer.
[4] Dvorkin, E. N. and Goldschmit, M. B. (2005) *Nonlinear Continua*. New York, Springer.
[5] Fleisch, D. (2012) *A Student's Guide to Vectors and Tensors*. New York, Cambridge University Press.
[6] Kusse, B. R. and Westwig, E. A. (2006) *Mathematical Physics: Applied Mathematics for Scientists and Engineers*, 2nd edn. Wiley-VCH Verlag GmbH & Co.
[7] Levi-Civita, T. (2013) *The Absolute Differential Calculus (Calculus of Tensors)*. New York, Dover Publications.
[8] Matthews, P. C. (1998) *Vector Calculus*. London, Springer.
[9] Gurtin, M. E. (1982) *An Introduction to Continuum Mechanics*. Oxford, Elsevier Science.
[10] Gurtin, M. E., (1983) The linear theory of elasticity. In Flügge, Siegfried; Truesdell, Clifford A., *Festkörpermechanik/Mechanics of Solids, Handbuch der Physik (Encyclopedia of Physics)*. Springer-Verlag.
[11] Truesdell, C. (1992) *A First Course in Rational Continuum Mechanics*. Oxford, Elsevier Science.
[12] Truesdell, C. (1985) *The Elements of Continuum Mechanics*. Berlin, Springer-Verlag.

12

Second Order Tensors in 3D

12.1 Stress Tensor in 3D

Homogeneous Internal Forces. A crack has appeared in a solid 3D concrete block under uniform load as shown in Figure 12.1. The crack shown is a flat surface described by the surface vector **a**, where **a** is a vector of magnitude equal to the surface area and with a direction orthogonal to the surface plane.

The question is: What was the force **f** that caused the concrete material to break and the crack to appear?

Here, force **f** is defined as the internal force that has kept the two crack walls together. This internal force is in essence the resultant of the inter-atomic forces across the crack walls. In the example shown in Figure 12.1, the internal forces are uniformly distributed over any given surface. The resultant force is therefore proportional to the surface area; if one doubles the surface area, the force will double as well. In other words,

$$\mathbf{f}(s\mathbf{a}) = s\mathbf{f}(\mathbf{a}), \tag{12.1}$$

where s is a real number. In a similar manner it follows that the force on a sum of surfaces **b** and **c**

$$\mathbf{a} = \mathbf{b} + \mathbf{c} \tag{12.2}$$

is equal to the sum of the forces on each of the surfaces

$$\mathbf{f}(\mathbf{b} + \mathbf{c}) = \mathbf{f}(\mathbf{b}) + \mathbf{f}(\mathbf{c}) \tag{12.3}$$

Thus,

$$\mathbf{f}(s_b\mathbf{b} + s_c\mathbf{c}) = s_b\mathbf{f}(\mathbf{b}) + s_c\mathbf{f}(\mathbf{c}). \tag{12.4}$$

Large Strain Finite Element Method: A Practical Course, First Edition. Antonio Munjiza, Esteban Rougier and Earl E. Knight.

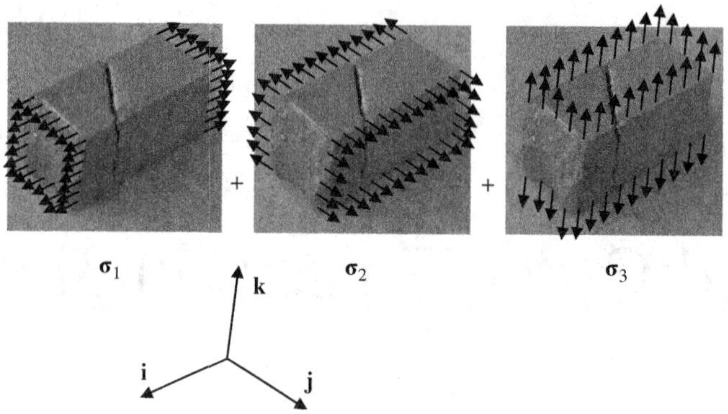

Figure 12.1 A brick under homogeneous internal forces together with the crack walls being shown

Nonhomogeneous Internal Forces. In general, the internal forces are not homogeneous and for a given point $P = (x, y, z)$ they are a function of (x, y, z). However, in the limit where the size of surface **a** becomes infinitesimally small, one can always consider that in an infinitesimally close vicinity of any point P, the internal forces are in fact homogeneous.

From the above, it can be concluded that for an internal force on any surface in the infinitesimal vicinity of a given point P there exists a linear mapping that maps each surface **a** into a corresponding force

$$\mathbf{f} = \mathbf{f}(\mathbf{a}) = \mathbf{F}\mathbf{a}. \tag{12.5}$$

This is a linear mapping of 3D surfaces into 3D internal forces and, as such, is called *the stress tensor*. For a given point P, the stress tensor describes the state of internal forces in the infinitesimal vicinity of that point.

If the internal forces are too large, the material may be broken at point P. One way to break a material is to produce a discrete crack through point P. The problem is that for the same internal forces one can assume an infinite number of possible cracks. In theory one would need to test the strength of the material for all these possible cracks and find the critical one. The issue that one encounters is that internal forces across different cracks are also different. Nevertheless, they can be described by the stress tensor, i.e., by a linear mapping of surface vectors into force vectors.

Surfaces in 3D. The best way to describe surfaces in 3D is by using the global orthonormal base of vectors $[\mathbf{i}\ \ \mathbf{j}\ \ \mathbf{k}]$ such that any surface **a** is given by

$$\begin{aligned} \mathbf{a} &= a_i\mathbf{i} + a_j\mathbf{j} + a_k\mathbf{k} \\ &= [\mathbf{i}\ \ \mathbf{j}\ \ \mathbf{k}] \begin{bmatrix} a_i \\ a_j \\ a_k \end{bmatrix} \end{aligned} \tag{12.6}$$

or simply

$$\mathbf{a} = \begin{bmatrix} a_i \\ a_j \\ a_k \end{bmatrix}, \tag{12.7}$$

where the equal sign does not mean that the vector **a** is equal to the matrix

$$\begin{bmatrix} a_i \\ a_j \\ a_k \end{bmatrix}, \tag{12.8}$$

but that the vector **a** is represented by the matrix (of its components)

$$\begin{bmatrix} a_i \\ a_j \\ a_k \end{bmatrix}, \tag{12.9}$$

using the base $[\mathbf{i}\ \ \mathbf{j}\ \ \mathbf{k}]$.

Forces in 3D. The resultant internal force on surface **i** can be represented by the matrix

$$\mathbf{f_i} = [\mathbf{i}\ \ \mathbf{j}\ \ \mathbf{k}] \begin{bmatrix} f_{ii} \\ f_{ji} \\ f_{ki} \end{bmatrix} = \begin{bmatrix} f_{ii} \\ f_{ji} \\ f_{ki} \end{bmatrix}. \tag{12.10}$$

The force on surface **j** can be represented by the matrix

$$\mathbf{f_j} = [\mathbf{i}\ \ \mathbf{j}\ \ \mathbf{k}] \begin{bmatrix} f_{ij} \\ f_{jj} \\ f_{kj} \end{bmatrix} = \begin{bmatrix} f_{ij} \\ f_{jj} \\ f_{kj} \end{bmatrix}. \tag{12.11}$$

The force on surface **k** can be represented by the matrix

$$\mathbf{f_k} = [\mathbf{i}\ \ \mathbf{j}\ \ \mathbf{k}] \begin{bmatrix} f_{ik} \\ f_{jk} \\ f_{kk} \end{bmatrix} = \begin{bmatrix} f_{ik} \\ f_{jk} \\ f_{kk} \end{bmatrix}, \tag{12.12}$$

where the global base $[\mathbf{i}\ \ \mathbf{j}\ \ \mathbf{k}]$ has been used to represent all three forces.

The force on any other surface **a** from Equation (12.7), is by definition given by

$$\begin{aligned} \mathbf{f}(\mathbf{a}) &= a_i\mathbf{f}(\mathbf{i}) + a_j\mathbf{f}(\mathbf{j}) + a_k\mathbf{f}(\mathbf{k}) \\ &= a_i\mathbf{f_i} + a_j\mathbf{f_j} + a_k\mathbf{f_k} \\ &= [\mathbf{f_i}\ \ \mathbf{f_j}\ \ \mathbf{f_k}] \begin{bmatrix} a_i \\ a_j \\ a_k \end{bmatrix}. \end{aligned} \tag{12.13}$$

The Stress Tensor in 3D. Equation (12.13) conveniently describes the linear mapping of a space of surfaces into a space of forces, i.e., it describes the stress tensor. As it maps vectors into vectors it is considered a second order tensor as opposed to a first order tensor (which maps vectors into scalars). As both forces and surfaces are vectors in 3D, the above tensor is a second order 3D tensor. In the case that surfaces are represented by the global base $[\mathbf{i}\ \ \mathbf{j}\ \ \mathbf{k}]$, the second order 3D stress tensor **F** is best represented by the matrix of three vectors

$$\mathbf{F} = [\mathbf{f_i}\ \ \mathbf{f_j}\ \ \mathbf{f_k}]. \tag{12.14}$$

Forces $\mathbf{f_i}$, $\mathbf{f_j}$ and $\mathbf{f_k}$ can be represented using any vector base and they do not have to be the same base used for the surfaces. In the case that the global base $[\mathbf{i}\ \ \mathbf{j}\ \ \mathbf{k}]$ is used to represent the forces, the stress tensor is represented by the matrix

$$\mathbf{F} = \begin{bmatrix} f_{ii} & f_{ij} & f_{ik} \\ f_{ji} & f_{jj} & f_{jk} \\ f_{ki} & f_{kj} & f_{kk} \end{bmatrix}, \tag{12.15}$$

where

$$\mathbf{f_i} = \begin{bmatrix} f_{ii} \\ f_{ji} \\ f_{ki} \end{bmatrix}; \ \mathbf{f_j} = \begin{bmatrix} f_{ij} \\ f_{jj} \\ f_{kj} \end{bmatrix}; \ \mathbf{f_k} = \begin{bmatrix} f_{ik} \\ f_{jk} \\ f_{kk} \end{bmatrix} \tag{12.16}$$

are the matrix representations of forces on surfaces **i**, **j** and **k** respectively. In Equation (12.15) the first index represents the direction of the force and the second index represents the direction of the surface, such that

$$f_{ij} \tag{12.17}$$

is the force in the direction **i** on the surface **j** as illustrated in Figure 12.2. In a similar manner, the force f_{ki} is the force in the direction **k** on the surface **i**.

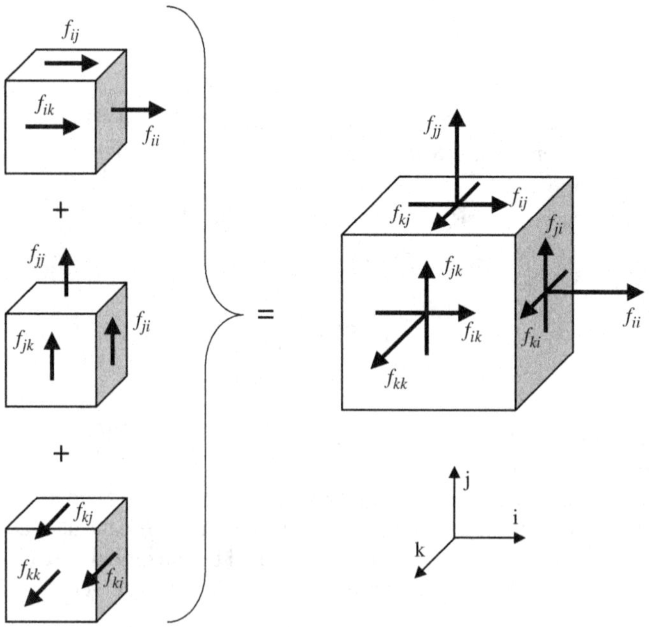

Figure 12.2 Graphical representation of a 3D stress tensor matrix

12.2 General Base for Surfaces

It is very often convenient to use one base for surfaces, called the surface base

$$[\boldsymbol{\alpha} \ \ \boldsymbol{\beta} \ \ \boldsymbol{\gamma}] \tag{12.18}$$

and another base for forces, called the force base

$$[\boldsymbol{\xi} \ \ \boldsymbol{\eta} \ \ \boldsymbol{\zeta}], \tag{12.19}$$

where

$$[\boldsymbol{\alpha} \ \ \boldsymbol{\beta} \ \ \boldsymbol{\gamma}] = \begin{bmatrix} \alpha_i & \beta_i & \gamma_i \\ \alpha_j & \beta_j & \gamma_j \\ \alpha_k & \beta_k & \gamma_k \end{bmatrix} \tag{12.20}$$

and

$$[\boldsymbol{\xi} \ \ \boldsymbol{\eta} \ \ \boldsymbol{\zeta}] = \begin{bmatrix} \xi_i & \eta_i & \zeta_i \\ \xi_j & \eta_j & \zeta_j \\ \xi_k & \eta_k & \zeta_k \end{bmatrix} \tag{12.21}$$

represent base matrices obtained by using the global base $[\mathbf{i} \ \ \mathbf{j} \ \ \mathbf{k}]$, such that the matrices of the base vectors are given by

$$\boldsymbol{\alpha} = \begin{bmatrix} \alpha_i \\ \alpha_j \\ \alpha_k \end{bmatrix}; \ \boldsymbol{\beta} = \begin{bmatrix} \beta_i \\ \beta_j \\ \beta_k \end{bmatrix}; \ \boldsymbol{\gamma} = \begin{bmatrix} \gamma_i \\ \gamma_j \\ \gamma_k \end{bmatrix} \tag{12.22}$$

$$\boldsymbol{\xi} = \begin{bmatrix} \xi_i \\ \xi_j \\ \xi_k \end{bmatrix}; \ \boldsymbol{\eta} = \begin{bmatrix} \eta_i \\ \eta_j \\ \eta_k \end{bmatrix}; \ \boldsymbol{\zeta} = \begin{bmatrix} \zeta_i \\ \zeta_j \\ \zeta_k \end{bmatrix}. \tag{12.23}$$

Any given surface vector $\mathbf{a}$ can be expressed using the surface base

$$\begin{aligned} \mathbf{a} &= a_\alpha \boldsymbol{\alpha} + a_\beta \boldsymbol{\beta} + a_\gamma \boldsymbol{\gamma} \\ &= [\boldsymbol{\alpha} \ \ \boldsymbol{\beta} \ \ \boldsymbol{\gamma}] \begin{bmatrix} a_\alpha \\ a_\beta \\ a_\gamma \end{bmatrix} = \begin{bmatrix} a_\alpha \\ a_\beta \\ a_\gamma \end{bmatrix}, \end{aligned} \tag{12.24}$$

where

$$\begin{aligned} \mathbf{a} = a_i\mathbf{i} + a_j\mathbf{j} + a_k\mathbf{k} = [\mathbf{i} \ \ \mathbf{j} \ \ \mathbf{k}] \begin{bmatrix} a_i \\ a_j \\ a_k \end{bmatrix} = \begin{bmatrix} a_i \\ a_j \\ a_k \end{bmatrix} = \begin{bmatrix} \alpha_i & \beta_i & \gamma_i \\ \alpha_j & \beta_j & \gamma_j \\ \alpha_k & \beta_k & \gamma_k \end{bmatrix} \begin{bmatrix} a_\alpha \\ a_\beta \\ a_\gamma \end{bmatrix} \\ = a_\alpha \begin{bmatrix} \alpha_i \\ \alpha_j \\ \alpha_k \end{bmatrix} + a_\beta \begin{bmatrix} \beta_i \\ \beta_j \\ \beta_k \end{bmatrix} + a_\gamma \begin{bmatrix} \gamma_i \\ \gamma_j \\ \gamma_k \end{bmatrix}. \end{aligned} \tag{12.25}$$

The stress **f** on any surface **a** is by definition given by

$$\begin{aligned}\mathbf{f} &= \mathbf{f_i} a_i + \mathbf{f_j} a_j + \mathbf{f_k} a_k = [\mathbf{f_i} \;\; \mathbf{f_j} \;\; \mathbf{f_k}] \begin{bmatrix} a_i \\ a_j \\ a_k \end{bmatrix} \\ &= \begin{bmatrix} f_i \\ f_j \\ f_k \end{bmatrix} = \begin{bmatrix} f_{ii} & f_{ij} & f_{ik} \\ f_{ji} & f_{jj} & f_{jk} \\ f_{ki} & f_{kj} & f_{kk} \end{bmatrix} \begin{bmatrix} a_i \\ a_j \\ a_k \end{bmatrix} \\ &= a_i \begin{bmatrix} f_{ii} \\ f_{ji} \\ f_{ki} \end{bmatrix} + a_j \begin{bmatrix} f_{ij} \\ f_{jj} \\ f_{kj} \end{bmatrix} + a_k \begin{bmatrix} f_{ik} \\ f_{jk} \\ f_{kk} \end{bmatrix}.\end{aligned} \tag{12.26}$$

By substituting Equation (12.25) into Equation (12.26)

$$\begin{aligned}\mathbf{f} = \begin{bmatrix} f_i \\ f_j \\ f_k \end{bmatrix} &= \begin{bmatrix} f_{ii} & f_{ij} & f_{ik} \\ f_{ji} & f_{jj} & f_{jk} \\ f_{ki} & f_{kj} & f_{kk} \end{bmatrix} \left(\begin{bmatrix} \alpha_i & \beta_i & \gamma_i \\ \alpha_j & \beta_j & \gamma_j \\ \alpha_k & \beta_k & \gamma_k \end{bmatrix} \begin{bmatrix} a_\alpha \\ a_\beta \\ a_\gamma \end{bmatrix} \right) \\ &= \left(\begin{bmatrix} f_{ii} & f_{ij} & f_{ik} \\ f_{ji} & f_{jj} & f_{jk} \\ f_{ki} & f_{kj} & f_{kk} \end{bmatrix} \begin{bmatrix} \alpha_i & \beta_i & \gamma_i \\ \alpha_j & \beta_j & \gamma_j \\ \alpha_k & \beta_k & \gamma_k \end{bmatrix} \right) \begin{bmatrix} a_\alpha \\ a_\beta \\ a_\gamma \end{bmatrix} \\ &= \begin{bmatrix} f_{i\alpha} & f_{i\beta} & f_{i\gamma} \\ f_{j\alpha} & f_{j\beta} & f_{j\gamma} \\ f_{k\alpha} & f_{k\beta} & f_{k\gamma} \end{bmatrix} \begin{bmatrix} a_\alpha \\ a_\beta \\ a_\gamma \end{bmatrix},\end{aligned} \tag{12.27}$$

where

$$\begin{bmatrix} f_{i\alpha} & f_{i\beta} & f_{i\gamma} \\ f_{j\alpha} & f_{j\beta} & f_{j\gamma} \\ f_{k\alpha} & f_{k\beta} & f_{k\gamma} \end{bmatrix} = \begin{bmatrix} f_{ii} & f_{ij} & f_{ik} \\ f_{ji} & f_{jj} & f_{jk} \\ f_{ki} & f_{kj} & f_{kk} \end{bmatrix} \begin{bmatrix} \alpha_i & \beta_i & \gamma_i \\ \alpha_j & \beta_j & \gamma_j \\ \alpha_k & \beta_k & \gamma_k \end{bmatrix}. \tag{12.28}$$

Equation (12.27) uses the surface base $[\boldsymbol{\alpha} \;\; \boldsymbol{\beta} \;\; \boldsymbol{\gamma}]$ for surfaces and the global base for forces, as shown in Figure 12.3. Equation (12.28) can also be written as

$$\begin{bmatrix} f_{i\alpha} \\ f_{j\alpha} \\ f_{k\alpha} \end{bmatrix} = \begin{bmatrix} f_{ii} & f_{ij} & f_{ik} \\ f_{ji} & f_{jj} & f_{jk} \\ f_{ki} & f_{kj} & f_{kk} \end{bmatrix} \begin{bmatrix} \alpha_i \\ \alpha_j \\ \alpha_k \end{bmatrix} = \alpha_i \begin{bmatrix} f_{ii} \\ f_{ji} \\ f_{ki} \end{bmatrix} + \alpha_j \begin{bmatrix} f_{ij} \\ f_{jj} \\ f_{kj} \end{bmatrix} + \alpha_k \begin{bmatrix} f_{ik} \\ f_{jk} \\ f_{kk} \end{bmatrix} \tag{12.29}$$

$$\begin{bmatrix} f_{i\beta} \\ f_{j\beta} \\ f_{k\beta} \end{bmatrix} = \begin{bmatrix} f_{ii} & f_{ij} & f_{ik} \\ f_{ji} & f_{jj} & f_{jk} \\ f_{ki} & f_{kj} & f_{kk} \end{bmatrix} \begin{bmatrix} \beta_i \\ \beta_j \\ \beta_k \end{bmatrix} = \beta_i \begin{bmatrix} f_{ii} \\ f_{ji} \\ f_{ki} \end{bmatrix} + \beta_j \begin{bmatrix} f_{ij} \\ f_{jj} \\ f_{kj} \end{bmatrix} + \beta_k \begin{bmatrix} f_{ik} \\ f_{jk} \\ f_{kk} \end{bmatrix} \tag{12.30}$$

$$\begin{bmatrix} f_{i\gamma} \\ f_{j\gamma} \\ f_{k\gamma} \end{bmatrix} = \begin{bmatrix} f_{ii} & f_{ij} & f_{ik} \\ f_{ji} & f_{jj} & f_{jk} \\ f_{ki} & f_{kj} & f_{kk} \end{bmatrix} \begin{bmatrix} \gamma_i \\ \gamma_j \\ \gamma_k \end{bmatrix} = \gamma_i \begin{bmatrix} f_{ii} \\ f_{ji} \\ f_{ki} \end{bmatrix} + \gamma_j \begin{bmatrix} f_{ij} \\ f_{jj} \\ f_{kj} \end{bmatrix} + \gamma_k \begin{bmatrix} f_{ik} \\ f_{jk} \\ f_{kk} \end{bmatrix}, \tag{12.31}$$

which is basically the straightforward definition of the stress tensor itself.

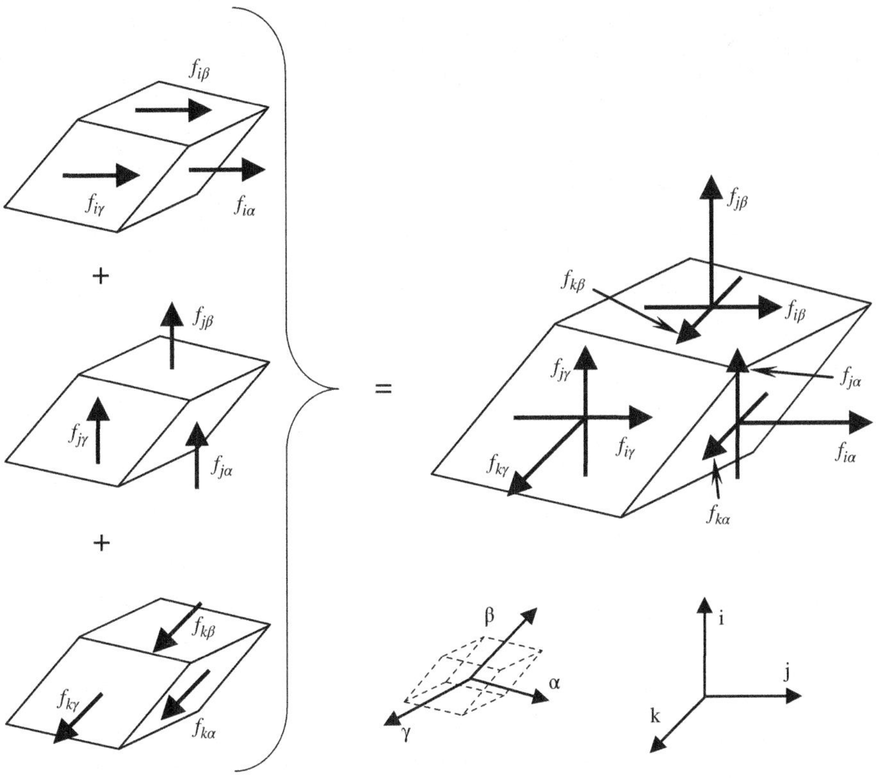

Figure 12.3 Graphical representation of the first Piola-Kirchhoff stress tensor matrix

Equation (12.27) can be written directly from the definition of the stress tensor where the stress tensor **F** is a linear mapping of surfaces into forces described by

$$\begin{aligned}\mathbf{F} &= a_\alpha \mathbf{f}_\alpha + a_\beta \mathbf{f}_\beta + a_\gamma \mathbf{f}_\gamma = a_\alpha \begin{bmatrix} f_{i\alpha} \\ f_{j\alpha} \\ f_{k\alpha} \end{bmatrix} + a_\beta \begin{bmatrix} f_{i\beta} \\ f_{j\beta} \\ f_{k\beta} \end{bmatrix} + a_\gamma \begin{bmatrix} f_{i\gamma} \\ f_{j\gamma} \\ f_{k\gamma} \end{bmatrix} \\ &= [\mathbf{f}_\alpha \;\; \mathbf{f}_\beta \;\; \mathbf{f}_\gamma] \begin{bmatrix} a_\alpha \\ a_\beta \\ a_\gamma \end{bmatrix} = \begin{bmatrix} f_{i\alpha} & f_{i\beta} & f_{i\gamma} \\ f_{j\alpha} & f_{j\beta} & f_{j\gamma} \\ f_{k\alpha} & f_{k\beta} & f_{k\gamma} \end{bmatrix} \begin{bmatrix} a_\alpha \\ a_\beta \\ a_\gamma \end{bmatrix}.\end{aligned} \tag{12.32}$$

The equal sign in Equation (12.32) does not mean that the stress tensor **F** is equal to a matrix. The matrix contains stress components for surfaces expressed using the surface base $[\boldsymbol{\alpha} \;\; \boldsymbol{\beta} \;\; \boldsymbol{\gamma}]$ and for forces using the global base $[\mathbf{i} \;\; \mathbf{j} \;\; \mathbf{k}]$. One can alternatively say that the stress tensor **F** is described by its matrix, i.e.,

$$\mathbf{F} = \begin{bmatrix} f_{i\alpha} & f_{i\beta} & f_{i\gamma} \\ f_{j\alpha} & f_{j\beta} & f_{j\gamma} \\ f_{k\alpha} & f_{k\beta} & f_{k\gamma} \end{bmatrix}. \tag{12.33}$$

The matrix components shown on the right hand side of Equation (12.32) have a physical meaning: $f_{k\beta}$ is the force in the global direction **k** on base surface $\boldsymbol{\beta}$; $f_{j\gamma}$ is the force in the global direction **j** on base surface $\boldsymbol{\gamma}$, etc., see Figure 12.3.

12.3 General Base for Forces

The force **f** can also be represented using the general force base

$$[\boldsymbol{\xi} \quad \boldsymbol{\eta} \quad \boldsymbol{\zeta}] \tag{12.34}$$

so that

$$\begin{aligned} \mathbf{f} &= f_\xi \boldsymbol{\xi} + f_\eta \boldsymbol{\eta} + f_\zeta \boldsymbol{\zeta} \\ &= [\boldsymbol{\xi} \quad \boldsymbol{\eta} \quad \boldsymbol{\zeta}] \begin{bmatrix} f_\xi \\ f_\eta \\ f_\zeta \end{bmatrix}. \end{aligned} \tag{12.35}$$

or simply

$$\mathbf{f} = \begin{bmatrix} f_\xi \\ f_\eta \\ f_\zeta \end{bmatrix}, \tag{12.36}$$

where

$$\begin{bmatrix} f_\xi \\ f_\eta \\ f_\zeta \end{bmatrix} \tag{12.37}$$

is the matrix that represents vector **f** in the force base

$$[\boldsymbol{\xi} \quad \boldsymbol{\eta} \quad \boldsymbol{\zeta}]. \tag{12.38}$$

As the base vectors of the force base are described using the global base, one can write

$$\begin{aligned} \mathbf{f} = \begin{bmatrix} f_i \\ f_j \\ f_k \end{bmatrix} &= f_\xi \boldsymbol{\xi} + f_\eta \boldsymbol{\eta} + f_\zeta \boldsymbol{\zeta} \\ &= [\boldsymbol{\xi} \quad \boldsymbol{\eta} \quad \boldsymbol{\zeta}] \begin{bmatrix} f_\xi \\ f_\eta \\ f_\zeta \end{bmatrix} = f_\xi \begin{bmatrix} \xi_i \\ \xi_j \\ \xi_k \end{bmatrix} + f_\eta \begin{bmatrix} \eta_i \\ \eta_j \\ \eta_k \end{bmatrix} + f_\zeta \begin{bmatrix} \zeta_i \\ \zeta_j \\ \zeta_k \end{bmatrix} \\ &= \begin{bmatrix} \xi_i & \eta_i & \zeta_i \\ \xi_j & \eta_j & \zeta_j \\ \xi_k & \eta_k & \zeta_k \end{bmatrix} \begin{bmatrix} f_\xi \\ f_\eta \\ f_\zeta \end{bmatrix}. \end{aligned} \tag{12.39}$$

By multiplying from the left by

$$\begin{bmatrix} \xi_i & \eta_i & \zeta_i \\ \xi_j & \eta_j & \zeta_j \\ \xi_k & \eta_k & \zeta_k \end{bmatrix}^{-1}, \tag{12.40}$$

one obtains

$$\mathbf{f} = \begin{bmatrix} f_\xi \\ f_\eta \\ f_\zeta \end{bmatrix} = \begin{bmatrix} i_\xi & j_\xi & k_\xi \\ i_\eta & j_\eta & k_\eta \\ i_\zeta & j_\zeta & k_\zeta \end{bmatrix} \begin{bmatrix} f_i \\ f_j \\ f_k \end{bmatrix}, \tag{12.41}$$

where

$$\begin{bmatrix} i_\xi & j_\xi & k_\xi \\ i_\eta & j_\eta & k_\eta \\ i_\zeta & j_\zeta & k_\zeta \end{bmatrix} = \begin{bmatrix} \xi_i & \eta_i & \zeta_i \\ \xi_j & \eta_j & \zeta_j \\ \xi_k & \eta_k & \zeta_k \end{bmatrix}^{-1}. \tag{12.42}$$

This means that

$$\begin{aligned} \mathbf{f} &= [\boldsymbol{\xi} \;\; \boldsymbol{\eta} \;\; \boldsymbol{\zeta}] \begin{bmatrix} f_\xi \\ f_\eta \\ f_\zeta \end{bmatrix} \\ &= f_i \begin{bmatrix} i_\xi \\ i_\eta \\ i_\zeta \end{bmatrix} + f_j \begin{bmatrix} j_\xi \\ j_\eta \\ j_\zeta \end{bmatrix} + f_k \begin{bmatrix} k_\xi \\ k_\eta \\ k_\zeta \end{bmatrix} = \begin{bmatrix} i_\xi & j_\xi & k_\xi \\ i_\eta & j_\eta & k_\eta \\ i_\zeta & j_\zeta & k_\zeta \end{bmatrix} \begin{bmatrix} f_i \\ f_j \\ f_k \end{bmatrix}. \end{aligned} \tag{12.43}$$

By substituting

$$\begin{bmatrix} f_i \\ f_j \\ f_k \end{bmatrix} = \begin{bmatrix} f_{ii} & f_{ij} & f_{ik} \\ f_{ji} & f_{jj} & f_{jk} \\ f_{ki} & f_{kj} & f_{kk} \end{bmatrix} \begin{bmatrix} a_i \\ a_j \\ a_k \end{bmatrix} \tag{12.44}$$

and

$$\begin{bmatrix} i_\xi & j_\xi & k_\xi \\ i_\eta & j_\eta & k_\eta \\ i_\zeta & j_\zeta & k_\zeta \end{bmatrix} = \begin{bmatrix} \xi_i & \eta_i & \zeta_i \\ \xi_j & \eta_j & \zeta_j \\ \xi_k & \eta_k & \zeta_k \end{bmatrix}^{-1}, \tag{12.45}$$

one obtains

$$\begin{aligned} \mathbf{f} = \begin{bmatrix} f_\xi \\ f_\eta \\ f_\zeta \end{bmatrix} &= \begin{bmatrix} \xi_i & \eta_i & \zeta_i \\ \xi_j & \eta_j & \zeta_j \\ \xi_k & \eta_k & \zeta_k \end{bmatrix}^{-1} \begin{bmatrix} f_{ii} & f_{ij} & f_{ik} \\ f_{ji} & f_{jj} & f_{jk} \\ f_{ki} & f_{kj} & f_{kk} \end{bmatrix} \begin{bmatrix} a_i \\ a_j \\ a_k \end{bmatrix} \\ &= \begin{bmatrix} f_{\xi i} & f_{\xi j} & f_{\xi k} \\ f_{\eta i} & f_{\eta j} & f_{\eta k} \\ f_{\zeta i} & f_{\zeta j} & f_{\zeta k} \end{bmatrix} \begin{bmatrix} a_i \\ a_j \\ a_k \end{bmatrix}, \end{aligned} \tag{12.46}$$

where

$$\begin{bmatrix} f_{\xi i} & f_{\xi j} & f_{\xi k} \\ f_{\eta i} & f_{\eta j} & f_{\eta k} \\ f_{\zeta i} & f_{\zeta j} & f_{\zeta k} \end{bmatrix} = \begin{bmatrix} \xi_i & \eta_i & \zeta_i \\ \xi_j & \eta_j & \zeta_j \\ \xi_k & \eta_k & \zeta_k \end{bmatrix}^{-1} \begin{bmatrix} f_{ii} & f_{ij} & f_{ik} \\ f_{ji} & f_{jj} & f_{jk} \\ f_{ki} & f_{kj} & f_{kk} \end{bmatrix}. \quad (12.47)$$

12.4 General Base for Forces and Surfaces

By substituting Equation (12.27) into Equation (12.46) one obtains

$$\begin{aligned} \mathbf{f} = \begin{bmatrix} f_\xi \\ f_\eta \\ f_\zeta \end{bmatrix} &= \begin{bmatrix} \xi_i & \eta_i & \zeta_i \\ \xi_j & \eta_j & \zeta_j \\ \xi_k & \eta_k & \zeta_k \end{bmatrix}^{-1} \left(\begin{bmatrix} f_{ii} & f_{ij} & f_{ik} \\ f_{ji} & f_{jj} & f_{jk} \\ f_{ki} & f_{kj} & f_{kk} \end{bmatrix} \begin{bmatrix} \alpha_i & \beta_i & \gamma_i \\ \alpha_j & \beta_j & \gamma_j \\ \alpha_k & \beta_k & \gamma_k \end{bmatrix} \begin{bmatrix} a_\alpha \\ a_\beta \\ a_\gamma \end{bmatrix} \right) \\ &= \left(\begin{bmatrix} \xi_i & \eta_i & \zeta_i \\ \xi_j & \eta_j & \zeta_j \\ \xi_k & \eta_k & \zeta_k \end{bmatrix}^{-1} \begin{bmatrix} f_{ii} & f_{ij} & f_{ik} \\ f_{ji} & f_{jj} & f_{jk} \\ f_{ki} & f_{kj} & f_{kk} \end{bmatrix} \begin{bmatrix} \alpha_i & \beta_i & \gamma_i \\ \alpha_j & \beta_j & \gamma_j \\ \alpha_k & \beta_k & \gamma_k \end{bmatrix} \right) \begin{bmatrix} a_\alpha \\ a_\beta \\ a_\gamma \end{bmatrix}. \end{aligned} \quad (12.48)$$

This is a general description of the stress tensor **F** using the general base $[\boldsymbol{\xi} \ \boldsymbol{\eta} \ \boldsymbol{\zeta}]$ for forces and the general base $[\boldsymbol{\alpha} \ \boldsymbol{\beta} \ \boldsymbol{\gamma}]$ for surfaces; the corresponding matrix of the stress tensor is

$$\mathbf{F} = \begin{bmatrix} f_{\xi\alpha} & f_{\xi\beta} & f_{\xi\gamma} \\ f_{\eta\alpha} & f_{\eta\beta} & f_{\eta\gamma} \\ f_{\zeta\alpha} & f_{\zeta\beta} & f_{\zeta\gamma} \end{bmatrix}, \quad (12.49)$$

where

$$\begin{bmatrix} f_{\xi\alpha} & f_{\xi\beta} & f_{\xi\gamma} \\ f_{\eta\alpha} & f_{\eta\beta} & f_{\eta\gamma} \\ f_{\zeta\alpha} & f_{\zeta\beta} & f_{\zeta\gamma} \end{bmatrix} = \begin{bmatrix} i_\xi & j_\xi & k_\xi \\ i_\eta & j_\eta & k_\eta \\ i_\zeta & j_\zeta & k_\zeta \end{bmatrix} \begin{bmatrix} f_{ii} & f_{ij} & f_{ik} \\ f_{ji} & f_{jj} & f_{jk} \\ f_{ki} & f_{kj} & f_{kk} \end{bmatrix} \begin{bmatrix} \alpha_i & \beta_i & \gamma_i \\ \alpha_j & \beta_j & \gamma_j \\ \alpha_k & \beta_k & \gamma_k \end{bmatrix} \quad (12.50)$$

or

$$\begin{bmatrix} f_{\xi\alpha} & f_{\xi\beta} & f_{\xi\gamma} \\ f_{\eta\alpha} & f_{\eta\beta} & f_{\eta\gamma} \\ f_{\zeta\alpha} & f_{\zeta\beta} & f_{\zeta\gamma} \end{bmatrix} = \begin{bmatrix} \xi_i & \eta_i & \zeta_i \\ \xi_j & \eta_j & \zeta_j \\ \xi_k & \eta_k & \zeta_k \end{bmatrix}^{-1} \begin{bmatrix} f_{ii} & f_{ij} & f_{ik} \\ f_{ji} & f_{jj} & f_{jk} \\ f_{ki} & f_{kj} & f_{kk} \end{bmatrix} \begin{bmatrix} \alpha_i & \beta_i & \gamma_i \\ \alpha_j & \beta_j & \gamma_j \\ \alpha_k & \beta_k & \gamma_k \end{bmatrix}. \quad (12.51)$$

Of course, the matrix is not the stress tensor itself; it is merely the matrix representing the stress tensor (linear mapping of surfaces into internal forces) obtained using the bases

$$[\boldsymbol{\alpha} \ \boldsymbol{\beta} \ \boldsymbol{\gamma}] \quad (12.52)$$

and

$$[\boldsymbol{\xi} \ \boldsymbol{\eta} \ \boldsymbol{\zeta}], \quad (12.53)$$

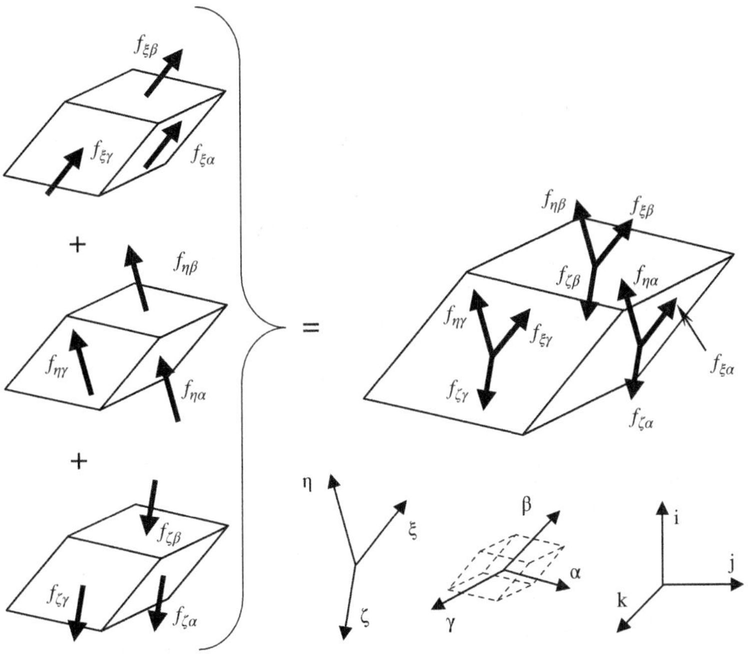

Figure 12.4 Physical meaning of the components of stress tensor matrix obtained using base $[\boldsymbol{\alpha}\ \boldsymbol{\beta}\ \boldsymbol{\gamma}]$ for surfaces and base $[\boldsymbol{\xi}\ \boldsymbol{\eta}\ \boldsymbol{\zeta}]$ for forces

for surfaces and forces respectively. Each of the components of the stress tensor matrix has a physical meaning, as shown in Figure 12.4, where for example

$$f_{\xi\beta} \tag{12.54}$$

is the force in $\boldsymbol{\xi}$ direction on the surface $\boldsymbol{\beta}$.

One should note that Equation (12.51) can be written directly from the definition of the stress tensor by simply considering that

$$\begin{bmatrix} f_{i\alpha} \\ f_{j\alpha} \\ f_{k\alpha} \end{bmatrix} = \alpha_i \begin{bmatrix} f_{ii} \\ f_{ji} \\ f_{ki} \end{bmatrix} + \alpha_j \begin{bmatrix} f_{ij} \\ f_{jj} \\ f_{kj} \end{bmatrix} + \alpha_k \begin{bmatrix} f_{ik} \\ f_{jk} \\ f_{kk} \end{bmatrix} \tag{12.55}$$

and also that

$$f_{\xi\alpha} = i_\xi f_{i\alpha} + j_\xi f_{j\alpha} + k_\xi f_{k\alpha} = \begin{bmatrix} i_\xi & j_\xi & k_\xi \end{bmatrix} \begin{bmatrix} f_{i\alpha} \\ f_{j\alpha} \\ f_{k\alpha} \end{bmatrix}. \tag{12.56}$$

Very often, in the relevant literature, the above equations are referred to as the "tensor transformation." For this book, this term is deliberately avoided in light of the definition of the second order tensor itself: the stress tensor is simply a mapping of surfaces into forces. As such, it cannot be transformed or changed by changing the corresponding bases used to represent the vectors. It is a well-defined physical reality that describes the state of the internal forces and of course, any change of the vector base has no impact on these forces. Equation (12.56) uses the tensor definition directly by answering the question: what is the force in $\boldsymbol{\xi}$ direction on the surface $\boldsymbol{\alpha}$?

What does change in the above equations is the matrix representing the tensor itself. Subsequently, this has a perfect analogy with vectors; a force vector does not change when one changes the vector base: the matrix representing it changes. It is important to keep a clear distinction between matrices and tensors. Tensors are physical quantities representing for example stress at a given point P, much like scalars represent for example density at the same point P. Real numbers are a convenient way for expressing scalars. Matrices are a convenient way of expressing tensors. Both scalars and tensors can be expressed in alternative ways as is well illustrated by the multiple approaches used in some stress analysis books from something as simple as Mohr's circle (see Timoshenko).

Here, the stress tensor is used to explain the 3D second order tensor in general. There are many other physical quantities that are described by second order tensors.

12.5 The Cauchy Stress Tensor Matrix in 3D

By choosing the global orthonormal base $[\mathbf{i} \;\; \mathbf{j} \;\; \mathbf{k}]$ for both surface vectors and force vectors, the stress tensor can be represented by the following matrix

$$\mathbf{F} = \begin{bmatrix} f_{ii} & f_{ij} & f_{ik} \\ f_{ji} & f_{jj} & f_{jk} \\ f_{ki} & f_{kj} & f_{kk} \end{bmatrix}, \tag{12.57}$$

where the matrix component f_{ij} represent a force on surface $\mathbf{j}$ in the $\mathbf{i}$ direction. For historical reasons this matrix is called the Cauchy stress tensor matrix, and is therefore noted as

$$\mathbf{C} = \begin{bmatrix} c_{ii} & c_{ij} & c_{ik} \\ c_{ji} & c_{jj} & c_{jk} \\ c_{ki} & c_{kj} & c_{kk} \end{bmatrix}. \tag{12.58}$$

12.6 The First Piola-Kirchhoff Stress Tensor Matrix in 3D

For the general surface base $[\boldsymbol{\alpha} \;\; \boldsymbol{\beta} \;\; \boldsymbol{\gamma}]$ shown in Figure 12.5 one can represent the stress tensor as follows

$$\mathbf{F} = \begin{bmatrix} f_{i\alpha} & f_{i\beta} & f_{i\gamma} \\ f_{j\alpha} & f_{j\beta} & f_{j\gamma} \\ f_{k\alpha} & f_{k\beta} & f_{k\gamma} \end{bmatrix}, \tag{12.59}$$

where the matrix

$$\mathbf{P} = \begin{bmatrix} f_{i\alpha} & f_{i\beta} & f_{i\gamma} \\ f_{j\alpha} & f_{j\beta} & f_{j\gamma} \\ f_{k\alpha} & f_{k\beta} & f_{k\gamma} \end{bmatrix} = \begin{bmatrix} p_{i\alpha} & p_{i\beta} & p_{i\gamma} \\ p_{j\alpha} & p_{j\beta} & p_{j\gamma} \\ p_{k\alpha} & p_{k\beta} & p_{k\gamma} \end{bmatrix} \tag{12.60}$$

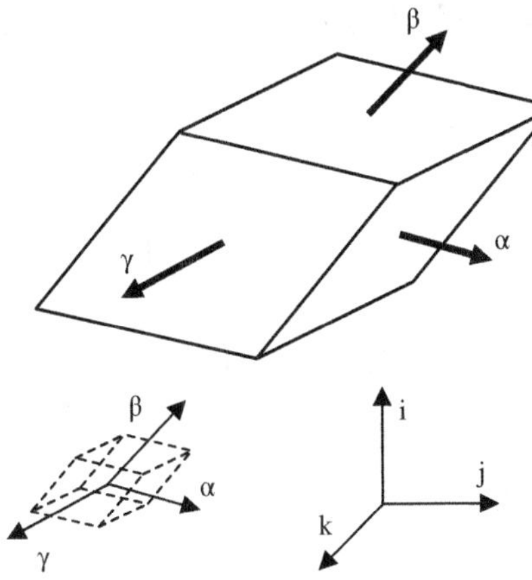

Figure 12.5 General surface base in 3D

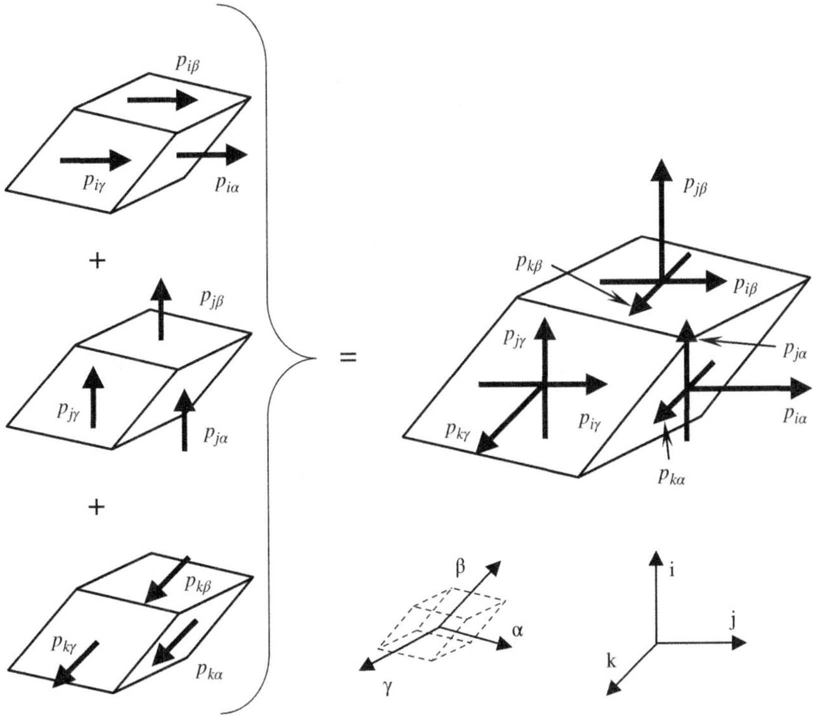

Figure 12.6 Graphical representation of the first Piola-Kirchhoff matrix of the stress tensor in 3D

is for historical reasons called the first Piola-Kirchhoff stress tensor matrix. The component $f_{i\alpha}$ of this matrix represents a force in the **i** direction (of the global orthonormal base $[\mathbf{i}\ \ \mathbf{j}\ \ \mathbf{k}]$) on the surface $\boldsymbol{\alpha}$ of the general (nonorthonormal) base $[\boldsymbol{\alpha}\ \ \boldsymbol{\beta}\ \ \boldsymbol{\gamma}]$, Figure 12.6.

In other words, the first Piola-Kirchhoff matrix answers the question: what is the resultant internal force in the **i** direction on surface $\boldsymbol{\alpha}$? The answer is $p_{i\alpha}$.

12.7 The Second Piola-Kirchhoff Stress Tensor Matrix in 3D

The components of the second Piola-Kirchhoff stress tensor matrix are given by

$$\mathbf{K} = \begin{bmatrix} k_{\xi\alpha} & k_{\xi\beta} & k_{\xi\gamma} \\ k_{\eta\alpha} & k_{\eta\beta} & k_{\eta\gamma} \\ k_{\zeta\alpha} & k_{\zeta\beta} & k_{\zeta\gamma} \end{bmatrix}. \tag{12.61}$$

Each of the components $k_{\xi\alpha}$ represent the ξ-component of the internal force on the $\boldsymbol{\alpha}$-surface.

The second Piola-Kirchhoff stress tensor matrix is calculated from the first Piola-Kirchhoff stress matrix, which in turn is calculated from the Cauchy stress matrix

$$\mathbf{K} = \begin{bmatrix} k_{\xi\alpha} & k_{\xi\beta} & k_{\xi\gamma} \\ k_{\eta\alpha} & k_{\eta\beta} & k_{\eta\gamma} \\ k_{\zeta\alpha} & k_{\zeta\beta} & k_{\zeta\gamma} \end{bmatrix} = \begin{bmatrix} i_\xi & j_\xi & k_\xi \\ i_\eta & j_\eta & k_\eta \\ i_\zeta & j_\zeta & k_\zeta \end{bmatrix} \begin{bmatrix} p_{i\alpha} & p_{i\beta} & p_{i\gamma} \\ p_{j\alpha} & p_{j\beta} & p_{j\gamma} \\ p_{k\alpha} & p_{k\beta} & p_{k\gamma} \end{bmatrix} \tag{12.62}$$

where

$$\begin{bmatrix} i_\xi & j_\xi & k_\xi \\ i_\eta & j_\eta & k_\eta \\ i_\zeta & j_\zeta & k_\zeta \end{bmatrix} = \begin{bmatrix} \xi_i & \eta_i & \zeta_i \\ \xi_j & \eta_j & \zeta_j \\ \xi_k & \eta_k & \zeta_k \end{bmatrix}^{-1}, \tag{12.63}$$

In addition,

$$\begin{bmatrix} p_{i\alpha} & p_{i\beta} & p_{i\gamma} \\ p_{j\alpha} & p_{j\beta} & p_{j\gamma} \\ p_{k\alpha} & p_{k\beta} & p_{k\gamma} \end{bmatrix} = \begin{bmatrix} c_{ii} & c_{ij} & c_{ik} \\ c_{ji} & c_{jj} & c_{jk} \\ c_{ki} & c_{kj} & c_{kk} \end{bmatrix} \begin{bmatrix} \alpha_i & \beta_i & \gamma_i \\ \alpha_j & \beta_l & \gamma_j \\ \alpha_k & \beta_k & \gamma_k \end{bmatrix} \tag{12.64}$$

where

$$\begin{bmatrix} \alpha_i & \beta_i & \gamma_i \\ \alpha_j & \beta_l & \gamma_j \\ \alpha_k & \beta_k & \gamma_k \end{bmatrix} = g \begin{bmatrix} \xi_i & \eta_i & \zeta_i \\ \xi_j & \eta_j & \zeta_j \\ \xi_k & \eta_k & \zeta_k \end{bmatrix}^{-\mathrm{T}}. \tag{12.65}$$

These yield

$$\begin{aligned} \mathbf{K} &= \begin{bmatrix} k_{\xi\alpha} & k_{\xi\beta} & k_{\xi\gamma} \\ k_{\eta\alpha} & k_{\eta\beta} & k_{\eta\gamma} \\ k_{\zeta\alpha} & k_{\zeta\beta} & k_{\zeta\gamma} \end{bmatrix} \\ &= g \begin{bmatrix} \xi_i & \eta_i & \zeta_i \\ \xi_j & \eta_j & \zeta_j \\ \xi_k & \eta_k & \zeta_k \end{bmatrix}^{-1} \begin{bmatrix} c_{ii} & c_{ij} & c_{ik} \\ c_{ji} & c_{jj} & c_{jk} \\ c_{ki} & c_{kj} & c_{kk} \end{bmatrix} \begin{bmatrix} \xi_i & \eta_i & \zeta_i \\ \xi_j & \eta_j & \zeta_j \\ \xi_k & \eta_k & \zeta_k \end{bmatrix}^{-\mathrm{T}}. \end{aligned} \tag{12.66}$$

The components of the second Piola-Kirchhoff stress tensor matrix answer the question: what is the resultant internal force in $\boldsymbol{\xi}$ direction on surface $\boldsymbol{\alpha}$? The answer is $k_{\xi\alpha}$, see Figure 12.7.

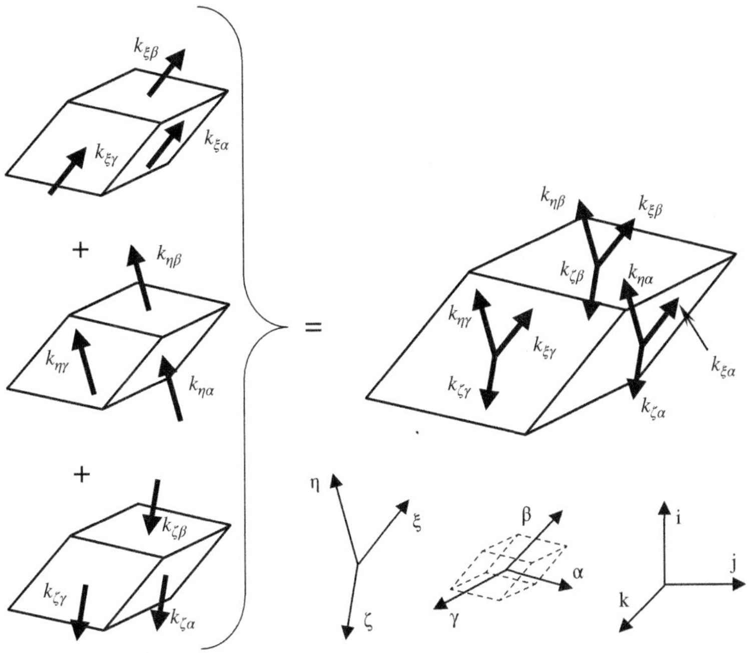

Figure 12.7 Graphical representation of the second Piola-Kirchhoff matrix of the stress tensor in 3D

12.8 Summary

The second order tensor in 3D has been explained using the stress tensor. A complete analogy with the 2D stress tensor chapter is deliberatively observed, often involving some repetition; these are for didactical reasons.

The reader is encouraged never to remember any formula, but to always start from the first principles and derive all the necessary formulas, even when solving a particular example.

Remember, if 2D is properly understood, the reader should ideally be able to derive 3D without even looking at this chapter at all.

Further Reading

[1] Abraham, R., Marsden, J. E. and Ratiu, T. S. (1988) *Manifolds, Tensor Analysis, and Applications.* New York, Springer-Verlag.

[2] Aris, R. (1962) *Vectors, Tensors and the Basic Equations of Fluid Mechanics.* New Jersey, Prentice Hall.

[3] Bronstein, I. N. and Semendyayev, S. (2007) *Handbook of Mathematics.* Springer, New York.

[4] Dvorkin, E. N. and Goldschmit, M. B. (2005) *Nonlinear Continua.* New York, Springer.

[5] Fleisch, D. (2012) *A Student's Guide to Vectors and Tensors.* New York, Cambridge University Press.

[6] Kusse, B. R. and Westwig, E. A. (2006) *Mathematical Physics: Applied Mathematics for Scientists and Engineers*, 2nd edn. Wiley-VCH Verlag GmbH & Co.

[7] Levi-Civita, T. (2013) *The Absolute Differential Calculus (Calculus of Tensors).* New York, Dover Publications.

[8] Matthews, P. C. (1998) *Vector Calculus.* London, Springer.

13

Second Order Tensors in nD

13.1 Second Order Tensor in n-Dimensions

Surfaces in 4D. Extending the "surfaces" into the fourth dimension is straight forward

$$\mathbf{a} = a_\alpha \boldsymbol{\alpha} + a_\beta \boldsymbol{\beta} + a_\gamma \boldsymbol{\gamma} + a_\delta \boldsymbol{\delta}, \tag{13.1}$$

where $\boldsymbol{\alpha}$, $\boldsymbol{\beta}$, $\boldsymbol{\gamma}$ and $\boldsymbol{\delta}$ are base vectors of the "surface" base

$$\begin{bmatrix} \boldsymbol{\alpha} & \boldsymbol{\beta} & \boldsymbol{\gamma} & \boldsymbol{\delta} \end{bmatrix}. \tag{13.2}$$

These can be expressed using the global orthonormal base

$$\begin{bmatrix} \mathbf{i} & \mathbf{j} & \mathbf{k} & \mathbf{t} \end{bmatrix} \tag{13.3}$$

so that

$$\begin{aligned}
\boldsymbol{\alpha} &= \alpha_i \mathbf{i} + \alpha_j \mathbf{j} + \alpha_k \mathbf{k} + \alpha_t \mathbf{t} \\
\boldsymbol{\beta} &= \beta_i \mathbf{i} + \beta_j \mathbf{j} + \beta_k \mathbf{k} + \beta_t \mathbf{t} \\
\boldsymbol{\gamma} &= \gamma_i \mathbf{i} + \gamma_j \mathbf{j} + \gamma_k \mathbf{k} + \gamma_t \mathbf{t} \\
\boldsymbol{\delta} &= \delta_i \mathbf{i} + \delta_j \mathbf{j} + \delta_k \mathbf{k} + \delta_t \mathbf{t}.
\end{aligned} \tag{13.4}$$

Large Strain Finite Element Method: A Practical Course, First Edition. Antonio Munjiza, Esteban Rougier and Earl E. Knight.

or simply

$$[\boldsymbol{\alpha}\ \boldsymbol{\beta}\ \boldsymbol{\gamma}\ \boldsymbol{\delta}] = [\mathbf{i}\ \mathbf{j}\ \mathbf{k}\ \mathbf{t}] \begin{bmatrix} \alpha_i & \beta_i & \gamma_i & \delta_i \\ \alpha_j & \beta_j & \gamma_j & \delta_j \\ \alpha_k & \beta_k & \gamma_k & \delta_k \\ \alpha_t & \beta_t & \gamma_t & \delta_t \end{bmatrix}, \tag{13.5}$$

which can be shortened to

$$[\boldsymbol{\alpha}\ \boldsymbol{\beta}\ \boldsymbol{\gamma}\ \boldsymbol{\delta}] = \begin{bmatrix} \alpha_i & \beta_i & \gamma_i & \delta_i \\ \alpha_j & \beta_j & \gamma_j & \delta_j \\ \alpha_k & \beta_k & \gamma_k & \delta_k \\ \alpha_t & \beta_t & \gamma_t & \delta_t \end{bmatrix}, \tag{13.6}$$

where the equal sign means that the base vectors are represented by the corresponding columns of the above matrix.

Visualizing these surfaces is not easy for the human brain is "tailor made" for visualizing 3D spaces. Nevertheless, one can proceed with the straight forward analogy to the 3D, where it is not always easy to plot vectors in 4D.

Forces in 4D. In 4D "forces" can be represented as

$$\mathbf{f} = f_\xi \boldsymbol{\xi} + f_\eta \boldsymbol{\eta} + f_\zeta \boldsymbol{\zeta} + f_\tau \boldsymbol{\tau}, \tag{13.7}$$

where

$$\begin{aligned} \boldsymbol{\xi} &= \xi_i \mathbf{i} + \xi_j \mathbf{j} + \xi_k \mathbf{k} + \xi_t \mathbf{t} \\ \boldsymbol{\eta} &= \eta_i \mathbf{i} + \eta_j \mathbf{j} + \eta_k \mathbf{k} + \eta_t \mathbf{t} \\ \boldsymbol{\zeta} &= \zeta_i \mathbf{i} + \zeta_j \mathbf{j} + \zeta_k \mathbf{k} + \zeta_t \mathbf{t} \\ \boldsymbol{\tau} &= \tau_i \mathbf{i} + \tau_j \mathbf{j} + \tau_k \mathbf{k} + \tau_t \mathbf{t} \end{aligned} \tag{13.8}$$

are base vectors of the "force" base in 4D

$$[\boldsymbol{\xi}\ \boldsymbol{\eta}\ \boldsymbol{\zeta}\ \boldsymbol{\tau}]. \tag{13.9}$$

Stress in 4D. The "stress" tensor in 4D can be defined as a linear mapping of "surface" vectors in 4D to "force" vectors in 4D such that

$$\begin{aligned} \mathbf{f}(s\mathbf{a}) &= s\mathbf{f}(\mathbf{a}) \\ \mathbf{f}(\mathbf{a}+\mathbf{b}) &= \mathbf{f}(\mathbf{a}) + \mathbf{f}(\mathbf{b}). \end{aligned} \tag{13.10}$$

By definition

$$\begin{aligned} \mathbf{f}(\mathbf{a}) &= f_i \mathbf{i} + f_j \mathbf{j} + f_k \mathbf{k} + f_t \mathbf{t} \\ &= \mathbf{f_i} a_i + \mathbf{f_j} a_j + \mathbf{f_k} a_k + \mathbf{f_t} a_t. \end{aligned} \tag{13.11}$$

By substituting expressions for the surface base vectors in terms of the global base vectors, i.e.

$$\mathbf{a} = a_\alpha \boldsymbol{\alpha} + a_\beta \boldsymbol{\beta} + a_\gamma \boldsymbol{\gamma} + a_\delta \boldsymbol{\delta} \tag{13.12}$$

and

$$\mathbf{a} = a_i\mathbf{i} + a_j\mathbf{j} + a_k\mathbf{k} + a_t\mathbf{t}, \tag{13.13}$$

one then obtains

$$\begin{aligned}\mathbf{a} = {} & a_\alpha(\alpha_i\mathbf{i} + \alpha_j\mathbf{j} + \alpha_k\mathbf{k} + \alpha_t\mathbf{t}) \\ & + a_\beta(\beta_i\mathbf{i} + \beta_j\mathbf{j} + \beta_k\mathbf{k} + \beta_t\mathbf{t}) \\ & + a_\gamma(\gamma_i\mathbf{i} + \gamma_j\mathbf{j} + \gamma_k\mathbf{k} + \gamma_t\mathbf{t}) \\ & + a_\delta(\delta_i\mathbf{i} + \delta_j\mathbf{j} + \delta_k\mathbf{k} + \delta_t\mathbf{t})\end{aligned} \tag{13.14}$$

or, in a matrix form

$$\begin{aligned}\begin{bmatrix} a_i \\ a_j \\ a_k \\ a_t \end{bmatrix} &= a_\alpha \begin{bmatrix} \alpha_i \\ \alpha_j \\ \alpha_k \\ \alpha_t \end{bmatrix} + a_\beta \begin{bmatrix} \beta_i \\ \beta_j \\ \beta_k \\ \beta_t \end{bmatrix} + a_\gamma \begin{bmatrix} \gamma_i \\ \gamma_j \\ \gamma_k \\ \gamma_t \end{bmatrix} + a_\delta \begin{bmatrix} \delta_i \\ \delta_j \\ \delta_k \\ \delta_t \end{bmatrix} \\ &= \begin{bmatrix} \alpha_i & \beta_i & \gamma_i & \delta_i \\ \alpha_j & \beta_j & \gamma_j & \delta_j \\ \alpha_k & \beta_k & \gamma_k & \delta_k \\ \alpha_t & \beta_t & \gamma_t & \delta_t \end{bmatrix} \begin{bmatrix} a_\alpha \\ a_\beta \\ a_\gamma \\ a_\delta \end{bmatrix}.\end{aligned} \tag{13.15}$$

Combining Equation (13.15) and Equation (13.11) yields

$$\begin{aligned}\mathbf{f} = f_i\mathbf{i} + f_j\mathbf{j} + f_k\mathbf{k} + f_t\mathbf{t} &= \begin{bmatrix} f_i \\ f_j \\ f_k \\ f_t \end{bmatrix} = [\mathbf{f_i} \;\; \mathbf{f_j} \;\; \mathbf{f_k} \;\; \mathbf{f_t}] \begin{bmatrix} a_i \\ a_j \\ a_k \\ a_t \end{bmatrix} \\ &= a_i \begin{bmatrix} f_{ii} \\ f_{ji} \\ f_{ki} \\ f_{ti} \end{bmatrix} + a_j \begin{bmatrix} f_{ij} \\ f_{jj} \\ f_{kj} \\ f_{tj} \end{bmatrix} + a_k \begin{bmatrix} f_{ik} \\ f_{jk} \\ f_{kk} \\ f_{tk} \end{bmatrix} + a_t \begin{bmatrix} f_{it} \\ f_{jt} \\ f_{kt} \\ f_{tt} \end{bmatrix} \\ &= \begin{bmatrix} f_{ii} & f_{ij} & f_{ik} & f_{it} \\ f_{ji} & f_{jj} & f_{jk} & f_{jt} \\ f_{ki} & f_{kj} & f_{kk} & f_{kt} \\ f_{ti} & f_{tj} & f_{tk} & f_{tt} \end{bmatrix} \begin{bmatrix} a_i \\ a_j \\ a_k \\ a_t \end{bmatrix} \\ &= \begin{bmatrix} f_{ii} & f_{ij} & f_{ik} & f_{it} \\ f_{ji} & f_{jj} & f_{jk} & f_{jt} \\ f_{ki} & f_{kj} & f_{kk} & f_{kt} \\ f_{ti} & f_{tj} & f_{tk} & f_{tt} \end{bmatrix} \left(\begin{bmatrix} \alpha_i & \beta_i & \gamma_i & \delta_i \\ \alpha_j & \beta_j & \gamma_j & \delta_j \\ \alpha_k & \beta_k & \gamma_k & \delta_k \\ \alpha_t & \beta_t & \gamma_t & \delta_t \end{bmatrix} \begin{bmatrix} a_\alpha \\ a_\beta \\ a_\gamma \\ a_\delta \end{bmatrix} \right) \\ &= \left(\begin{bmatrix} f_{ii} & f_{ij} & f_{ik} & f_{it} \\ f_{ji} & f_{jj} & f_{jk} & f_{jt} \\ f_{ki} & f_{kj} & f_{kk} & f_{kt} \\ f_{ti} & f_{tj} & f_{tk} & f_{tt} \end{bmatrix} \begin{bmatrix} \alpha_i & \beta_i & \gamma_i & \delta_i \\ \alpha_j & \beta_j & \gamma_j & \delta_j \\ \alpha_k & \beta_k & \gamma_k & \delta_k \\ \alpha_t & \beta_t & \gamma_t & \delta_t \end{bmatrix} \right) \begin{bmatrix} a_\alpha \\ a_\beta \\ a_\gamma \\ a_\delta \end{bmatrix}.\end{aligned} \tag{13.16}$$

In the above equation the 4D force **f** is still expressed using the base

$$[\mathbf{i}\ \ \mathbf{j}\ \ \mathbf{k}\ \ \mathbf{t}]. \tag{13.17}$$

Since

$$\begin{aligned}\mathbf{f} &= f_i\mathbf{i}+f_j\mathbf{j}+f_k\mathbf{k}+f_t\mathbf{t}\\ &= f_\xi\boldsymbol{\xi}+f_\eta\boldsymbol{\eta}+f_\zeta\boldsymbol{\zeta}+f_\tau\boldsymbol{\tau}.\end{aligned} \tag{13.18}$$

and

$$\begin{aligned}\begin{bmatrix} f_i \\ f_j \\ f_k \\ f_t \end{bmatrix} &= [\boldsymbol{\xi}\ \ \boldsymbol{\eta}\ \ \boldsymbol{\zeta}\ \ \boldsymbol{\tau}]\begin{bmatrix} f_\xi \\ f_\eta \\ f_\zeta \\ f_\tau \end{bmatrix}\\ &= \begin{bmatrix} \xi_i & \eta_i & \zeta_i & \delta_i \\ \xi_j & \eta_j & \zeta_j & \delta_j \\ \xi_k & \eta_k & \zeta_k & \delta_k \\ \xi_t & \eta_t & \zeta_t & \delta_t \end{bmatrix}\begin{bmatrix} f_\xi \\ f_\eta \\ f_\zeta \\ f_\tau \end{bmatrix},\end{aligned} \tag{13.19}$$

it follows that

$$\begin{bmatrix} f_\xi \\ f_\eta \\ f_\zeta \\ f_\tau \end{bmatrix} = \begin{bmatrix} \xi_i & \eta_i & \zeta_i & \delta_i \\ \xi_j & \eta_j & \zeta_j & \delta_j \\ \xi_k & \eta_k & \zeta_k & \delta_k \\ \xi_t & \eta_t & \zeta_t & \delta_t \end{bmatrix}^{-1}\begin{bmatrix} f_i \\ f_j \\ f_k \\ f_t \end{bmatrix}. \tag{13.20}$$

By substituting Equation (13.16) into Equation (13.20) one obtains

$$\begin{aligned}\begin{bmatrix} f_\xi \\ f_\eta \\ f_\zeta \\ f_\tau \end{bmatrix} &= \begin{bmatrix} \xi_i & \eta_i & \zeta_i & \delta_i \\ \xi_j & \eta_j & \zeta_j & \delta_j \\ \xi_k & \eta_k & \zeta_k & \delta_k \\ \xi_t & \eta_t & \zeta_t & \delta_t \end{bmatrix}^{-1}\left(\begin{bmatrix} f_{ii} & f_{ij} & f_{ik} & f_{it} \\ f_{ji} & f_{jj} & f_{jk} & f_{jt} \\ f_{ki} & f_{kj} & f_{kk} & f_{kt} \\ f_{ti} & f_{tj} & f_{tk} & f_{tt} \end{bmatrix}\begin{bmatrix} \alpha_i & \beta_i & \gamma_i & \delta_i \\ \alpha_j & \beta_j & \gamma_j & \delta_j \\ \alpha_k & \beta_k & \gamma_k & \delta_k \\ \alpha_t & \beta_t & \gamma_t & \delta_t \end{bmatrix}\begin{bmatrix} a_\alpha \\ a_\beta \\ a_\gamma \\ a_\delta \end{bmatrix}\right)\\ &= \left(\begin{bmatrix} \xi_i & \eta_i & \zeta_i & \delta_i \\ \xi_j & \eta_j & \zeta_j & \delta_j \\ \xi_k & \eta_k & \zeta_k & \delta_k \\ \xi_t & \eta_t & \zeta_t & \delta_t \end{bmatrix}^{-1}\begin{bmatrix} f_{ii} & f_{ij} & f_{ik} & f_{it} \\ f_{ji} & f_{jj} & f_{jk} & f_{jt} \\ f_{ki} & f_{kj} & f_{kk} & f_{kt} \\ f_{ti} & f_{tj} & f_{tk} & f_{tt} \end{bmatrix}\begin{bmatrix} \alpha_i & \beta_i & \gamma_i & \delta_i \\ \alpha_j & \beta_j & \gamma_j & \delta_j \\ \alpha_k & \beta_k & \gamma_k & \delta_k \\ \alpha_t & \beta_t & \gamma_t & \delta_t \end{bmatrix}\right)\begin{bmatrix} a_\alpha \\ a_\beta \\ a_\gamma \\ a_\delta \end{bmatrix},\end{aligned} \tag{13.21}$$

where tensor matrix

$$\begin{bmatrix} f_{\xi\alpha} & f_{\xi\beta} & f_{\xi\gamma} & f_{\xi\delta} \\ f_{\eta\alpha} & f_{\eta\beta} & f_{\eta\gamma} & f_{\eta\delta} \\ f_{\zeta\alpha} & f_{\zeta\beta} & f_{\zeta\gamma} & f_{\zeta\delta} \\ f_{\tau\alpha} & f_{\tau\beta} & f_{\tau\gamma} & f_{\tau\delta} \end{bmatrix} = \begin{bmatrix} \xi_i & \eta_i & \zeta_i & \delta_i \\ \xi_j & \eta_j & \zeta_j & \delta_j \\ \xi_k & \eta_k & \zeta_k & \delta_k \\ \xi_t & \eta_t & \zeta_t & \delta_t \end{bmatrix}^{-1}\begin{bmatrix} f_{ii} & f_{ij} & f_{ik} & f_{it} \\ f_{ji} & f_{jj} & f_{jk} & f_{jt} \\ f_{ki} & f_{kj} & f_{kk} & f_{kt} \\ f_{ti} & f_{tj} & f_{tk} & f_{tt} \end{bmatrix}\begin{bmatrix} \alpha_i & \beta_i & \gamma_i & \delta_i \\ \alpha_j & \beta_j & \gamma_j & \delta_j \\ \alpha_k & \beta_k & \gamma_k & \delta_k \\ \alpha_t & \beta_t & \gamma_t & \delta_t \end{bmatrix} \tag{13.22}$$

is obtained using the base

$$[\boldsymbol{\alpha}\ \ \boldsymbol{\beta}\ \ \boldsymbol{\gamma}\ \ \boldsymbol{\delta}] \tag{13.23}$$

for the "surfaces" and the base

$$[\boldsymbol{\xi}\ \ \boldsymbol{\eta}\ \ \boldsymbol{\zeta}\ \ \boldsymbol{\tau}] \tag{13.24}$$

for the "forces."

It is worth noting that in Equation (13.22)

$$\begin{bmatrix} \xi_i & \eta_i & \zeta_i & \delta_i \\ \xi_j & \eta_j & \zeta_j & \delta_j \\ \xi_k & \eta_k & \zeta_k & \delta_k \\ \xi_t & \eta_t & \zeta_t & \delta_t \end{bmatrix}^{-1} = \begin{bmatrix} i_\xi & j_\xi & k_\xi & t_\xi \\ i_\eta & j_\eta & k_\eta & t_\eta \\ i_\zeta & j_\zeta & k_\zeta & t_\zeta \\ i_\delta & j_\delta & k_\delta & t_\delta \end{bmatrix}. \tag{13.25}$$

This means that

$$\begin{bmatrix} f_{\xi\alpha} & f_{\xi\beta} & f_{\xi\gamma} & f_{\xi\delta} \\ f_{\eta\alpha} & f_{\eta\beta} & f_{f\gamma} & f_{\eta\delta} \\ f_{\zeta\alpha} & f_{\zeta\beta} & f_{\zeta\gamma} & f_{\zeta\delta} \\ f_{\tau\alpha} & f_{\tau\beta} & f_{\tau\gamma} & f_{\tau\delta} \end{bmatrix} = \begin{bmatrix} i_\xi & j_\xi & k_\xi & t_\xi \\ i_\eta & j_\eta & k_\eta & t_\eta \\ i_\zeta & j_\zeta & k_\zeta & t_\zeta \\ i_\delta & j_\delta & k_\delta & t_\delta \end{bmatrix} \begin{bmatrix} f_{ii} & f_{ij} & f_{ik} & f_{it} \\ f_{ji} & f_{jj} & f_{jk} & f_{jt} \\ f_{ki} & f_{kj} & f_{kk} & f_{kt} \\ f_{ti} & f_{tj} & f_{tk} & f_{tt} \end{bmatrix} \begin{bmatrix} \alpha_i & \beta_i & \gamma_i & \delta_i \\ \alpha_j & \beta_j & \gamma_j & \delta_j \\ \alpha_k & \beta_k & \gamma_k & \delta_k \\ \alpha_t & \beta_t & \gamma_t & \delta_t \end{bmatrix}. \tag{13.26}$$

In this manner, one can write Equation (13.22) straight from the definition without any need for a lengthy process of deriving the equation.

Of course the term "surface" in a particular problem is replaced by the terminology defined by the nature of the problem. In a similar manner, the term "force" is replaced by the terminology that a particular problem dictates.

Second Order Tensor in *n*-D. Extension of the concept of the second order tensor to *n*-dimensional space is obtained by using the same steps described above to extend the tensor to 4D space.

Again, the best way to aid understanding and to emphasize that the tensor is a physical quantity is to introduce the concept of "stress" in *n*-D space. In order to do that, first a concept of the "surface" in *n*-D space is introduced.

$$\mathbf{a} = a_1\boldsymbol{\alpha}_1 + a_2\boldsymbol{\alpha}_2 + a_3\boldsymbol{\alpha}_3 + \cdots + a_n\boldsymbol{\alpha}_n \tag{13.27}$$

where

$$[\boldsymbol{\alpha}_1\ \ \boldsymbol{\alpha}_2\ \ \boldsymbol{\alpha}_3\ \ \cdots\ \ \boldsymbol{\alpha}_n] \tag{13.28}$$

is the general surface base in *n*-D space. The individual vectors of this base can be expressed using the general orthonormal base

$$[\mathbf{i}_1\ \ \mathbf{i}_2\ \ \mathbf{i}_3\ \ \cdots\ \ \mathbf{i}_n] \tag{13.29}$$

so that

$$[\boldsymbol{\alpha}_1 \;\; \boldsymbol{\alpha}_2 \;\; \boldsymbol{\alpha}_3 \;\; \cdots \;\; \boldsymbol{\alpha}_n] = \begin{bmatrix} \alpha_{i_1 1} & \alpha_{i_1 2} & \alpha_{i_1 3} & \cdots & \alpha_{i_1 n} \\ \alpha_{i_2 1} & \alpha_{i_2 2} & \alpha_{i_2 3} & \cdots & \alpha_{i_2 n} \\ \alpha_{i_3 1} & \alpha_{i_3 2} & \alpha_{i_3 3} & \cdots & \alpha_{i_3 n} \\ \vdots & \vdots & \vdots & \ddots & \vdots \\ \alpha_{i_n 1} & \alpha_{i_n 2} & \alpha_{i_n 3} & \cdots & \alpha_{i_n n} \end{bmatrix}. \tag{13.30}$$

In a similar manner, forces in n-D space are introduced

$$\mathbf{f} = f_1\boldsymbol{\xi}_1 + f_2\boldsymbol{\xi}_2 + f_3\boldsymbol{\xi}_3 + \cdots + f_n\boldsymbol{\xi}_n, \tag{13.31}$$

where

$$[\boldsymbol{\xi}_1 \;\; \boldsymbol{\xi}_2 \;\; \boldsymbol{\xi}_3 \;\; \cdots \;\; \boldsymbol{\xi}_n] \tag{13.32}$$

is the general force base in n-D space defined by using the global orthonormal base as follows

$$[\boldsymbol{\xi}_1 \;\; \boldsymbol{\xi}_2 \;\; \boldsymbol{\xi}_3 \;\; \cdots \;\; \boldsymbol{\xi}_n] = [\mathbf{i}_1 \;\; \mathbf{i}_2 \;\; \mathbf{i}_3 \;\; \cdots \;\; \mathbf{i}_n] \begin{bmatrix} \xi_{i_1 1} & \xi_{i_1 2} & \xi_{i_1 3} & \cdots & \xi_{i_1 n} \\ \xi_{i_2 1} & \xi_{i_2 2} & \xi_{i_2 3} & \cdots & \xi_{i_2 n} \\ \xi_{i_3 1} & \xi_{i_3 2} & \xi_{i_3 3} & \cdots & \xi_{i_3 n} \\ \vdots & \vdots & \vdots & \ddots & \vdots \\ \xi_{i_n 1} & \xi_{i_n 2} & \xi_{i_n 3} & \cdots & \xi_{i_n n} \end{bmatrix} \tag{13.33}$$

or simply

$$[\boldsymbol{\xi}_1 \;\; \boldsymbol{\xi}_2 \;\; \boldsymbol{\xi}_3 \;\; \cdots \;\; \boldsymbol{\xi}_n] = \begin{bmatrix} \xi_{i_1 1} & \xi_{i_1 2} & \xi_{i_1 3} & \cdots & \xi_{i_1 n} \\ \xi_{i_2 1} & \xi_{i_2 2} & \xi_{i_2 3} & \cdots & \xi_{i_2 n} \\ \xi_{i_3 1} & \xi_{i_3 2} & \xi_{i_3 3} & \cdots & \xi_{i_3 n} \\ \vdots & \vdots & \vdots & \ddots & \vdots \\ \xi_{i_n 1} & \xi_{i_n 2} & \xi_{i_n 3} & \cdots & \xi_{i_n n} \end{bmatrix}. \tag{13.34}$$

The stress in n-D can be defined as linear mapping of surfaces in n-D to forces in n-D such that

$$\begin{aligned} \mathbf{f}(s\mathbf{a}) &= s\mathbf{f}(\mathbf{a}) \\ \mathbf{f}(\mathbf{a}+\mathbf{b}) &= \mathbf{f}(\mathbf{a}) + \mathbf{f}(\mathbf{b}). \end{aligned} \tag{13.35}$$

By using the global base for both surfaces and forces one obtains

$$\begin{aligned} \mathbf{f} &= \mathbf{f}_{i_1}a_{i_1} + \mathbf{f}_{i_2}a_{i_2} + \mathbf{f}_{i_3}a_{i_3} + \cdots + \mathbf{f}_{i_n}a_{i_n} \\ &= [\mathbf{f}_{i_1} \;\; \mathbf{f}_{i_2} \;\; \mathbf{f}_{i_3} \;\; \cdots \;\; \mathbf{f}_{i_n}] \begin{bmatrix} a_{i_1} \\ a_{i_2} \\ a_{i_3} \\ \vdots \\ a_{i_n} \end{bmatrix}. \end{aligned} \tag{13.36}$$

where

$$\begin{bmatrix} \mathbf{f}_{i_1} & \mathbf{f}_{i_2} & \mathbf{f}_{i_3} & \cdots & \mathbf{f}_{i_n} \end{bmatrix} \tag{13.37}$$

are the forces on the base surfaces

$$\begin{bmatrix} \mathbf{i}_1 & \mathbf{i}_2 & \mathbf{i}_3 & \cdots & \mathbf{i}_n \end{bmatrix} \tag{13.38}$$

and

$$\mathbf{a} = a_{i_1}\mathbf{i}_1 + a_{i_2}\mathbf{i}_2 + a_{i_3}\mathbf{i}_3 + \cdots + a_{i_n}\mathbf{i}_n \tag{13.39}$$

is a surface in n-D space. Thus,

$$\mathbf{f} = \begin{bmatrix} f_{i_1} \\ f_{i_2} \\ f_{i_3} \\ \vdots \\ f_{i_n} \end{bmatrix} = a_{i_1}\begin{bmatrix} f_{i_1 i_1} \\ f_{i_2 i_1} \\ f_{i_3 i_1} \\ \vdots \\ f_{i_n i_1} \end{bmatrix} + a_{i_2}\begin{bmatrix} f_{i_1 i_2} \\ f_{i_2 i_2} \\ f_{i_3 i_2} \\ \vdots \\ f_{i_n i_2} \end{bmatrix} + a_{i_3}\begin{bmatrix} f_{i_1 i_3} \\ f_{i_2 i_3} \\ f_{i_3 i_3} \\ \vdots \\ f_{i_n i_3} \end{bmatrix} + \ldots + a_{i_n}\begin{bmatrix} f_{i_1 i_n} \\ f_{i_2 i_n} \\ f_{i_3 i_n} \\ \vdots \\ f_{i_n i_n} \end{bmatrix}$$
$$= \begin{bmatrix} f_{i_1 i_1} & f_{i_1 i_2} & f_{i_1 i_3} & \cdots & f_{i_1 i_n} \\ f_{i_2 i_1} & f_{i_2 i_2} & f_{i_2 i_3} & \cdots & f_{i_2 i_n} \\ f_{i_3 i_1} & f_{i_3 i_2} & f_{i_3 i_3} & \cdots & f_{i_3 i_n} \\ \vdots & \vdots & \vdots & \ddots & \vdots \\ f_{i_n i_1} & f_{i_n i_2} & f_{i_n i_3} & \cdots & f_{i_n i_n} \end{bmatrix} \begin{bmatrix} a_{i_1} \\ a_{i_2} \\ a_{i_3} \\ \vdots \\ a_{i_n} \end{bmatrix}. \tag{13.40}$$

The matrix of the surface vector is given by

$$\begin{bmatrix} a_{i_1} \\ a_{i_2} \\ a_{i_3} \\ \vdots \\ a_{i_n} \end{bmatrix} = \begin{bmatrix} \alpha_{i_1 1} & \alpha_{i_1 2} & \alpha_{i_1 3} & \cdots & \alpha_{i_1 n} \\ \alpha_{i_2 1} & \alpha_{i_2 2} & \alpha_{i_2 3} & \cdots & \alpha_{i_2 n} \\ \alpha_{i_3 1} & \alpha_{i_3 2} & \alpha_{i_3 3} & \cdots & \alpha_{i_3 n} \\ \vdots & \vdots & \vdots & \ddots & \vdots \\ \alpha_{i_n 1} & \alpha_{i_n 2} & \alpha_{i_n 3} & \cdots & \alpha_{i_n n} \end{bmatrix} \begin{bmatrix} a_{\alpha_1} \\ a_{\alpha_2} \\ a_{\alpha_3} \\ \vdots \\ a_{\alpha_n} \end{bmatrix}. \tag{13.41}$$

And the matrix of the force vector is given by

$$\begin{bmatrix} f_{i_1} \\ f_{i_2} \\ f_{i_3} \\ \vdots \\ f_{i_n} \end{bmatrix} = \begin{bmatrix} \xi_{i_1 1} & \xi_{i_1 2} & \xi_{i_1 3} & \cdots & \xi_{i_1 n} \\ \xi_{i_2 1} & \xi_{i_2 2} & \xi_{i_2 3} & \cdots & \xi_{i_2 n} \\ \xi_{i_3 1} & \xi_{i_3 2} & \xi_{i_3 3} & \cdots & \xi_{i_3 n} \\ \vdots & \vdots & \vdots & \ddots & \vdots \\ \xi_{i_n 1} & \xi_{i_n 2} & \xi_{i_n 3} & \cdots & \xi_{i_n n} \end{bmatrix} \begin{bmatrix} f_{\xi_1} \\ f_{\xi_2} \\ f_{\xi_3} \\ \vdots \\ f_{\xi_n} \end{bmatrix}. \tag{13.42}$$

By substituting Equations (13.41) and (13.42) into Equation (13.40) one obtains

$$\mathbf{f} = f_{\xi_1}\xi_1 + f_{\xi_2}\xi_2 + f_{\xi_3}\xi_3 + \cdots + f_{\xi_n}\xi_n = \begin{bmatrix} f_{\xi_1} \\ f_{\xi_2} \\ f_{\xi_3} \\ \vdots \\ f_{\xi_n} \end{bmatrix}$$

$$= \begin{bmatrix} \xi_{i_1 1} & \xi_{i_1 2} & \xi_{i_1 3} & \cdots & \xi_{i_1 n} \\ \xi_{i_2 1} & \xi_{i_2 2} & \xi_{i_2 3} & \cdots & \xi_{i_2 n} \\ \xi_{i_3 1} & \xi_{i_3 2} & \xi_{i_3 3} & \cdots & \xi_{i_3 n} \\ \vdots & \vdots & \vdots & \ddots & \vdots \\ \xi_{i_n 1} & \xi_{i_n 2} & \xi_{i_n 3} & \cdots & \xi_{i_n n} \end{bmatrix}^{-1} \begin{bmatrix} f_{i_1 i_1} & f_{i_1 i_2} & f_{i_1 i_3} & \cdots & f_{i_1 i_n} \\ f_{i_2 i_1} & f_{i_2 i_2} & f_{i_2 i_3} & \cdots & f_{i_2 i_n} \\ f_{i_3 i_1} & f_{i_3 i_2} & f_{i_3 i_3} & \cdots & f_{i_3 i_n} \\ \vdots & \vdots & \vdots & \ddots & \vdots \\ f_{i_n i_1} & f_{i_n i_2} & f_{i_n i_3} & \cdots & f_{i_n i_n} \end{bmatrix}$$

$$\begin{bmatrix} \alpha_{i_1 1} & \alpha_{i_1 2} & \alpha_{i_1 3} & \cdots & \alpha_{i_1 n} \\ \alpha_{i_2 1} & \alpha_{i_2 2} & \alpha_{i_2 3} & \cdots & \alpha_{i_2 n} \\ \alpha_{i_3 1} & \alpha_{i_3 2} & \alpha_{i_3 3} & \cdots & \alpha_{i_3 n} \\ \vdots & \vdots & \vdots & \ddots & \vdots \\ \alpha_{i_n 1} & \alpha_{i_n 2} & \alpha_{i_n 3} & \cdots & \alpha_{i_n n} \end{bmatrix} \begin{bmatrix} a_{\alpha_1} \\ a_{\alpha_2} \\ a_{\alpha_3} \\ \vdots \\ a_{\alpha_n} \end{bmatrix} \tag{13.43}$$

$$= \left(\begin{bmatrix} \xi_{i_1 1} & \xi_{i_1 2} & \xi_{i_1 3} & \cdots & \xi_{i_1 n} \\ \xi_{i_2 1} & \xi_{i_2 2} & \xi_{i_2 3} & \cdots & \xi_{i_2 n} \\ \xi_{i_3 1} & \xi_{i_3 2} & \xi_{i_3 3} & \cdots & \xi_{i_3 n} \\ \vdots & \vdots & \vdots & \ddots & \vdots \\ \xi_{i_n 1} & \xi_{i_n 2} & \xi_{i_n 3} & \cdots & \xi_{i_n n} \end{bmatrix}^{-1} \begin{bmatrix} f_{i_1 i_1} & f_{i_1 i_2} & f_{i_1 i_3} & \cdots & f_{i_1 i_n} \\ f_{i_2 i_1} & f_{i_2 i_2} & f_{i_2 i_3} & \cdots & f_{i_2 i_n} \\ f_{i_3 i_1} & f_{i_3 i_2} & f_{i_3 i_3} & \cdots & f_{i_3 i_n} \\ \vdots & \vdots & \vdots & \ddots & \vdots \\ f_{i_n i_1} & f_{i_n i_2} & f_{i_n i_3} & \cdots & f_{i_n i_n} \end{bmatrix} \right.$$

$$\left. \begin{bmatrix} \alpha_{i_1 1} & \alpha_{i_1 2} & \alpha_{i_1 3} & \cdots & \alpha_{i_1 n} \\ \alpha_{i_2 1} & \alpha_{i_2 2} & \alpha_{i_2 3} & \cdots & \alpha_{i_2 n} \\ \alpha_{i_3 1} & \alpha_{i_3 2} & \alpha_{i_3 3} & \cdots & \alpha_{i_3 n} \\ \vdots & \vdots & \vdots & \ddots & \vdots \\ \alpha_{i_n 1} & \alpha_{i_n 2} & \alpha_{i_n 3} & \cdots & \alpha_{i_n n} \end{bmatrix} \right) \begin{bmatrix} a_{\alpha_1} \\ a_{\alpha_2} \\ a_{\alpha_3} \\ \vdots \\ a_{\alpha_n} \end{bmatrix},$$

where the forces are expressed using the force's base and the surfaces are expressed using the surface's base and the tensor is represented by the tensor matrix

$$\mathbf{F}=\begin{bmatrix} f_{\xi_1\alpha_1} & f_{\xi_1\alpha_2} & f_{\xi_1\alpha_3} & \cdots & f_{\xi_1\alpha_n} \\ f_{\xi_2\alpha_1} & f_{\xi_2\alpha_2} & f_{\xi_2\alpha_3} & \cdots & f_{\xi_2\alpha_n} \\ f_{\xi_3\alpha_1} & f_{\xi_3\alpha_2} & f_{\xi_3\alpha_3} & \cdots & f_{\xi_3\alpha_n} \\ \vdots & \vdots & \vdots & \ddots & \vdots \\ f_{\xi_n\alpha_1} & f_{\xi_n\alpha_2} & f_{\xi_n\alpha_3} & \cdots & f_{\xi_n\alpha_n} \end{bmatrix}$$

$$=\begin{bmatrix} \xi_{i_11} & \xi_{i_12} & \xi_{i_13} & \cdots & \xi_{i_1n} \\ \xi_{i_21} & \xi_{i_22} & \xi_{i_23} & \cdots & \xi_{i_2n} \\ \xi_{i_31} & \xi_{i_32} & \xi_{i_33} & \cdots & \xi_{i_3n} \\ \vdots & \vdots & \vdots & \ddots & \vdots \\ \xi_{i_n1} & \xi_{i_n2} & \xi_{i_n3} & \cdots & \xi_{i_nn} \end{bmatrix}^{-1} \begin{bmatrix} f_{i_1i_1} & f_{i_1i_2} & f_{i_1i_3} & \cdots & f_{i_1i_n} \\ f_{i_2i_1} & f_{i_2i_2} & f_{i_2i_3} & \cdots & f_{i_2i_n} \\ f_{i_3i_1} & f_{i_3i_2} & f_{i_3i_3} & \cdots & f_{i_3i_n} \\ \vdots & \vdots & \vdots & \ddots & \vdots \\ f_{i_ni_1} & f_{i_ni_2} & f_{i_ni_3} & \cdots & f_{i_ni_n} \end{bmatrix} \begin{bmatrix} \alpha_{i_11} & \alpha_{i_12} & \alpha_{i_13} & \cdots & \alpha_{i_1n} \\ \alpha_{i_21} & \alpha_{i_22} & \alpha_{i_23} & \cdots & \alpha_{i_2n} \\ \alpha_{i_31} & \alpha_{i_32} & \alpha_{i_33} & \cdots & \alpha_{i_3n} \\ \vdots & \vdots & \vdots & \ddots & \vdots \\ \alpha_{i_n1} & \alpha_{i_n2} & \alpha_{i_n3} & \cdots & \alpha_{i_nn} \end{bmatrix} \tag{13.44}$$

in the global bases $[\boldsymbol{\alpha}_1 \;\; \boldsymbol{\alpha}_2 \;\; \boldsymbol{\alpha}_3 \;\; \cdots \;\; \boldsymbol{\alpha}_n]$ and $[\boldsymbol{\xi}_1 \;\; \boldsymbol{\xi}_2 \;\; \boldsymbol{\xi}_3 \;\; \cdots \;\; \boldsymbol{\xi}_n]$ for surfaces and forces respectively.

It is worth noting that

$$\begin{bmatrix} \xi_{i_11} & \xi_{i_12} & \xi_{i_13} & \cdots & \xi_{i_1n} \\ \xi_{i_21} & \xi_{i_22} & \xi_{i_23} & \cdots & \xi_{i_2n} \\ \xi_{i_31} & \xi_{i_32} & \xi_{i_33} & \cdots & \xi_{i_3n} \\ \vdots & \vdots & \vdots & \ddots & \vdots \\ \xi_{i_n1} & \xi_{i_n2} & \xi_{i_n3} & \cdots & \xi_{i_nn} \end{bmatrix}^{-1} = \begin{bmatrix} i_{\xi_11} & i_{\xi_12} & i_{\xi_13} & \cdots & i_{\xi_1n} \\ i_{\xi_21} & i_{\xi_22} & i_{\xi_23} & \cdots & i_{\xi_2n} \\ i_{\xi_31} & i_{\xi_32} & i_{\xi_33} & \cdots & i_{\xi_3n} \\ \vdots & \vdots & \vdots & \ddots & \vdots \\ i_{\xi_n1} & i_{\xi_n2} & i_{\xi_n3} & \cdots & i_{\xi_nn} \end{bmatrix}, \tag{13.45}$$

which yields

$$\mathbf{F}=\begin{bmatrix} f_{\xi_1\alpha_1} & f_{\xi_1\alpha_2} & f_{\xi_1\alpha_3} & \cdots & f_{\xi_1\alpha_n} \\ f_{\xi_2\alpha_1} & f_{\xi_2\alpha_2} & f_{\xi_2\alpha_3} & \cdots & f_{\xi_2\alpha_n} \\ f_{\xi_3\alpha_1} & f_{\xi_3\alpha_2} & f_{\xi_3\alpha_3} & \cdots & f_{\xi_3\alpha_n} \\ \vdots & \vdots & \vdots & \ddots & \vdots \\ f_{\xi_n\alpha_1} & f_{\xi_n\alpha_2} & f_{\xi_n\alpha_3} & \cdots & f_{\xi_n\alpha_n} \end{bmatrix}$$

$$=\begin{bmatrix} i_{\xi_11} & i_{\xi_12} & i_{\xi_13} & \cdots & i_{\xi_1n} \\ i_{\xi_21} & i_{\xi_22} & i_{\xi_23} & \cdots & i_{\xi_2n} \\ i_{\xi_31} & i_{\xi_32} & i_{\xi_33} & \cdots & i_{\xi_3n} \\ \vdots & \vdots & \vdots & \ddots & \vdots \\ i_{\xi_n1} & i_{\xi_n2} & i_{\xi_n3} & \cdots & i_{\xi_nn} \end{bmatrix} \begin{bmatrix} f_{i_1i_1} & f_{i_1i_2} & f_{i_1i_3} & \cdots & f_{i_1i_n} \\ f_{i_2i_1} & f_{i_2i_2} & f_{i_2i_3} & \cdots & f_{i_2i_n} \\ f_{i_3i_1} & f_{i_3i_2} & f_{i_3i_3} & \cdots & f_{i_3i_n} \\ \vdots & \vdots & \vdots & \ddots & \vdots \\ f_{i_ni_1} & f_{i_ni_2} & f_{i_ni_3} & \cdots & f_{i_ni_n} \end{bmatrix} \begin{bmatrix} \alpha_{i_11} & \alpha_{i_12} & \alpha_{i_13} & \cdots & \alpha_{i_1n} \\ \alpha_{i_21} & \alpha_{i_22} & \alpha_{i_23} & \cdots & \alpha_{i_2n} \\ \alpha_{i_31} & \alpha_{i_32} & \alpha_{i_33} & \cdots & \alpha_{i_3n} \\ \vdots & \vdots & \vdots & \ddots & \vdots \\ \alpha_{i_n1} & \alpha_{i_n2} & \alpha_{i_n3} & \cdots & \alpha_{i_nn} \end{bmatrix}. \tag{13.46}$$

13.2 Summary

The second order tensor has first been extended to 4D space. This is followed by the extension to n-dimensional space. A complete analogy with 2D and 3D spaces has been deliberately exploited, thus making the mental jump to n-dimensional space as simple and natural as possible.

Further Reading

[1] Abraham, R., Marsden, J. E. and Ratiu, T. S. (1988) *Manifolds, Tensor Analysis, and Applications*. New York, Springer-Verlag.

[2] Aris, R. (1962) *Vectors, Tensors and the Basic Equations of Fluid Mechanics*. New Jersey, Prentice Hall.

[3] Bronstein, I. N. and Semendyayev, S. (2007) *Handbook of Mathematics*. New York, Springer.

[4] Dvorkin, E. N. and Goldschmit, M. B. (2005) *Nonlinear Continua*. New York, Springer.

[5] Fleisch, D. (2012) *A Student's Guide to Vectors and Tensors*. New York, Cambridge University Press.

[6] Kusse, B. R. and Westwig, E. A. (2006) *Mathematical Physics: Applied Mathematics for Scientists and Engineers*, 2nd edn. Wiley-VCH Verlag GmbH & Co.

[7] Levi-Civita, T. (2013) *The Absolute Differential Calculus (Calculus of Tensors)*. New York, Dover Publications.

[8] Matthews, P. C. (1998) *Vector Calculus*. London, Springer.

Part Three

Deformability and Material Modeling

14

Kinematics of Deformation in 1D

14.1 Geometric Nonlinearity in General

When the material points of a solid body move in space, the solid body deforms, i.e., it translates, rotates and changes its shape and size. If the rotations and translations are relatively large, then a large displacements problem is being solved. On the other hand, if the solid stretches or flows a lot, such as rubber, plasticine, soft clay, hot metals, high strength fibers, biological tissue, etc., then a finite (large) strain problem is being solved.

Geometric nonlinearity is the terminology associated with both large displacements and large strains.

In the finite element literature, geometric nonlinearity is often considered by exploiting the existing small strain, small displacement approaches (already implemented in finite element packages) and redeploying them through a co-rotational formulation. The co-rotational formulation takes into account finite rotations by explicitly introducing the co-rotated configuration (position). As the solid deforms, the small strain material laws are implemented in an incremental fashion. This is called additive decomposition. Such an approach requires a continuous correction of the Cauchy stress matrix, which is often done by using the Jaumann rate.

These approaches are well defined in the finite element literature and can be found in a myriad of related text books. That stated, they do fall short of implementing the finite strain deformability theory consistently (see Gurtin, 1981; Truesdell, 1955) and are difficult to adopt for anisotropic materials.

In order to implement the original finite strain theory consistently, it is necessary to employ the multiplicative decomposition of both rotations (displacements) and stretches (strains). Early researchers in the field have realized this fact (see Simo, Peric, Munjiza). However, the problem with incorporating this approach is that it would require a complete re-development (and re-writing) of the existing element libraries found in many finite element packages. In addition, the formulation has to be, by default, explicit (for dynamic problems) or explicitly iterative

Large Strain Finite Element Method: A Practical Course, First Edition. Antonio Munjiza, Esteban Rougier and Earl E. Knight.

(for static problems). This is due to the fact that it is not always easy or even possible to obtain a consistent tangential stiffness matrix. This is not unique to finite strain finite element problems, for it has also been encountered with some other nonlinear problems. For example, in conducting finite element analysis of contact problems one encounters a discontinuity (a jump) with contact-make and contact-release. In these cases, the tangential stiffness matrix is not even unique apart from the reality that it is difficult to calculate.

In this book only the consistent multiplicative decomposition-based formulation is considered in the context of explicit or explicitly iterative formulations.

For didactical reasons it is best to start by studying the deformation of solids in 1D. In Figure 14.1a, a piece of rubber is shown with a linear scale sketched on the rubber.

The material point P has been marked on the rubber strip. In Figure 14.1a the initial position of the material point P is described by its initial coordinate

$$\bar{x}. \tag{14.1}$$

In Figure 14.1b the same material point P has moved in space. Since this is a 1D problem, it has moved along coordinate axis x. Its current position is therefore $\tilde{x}$, where

$$\tilde{x} = \tilde{\mathrm{x}}(\bar{x}). \tag{14.2}$$

Note that $\bar{x}$ and $\tilde{x}$ are scalar variables (1D space), thus they are written using italic font and $\tilde{\mathrm{x}}$ is a function, thus it is written using a nonslanted font. This is the standard convention used throughout this book.

The position of the solid body shown in Figure 14.1a is sometimes called the reference configuration. It is also called the initial configuration or the initial position. Sometimes it is called the nonstretched configuration. The position of the solid body in Figure 14.1b is called the current configuration, current position, deformed position, or stretched position.

From this point on, the position of the solid body in Figure 14.1a will be referred to as the initial position, where each material point is defined by its initial coordinate $\bar{x}$ (x-bar, where the bar stands for initial). In a similar manner, the position of the solid body shown in Figure 14.1b will be referred to as the current position, where each material point of the solid body is defined by its current coordinate $\tilde{x}$ (x-tilde, where the tilde stands for current).

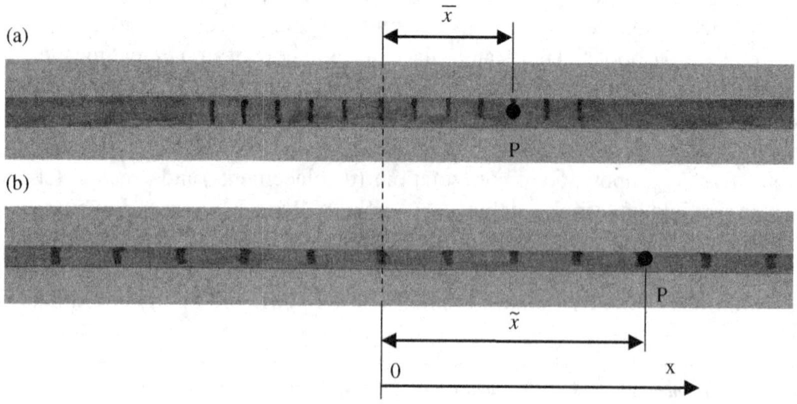

Figure 14.1 Stretching a piece of rubber: (a) initial positions, (b) deformed positions

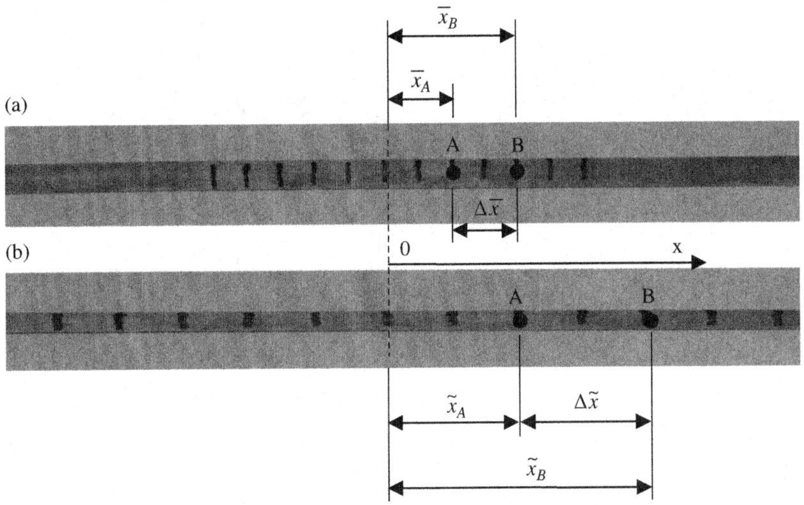

Figure 14.2 Initial (a) and current (b) distance between two material points

In short, any material point P of the rubber strip shown in Figure 14.1 is described by its initial coordinate $\bar{x}$ and its current coordinate $\tilde{x}$, where

$$\tilde{x} = \tilde{x}(\bar{x}). \tag{14.3}$$

In Figure 14.2, the same rubber strip is shown, however, this time two material points A and B are marked.

By observation it can be seen that the rubber element of initial length

$$\Delta\bar{x} = \bar{x}_B - \bar{x}_A \tag{14.4}$$

has changed into the rubber element of current length

$$\Delta\tilde{x} = \tilde{x}_B - \tilde{x}_A. \tag{14.5}$$

14.2 Stretch

In physical terms, the rubber element in Figure 14.2 has been subjected to a stretch defined by

$$s = \frac{\Delta\tilde{x}}{\Delta\bar{x}}. \tag{14.6}$$

The best way to understand the concept of stretch is to take a rubber element of initial length

$$\Delta\bar{x} = 1 \tag{14.7}$$

and stretch it, thus producing the current length

$$\Delta\tilde{x} = s \cdot \Delta\bar{x} = s \cdot 1 = s. \tag{14.8}$$

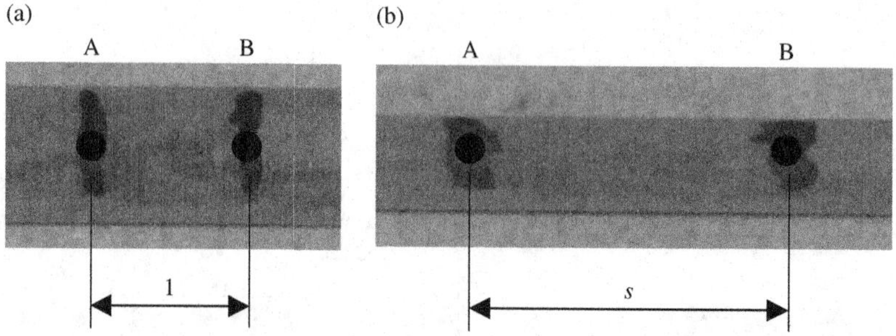

Figure 14.3 Stretching a unit-length element of rubber: (a) initial positions, (b) deformed positions

Stretch s turns a rubber element of unit initial length into a rubber element of current length s, Figure 14.3.

It is also useful to consider what happens to the stretch s as the initial length of the rubber element is chosen to be smaller and smaller. In other words, if initially points A and B are very close to each other. One can think of the initial rubber element being infinitesimally small

$$\Delta\bar{x} \rightarrow 0. \tag{14.9}$$

"Infinitesimally small" simply means smaller than any number that is greater than zero; by this definition $\Delta\bar{x}$ is always greater than zero.

While equation (14.6) gives the average stretch between points A and B, the following equation

$$s = \lim_{\Delta\bar{x}\rightarrow 0} \frac{\Delta\tilde{x}}{\Delta\bar{x}} \tag{14.10}$$

gives the stretch between two points A and B that are virtually identical; i.e., it gives the stretch s at a point. This is best illustrated by the following two examples.

Example 1. Take

$$\tilde{x} = 2\bar{x} \tag{14.11}$$

in this case the stretch s is given by

$$\begin{aligned} s &= \frac{\tilde{x}(\bar{x}+\Delta\bar{x}) - \tilde{x}(\bar{x})}{\Delta\bar{x}} \\ &= \frac{[2(\bar{x}+\Delta\bar{x})] - [2(\bar{x})]}{\Delta\bar{x}} \\ &= \frac{2\Delta\bar{x}}{\Delta\bar{x}}. \end{aligned} \tag{14.12}$$

At $\bar{x}=1$, for $\Delta\bar{x}=1$ Equation (14.12) gives 2; for $\Delta\bar{x}=1/2$ Equation (14.12) gives 2; for $\Delta\bar{x}=1/4$ Equation (14.12) gives 2; etc. For $\Delta\bar{x}$ infinitesimally small Equation (14.12) also gives 2. Thus

$$\begin{aligned} s &= \lim_{\Delta\bar{x}\to 0}\frac{\tilde{x}(\bar{x}+\Delta\bar{x})-\tilde{x}(\bar{x})}{\Delta\bar{x}} \\ &= \frac{d\tilde{x}}{d\bar{x}} \\ &= \frac{d(2\bar{x})}{d\bar{x}} \\ &= 2. \end{aligned} \tag{14.13}$$

Equation (14.13) simply states that any infinitesimally small rubber element can be stretched to twice its original length. Again it is convenient to think that the original length of the rubber element is equal to 1, Figure 14.4, i.e., equal to the unit length

$$\Delta\bar{x} = 1\,\mathrm{im}, \tag{14.14}$$

where the unit length itself is not m, nor μm, but it is infinitesimally small. In further text, such an infinitesimally small unit for length is called the infimeter. After stretching, the rubber element becomes of length s, where

$$s = \frac{d\tilde{x}}{d\bar{x}} \tag{14.15}$$

Example 2. Say that

$$\tilde{x} = \bar{x}^2. \tag{14.16}$$

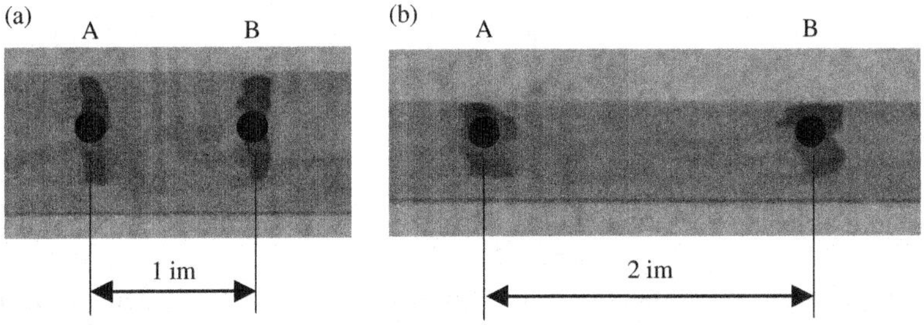

Figure 14.4 Stretching a infinitesimally short unit-length element of rubber

In this case

$$
\begin{aligned}
s &= \frac{\tilde{x}(\bar{x}+\Delta\bar{x})-\tilde{x}(\bar{x})}{\Delta\bar{x}} \\
&= \frac{(\bar{x}+\Delta\bar{x})^2-(\bar{x})^2}{\Delta\bar{x}} \\
&= \frac{\bar{x}^2+2\bar{x}\Delta\bar{x}+(\Delta\bar{x})^2-\bar{x}^2}{\Delta\bar{x}} \\
&= \frac{2\bar{x}\Delta\bar{x}+\Delta\bar{x}\Delta\bar{x}}{\Delta\bar{x}} = 2\bar{x}+\Delta\bar{x}.
\end{aligned}
\tag{14.17}
$$

At $\bar{x}=1$, for $\Delta\bar{x}=1$ Equation (14.17) gives 3; for $\Delta\bar{x}=1/10$ Equation (14.17) gives 2.01; for $\Delta\bar{x}=1/100$ Equation (14.17) gives 2.0001; etc. for $\Delta\bar{x}$ infinitesimally small Equation (14.17) gives 2. Thus, by gradually decreasing $\Delta\bar{x}$ one obtains the stretch at point A, as opposed to the average stretch. In the limit when $\Delta\bar{x}\to 0$

$$s=2\bar{x}. \tag{14.18}$$

or in general

$$s=\frac{d\tilde{x}}{d\bar{x}}=2\bar{x}. \tag{14.19}$$

14.3 Material Element and Continuum Assumption

In the previous section a material element was intuitively used. A piece of rubber of unit length was marked on the material itself. As the rubber deformed, it was observed that the piece of rubber stretched in such a way that its length became equal to s, where

$$s=\frac{\Delta\tilde{x}}{\Delta\bar{x}}. \tag{14.20}$$

Using this approach, an average stretch of a finite length of rubber band was obtained. In order to obtain the stretch at a point, it was necessary to progressively reduce the initial length $\Delta\bar{x}$ of the solid element. Since

$$\Delta\bar{x}=(\text{real number})\cdot\text{unit}, \tag{14.21}$$

this can be done in two ways (Figure 14.5):

a. reduce the real number to $d\bar{x}$ (infinitesimally small);
b. reduce the unit to infinitesimally small – from m to mm to μm, etc…, to im (infi-meter).

Option (b) often aids understanding, while option (a) is convenient when writing equations.

It is worth mentioning that by reducing the length of the solid (rubber) element, it is implicitly assumed that it is still the same macroscopically observed rubber with the same macroscopically observed properties.

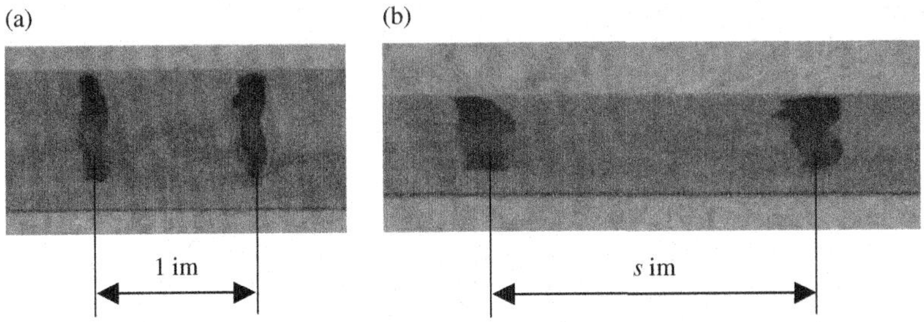

Figure 14.5 Stretching a piece of rubber: (a) initial position, (b) deformed position

This assumption is called the assumption of continuum. It is of essential importance in all problems where differential equations, and by default, where infinitesimal length and time scales are used.

Strictly speaking, the assumption of continuum is physically never true in either space or time for, as the size of the solid element reduces, one hits crystals, molecules, atoms, etc. However, its enforcement makes it possible to use differential calculus in modeling the macroscopic behavior. In many applications, the assumption of continuum does not affect the validity of the formulations as long as only macroscopic phenomena are being described.

That said, in many cases the assumption of continuum is not valid and one has to explicitly consider discontinuum as such. The applied mechanics dealing with such problems is called Mechanics of Discontinua. Continuum mechanics applies to problems where the assumption of continuum is valid. In contrast, Mechanics of Discontinua applies to problems where the discontinuous nature of solids is a-priori assumed, for more details see Computational Mechanics of Discontinua [1].

14.4 Strain

As a material stretches, its microstructure is strained and there can be instances where the microstructure starts disintegrating to the point that the material cracks, fractures, fails, yields, flows or undergoes other similar phenomena.

In order to quantify the straining of the stretched solid material, a strain measure is introduced. The simplest possible strain measure is

$$e = s - 1, \tag{14.22}$$

where s is the stretch. This strain measure is called the engineering strain, for it typically is used in many engineering applications.

Example 3. A rubber band deforms in such a way that its current position is described by

$$\tilde{x} = \tilde{x}(\bar{x}) = \frac{1}{200}\bar{x}^2 + \bar{x}. \tag{14.23}$$

Calculate the engineering strain at point P which has the initial position given by

$$\bar{x} = 3. \tag{14.24}$$

Solution:
Step 1: Stretch s is by definition given by

$$
\begin{aligned}
s = \frac{d\tilde{x}}{d\bar{x}} &= \frac{d\left(\frac{1}{200}\bar{x}^2 + \bar{x}\right)}{d\bar{x}} \\
&= \frac{1}{100}\bar{x} + 1.
\end{aligned}
\tag{14.25}
$$

At point P the stretch is given by

$$
s = \frac{1}{100}\bar{x} + 1 = \frac{1}{100}3 + 1 = \frac{3}{100} + 1. \tag{14.26}
$$

By definition, the engineering strain is given by

$$
e = s - 1 = \frac{3}{100} + 1 - 1 = \frac{3}{100} = 3\%. \tag{14.27}
$$

In other words, in the infinitesimal vicinity of point P the material has elongated by 3%. For some materials such as glass, a strain of this size is enough to break the material. For other materials, such as aluminum, this strain is big enough to cause yielding (flow) of the material. In both cases the microstructure starts changing. For materials such as rubber, this is an insignificant strain that will cause no damage, flow or change to the microstructure of the material.

There are many other strain measures used in science and engineering, for example:

a. Logarithmic strain, also known as true strain or natural strain

$$
e = \ln(s). \tag{14.28}
$$

b. Euler-Almansi strain

$$
e = \frac{1}{2}\left(1 - \frac{1}{s^2}\right). \tag{14.29}
$$

c. Green-Lagrange strain

$$
e = \frac{1}{2}(s^2 - 1). \tag{14.30}
$$

Example 4. The stretch at point P of a solid medium is given by

$$
s = 1.0001. \tag{14.31}
$$

Calculate at least three different strain measures.

Solution:
Engineering strain

$$
e = s - 1 = 1.0001 - 1 = 1.0000\text{e}{-4} = 1\text{e}{-4}. \tag{14.32}
$$

Logarithmic strain

$$e = \ln(s) = \ln(1.0001) = 0.000099995 = 0.99995\text{e}-4. \tag{14.33}$$

Euler-Almansi strain

$$\begin{aligned} e &= \frac{1}{2}\left(1-\frac{1}{s^2}\right) = \frac{1}{2}\left(1-\frac{1}{(1.0001)^2}\right) \\ &= 0.000099985 = 0.99985\text{e}-4. \end{aligned} \tag{14.34}$$

Green-Lagrange strain

$$\begin{aligned} e &= \frac{1}{2}(s^2-1) = \frac{1}{2}\left((1.0001)^2-1\right) \\ &= 0.000100005 = 1.00005\text{e}-4. \end{aligned} \tag{14.35}$$

Note that in this example all of the strain measures are nearly identical.

Example 5. The stretch at point P of a solid medium is given by

$$s = 0.9999. \tag{14.36}$$

Calculate the different strain measures.

Solution:
Engineering strain

$$e = s-1 = 0.9999-1 = -0.0001 = -1\text{e}-4. \tag{14.37}$$

Logarithmic strain

$$e = \ln(s) = \ln(0.9999) = -0.000100005 = -1.00005\text{e}-4. \tag{14.38}$$

Euler-Almansi strain

$$\begin{aligned} e &= \frac{1}{2}\left(1-\frac{1}{s^2}\right) = \frac{1}{2}\left(1-\frac{1}{(0.9999)^2}\right) \\ &= -0.000100015 = -1.00015\text{e}-4 \end{aligned} \tag{14.39}$$

Green-Lagrange strain

$$\begin{aligned} e &= \frac{1}{2}(s^2-1) = \frac{1}{2}\left((0.9999)^2-1\right) \\ &= 0.000099995 = -0.99995\text{e}-4. \end{aligned} \tag{14.40}$$

Again, all the strain measures are close to each other.

Example 6. The stretch at point P of a solid medium is given by

$$s = 0.1. \tag{14.41}$$

Calculate the different strain measures.

Solution:
Engineering strain

$$e = s - 1 = 0.1 - 1 = -9.0\mathrm{e}-1. \tag{14.42}$$

Logarithmic strain

$$e = \ln(s) = \ln(0.1) = -2.3026. \tag{14.43}$$

Euler-Almansi strain

$$e = \frac{1}{2}\left(1 - \frac{1}{s^2}\right) = \frac{1}{2}\left(1 - \frac{1}{(0.1)^2}\right) = -4.95. \tag{14.44}$$

Green-Lagrange strain

$$e = \frac{1}{2}(s^2 - 1) = \frac{1}{2}\left((0.1)^2 - 1\right) = -0.495 = -4.95\mathrm{e}-1. \tag{14.45}$$

The stretch in this example is far from 1 and different strain measures are quite different from each other.

Example 7. The stretch at point P of a solid medium is given by

$$s = 2.6. \tag{14.46}$$

Calculate the different strain measures.

Solution:
Engineering strain

$$e = s - 1 = 2.6 - 1 = 1.6. \tag{14.47}$$

Logarithmic strain

$$e = \ln(s) = \ln(2.6) = 0.95551. \tag{14.48}$$

Euler-Almansi strain

$$e = \frac{1}{2}\left(1 - \frac{1}{s^2}\right) = \frac{1}{2}\left(1 - \frac{1}{(2.6)^2}\right) = 0.426036. \tag{14.49}$$

Green-Lagrange strain

$$e = \frac{1}{2}(s^2 - 1) = \frac{1}{2}\left((2.6)^2 - 1\right) = 2.8. \tag{14.50}$$

By comparing the above examples it can be observed that for a value of stretch that is close to one all strain measures give nearly the same result. For larger stretches they all give quite different results. Examples four and five correspond to the case of small strains. The sixth and seventh examples correspond to the case of large strains. Large strains are often called finite strains. Small strains are sometimes called infinitesimal strains.

14.5 Stress

From strains, stresses can be calculated using a constitutive law (material equations). The simplest of these is the Hooke's law; in 1D the stress, σ, is simply

$$\sigma = \mathrm{E}e, \tag{14.51}$$

where E (note the nonitalic font) is the modulus of elasticity (approximately: 72 GPa for aluminum, 210 GPa for steel, 180 GPa for carbon fiber, 70 GPa for Kevlar fiber, 80 GPa for glass fiber, 1 GPa for soft wood, etc.). Hooke's law describes the behavior of linear elastic materials. However, the behavior of some other materials is best described by nonlinear elasticity, Figure 14.6.

In general, the term elastic means that the material goes back to its original shape when the load is removed, while linear or nonlinear refers to the stress-strain relationship.

In many cases, the material yields (flows) i.e., behaves plastically (steel, aluminum, etc.), Figure 14.7.

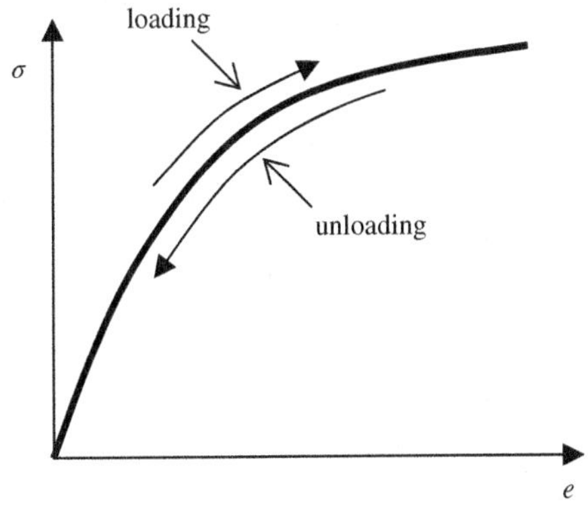

Figure 14.6 Stress-strain behavior for a nonlinear elastic material

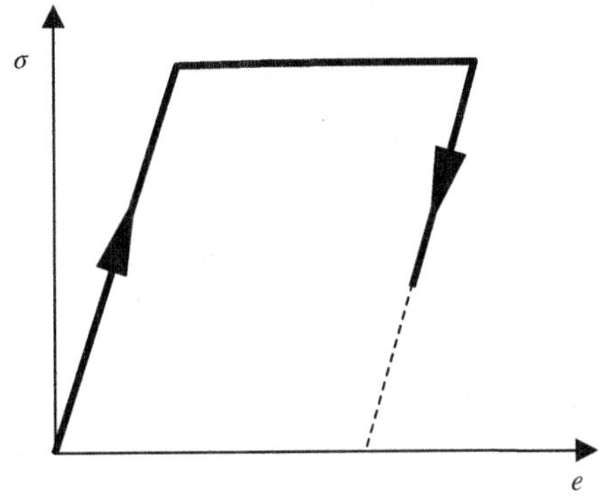

Figure 14.7 Stress-strain behavior for a plastic material

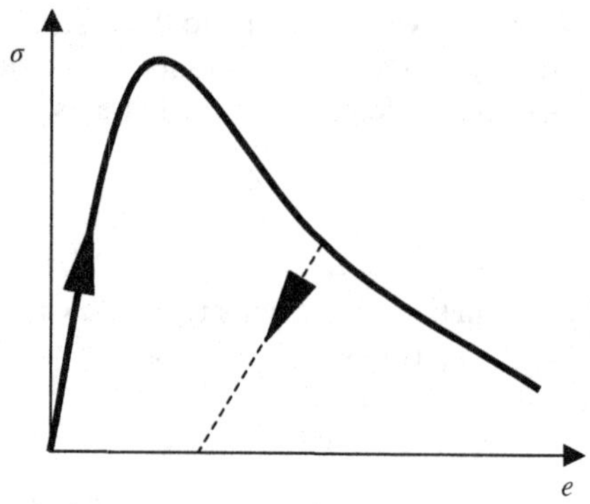

Figure 14.8 Stress-strain behavior for a nonlinear softening material: instead of e, a localized displacement variable, such as crack opening is often used

In some cases, the material yields through damage (rock, wood, composites, paper, etc.), Figure 14.8.

14.6 Summary

In this chapter some key aspects of deformation have been demonstrated using 1D solids. In the process the following points were covered:

- Initial (position) configuration has been explained.
- Current position (configuration) has been introduced.
- The concept of stretch has been presented.
- The role of the material element has been emphasized together with the role of differential calculus.
- The assumption of continuum has been introduced.
- Small and large strains have been explained.

Further Reading

[1] Bathe, K. J. (1982) *Finite Element Procedures Engineering Analysis.* New Jersey, Prentice-Hall.
[2] Bathe, K. J. (1996) *Finite Element Procedures.* New Jersey, Prentice-Hall.
[3] Dvorkin, E. N. and Goldschmit, M. B. (2005) *Nonlinear Continua.* New York, Springer.
[4] Gurtin, M. E. (1981) *An Introduction to Continuum Mechanics.* New York, Academic Press.
[5] Gurtin, M. E. and K. Spear (1983) On the relationship between the logarithmic strain rate and the stretching tensor. *Int. J. Solids Struct.*, **19**(5), 437–44.
[6] Hill, R. (1979) *Aspects of Invariance in Solid Mechanics*, ed. Y. Chia-Shun, Elsevier, pp. 1–75.
[7] Malvern, L. E. (1969) *Introduction to the Mechanics of a Continuous Medium.* New-Jersey, Prentice Hall.
[8] Munjiza, A. (2004) *The Combined Finite-Discrete Element Method.* Chichester, John Wiley & Sons, Ltd.
[9] Peric, D. (1992) On consistent stress rates in solid mechanics – computational implications. *International Journal for Numerical Methods in Engineering*, **33**(4), 799–817.
[10] Peric, D. and Owen, D. R. J. (1998) Finite-element applications to the nonlinear mechanics of solids. *Reports on Progress in Physics*, **61**(11), 1495–1574.
[11] Simo, J. C. (1985) On the computational significance of the intermediate configuration and hyperelastic stress relations in finite deformation elastoplasticity. *Mech. Mater.*, **4**(3–4), 439–51.
[12] Simo, J. C. (1988) A framework for finite strain elastoplasticity based on maximum plastic dissipation and the multiplicative decomposition. 1. Continuum formulation. *Comput. Method Appl. M.*, **66**(2), 199–219.

[13] Simo, J. C. (1988) A framework for finite strain elastoplasticity based on maximum plastic dissipation and the multiplicative decomposition. 2. Computational aspects, *Comput. Method Appl. M.*, **68**(1), 1–31.
[14] Simo, J. C. and Ortiz, M. (1985) A unified approach to finite deformation elastoplastic analysis based on the use of hyperelastic constitutive equations. *Comput. Method Appl. M.*, **49**(2), 221–45.
[15] Simo, J. C. and Ju, J. W. (1989) Finite deformation damage-elastoplasticity: a non conventional framework, *International Journal of Computational Mechanics*, **5**, 375–400.
[16] Simo, J. C. and Hughes, T. J. R. (1998) *Computational Inelasticity.* New York, Springer.
[17] Simo, J. C., Armero, F. and Taylor, R. L. (1993) Improved versions of assumed enhanced strain tri-linear elements for 3D finite deformation problems. *Comput Method Appl M*, **110**(3–4), 359–86.
[18] Truesdell, C. (1955), Hypo-elasticity. *J. Ration. Mech. Anal.*, **4**(1), 83–131.
[19] Truesdell, C. (1955) The simplest rate theory of pure elasticity, *Communications on Pure and Applied Mathematics*, **8**(1), 123–32.
[20] Truesdell, C. and W. Noll (1965) The non-linear field theories of mechanics. In *Hanbuch der Physik*, ed. S. Fluegge, Springer-Verlag, Berlin.
[21] Zienkiewicz, O. C. (1971) *The Finite Element Method in Engineering Science.* New York, McGraw-Hill.
[22] Zienkiewicz, O. C. and Taylor, R. L. (2005) *The Finite Element Method Set.* Oxford, Elsevier Science.

15

Kinematics of Deformation in 2D

15.1 Isotropic Solids

A solid that has properties that are the same in all directions is called an isotropic solid. One way to explain what an isotropic solid is would be to take a flat 2D solid sheet and cut a strip out of it. The properties of the newly cut strip will not depend on which direction the sheet has been cut, as shown in Figure 15.1.

15.2 Homogeneous Solids

A solid is homogeneous if its solid properties are the same regardless of the place where the properties have been measured. For example, two squares cut out of a solid sheet should be exactly the same, property wise, regardless of he fact that they have been taken from different locations, Figure 15.2.

More often than not solids are not homogeneous. However, in the infinitesimally close vicinity of a given material point P (note the nonitalic fonts for points) even a heterogeneous solid can be considered locally homogeneous (but not necessarily isotropic), regardless of the location or the orientation.

15.3 Homogeneous and Isotropic Solids

A solid with properties that do not change with either direction or location is considered to be both homogeneous and isotropic. This is illustrated by three identical squares cut from a homogeneous isotropic solid sheet, Figure 15.3. The physical properties of all three squares are the same.

Large Strain Finite Element Method: A Practical Course, First Edition. Antonio Munjiza, Esteban Rougier and Earl E. Knight.

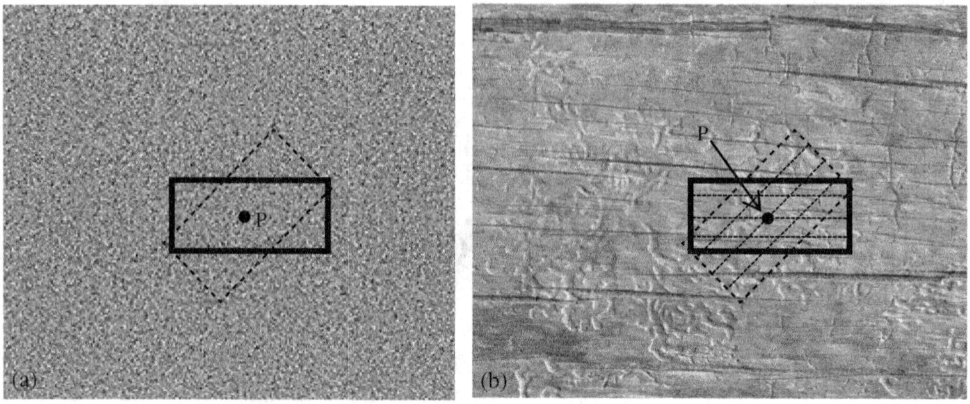

Figure 15.1 a) and b) Isotropic and anisotropic solid: (a) isotropic material; no matter in which direction the strip is cut, its material properties are the same, (b) anisotropic material; the two strips shown have different properties

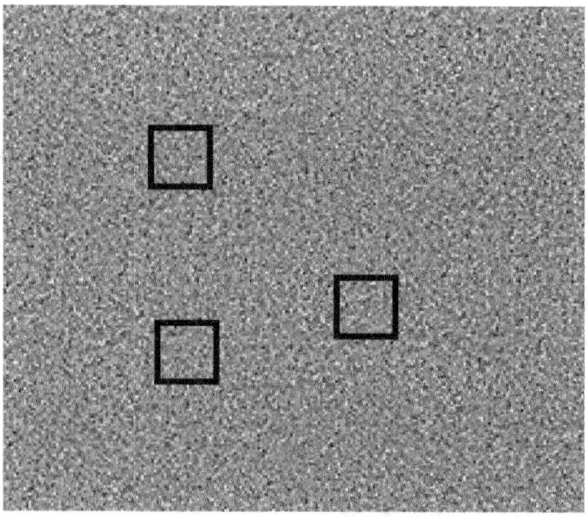

Figure 15.2 Homogeneous solid: no matter where the squares are taken from, their material properties are the same

15.4 Nonhomogeneous and Anisotropic Solids

In the most general case, solids are both nonhomogeneous and nonisotropic. Let us consider the layered rock shown in Figure 15.4. The lines show the boundaries of the individual layers. Within each layer the rock may often be considered not to be homogeneous. In addition, the properties of the rock are not the same in all directions. The dotted lines in Figure 15.5 show some preferred material directions (orientations, axes).

In the general case, the material directions do not have to be orthogonal to each other. Take for example, a sheet of glass fiber reinforced plastics as shown in Figure 15.6.

Two layers of glass fibers have been introduced defining the material element as a parallelogram defined by the material axes $\boldsymbol{\alpha}$ and $\boldsymbol{\beta}$, where in 2D $\boldsymbol{\alpha}$ and $\boldsymbol{\beta}$ are in general nonunit nonorthogonal vectors that form the material vector base, also called material embedded vector base

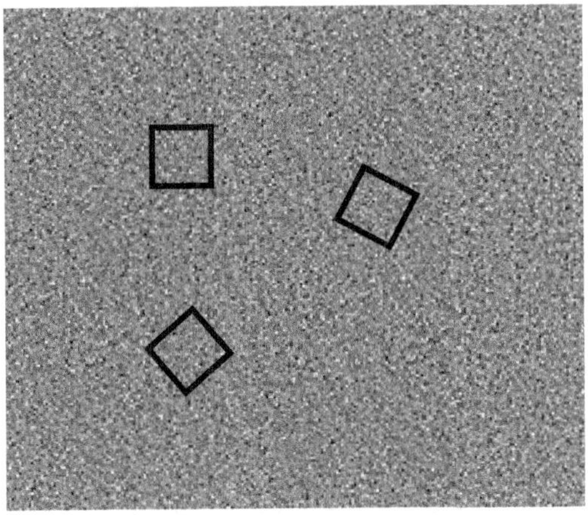

Figure 15.3 Homogeneous and isotropic solid: no matter from which part of the solid the squares are cut from or what their orientation is, their material properties are the same

Figure 15.4 Layered rock

$$[\boldsymbol{\alpha} \ \ \boldsymbol{\beta}]. \tag{15.1}$$

As the material is also nonhomogeneous these material axes change when one moves to a different material point, Figure 15.7.

The material vector base

$$[\boldsymbol{\alpha} \ \ \boldsymbol{\beta}] \tag{15.2}$$

is described using the global orthonormal base

$$[\mathbf{i} \ \ \mathbf{j}], \tag{15.3}$$

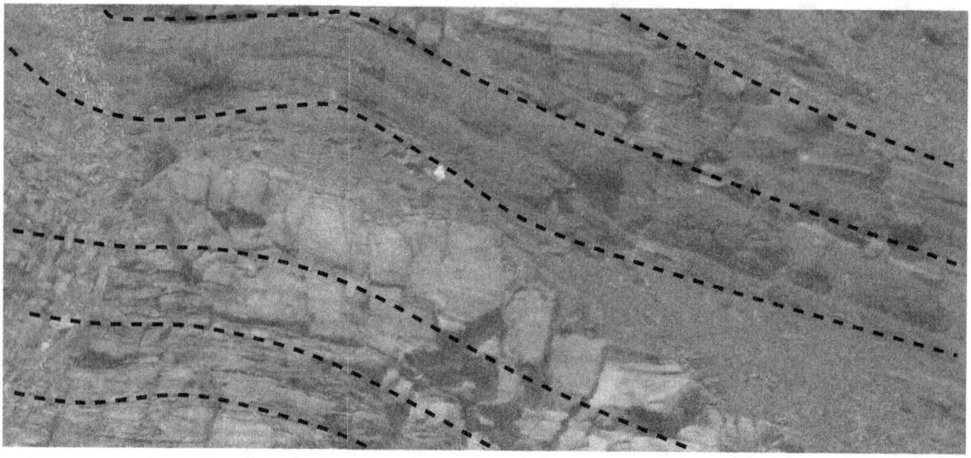

Figure 15.5 Material orientation

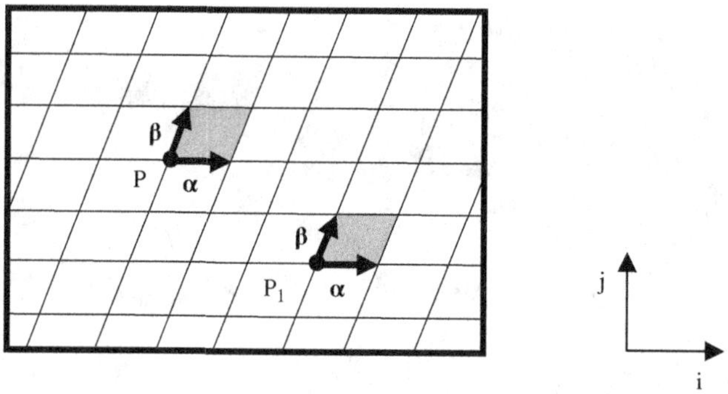

Figure 15.6 Anisotropic material elements for fiber reinforced plastics (solid lines indicate the direction of fibers)

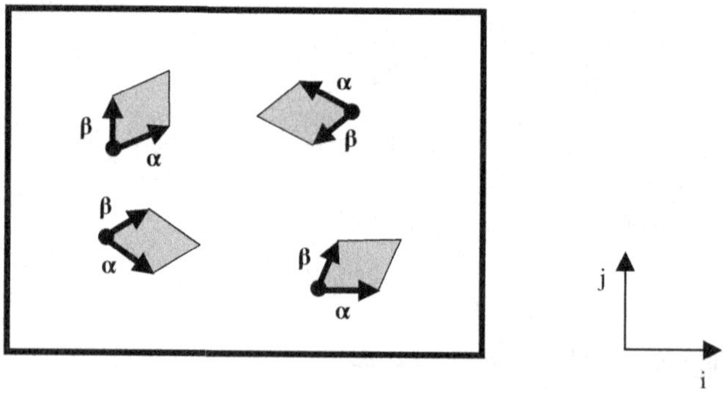

Figure 15.7 Anisotropic material elements at different locations of nonhomogeneous anisotropic solid

so that

$$\boldsymbol{\alpha} = \alpha_i \mathbf{i} + \alpha_j \mathbf{j} = \begin{bmatrix} \alpha_i \\ \alpha_j \end{bmatrix}$$
$$\boldsymbol{\beta} = \beta_i \mathbf{i} + \beta_j \mathbf{j} = \begin{bmatrix} \beta_i \\ \beta_j \end{bmatrix} \tag{15.4}$$

or simply

$$[\boldsymbol{\alpha} \;\; \boldsymbol{\beta}] = [\mathbf{i} \;\; \mathbf{j}] \begin{bmatrix} \alpha_i & \beta_i \\ \alpha_j & \beta_j \end{bmatrix}, \tag{15.5}$$

which is often written as

$$[\boldsymbol{\alpha} \;\; \boldsymbol{\beta}] = \begin{bmatrix} \alpha_i & \beta_i \\ \alpha_j & \beta_j \end{bmatrix} \tag{15.6}$$

where the equal sign means that the material vector base is represented by the corresponding matrix

$$\begin{bmatrix} \alpha_i & \beta_i \\ \alpha_j & \beta_j \end{bmatrix}. \tag{15.7}$$

15.5 Material Element Deformation

In order to define solid deformation in the vicinity of a given material point P, it has been customary in the literature to consider an infinitesimally small square-shaped material element at point P, Figure 15.8. The edges of this material element are defined by vectors

$$\mathbf{i} \text{ im} \quad \text{and} \quad \mathbf{j} \text{ im}, \tag{15.8}$$

where, as noted before, an infinitesimally small unit for length is used ("infi-meter", im) to indicate that the material element is infinitesimally small. The problem with this material element is that it does not represent the best way to describe anisotropic materials.

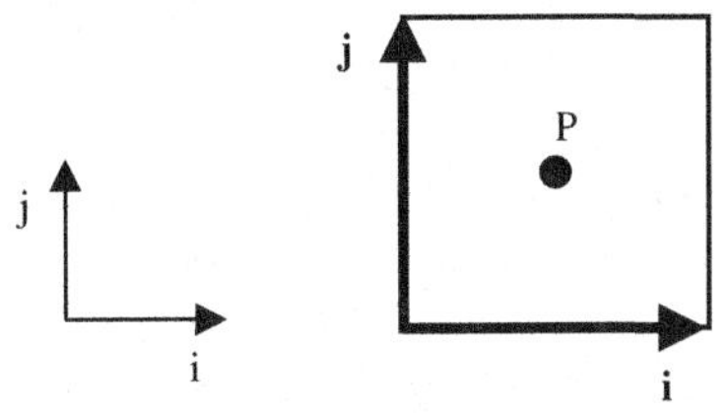

Figure 15.8 The infinitesimally small square-shaped material element at point P

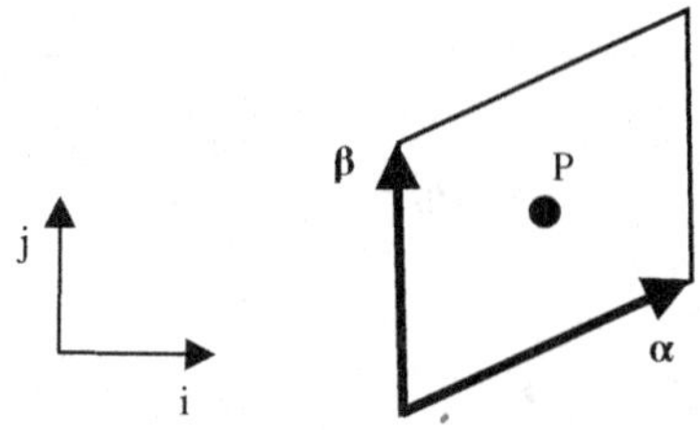

Figure 15.9 The infinitesimally small parallelogram-shaped material element at point P. This element is called the generalized material element, or simply the material element

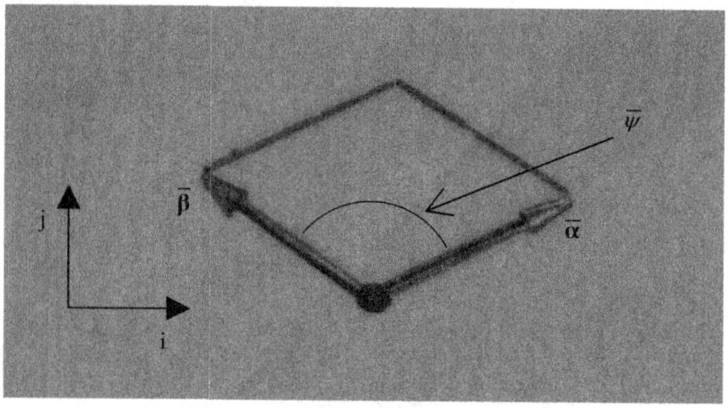

Figure 15.10 A photograph of a rubber sheet with the generalized infinitesimal material element plotted on it

In order to better represent the anisotropic material, it is convenient to introduce a parallelogram-shaped infinitesimal material element, Figure 15.9. In order to distinguish between the two material elements, the parallelogram-shaped material element is called the Munjiza generalized material element. For brevity, hereafter, the book refers to this element simply as the generalized material element. The square-shaped material element will be referred to explicitly as the square-shaped material element.

In Figure 15.10 a rubber sheet with an infinitesimal generalized material element plotted on it is shown. The material element is defined by its edges that are $|\bar{\alpha}|$ and $|\bar{\beta}|$ (im) long. One should note that these edges are not orthogonal to each other. They simply define an infinitesimally small volume of material cut in such a way that its shape has some physical meaning. For example, for layered rock these edges may follow the orientation of the rock joints; for fiber reinforced composites these edges may define the direction of the individual fibers.

In Figure 15.11 the rubber sheet from Figure 15.10 has been first rotated and then stretched.

In a similar manner, in Figure 15.12 the same rubber sheet has first been stretched and then rotated.

The end result in both cases is the same. The generalized material element has changed in shape and in size. In addition, it has been rotated. In short, it has been deformed (translated plus rotated plus stretched).

Of course, rotation and translation do not change the shape of the material element. They only change the orientation and position of the material element. As such, they do not strain the material element in any way.

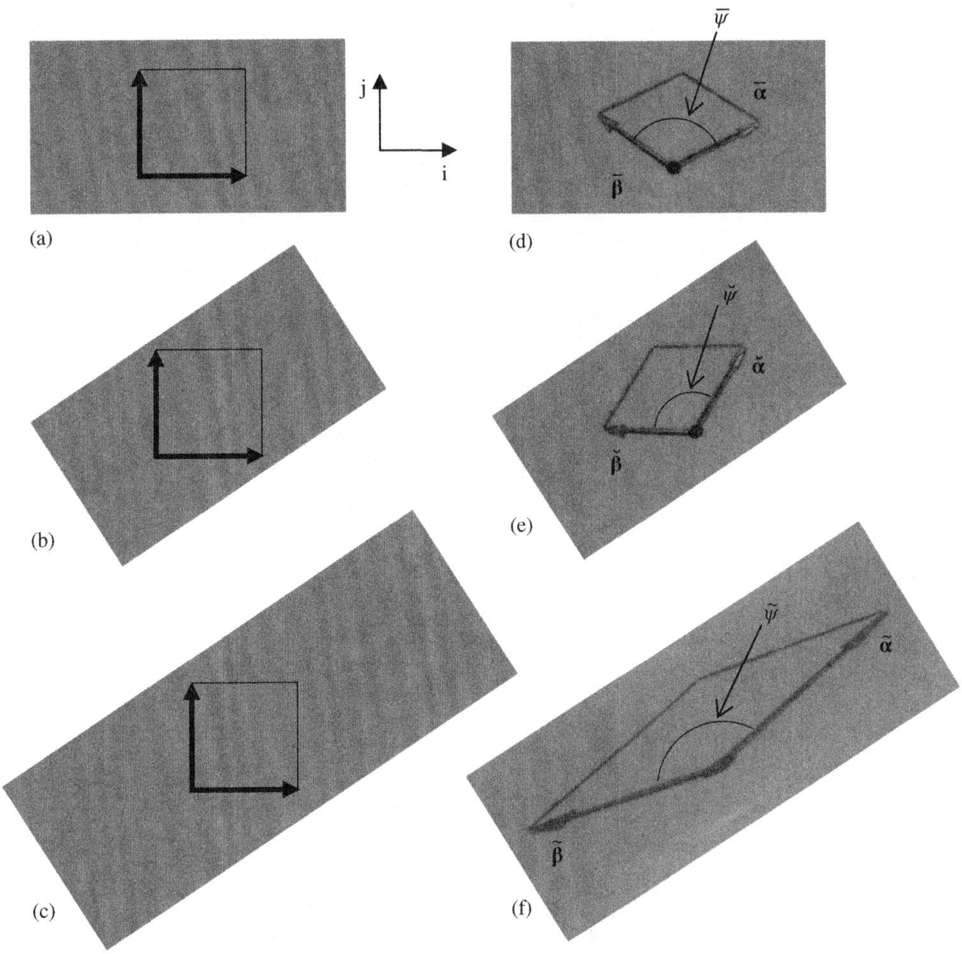

Figure 15.11 Left: A pictograph of the deformation of a rubber sheet with a square-shaped material element plotted on it: (a) initial position, (b) rotation, (c) stretch; Right: A pictograph of deformation of a rubber sheet with the generalized material element plotted on it: (d) initial position, (e) rotation, (f) stretch

As opposed to this, stretching changes both the shape and the size of the material element in such a way that:

a. it stays a parallelogram;
b. its edges become $\alpha = |\tilde{\boldsymbol{\alpha}}|$ and $\beta = |\tilde{\boldsymbol{\beta}}|$ long;
c. the angle between its edges becomes $\tilde{\psi}$.

The original material element defined by edges

$$\begin{bmatrix} \bar{\boldsymbol{\alpha}} & \bar{\boldsymbol{\beta}} \end{bmatrix} = \begin{bmatrix} \bar{\alpha}_i & \bar{\beta}_i \\ \bar{\alpha}_j & \bar{\beta}_j \end{bmatrix} \tag{15.9}$$

is therefore turned into the deformed solid element defined by edges

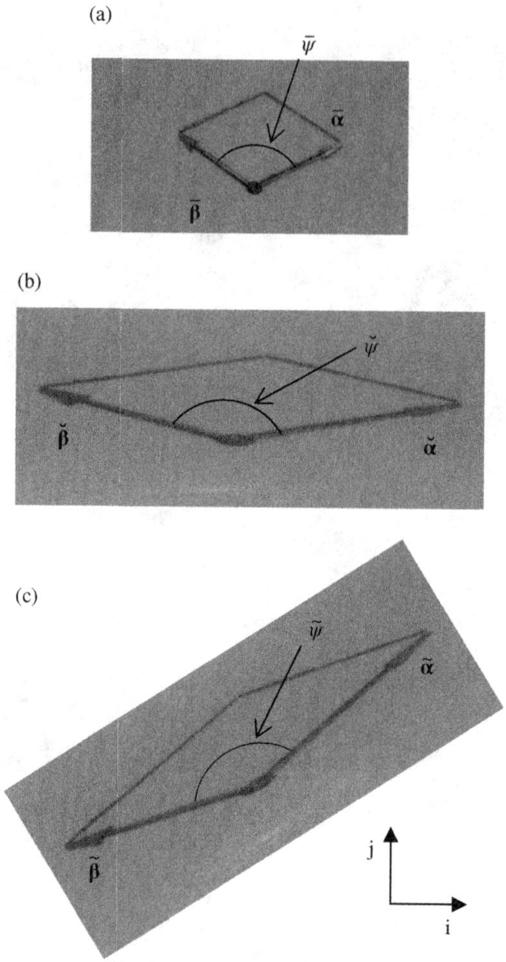

Figure 15.12 A pictograph of the deformation of a rubber sheet with the generalized material element being stretched first followed by a rotation: (a) initial position, (b) stretch, (c) rotation

$$\begin{bmatrix} \tilde{\boldsymbol{\alpha}} & \tilde{\boldsymbol{\beta}} \end{bmatrix} = \begin{bmatrix} \tilde{\alpha}_i & \tilde{\beta}_i \\ \tilde{\alpha}_j & \tilde{\beta}_j \end{bmatrix}. \tag{15.10}$$

Both $\bar{\boldsymbol{\alpha}}$ and $\tilde{\boldsymbol{\alpha}}$ go through the exact same material points of the solid; also $\bar{\boldsymbol{\beta}}$ and $\tilde{\boldsymbol{\beta}}$ coincide with the exact same material points of the solid. One can envision that they are plotted on the initial solid and as such, they move with the material points of the solid. The difference is that $\bar{\boldsymbol{\alpha}}$ and $\bar{\boldsymbol{\beta}}$ were plotted on the solid before it was deformed and subsequently they have moved with the material points of the deformed solid eventually turning (being transfigured) into $\tilde{\boldsymbol{\alpha}}$ and $\tilde{\boldsymbol{\beta}}$.

Vectors $\tilde{\boldsymbol{\alpha}}$ and $\tilde{\boldsymbol{\beta}}$ will stay as straight lines after deformation. The reason for this is that the generalized material element is infinitesimally small; as a consequence, any smooth curve of infinitesimally short length is a straight line. For the same reason the parallel edges of the generalized material element stay parallel after deformation.

15.6 Cauchy Stress Matrix for the Solid Element

The stress on the infinitesimal material element is defined as a linear mapping of surfaces to internal forces. In Figure 15.13 the current shape of the generalized material element is shown.

The current position and shape of the generalized material element are uniquely defined by edges $\tilde{\boldsymbol{\alpha}}$ and $\tilde{\boldsymbol{\beta}}$. The infinitesimal edge $\tilde{\boldsymbol{\alpha}}$ is obtained by rotating, translating and stretching the initial vector $\bar{\boldsymbol{\alpha}}$. The infinitesimal edge $\tilde{\boldsymbol{\beta}}$ is obtained by rotating, translating and stretching the initial vector $\bar{\boldsymbol{\beta}}$. The current surfaces $\tilde{\mathbf{a}}$ and $\tilde{\mathbf{b}}$ of the generalized material element are shown in Figure 15.14.

In the same figure, a square-shaped material element has been plotted on the material points in their current position. This square-shaped material element is different from the initially square-shaped material element shown in Figure 15.11. The square-shaped element from Figure 15.11 was initially square-shaped and has changed its shape due to deformation. In order to distinguish between the two, the square-shaped material element from Figure 15.14 is called the Cauchy material element. The Cauchy material element does not change with the deformation of the solid. As such it is not fixed to the material points and generally consists of different material points as the current position keeps changing, nevertheless the same material point P is always in the center.

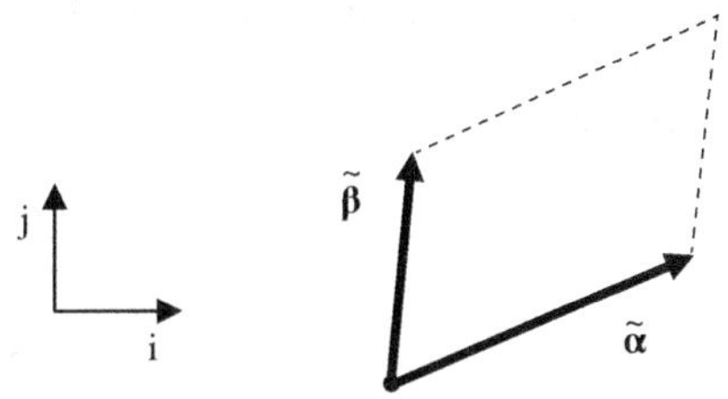

Figure 15.13 The current position of the generalized material element

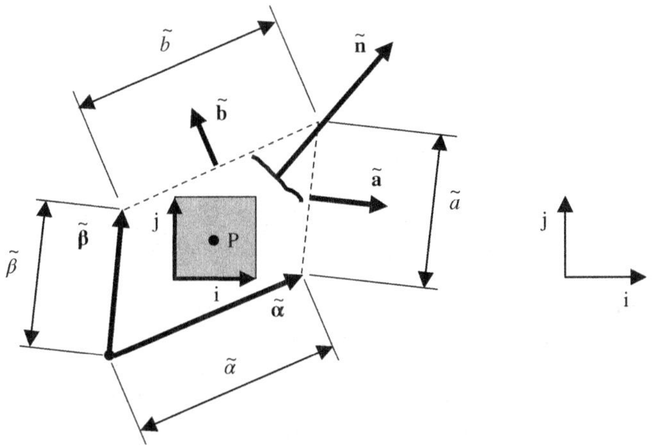

Figure 15.14 Current surfaces $\tilde{\mathbf{a}}$ and $\tilde{\mathbf{b}}$ of the generalized material element with the Cauchy material element also being shown for comparison purposes

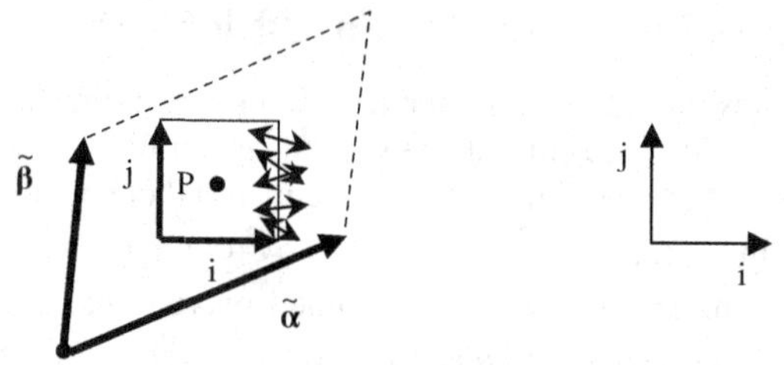

Figure 15.15 Components of the Cauchy stress matrix obtained as a resultant of internal forces across the surfaces of the Cauchy material element

The internal forces on (i.e., across) surfaces **i** and **j** of the Cauchy material element are given by

$$\mathbf{c_i} = c_{ii}\mathbf{i} + c_{ij}\mathbf{j}$$
$$\mathbf{c_j} = c_{ji}\mathbf{i} + c_{jj}\mathbf{j}. \tag{15.11}$$

These forces are in actual fact the resultant of all inter-atomic forces across the base surfaces **i** and **j**, Figure 15.15.

The resultant internal force on any other surface $\tilde{\mathbf{n}}$

$$\tilde{\mathbf{n}} = \tilde{n}_i\mathbf{i} + \tilde{n}_j\mathbf{j} \tag{15.12}$$

of the deformed solid is by definition of the stress tensor given by

$$\begin{aligned}\mathbf{f} &= \mathbf{c_i}\tilde{n}_i + \mathbf{c_j}\tilde{n}_j \\ &= \begin{bmatrix}\mathbf{c_i} & \mathbf{c_j}\end{bmatrix}\begin{bmatrix}\tilde{n}_i \\ \tilde{n}_j\end{bmatrix} \\ &= \begin{bmatrix}c_{ii} & c_{ij} \\ c_{ji} & c_{jj}\end{bmatrix}\begin{bmatrix}\tilde{n}_i \\ \tilde{n}_j\end{bmatrix} \\ &= \mathbf{C}\tilde{\mathbf{n}},\end{aligned} \tag{15.13}$$

where $\mathbf{C}$ is the Cauchy stress tensor matrix, $\mathbf{c_i}$ and $\mathbf{c_j}$ are the internal forces on surfaces **i** and **j** respectively, and $\tilde{\mathbf{n}}$ is a general surface plotted over material points in their current deformed position (thus ~).

Each of the components c_{ij} of the matrix $\mathbf{C}$ defines the internal force in the **i** direction on the **j** surface of the deformed solid. This is illustrated in Figure 15.16, where it can be observed that first the piece of a solid (say rubber) has been deformed. Then on the deformed rubber sheet, the solid element $[\mathbf{i} \;\; \mathbf{j}]$ has been plotted. After that, the surfaces **i** and **j** have been identified. Finally, the resultant internal forces on these surfaces is shown.

From the above, it is evident that the resultant internal force that the Cauchy stress matrix shows is the force on the deformed surface expressed using the global orthonormal base $[\mathbf{i} \;\; \mathbf{j}]$.

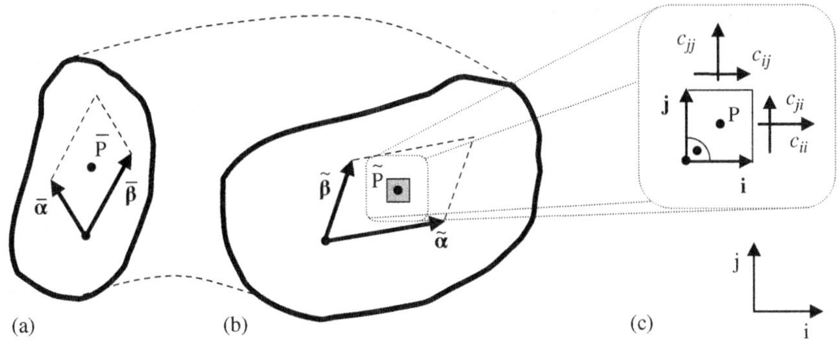

Figure 15.16 The Cauchy stress tensor matrix: (a) the initial position, (b) the current position, (c) a zoom-in of the Cauchy material element with components of the Cauchy matrix of the stress tensor

As **i** and **j** are infinitesimally small unit surfaces, the components of the Cauchy stress matrix show the internal forces per real unit surface of the deformed solid.

15.7 Coordinate Systems in 2D

In order to describe the position of solid material points in 2D space, it is convenient to introduce the Cartesian coordinate system, which is defined by two straight-line orthogonal axes (x, y), Figure 15.17.

Apart from the global Cartesian coordinate system, it is also often convenient to introduce the solid-embedded coordinate system

$$(\xi, \eta), \tag{15.14}$$

such that the coordinate lines η and ξ are given by

$$\begin{aligned} \xi &= \text{const} \\ \eta &= \text{const} \end{aligned} \tag{15.15}$$

respectively. These coordinate lines are fixed to the material points of the solid, Figure 15.17.

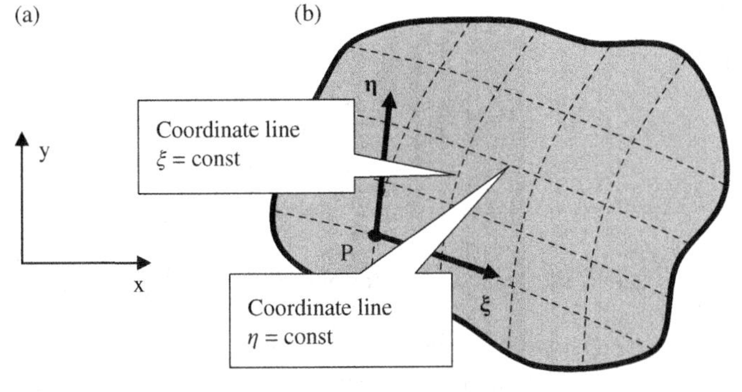

Figure 15.17 The Cartesian coordinate system (a) and the solid-embedded coordinate system (b)

15.8 The Solid- and the Material-Embedded Vector Bases

At any material point P, it is possible to define a set of infinitesimally short base vectors

$$[\boldsymbol{\xi} \quad \boldsymbol{\eta}] \tag{15.16}$$

that are tangential to the coordinate lines ξ and η respectively, Figure 15.17.

These vectors are given by

$$\begin{aligned} \boldsymbol{\xi} &= \frac{\partial x}{\partial \xi}\mathbf{i} + \frac{\partial y}{\partial \xi}\mathbf{j} \\ \boldsymbol{\eta} &= \frac{\partial x}{\partial \eta}\mathbf{i} + \frac{\partial y}{\partial \eta}\mathbf{j}. \end{aligned} \tag{15.17}$$

These can be interpreted as:

a. How much does one move $\boldsymbol{\xi}$ if ξ is changed by an infinitesimally small amount and η is kept constant? The answer is that one moves tangentially along the corresponding line $\eta = \text{const}$;
b. How much does one move $\boldsymbol{\eta}$ if η is changed by an infinitesimally small amount and ξ is kept constant? The answer is that one moves tangentially along the corresponding line $\xi = \text{const}$.

Equation (15.17) can be written in a matrix form

$$\begin{aligned} \boldsymbol{\xi} &= [\mathbf{i} \quad \mathbf{j}] \begin{bmatrix} \dfrac{\partial x}{\partial \xi} \\ \dfrac{\partial y}{\partial \xi} \end{bmatrix} \\ \boldsymbol{\eta} &= [\mathbf{i} \quad \mathbf{j}] \begin{bmatrix} \dfrac{\partial x}{\partial \eta} \\ \dfrac{\partial y}{\partial \eta} \end{bmatrix} \end{aligned} \tag{15.18}$$

or simply

$$[\boldsymbol{\xi} \quad \boldsymbol{\eta}] = [\mathbf{i} \quad \mathbf{j}] \begin{bmatrix} \dfrac{\partial x}{\partial \xi} & \dfrac{\partial x}{\partial \eta} \\ \dfrac{\partial y}{\partial \xi} & \dfrac{\partial y}{\partial \eta} \end{bmatrix} = \begin{bmatrix} \dfrac{\partial x}{\partial \xi} & \dfrac{\partial x}{\partial \eta} \\ \dfrac{\partial y}{\partial \xi} & \dfrac{\partial y}{\partial \eta} \end{bmatrix}. \tag{15.19}$$

This vector base is called the solid-embedded vector base, for its vectors deform with the material of the solid.

Since both, the solid-embedded and the material-embedded vector bases are fixed to the materials points of the solid, there is a deformation independent (invariant) relationship between these two bases

$$[\boldsymbol{\alpha}\ \ \boldsymbol{\beta}] = [\alpha_\xi \boldsymbol{\xi} + \alpha_\eta \boldsymbol{\eta}\ \ \beta_\xi \boldsymbol{\xi} + \beta_\eta \boldsymbol{\eta}] = [\boldsymbol{\xi}\ \ \boldsymbol{\eta}] \begin{bmatrix} \alpha_\xi & \beta_\xi \\ \alpha_\eta & \beta_\eta \end{bmatrix} \tag{15.20}$$

or simply

$$[\boldsymbol{\alpha}\ \ \boldsymbol{\beta}] = \begin{bmatrix} \alpha_\xi & \beta_\xi \\ \alpha_\eta & \beta_\eta \end{bmatrix}. \tag{15.21}$$

The matrix

$$\begin{bmatrix} \alpha_\xi & \beta_\xi \\ \alpha_\eta & \beta_\eta \end{bmatrix} \tag{15.22}$$

stays constant regardless of the solid deformation and as such, is the same in the initial and in the current position of the material points.

15.9 Kinematics of 2D Deformation

As the solid deforms, its material points move in space. In this manner, the global coordinates (x, y) of any given material point change. Of particular interest are the initial, the previous and the current positions of a given material point P. These are respectively given by

$$\bar{\mathrm{P}} = (\bar{x},\ \bar{y})\,;\ \hat{\mathrm{P}} = (\hat{x},\ \hat{y}) \quad \text{and} \quad \tilde{\mathrm{P}} = (\tilde{x},\ \tilde{y}), \tag{15.23}$$

where "–" stand for "initial", "^" stands for "previous" and "~" stands for "current."

In contrast to the global coordinates of the material points (which change with solid's deformation), the local coordinates of any material point P do not change with the solid's deformation. This means that the initial, previous and current positions of any given material point P are respectively

$$\begin{aligned} \bar{\mathrm{P}} &= (\bar{\xi}, \bar{\eta}) = (\xi, \eta) \\ \hat{\mathrm{P}} &= (\hat{\xi}, \hat{\eta}) = (\xi, \eta) \\ \tilde{\mathrm{P}} &= (\tilde{\xi}, \tilde{\eta}) = (\xi, \eta). \end{aligned} \tag{15.24}$$

However, the relationships between the global and local coordinates change as the solid deforms

$$\begin{aligned} x &= \mathrm{x}(\xi, \eta) \\ y &= \mathrm{y}(\xi, \eta) \\ z &= \mathrm{z}(\xi, \eta), \end{aligned} \tag{15.25}$$

thus giving the initial, the previous, and the current coordinates as a function of ξ and η.

$$\begin{aligned} \bar{x} &= \bar{\mathrm{x}}(\xi, \eta) \\ \bar{y} &= \bar{\mathrm{y}}(\xi, \eta) \\ \bar{z} &= \bar{\mathrm{z}}(\xi, \eta) \end{aligned} \tag{15.26}$$

$$\begin{aligned}\hat{x} &= \hat{\mathrm{x}}(\xi, \eta)\\ \hat{y} &= \hat{\mathrm{y}}(\xi, \eta)\\ \hat{z} &= \hat{\mathrm{z}}(\xi, \eta)\end{aligned} \tag{15.27}$$

$$\begin{aligned}\tilde{x} &= \tilde{\mathrm{x}}(\xi, \eta)\\ \tilde{y} &= \tilde{\mathrm{y}}(\xi, \eta)\\ \tilde{z} &= \tilde{\mathrm{z}}(\xi, \eta).\end{aligned} \tag{15.28}$$

As the solid deforms, the solid-embedded vector base $[\boldsymbol{\xi}\ \ \boldsymbol{\eta}]$ changes giving the initial, the previous, and the current positions of the vector base $[\boldsymbol{\xi}\ \ \boldsymbol{\eta}]$

$$\begin{aligned}\bar{\boldsymbol{\xi}} &= \frac{\partial \bar{x}}{\partial \xi}\mathbf{i} + \frac{\partial \bar{y}}{\partial \xi}\mathbf{j}\\ \bar{\boldsymbol{\eta}} &= \frac{\partial \bar{x}}{\partial \eta}\mathbf{i} + \frac{\partial \bar{y}}{\partial \eta}\mathbf{j}\end{aligned} \tag{15.29}$$

$$\begin{aligned}\hat{\boldsymbol{\xi}} &= \frac{\partial \hat{x}}{\partial \xi}\mathbf{i} + \frac{\partial \hat{y}}{\partial \xi}\mathbf{j}\\ \hat{\boldsymbol{\eta}} &= \frac{\partial \hat{x}}{\partial \eta}\mathbf{i} + \frac{\partial \hat{y}}{\partial \eta}\mathbf{j}\end{aligned} \tag{15.30}$$

$$\begin{aligned}\tilde{\boldsymbol{\xi}} &= \frac{\partial \tilde{x}}{\partial \xi}\mathbf{i} + \frac{\partial \tilde{y}}{\partial \xi}\mathbf{j}\\ \tilde{\boldsymbol{\eta}} &= \frac{\partial \tilde{x}}{\partial \eta}\mathbf{i} + \frac{\partial \tilde{y}}{\partial \eta}\mathbf{j}.\end{aligned} \tag{15.31}$$

The above equations are written using matrices as follows:

$$[\bar{\boldsymbol{\xi}}\ \ \bar{\boldsymbol{\eta}}] = [\mathbf{i}\ \ \mathbf{j}]\begin{bmatrix}\frac{\partial \bar{x}}{\partial \xi} & \frac{\partial \bar{y}}{\partial \xi}\\ \frac{\partial \bar{x}}{\partial \eta} & \frac{\partial \bar{y}}{\partial \eta}\end{bmatrix} = \begin{bmatrix}\frac{\partial \bar{x}}{\partial \xi} & \frac{\partial \bar{y}}{\partial \xi}\\ \frac{\partial \bar{x}}{\partial \eta} & \frac{\partial \bar{y}}{\partial \eta}\end{bmatrix} \tag{15.32}$$

$$[\hat{\boldsymbol{\xi}}\ \ \hat{\boldsymbol{\eta}}] = [\mathbf{i}\ \ \mathbf{j}]\begin{bmatrix}\frac{\partial \hat{x}}{\partial \xi} & \frac{\partial \hat{y}}{\partial \xi}\\ \frac{\partial \hat{x}}{\partial \eta} & \frac{\partial \hat{y}}{\partial \eta}\end{bmatrix} = \begin{bmatrix}\frac{\partial \hat{x}}{\partial \xi} & \frac{\partial \hat{y}}{\partial \xi}\\ \frac{\partial \hat{x}}{\partial \eta} & \frac{\partial \hat{y}}{\partial \eta}\end{bmatrix} \tag{15.33}$$

$$[\tilde{\boldsymbol{\xi}}\ \ \tilde{\boldsymbol{\eta}}] = [\mathbf{i}\ \ \mathbf{j}]\begin{bmatrix}\frac{\partial \tilde{x}}{\partial \xi} & \frac{\partial \tilde{y}}{\partial \xi}\\ \frac{\partial \tilde{x}}{\partial \eta} & \frac{\partial \tilde{y}}{\partial \eta}\end{bmatrix} = \begin{bmatrix}\frac{\partial \tilde{x}}{\partial \xi} & \frac{\partial \tilde{y}}{\partial \xi}\\ \frac{\partial \tilde{x}}{\partial \eta} & \frac{\partial \tilde{y}}{\partial \eta}\end{bmatrix}. \tag{15.34}$$

These three sets of base vectors define the initial, the previous, and the current position of the solid-embedded vector base and consequently the initial, the current, and the previous shapes of the generalized material element.

The initial, the previous, and the current positions of the material-embedded vector base are obtained from the initial, the previous, and the current position of the solid-embedded vector base using Equation (15.21):

$$\begin{bmatrix}\bar{\boldsymbol{\alpha}} & \bar{\boldsymbol{\beta}}\end{bmatrix} = \begin{bmatrix}\bar{\boldsymbol{\xi}} & \bar{\boldsymbol{\eta}}\end{bmatrix} \begin{bmatrix}\alpha_\xi & \beta_\xi \\ \alpha_\eta & \beta_\eta\end{bmatrix}$$

$$= \begin{bmatrix}\dfrac{\partial \bar{x}}{\partial \xi} & \dfrac{\partial \bar{x}}{\partial \eta} \\ \dfrac{\partial \bar{y}}{\partial \xi} & \dfrac{\partial \bar{y}}{\partial \eta}\end{bmatrix} \begin{bmatrix}\alpha_\xi & \beta_\xi \\ \alpha_\eta & \beta_\eta\end{bmatrix} = \begin{bmatrix}\bar{\xi}_i & \bar{\eta}_i \\ \bar{\xi}_j & \bar{\eta}_j\end{bmatrix} \begin{bmatrix}\alpha_\xi & \beta_\xi \\ \alpha_\eta & \beta_\eta\end{bmatrix} \tag{15.35}$$

$$\begin{bmatrix}\hat{\boldsymbol{\alpha}} & \hat{\boldsymbol{\beta}}\end{bmatrix} = \begin{bmatrix}\hat{\boldsymbol{\xi}} & \hat{\boldsymbol{\eta}}\end{bmatrix} \begin{bmatrix}\alpha_\xi & \beta_\xi \\ \alpha_\eta & \beta_\eta\end{bmatrix}$$

$$= \begin{bmatrix}\dfrac{\partial \hat{x}}{\partial \xi} & \dfrac{\partial \hat{x}}{\partial \eta} \\ \dfrac{\partial \hat{y}}{\partial \xi} & \dfrac{\partial \hat{y}}{\partial \eta}\end{bmatrix} \begin{bmatrix}\alpha_\xi & \beta_\xi \\ \alpha_\eta & \beta_\eta\end{bmatrix} = \begin{bmatrix}\hat{\xi}_i & \hat{\eta}_i \\ \hat{\xi}_j & \hat{\eta}_j\end{bmatrix} \begin{bmatrix}\alpha_\xi & \beta_\xi \\ \alpha_\eta & \beta_\eta\end{bmatrix} \tag{15.36}$$

$$\begin{bmatrix}\tilde{\boldsymbol{\alpha}} & \tilde{\boldsymbol{\beta}}\end{bmatrix} = \begin{bmatrix}\tilde{\boldsymbol{\xi}} & \tilde{\boldsymbol{\eta}}\end{bmatrix} \begin{bmatrix}\alpha_\xi & \beta_\xi \\ \alpha_\eta & \beta_\eta\end{bmatrix}$$

$$= \begin{bmatrix}\dfrac{\partial \tilde{x}}{\partial \xi} & \dfrac{\partial \tilde{x}}{\partial \eta} \\ \dfrac{\partial \tilde{y}}{\partial \xi} & \dfrac{\partial \tilde{y}}{\partial \eta}\end{bmatrix} \begin{bmatrix}\alpha_\xi & \beta_\xi \\ \alpha_\eta & \beta_\eta\end{bmatrix} = \begin{bmatrix}\tilde{\xi}_i & \tilde{\eta}_i \\ \tilde{\xi}_j & \tilde{\eta}_j\end{bmatrix} \begin{bmatrix}\alpha_\xi & \beta_\xi \\ \alpha_\eta & \beta_\eta\end{bmatrix}. \tag{15.37}$$

These six vectors $\begin{bmatrix}\bar{\boldsymbol{\alpha}} & \bar{\boldsymbol{\beta}}\end{bmatrix}, \begin{bmatrix}\hat{\boldsymbol{\alpha}} & \hat{\boldsymbol{\beta}}\end{bmatrix}, \begin{bmatrix}\tilde{\boldsymbol{\alpha}} & \tilde{\boldsymbol{\beta}}\end{bmatrix}$ define the initial, the previous and the current position of the generalized material element. As such, they uniquely define the deformation-induced internal forces over any surface including the surfaces of the Cauchy material element, i.e., the Cauchy stress tensor matrix.

15.10 2D Equilibrium Using the Virtual Work of Internal Forces

The geometry of this material element uniquely defines the solid's deformation state in the infinitesimal vicinity of a given material point P. By employing the constitutive law one calculates the Cauchy stress tensor matrix.

$$\mathbf{C} = \begin{bmatrix}c_{ii} & c_{ij} \\ c_{ji} & c_{jj}\end{bmatrix}. \tag{15.38}$$

Here, c_{ij} represents the internal force on surface **j** in the global **i** direction. The first row of matrix **C** represents the "**i**" forces (**i** components of the internal forces), the second row represents the "**j**" forces.

These internal forces change from material point to material point and in general, are a function of the current global coordinates, as shown in Figure 15.18. In other words, the internal force on two parallel surfaces of the Cauchy material element are in general not the same

Nevertheless, the Cauchy material element can be considered to be in a state of static or dynamic equilibrium. In order to consider the equilibrium of the Cauchy material element, it is often convenient to use the principle of virtual work.

Virtual Displacement in i Direction. For equilibrium of the forces in the **i** direction (**i** forces) an infinitesimally small virtual displacement $u > 0$ is supplied, Figure 15.19. This virtual displacement is imposed on the current position of the material points, thus moving the material points in **i** direction by

$$u = \mathrm{u}(\tilde{x}, \tilde{y}). \tag{15.39}$$

Since the virtual displacements in **j** direction are zero, the work of all **j** forces is zero. The only internal forces that do work are the **i** forces: the virtual work of the internal forces in

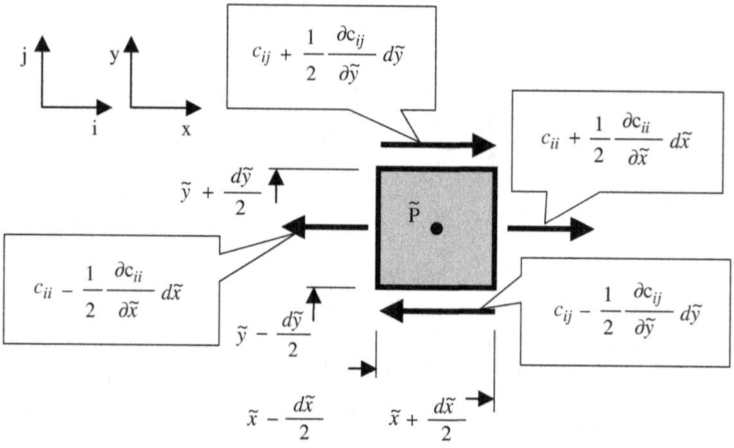

Figure 15.18 The Cauchy material element with **i** components of the internal forces shown

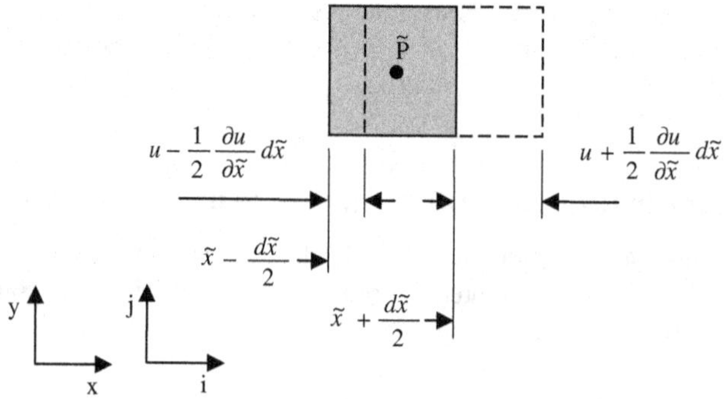

Figure 15.19 Virtual displacement of the Cauchy material element in **i** direction

i direction, dW_i, is obtained by multiplying the internal forces by the corresponding virtual displacements:

$$\begin{aligned} dW_i = &-\left(c_{ii}-\frac{1}{2}\frac{\partial c_{ii}}{\partial \tilde{x}}d\tilde{x}\right)d\tilde{y}\left(u-\frac{1}{2}\frac{\partial u}{\partial \tilde{x}}d\tilde{x}\right)\\ &+\left(c_{ii}+\frac{1}{2}\frac{\partial c_{ii}}{\partial \tilde{x}}d\tilde{x}\right)d\tilde{y}\left(u+\frac{1}{2}\frac{\partial u}{\partial \tilde{x}}d\tilde{x}\right)\\ &-\left(c_{ij}-\frac{1}{2}\frac{\partial c_{ij}}{\partial \tilde{y}}d\tilde{y}\right)d\tilde{x}\left(u-\frac{1}{2}\frac{\partial u}{\partial \tilde{y}}d\tilde{y}\right)\\ &+\left(c_{ij}+\frac{1}{2}\frac{\partial c_{ij}}{\partial \tilde{y}}d\tilde{y}\right)d\tilde{x}\left(u+\frac{1}{2}\frac{\partial u}{\partial \tilde{y}}d\tilde{y}\right). \end{aligned} \tag{15.40}$$

Since the products $d\tilde{x}d\tilde{x}d\tilde{y}$ and $d\tilde{x}d\tilde{y}d\tilde{y}$ are infinitesimally small in comparison with the products $d\tilde{x}d\tilde{y}$, the corresponding virtual work contributions are zero and one is left with the following expression for the work of the internal forces on the virtual displacements in the **i** direction

$$\begin{aligned} dW_i &= \left(c_{ii}\frac{\partial u}{\partial \tilde{x}}+c_{ij}\frac{\partial u}{\partial \tilde{y}}+\frac{\partial c_{ii}}{\partial \tilde{x}}u+\frac{\partial c_{ij}}{\partial \tilde{y}}u\right)d\tilde{x}d\tilde{y}\\ &= \left(\begin{bmatrix} c_{ii} & c_{ij}\end{bmatrix}\begin{bmatrix}\dfrac{\partial u}{\partial \tilde{x}}\\[2mm] \dfrac{\partial u}{\partial \tilde{y}}\end{bmatrix}+\begin{bmatrix}\dfrac{\partial c_{ii}}{\partial \tilde{x}} & \dfrac{\partial c_{ij}}{\partial \tilde{y}}\end{bmatrix}\begin{bmatrix}u\\ u\end{bmatrix}\right)d\tilde{x}d\tilde{y}. \end{aligned} \tag{15.41}$$

The term

$$\frac{\partial c_{ii}}{\partial \tilde{x}}u+\frac{\partial c_{ij}}{\partial \tilde{y}}u$$

represents the external forces that correspond to the given stress state. These are in general different from the external loads, thus the so-called weak or approximate formulation of equilibrium. In the case that the external forces are equal to the external loads, the strong (exact) formulation of equilibrium is recovered. When calculating the virtual work of the external loads **p**, only the residual external load is considered, i.e.,

$$p_x-\left(\frac{\partial c_{ii}}{\partial \tilde{x}}+\frac{\partial c_{ij}}{\partial \tilde{y}}\right)$$

Virtual Displacement in j Direction. The work done by the internal forces, Figure 15.20, due to the virtual displacements, Figure 15.21, is obtained by multiplying the internal forces in the **j** direction by the corresponding virtual displacements

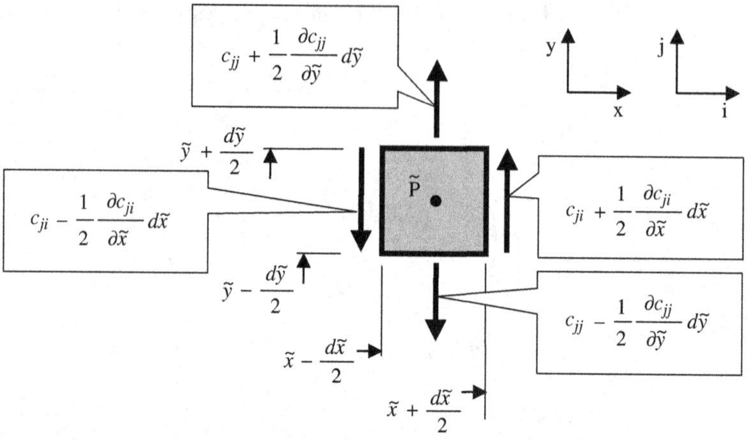

Figure 15.20 The Cauchy material element with **j** components of the internal forces shown

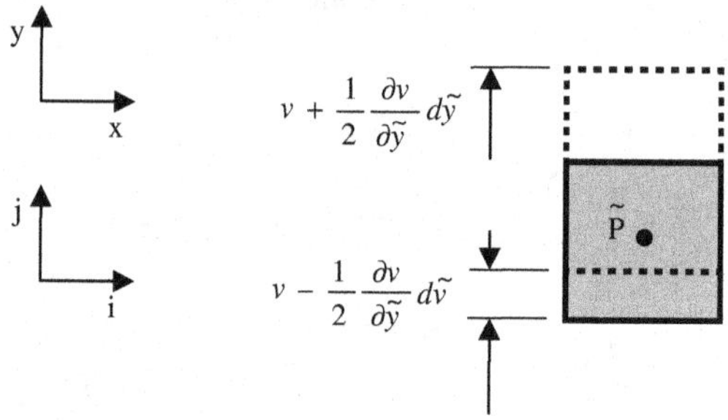

Figure 15.21 Virtual displacement of Cauchy material element in **j** direction

$$
\begin{aligned}
dW_j = &-\left(c_{ji} - \frac{1}{2}\frac{\partial c_{ji}}{\partial \tilde{x}} d\tilde{x}\right) d\tilde{y} \left(v - \frac{1}{2}\frac{\partial v}{\partial \tilde{x}} d\tilde{x}\right) \\
&+ \left(c_{ji} + \frac{1}{2}\frac{\partial c_{ji}}{\partial \tilde{x}} d\tilde{x}\right) d\tilde{y} \left(v + \frac{1}{2}\frac{\partial v}{\partial \tilde{x}} d\tilde{x}\right) \\
&- \left(c_{jj} - \frac{1}{2}\frac{\partial c_{jj}}{\partial \tilde{y}} d\tilde{y}\right) d\tilde{x} \left(v - \frac{1}{2}\frac{\partial v}{\partial \tilde{y}} d\tilde{y}\right) \\
&+ \left(c_{jj} + \frac{1}{2}\frac{\partial c_{jj}}{\partial \tilde{y}} d\tilde{y}\right) d\tilde{x} \left(v + \frac{1}{2}\frac{\partial v}{\partial \tilde{y}} d\tilde{y}\right).
\end{aligned}
\tag{15.42}
$$

The terms $d\tilde{x}d\tilde{x}d\tilde{y}$ and $d\tilde{x}d\tilde{y}d\tilde{y}$ are small relative to the terms $d\tilde{x}d\tilde{y}$ and the corresponding virtual work is therefore zero, which yields the virtual work of **j** forces on the v virtual displacement

$$dW_j = \left(c_{ji}\frac{\partial v}{\partial \tilde{x}} + c_{jj}\frac{\partial v}{\partial \tilde{y}} + \frac{\partial c_{ji}}{\partial \tilde{x}}v + \frac{\partial c_{jj}}{\partial \tilde{y}}v \right) d\tilde{x}d\tilde{y}$$

$$= \left(\begin{bmatrix} c_{ji} & c_{jj} \end{bmatrix} \begin{bmatrix} \frac{\partial v}{\partial \tilde{x}} \\ \frac{\partial v}{\partial \tilde{y}} \end{bmatrix} + \begin{bmatrix} \frac{\partial c_{ji}}{\partial \tilde{x}} & \frac{\partial c_{jj}}{\partial \tilde{y}} \end{bmatrix} \begin{bmatrix} v \\ v \end{bmatrix} \right) d\tilde{x}d\tilde{y}. \tag{15.43}$$

For a general virtual displacement

$$\begin{bmatrix} u \\ v \end{bmatrix} = \begin{bmatrix} \mathrm{u}(\tilde{x},\tilde{y}) \\ \mathrm{v}(\tilde{x},\tilde{y}) \end{bmatrix} \tag{15.44}$$

the total virtual work of the internal forces is obtained by integration over the solid in its current position,

$$W = \iint_{V_e} \left(\begin{bmatrix} c_{ii} & c_{ij} \end{bmatrix} \begin{bmatrix} \frac{\partial u}{\partial \tilde{x}} \\ \frac{\partial u}{\partial \tilde{y}} \end{bmatrix} + \begin{bmatrix} \frac{\partial c_{ii}}{\partial \tilde{x}} & \frac{\partial c_{ij}}{\partial \tilde{y}} \end{bmatrix} \begin{bmatrix} u \\ u \end{bmatrix} \right) d\tilde{x}d\tilde{y}$$
$$+ \iint_{V_e} \left(\begin{bmatrix} c_{ji} & c_{jj} \end{bmatrix} \begin{bmatrix} \frac{\partial v}{\partial \tilde{x}} \\ \frac{\partial v}{\partial \tilde{y}} \end{bmatrix} + \begin{bmatrix} \frac{\partial c_{ji}}{\partial \tilde{x}} & \frac{\partial c_{jj}}{\partial \tilde{y}} \end{bmatrix} \begin{bmatrix} v \\ v \end{bmatrix} \right) d\tilde{x}d\tilde{y}. \tag{15.45}$$

By considering in Equation (15.45) the current position $(\tilde{x}, \tilde{y})$ of the solid's material points, it is ensured that the equilibrium is written for the deformed solid, thus representing the large displacements (geometric nonlinearity) exactly.

15.11 Examples

Example 1. A solid body of square shape (initially), see Figure 15.22, has moved (deformed) in space. The current position of the body is described by:

$$\tilde{x} = 4 + 2\bar{x}$$
$$\tilde{y} = 3 + \bar{y}. \tag{15.46}$$

Sketch the deformed shape of the body.

Solution: the solution is shown in Figure 15.23.

Example 2. The Cauchy stress is constant over the solid body from example 1 and is given by

$$\mathbf{C} = \begin{bmatrix} 100 & 0 \\ 0 & 0 \end{bmatrix}. \tag{15.47}$$

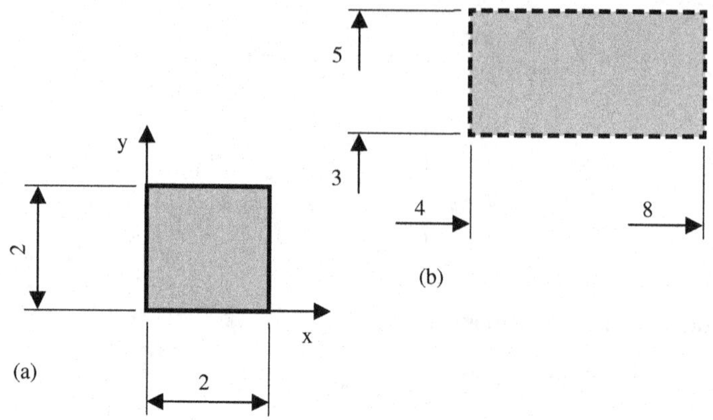

Figure 15.22 The initial position of a square-shaped solid body

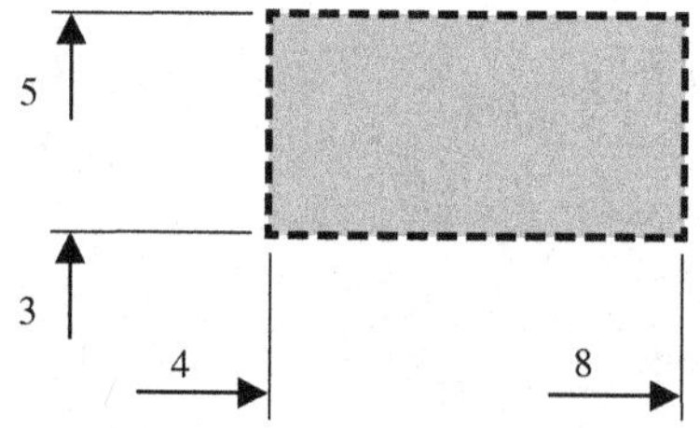

Figure 15.23 The current position of the initially square-shaped solid body

Calculate the virtual work of the internal forces on a virtual displacement given by

$$u=\tilde{x}\ \text{im}, \tag{15.48}$$

where im stands for a infinitesimally small length unit.

Solution: The virtual work is given by (see Equation (15.45))

$$\begin{aligned} dW_i &= \left(c_{ii}\frac{\partial u}{\partial \tilde{x}} + c_{ij}\frac{\partial u}{\partial \tilde{y}} + \frac{\partial c_{ii}}{\partial \tilde{x}}u + \frac{\partial c_{ij}}{\partial \tilde{y}}u \right) d\tilde{x}d\tilde{y} \\ &= \left(100\frac{\partial(\tilde{x})}{\partial \tilde{x}} + 0\frac{\partial(\tilde{x})}{\partial \tilde{y}} + \frac{\partial 100}{\partial \tilde{x}}(\tilde{x}) + \frac{\partial 0}{\partial \tilde{y}}(\tilde{x}) \right) d\tilde{x}d\tilde{y} \\ &= (100\cdot 1 + 0\cdot 0 + 0\cdot \tilde{x} + 0\cdot \tilde{x})d\tilde{x}d\tilde{y} \\ &= 100\, d\tilde{x}d\tilde{y}. \end{aligned} \tag{15.49}$$

The total virtual work is given by Equation (15.45)

$$W = \iint_{V_e} dW_i \, d\tilde{x} d\tilde{y}$$
$$= \int_{\tilde{y}=3}^{\tilde{y}=5} \int_{\tilde{x}=4}^{\tilde{x}=8} 100 \, d\tilde{x} d\tilde{y} = 100 \cdot 4 \cdot 2 = 800. \tag{15.50}$$

Example 3. For the Cauchy stress given by Equation (15.47) calculate the virtual work of the internal forces on virtual displacements

$$u = 1 \text{ im}. \tag{15.51}$$

Solution: The virtual work is given by (see Equation (15.45))

$$dW_i = \left(c_{ii} \frac{\partial u}{\partial \tilde{x}} + c_{ij} \frac{\partial u}{\partial \tilde{y}} + \frac{\partial c_{ii}}{\partial \tilde{x}} u + \frac{\partial c_{ij}}{\partial \tilde{y}} u \right) d\tilde{x} d\tilde{y}$$
$$= \left(100 \frac{\partial 1}{\partial \tilde{x}} + 0 \frac{\partial 1}{\partial \tilde{y}} + \frac{\partial 100}{\partial \tilde{x}} 1 + \frac{\partial 0}{\partial \tilde{y}} 1 \right) d\tilde{x} d\tilde{y} \tag{15.52}$$
$$= (100 \cdot 0 + 0 \cdot 0 + 0 \cdot 1 + 0 \cdot 1) d\tilde{x} d\tilde{y}$$
$$= 0.$$

The total virtual work is given by Equation (15.45)

$$W = \iint_{V_e} 0 \, d\tilde{x} d\tilde{y}$$
$$= \int_{\tilde{y}=3}^{\tilde{y}=5} \int_{\tilde{x}=4}^{\tilde{x}=8} 0 \, d\tilde{x} d\tilde{y} = 0 \cdot 4 \cdot 2 = 0. \tag{15.53}$$

Example 4. The Cauchy stress over the solid body is given by

$$\mathbf{C} = \begin{bmatrix} 100\tilde{x} & 0 \\ 0 & 0 \end{bmatrix}. \tag{15.54}$$

Calculate the virtual work of the internal forces on a virtual displacement

$$u = 1 \text{ im}. \tag{15.55}$$

Solution: The virtual work is given by (see Equation (15.45))

$$dW_i = \left(c_{ii}\frac{\partial u}{\partial \tilde{x}} + c_{ij}\frac{\partial u}{\partial \tilde{y}} + \frac{\partial c_{ii}}{\partial \tilde{x}}u + \frac{\partial c_{ij}}{\partial \tilde{y}}u \right) d\tilde{x}d\tilde{y}$$

$$= \left((100\tilde{x})\frac{\partial 1}{\partial \tilde{x}} + 0\frac{\partial 1}{\partial \tilde{y}} + \frac{\partial (100\tilde{x})}{\partial \tilde{x}}1 + \frac{\partial 0}{\partial \tilde{y}}1 \right) d\tilde{x}d\tilde{y} \tag{15.56}$$

$$= (100\tilde{x}\cdot 0 + 0\cdot 0 + 100\cdot 1 + 0\cdot 1)d\tilde{x}d\tilde{y}$$

$$= 100\, d\tilde{x}d\tilde{y}.$$

The total virtual work is given by Equation (15.45)

$$W = \iint_{V_e} 100\, d\tilde{x}d\tilde{y}$$

$$= \int_{\tilde{y}=3}^{\tilde{y}=5}\int_{\tilde{x}=4}^{\tilde{x}=8} 100\, d\tilde{x}d\tilde{y} = 100\cdot 4\cdot 2 = 800. \tag{15.57}$$

15.12 Summary

In this chapter, first nonhomogeneous and anisotropic solids were introduced. In order to describe the kinematics of solid deformation in the infinitesimal vicinity of a given point P a nonsquare shaped infinitesimal material element (generalized material element) has been introduced.

Through the generalized material element, the kinematics of the material deformation is described by using the initial, the current and the previous positions of the solid-embedded vector base and the material-embedded vector base.

In order to formulate the equilibrium, the Cauchy material element has been introduced (consisting of material points in their current position) with the Cauchy stress tensor matrix. This has been combined with virtual displacements, thus producing the virtual work of internal forces.

This virtual work is ready to be used in any equilibrium considerations of the deformed solid. As such, it is used in the rest of the book to consider equilibrium of finite elements.

Further Reading

[1] Bathe, K. J. (1982) *Finite Element Procedures Engineering Analysis*. New Jersey, Prentice-Hall.

[2] Bathe, K. J. (1996) *Finite Element Procedures*. New Jersey, Prentice-Hall.

[3] Bonet, J. and Burton, A. J. (1998) A simple orthotropic, transversely isotropic hyperelastic constitutive equation for large strain computations. *Comput. Method Appl. M.*, **162**(1–4): 151–64.

[4] Bonet, J. and Wood, R. D. (1997) *Nonlinear Continuum Mechanics for Finite Element Analysis*. Cambridge, Cambridge University Press.

[5] de Souza Neto, E. A. and Peric, D. (1996) A computational framework for a class of fully coupled models for elastoplastic damage at finite strains with reference to the linearization aspects. *Comput. Method Appl. M.*, **130**(1–2):179–93.

[6] de Souza Neto, E. A., Pires, F. M. A. and Owen, D. R. J. (2005) F-bar-based linear triangles and tetrahedra for finite strain analysis of nearly incompressible solids. Part I: formulation and benchmarking. *International Journal for Numerical Methods in Engineering*, **62**(3), 353–83.
[7] de Souza Neto, E. A., Peric, D., Huang, G. C. and Owen, D. R. J. (1995) Remarks on the stability of enhanced strain elements in finite elasticity and elastoplasticity. *Communications in Numerical Methods in Engineering*, **11**(11): 951–61.
[8] Dvorkin, E. N. and Goldschmit, M. B. (2005) *Nonlinear Continua*, New York, Springer.
[9] Green, A. E. and Naghdi, P. M. (1971) Some remarks on elastic-plastic deformation at finite strain. *International Journal of Engineering Science*, **9**(12), 1219–29.
[10] Gurtin, M. E. (1981) *An Introduction to Continuum Mechanics*. New York, Academic Press.
[11] Hill, R. (1979) *Aspects of Invariance in Solid Mechanics*, ed. Y. Chia-Shun, Elsevier, pp. 1–75.
[12] Hughes, T.J.R. (2000) *The Finite Element Method: Linear Static and Dynamic Finite Element Analysis*. New York, Dover Publications.
[13] Malvern, L. E. (1969) *Introduction to the Mechanics of a Continuous Medium*. New York, Prentice Hall.
[14] Munjiza, A. (2004) *The Combined Finite-Discrete Element Method*. Chichester, John Wiley & Sons, Ltd.
[15] Peric, D. (1992) On consistent stress rates in solid mechanics – computational implications. *International Journal for Numerical Methods in Engineering*, **33**(4), 799–817.
[16] Peric, D. and Owen, D. R. J. (1998) Finite-element applications to the nonlinear mechanics of solids. *Reports on Progress in Physics*, **61**(11), 1495–1574.
[17] Simo, J. C. (1985) On the computational significance of the intermediate configuration and hyperelastic stress relations in finite deformation elastoplasticity. *Mech. Mater.*, **4**(3–4), 439–51.
[18] Simo, J. C. (1988) A framework for finite strain elastoplasticity based on maximum plastic dissipation and the multiplicative decomposition. 1. Continuum formulation. *Comput. Method Appl. M.*, **66**(2), 199–219.
[19] Simo, J. C. (1988), A framework for finite strain elastoplasticity based on maximum plastic dissipation and the multiplicative decomposition. 2. Computational aspects. *Comput. Method Appl. M.*, **68**(1), 1–31.
[20] Simo, J. C. and Ortiz, M. (1985) A unified approach to finite deformation elastoplastic analysis based on the use of hyperelastic constitutive equations. *Comput. Method Appl. M.*, **49**(2), 221–45.
[21] Simo, J. C. and Ju, J. W. (1989) Finite deformation damage-elastoplasticity: a non conventional framework. *International Journal of Computational Mechanics*, **5**, 375–400.
[22] Simo, J. C. and Hughes, T. J. R. (1998) *Computational Inelasticity*. New York, Springer.
[23] Simo, J. C., Armero, F. and Taylor, R. L. (1993) Improved versions of assumed enhanced strain tri-linear elements for 3D finite deformation problems. *Comput. Method Appl. M.*, **110**(3–4) 359–86.
[24] Zienkiewicz, O. C. (1971) *The Finite Element Method in Engineering Science*. New York, McGraw-Hill.
[25] Zienkiewicz, O. C. and Taylor, R. L. (2005) *The Finite Element Method Set*. Oxford, Elsevier Science.

16

Kinematics of Deformation in 3D

16.1 The Cartesian Coordinate System in 3D

The Cartesian coordinate system in 3D is defined by three straight-line orthogonal axes (x, y, z), Figure 16.1.

For the right hand system, the thumb of the right hand faces in the z direction when the index finger of the right hand faces in the x direction and the surface of the palm of the right hand is perpendicular to the y direction.

For the left hand system the situation is the same, but utilizing the left hand, see Figure 16.2.

16.2 The Solid-Embedded Coordinate System

Apart from the global Cartesian coordinate system, it is often convenient to introduce another coordinate system called the solid-embedded coordinate system

$$(\xi, \eta, \zeta), \tag{16.1}$$

such that the coordinate surfaces

$$\begin{aligned} \xi &= \text{const} \\ \eta &= \text{const} \\ \zeta &= \text{const}. \end{aligned} \tag{16.2}$$

are fixed to the solid's material points and as such, they move with the solid, Figure 16.3.

Large Strain Finite Element Method: A Practical Course, First Edition. Antonio Munjiza, Esteban Rougier and Earl E. Knight.

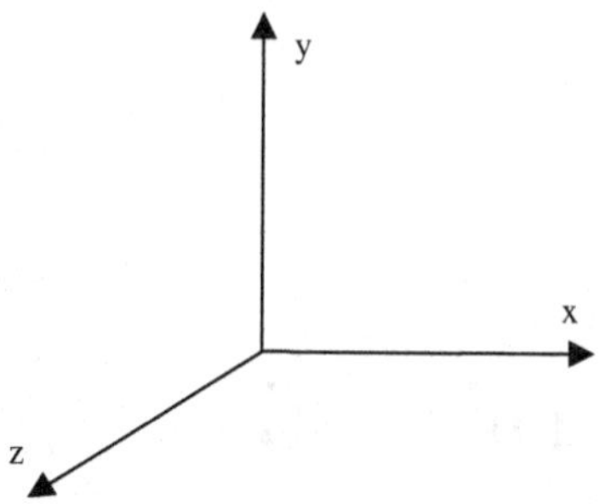

Figure 16.1 The right hand Cartesian coordinate system

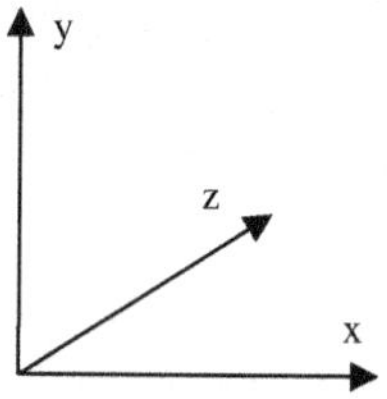

Figure 16.2 The left hand Cartesian coordinate system

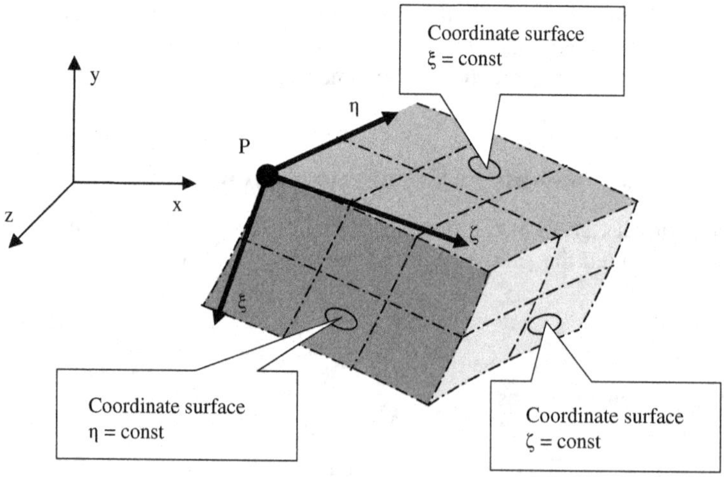

Figure 16.3 The solid-embedded coordinate system, with both coordinate surfaces and coordinate lines being shown

The coordinate lines ξ, η, and ζ are obtained by respectively setting

$$
\begin{aligned}
&\eta = \text{const}, \zeta = \text{const for coordinate line } \xi \\
&\xi = \text{const}, \zeta = \text{const for coordinate line } \eta \\
&\xi = \text{const}, \eta = \text{const for coordinate line } \zeta.
\end{aligned}
\tag{16.3}
$$

The coordinate lines are also fixed to the material points of the solid. As such, they move (deform) with the solid.

16.3 The Global and the Solid-Embedded Vector Bases

In order to represent vectors in 3D, the global $[\mathbf{i} \ \mathbf{j} \ \mathbf{k}]$ vector base is used. In addition, for a given material point P, it is convenient to introduce a local infinitesimal solid-embedded vector base

$$[\boldsymbol{\xi} \ \boldsymbol{\eta} \ \boldsymbol{\zeta}](\text{im}). \tag{16.4}$$

This vector base consists of three infinitesimally small (thus "infi-meter", im) base vectors $\boldsymbol{\xi}$, $\boldsymbol{\eta}$ and $\boldsymbol{\zeta}$ that are fixed to the material points in the infinitesimal vicinity of a given material point

$$\text{P} = (\xi, \eta, \zeta). \tag{16.5}$$

Vectors $\boldsymbol{\xi}$, $\boldsymbol{\eta}$ and $\boldsymbol{\zeta}$ are respectively made to be tangential to the coordinate lines ξ, η, and ζ through point P, Figure 16.4.

As the coordinate lines are fixed to the solid's material points, it follows that the tangents to these lines are also fixed to the material points in the infinitesimal vicinity of point P. One way to define these vectors is:

$$\begin{aligned}
\boldsymbol{\xi} &= \left(\frac{\partial x}{\partial \xi}\mathbf{i} + \frac{\partial y}{\partial \xi}\mathbf{j} + \frac{\partial z}{\partial \xi}\mathbf{k}\right) \text{im} \\
\boldsymbol{\eta} &= \left(\frac{\partial x}{\partial \eta}\mathbf{i} + \frac{\partial y}{\partial \eta}\mathbf{j} + \frac{\partial z}{\partial \eta}\mathbf{k}\right) \text{im} \\
\boldsymbol{\zeta} &= \left(\frac{\partial x}{\partial \zeta}\mathbf{i} + \frac{\partial y}{\partial \zeta}\mathbf{j} + \frac{\partial z}{\partial \zeta}\mathbf{k}\right) \text{im},
\end{aligned} \tag{16.6}$$

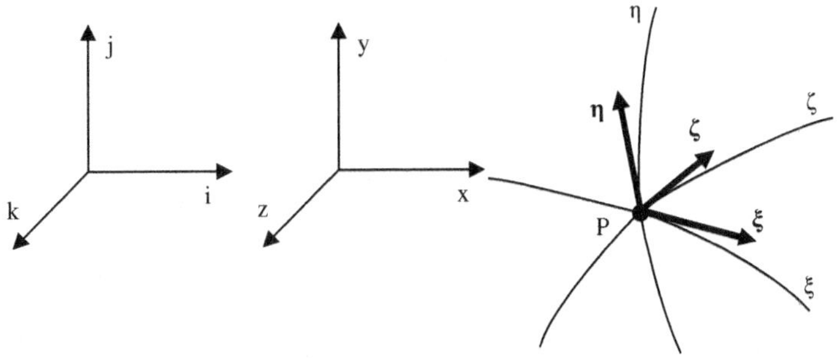

Figure 16.4 Vector base fixed to the material points in the infinitesimal vicinity of point P. This vector base is called the solid-embedded vector base

which for $\boldsymbol{\xi}$ can be interpreted as the displacement from point P due to an infinitesimal change in ξ only. In a similar manner, for $\boldsymbol{\eta}$ and $\boldsymbol{\zeta}$ these also mean that the displacement from point P is due to infinitesimal changes of η and ζ respectively.

Equation (16.6) can be written in a matrix form,

$$[\boldsymbol{\xi}\ \boldsymbol{\eta}\ \boldsymbol{\zeta}]=[\mathbf{i}\ \mathbf{j}\ \mathbf{k}]\begin{bmatrix}\xi_i & \eta_i & \zeta_i\\ \xi_j & \eta_j & \zeta_j\\ \xi_k & \eta_k & \zeta_k\end{bmatrix}=\begin{bmatrix}\xi_i & \eta_i & \zeta_i\\ \xi_j & \eta_j & \zeta_j\\ \xi_k & \eta_k & \zeta_k\end{bmatrix}$$

$$=\begin{bmatrix}\frac{\partial x}{\partial \xi} & \frac{\partial x}{\partial \eta} & \frac{\partial x}{\partial \zeta}\\ \frac{\partial y}{\partial \xi} & \frac{\partial y}{\partial \eta} & \frac{\partial y}{\partial \zeta}\\ \frac{\partial z}{\partial \xi} & \frac{\partial z}{\partial \eta} & \frac{\partial z}{\partial \zeta}\end{bmatrix}, \tag{16.7}$$

where each column of the matrix represents one base vector.

16.4 Deformation of the Solid

As the solid deforms, its material points move in space. In this manner, the global coordinates (x, y, z) of any given material point change with time. Of particular interest are the initial, previous and current positions of a given material point P. These positions are respectively given by

$$\bar{\mathrm{P}}=(\bar{x}, \bar{y}, \bar{z})\ ;\quad \hat{\mathrm{P}}=(\hat{x}, \hat{y}, \hat{z})\quad \text{and}\quad \tilde{\mathrm{P}}=(\tilde{x}, \tilde{y}, \tilde{z}). \tag{16.8}$$

In contrast to the global coordinates of the material points (which change with deformation of the solid), the local coordinates of any material point P do not change with the deformation of the solid. As a result, the initial, previous, and current positions of any given material point P are

$$\begin{aligned}\bar{\mathrm{P}}&=(\bar{\xi}, \bar{\eta}, \bar{\zeta})=(\xi, \eta, \zeta)\\ \hat{\mathrm{P}}&=(\hat{\xi}, \hat{\eta}, \hat{\zeta})=(\xi, \eta, \zeta)\\ \tilde{\mathrm{P}}&=(\tilde{\xi}, \tilde{\eta}, \tilde{\zeta})=(\xi, \eta, \zeta).\end{aligned} \tag{16.9}$$

Nevertheless, the relationship between the global and the local coordinates changes as the solid deforms, thus

$$\begin{aligned}x&=\mathrm{x}(\xi, \eta, \zeta)\\ y&=\mathrm{y}(\xi, \eta, \zeta)\\ z&=\mathrm{z}(\xi, \eta, \zeta)\ ,\end{aligned} \tag{16.10}$$

which yields the initial, previous, and current coordinates as a function of ξ, η and ζ:

$$\begin{aligned}\bar{x} &= \bar{\mathrm{x}}(\xi, \eta, \zeta)\\ \bar{y} &= \bar{\mathrm{y}}(\xi, \eta, \zeta)\\ \bar{z} &= \bar{\mathrm{z}}(\xi, \eta, \zeta)\end{aligned} \tag{16.11}$$

$$\begin{aligned}\hat{x} &= \hat{\mathrm{x}}(\xi, \eta, \zeta)\\ \hat{y} &= \hat{\mathrm{y}}(\xi, \eta, \zeta)\\ \hat{z} &= \hat{\mathrm{z}}(\xi, \eta, \zeta)\end{aligned} \tag{16.12}$$

$$\begin{aligned}\tilde{x} &= \tilde{\mathrm{x}}(\xi, \eta, \zeta)\\ \tilde{y} &= \tilde{\mathrm{y}}(\xi, \eta, \zeta)\\ \tilde{z} &= \tilde{\mathrm{z}}(\xi, \eta, \zeta)\ .\end{aligned} \tag{16.13}$$

In the above equations "–" stands for "initial", "^" stands for "previous" and "~" stands for current.

The solid-embedded vector base $[\boldsymbol{\xi}\ \ \boldsymbol{\eta}\ \ \boldsymbol{\zeta}]$ changes with the solid's deformation, giving the initial position of the vector base $[\bar{\boldsymbol{\xi}}\ \ \bar{\boldsymbol{\eta}}\ \ \bar{\boldsymbol{\zeta}}]$

$$\begin{aligned}\bar{\boldsymbol{\xi}} &= \frac{\partial \bar{x}}{\partial \xi}\mathbf{i} + \frac{\partial \bar{y}}{\partial \xi}\mathbf{j} + \frac{\partial \bar{z}}{\partial \xi}\mathbf{k}\\ \bar{\boldsymbol{\eta}} &= \frac{\partial \bar{x}}{\partial \eta}\mathbf{i} + \frac{\partial \bar{y}}{\partial \eta}\mathbf{j} + \frac{\partial \bar{z}}{\partial \eta}\mathbf{k}\\ \bar{\boldsymbol{\zeta}} &= \frac{\partial \bar{x}}{\partial \zeta}\mathbf{i} + \frac{\partial \bar{y}}{\partial \zeta}\mathbf{j} + \frac{\partial \bar{z}}{\partial \zeta}\mathbf{k},\end{aligned} \tag{16.14}$$

which can be written in a matrix form

$$[\bar{\boldsymbol{\xi}}\ \ \bar{\boldsymbol{\eta}}\ \ \bar{\boldsymbol{\zeta}}] = [\mathbf{i}\ \ \mathbf{j}\ \ \mathbf{k}]\begin{bmatrix}\frac{\partial \bar{x}}{\partial \xi} & \frac{\partial \bar{x}}{\partial \eta} & \frac{\partial \bar{x}}{\partial \zeta}\\ \frac{\partial \bar{y}}{\partial \xi} & \frac{\partial \bar{y}}{\partial \eta} & \frac{\partial \bar{y}}{\partial \zeta}\\ \frac{\partial \bar{z}}{\partial \xi} & \frac{\partial \bar{z}}{\partial \eta} & \frac{\partial \bar{z}}{\partial \zeta}\end{bmatrix} = \begin{bmatrix}\frac{\partial \bar{x}}{\partial \xi} & \frac{\partial \bar{x}}{\partial \eta} & \frac{\partial \bar{x}}{\partial \zeta}\\ \frac{\partial \bar{y}}{\partial \xi} & \frac{\partial \bar{y}}{\partial \eta} & \frac{\partial \bar{y}}{\partial \zeta}\\ \frac{\partial \bar{z}}{\partial \xi} & \frac{\partial \bar{z}}{\partial \eta} & \frac{\partial \bar{z}}{\partial \zeta}\end{bmatrix}. \tag{16.15}$$

In a similar manner, the previous position of the solid-embedded base, $[\hat{\boldsymbol{\xi}}\ \ \hat{\boldsymbol{\eta}}\ \ \hat{\boldsymbol{\zeta}}]$, is given by

$$[\hat{\boldsymbol{\xi}}\ \ \hat{\boldsymbol{\eta}}\ \ \hat{\boldsymbol{\zeta}}] = [\mathbf{i}\ \ \mathbf{j}\ \ \mathbf{k}]\begin{bmatrix}\frac{\partial \hat{x}}{\partial \xi} & \frac{\partial \hat{x}}{\partial \eta} & \frac{\partial \hat{x}}{\partial \zeta}\\ \frac{\partial \hat{y}}{\partial \xi} & \frac{\partial \hat{y}}{\partial \eta} & \frac{\partial \hat{y}}{\partial \zeta}\\ \frac{\partial \hat{z}}{\partial \xi} & \frac{\partial \hat{z}}{\partial \eta} & \frac{\partial \hat{z}}{\partial \zeta}\end{bmatrix} = \begin{bmatrix}\frac{\partial \hat{x}}{\partial \xi} & \frac{\partial \hat{x}}{\partial \eta} & \frac{\partial \hat{x}}{\partial \zeta}\\ \frac{\partial \hat{y}}{\partial \xi} & \frac{\partial \hat{y}}{\partial \eta} & \frac{\partial \hat{y}}{\partial \zeta}\\ \frac{\partial \hat{z}}{\partial \xi} & \frac{\partial \hat{z}}{\partial \eta} & \frac{\partial \hat{z}}{\partial \zeta}\end{bmatrix}. \tag{16.16}$$

Also, the current position of the solid-embedded vector base is obtained from the current position of the material points

$$\begin{bmatrix}\tilde{\boldsymbol{\xi}} & \tilde{\boldsymbol{\eta}} & \tilde{\boldsymbol{\zeta}}\end{bmatrix} = \begin{bmatrix}\mathbf{i} & \mathbf{j} & \mathbf{k}\end{bmatrix}\begin{bmatrix}\frac{\partial\tilde{x}}{\partial\xi} & \frac{\partial\tilde{x}}{\partial\eta} & \frac{\partial\tilde{x}}{\partial\zeta}\\ \frac{\partial\tilde{y}}{\partial\xi} & \frac{\partial\tilde{y}}{\partial\eta} & \frac{\partial\tilde{y}}{\partial\zeta}\\ \frac{\partial\tilde{z}}{\partial\xi} & \frac{\partial\tilde{z}}{\partial\eta} & \frac{\partial\tilde{z}}{\partial\zeta}\end{bmatrix} = \begin{bmatrix}\frac{\partial\tilde{x}}{\partial\xi} & \frac{\partial\tilde{x}}{\partial\eta} & \frac{\partial\tilde{x}}{\partial\zeta}\\ \frac{\partial\tilde{y}}{\partial\xi} & \frac{\partial\tilde{y}}{\partial\eta} & \frac{\partial\tilde{y}}{\partial\zeta}\\ \frac{\partial\tilde{z}}{\partial\xi} & \frac{\partial\tilde{z}}{\partial\eta} & \frac{\partial\tilde{z}}{\partial\zeta}\end{bmatrix}. \tag{16.17}$$

16.5 Generalized Material Element

In order to define the properties of an anisotropic material, for each material point P, a local infinitesimal material element is defined. This element has the shape of a parallelepiped of infinitesimally small size.

When compared to the familiar cube-shaped material element, this element is defined by three nonorthogonal nonunit vectors $[\boldsymbol{\alpha}\ \ \boldsymbol{\beta}\ \ \boldsymbol{\gamma}]$. In order to distinguish the two, the second is called the generalized material element, while the first one is always explicitly referred to as the cube-shaped material element, Figure 16.5.

The geometry of the generalized material element is uniquely defined by three vectors $\boldsymbol{\alpha}$, $\boldsymbol{\beta}$ and $\boldsymbol{\gamma}$ forming a vector base

$$[\boldsymbol{\alpha}\ \ \boldsymbol{\beta}\ \ \boldsymbol{\gamma}](\mathrm{im}). \tag{16.18}$$

The vectors are fixed (glued) to the solid's material points in the vicinity of a given material point P and as such they deform with the material.

In a sense, it is like plotting an infinitesimal (very small) parallelepiped at point P. As the material points at point P move, so does the parallelepiped. In other words, this parallelepiped translates, rotates and stretches with the material in an infinitesimally close vicinity to point P, i.e., it "measures" (characterizes) the solid's deformation at point P. Because the size of the generalized material element is infinitesimally small, its surfaces stay flat and parallel to each other; i.e., its shape is always a parallelepiped.

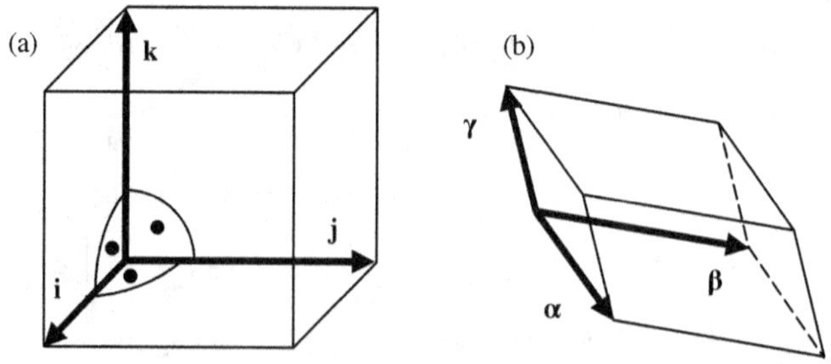

Figure 16.5 (a) Cube-shaped material element; (b) generalized material element

The generalized material-fixed vector base, $[\boldsymbol{\alpha}\ \ \boldsymbol{\beta}\ \ \boldsymbol{\gamma}]$, is supplied as input to the analysis. For example, a fiber reinforced composite may have the generalized material-embedded vector base in the direction of the fibers.

It is worth noting that for non-homogeneous materials, the generalized material element will change from one material point to the next.

16.6 Kinematic of Deformation in 3D

Generalized material-embedded vectors are fixed to the material points in the infinitesimal vicinity of a given material point P. At the same time, the solid-embedded vector base is also fixed to the material points in the infinitesimal vicinity of the same point P. In other words, bases

$$[\boldsymbol{\alpha}\ \ \boldsymbol{\beta}\ \ \boldsymbol{\gamma}] \tag{16.19}$$

and

$$[\boldsymbol{\xi}\ \ \boldsymbol{\eta}\ \ \boldsymbol{\zeta}] \tag{16.20}$$

are both fixed to the material points in the infinitesimal vicinity of point P and they both deform (move) with the material points in the infinitesimal vicinity of point P.

As a consequence, it is possible to express vectors of one base with vectors of the other base, as follows

$$\begin{aligned}
\boldsymbol{\alpha} &= \alpha_\xi \boldsymbol{\xi} + \alpha_\eta \boldsymbol{\eta} + \alpha_\zeta \boldsymbol{\zeta} \\
\boldsymbol{\beta} &= \beta_\xi \boldsymbol{\xi} + \beta_\eta \boldsymbol{\eta} + \beta_\zeta \boldsymbol{\zeta} \\
\boldsymbol{\gamma} &= \gamma_\xi \boldsymbol{\xi} + \gamma_\eta \boldsymbol{\eta} + \gamma_\zeta \boldsymbol{\zeta}.
\end{aligned} \tag{16.21}$$

The above equations can be written in matrix form

$$[\boldsymbol{\alpha}\ \ \boldsymbol{\beta}\ \ \boldsymbol{\gamma}] = [\boldsymbol{\xi}\ \ \boldsymbol{\eta}\ \ \boldsymbol{\zeta}] \begin{bmatrix} \alpha_\xi & \beta_\xi & \gamma_\xi \\ \alpha_\eta & \beta_\eta & \gamma_\eta \\ \alpha_\zeta & \beta_\zeta & \gamma_\zeta \end{bmatrix}, \tag{16.22}$$

or simply

$$[\boldsymbol{\alpha}\ \ \boldsymbol{\beta}\ \ \boldsymbol{\gamma}] = \begin{bmatrix} \alpha_\xi & \beta_\xi & \gamma_\xi \\ \alpha_\eta & \beta_\eta & \gamma_\eta \\ \alpha_\zeta & \beta_\zeta & \gamma_\zeta \end{bmatrix}, \tag{16.23}$$

where, in Equation (16.23), the solid embedded base

$$[\boldsymbol{\xi}\ \ \boldsymbol{\eta}\ \ \boldsymbol{\zeta}] \tag{16.24}$$

is implicitly assumed.

From the above, it follows that the initial position of the generalized material-embedded vectors is given by

$$\begin{aligned}\begin{bmatrix}\bar{\boldsymbol{\alpha}} & \bar{\boldsymbol{\beta}} & \bar{\boldsymbol{\gamma}}\end{bmatrix} &= \begin{bmatrix}\mathbf{i} & \mathbf{j} & \mathbf{k}\end{bmatrix}\begin{bmatrix}\bar{\alpha}_i & \bar{\beta}_i & \bar{\gamma}_i \\ \bar{\alpha}_j & \bar{\beta}_j & \bar{\gamma}_j \\ \bar{\alpha}_k & \bar{\beta}_k & \bar{\gamma}_k\end{bmatrix} \\ &= \begin{bmatrix}\bar{\boldsymbol{\xi}} & \bar{\boldsymbol{\eta}} & \bar{\boldsymbol{\zeta}}\end{bmatrix}\begin{bmatrix}\alpha_\xi & \beta_\xi & \gamma_\xi \\ \alpha_\eta & \beta_\eta & \gamma_\eta \\ \alpha_\zeta & \beta_\zeta & \gamma_\zeta\end{bmatrix}.\end{aligned} \tag{16.25}$$

After substitution of the solid-embedded vector base's initial position

$$\begin{bmatrix}\bar{\boldsymbol{\xi}} & \bar{\boldsymbol{\eta}} & \bar{\boldsymbol{\zeta}}\end{bmatrix} = \begin{bmatrix}\mathbf{i} & \mathbf{j} & \mathbf{k}\end{bmatrix}\begin{bmatrix}\bar{\xi}_i & \bar{\eta}_i & \bar{\zeta}_i \\ \bar{\xi}_j & \bar{\eta}_j & \bar{\zeta}_j \\ \bar{\xi}_k & \bar{\eta}_k & \bar{\zeta}_k\end{bmatrix}, \tag{16.26}$$

one obtains

$$\begin{bmatrix}\bar{\alpha}_i & \bar{\beta}_i & \bar{\gamma}_i \\ \bar{\alpha}_j & \bar{\beta}_j & \bar{\gamma}_j \\ \bar{\alpha}_k & \bar{\beta}_k & \bar{\gamma}_k\end{bmatrix} = \begin{bmatrix}\bar{\xi}_i & \bar{\eta}_i & \bar{\zeta}_i \\ \bar{\xi}_j & \bar{\eta}_j & \bar{\zeta}_j \\ \bar{\xi}_k & \bar{\eta}_k & \bar{\zeta}_k\end{bmatrix}\begin{bmatrix}\alpha_\xi & \beta_\xi & \gamma_\xi \\ \alpha_\eta & \beta_\eta & \gamma_\eta \\ \alpha_\zeta & \beta_\zeta & \gamma_\zeta\end{bmatrix}. \tag{16.27}$$

Since

$$\begin{bmatrix}\bar{\alpha}_i & \bar{\beta}_i & \bar{\gamma}_i \\ \bar{\alpha}_j & \bar{\beta}_j & \bar{\gamma}_j \\ \bar{\alpha}_k & \bar{\beta}_k & \bar{\gamma}_k\end{bmatrix} \tag{16.28}$$

is provided as an input to the analysis, it is known and is therefore used to calculate

$$\begin{bmatrix}\alpha_\xi & \beta_\xi & \gamma_\xi \\ \alpha_\eta & \beta_\eta & \gamma_\eta \\ \alpha_\zeta & \beta_\zeta & \gamma_\zeta\end{bmatrix} = \begin{bmatrix}\bar{\xi}_i & \bar{\eta}_i & \bar{\zeta}_i \\ \bar{\xi}_j & \bar{\eta}_j & \bar{\zeta}_j \\ \bar{\xi}_k & \bar{\eta}_k & \bar{\zeta}_k\end{bmatrix}^{-1}\begin{bmatrix}\bar{\alpha}_i & \bar{\beta}_i & \bar{\gamma}_i \\ \bar{\alpha}_j & \bar{\beta}_j & \bar{\gamma}_j \\ \bar{\alpha}_k & \bar{\beta}_k & \bar{\gamma}_k\end{bmatrix}. \tag{16.29}$$

By using Equation (16.29), the previous position of the generalized material-embedded base vectors is then obtained as follows

$$\begin{aligned}\begin{bmatrix}\hat{\boldsymbol{\alpha}} & \hat{\boldsymbol{\beta}} & \hat{\boldsymbol{\gamma}}\end{bmatrix} &= \begin{bmatrix}\mathbf{i} & \mathbf{j} & \mathbf{k}\end{bmatrix}\begin{bmatrix}\hat{\alpha}_i & \hat{\beta}_i & \hat{\gamma}_i \\ \hat{\alpha}_j & \hat{\beta}_j & \hat{\gamma}_j \\ \hat{\alpha}_k & \hat{\beta}_k & \hat{\gamma}_k\end{bmatrix} \\ &= \begin{bmatrix}\hat{\alpha}_i & \hat{\beta}_i & \hat{\gamma}_i \\ \hat{\alpha}_j & \hat{\beta}_j & \hat{\gamma}_j \\ \hat{\alpha}_k & \hat{\beta}_k & \hat{\gamma}_k\end{bmatrix} = \begin{bmatrix}\hat{\xi}_i & \hat{\eta}_i & \hat{\zeta}_i \\ \hat{\xi}_j & \hat{\eta}_j & \hat{\zeta}_j \\ \hat{\xi}_k & \hat{\eta}_k & \hat{\zeta}_k\end{bmatrix}\begin{bmatrix}\alpha_\xi & \beta_\xi & \gamma_\xi \\ \alpha_\eta & \beta_\eta & \gamma_\eta \\ \alpha_\zeta & \beta_\zeta & \gamma_\zeta\end{bmatrix}.\end{aligned} \tag{16.30}$$

In a similar manner, the current position of the generalized material-embedded vector base is given by

$$\begin{aligned}
[\tilde{\boldsymbol{\alpha}}\ \tilde{\boldsymbol{\beta}}\ \tilde{\boldsymbol{\gamma}}] &= [\mathbf{i}\ \mathbf{j}\ \mathbf{k}]\begin{bmatrix} \tilde{\alpha}_i & \tilde{\beta}_i & \tilde{\gamma}_i \\ \tilde{\alpha}_j & \tilde{\beta}_j & \tilde{\gamma}_j \\ \tilde{\alpha}_k & \tilde{\beta}_k & \tilde{\gamma}_k \end{bmatrix} \\
&= \begin{bmatrix} \tilde{\alpha}_i & \tilde{\beta}_i & \tilde{\gamma}_i \\ \tilde{\alpha}_j & \tilde{\beta}_j & \tilde{\gamma}_j \\ \tilde{\alpha}_k & \tilde{\beta}_k & \tilde{\gamma}_k \end{bmatrix} = \begin{bmatrix} \tilde{\xi}_i & \tilde{\eta}_i & \tilde{\zeta}_i \\ \tilde{\xi}_j & \tilde{\eta}_j & \tilde{\zeta}_j \\ \tilde{\xi}_k & \tilde{\eta}_k & \tilde{\zeta}_k \end{bmatrix} \begin{bmatrix} \alpha_\xi & \beta_\xi & \gamma_\xi \\ \alpha_\eta & \beta_\eta & \gamma_\eta \\ \alpha_\zeta & \beta_\zeta & \gamma_\zeta \end{bmatrix}.
\end{aligned} \tag{16.31}$$

These three vector bases (Equations (16.28), (16.30) and (16.31)) uniquely define the initial, the previous, and the current positions of the generalized material element. One can say that they define the kinematics of deformation in 3D. As such, they can be directly used in calculating the stress.

16.7 The Virtual Work of Internal Forces

At this point, it is convenient to introduce the Cauchy material element in 3D, which is an infinitesimally small cube taken in the vicinity of a given material point P by considering the solid's current position, Figure 16.6.

The deformation kinematics obtained in Section 16.6 induces internal forces between microstructural elements of the material (say atoms). These are represented by the stress tensor. The stress tensor is obtained from the deformation kinematics using the constitutive law (material law).

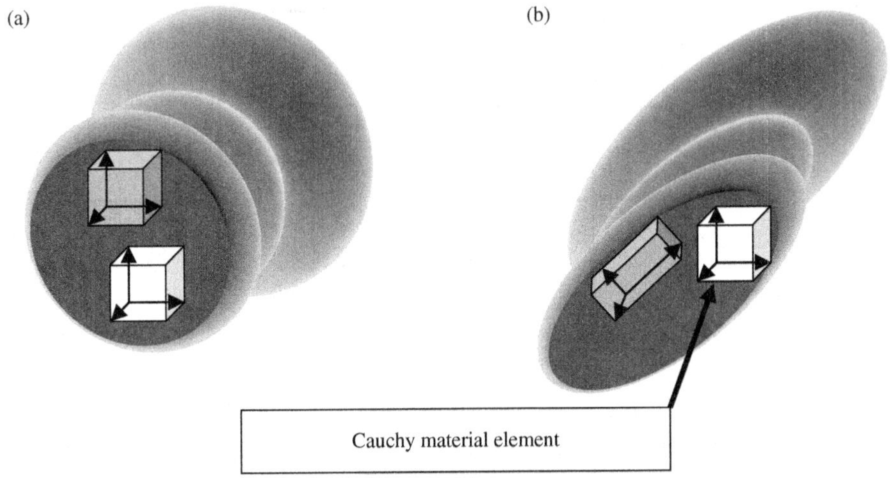

Figure 16.6 (a) Cube shaped material element taken at initial position. The same element has been deformed considerably in the current position. (b) The Cauchy material element is taken in the current position and is therefore a cube

The constitutive law takes as input the deformation kinematics as derived in the previous section and produces the internal forces on the surfaces of the Cauchy material element, i.e., it produces the Cauchy stress tensor matrix

$$\mathbf{C} = \begin{bmatrix} c_{ii} & c_{ij} & c_{ik} \\ c_{ji} & c_{jj} & c_{jk} \\ c_{ki} & c_{kj} & c_{kk} \end{bmatrix}. \tag{16.32}$$

It is worth noting that these are functions of the material point's current coordinates, i.e.,

$$\mathbf{C} = \begin{bmatrix} c_{ii}(\tilde{x},\tilde{y},\tilde{z}) & c_{ij}(\tilde{x},\tilde{y},\tilde{z}) & c_{ik}(\tilde{x},\tilde{y},\tilde{z}) \\ c_{ji}(\tilde{x},\tilde{y},\tilde{z}) & c_{jj}(\tilde{x},\tilde{y},\tilde{z}) & c_{jk}(\tilde{x},\tilde{y},\tilde{z}) \\ c_{ki}(\tilde{x},\tilde{y},\tilde{z}) & c_{kj}(\tilde{x},\tilde{y},\tilde{z}) & c_{kk}(\tilde{x},\tilde{y},\tilde{z}) \end{bmatrix}. \tag{16.33}$$

In order to consider equilibrium of the internal and external forces (and possibly inertia forces for dynamic problems) on the deformed solid, it is often convenient to use the principle of virtual work. Using the principle of virtual work involves the evaluation of the work of both internal and external (and inertia) forces on virtual displacements.

Work of Internal Forces on Virtual Displacement in the i Direction. If the virtual displacements have only **i** components, then only forces in the **i** direction will produce virtual work. The **i** force components are illustrated in Figure 16.7 where it can be seen that the internal forces change between two parallel surfaces of the Cauchy material element.

As explained above, one has to consider the equilibrium of internal and external forces. Equilibrium has, by definition, to be considered at current positions of the material points $(\tilde{x},\tilde{y},\tilde{z})$. The most often used equilibrium formulation is virtual work, which states that the total

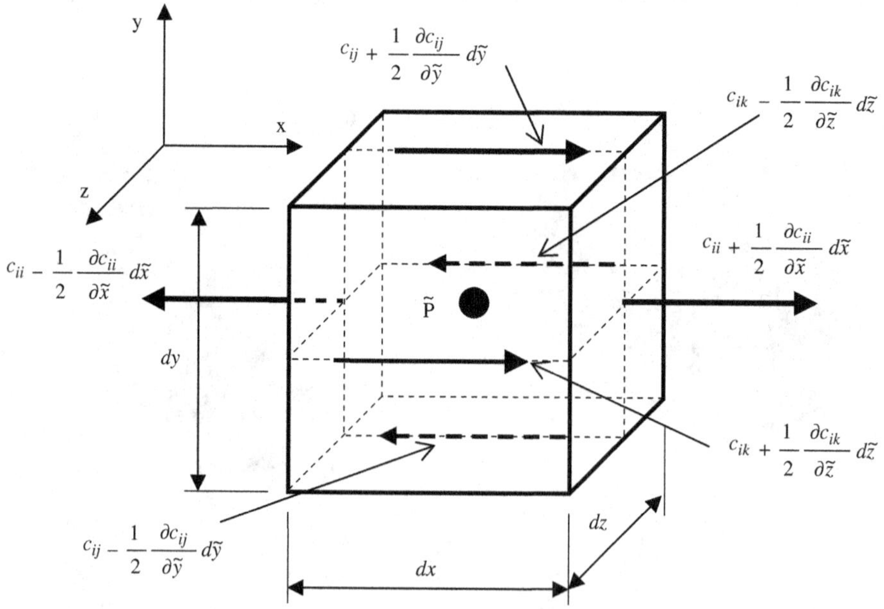

Figure 16.7 Cauchy stress tensor matrix – only the **i** components of internal forces on the surfaces of the Cauchy material element are shown

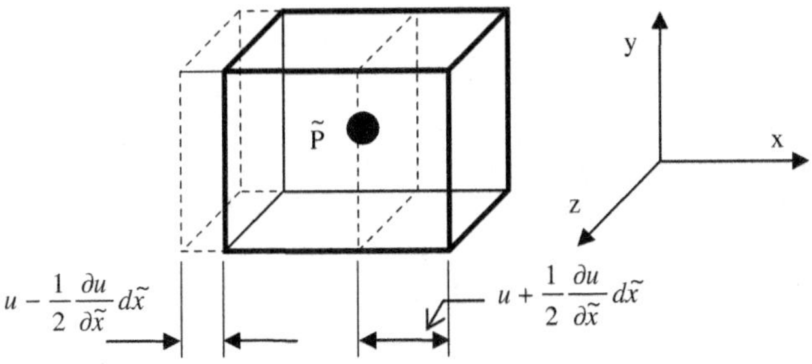

Figure 16.8 Virtual displacement of Cauchy material element in **i** direction

work of all forces on virtual displacements should be zero. This requires the evaluation of the work of all internal forces on the given virtual displacements. Virtual displacements (in the vicinity of point $\tilde{P}$) in the **i** direction are shown in Figure 16.8.

By inspection, the work done by the internal forces due to a virtual displacement in the **i** direction, $u=u(\tilde{x},\tilde{y},\tilde{z})$, Figure 16.8, is given by

$$\begin{aligned} dW_i = &-\left(c_{ii}-\frac{1}{2}\frac{\partial c_{ii}}{\partial \tilde{x}}d\tilde{x}\right)d\tilde{y}d\tilde{z}\left(u-\frac{1}{2}\frac{\partial u}{\partial \tilde{x}}d\tilde{x}\right)+\left(c_{ii}+\frac{1}{2}\frac{\partial c_{ii}}{\partial \tilde{x}}d\tilde{x}\right)d\tilde{y}d\tilde{z}\left(u+\frac{1}{2}\frac{\partial u}{\partial \tilde{x}}d\tilde{x}\right)\\ &-\left(c_{ij}-\frac{1}{2}\frac{\partial c_{ij}}{\partial \tilde{y}}d\tilde{y}\right)d\tilde{x}d\tilde{z}\left(u-\frac{1}{2}\frac{\partial u}{\partial \tilde{y}}d\tilde{y}\right)+\left(c_{ij}+\frac{1}{2}\frac{\partial c_{ij}}{\partial \tilde{y}}d\tilde{y}\right)d\tilde{x}d\tilde{z}\left(u+\frac{1}{2}\frac{\partial u}{\partial \tilde{y}}d\tilde{y}\right)\\ &-\left(c_{ik}-\frac{1}{2}\frac{\partial c_{ik}}{\partial \tilde{z}}d\tilde{z}\right)d\tilde{x}d\tilde{y}\left(u-\frac{1}{2}\frac{\partial u}{\partial \tilde{z}}d\tilde{z}\right)+\left(c_{ik}+\frac{1}{2}\frac{\partial c_{ik}}{\partial \tilde{z}}d\tilde{z}\right)d\tilde{x}d\tilde{y}\left(u+\frac{1}{2}\frac{\partial u}{\partial \tilde{z}}d\tilde{z}\right). \end{aligned} \tag{16.34}$$

Since the terms $d\tilde{x}d\tilde{x}d\tilde{y}d\tilde{z}$, $d\tilde{x}d\tilde{y}d\tilde{y}d\tilde{z}$ and $d\tilde{x}d\tilde{y}d\tilde{z}d\tilde{z}$ are infinitesimally small in comparison to the terms $d\tilde{x}d\tilde{y}d\tilde{z}$, they produce zero work and one is left with the following expression for the work of the internal forces on the virtual displacements in the **i** direction

$$\begin{aligned} dW_i &= \left(c_{ii}\frac{\partial u}{\partial \tilde{x}}+c_{ij}\frac{\partial u}{\partial \tilde{y}}+c_{ik}\frac{\partial u}{\partial \tilde{z}}+\frac{\partial c_{ii}}{\partial \tilde{x}}u+\frac{\partial c_{ij}}{\partial \tilde{y}}u+\frac{\partial c_{ik}}{\partial \tilde{z}}u\right)d\tilde{x}d\tilde{y}d\tilde{z}\\ &= \begin{bmatrix} c_{ii} & c_{ij} & c_{ik}\end{bmatrix}\begin{bmatrix}\frac{\partial u}{\partial \tilde{x}}\\ \frac{\partial u}{\partial \tilde{y}}\\ \frac{\partial u}{\partial \tilde{z}}\end{bmatrix}d\tilde{x}d\tilde{y}d\tilde{z}+\begin{bmatrix}\frac{\partial c_{ii}}{\partial \tilde{x}} & \frac{\partial c_{ij}}{\partial \tilde{y}} & \frac{\partial c_{ik}}{\partial \tilde{z}}\end{bmatrix}\begin{bmatrix}u\\ u\\ u\end{bmatrix}d\tilde{x}d\tilde{y}d\tilde{z}. \end{aligned} \tag{16.35}$$

It is worth noting that

$$\begin{bmatrix}\frac{\partial c_{ii}}{\partial \tilde{x}} & \frac{\partial c_{ij}}{\partial \tilde{y}} & \frac{\partial c_{ik}}{\partial \tilde{z}}\end{bmatrix}$$

corresponds to distributed forces due to the stress field. In the ideal case, these should be equal to the distributed external load **p**. In practical applications, this is not the case; thus when considering the virtual work of the external forces, one has to consider the residual

$$r_x = p_x - \left(\frac{\partial c_{ii}}{\partial \tilde{x}} + \frac{\partial c_{ij}}{\partial \tilde{y}} + \frac{\partial c_{ik}}{\partial \tilde{z}} \right)$$

where p_x is the distributed external load in the x direction.

Work of Internal Forces on Virtual Displacement in the j Direction. By providing a virtual displacement, $v = v(\tilde{x}, \tilde{y}, \tilde{z})$ in the global **j** direction, Figure 16.10, and by combining these with the corresponding internal forces in the **j** direction as shown in Figure 16.9, one obtains

$$\begin{aligned} dW_j = & -\left(c_{ji} - \frac{1}{2}\frac{\partial c_{ji}}{\partial \tilde{x}} d\tilde{x} \right) d\tilde{y} d\tilde{z} \left(v - \frac{1}{2}\frac{\partial v}{\partial \tilde{x}} d\tilde{x} \right) + \left(c_{ji} + \frac{1}{2}\frac{\partial c_{ji}}{\partial \tilde{x}} d\tilde{x} \right) d\tilde{y} d\tilde{z} \left(v + \frac{1}{2}\frac{\partial v}{\partial \tilde{x}} d\tilde{x} \right) \\ & -\left(c_{jj} - \frac{1}{2}\frac{\partial c_{jj}}{\partial \tilde{y}} d\tilde{y} \right) d\tilde{x} d\tilde{z} \left(v - \frac{1}{2}\frac{\partial v}{\partial \tilde{y}} d\tilde{y} \right) + \left(c_{jj} + \frac{1}{2}\frac{\partial c_{jj}}{\partial \tilde{y}} d\tilde{y} \right) d\tilde{x} d\tilde{z} \left(v + \frac{1}{2}\frac{\partial v}{\partial \tilde{y}} d\tilde{y} \right) \\ & -\left(c_{jk} - \frac{1}{2}\frac{\partial c_{jk}}{\partial \tilde{z}} d\tilde{z} \right) d\tilde{x} d\tilde{y} \left(v - \frac{1}{2}\frac{\partial v}{\partial \tilde{z}} d\tilde{z} \right) + \left(c_{jk} + \frac{1}{2}\frac{\partial c_{jk}}{\partial \tilde{z}} d\tilde{z} \right) d\tilde{x} d\tilde{y} \left(v + \frac{1}{2}\frac{\partial v}{\partial \tilde{z}} d\tilde{z} \right). \end{aligned} \tag{16.36}$$

Since the terms on $d\tilde{x}d\tilde{y}d\tilde{y}d\tilde{z}$, $d\tilde{x}d\tilde{x}d\tilde{y}d\tilde{z}$ and $d\tilde{x}d\tilde{y}d\tilde{z}d\tilde{z}$ are infinitesimally small in comparison with the terms on $d\tilde{x}d\tilde{y}d\tilde{z}$, they produce zero virtual work and one is left with the following expression for the work of the Cauchy stress on the **j** virtual displacements

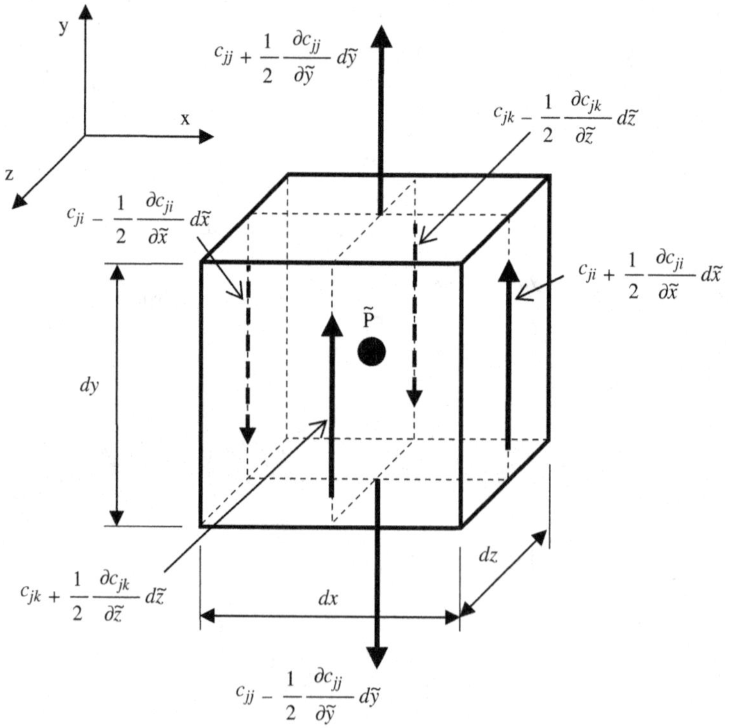

Figure 16.9 Cauchy stress tensor matrix – only the **j** components of internal forces produce work on the virtual displacements that are in **j** direction, thus only the **j** forces on the surfaces of the Cauchy material element are shown

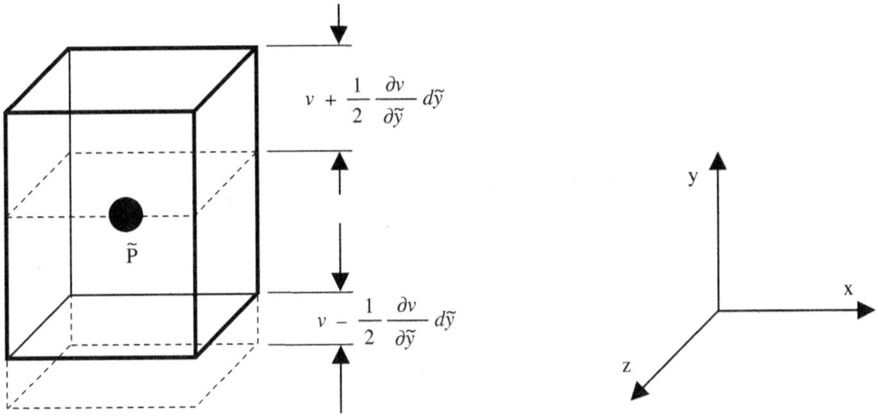

Figure 16.10 Virtual displacement of Cauchy material element in the **j** direction

$$dW_j = \left(c_{ji}\frac{\partial v}{\partial \tilde{x}} + c_{jj}\frac{\partial v}{\partial \tilde{y}} + c_{jk}\frac{\partial v}{\partial \tilde{z}} + \frac{\partial c_{ji}}{\partial \tilde{x}}v + \frac{\partial c_{jj}}{\partial \tilde{y}}v + \frac{\partial c_{jk}}{\partial \tilde{z}}v \right) d\tilde{x}d\tilde{y}d\tilde{z}$$

$$= \begin{bmatrix} c_{ji} & c_{jj} & c_{jk} \end{bmatrix} \begin{bmatrix} \frac{\partial v}{\partial \tilde{x}} \\ \frac{\partial v}{\partial \tilde{y}} \\ \frac{\partial v}{\partial \tilde{z}} \end{bmatrix} d\tilde{x}d\tilde{y}d\tilde{z} + \begin{bmatrix} \frac{\partial c_{ji}}{\partial \tilde{x}} & \frac{\partial c_{jj}}{\partial \tilde{y}} & \frac{\partial c_{jk}}{\partial \tilde{z}} \end{bmatrix} \begin{bmatrix} v \\ v \\ v \end{bmatrix} d\tilde{x}d\tilde{y}d\tilde{z}. \quad (16.37)$$

Work of Internal Forces on Virtual Displacement in the k Direction. In a similar manner, the work of internal forces in the **k** direction is obtained by giving a virtual displacement $w = w(\tilde{x}, \tilde{y}, \tilde{z})$ in the **k** direction (Figure 16.12) and multiplying these with the corresponding internal forces, Figure 16.11.

This yields the virtual work

$$dW_k = \left(c_{ki}\frac{\partial w}{\partial \tilde{x}} + c_{kj}\frac{\partial w}{\partial \tilde{y}} + c_{kk}\frac{\partial w}{\partial \tilde{z}} + \frac{\partial c_{ki}}{\partial \tilde{x}}w + \frac{\partial c_{kj}}{\partial \tilde{y}}w + \frac{\partial c_{kk}}{\partial \tilde{z}}w \right) d\tilde{x}d\tilde{y}d\tilde{z}$$

$$= \begin{bmatrix} c_{ki} & c_{kj} & c_{kk} \end{bmatrix} \begin{bmatrix} \frac{\partial w}{\partial \tilde{x}} \\ \frac{\partial w}{\partial \tilde{y}} \\ \frac{\partial w}{\partial \tilde{z}} \end{bmatrix} d\tilde{x}d\tilde{y}d\tilde{z} + \begin{bmatrix} \frac{\partial c_{ki}}{\partial \tilde{x}} & \frac{\partial c_{kj}}{\partial \tilde{y}} & \frac{\partial c_{kk}}{\partial \tilde{z}} \end{bmatrix} \begin{bmatrix} w \\ w \\ w \end{bmatrix} d\tilde{x}d\tilde{y}d\tilde{z}. \quad (16.38)$$

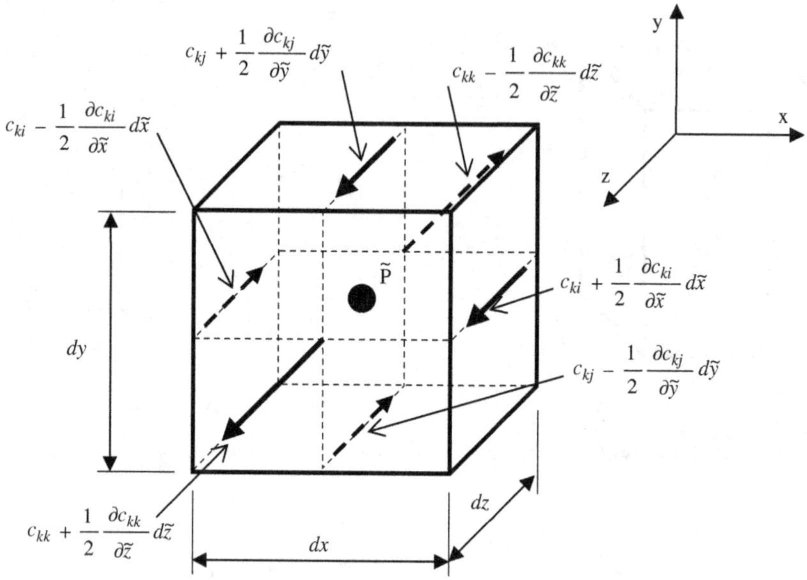

Figure 16.11 The **k** components of internal forces on the surfaces of the Cauchy material element

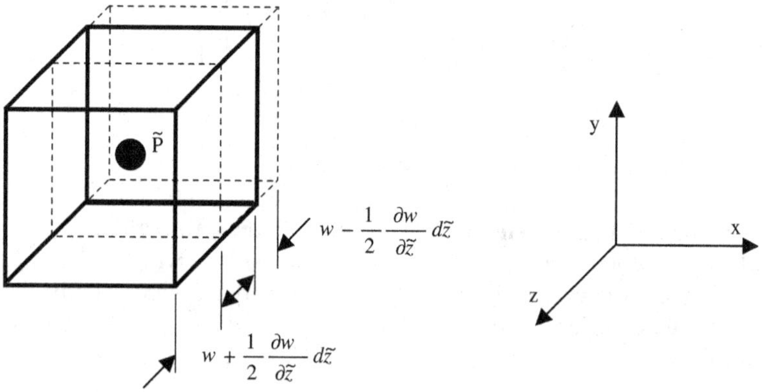

Figure 16.12 The **k** components the virtual displacement of Cauchy material element

Work of Internal Forces on General Virtual Displacements. The total virtual work of the internal forces on a general virtual displacement

$$\begin{bmatrix} u \\ v \\ w \end{bmatrix} = \begin{bmatrix} \mathrm{u}(\tilde{x},\tilde{y},\tilde{z}) \\ \mathrm{v}(\tilde{x},\tilde{y},\tilde{z}) \\ \mathrm{w}(\tilde{x},\tilde{y},\tilde{z}) \end{bmatrix} \tag{16.39}$$

is obtained by combining Equations (16.35), (16.37) and (16.38) and integrating them over the solid in its current position,

$$
\begin{aligned}
W = & \iiint_{\tilde{V}_e} \left(\begin{bmatrix} c_{ii} & c_{ij} & c_{ik} \end{bmatrix} \begin{bmatrix} \dfrac{\partial u}{\partial \tilde{x}} \\ \dfrac{\partial u}{\partial \tilde{y}} \\ \dfrac{\partial u}{\partial \tilde{z}} \end{bmatrix} + \begin{bmatrix} \dfrac{\partial c_{ii}}{\partial \tilde{x}} & \dfrac{\partial c_{ij}}{\partial \tilde{y}} & \dfrac{\partial c_{ik}}{\partial \tilde{z}} \end{bmatrix} \begin{bmatrix} u \\ u \\ u \end{bmatrix} \right) d\tilde{x} d\tilde{y} d\tilde{z} \\
& + \iiint_{\tilde{V}_e} \left(\begin{bmatrix} c_{ji} & c_{jj} & c_{jk} \end{bmatrix} \begin{bmatrix} \dfrac{\partial v}{\partial \tilde{x}} \\ \dfrac{\partial v}{\partial \tilde{y}} \\ \dfrac{\partial v}{\partial \tilde{z}} \end{bmatrix} + \begin{bmatrix} \dfrac{\partial c_{ji}}{\partial \tilde{x}} & \dfrac{\partial c_{jj}}{\partial \tilde{y}} & \dfrac{\partial c_{jk}}{\partial \tilde{z}} \end{bmatrix} \begin{bmatrix} v \\ v \\ v \end{bmatrix} \right) d\tilde{x} d\tilde{y} d\tilde{z} \\
& + \iiint_{\tilde{V}_e} \left(\begin{bmatrix} c_{ki} & c_{kj} & c_{kk} \end{bmatrix} \begin{bmatrix} \dfrac{\partial w}{\partial \tilde{x}} \\ \dfrac{\partial w}{\partial \tilde{y}} \\ \dfrac{\partial w}{\partial \tilde{z}} \end{bmatrix} + \begin{bmatrix} \dfrac{\partial c_{ki}}{\partial \tilde{x}} & \dfrac{\partial c_{kj}}{\partial \tilde{y}} & \dfrac{\partial c_{kk}}{\partial \tilde{z}} \end{bmatrix} \begin{bmatrix} w \\ w \\ w \end{bmatrix} \right) d\tilde{x} d\tilde{y} d\tilde{z}.
\end{aligned}
\tag{16.40}
$$

16.8 Summary

In this chapter a logical extension of deformation kinematics from 2D space to 3D space is given. The formulation presented considers the deformation kinematics of anisotropic materials. This is done through the introduction of the Munjiza generalized material element defined by three base vectors. These base vectors move (deform) with the material in the infinitesimal vicinity of a given material point P. As such, they describe the deformation and stretching of the material in the infinitesimal vicinity of point P. Consequently, in conjunction with the material law, they describe the internal forces (stress) in the infinitesimal vicinity of point P. Using this approach, any type of anisotropic material law can be considered.

The internal forces (stress) are represented by the internal forces on the surfaces of the Cauchy material element (an infinitesimal cube taken in the infinitesimal vicinity of point P after deformation), i.e., the Cauchy stress tensor matrix.

The Cauchy stress tensor matrix is used in formulating the static or dynamic equilibrium for the solid's deformed position. Both static and dynamic equilibrium considers equilibrium of all forces (including the internal forces) in conjunction with the current (deformed) geometry. As such, the deformation kinematics formulation presented in this chapter covers both large displacements and large strains.

The dynamic equilibrium, by its nature, also includes inertia forces, which are not present in static problems.

In this chapter, equilibrium using virtual work has also been described in detail.

Further Reading

[1] Bathe, K. J. (1982) *Finite Element Procedures Engineering Analysis.* New Jersey, Prentice-Hall.

[2] Bathe, K. J. (1996) *Finite Element Procedures.* New Jersey, Prentice-Hall.

[3] Bonet, J. and Burton, A. J. (1998) A simple orthotropic, transversely isotropic hyperelastic constitutive equation for large strain computations. *Comput. Method Appl. M.*, **162**(1–4): 151–64.

[4] Bonet, J. and Wood, R. D. (1997) *Nonlinear Continuum Mechanics for Finite Element Analysis.* Cambridge, Cambridge University Press.

[5] de Souza Neto, E. A. and Peric, D. (1996) A computational framework for a class of fully coupled models for elastoplastic damage at finite strains with reference to the linearization aspects. *Comput. Method Appl. M.*, **130**(1–2): 179–93.

[6] de Souza Neto, E. A., Pires, F. M. A. and Owen, D. R. J. (2005) F-bar-based linear triangles and tetrahedra for finite strain analysis of nearly incompressible solids. Part I: formulation and benchmarking. *International Journal for Numerical Methods in Engineering*, **62**(3): 353–83.

[7] de Souza Neto, E. A., Peric, D., Huang, G. C. and Owen, D. R. J. (1995) Remarks on the stability of enhanced strain elements in finite elasticity and elastoplasticity. *Communications in Numerical Methods in Engineering*, **11**(11): 951–61.

[8] Dvorkin, E. N. and Goldschmit, M. B. (2005) *Nonlinear Continua.* New York, Springer.

[9] Green, A. E. and Naghdi, P. M. (1971) Some remarks on elastic-plastic deformation at finite strain. *International Journal of Engineering Science*, **9**(12): 1219–29.

[10] Gurtin, M. E. (1981) *An Introduction to Continuum Mechanics.* New York, Academic Press.

[11] Hill, R. (1979) *Aspects of Invariance in Solid Mechanics*, ed. Y. Chia-Shun, Elsevier, pp. 1–75.

[12] Hughes, T.J.R. (2000) *The Finite Element Method: Linear Static and Dynamic Finite Element Analysis.* New York, Dover Publications.

[13] Malvern, L. E. (1969) *Introduction to the Mechanics of a Continuous Medium.* New Jersey, Prentice Hall.

[14] Munjiza, A. (2004) *The Combined Finite-Discrete Element Method.* Chichester, John Wiley & Sons, Ltd.

[15] Peric, D. (1992) On consistent stress rates in solid mechanics – computational implications. *International Journal for Numerical Methods in Engineering*, **33**(4): 799–817.

[16] Peric, D. and Owen, D. R. J. (1998) Finite-element applications to the nonlinear mechanics of solids. *Reports on Progress in Physics*, **61**(11): 1495–1574.

[17] Simo, J. C. (1985) On the computational significance of the intermediate configuration and hyperelastic stress relations in finite deformation elastoplasticity, *Mech. Mater.*, **4**(3–4): 439–51.

[18] Simo, J. C. (1988) A framework for finite strain elastoplasticity based on maximum plastic dissipation and the multiplicative decomposition. 1. Continuum formulation, *Comput. Method Appl. M.*, **66**(2): 199–219.

[19] Simo, J. C. (1988) A framework for finite strain elastoplasticity based on maximum plastic dissipation and the multiplicative decomposition. 2. Computational aspects, *Comput Method Appl M*, **68**(1): 1–31.

[20] Simo, J. C. and Ortiz, M. (1985) A unified approach to finite deformation elastoplastic analysis based on the use of hyperelastic constitutive equations. *Comput Method Appl M*, **49**(2): 221–45.

[21] Simo, J. C. and Ju, J. W. (1989) Finite deformation damage-elastoplasticity: a non conventional framework. *International Journal of Computational Mechanics*, **5**: 375–400.

[22] Simo, J. C. and Hughes, T. J. R. (1998) *Computational Inelasticity*, New York, Springer.

[23] Simo, J. C., Armero, F. and Taylor, R. L. (1993) Improved versions of assumed enhanced strain tri-linear elements for 3D finite deformation problems. *Comput. Method Appl. M.*, **110**(3–4): 359–86.

[24] Zienkiewicz, O. C. (1971) *The Finite Element Method in Engineering Science.* New York, McGraw-Hill.

[25] Zienkiewicz, O. C. and Taylor, R. L. (2005) *The Finite Element Method Set*, Oxford, Elsevier Science.

17

The Unified Constitutive Approach in 2D

17.1 Introduction

The initial configuration of the 2D generalized material element is defined by vectors $\bar{\boldsymbol{\alpha}}$ and $\bar{\boldsymbol{\beta}}$. The current (deformed) configuration (position) of the generalized material element is defined by vectors $\tilde{\boldsymbol{\alpha}}$ and $\tilde{\boldsymbol{\beta}}$ where $\tilde{\boldsymbol{\alpha}}$ is the deformed (rotated and stretched) vector $\bar{\boldsymbol{\alpha}}$ and $\tilde{\boldsymbol{\beta}}$ is the deformed vector $\bar{\boldsymbol{\beta}}$.

By observing these two vectors, it is possible to characterize the deformation of the generalized material element. The deformation is made up of translation followed by rotation, and then followed by stretch. The generalized material element's stretch can be defined as:

a. stretching of the vector $\bar{\boldsymbol{\alpha}}$ so that

$$s_\alpha = \frac{\tilde{\alpha}}{\bar{\alpha}}, \tag{17.1}$$

where $\tilde{\alpha}$ is the magnitude of vector $\tilde{\boldsymbol{\alpha}}$ and $\bar{\alpha}$ is the magnitude of vector $\bar{\boldsymbol{\alpha}}$;

b. stretching of the vector $\bar{\boldsymbol{\beta}}$ so that

$$s_\beta = \frac{\tilde{\beta}}{\bar{\beta}}, \tag{17.2}$$

where $\tilde{\beta}$ is the magnitude of vector $\tilde{\boldsymbol{\beta}}$ and $\bar{\beta}$ is the magnitude of vector $\bar{\boldsymbol{\beta}}$;

c. change of angle between vectors $\bar{\boldsymbol{\alpha}}$ and $\bar{\boldsymbol{\beta}}$

$$s_{\alpha\beta} = \frac{\tilde{\psi}_{\alpha\beta}}{\bar{\psi}_{\alpha\beta}}; \tag{17.3}$$

Large Strain Finite Element Method: A Practical Course, First Edition. Antonio Munjiza, Esteban Rougier and Earl E. Knight.

d. change of volume of the solid element

$$s_v = \frac{\tilde{v}}{\bar{v}}. \tag{17.4}$$

This is shown in Figure 17.1.

One can say that the solid element resists stretching via the internal forces between atoms that keep the solid in one piece. The resultant internal forces over a given surface for a given point P are best described using the stress tensor. The stress tensor is a linear mapping that for a given surface vector returns the corresponding resultant internal force. As such, this linear mapping is best described using matrices such as the Cauchy stress tensor matrix, the First Piola-Kirchhoff matrix, or the Second Piola-Kirchhoff stress tensor matrix.

The relationship between stress and stretching is called the constitutive law. Among the constitutive laws there exists a whole class of hypo-elastic constitutive laws in which the material is stretched in small increments and the stresses are accumulated. Another approach is called the hyper-elastic approach, in which the stresses are calculated in one increment by considering only the initial and the current shapes of the material element.

Both hypo-elastic and hyper-elastic formulations are difficult to generalize in a sense that:

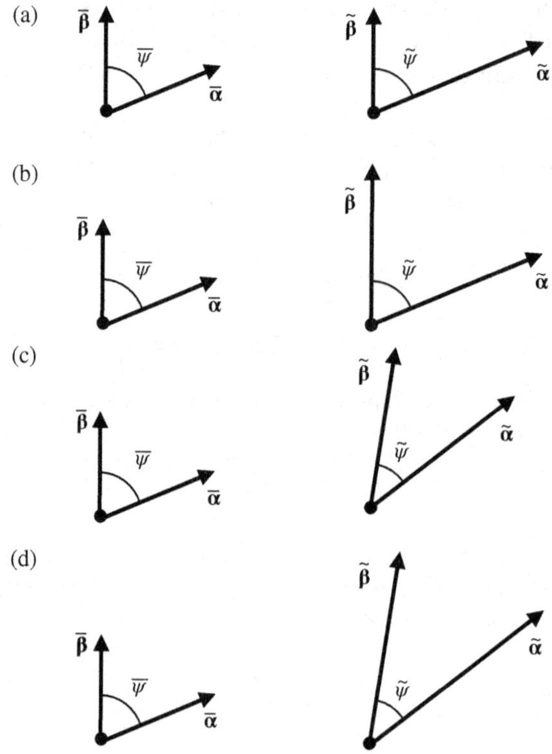

Figure 17.1 (a) Stretching of the vector **α**; (b) stretching of vector **β**; (c) shearing of the generalized material element; (d) stretching of the generalized material element

a. anisotropic materials do not fit naturally into either hypo-elastic or hyper-elastic formulations;
b. when it comes to describing inelastic behavior caused by something like kinematic hardening, anisotropy is introduced by default and the additive decomposition of strains leads to serious formulation problems.

These issues were first explained by the theoretical founders of modern Continuum Mechanics (Gurtin, Truesdell). In the context of Computational Mechanics, it was pointed out by Simo, Munjiza and Peric, who introduced the term multiplicative decomposition. The idea is relatively simple: at each increment the previous configuration is considered as the initial configuration and stretches are subsequently calculated. However, even with this formulation the problems associated with the additive decomposition of the stress remains and the stress is still corrected using, for example, the Jaumann rate.

Recently within the combined finite-discrete element method, the unified hyper-elastic hypo-elastic formulation has been developed (Munjiza, 2004). This approach generalizes both the hypo-elastic and the hyper-elastic formulations together with the multiplicative decomposition, resulting in no need for any type of stress correction, i.e., Jaumann rate, Truesdell rate, Green-Naghdi rate, etc. Because of its engineering simplicity and its general applicability to both elastic and inelastic materials and to isotropic and anisotropic materials, it is the only constitutive law formulation described in detail in this book. It is worth noting that a number of finite strain computer programs have already adopted this formulation, for example the Los Alamos National Laboratory's MUNROU finite and finite-discrete element package.

Alternative formulations have been described many times in different publications and there is no need to repeat them here, especially given the fact that they have, in many cases, become obsolete with the introduction of the unified approach.

17.2 Material Axes

Consider the 2D material element shown in Figure 17.2, where the initial, the previous and the current positions of an infinitesimal solid element with edges

$$\begin{array}{c}\bar{\boldsymbol{\alpha}}(\text{im})\\ \bar{\boldsymbol{\beta}}(\text{im})\end{array} \tag{17.5}$$

are shown. This element is called the generalized material element. The symbol im represents an infinitesimally small unit for length (infi-meter). As the generalized material element deforms, its edges stretch in such a way that the edge $\bar{\boldsymbol{\alpha}}$ becomes $\tilde{\boldsymbol{\alpha}}$ and the edge $\bar{\boldsymbol{\beta}}$ becomes $\tilde{\boldsymbol{\beta}}$.

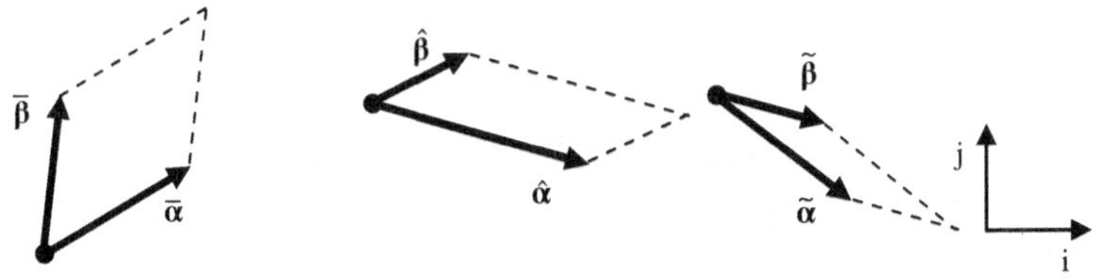

Figure 17.2 The initial, the previous, and the current positions of a generalized material element

The best way to observe these changes is to draw $\bar{\boldsymbol{\alpha}}$ and $\bar{\boldsymbol{\beta}}$ on the undeformed solid and then deform the solid in order to observe how these two vectors change both their position and magnitude to become $\tilde{\boldsymbol{\alpha}}$ and $\tilde{\boldsymbol{\beta}}$.

In addition to the initial position of the generalized material element, it is also useful to know the previous position of the material element.

All six vectors can be defined using the global vector base

$$[\mathbf{i}\ \mathbf{j}], \tag{17.6}$$

so that

$$[\bar{\boldsymbol{\alpha}}\ \bar{\boldsymbol{\beta}}] = [\mathbf{i}\ \mathbf{j}]\begin{bmatrix} \bar{\alpha}_i & \bar{\beta}_i \\ \bar{\alpha}_j & \bar{\beta}_j \end{bmatrix} = \begin{bmatrix} \bar{\alpha}_i & \bar{\beta}_i \\ \bar{\alpha}_j & \bar{\beta}_j \end{bmatrix}, \tag{17.7}$$

$$[\hat{\boldsymbol{\alpha}}\ \hat{\boldsymbol{\beta}}] = [\mathbf{i}\ \mathbf{j}]\begin{bmatrix} \hat{\alpha}_i & \hat{\beta}_i \\ \hat{\alpha}_j & \hat{\beta}_j \end{bmatrix} = \begin{bmatrix} \hat{\alpha}_i & \hat{\beta}_i \\ \hat{\alpha}_j & \hat{\beta}_j \end{bmatrix} \tag{17.8}$$

and

$$[\tilde{\boldsymbol{\alpha}}\ \tilde{\boldsymbol{\beta}}] = [\mathbf{i}\ \mathbf{j}]\begin{bmatrix} \tilde{\alpha}_i & \tilde{\beta}_i \\ \tilde{\alpha}_j & \tilde{\beta}_j \end{bmatrix} = \begin{bmatrix} \tilde{\alpha}_i & \tilde{\beta}_i \\ \tilde{\alpha}_j & \tilde{\beta}_j \end{bmatrix}. \tag{17.9}$$

Vectors $\bar{\boldsymbol{\alpha}}$ and $\bar{\boldsymbol{\beta}}$ are called the generalized material-embedded base vectors. For nonhomogeneous anisotropic materials they change from one material point $\bar{\mathrm{P}} = (\bar{x}, \bar{y})$ to another.

These vectors have been calculated in Chapter 15 using the deformation kinematics. As such, they are readily available as an input to the constitutive law calculations.

17.3 Micromechanical Aspects and Homogenization

When considering a constitutive law for a given material, the material parameters are not measured on an infinitesimal material element. For obvious reasons they are measured on a finite size material element, that is referred to as the sample. For example, in Figure 17.3 a concrete block compression test experiment is shown.

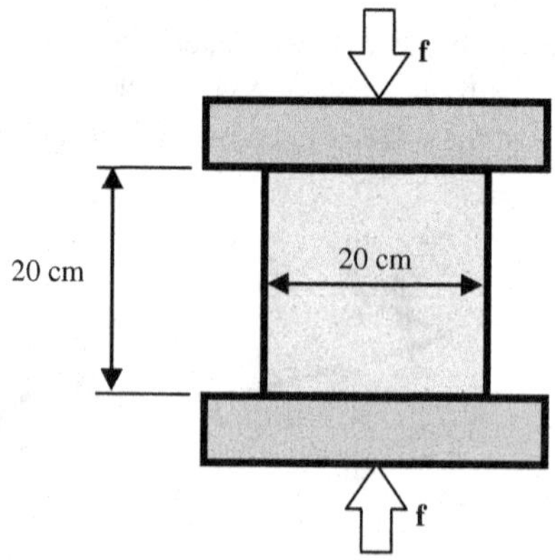

Figure 17.3 Concrete sample under compressive testing

For modeling purposes, the material element is conveniently replaced with what is called a representative volume. This is done because of the assumption of continuum on which the whole theory of Continuum Mechanics rests.

The problem, however, is that real materials are not continuum. They have a quite elaborate microstructure. In the case of volcanic rock, this is illustrated in Figure 17.4a. In the case of wood this is illustrated in Figure 17.4b. For the case of bone-graft substitute, this is shown in Figure 17.4c.

In terms of transmitting internal forces, these materials are made of structural blocks connected together in a manner similar to that of a framed building which has beams and columns that form its structural skeleton.

The difference is that the individual structural elements for concrete vary in size from a fraction of a millimeter to a couple of centimeters. Individual fibers for tetrapack are even smaller, which is also the case for bone. For this reason, these structural elements form the structure on a micro level, thus the name microstructure.

Take for example the unidirectional glass fiber reinforced plastic shown in Figure 17.5.

For a tension test, the experimentally measured force **f** is the sum of forces in the individual glass fibers, $\mathbf{f}_i$. In practice, the fibers are never perfectly spaced. Thus, there is a random component to the exact position of the individual internal forces $\mathbf{f}_i$. In addition, there may be some

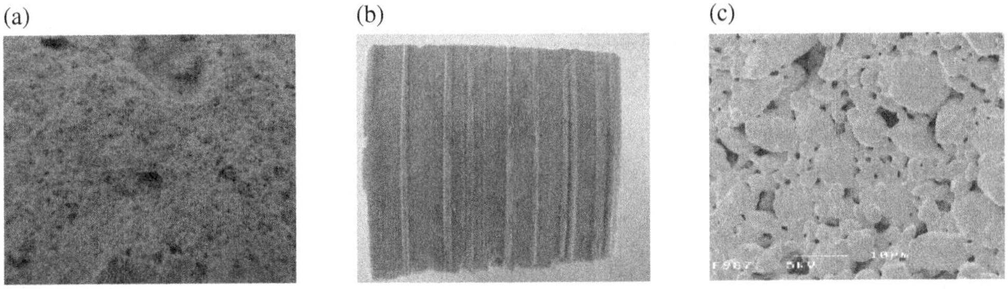

Figure 17.4 Microstructure of different materials. (a) Volcanic rock, (b) wood, (c) calcium phosphate bone-graft substitute (courtesy of Karin Hing)

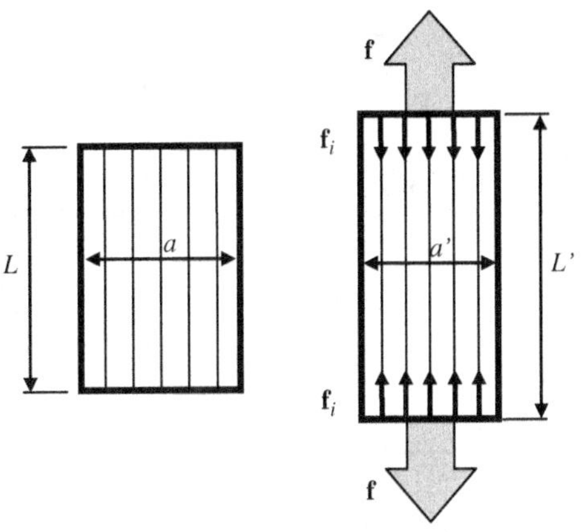

Figure 17.5 Force transmission in glass fiber reinforced plastic

difference in the forces in the individual fibers. For this reason, the measured results for force **f** are a function of the size of the sample, a. For a very small value of a, the measured force **f** will change significantly from sample to sample. As the value of a is increased, the scatter on the values of the measured force **f** will get smaller and smaller.

Figure 17.6 depicts a sample size a_r that delineates a size wherein random changes in the sample's microstructure do not have much effect on the tensile force. Consequently, any sample of size

$$a > a_r \tag{17.10}$$

can be considered a representative sample in the sense that it properly represents the microstructure of the material. For different materials different representative samples are required.

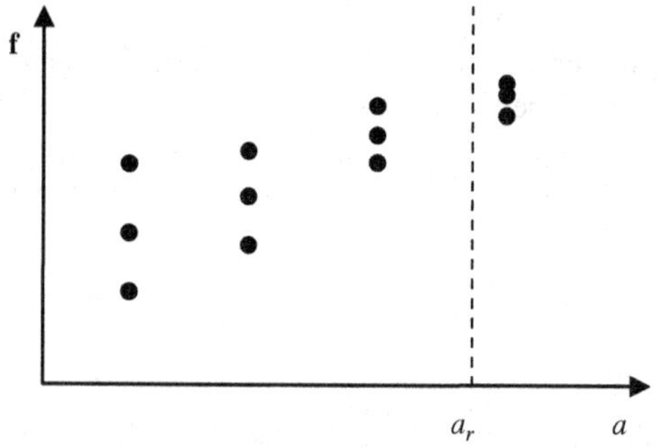

Figure 17.6 A typical force **f** results for samples of different sizes

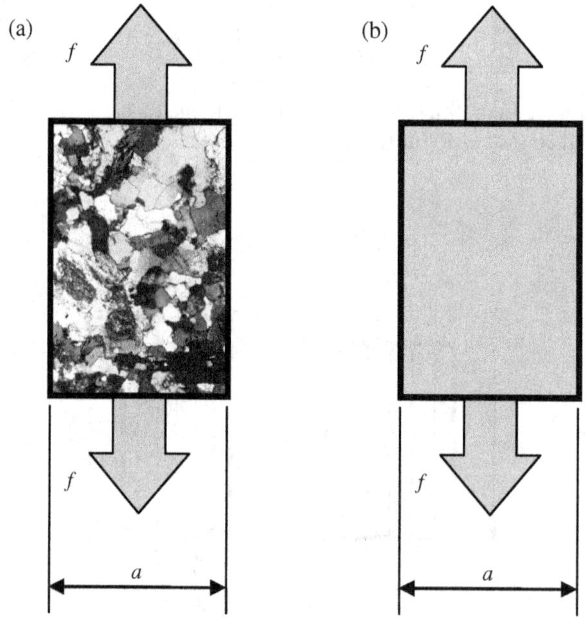

Figure 17.7 Homogenization: a sample of the real material (courtesy of Jennifer Wilson) (a) and a model of a fictitious continuum (b) produce the same force

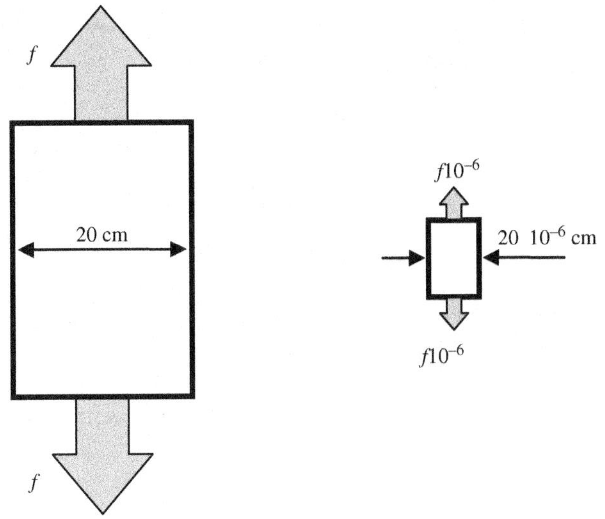

Figure 17.8 Making an infinitesimal solid element through the assumption of Continuum

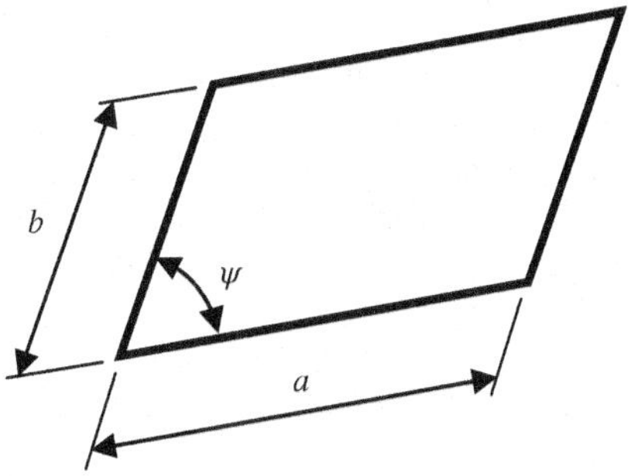

Figure 17.9 The generalized representative sample

When measuring the force **f** in an experiment one can ignore the microstructure of the material, and assume that the sample is made of a homogeneous continuum, which when stretched produces force **f**, i.e., the same force as the real material, Figure 17.7.

One can now easily move from a continuum sample of say $a = 20$ cm to say $a = 20\ 10^{-6}$ cm, Figure 17.8.

17.4 Generalized Homogenization

The representative sample is usually assumed to be a square of edge a in 2D. This assumption is fine when dealing with isotropic or orthotropic materials. However, when dealing with general anisotropic materials, the representative sample needs to be a parallelogram, as shown in Figure 17.9. This sample is called the generalized representative sample.

Through the homogenization process, one can measure forces due to different modes of stretching via experiments or other means while, keeping a complete analogy to the square sample. The forces obtained refer to a finite size representative volume. These are then transformed

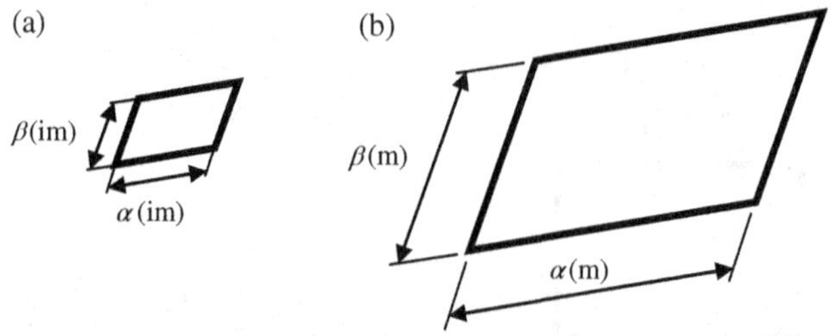

Figure 17.10 Generalized infinitesimal anisotropic material element (a) obtained through the homogenization of the generalized representative sample (b) and the assumption of Continuum

into an infinitesimal anisotropic material element, called the generalized material element, Figure 17.10.

17.5 The Material Package

Modern computer software dealing with constitutive laws usually have a material package as one of the main components. The material package has a whole library of materials available (constitutive laws). These include linear elastic, nonlinear elastic, plastic materials, etc.

Some of these different material packages are implemented using the hypo-elastic formulation. Others are implemented using the hyper-elastic formulation. The material package is, in general, developed separately from the finite element package. For the material package to be truly portable it is important that it can be shared between different types of finite element packages.

As material models are often developed by specialized teams depending on the application, an ideal situation for material model developers looks like this:

Use your finite element software but with my material package.

In order to achieve this, it is necessary to divide the work between the material package and the rest of the finite element software. The division line is shown in Figure 17.11.

The finite element package deals with deformation kinematics. For each material point it produces six vectors that are passed to the material package

$$\begin{bmatrix}\bar{\boldsymbol{\alpha}} & \bar{\boldsymbol{\beta}}\end{bmatrix} \begin{bmatrix}\tilde{\boldsymbol{\alpha}} & \tilde{\boldsymbol{\beta}}\end{bmatrix} \begin{bmatrix}\hat{\boldsymbol{\alpha}} & \hat{\boldsymbol{\beta}}\end{bmatrix}, \tag{17.11}$$

usually as matrices

$$\begin{bmatrix}\bar{\alpha}_i & \bar{\beta}_i \\ \bar{\alpha}_j & \bar{\beta}_j\end{bmatrix}; \begin{bmatrix}\tilde{\alpha}_i & \tilde{\beta}_i \\ \tilde{\alpha}_j & \tilde{\beta}_j\end{bmatrix} \text{ and } \begin{bmatrix}\hat{\alpha}_i & \hat{\beta}_i \\ \hat{\alpha}_j & \hat{\beta}_j\end{bmatrix}. \tag{17.12}$$

These matrices are all that the material package needs in terms of deformation kinematics. From these, it should be able to process any material model based on both hypo and hyper-elastic formulations.

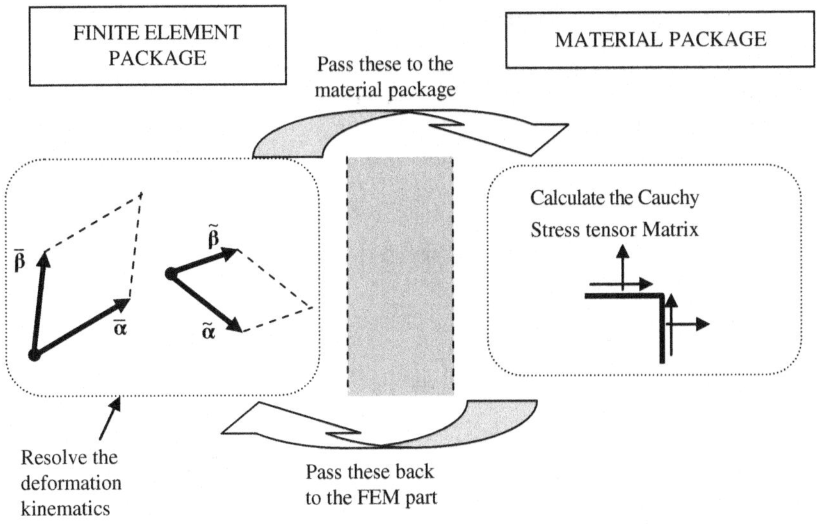

Figure 17.11 The boundary between the material package and the rest of the finite element method

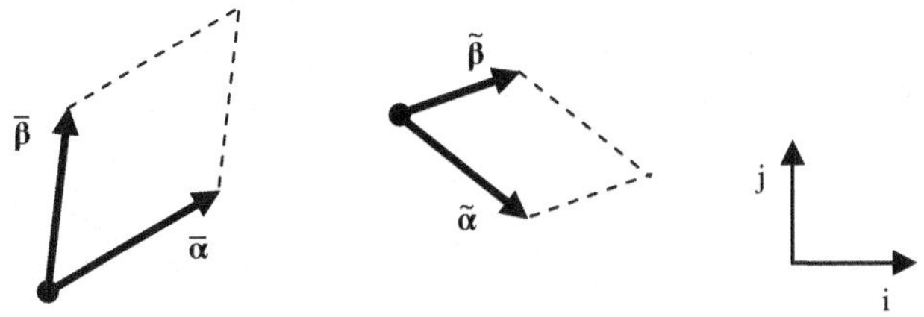

Figure 17.12 The hyper-elastic formulation: the deformation occurs in a single (big) step with the initial (undeformed), the previous, and the current (deformed) shape of the material element being available to the material package

17.6 Hyper-Elastic Constitutive Law

For the purposes of understanding the most general case of hyper-elastic formulation for an anisotropic material element, three configurations are known (see Figure 17.12):

a. the initial configuration of the generalized material element described by vectors

$$\left[\bar{\boldsymbol{\alpha}}\ \bar{\boldsymbol{\beta}}\right], \tag{17.13}$$

b. the current configuration of the generalized material element described by vectors

$$\left[\tilde{\boldsymbol{\alpha}}\ \tilde{\boldsymbol{\beta}}\right], \tag{17.14}$$

c. the previous configuration of the generalized material element described by vectors

$$\left[\hat{\boldsymbol{\alpha}}\ \hat{\boldsymbol{\beta}}\right]. \tag{17.15}$$

17.7 Hypo-Elastic Constitutive Law

The essence of the hypo-elastic formulation is that the deformation and stretching of the solid element proceeds in many very small increments, as illustrated in Figure 17.13.

The easiest way to incorporate these increments is to consider the situation shown in Figure 17.14, where the current and previous positions of the solid element are shown, and simply assume that

$$\begin{bmatrix} \tilde{\boldsymbol{\alpha}}_7 & \tilde{\boldsymbol{\beta}}_7 \end{bmatrix} \tag{17.16}$$

is temporarily the initial position, while

$$\begin{bmatrix} \tilde{\boldsymbol{\alpha}}_8 & \tilde{\boldsymbol{\beta}}_8 \end{bmatrix} \tag{17.17}$$

is temporarily the current position.

Using the hypo-elastic formulation from the material package, the stress increment is calculated, added to the previous stress and remembered inside the material package in a deformation-invariant format. The material package calculates the Cauchy stress matrix and passes it back to the finite element package.

The above process is repeated for the rest of the increments.

$$\begin{array}{l} \begin{bmatrix} \tilde{\boldsymbol{\alpha}}_8 & \tilde{\boldsymbol{\beta}}_8 \end{bmatrix} \rightarrow \begin{bmatrix} \tilde{\boldsymbol{\alpha}}_9 & \tilde{\boldsymbol{\beta}}_9 \end{bmatrix} \\ \begin{bmatrix} \tilde{\boldsymbol{\alpha}}_9 & \tilde{\boldsymbol{\beta}}_9 \end{bmatrix} \rightarrow \begin{bmatrix} \tilde{\boldsymbol{\alpha}}_{10} & \tilde{\boldsymbol{\beta}}_{10} \end{bmatrix} \\ \text{etc.} \end{array} \tag{17.18}$$

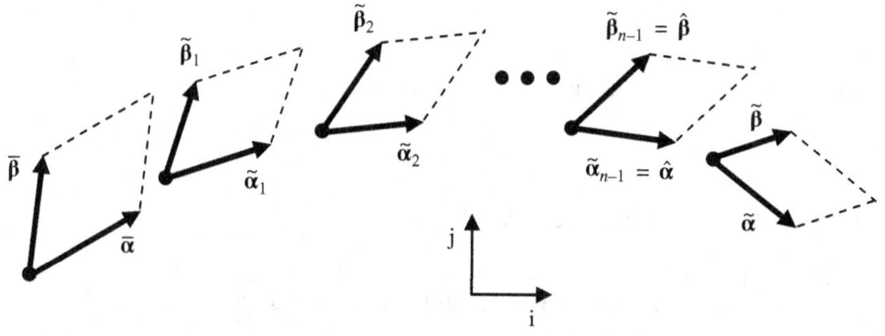

Figure 17.13 Hypo-elastic formulation. The deformation occurs in small increments and the constitutive law is resolved at each increment

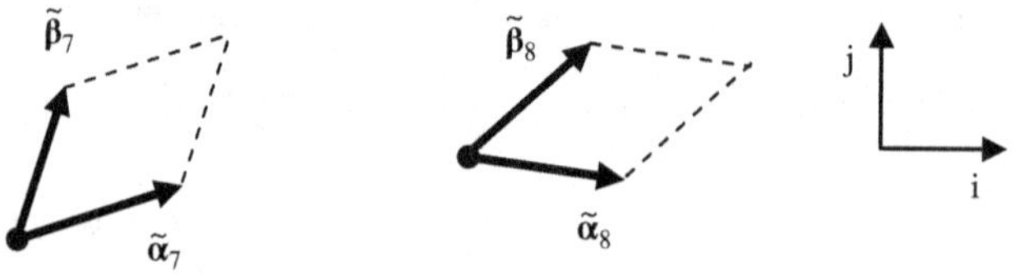

Figure 17.14 Multiplicative decomposition

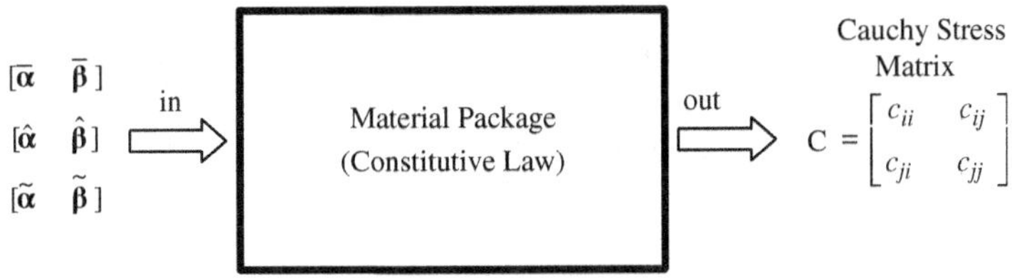

Figure 17.15 Input and output scenario for the material package

This process is called multiplicative decomposition. It has no need for stress corrections, such as Jaumann rate, Truesdell rate or Green-Naghdi rate type of corrections. This is because it is not the Cauchy stress tensor matrix that is remembered within the material package, but one of the many possible deformation-independent matrices representing the stress tensor and therefore, the internal forces. In addition, it basically uses the same kinematics as the hyper-elastic formulation.

17.8 A Unified Framework for Developing Anisotropic Material Models in 2D

On the material package side, the developer faces the situation shown in Figure 17.15.

The kinematics of deformation serves as the input for the material package. The kinematics of deformation consists of the following vectors:

a. $\begin{bmatrix} \bar{\boldsymbol{\alpha}} & \bar{\boldsymbol{\beta}} \end{bmatrix}$ describing the initial (undeformed) generalized material element
b. $\begin{bmatrix} \tilde{\boldsymbol{\alpha}} & \tilde{\boldsymbol{\beta}} \end{bmatrix}$ describing the current (deformed) generalized material element
c. $\begin{bmatrix} \hat{\boldsymbol{\alpha}} & \hat{\boldsymbol{\beta}} \end{bmatrix}$ describing the deformed generalized material element from the previous step or increment; it is therefore called the previous generalized material element

Based on these, different constitutive laws (material models) are developed. In the case that time effects, viscosity, etc. need to be taken into account, one may have to provide the time step size as an input to the material package. In some special cases temperature, energy, chemical reactivity or similar field variables may also need to be provided. However, in the most straightforward case, the material model has the initial position, the current position and the previous position of the generalized material element as input and with this information it produces the Cauchy stress matrix **C** as output.

17.9 Generalized Hyper-Elastic Material

The Deformation Kinematics. As described in the previous section, the initial position of the generalized material element is provided as an input to the material package and is described by two vectors

$$\bar{\boldsymbol{\alpha}} = [\mathbf{i} \;\; \mathbf{j}] \begin{bmatrix} \bar{\alpha}_i \\ \bar{\alpha}_j \end{bmatrix} = \begin{bmatrix} \bar{\alpha}_i \\ \bar{\alpha}_j \end{bmatrix}, \tag{17.19}$$

$$\bar{\boldsymbol{\beta}} = [\mathbf{i}\ \ \mathbf{j}]\begin{bmatrix}\bar{\beta}_i \\ \bar{\beta}_j\end{bmatrix} = \begin{bmatrix}\bar{\beta}_i \\ \bar{\beta}_j\end{bmatrix}, \tag{17.20}$$

or simply

$$\begin{bmatrix}\bar{\boldsymbol{\alpha}} & \bar{\boldsymbol{\beta}}\end{bmatrix} = \begin{bmatrix}\bar{\alpha}_i & \bar{\beta}_i \\ \bar{\alpha}_j & \bar{\beta}_j\end{bmatrix}. \tag{17.21}$$

The current position of the generalized material element is described by vectors

$$\tilde{\boldsymbol{\alpha}} = [\mathbf{i}\ \ \mathbf{j}]\begin{bmatrix}\tilde{\alpha}_i \\ \tilde{\alpha}_j\end{bmatrix} = \begin{bmatrix}\tilde{\alpha}_i \\ \tilde{\alpha}_j\end{bmatrix}, \tag{17.22}$$

$$\tilde{\boldsymbol{\beta}} = [\mathbf{i}\ \ \mathbf{j}]\begin{bmatrix}\tilde{\beta}_i \\ \tilde{\beta}_j\end{bmatrix} = \begin{bmatrix}\tilde{\beta}_i \\ \tilde{\beta}_j\end{bmatrix}, \tag{17.23}$$

which can be written in the matrix form

$$\begin{bmatrix}\tilde{\boldsymbol{\alpha}} & \tilde{\boldsymbol{\beta}}\end{bmatrix} = \begin{bmatrix}\tilde{\alpha}_i & \tilde{\beta}_i \\ \tilde{\alpha}_j & \tilde{\beta}_j\end{bmatrix}, \tag{17.24}$$

inside the material package.

Volumetric Stretch. The initial volume of the generalized material element is calculated as

$$\begin{aligned}\bar{v} &= \det\begin{bmatrix}\bar{\alpha}_i & \bar{\alpha}_j \\ \bar{\beta}_i & \bar{\beta}_j\end{bmatrix} \\ &= \bar{\alpha}_i\bar{\beta}_j - \bar{\alpha}_j\bar{\beta}_i.\end{aligned} \tag{17.25}$$

The current volume of the generalized material element is given by

$$\begin{aligned}\tilde{v} &= \det\begin{bmatrix}\tilde{\alpha}_i & \tilde{\alpha}_j \\ \tilde{\beta}_i & \tilde{\beta}_j\end{bmatrix} \\ &= \tilde{\alpha}_i\tilde{\beta}_j - \tilde{\alpha}_j\tilde{\beta}_i.\end{aligned} \tag{17.26}$$

The stretching of the generalized material element's volume (volumetric stretch) is defined by

$$s_v = \frac{\tilde{v}}{\bar{v}}. \tag{17.27}$$

It describes the current volume of what was initially a unit volume, so that

$$s_v = 1 \tag{17.28}$$

means no volume change,

$$s_v < 1 \tag{17.29}$$

means a contraction of the volume and

$$s_v > 1 \tag{17.30}$$

means an expansion of the volume. The volumetric stretch quantifies the volume change due to deformation of the generalized material element.

Edge Stretches. From the initial position of the generalized material element, the individual edge's lengths are obtained as follows

$$\begin{aligned} \bar{\alpha} &= |\bar{\boldsymbol{\alpha}}| = \sqrt{\bar{\alpha}_i^2 + \bar{\alpha}_j^2} \\ \bar{\beta} &= |\bar{\boldsymbol{\beta}}| = \sqrt{\bar{\beta}_i^2 + \bar{\beta}_j^2}. \end{aligned} \tag{17.31}$$

In a similar manner, the lengths of the edges of the current solid element are given by

$$\begin{aligned} \tilde{\alpha} &= |\tilde{\boldsymbol{\alpha}}| = \sqrt{\tilde{\alpha}_i^2 + \tilde{\alpha}_j^2} \\ \tilde{\beta} &= |\tilde{\boldsymbol{\beta}}| = \sqrt{\tilde{\beta}_i^2 + \tilde{\beta}_j^2}. \end{aligned} \tag{17.32}$$

Stretching of the two edges of the generalized material element (edge stretches) are defined by

$$\begin{aligned} s_\alpha &= \frac{\tilde{\alpha}}{\bar{\alpha}} \\ s_\beta &= \frac{\tilde{\beta}}{\bar{\beta}}. \end{aligned} \tag{17.33}$$

These describe the edges' length change, with

$$s_\alpha = 1; \quad s_\beta = 1 \tag{17.34}$$

meaning no length change,

$$s_\alpha > 1; \quad s_\beta > 1 \tag{17.35}$$

representing an increase in the lengths (extension) and

$$s_\alpha < 1; \quad s_\beta < 1 \tag{17.36}$$

representing a reduction in the lengths (contraction).

Angle Stretch. The initial angle between the edges $\bar{\boldsymbol{\alpha}}$ and $\bar{\boldsymbol{\beta}}$ is given by

$$\bar{\psi} = \arccos\left(\frac{\bar{\boldsymbol{\alpha}} \cdot \bar{\boldsymbol{\beta}}}{\bar{\alpha}\bar{\beta}}\right). \tag{17.37}$$

The current angle between the edges $\tilde{\boldsymbol{\alpha}}$ and $\tilde{\boldsymbol{\beta}}$ is

$$\tilde{\psi} = \arccos\left(\frac{\tilde{\boldsymbol{\alpha}} \cdot \tilde{\boldsymbol{\beta}}}{\tilde{\alpha}\tilde{\beta}}\right), \tag{17.38}$$

where both $\bar{\psi}$ and $\tilde{\psi}$ range between an interval of 0 and π. The stretching of the angle between the solid element's edges (angle stretch) is defined as

$$s_\psi = \frac{\tilde{\psi}}{\bar{\psi}}, \tag{17.39}$$

where

$$s_\psi = 1 \tag{17.40}$$

means no angle change,

$$s_\psi < 1 \tag{17.41}$$

means the angle is contracting, and

$$s_\psi > 1 \tag{17.42}$$

means the angle is expanding. For example, if

$$\bar{\psi} = \frac{\pi}{2} \text{ and } \tilde{\psi} = \frac{\pi}{2} - \frac{1}{100}\frac{\pi}{2}, \tag{17.43}$$

one obtains

$$s_\psi = \frac{\frac{\pi}{2} - \frac{1}{100}\frac{\pi}{2}}{\frac{\pi}{2}} = 1 - \frac{1}{100} = 0.99. \tag{17.44}$$

In a similar manner, for

$$\bar{\psi} = \frac{\pi}{2} \text{ and } \tilde{\psi} = \frac{\pi}{2} + \frac{1}{10}\frac{\pi}{2}, \tag{17.45}$$

one obtains

$$s_\psi = \frac{\frac{\pi}{2} + \frac{1}{10}\frac{\pi}{2}}{\frac{\pi}{2}} = 1 + \frac{1}{10} = 1.1. \tag{17.46}$$

For

$$\bar{\psi} = \frac{\pi}{2} \text{ and } \tilde{\psi} = \frac{\pi}{4}, \tag{17.47}$$

one obtains

$$s_{\psi} = \frac{\pi/4}{\pi/2} = 0.5. \tag{17.48}$$

It is evident that for $\bar{\psi} = \pi/2$ and infinitesimal angle changes

$$s_{\psi} - 1 = 2\varepsilon_{xy}, \tag{17.49}$$

i.e., the small strain formulation for engineering shear strain ε_{xy} is recovered.

From the above, three modes of stretching for the generalized material element can be identified. The first mode is stretching in the $\boldsymbol{\alpha}$ direction, Figure 17.16.

The second mode is stretching in the $\boldsymbol{\beta}$ direction, Figure 17.17.

The third mode is shearing where changes in the angle between $\boldsymbol{\alpha}$ and $\boldsymbol{\beta}$ occur, Figure 17.18.

The above modes produce the stretches shown in Table 17.1, where it can be seen that all three modes also produce volumetric stretch.

In other words, each of the modes produces a nonzero stretch in its corresponding direction, plus a nonzero volumetric stretch combined with all other stretches being zero.

Internal Forces. Using these stretches, internal forces on the surfaces of the solid element (i.e. stress) can be calculated.

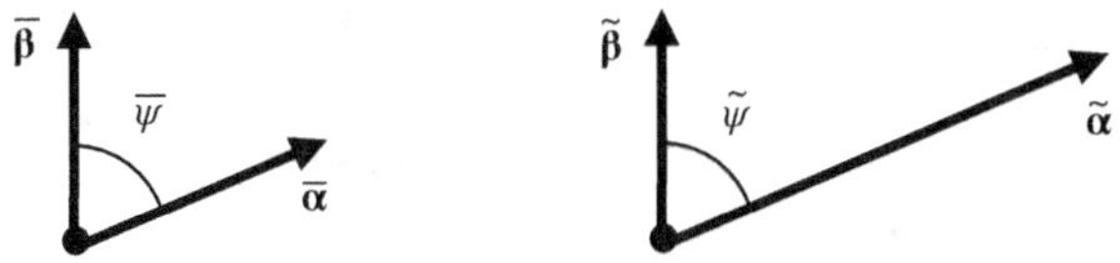

Figure 17.16 Mode α: $\tilde{\beta} = \bar{\beta}$; $\tilde{\psi} = \bar{\psi}$

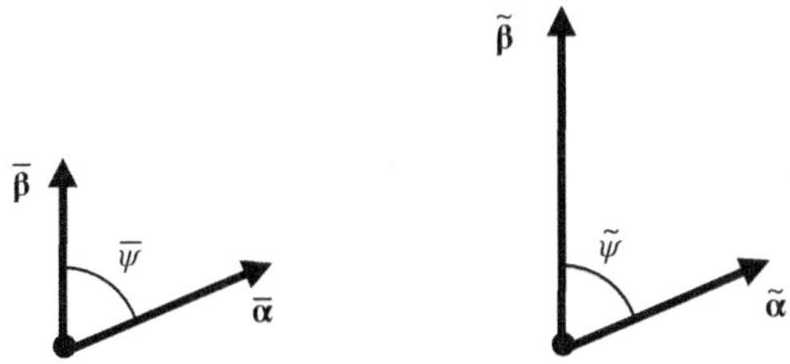

Figure 17.17 Mode β: $\tilde{\alpha} = \bar{\alpha}$; $\tilde{\psi} = \bar{\psi}$

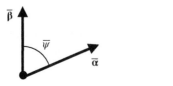

Figure 17.18 Mode ψ: $\tilde{\alpha} = \bar{\alpha}$; $\tilde{\beta} = \bar{\beta}$; $\tilde{\psi} \neq \bar{\psi}$

Table 17.1 Stretches created by the three modes of solid element deformation

	s_α	s_β	s_ψ	s_v
Mode α	●			●
Mode β		●		●
Mode ψ			●	●

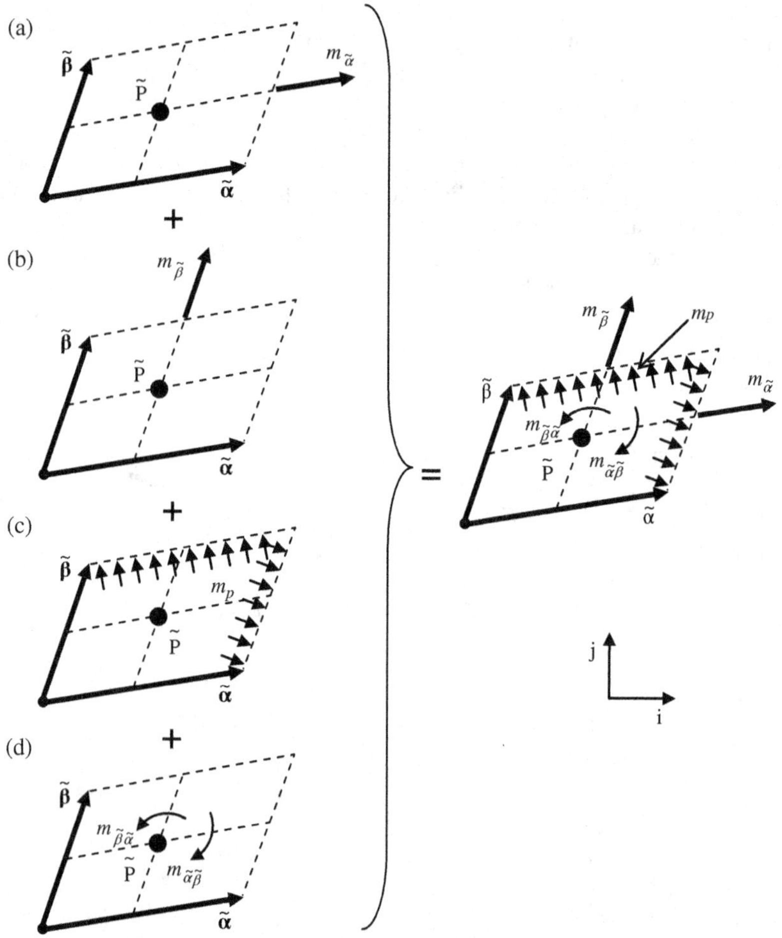

Figure 17.19 Components of the Munjiza stress tensor matrix

For anisotropic material it is convenient to use the Munjiza stress tensor matrix, Figure 17.19. First there is the internal force $m_{\tilde{\alpha}}$ in the direction of the vector $\tilde{\alpha}$. This force is produced solely by edge stretch s_α. This can, for example, be a force produced by a carbon fiber in the $\tilde{\alpha}$ direction, Figure 17.20. Second, there is force $m_{\tilde{\beta}}$, which is parallel to the vector $\tilde{\beta}$. It is produced by homogenization of a representative volume of say carbon fibers that are parallel to vector $\tilde{\beta}$. Stretching of these fibers produces internal forces, Figure 17.21. Third, there is internal pressure m_p. This can be illustrated by a fluidlike material trapped inside the

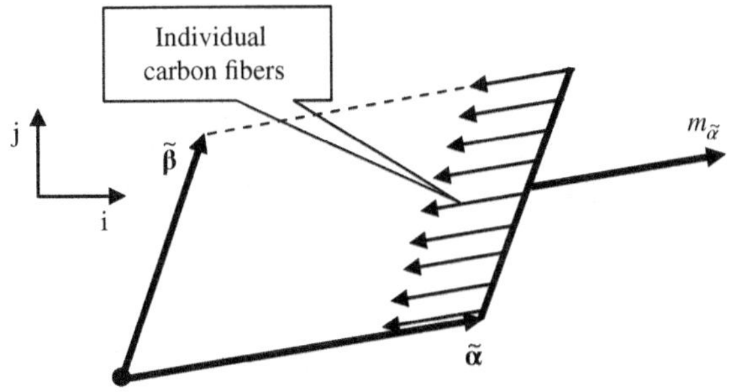

Figure 17.20 The origin of force $m_{\tilde{\alpha}}$ obtained from a representative volume using homogenization

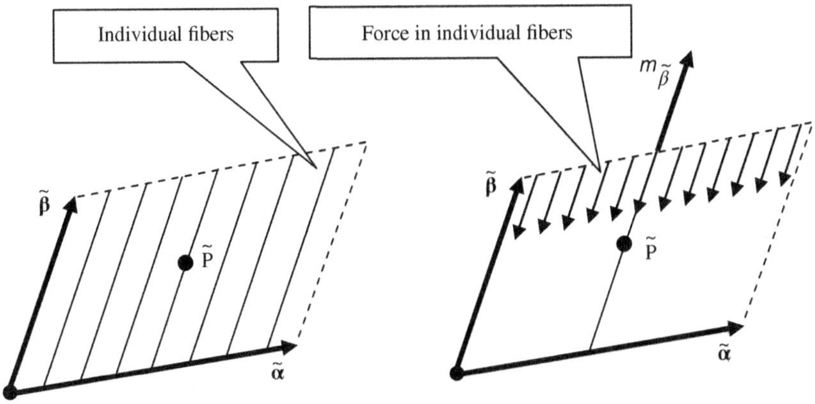

Figure 17.21 Representative volume with carbon fibers in $\tilde{\beta}$ direction

representative volume. As the volume increases or decreases, this fluid will produce pressure on the representative volume's surfaces. The internal forces on the surface of the representative volume due to this pressure are parallel to the surface vectors. The resultant forces due to this pressure are parallel to the surface vectors and are proportional to the surface area (the magnitude of the surface vector), Figure 17.22.

Finally, there is a moment on each of the surfaces of the generalized material volume. These moments resist any changes to the generalized material element's initial angle $\bar{\psi}$. They can easily be replaced by equivalent force couples as shown in Figure 17.23.

As $m_{\tilde{\alpha}}, m_{\tilde{\beta}}$ and m_p produce no moments about point P, it follows from the equilibrium of moments about point P that

$$m_{\tilde{\alpha}\tilde{\beta}} = m_{\tilde{\beta}\tilde{\alpha}}. \tag{17.50}$$

These moments can be physically illustrated (visualized) by a representative volume that contains two solid bars linked by a torsional pin, Figure 17.24. The torsional pin resists any change of the angle between the two bars, thus producing a concentrated point moments at each of the bars. The moments are equal in size and opposite in direction. For each of the bars, the corresponding moment is simply replaced by an equivalent force couple with the forces being perpendicular to the bar, Figure 17.24.

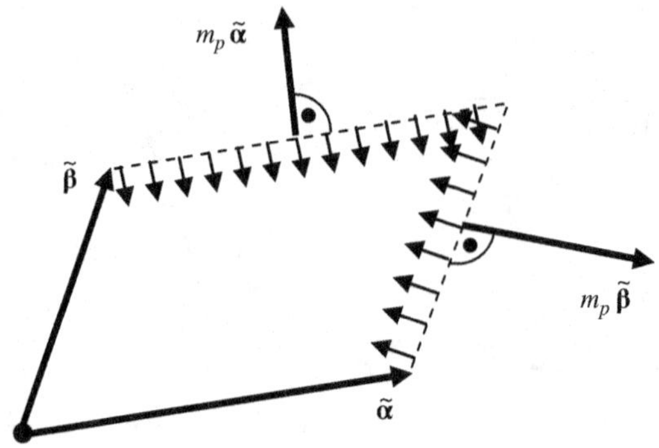

Figure 17.22 The resultant internal forces due to the Munjiza pressure

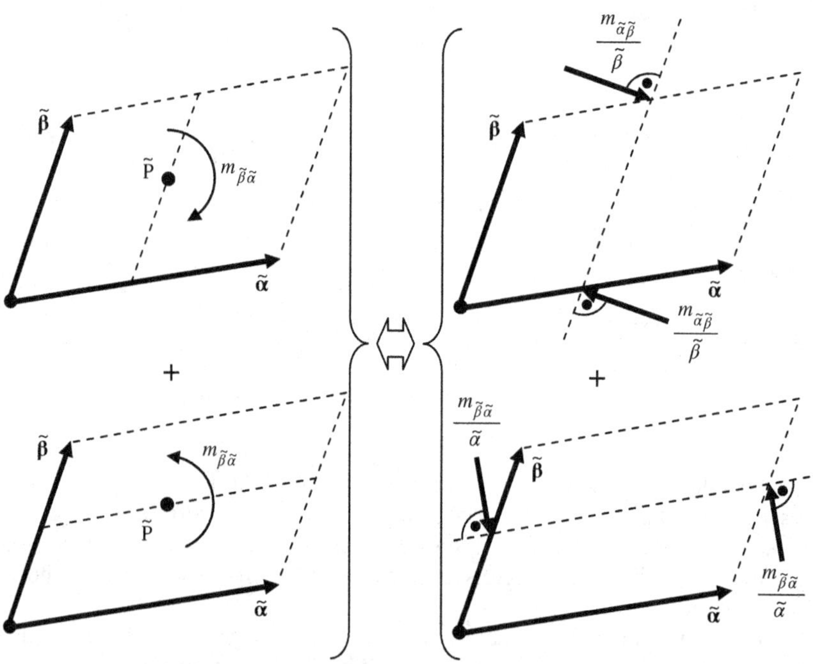

Figure 17.23 The forces due to the Munjiza moments

By the above procedure the generalized material element has been turned into a mechanical system consisting of two bars, each of which can be stretched: the bars can also shear relative to each other. In addition, the material element's volume change will produce pressure.

17.10 Converting the Munjiza Stress Matrix to the Cauchy Stress Matrix

The Munjiza stress matrix is independent of deformation. As such, it is stored within the material package, where the Cauchy stress matrix is calculated from the Munjiza stress matrix, when needed. It is not remembered within the material package, but is passed to the

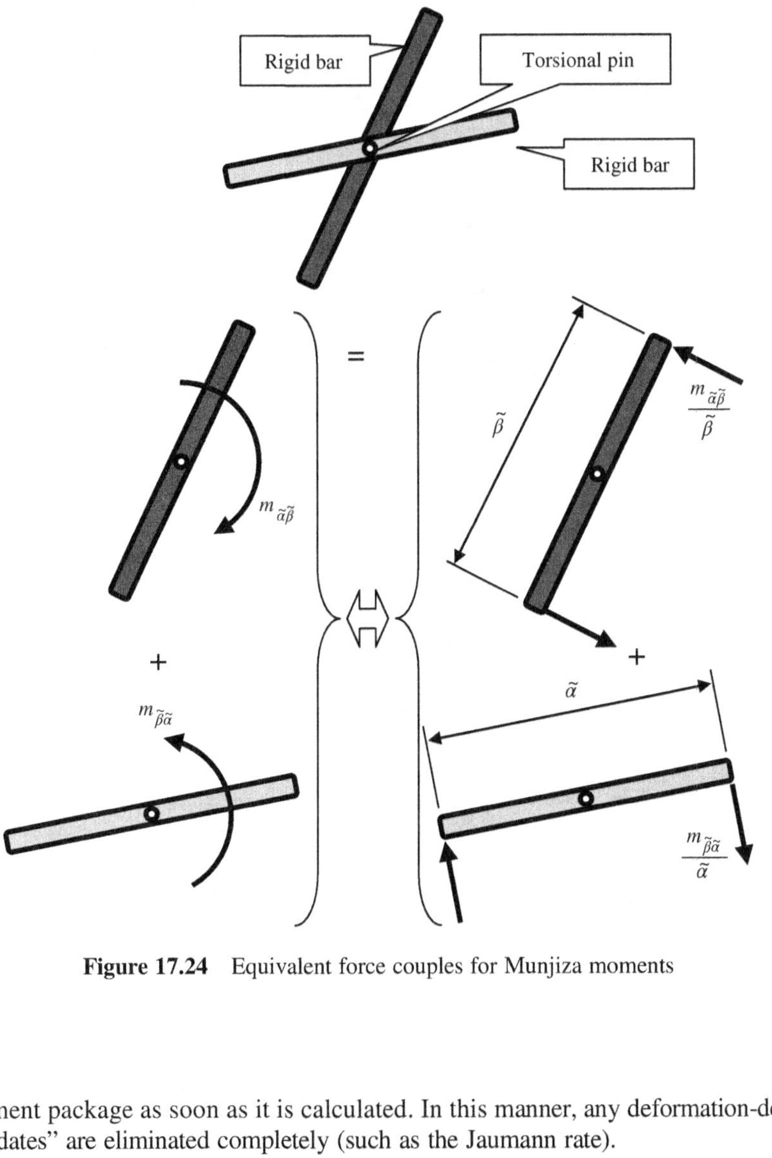

Figure 17.24 Equivalent force couples for Munjiza moments

finite element package as soon as it is calculated. In this manner, any deformation-dependent "stress updates" are eliminated completely (such as the Jaumann rate).

First Piola-Kirchhoff Stress Matrix. Generalized stress matrix components can be easily converted into the first Piola-Kirchhoff stress matrix. As noted earlier, the first Piola-Kirchhoff matrix components describe the forces on the deformed generalized material element's surfaces. Most important, these forces are conveniently placed in the global **i** and **j** directions, Figure 17.25.

The Surfaces of the Generalized Material Element. The surfaces of the generalized material element are shown in Figure 17.26.

The generalized material element's current volume is calculated from the current position

$$\tilde{v} = \det\begin{bmatrix} \tilde{\alpha}_i & \tilde{\beta}_i \\ \tilde{\alpha}_j & \tilde{\beta}_j \end{bmatrix} = \tilde{\alpha}_i\tilde{\beta}_j - \tilde{\beta}_i\tilde{\alpha}_j. \tag{17.51}$$

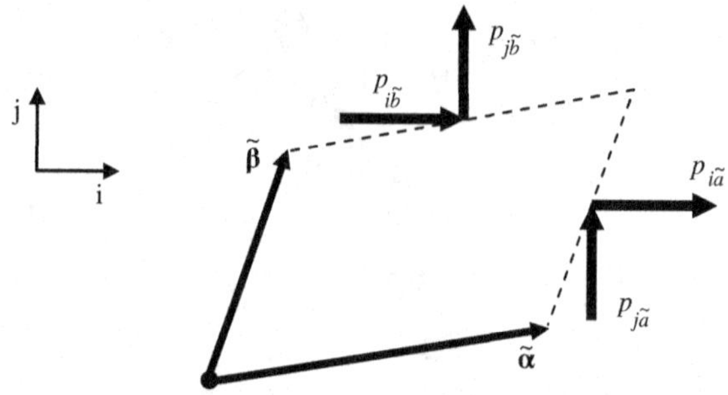

Figure 17.25 The first Piola-Kirchhoff stress matrix components for the generalized material element: internal forces on surfaces $\tilde{\mathbf{a}}$ and $\tilde{\mathbf{b}}$ are expressed using **i** and **j** directions

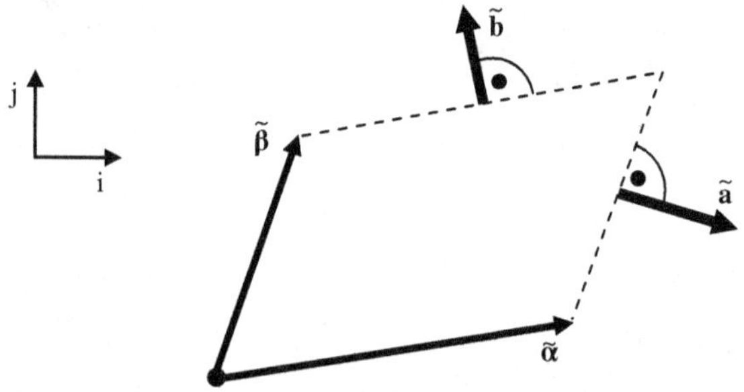

Figure 17.26 Surfaces of the current deformed generalized material element

The solid element's surfaces (vectors $\tilde{\mathbf{a}}$ and $\tilde{\mathbf{b}}$) are given by

$$\begin{bmatrix} \tilde{a}_i & \tilde{b}_i \\ \tilde{a}_j & \tilde{b}_j \end{bmatrix} = \tilde{v} \begin{bmatrix} \tilde{\alpha}_i & \tilde{\beta}_i \\ \tilde{\alpha}_j & \tilde{\beta}_j \end{bmatrix}^{-T}. \tag{17.52}$$

Equation (17.52) is obtained directly by simply considering that the scalar products

$$\tilde{\mathbf{a}} \bullet \tilde{\boldsymbol{\alpha}} = \tilde{v} \quad \text{and} \quad \underline{\tilde{\boldsymbol{\alpha}}} \bullet \tilde{\boldsymbol{\alpha}} = 1, \tag{17.53}$$

$$\tilde{\mathbf{a}} \bullet \tilde{\boldsymbol{\beta}} = 0 \quad \text{and} \quad \tilde{\mathbf{b}} \bullet \tilde{\boldsymbol{\alpha}} = 0. \tag{17.54}$$

Munjiza Pressure. The First Piola-Kirchhoff internal forces (components of the First Piola-Kirchhoff stress matrix) due to Munjiza pressure (pressure component of the Munjiza stress matrix) are given by

$$\begin{aligned} m_p\tilde{\mathbf{a}} &= \begin{bmatrix} m_p\tilde{a}_i \\ m_p\tilde{a}_j \end{bmatrix} \\ m_p\tilde{\mathbf{b}} &= \begin{bmatrix} m_p\tilde{b}_i \\ m_p\tilde{b}_j \end{bmatrix}. \end{aligned} \tag{17.55}$$

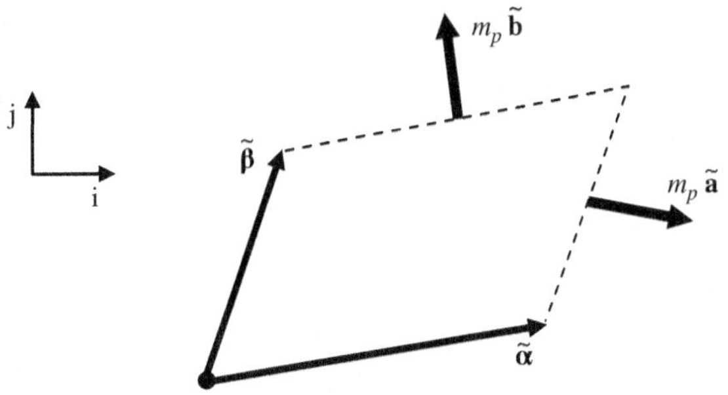

Figure 17.27 First Piola-Kirchhoff components due to Munjiza pressure

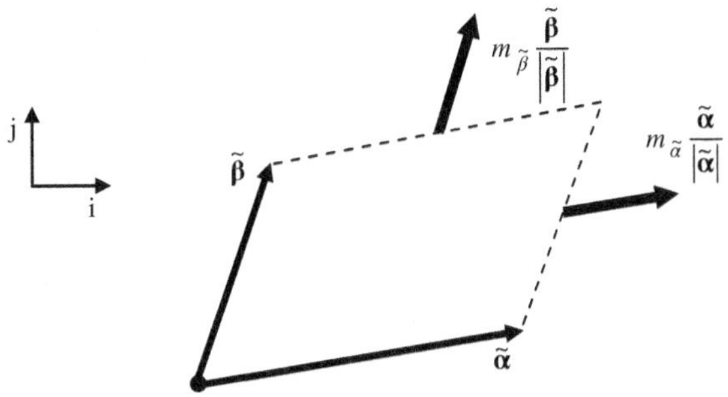

Figure 17.28 First Piola-Kirchhoff internal forces due to $m_{\tilde{\alpha}}$ and $m_{\tilde{\beta}}$ Munjiza forces

as shown in Figure 17.27.

Edge Stretches. The First Piola-Kirchhoff internal forces due to $m_{\tilde{\alpha}}$ and $m_{\tilde{\beta}}$ components of the Munjiza stress matrix are by definition given by

$$m_{\tilde{\alpha}}\frac{\tilde{\boldsymbol{\alpha}}}{|\tilde{\boldsymbol{\alpha}}|} = \begin{bmatrix} \frac{m_{\tilde{\alpha}}}{|\tilde{\boldsymbol{\alpha}}|}\tilde{\alpha}_i \\ \frac{m_{\tilde{\alpha}}}{|\tilde{\boldsymbol{\alpha}}|}\tilde{\alpha}_j \end{bmatrix}$$
$$m_{\tilde{\beta}}\frac{\tilde{\boldsymbol{\beta}}}{|\tilde{\boldsymbol{\beta}}|} = \begin{bmatrix} \frac{m_{\tilde{\beta}}}{|\tilde{\boldsymbol{\beta}}|}\tilde{\beta}_i \\ \frac{m_{\tilde{\beta}}}{|\tilde{\boldsymbol{\beta}}|}\tilde{\beta}_j \end{bmatrix}, \tag{17.56}$$

as shown in Figure 17.28. It is worth noting that the Munjiza forces $m_{\tilde{\alpha}}$ and $m_{\tilde{\beta}}$ do not change with the generalized material element's geometry. In contrast, the First Piola-Kirchhoff forces

$$m_{\tilde{\alpha}}\frac{\tilde{\boldsymbol{\alpha}}}{|\tilde{\boldsymbol{\alpha}}|} \quad \text{and} \quad m_{\tilde{\beta}}\frac{\tilde{\boldsymbol{\beta}}}{|\tilde{\boldsymbol{\beta}}|} \tag{17.57}$$

depend on $\tilde{\boldsymbol{\alpha}}$ and $\tilde{\boldsymbol{\beta}}$. As such, they are never stored in the material package, for they are deformation-dependent.

Munjiza Moments. The First Piola-Kirchhoff forces due to the Munjiza moments $m_{\tilde{\alpha}\tilde{\beta}}$ and $m_{\tilde{\beta}\tilde{\alpha}}$ are calculated by first replacing the moments with the equivalent force couples, which are perpendicular to vectors $\tilde{\boldsymbol{\alpha}}$ and $\tilde{\boldsymbol{\beta}}$.

The corresponding first Piola-Kirchhoff forces are therefore

$$\frac{m_{\tilde{\alpha}\tilde{\beta}}}{|\tilde{\boldsymbol{\alpha}}|}\frac{\tilde{\mathbf{b}}}{|\tilde{\mathbf{b}}|} \tag{17.58}$$

and

$$\frac{m_{\tilde{\beta}\tilde{\alpha}}}{|\tilde{\boldsymbol{\beta}}|}\frac{\tilde{\mathbf{a}}}{|\tilde{\mathbf{a}}|}, \tag{17.59}$$

as shown in Figure 17.29.

The First Piola-Kirchhoff Stress Tensor Matrix. The First Piola-Kirchhoff stress tensor matrix is obtained by adding each of the above calculated contributions. This is shown in Figure 17.30.

From the first Piola-Kirchhoff stress matrix **P**, the Cauchy stress matrix **C** is obtained from the fact that

$$\begin{bmatrix} p_{i\tilde{a}} & p_{i\tilde{b}} \\ p_{j\tilde{a}} & p_{j\tilde{b}} \end{bmatrix} = \begin{bmatrix} c_{ii} & c_{ij} \\ c_{ji} & c_{jj} \end{bmatrix} \begin{bmatrix} \tilde{a}_i & \tilde{b}_i \\ \tilde{a}_j & \tilde{b}_j \end{bmatrix}, \tag{17.60}$$

which yields

$$\begin{bmatrix} c_{ii} & c_{ij} \\ c_{ji} & c_{jj} \end{bmatrix} = \begin{bmatrix} p_{i\tilde{a}} & p_{i\tilde{b}} \\ p_{j\tilde{a}} & p_{j\tilde{b}} \end{bmatrix} \begin{bmatrix} \tilde{a}_i & \tilde{b}_i \\ \tilde{a}_j & \tilde{b}_j \end{bmatrix}^{-1}, \tag{17.61}$$

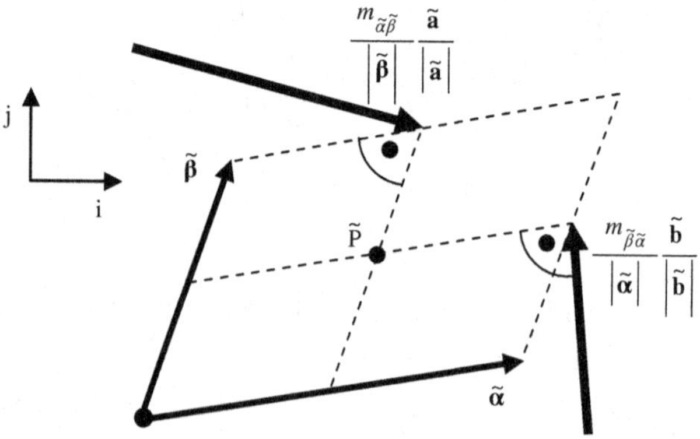

Figure 17.29 Second Piola-Kirchhoff forces due to Munjiza moments

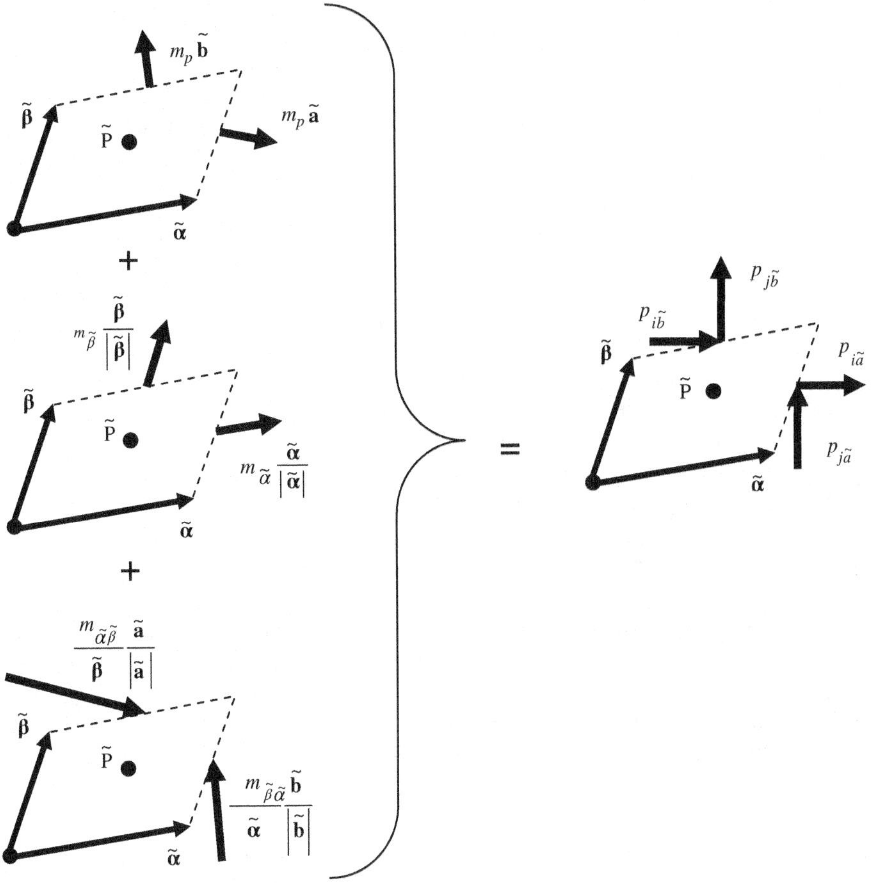

Figure 17.30 Adding all the internal forces in order to obtain the First Piola-Kirchhoff stress matrix components

where (see Equation (17.52))

$$\begin{bmatrix} \tilde{a}_i & \tilde{b}_i \\ \tilde{a}_j & \tilde{b}_j \end{bmatrix} = \begin{bmatrix} \tilde{\alpha}_i & \tilde{\beta}_i \\ \tilde{\alpha}_j & \tilde{\beta}_j \end{bmatrix}^{-T} \det\left(\begin{bmatrix} \tilde{\alpha}_i & \tilde{\beta}_i \\ \tilde{\alpha}_j & \tilde{\beta}_j \end{bmatrix} \right). \quad (17.62)$$

17.11 Developing Constitutive Laws

The general procedure of obtaining the Cauchy stress tensor can be summarized as follows:

a. Take the initial and the current generalized material elements. These are obtained by considering the deformation kinematics, Chapter 15 and are passed as input to the material package

$$\text{initial}: \begin{bmatrix} \bar{\alpha}_i & \bar{\beta}_i \\ \bar{\alpha}_j & \bar{\beta}_j \end{bmatrix} \quad \text{current}: \begin{bmatrix} \tilde{\alpha}_i & \tilde{\beta}_i \\ \tilde{\alpha}_j & \tilde{\beta}_j \end{bmatrix}. \tag{17.63}$$

b. Calculate the generalized stretches (using Equations (17.27), (17.33) and (17.39))

$$s_v \; ; \; s_\alpha \; ; \; s_\beta \; ; \; s_\psi. \tag{17.64}$$

c. Calculate the Munjiza stress matrix components and save them together with the state variables (if any)

$$m_p \; ; \; m_\alpha \; ; \; m_\beta \; ; \; m_\psi. \tag{17.65}$$

d. From the Munjiza stress matrix components calculate the First Piola-Kirchhoff stress matrix components, as explained in Figure 17.30. Do not save the First Piola-Kirchhoff stress matrix for it changes as the solid rotates, i.e., it is not objective.
e. From the First Piola-Kirchhoff stress matrix components calculate the Cauchy stress matrix components, Equation (17.61). Do not remember this matrix for it also changes with solid rotation, i.e., it is not objective.
f. Return the Cauchy stress matrix to the finite element package.

It is worth noting that the only step that depends on the actual material type is step (c). All other steps are the same irrespective of what material model is used. This makes developing a new material model as simple as writing step (c). This is illustrated by the examples listed below.

An Example of an Elastic Anisotropic Solid. As explained above, only step (c) needs to be implemented to create a new material. For a simple hyper-elastic solid, the following procedure can be adopted:

a. Starting from the generalized stretches, calculate the logarithmic strains:

$$\begin{aligned} e_v &= \ln s_v \\ e_\alpha &= \ln s_\alpha \\ e_\beta &= \ln s_\beta \\ e_\psi &= \ln s_\psi. \end{aligned} \tag{17.66}$$

b. Calculate the Munjiza stresses components:

$$\begin{aligned} m_v &= \mathrm{M}_v e_v \\ m_\alpha &= \mathrm{M}_\alpha e_\alpha \\ m_\beta &= \mathrm{M}_\beta e_\beta \\ m_\psi &= \mathrm{M}_\psi e_\psi, \end{aligned} \tag{17.67}$$

where M_v, M_α, M_β and M_ψ are the so-called Munjiza elastic constants defining the anisotropic material.

The above model is fully anisotropic. An isotropic solid is obtained by simply supplying the material constants corresponding to the isotropic solid. First the generalized material element is of square shape, Figure 17.31.

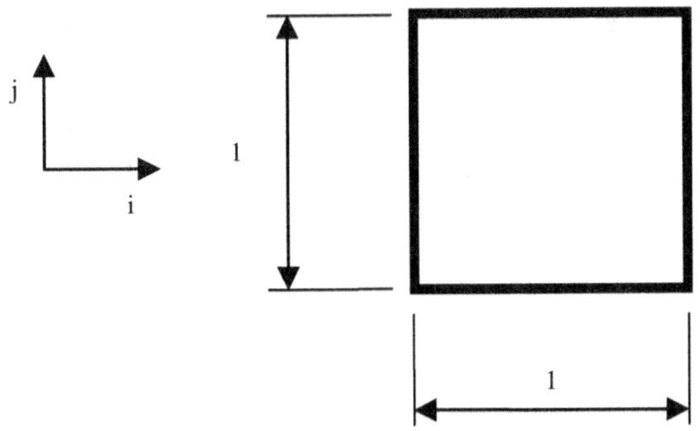

Figure 17.31 Generalized material element for isotropic material

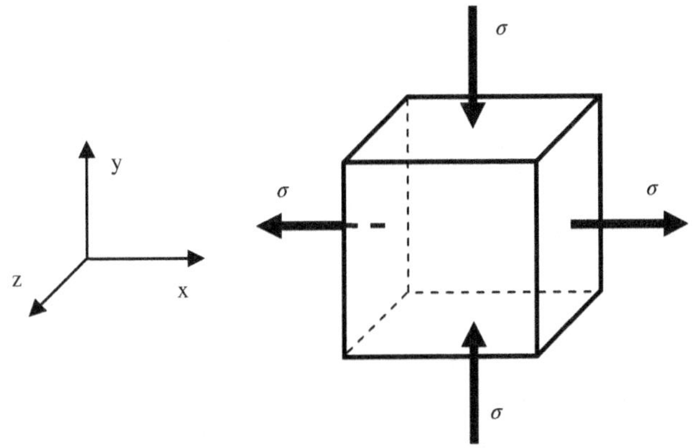

Figure 17.32 A small strain elasticity problem with $\sigma_z = 0$

Plane Stress. Plane stress 2D deformation is characterized by zero stress in the z direction. Consider a small strain 3D cube-shaped material element subject to stress, as shown in Figure 17.32. From Hooke's law it follows that

$$\begin{aligned}\varepsilon_x &= \frac{\sigma}{E} + \nu\frac{\sigma}{E}\\ \varepsilon_y &= -\frac{\sigma}{E} - \nu\frac{\sigma}{E}\\ \varepsilon_z &= -\nu\frac{\sigma}{E} + \nu\frac{\sigma}{E} = 0.\end{aligned} \tag{17.68}$$

The volumetric strain (for the small strains case) is given by

$$\varepsilon_v = \varepsilon_x + \varepsilon_y + \varepsilon_z = 0. \tag{17.69}$$

By substituting ε_x for e_α, ε_y for e_β and ε_v for e_v in Equation (17.67) one obtains

$$m_v = \mathrm{M}_v\varepsilon_v = \mathrm{M}_v 0 = 0$$
$$m_\alpha = \mathrm{M}_\alpha e_\alpha = \mathrm{M}_\alpha \varepsilon_x = \mathrm{M}_\alpha\left(\frac{\sigma}{\mathrm{E}} + \nu\frac{\sigma}{\mathrm{E}}\right). \tag{17.70}$$

In order for Equation (17.67) to converge to Equation (17.68) for small strains, it is necessary that

$$m_\alpha = \sigma. \tag{17.71}$$

From this, it follows that

$$\sigma = \mathrm{M}_\alpha\left(\frac{\sigma}{\mathrm{E}} + \nu\frac{\sigma}{\mathrm{E}}\right), \tag{17.72}$$

which yields

$$\mathrm{M}_\alpha = \frac{\mathrm{E}}{1+\nu}. \tag{17.73}$$

By analogy

$$\mathrm{M}_\beta = \frac{\mathrm{E}}{1+\nu}. \tag{17.74}$$

In a similar manner, by considering the problem shown in Figure 17.33. By employing Hooke's law, one obtains the following small strains

$$\varepsilon_x = \frac{\sigma}{\mathrm{E}}$$
$$\varepsilon_y = -\nu\frac{\sigma}{\mathrm{E}} \tag{17.75}$$
$$\varepsilon_z = -\nu\frac{\sigma}{\mathrm{E}}.$$

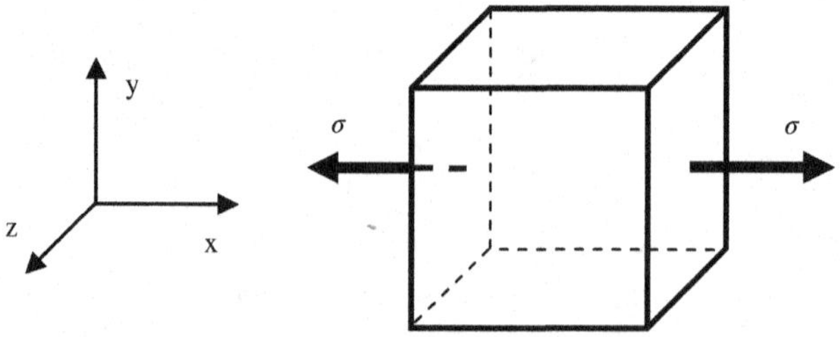

Figure 17.33 A small strain elasticity problem with $\sigma_y = \sigma_z = 0$

The volumetric strain in the case of small strains is given by

$$\begin{aligned}\varepsilon_v &= \varepsilon_x + \varepsilon_y + \varepsilon_z \\ &= \frac{\sigma}{E}(1-\nu-\nu) = \frac{\sigma}{E}(1-2\nu).\end{aligned} \tag{17.76}$$

However, in order to obtain the 2D plane stress formulation, the volumetric strain is redefined as 2D volumetric strain, such that

$$\begin{aligned}\varepsilon_v &= \varepsilon_x + \varepsilon_y \\ &= \frac{\sigma}{E}(1-\nu).\end{aligned} \tag{17.77}$$

The stress obtained using Equation (17.67) should, for infinitesimal strains, converge to

$$m_v + m_\alpha = \sigma, \tag{17.78}$$

where

$$\begin{aligned}m_v &= M_v e_v = M_v \varepsilon_v = M_v \left[\frac{\sigma}{E}(1-\nu)\right] \\ m_\alpha &= M_\alpha e_\alpha = M_\alpha \varepsilon_x = M_\alpha \left(\frac{\sigma}{E}\right).\end{aligned} \tag{17.79}$$

By substituting M_α from Equation (17.73) one obtains

$$\begin{aligned}m_v &= M_v e_v = M_v \varepsilon_v = M_v \left[\frac{\sigma}{E}(1-\nu)\right] \\ m_\alpha &= \frac{E}{1+\nu}\left(\frac{\sigma}{E}\right) = \frac{\sigma}{1+\nu}.\end{aligned} \tag{17.80}$$

After substitution of these into Equation (17.78), one obtains

$$M_v \left[\frac{\sigma}{E}(1-\nu)\right] + \frac{\sigma}{1+\nu} = \sigma, \tag{17.81}$$

which yields

$$M_v \frac{1-\nu}{E} = 1 - \frac{1}{1+\nu} \tag{17.82}$$

or

$$M_v \frac{(1-\nu)(1+\nu)}{E} = \nu \tag{17.83}$$

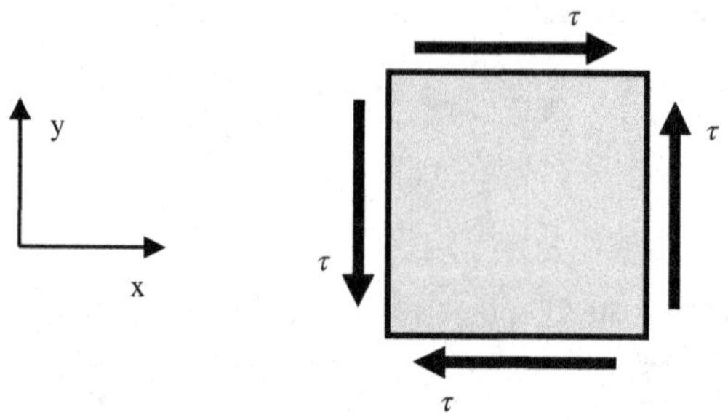

Figure 17.34 A pure shear problem in 2D

and

$$M_v = \frac{\nu E}{1-\nu^2}. \tag{17.84}$$

By considering the small strain pure shear problem shown in Figure 17.34 and applying Hooke's law to it, one obtains

$$\varepsilon_{xy} = \frac{\tau}{2G}. \tag{17.85}$$

For infinitesimally small strains, Equation (17.67) has to produce the same result, which yields

$$m_\psi = \tau = M_\psi e_\psi = M_\psi (2\varepsilon_{xy}) = M_\psi \frac{2\tau}{2G}. \tag{17.86}$$

This yields

$$M_\psi = G. \tag{17.87}$$

From the above considerations, it follows that the material constants from Equation (17.67) are, for the plane stress case, given by

$$\begin{aligned} M_v &= \frac{\nu E}{(1-\nu^2)} \\ M_\alpha &= \frac{E}{(1+\nu)} \\ M_\beta &= \frac{E}{(1+\nu)} \\ M_\psi &= \frac{E}{2(1+\nu)}. \end{aligned} \tag{17.88}$$

Plane Strain Isotropic Material. Elastic constants from Equation (17.67) can be derived from Hooke's law in such a way that in the limit when strains are small, the plane strain

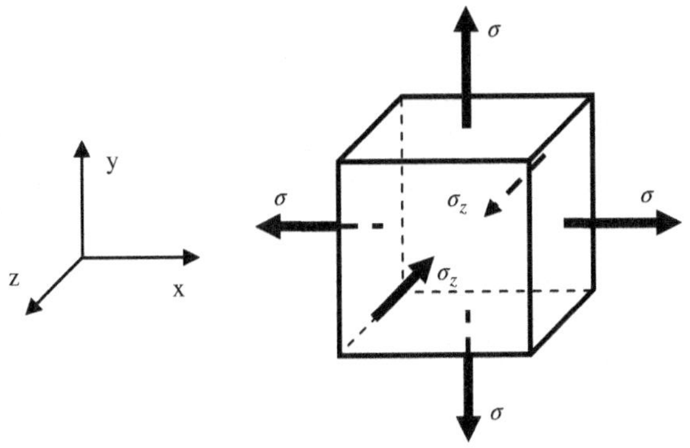

Figure 17.35 A small strain elasticity problem with $\sigma_z \neq 0$

constitutive law for a small strain isotropic material is recovered. Consider the small strain elastic problem shown in Figure 17.35.

By employing Hooke's law, one obtains the following strains

$$\begin{aligned}
\varepsilon_x &= \frac{\sigma}{E}(1-\nu) - \nu\frac{\sigma_z}{E} \\
\varepsilon_y &= \frac{\sigma}{E}(1-\nu) - \nu\frac{\sigma_z}{E} \\
\varepsilon_z &= \frac{\sigma_z}{E} - 2\nu\frac{\sigma}{E}.
\end{aligned} \tag{17.89}$$

From the definition of plane strain it follows that

$$\varepsilon_z = 0, \tag{17.90}$$

which means that

$$\varepsilon_z = \frac{\sigma_z}{E} - 2\nu\frac{\sigma}{E} = 0 \tag{17.91}$$

and

$$\sigma_z = 2\nu\sigma. \tag{17.92}$$

When substituted into Equation (17.89) this yields

$$\begin{aligned}
\varepsilon_x &= \frac{\sigma}{E}(1-\nu) - \nu\frac{2\nu\sigma}{E} \\
\varepsilon_y &= \frac{\sigma}{E}(1-\nu) - \nu\frac{2\nu\sigma}{E} \\
\varepsilon_z &= \frac{2\nu\sigma}{E} - \frac{2\nu\sigma}{E} = 0.
\end{aligned} \tag{17.93}$$

As the strains are small, the volumetric strain is given by

$$\varepsilon_v = \varepsilon_x + \varepsilon_y = 2\frac{\sigma}{E}\left(1-\nu-2\nu^2\right). \tag{17.94}$$

The stress m_v should, in the limit for small strains, converge to stress σ_z, thus

$$m_v = \sigma_z \tag{17.95}$$

or

$$M_v \varepsilon_v = \sigma_z, \tag{17.96}$$

which after substitution for ε_v from Equation (17.94) and for σ_z from Equation (17.92) yields

$$M_v\left(2\frac{\sigma}{E}\left(1-\nu-2\nu^2\right)\right) = 2\nu\sigma, \tag{17.97}$$

which means that

$$M_v = \frac{\nu E}{(1+\nu)(1-2\nu)}. \tag{17.98}$$

In a similar manner, by considering the problem shown in Figure 17.36, the strains are obtained directly from Hooke's law

$$\begin{aligned} \varepsilon_x &= \frac{\sigma}{E} + \nu\frac{\sigma}{E} \\ \varepsilon_y &= -\frac{\sigma}{E} - \nu\frac{\sigma}{E} \\ \varepsilon_z &= -\nu\frac{\sigma}{E} + \nu\frac{\sigma}{E} = 0. \end{aligned} \tag{17.99}$$

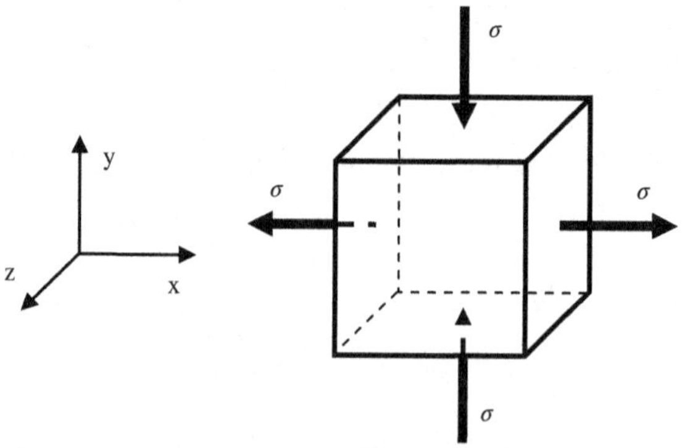

Figure 17.36 A zero volume change deformation

The volume change is therefore equal to zero and so is the volumetric strain

$$\varepsilon_v = \varepsilon_x + \varepsilon_y + \varepsilon_z = 0, \tag{17.100}$$

By substituting these into Equation (17.67), one obtains

$$\begin{aligned} m_v &= \mathrm{M_v} e_v = \mathrm{M_v}\varepsilon_v = 0 \\ m_\alpha &= \mathrm{M_\alpha} e_\alpha = \mathrm{M_\alpha}\varepsilon_\alpha = \mathrm{M_\alpha}\left(\frac{\sigma}{\mathrm{E}} + \nu\frac{\sigma}{\mathrm{E}}\right). \end{aligned} \tag{17.101}$$

In the limit for small strains

$$\sigma = m_v + m_\alpha. \tag{17.102}$$

After substituting m_v and m_α from Equation (17.101) one obtains

$$\sigma = 0 + \mathrm{M_\alpha}\left(\frac{\sigma}{\mathrm{E}} + \nu\frac{\sigma}{\mathrm{E}}\right) \tag{17.103}$$

or

$$\mathrm{M_\alpha} = \frac{\mathrm{E}}{1+\nu} = 2\left(\frac{\mathrm{E}}{2(1+\nu)}\right) = 2\mathrm{G}. \tag{17.104}$$

In a similar manner

$$\mathrm{M_\beta} = \frac{\mathrm{E}}{1+\nu} = 2\mathrm{G}. \tag{17.105}$$

The elastic constant $\mathrm{M_\psi}$ is obtained by considering a pure shear case, Figure 17.34. Hooke's law gives

$$\tau = 2\mathrm{G}\varepsilon_{xy}. \tag{17.106}$$

In the limit τ and m_ψ should be the same, which yields

$$\mathrm{M_\psi} = \mathrm{G} = \frac{\mathrm{E}}{2(1+\nu)}. \tag{17.107}$$

In summary, the plane strain material constants from Equation (17.67) are given by

$$\begin{aligned} \mathrm{M_v} &= \frac{\nu \mathrm{E}}{(1+\nu)(1-2\nu)} \\ \mathrm{M_\alpha} &= \frac{\mathrm{E}}{(1+\nu)} \\ \mathrm{M_\beta} &= \frac{\mathrm{E}}{(1+\nu)} \\ \mathrm{M_\psi} &= \frac{\mathrm{E}}{2(1+\nu)}. \end{aligned} \tag{17.108}$$

where E is the elastic modulus (also known as Young's modulus) and ν is Poisson's ratio.

It is worth noting that the only difference between the 2D plane stress and the 2D plane strain problem is in the elastic constants as given by Equations (17.88) and (17.108). Otherwise, the deformation kinematics is exactly the same. This means that the same finite element formulation is used for both plane stress and plane strain.

17.12 Generalized Hypo-Elastic Material

The current position of the generalized material element is described by the vectors

$$\tilde{\boldsymbol{\alpha}} = [\mathbf{i}\ \ \mathbf{j}]\begin{bmatrix}\tilde{\alpha}_i \\ \tilde{\alpha}_j\end{bmatrix} = \begin{bmatrix}\tilde{\alpha}_i \\ \tilde{\alpha}_j\end{bmatrix}, \tag{17.109}$$

$$\tilde{\boldsymbol{\beta}} = [\mathbf{i}\ \ \mathbf{j}]\begin{bmatrix}\tilde{\beta}_i \\ \tilde{\beta}_j\end{bmatrix} = \begin{bmatrix}\tilde{\beta}_i \\ \tilde{\beta}_j\end{bmatrix}. \tag{17.110}$$

The previous position of the generalized material element is described by vectors

$$\hat{\boldsymbol{\alpha}} = [\mathbf{i}\ \ \mathbf{j}]\begin{bmatrix}\hat{\alpha}_i \\ \hat{\alpha}_j\end{bmatrix} = \begin{bmatrix}\hat{\alpha}_i \\ \hat{\alpha}_j\end{bmatrix}, \tag{17.111}$$

$$\hat{\boldsymbol{\beta}} = [\mathbf{i}\ \ \mathbf{j}]\begin{bmatrix}\hat{\beta}_i \\ \hat{\beta}_j\end{bmatrix} = \begin{bmatrix}\hat{\beta}_i \\ \hat{\beta}_j\end{bmatrix}. \tag{17.112}$$

Multiplicative Decomposition of the Volumetric Stretch. The previous volume of the generalized material element is given by

$$\hat{v} = \det\begin{bmatrix}\hat{\alpha}_i & \hat{\beta}_i \\ \hat{\alpha}_j & \hat{\beta}_j\end{bmatrix} = \hat{\alpha}_i\hat{\beta}_j - \hat{\alpha}_j\hat{\beta}_i. \tag{17.113}$$

The current volume of the generalized material element is given by

$$\tilde{v} = \det\begin{bmatrix}\tilde{\alpha}_i & \tilde{\beta}_i \\ \tilde{\alpha}_j & \tilde{\beta}_j\end{bmatrix} = \tilde{\alpha}_i\tilde{\beta}_j - \tilde{\alpha}_j\tilde{\beta}_i. \tag{17.114}$$

The incremental stretching (multiplicative decomposition) of the volume (volumetric stretch) is defined as

$$s_v = \frac{\tilde{v}}{\hat{v}}. \tag{17.115}$$

Given the step increment such as a time step h, the rate of the volumetric stretch is obtained as

$$\dot{s}_v = \frac{s_v}{h}. \tag{17.116}$$

Multiplicative Decomposition of the Edge Stretches. The edge lengths of the previous generalized material element are given by

$$\hat{\alpha} = |\hat{\boldsymbol{\alpha}}| = \sqrt{\hat{\alpha}_i^2 + \hat{\alpha}_j^2}, \tag{17.117}$$

$$\hat{\beta} = |\hat{\boldsymbol{\beta}}| = \sqrt{\hat{\beta}_i^2 + \hat{\beta}_j^2}. \tag{17.118}$$

In a similar manner, the edge lengths of the current generalized material element are given by

$$\tilde{\alpha} = |\tilde{\boldsymbol{\alpha}}| = \sqrt{\tilde{\alpha}_i^2 + \tilde{\alpha}_j^2}, \tag{17.119}$$

$$\tilde{\beta} = |\tilde{\boldsymbol{\beta}}| = \sqrt{\tilde{\beta}_i^2 + \tilde{\beta}_j^2}. \tag{17.120}$$

The incremental stretching of the two edges (edge stretches) of the generalized material element are given by

$$s_\alpha = \frac{\tilde{\alpha}}{\hat{\alpha}}, \tag{17.121}$$

$$s_\beta = \frac{\tilde{\beta}}{\hat{\beta}}, \tag{17.122}$$

where the stretch increment is given in terms of the multiplicative decomposition.

The rates of the edge stretches are

$$\dot{s}_\alpha = \frac{s_\alpha}{h}, \tag{17.123}$$

$$\dot{s}_\beta = \frac{s_\beta}{h}, \tag{17.124}$$

where h is the time step.

Multiplicative Decomposition of the Angle Stretch. The initial angle between edges $\bar{\boldsymbol{\alpha}}$ and $\bar{\boldsymbol{\beta}}$ is given by

$$\bar{\psi} = \arccos\left(\frac{\bar{\boldsymbol{\alpha}} \bullet \bar{\boldsymbol{\beta}}}{\bar{\alpha}\bar{\beta}}\right). \tag{17.125}$$

The current angle between edges $\tilde{\boldsymbol{\alpha}}$ and $\tilde{\boldsymbol{\beta}}$ is

$$\tilde{\psi} = \arccos\left(\frac{\tilde{\boldsymbol{\alpha}} \bullet \tilde{\boldsymbol{\beta}}}{\tilde{\alpha}\tilde{\beta}}\right). \tag{17.126}$$

The previous angle between edges $\hat{\boldsymbol{\alpha}}$ and $\hat{\boldsymbol{\beta}}$ is

$$\hat{\psi} = \arccos\left(\frac{\hat{\boldsymbol{\alpha}} \bullet \hat{\boldsymbol{\beta}}}{\hat{\alpha}\hat{\beta}}\right). \tag{17.127}$$

All three angles, $\bar{\psi}, \tilde{\psi}$ and $\hat{\psi}$ are from the interval between 0 and π.

The incremental stretch of the angle (angle stretch) is defined by

$$s_{\psi} = \frac{\tilde{\psi}}{\hat{\psi}}, \tag{17.128}$$

where the multiplicative decomposition of the stretch increment is implied by the definition of the stretch increment itself. The rate of the angle stretch is

$$\dot{s}_{\psi} = \frac{s_{\psi}}{h}, \tag{17.129}$$

where h is the time step.

From the above, three modes of generalized material element stretching can be identified: The first mode is stretching in the $\boldsymbol{\alpha}$ direction, the second mode is stretching in the $\boldsymbol{\beta}$ direction and the third mode is shearing, Figure 17.37.

The above modes also produce incremental volumetric stretch, as shown in Table 17.2. It is evident from the table that the stretch increments are independent from each other, except for the volumetric stretch increment, which accompanies all other stretch increments.

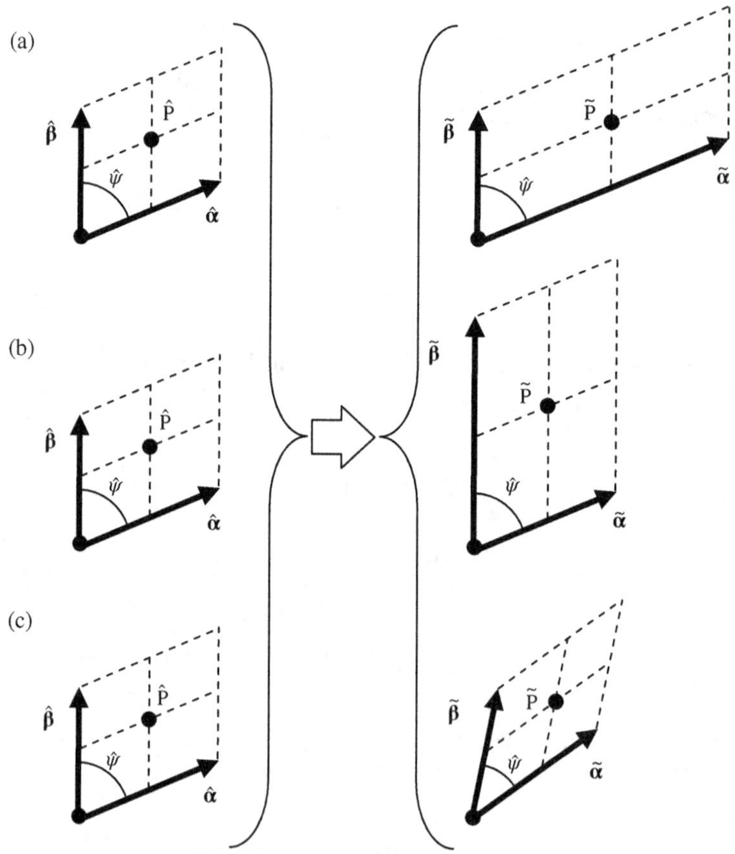

Figure 17.37 Three modes of incremental stretching the solid element: (a) Mode α - stretching in the $\boldsymbol{\alpha}$ direction; (b) mode β - stretching in the $\boldsymbol{\beta}$ direction and (c) mode ψ - shearing

Table 17.2 Stretches created by the three stretching modes

	s_α	s_β	s_ψ	s_ν
Mode α	•			•
Mode β		•		•
Mode ψ			•	•

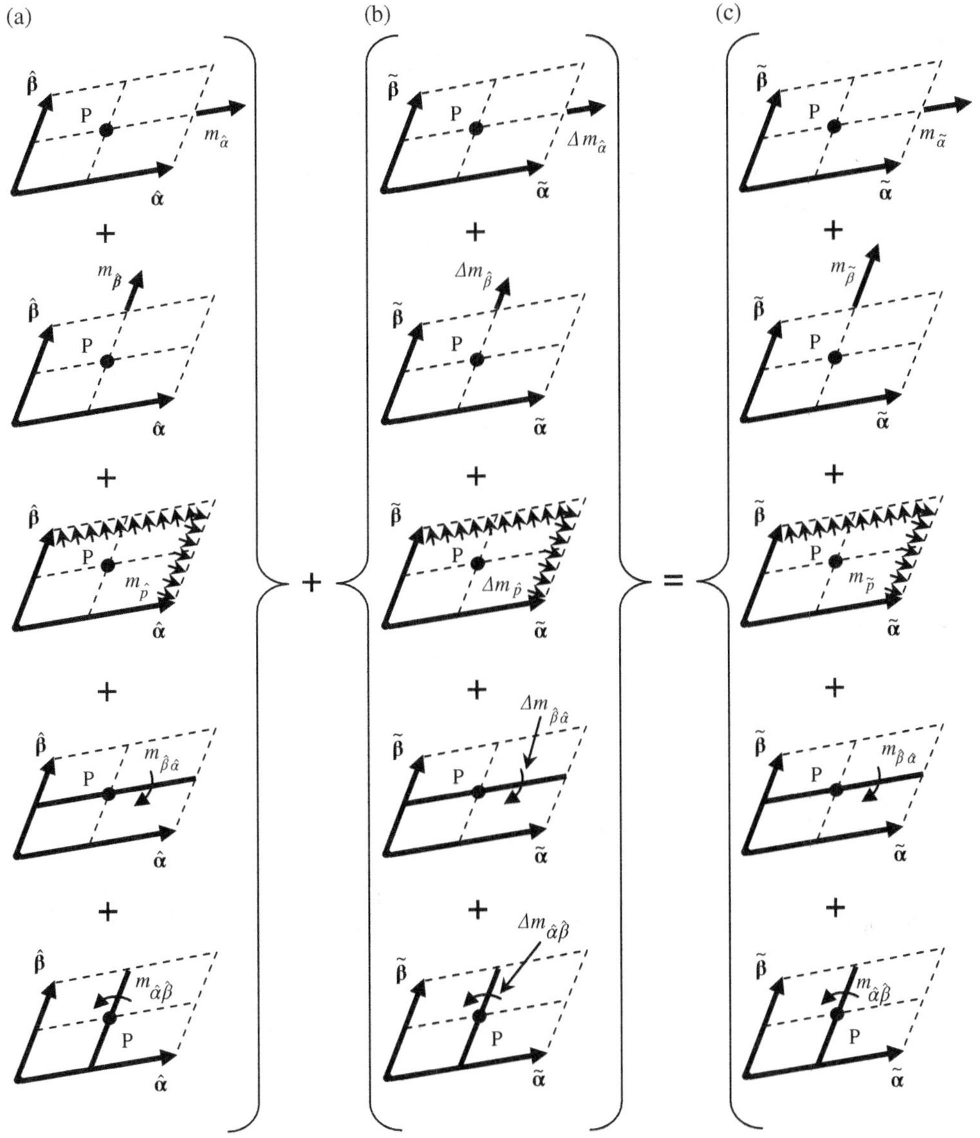

Figure 17.38 Multiplicative stress increments: (a) Total Munjiza forces at previous step. (b) Increment in the Munjiza forces due to deformation from the previous to the current step. (c) Total Munjiza forces at the current time step

Multiplicative Decomposition of the Munjiza Stress Tensor Matrix. From the incremental deformation kinematics, the increment in the internal forces is calculated and added to the already accumulated internal forces. One way of doing this is by using the Munjiza stress tensor matrix, Figure 17.38.

The incremental stress matrix components are added to the total stress matrix components. The total stress matrix components are stored as state variables inside the material package. In this way, the deformation kinematics does not have a need to know anything about: (a) what constitutive law is being used, (b) the different stress matrices being employed within the material package or (c) whether the formulation is hypo or hyper.

17.13 Unified Constitutive Approach for Strain Rate and Viscosity

Very often, constitutive laws involve strain rate and viscous internal forces. These are in general defined by the rate of the solid deformation. In order to consider the rate of stretching, it is enough to calculate the generalized stretch increments from the previous and current positions.

The stretch rate is then defined by simply dividing these increments by the time step h, i.e., time lapsed from the previous position to the current position

$$\begin{aligned}\dot{s}_v &= \frac{s_v}{h}\\ \dot{s}_\alpha &= \frac{s_\alpha}{h}\\ \dot{s}_\beta &= \frac{s_\beta}{h}\\ \dot{s}_\psi &= \frac{s_\psi}{h}.\end{aligned} \tag{17.130}$$

Once the stretch rates are known, the Munjiza stress matrix components are calculated using a suitable constitutive law formulation. The simplest law involves constants $\breve{\mathrm{M}}_\mathrm{v}, \breve{\mathrm{M}}_\alpha, \breve{\mathrm{M}}_\beta$, and $\breve{\mathrm{M}}_\psi$ such that

$$\begin{aligned}m_v &= \breve{\mathrm{M}}_\mathrm{v}\dot{s}_v\\ m_\alpha &= \breve{\mathrm{M}}_\alpha\dot{s}_\alpha\\ m_\beta &= \breve{\mathrm{M}}_\beta\dot{s}_\beta\\ m_\psi &= \breve{\mathrm{M}}_\psi\dot{s}_\psi,\end{aligned} \tag{17.131}$$

where m_v, m_α, m_β, and m_ψ are the resultant Munjiza forces due to viscosity-induced internal forces, and $\breve{\mathrm{M}}_\mathrm{v}, \breve{\mathrm{M}}_\alpha, \breve{\mathrm{M}}_\beta$ and $\breve{\mathrm{M}}_\psi$ are the viscosity related material constants.

An Example of Isotropic Material. In the case of isotropic plane stress, these constants can be obtained using an analogous approach as seen in the previous section, thus resulting in

$$\begin{aligned}\breve{\mathrm{M}}_\mathrm{v} &= \frac{\breve{\nu}\breve{\mathrm{E}}}{(1-\breve{\nu}^2)}\\ \breve{\mathrm{M}}_\alpha &= \frac{\breve{\mathrm{E}}}{(1+\breve{\nu})}\\ \breve{\mathrm{M}}_\beta &= \frac{\breve{\mathrm{E}}}{(1+\breve{\nu})}\\ \breve{\mathrm{M}}_\psi &= \frac{\breve{\mathrm{E}}}{2(1+\breve{\nu})}.\end{aligned} \tag{17.132}$$

An Example of Isotropic Plane Strain Material. In a similar manner to the plane strain case, one can obtain the material constants for the plane strain case.

$$\begin{aligned}
\breve{M}_v &= \frac{\breve{\nu}\breve{E}}{(1+\breve{\nu})(1-\breve{\nu})} \\
\breve{M}_\alpha &= \frac{\breve{E}}{(1+\breve{\nu})} \\
\breve{M}_\beta &= \frac{\breve{E}}{(1+\breve{\nu})} \\
\breve{M}_\psi &= \frac{\breve{E}}{2(1+\breve{\nu})}.
\end{aligned} \tag{17.133}$$

where $\breve{\nu}$ and $\breve{E}$ are given by

$$\begin{aligned}
\breve{\nu} &= 0 \\
\breve{E} &= 2\mu
\end{aligned} \tag{17.134}$$

and μ is the dynamic viscosity.

17.14 Summary

In this chapter the unified constitutive approach in 2D has been presented. The key idea is to store internal forces for the generalized material element inside the material package and never pass them outside of the material package. The stored internal forces must be such that they are rotation, stretch (deformation) invariant (i.e., objective). One example of such internal forces are the Munjiza internal forces.

Objectivity simply means that the already stored internal forces do not change with further say rotation of the generalized material element. This way, summing internal forces from different increments of hypo-deformation based formulation becomes trivial and involves no "stress correction" (as often employed in now outdated so called co-rotational formulations).

As a consequence, both hyper-deformation and hypo-deformation based constitutive law formulations become identical and converge to one so-called unified constitutive approach.

In this chapter the unified constitutive approach has been further demonstrated using examples of both anisotropic and isotropic materials.

Finally, it has been explained how only the Cauchy matrix of the stress tensor is passed outside of the material package, but is never stored within the material package – for it is not deformation-invariant.

Further Reading

[1] Anand, L. (1985) Constitutive equations for hot-working of metals. *Int. J. Plasticity*, **1**(3): 213–31.

[2] Argyris, J. H. and Kleiber, M. (1977) Incremental formulation in nonlinear mechanics and large strain elasto-plasticity – natural approach. Part 1. *Comput. Method Appl. M.*, **11**(2): 215–47.

[3] Argyris, J. H. and St. Doltsinis, J. (1979) On the large strain inelastic analysis in natural formulation. Part I: Quasistatic problems. *Comput. Method Appl. M.*, **20**(2): 213–51.

[4] Argyris, J. H. and St. Doltsinis,J. (1980) On the large strain inelastic analysis in natural formulation. Part II. Dynamic problems. *Comput. Method Appl. M.*, **21**(1): 91–126.

[5] Bathe, K. J. (1982) *Finite Element Procedures Engineering Analysis*. New Jersey, Prentice-Hall.
[6] Bathe, K. J. (1996) *Finite Element Procedures*. New Jersey, Prentice-Hall.
[7] Bonet, J. and Burton, A. J. (1998) A simple orthotropic, transversely isotropic hyperelastic constitutive equation for large strain computations. *Comput Method Appl M.*, **162**(1–4): 151–64.
[8] Bonet, J. and Wood, R. D. (1997) *Nonlinear Continuum Mechanics for Finite Element Analysis*. Cambridge, Cambridge University Press.
[9] de Souza Neto, E. A. and Peric, D. (1996) A computational framework for a class of fully coupled models for elastoplastic damage at finite strains with reference to the linearization aspects. *Comput. Method Appl. M.*, **130**(1–2): 179–93.
[10] de Souza Neto, E. A., Pires, F. M. A. and Owen, D. R. J. (2005) F-bar-based linear triangles and tetrahedra for finite strain analysis of nearly incompressible solids. Part I: formulation and benchmarking. *International Journal for Numerical Methods in Engineering*, **62**(3): 353–83.
[11] de Souza Neto, E. A. and Feijoo, R. A. (2008) On the equivalence between spatial and material volume averaging of stress in large strain multi-scale solid constitutive models. *Mech. Mater.*, **40**(10): 803–11.
[12] de Souza Neto, E. A., Peric, D., Huang, G. C. and Owen, D. R. J. (1995) Remarks on the stability of enhanced strain elements in finite elasticity and elastoplasticity. *Communications in Numerical Methods in Engineering*, **11**(11): 951–61.
[13] Green, A. E. and Naghdi, P. M. (1971) Some remarks on elastic-plastic deformation at finite strain. *International Journal of Engineering Science*, **9**(12): 1219–29.
[14] Gurtin, M. E. (1981) *An Introduction to Continuum Mechanics*. New York, Academic Press.
[15] Hill, R. (1979) *Aspects of Invariance in Solid Mechanics*, ed. Y. Chia-Shun, Elsevier, pp. 1–75.
[16] Hughes, T.J.R. (2000) *The Finite Element Method: Linear Static and Dynamic Finite Element Analysis*. New York, Dover Publications.
[17] Malvern, L. E. (1969) *Introduction to the Mechanics of a Continuous Medium*. New Jersey, Prentice Hall.
[18] Munjiza, A. (2004) *The Combined Finite-Discrete Element Method*. Chichester, John Wiley & Sons, Ltd.
[19] Peric, D. (1992) On consistent stress rates in solid mechanics – computational implications. *International Journal for Numerical Methods in Engineering*, **33**(4): 799–817.
[20] Peric, D. and Owen, D. R. J. (1998) Finite-element applications to the nonlinear mechanics of solids. *Reports on Progress in Physics*, **61**(11): 1495–1574.
[21] Peric, D., Owen, D. R. J. and Honnor, M. E. (1992) A model for finite strain elastoplasticity based on logarithmic strains – computational issues. *Comput. Method Appl. M.*, **94**(1): 35–61.
[22] Simo, J. C. (1985) On the computational significance of the intermediate configuration and hyperelastic stress relations in finite deformation elastoplasticity. *Mech. Mater.*, **4**(3–4): 439–51.
[23] Simo, J. C. (1988) A framework for finite strain elastoplasticity based on maximum plastic dissipation and the multiplicative decomposition. 1. Continuum formulation. *Comput. Method Appl. M.*, **66**(2): 199–219.
[24] Simo, J. C. (1988) A framework for finite strain elastoplasticity based on maximum plastic dissipation and the multiplicative decomposition. 2. Computational aspects. *Comput. Method Appl. M.*, **68**(1): 1–31.
[25] Simo, J. C. and Ortiz, M. (1985) A unified approach to finite deformation elastoplastic analysis based on the use of hyperelastic constitutive equations. *Comput. Method Appl. M.*, **49**(2): 221–45.
[26] Simo, J. C. and Ju, J. W. (1989) Finite deformation damage-elastoplasticity: a non conventional framework. *International Journal of Computational Mechanics*, **5**: 375–400.
[27] Simo, J. C. and Hughes, T. J. R. (1998) *Computational Inelasticity*. New York, Springer.
[28] Simo, J. C., Armero, F. and Taylor, R. L. (1993) Improved versions of assumed enhanced strain tri-linear elements for 3D finite deformation problems. *Comput. Method Appl. M.*, **110**(3–4): 359–86.
[29] Zienkiewicz, O. C. (1971) *The Finite Element Method in Engineering Science*. New York, McGraw-Hill.
[30] Zienkiewicz, O. C. and Taylor, R. L. (2005) *The Finite Element Method Set*. Oxford, Elsevier Science.

18

The Unified Constitutive Approach in 3D

18.1 Material Package Framework

In 3D space the developer of a constitutive law faces the situation shown in Figure 18.1. The initial generalized material element geometry is defined by vectors $\bar{\boldsymbol{\alpha}}$, $\bar{\boldsymbol{\beta}}$ and $\bar{\boldsymbol{\gamma}}$. The current generalized material element geometry is defined by vectors $\hat{\boldsymbol{\alpha}}$, $\hat{\boldsymbol{\beta}}$ and $\hat{\boldsymbol{\gamma}}$. The previous generalized material element geometry is defined by vectors $\tilde{\boldsymbol{\alpha}}$, $\tilde{\boldsymbol{\beta}}$ and $\tilde{\boldsymbol{\gamma}}$. These are obtained from the deformation kinematics and passed as an input to the material package.

18.2 Generalized Hyper-Elastic Material

The initial position of the generalized material element is provided as input to the material package and is described by three vectors (as explained in detail in Chapter 16)

$$\bar{\boldsymbol{\alpha}} = [\mathbf{i} \;\; \mathbf{j} \;\; \mathbf{k}] \begin{bmatrix} \bar{\alpha}_i \\ \bar{\alpha}_j \\ \bar{\alpha}_k \end{bmatrix} = \begin{bmatrix} \bar{\alpha}_i \\ \bar{\alpha}_j \\ \bar{\alpha}_k \end{bmatrix}, \tag{18.1}$$

$$\bar{\boldsymbol{\beta}} = [\mathbf{i} \;\; \mathbf{j} \;\; \mathbf{k}] \begin{bmatrix} \bar{\beta}_i \\ \bar{\beta}_j \\ \bar{\beta}_k \end{bmatrix} = \begin{bmatrix} \bar{\beta}_i \\ \bar{\beta}_j \\ \bar{\beta}_k \end{bmatrix}, \tag{18.2}$$

$$\bar{\boldsymbol{\gamma}} = [\mathbf{i} \;\; \mathbf{j} \;\; \mathbf{k}] \begin{bmatrix} \bar{\gamma}_i \\ \bar{\gamma}_j \\ \bar{\gamma}_k \end{bmatrix} = \begin{bmatrix} \bar{\gamma}_i \\ \bar{\gamma}_j \\ \bar{\gamma}_k \end{bmatrix}, \tag{18.3}$$

Large Strain Finite Element Method: A Practical Course, First Edition. Antonio Munjiza, Esteban Rougier and Earl E. Knight.

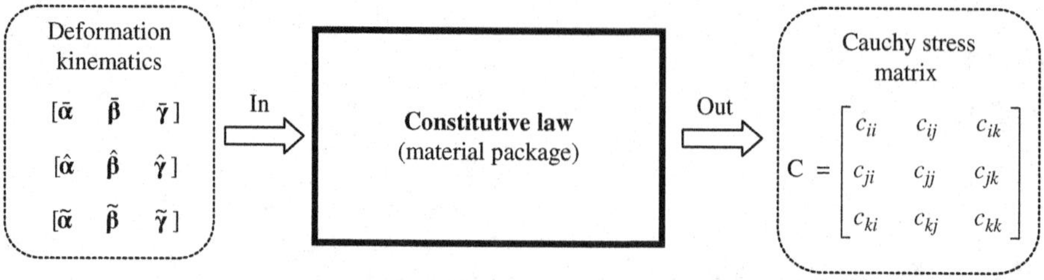

Figure 18.1 Input and output for the constitutive law calculation

or simply

$$\begin{bmatrix}\bar{\boldsymbol{\alpha}} & \bar{\boldsymbol{\beta}} & \bar{\boldsymbol{\gamma}}\end{bmatrix} = \begin{bmatrix}\bar{\alpha}_i & \bar{\beta}_i & \bar{\gamma}_i \\ \bar{\alpha}_j & \bar{\beta}_j & \bar{\gamma}_j \\ \bar{\alpha}_k & \bar{\beta}_k & \bar{\gamma}_k\end{bmatrix}. \tag{18.4}$$

The current position of the generalized material element is described by vectors

$$\tilde{\boldsymbol{\alpha}} = \begin{bmatrix}\mathbf{i} & \mathbf{j} & \mathbf{k}\end{bmatrix}\begin{bmatrix}\tilde{\alpha}_i \\ \tilde{\alpha}_j \\ \tilde{\alpha}_k\end{bmatrix} = \begin{bmatrix}\tilde{\alpha}_i \\ \tilde{\alpha}_j \\ \tilde{\alpha}_k\end{bmatrix}, \tag{18.5}$$

$$\tilde{\boldsymbol{\beta}} = \begin{bmatrix}\mathbf{i} & \mathbf{j} & \mathbf{k}\end{bmatrix}\begin{bmatrix}\tilde{\beta}_i \\ \tilde{\beta}_j \\ \tilde{\beta}_k\end{bmatrix} = \begin{bmatrix}\tilde{\beta}_i \\ \tilde{\beta}_j \\ \tilde{\beta}_k\end{bmatrix}, \tag{18.6}$$

$$\tilde{\boldsymbol{\gamma}} = \begin{bmatrix}\mathbf{i} & \mathbf{j} & \mathbf{k}\end{bmatrix}\begin{bmatrix}\tilde{\gamma}_i \\ \tilde{\gamma}_j \\ \tilde{\gamma}_k\end{bmatrix} = \begin{bmatrix}\tilde{\gamma}_i \\ \tilde{\gamma}_j \\ \tilde{\gamma}_k\end{bmatrix}, \tag{18.7}$$

or in matrix form,

$$\begin{bmatrix}\tilde{\boldsymbol{\alpha}} & \tilde{\boldsymbol{\beta}} & \tilde{\boldsymbol{\gamma}}\end{bmatrix} = \begin{bmatrix}\tilde{\alpha}_i & \tilde{\beta}_i & \tilde{\gamma}_i \\ \tilde{\alpha}_j & \tilde{\beta}_j & \tilde{\gamma}_j \\ \tilde{\alpha}_k & \tilde{\beta}_k & \tilde{\gamma}_k\end{bmatrix}. \tag{18.8}$$

Volumetric Stretch. The initial and current volumes of the generalized material element are calculated as

$$\begin{aligned}\bar{v} &= \det\begin{bmatrix}\bar{\alpha}_i & \bar{\beta}_i & \bar{\gamma}_i \\ \bar{\alpha}_j & \bar{\beta}_j & \bar{\gamma}_j \\ \bar{\alpha}_k & \bar{\beta}_k & \bar{\gamma}_k\end{bmatrix} \\ &= \bar{\alpha}_i\left(\bar{\beta}_j\bar{\gamma}_k - \bar{\beta}_k\bar{\gamma}_j\right) - \bar{\beta}_i\left(\bar{\alpha}_j\bar{\gamma}_k - \bar{\alpha}_k\bar{\gamma}_j\right) + \bar{\gamma}_i\left(\bar{\alpha}_j\bar{\beta}_k - \bar{\alpha}_k\bar{\beta}_j\right),\end{aligned} \tag{18.9}$$

$$\tilde{v} = \det \begin{bmatrix} \tilde{\alpha}_i & \tilde{\beta}_i & \tilde{\gamma}_i \\ \tilde{\alpha}_j & \tilde{\beta}_j & \tilde{\gamma}_j \\ \tilde{\alpha}_k & \tilde{\beta}_k & \tilde{\gamma}_k \end{bmatrix} = \tilde{\alpha}_i\left(\tilde{\beta}_j\tilde{\gamma}_k - \tilde{\beta}_k\tilde{\gamma}_j\right) - \tilde{\beta}_i\left(\tilde{\alpha}_j\tilde{\gamma}_k - \tilde{\alpha}_k\tilde{\gamma}_j\right) + \tilde{\gamma}_i\left(\tilde{\alpha}_j\tilde{\beta}_k - \tilde{\alpha}_k\tilde{\beta}_j\right). \tag{18.10}$$

The stretching of the volume (volumetric stretch) is defined by

$$s_v = \frac{\tilde{v}}{\bar{v}}. \tag{18.11}$$

Edge Stretches. From the initial position of the generalized material element the lengths of the three edges are obtained as follows

$$\bar{\alpha} = |\bar{\boldsymbol{\alpha}}| = \sqrt{\bar{\alpha}_i^2 + \bar{\alpha}_j^2 + \bar{\alpha}_j^2}, \tag{18.12}$$

$$\bar{\beta} = |\bar{\boldsymbol{\beta}}| = \sqrt{\bar{\beta}_i^2 + \bar{\beta}_j^2 + \bar{\beta}_j^2}, \tag{18.13}$$

$$\bar{\gamma} = |\bar{\boldsymbol{\gamma}}| = \sqrt{\bar{\gamma}_i^2 + \bar{\gamma}_j^2 + \bar{\gamma}_k^2}. \tag{18.14}$$

The lengths of the same deformed edges (of the current solid element) are given by

$$\tilde{\alpha} = |\tilde{\boldsymbol{\alpha}}| = \sqrt{\tilde{\alpha}_i^2 + \tilde{\alpha}_j^2 + \tilde{\alpha}_k^2}, \tag{18.15}$$

$$\tilde{\beta} = |\tilde{\boldsymbol{\beta}}| = \sqrt{\tilde{\beta}_i^2 + \tilde{\beta}_j^2 + \tilde{\beta}_k^2}, \tag{18.16}$$

$$\tilde{\gamma} = |\tilde{\boldsymbol{\gamma}}| = \sqrt{\tilde{\gamma}_i^2 + \tilde{\gamma}_j^2 + \tilde{\gamma}_k^2}. \tag{18.17}$$

Using these, the edge stretches (also called the axial stretches) are defined as follows

$$s_\alpha = \frac{\tilde{\alpha}}{\bar{\alpha}}, \tag{18.18}$$

$$s_\beta = \frac{\tilde{\beta}}{\bar{\beta}}, \tag{18.19}$$

$$s_\gamma = \frac{\tilde{\gamma}}{\bar{\gamma}}. \tag{18.20}$$

Angle Stretches. The shearing of the solid element is defined by the change of the initial angle between the edges $\bar{\boldsymbol{\alpha}}, \bar{\boldsymbol{\beta}}$ and $\bar{\boldsymbol{\gamma}}$

$$\bar{\psi}_{\alpha\beta} = \arccos\left(\frac{\bar{\boldsymbol{\alpha}} \bullet \bar{\boldsymbol{\beta}}}{\bar{\alpha}\bar{\beta}}\right) \tag{18.21}$$

and

$$\tilde{\psi}_{\alpha\beta} = \arccos\left(\frac{\tilde{\boldsymbol{\alpha}} \bullet \tilde{\boldsymbol{\beta}}}{\tilde{\alpha}\tilde{\beta}}\right), \tag{18.22}$$

where both $\bar{\psi}_{\alpha\beta}$ and $\tilde{\psi}_{\alpha\beta}$ range between the interval 0 and π. The angle stretch is defined as

$$s_{\alpha\beta} = \frac{\tilde{\psi}_{\alpha\beta}}{\bar{\psi}_{\alpha\beta}}. \tag{18.23}$$

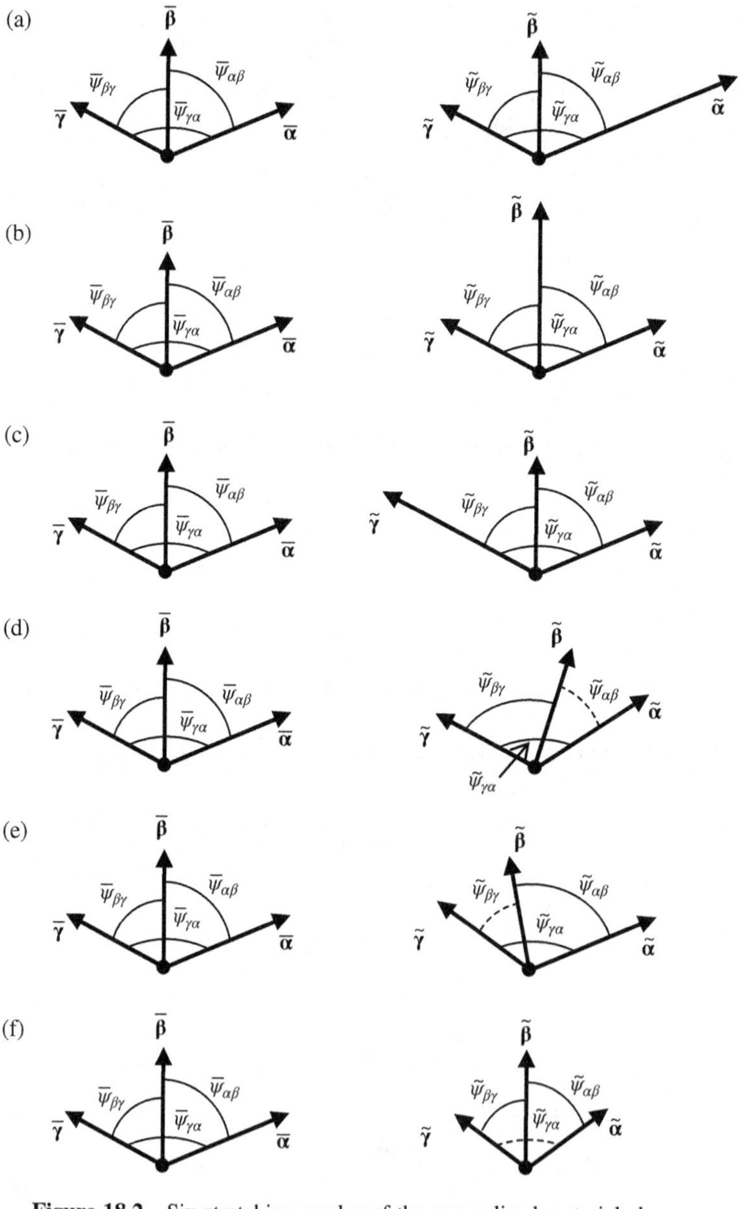

Figure 18.2 Six stretching modes of the generalized material element

In a similar manner, the initial and current angles between edges $\bar{\boldsymbol{\beta}}$ and $\bar{\boldsymbol{\gamma}}$ are

$$\bar{\psi}_{\beta\gamma} = \arccos\left(\frac{\bar{\boldsymbol{\beta}} \bullet \bar{\boldsymbol{\gamma}}}{\bar{\beta}\bar{\gamma}}\right), \tag{18.24}$$

$$\tilde{\psi}_{\beta\gamma} = \arccos\left(\frac{\tilde{\boldsymbol{\beta}} \bullet \tilde{\boldsymbol{\gamma}}}{\tilde{\beta}\tilde{\gamma}}\right), \tag{18.25}$$

resulting in an angle stretch of

$$s_{\beta\gamma} = \frac{\tilde{\psi}_{\beta\gamma}}{\bar{\psi}_{\beta\gamma}}. \tag{18.26}$$

The initial and current angles between edges $\bar{\boldsymbol{\gamma}}$ and $\bar{\boldsymbol{\alpha}}$ are

$$\bar{\psi}_{\gamma\alpha} = \arccos\left(\frac{\bar{\boldsymbol{\gamma}} \bullet \bar{\boldsymbol{\alpha}}}{\bar{\gamma}\bar{\alpha}}\right), \tag{18.27}$$

$$\tilde{\psi}_{\gamma\alpha} = \arccos\left(\frac{\tilde{\boldsymbol{\gamma}} \bullet \tilde{\boldsymbol{\alpha}}}{\tilde{\gamma}\tilde{\alpha}}\right), \tag{18.28}$$

producing an angle stretch of

$$s_{\gamma\alpha} = \frac{\tilde{\psi}_{\gamma\alpha}}{\bar{\psi}_{\gamma\alpha}}. \tag{18.29}$$

Stretching Modes. There are six modes of solid element stretching, as shown in Figure 18.2. From these stretches, using a constitutive law, the resultant internal forces on the solid element surfaces are calculated. They are stored within the material package. From this stored stress, the Cauchy stress matrix is calculated on demand.

In the case of an anisotropic solid, it is convenient to calculate and store the Munjiza internal forces on the solid element, which are shown in Figure 18.3.

These Munjiza resultant internal forces are only stored within the material package. When required, the internal forces on surfaces parallel to the global orthonormal base vectors $[\mathbf{i}\ \ \mathbf{j}\ \ \mathbf{k}]$ in the current deformed configuration are calculated and passed back to the finite element package; these are actually the components of the Cauchy stress matrix, Figure 18.4.

18.3 Generalized Hypo-Elastic Material

In the hypo-elastic formulation the generalized material element deforms in small multiplicative increments. By considering the kinematics of deformation (Chapter 16) one obtains the previous and current shape of the generalized material element respectively as defined by

$$[\hat{\boldsymbol{\alpha}}\ \ \hat{\boldsymbol{\beta}}\ \ \hat{\boldsymbol{\gamma}}], \tag{18.30}$$

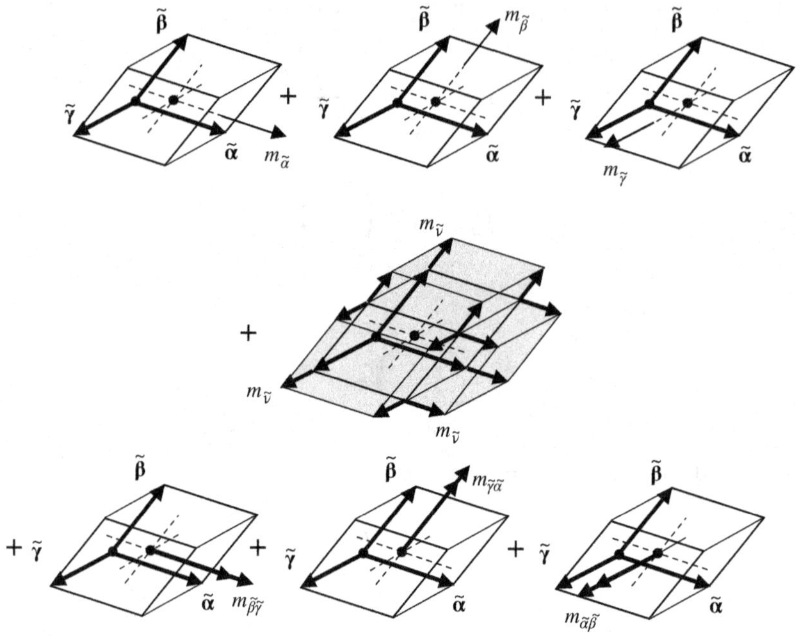

Figure 18.3 Internal forces corresponding to the stretching modes

$$\left[\tilde{\boldsymbol{\alpha}} \ \tilde{\boldsymbol{\beta}} \ \tilde{\boldsymbol{\gamma}}\right]. \tag{18.31}$$

Multiplicative Decomposition of Edge Stretches. The current and previous lengths of the generalized material element edges are calculated

$$\tilde{\alpha} = |\tilde{\boldsymbol{\alpha}}| = \sqrt{\tilde{\alpha}_i^2 + \tilde{\alpha}_j^2 + \tilde{\alpha}_k^2}, \tag{18.32}$$

$$\tilde{\beta} = |\tilde{\boldsymbol{\beta}}| = \sqrt{\tilde{\beta}_i^2 + \tilde{\beta}_j^2 + \tilde{\beta}_k^2}, \tag{18.33}$$

$$\tilde{\gamma} = |\tilde{\boldsymbol{\gamma}}| = \sqrt{\tilde{\gamma}_i^2 + \tilde{\gamma}_j^2 + \tilde{\gamma}_k^2} \tag{18.34}$$

and

$$\hat{\alpha} = |\hat{\boldsymbol{\alpha}}| = \sqrt{\hat{\alpha}_i^2 + \hat{\alpha}_j^2 + \hat{\alpha}_k^2}, \tag{18.35}$$

$$\hat{\beta} = |\hat{\boldsymbol{\beta}}| = \sqrt{\hat{\beta}_i^2 + \hat{\beta}_j^2 + \hat{\beta}_k^2}, \tag{18.36}$$

$$\hat{\gamma} = |\hat{\boldsymbol{\gamma}}| = \sqrt{\hat{\gamma}_i^2 + \hat{\gamma}_j^2 + \hat{\gamma}_k^2}. \tag{18.37}$$

The multiplicative increments of the stretch of the three edges of the solid element are given by

$$\tilde{s}_\alpha = \frac{\tilde{\alpha}}{\hat{\alpha}}, \tag{18.38}$$

$$\tilde{s}_\beta = \frac{\tilde{\beta}}{\hat{\beta}}, \tag{18.39}$$

$$\tilde{s}_\gamma = \frac{\tilde{\gamma}}{\hat{\gamma}}. \tag{18.40}$$

Multiplicative Decomposition of the Volumetric Stretch. In a similar manner, the multiplicative incremental stretch of the solid element volume is simply

$$\tilde{s}_v = \frac{\tilde{v}}{\hat{v}}, \tag{18.41}$$

where $\tilde{v}$ and $\hat{v}$ are the current and previous volumes of the solid element respectively.

Multiplicative Decomposition of Angle Stretches. The multiplicative increment of shear stretch is obtained as

$$\tilde{s}_{\psi_{\alpha\beta}} = \frac{\tilde{\psi}_{\alpha\beta}}{\hat{\psi}_{\alpha\beta}}, \tag{18.42}$$

where

$$\hat{\psi}_{\alpha\beta} = \arccos\left(\frac{\hat{\boldsymbol{\alpha}} \bullet \hat{\boldsymbol{\beta}}}{\hat{\alpha}\hat{\beta}}\right) \tag{18.43}$$

and

$$\tilde{\psi}_{\alpha\beta} = \arccos\left(\frac{\tilde{\boldsymbol{\alpha}} \bullet \tilde{\boldsymbol{\beta}}}{\tilde{\alpha}\tilde{\beta}}\right) \tag{18.44}$$

and the angles range between the interval 0 and π.

Strain Measures. From these stretches, the incremental strain measures are calculated, for example,

$$\begin{aligned}
\tilde{e}_v &= \ln \tilde{s}_v \\
\tilde{e}_\alpha &= \ln \tilde{s}_\alpha \\
\tilde{e}_\beta &= \ln \tilde{s}_\beta \\
\tilde{e}_\gamma &= \ln \tilde{s}_\gamma \\
\tilde{e}_{\psi_{\alpha\beta}} &= \ln \tilde{s}_{\psi_{\alpha\beta}} \\
\tilde{e}_{\psi_{\beta\gamma}} &= \ln \tilde{s}_{\psi_{\beta\gamma}} \\
\tilde{e}_{\psi_{\gamma\alpha}} &= \ln \tilde{s}_{\psi_{\gamma\alpha}}.
\end{aligned} \tag{18.45}$$

Munjiza Internal Forces. Depending on the particular material model being used, the increment in the Munjiza resultant internal forces is calculated from the above strain increments. For example

$$\begin{aligned}
\tilde{m}_v &= \mathrm{M}_v \tilde{e}_v \\
\tilde{m}_\alpha &= \mathrm{M}_\alpha \tilde{e}_\alpha \\
\tilde{m}_\beta &= \mathrm{M}_\beta \tilde{e}_\beta \\
\tilde{m}_\gamma &= \mathrm{M}_\gamma \tilde{e}_\gamma \\
\tilde{m}_{\psi_{\alpha\beta}} &= \mathrm{M}_{\psi_{\alpha\beta}} \tilde{e}_{\psi_{\alpha\beta}} \\
\tilde{m}_{\psi_{\beta\gamma}} &= \mathrm{M}_{\psi_{\beta\gamma}} \tilde{e}_{\psi_{\beta\gamma}} \\
\tilde{m}_{\psi_{\gamma\alpha}} &= \mathrm{M}_{\psi_{\gamma\alpha}} \tilde{e}_{\psi_{\gamma\alpha}},
\end{aligned} \tag{18.46}$$

where $\mathrm{M}_v, \mathrm{M}_\alpha, \mathrm{M}_\beta, \mathrm{M}_\gamma, \mathrm{M}_{\psi_{\alpha\beta}}, \mathrm{M}_{\psi_{\beta\gamma}}$, and $\mathrm{M}_{\psi_{\gamma\alpha}}$ are the elastic constants.

These are added to the accumulated total Munjiza internal forces from the previous step. The updated Munjiza internal forces are then stored within the material package for use in the next step.

Because the Munjiza internal forces do not depend on the orientation (or stretch) of the solid element, there is no need for any "transformation", "co-rotation" or "correction" of them from one step to the next.

In contrast, most FEM formulations use the co-rotational formulations (which are outside the scope of this book), at each step to create a "stress transformation" from one "configuration" to another. These "transformations" can, in some applications, lead to serious problems and errors and are, therefore not included in this book.

18.4 Developing Material Models

The modern approach to finite strain-finite displacement deformation as formulated by Gurtin and Thrusdell was first incorporated into the Finite Element Method by Simo and into the Combined Finite-Discrete Element Method by Munjiza. This approach is based on multiplicative decomposition, which is the exact finite-strain, finite-displacement approach that contains no truncation errors.

By introducing the Munjiza resultant internal forces, the development of material models has been further simplified and made objective in terms of both stress and internal variables (in the case of nonlinear materials, such as plastic materials).

18.5 Calculation of the Cauchy Stress Tensor Matrix

The surfaces of the current deformed solid element in the vicinity of point P are shown in Figure 18.4. On the deformed solid, a solid volume (Cauchy material element) with edges parallel to the global orthonormal base $[\mathbf{i} \;\; \mathbf{j} \;\; \mathbf{k}]$ is also shown in the figure.

The surfaces of the deformed generalized material element are defined by vectors $[\mathbf{a} \;\; \mathbf{b} \;\; \mathbf{c}]$ where the magnitude of each of these vectors defines the area of the corresponding surface. These quadrilateral shape surfaces face in the direction of their respective vectors.

In contrast, the surfaces of the Cauchy material element are defined by vectors $[\mathbf{i} \;\; \mathbf{j} \;\; \mathbf{k}]$ and are of unit area. The Cauchy material element is of cube shape, i.e., a 1 by 1 by 1 cube.

The surfaces

$$[\tilde{\mathbf{a}} \;\; \tilde{\mathbf{b}} \;\; \tilde{\mathbf{c}}] \tag{18.47}$$

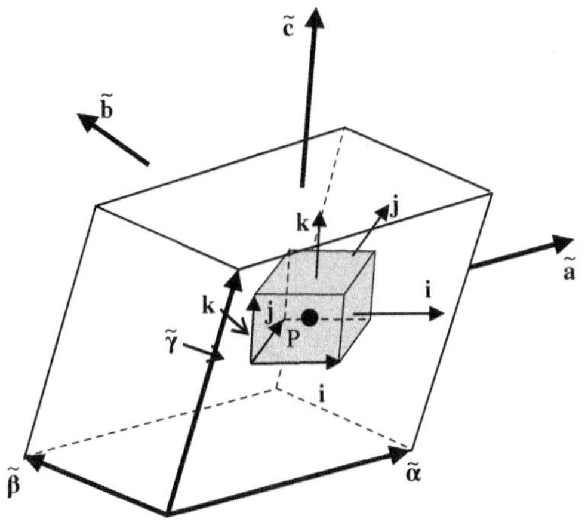

Figure 18.4 Surfaces of the generalized material element together with the Cauchy material element surfaces

are obtained from the following vector products

$$\begin{aligned}\tilde{\mathbf{a}} &= \tilde{\boldsymbol{\beta}} \times \tilde{\boldsymbol{\gamma}} \\ \tilde{\mathbf{b}} &= \tilde{\boldsymbol{\gamma}} \times \tilde{\boldsymbol{\alpha}} \\ \tilde{\mathbf{c}} &= \tilde{\boldsymbol{\alpha}} \times \tilde{\boldsymbol{\beta}}.\end{aligned} \tag{18.48}$$

These three surfaces can be written in a matrix form

$$\begin{aligned}\tilde{\mathbf{S}} = \begin{bmatrix}\tilde{\mathbf{a}} & \tilde{\mathbf{b}} & \tilde{\mathbf{c}}\end{bmatrix} &= \begin{bmatrix}\mathbf{i} & \mathbf{j} & \mathbf{k}\end{bmatrix}\begin{bmatrix}\tilde{a}_i & \tilde{b}_i & \tilde{c}_i \\ \tilde{a}_j & \tilde{b}_j & \tilde{c}_j \\ \tilde{a}_k & \tilde{b}_k & \tilde{c}_k\end{bmatrix} \\ &= \begin{bmatrix}\tilde{a}_i & \tilde{b}_i & \tilde{c}_i \\ \tilde{a}_j & \tilde{b}_j & \tilde{c}_j \\ \tilde{a}_k & \tilde{b}_k & \tilde{c}_k\end{bmatrix}.\end{aligned} \tag{18.49}$$

By multiplying both sides of Equation (18.49) from the right by

$$\begin{bmatrix}\tilde{a}_i & \tilde{b}_i & \tilde{c}_i \\ \tilde{a}_j & \tilde{b}_j & \tilde{c}_j \\ \tilde{a}_k & \tilde{b}_k & \tilde{c}_k\end{bmatrix}^{-1}, \tag{18.50}$$

one obtains

$$\begin{bmatrix}\mathbf{i} & \mathbf{j} & \mathbf{k}\end{bmatrix} = \begin{bmatrix}\tilde{\mathbf{a}} & \tilde{\mathbf{b}} & \tilde{\mathbf{c}}\end{bmatrix}\begin{bmatrix}\tilde{a}_i & \tilde{b}_i & \tilde{c}_i \\ \tilde{a}_j & \tilde{b}_j & \tilde{c}_j \\ \tilde{a}_k & \tilde{b}_k & \tilde{c}_k\end{bmatrix}^{-1}, \tag{18.51}$$

which can be written as

$$[\mathbf{i}\ \ \mathbf{j}\ \ \mathbf{k}] = [\tilde{\mathbf{a}}\ \ \tilde{\mathbf{b}}\ \ \tilde{\mathbf{c}}]\begin{bmatrix} i_{\tilde{a}} & j_{\tilde{a}} & k_{\tilde{a}} \\ i_{\tilde{b}} & j_{\tilde{b}} & k_{\tilde{b}} \\ i_{\tilde{c}} & j_{\tilde{c}} & k_{\tilde{c}} \end{bmatrix}, \tag{18.52}$$

where

$$\begin{bmatrix} i_{\tilde{a}} & j_{\tilde{a}} & k_{\tilde{a}} \\ i_{\tilde{b}} & j_{\tilde{b}} & k_{\tilde{b}} \\ i_{\tilde{c}} & j_{\tilde{c}} & k_{\tilde{c}} \end{bmatrix} = \begin{bmatrix} \tilde{a}_i & \tilde{b}_i & \tilde{c}_i \\ \tilde{a}_j & \tilde{b}_j & \tilde{c}_j \\ \tilde{a}_k & \tilde{b}_k & \tilde{c}_k \end{bmatrix}^{-1}. \tag{18.53}$$

Internal forces on surfaces $[\mathbf{a}\ \ \mathbf{b}\ \ \mathbf{c}]$ are calculated directly from the Munjiza internal forces as follows:

a. The internal force on surface **a** is, see Figure 18.5

$$\mathbf{p_a} = m_{\tilde{v}}\mathbf{a} + m_{\tilde{\alpha}}\frac{\tilde{\boldsymbol{\alpha}}}{|\tilde{\boldsymbol{\alpha}}|} + m_{\tilde{\beta}\tilde{\alpha}}\frac{1}{|\tilde{\boldsymbol{\alpha}}|}\mathbf{g}_{\tilde{\beta}\tilde{\alpha}} + m_{\tilde{\gamma}\tilde{\alpha}}\frac{1}{|\tilde{\boldsymbol{\alpha}}|}\mathbf{g}_{\tilde{\gamma}\tilde{\alpha}}, \tag{18.54}$$

where (see Figure 18.6)

$$\begin{aligned} \mathbf{g}_{\tilde{\beta}\tilde{\alpha}} &= \frac{(\tilde{\boldsymbol{\alpha}}\times\tilde{\boldsymbol{\beta}})\times\tilde{\boldsymbol{\alpha}}}{|(\tilde{\boldsymbol{\alpha}}\times\tilde{\boldsymbol{\beta}})\times\tilde{\boldsymbol{\alpha}}|} \\ \mathbf{g}_{\tilde{\gamma}\tilde{\alpha}} &= \frac{(\tilde{\boldsymbol{\alpha}}\times\tilde{\boldsymbol{\gamma}})\times\tilde{\boldsymbol{\alpha}}}{|(\tilde{\boldsymbol{\alpha}}\times\tilde{\boldsymbol{\gamma}})\times\tilde{\boldsymbol{\alpha}}|}. \end{aligned} \tag{18.55}$$

The individual components are shown in Figure 18.5, where the vector $\mathbf{g}_{\tilde{\beta}\tilde{\alpha}}$ is a unit vector inside the plane $\tilde{\boldsymbol{\alpha}}-\tilde{\boldsymbol{\beta}}$, such that it is perpendicular to $\tilde{\boldsymbol{\alpha}}$, Figure 18.6.

b. The internal force on surface **b** is

$$\mathbf{p_b} = m_{\tilde{v}}\mathbf{b} + m_{\tilde{\beta}}\frac{\tilde{\boldsymbol{\beta}}}{|\tilde{\boldsymbol{\beta}}|} + m_{\tilde{\alpha}\tilde{\beta}}\frac{1}{|\tilde{\boldsymbol{\beta}}|}\mathbf{g}_{\tilde{\alpha}\tilde{\beta}} + m_{\tilde{\gamma}\tilde{\beta}}\frac{1}{|\tilde{\boldsymbol{\beta}}|}\mathbf{g}_{\tilde{\gamma}\tilde{\beta}}, \tag{18.56}$$

where

$$\begin{aligned} \mathbf{g}_{\tilde{\alpha}\tilde{\beta}} &= \frac{(\tilde{\boldsymbol{\beta}}\times\tilde{\boldsymbol{\alpha}})\times\tilde{\boldsymbol{\beta}}}{|(\tilde{\boldsymbol{\beta}}\times\tilde{\boldsymbol{\alpha}})\times\tilde{\boldsymbol{\beta}}|} \\ \mathbf{g}_{\tilde{\gamma}\tilde{\beta}} &= \frac{(\tilde{\boldsymbol{\beta}}\times\tilde{\boldsymbol{\gamma}})\times\tilde{\boldsymbol{\beta}}}{|(\tilde{\boldsymbol{\beta}}\times\tilde{\boldsymbol{\gamma}})\times\tilde{\boldsymbol{\beta}}|}. \end{aligned} \tag{18.57}$$

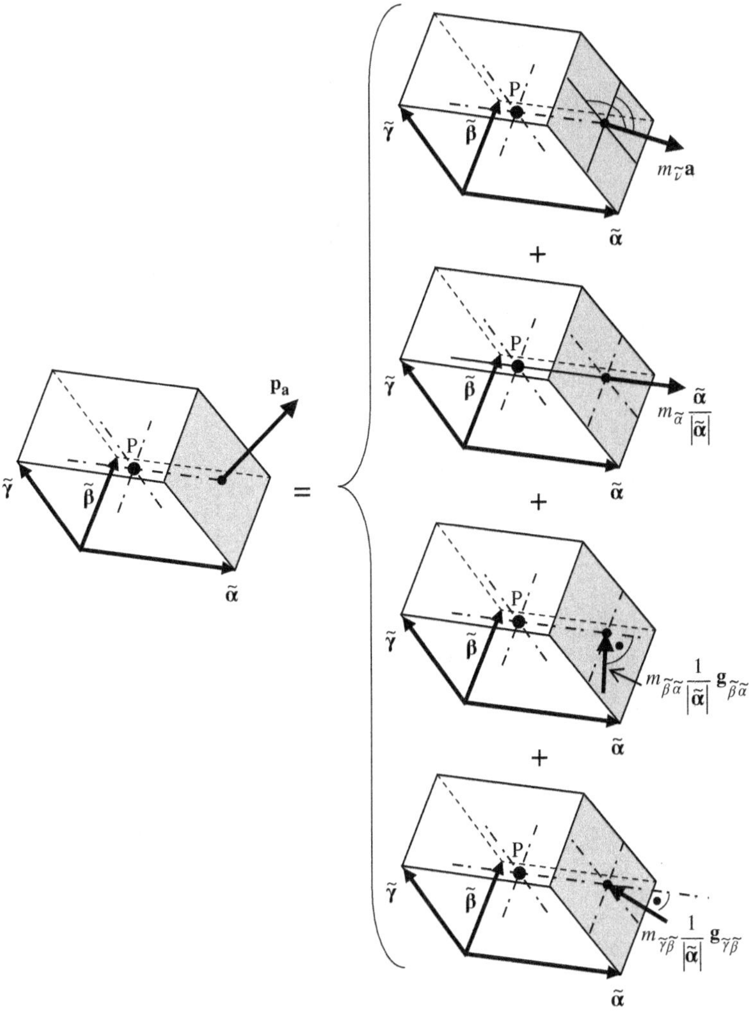

Figure 18.5 Graphical representation of the resultant internal forces on surface **a**, i.e., the components of the First Piola-Kirchhoff stress tensor matrix

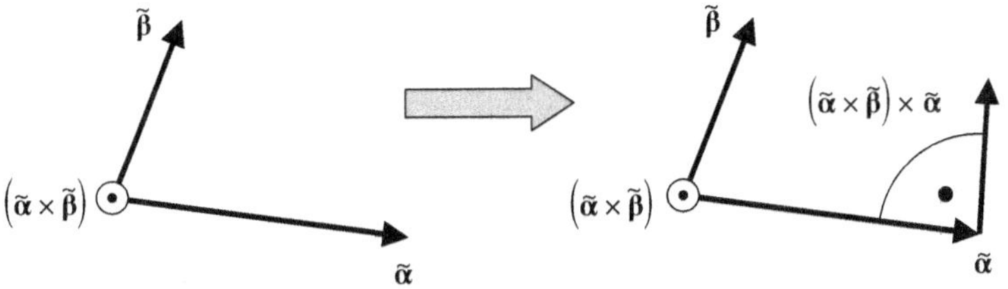

Figure 18.6 Graphical representation of vector $\mathbf{g}_{\tilde{\beta}\tilde{\alpha}} = (\tilde{\alpha} \times \tilde{\beta}) \times \tilde{\alpha}$

c. The internal force on surface **c** is

$$\mathbf{p_c} = m_{\tilde{\gamma}}\mathbf{c} + m_{\tilde{\gamma}}\frac{\tilde{\boldsymbol{\gamma}}}{|\tilde{\boldsymbol{\gamma}}|} + m_{\tilde{\alpha}\tilde{\gamma}}\frac{1}{|\tilde{\boldsymbol{\gamma}}|}\mathbf{g}_{\tilde{\alpha}\tilde{\gamma}} + m_{\tilde{\beta}\tilde{\gamma}}\frac{1}{|\tilde{\boldsymbol{\gamma}}|}\mathbf{g}_{\tilde{\beta}\tilde{\gamma}}, \tag{18.58}$$

where

$$\begin{aligned} \mathbf{g}_{\tilde{\alpha}\tilde{\gamma}} &= \frac{(\tilde{\boldsymbol{\gamma}} \times \tilde{\boldsymbol{\alpha}}) \times \tilde{\boldsymbol{\gamma}}}{|(\tilde{\boldsymbol{\gamma}} \times \tilde{\boldsymbol{\alpha}}) \times \tilde{\boldsymbol{\gamma}}|} \\ \mathbf{g}_{\tilde{\beta}\tilde{\gamma}} &= \frac{(\tilde{\boldsymbol{\gamma}} \times \tilde{\boldsymbol{\beta}}) \times \tilde{\boldsymbol{\gamma}}}{|(\tilde{\boldsymbol{\gamma}} \times \tilde{\boldsymbol{\beta}}) \times \tilde{\boldsymbol{\gamma}}|}. \end{aligned} \tag{18.59}$$

It is worth noting that

$$[\mathbf{p_a} \;\; \mathbf{p_b} \;\; \mathbf{p_c}] = \begin{bmatrix} p_{ia} & p_{ib} & p_{ic} \\ p_{ja} & p_{jb} & p_{jc} \\ p_{ka} & p_{kb} & p_{kc} \end{bmatrix} \tag{18.60}$$

are the components of the First Piola-Kirchhoff stress tensor matrix.

The internal forces on surfaces $[\mathbf{i} \;\; \mathbf{j} \;\; \mathbf{k}]$ of the Cauchy material element are by definition given by

$$\mathbf{c}_i = \begin{bmatrix} c_{ii} \\ c_{ji} \\ c_{ki} \end{bmatrix} = \mathbf{p_a} i_{\tilde{a}} + \mathbf{p_b} i_{\tilde{b}} + \mathbf{p_c} i_{\tilde{c}}, \tag{18.61}$$

$$\mathbf{c}_j = \begin{bmatrix} c_{ij} \\ c_{jj} \\ c_{kj} \end{bmatrix} = \mathbf{p_a} j_{\tilde{a}} + \mathbf{p_b} j_{\tilde{b}} + \mathbf{p_c} j_{\tilde{c}}, \tag{18.62}$$

$$\mathbf{c}_k = \begin{bmatrix} c_{ik} \\ c_{jk} \\ c_{kk} \end{bmatrix} = \mathbf{p_a} k_{\tilde{a}} + \mathbf{p_b} k_{\tilde{b}} + \mathbf{p_c} k_{\tilde{c}}. \tag{18.63}$$

These can be written in a matrix form

$$\begin{aligned} \begin{bmatrix} c_{ii} & c_{ij} & c_{ik} \\ c_{ji} & c_{jj} & c_{jk} \\ c_{ki} & c_{kj} & c_{kk} \end{bmatrix} &= [\mathbf{p_{\tilde{a}}} \;\; \mathbf{p_{\tilde{b}}} \;\; \mathbf{p_{\tilde{c}}}] \begin{bmatrix} i_{\tilde{a}} & j_{\tilde{a}} & k_{\tilde{a}} \\ i_{\tilde{b}} & j_{\tilde{b}} & k_{\tilde{b}} \\ i_{\tilde{c}} & j_{\tilde{c}} & k_{\tilde{c}} \end{bmatrix} \\ &= \begin{bmatrix} p_{i\tilde{a}} & p_{i\tilde{b}} & p_{i\tilde{c}} \\ p_{j\tilde{a}} & p_{j\tilde{b}} & p_{j\tilde{c}} \\ p_{k\tilde{a}} & p_{k\tilde{b}} & p_{k\tilde{c}} \end{bmatrix} \begin{bmatrix} i_{\tilde{a}} & j_{\tilde{a}} & k_{\tilde{a}} \\ i_{\tilde{b}} & j_{\tilde{b}} & k_{\tilde{b}} \\ i_{\tilde{c}} & j_{\tilde{c}} & k_{\tilde{c}} \end{bmatrix}. \end{aligned} \tag{18.64}$$

After substitution from Equation (18.51) one obtains

$$\begin{bmatrix} c_{ii} & c_{ij} & c_{ik} \\ c_{ji} & c_{jj} & c_{jk} \\ c_{ki} & c_{kj} & c_{kk} \end{bmatrix} = \begin{bmatrix} p_{i\tilde{a}} & p_{i\tilde{b}} & p_{i\tilde{c}} \\ p_{j\tilde{a}} & p_{j\tilde{b}} & p_{j\tilde{c}} \\ p_{k\tilde{a}} & p_{k\tilde{b}} & p_{k\tilde{c}} \end{bmatrix} \begin{bmatrix} \tilde{a}_i & \tilde{b}_i & \tilde{c}_i \\ \tilde{a}_j & \tilde{b}_j & \tilde{c}_j \\ \tilde{a}_k & \tilde{b}_k & \tilde{c}_k \end{bmatrix}^{-1} . \tag{18.65}$$

Starting from the volumetric stretches, angle stretches and edge stretches as derived in Sections 18.2 and 18.3, one can now proceed to develop material models for both isotropic and anisotropic solid. In this section, this process is explained using examples of an anisotropic elastic material followed by an example of an isotropic elastic material.

Anisotropic Elastic Material Example. For the generalized material element shown in Figure 18.7, by starting from the stretches one calculates the strain measures. One of the most often used strain measures are the logarithmic strains, given by

$$\begin{aligned} e_v &= \ln s_v \\ e_\alpha &= \ln s_\alpha \\ e_\beta &= \ln s_\beta \\ e_\gamma &= \ln s_\gamma \\ e_{\psi_{\alpha\beta}} &= \ln s_{\psi_{\alpha\beta}} \\ e_{\psi_{\beta\gamma}} &= \ln s_{\psi_{\beta\gamma}} \\ e_{\psi_{\gamma\alpha}} &= \ln s_{\psi_{\gamma\alpha}} . \end{aligned} \tag{18.66}$$

These strains can be used directly to calculate the Munjiza internal forces

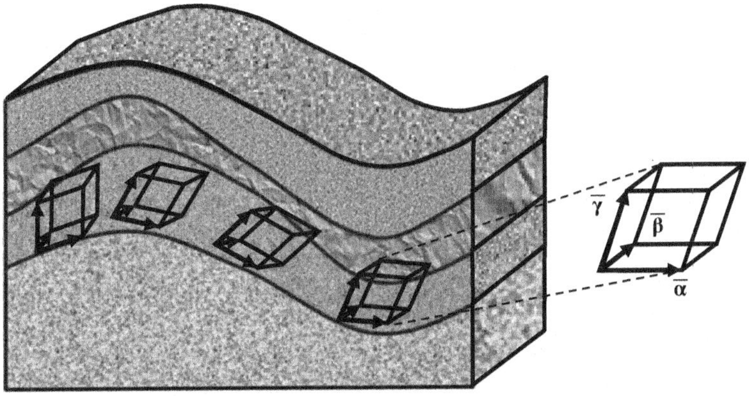

Figure 18.7 Generalized material element for anisotropic solid

$$
\begin{aligned}
m_v &= \mathrm{M}_v e_v \\
m_\alpha &= \mathrm{M}_\alpha e_\alpha \\
m_\beta &= \mathrm{M}_\beta e_\beta \\
m_\gamma &= \mathrm{M}_\gamma e_\gamma \\
m_{\psi_{\alpha\beta}} &= \mathrm{M}_{\psi_{\alpha\beta}} e_{\psi_{\alpha\beta}} \\
m_{\psi_{\beta\gamma}} &= \mathrm{M}_{\psi_{\beta\gamma}} e_{\psi_{\beta\gamma}} \\
m_{\psi_{\gamma\alpha}} &= \mathrm{M}_{\psi_{\gamma\alpha}} e_{\psi_{\gamma\alpha}},
\end{aligned}
\tag{18.67}
$$

where $\mathrm{M}_v, \mathrm{M}_\alpha, \mathrm{M}_\beta, \mathrm{M}_\gamma, \mathrm{M}_{\psi_{\alpha\beta}}, \mathrm{M}_{\psi_{\beta\gamma}}$ and $\mathrm{M}_{\psi_{\gamma\alpha}}$ are the generalized elastic constants that define the material's mechanical properties.

Isotropic Solid Example. In order to obtain the isotropic solid, the generalized material element has to be a cube of edge length equal to 1 (im), Figure 18.8.

Starting from the stretches, the strain for the cube shaped material element is obtained using the logarithmic strain measures calculated as

$$
\begin{aligned}
e_v &= \ln s_v \\
e_\alpha &= \ln s_\alpha \\
e_\beta &= \ln s_\beta \\
e_\gamma &= \ln s_\gamma \\
e_{\psi_{\alpha\beta}} &= \ln s_{\psi_{\alpha\beta}} \\
e_{\psi_{\beta\gamma}} &= \ln s_{\psi_{\beta\gamma}} \\
e_{\psi_{\gamma\alpha}} &= \ln s_{\psi_{\gamma\alpha}}.
\end{aligned}
\tag{18.68}
$$

The resultant Munjiza internal forces are obtained from these strains

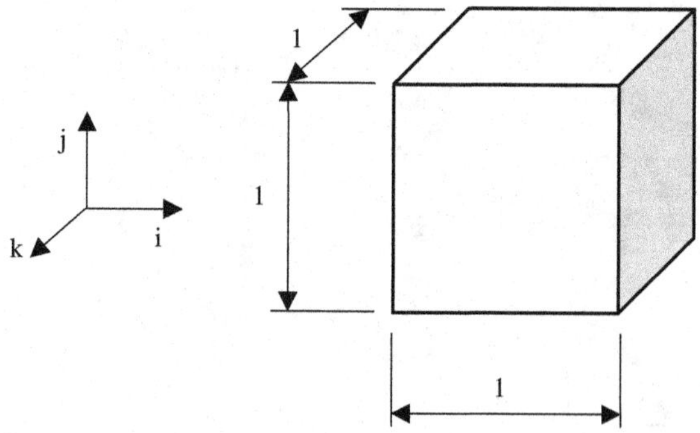

Figure 18.8 A generalized material element of cubical shape for an isotropic solid

$$
\begin{aligned}
m_v &= \mathrm{M}_v e_v \\
m_\alpha &= \mathrm{M}_\alpha e_\alpha \\
m_\beta &= \mathrm{M}_\beta e_\beta \\
m_\gamma &= \mathrm{M}_\gamma e_\gamma \\
m_{\psi_{\alpha\beta}} &= \mathrm{M}_{\psi_{\alpha\beta}} e_{\psi_{\alpha\beta}} \\
m_{\psi_{\beta\gamma}} &= \mathrm{M}_{\psi_{\beta\gamma}} e_{\psi_{\beta\gamma}} \\
m_{\psi_{\gamma\alpha}} &= \mathrm{M}_{\psi_{\gamma\alpha}} e_{\psi_{\gamma\alpha}},
\end{aligned}
\tag{18.69}
$$

using the generalized elastic constants $\mathrm{M}_v, \mathrm{M}_\alpha, \mathrm{M}_\beta, \mathrm{M}_\gamma, \mathrm{M}_{\psi_{\alpha\beta}}, \mathrm{M}_{\psi_{\beta\gamma}}$ and $\mathrm{M}_{\psi_{\gamma\alpha}}$. For the linear isotropic material these elastic constants can be obtained from Young's modulus, E, and Poisson's ratio, ν. This is done by considering that, in the simplest case of small strains, the constitutive law given by Equation (18.69) should reproduce Hooke's law. From this constraint, the elastic constants can be obtained as follows:

a. Consider the example shown in Figure 17.36.
   From Hooke's law, it follows that

$$
\begin{aligned}
\varepsilon_x &= \frac{\sigma}{\mathrm{E}} + \nu\frac{\sigma}{\mathrm{E}} \\
\varepsilon_y &= -\frac{\sigma}{\mathrm{E}} - \nu\frac{\sigma}{\mathrm{E}} \\
\varepsilon_z &= -\nu\frac{\sigma}{\mathrm{E}} + \nu\frac{\sigma}{\mathrm{E}} = 0.
\end{aligned}
\tag{18.70}
$$

The volumetric strain is therefore

$$
\varepsilon_v = \varepsilon_x + \varepsilon_y + \varepsilon_z = 0. \tag{18.71}
$$

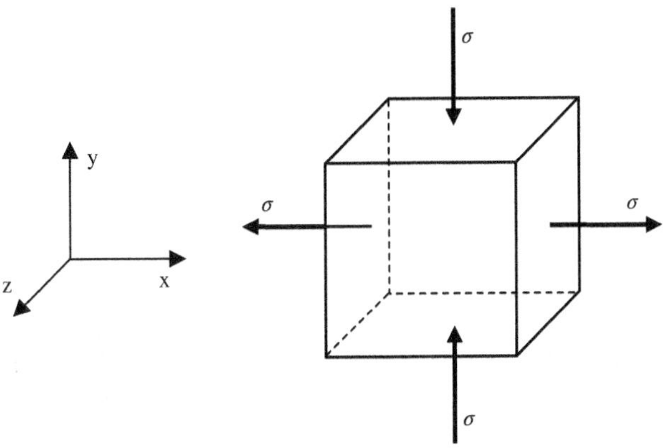

Figure 18.9 An example of small strain elasticity using Hooke's law

When substituted into Equation (18.69), this yields

$$m_v = M_v e_v = M_v \varepsilon_v = M_v 0 = 0$$
$$m_\alpha = M_\alpha e_\alpha = M_\alpha \varepsilon_x = M_\alpha \left[\frac{\sigma}{E} + \nu\frac{\sigma}{E}\right]. \tag{18.72}$$

In order that Equations (18.68) and (18.69) produce the same results, in the case of infinitesimally small strains, it follows from Equation (18.54) that

$$m_v + m_\alpha = \sigma, \tag{18.73}$$

which, when substituted into Equation (18.71) yields

$$M_\alpha \left[\frac{\sigma}{E} + \nu\frac{\sigma}{E}\right] = \sigma \tag{18.74}$$

or

$$M_\alpha = \frac{E}{(1+\nu)}. \tag{18.75}$$

By similar reasoning one obtains

$$M_\beta = \frac{E}{(1+\nu)} \text{ and } M_\gamma = \frac{E}{(1+\nu)}. \tag{18.76}$$

b. By considering a pure shear example shown in Figure 18.10 and by employing Hooke's law one obtains

$$\varepsilon_{xy} = \frac{\tau}{2G}, \tag{18.77}$$

where G is the shear modulus

$$G = \frac{E}{2(1+\nu)}. \tag{18.78}$$

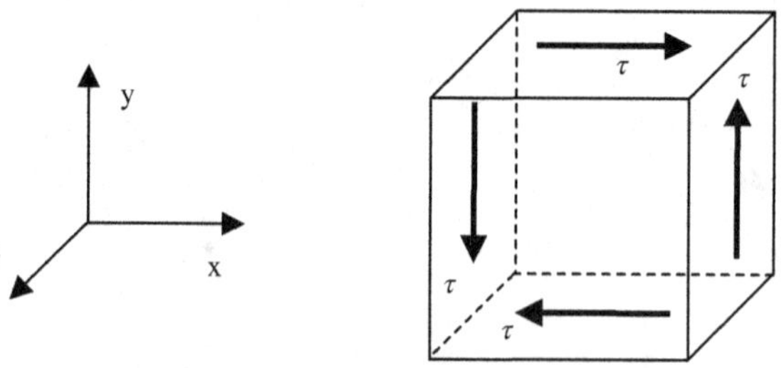

Figure 18.10 A pure shear example

Equation (18.67) for infinitesimal strains yields

$$e_{\psi} = 2\varepsilon_{xy} \text{ and } m_{\psi} = \tau, \tag{18.79}$$

which, when substituted into Equation (18.68) produces

$$m_{\psi} = \mathrm{M}_{\psi} e_{\psi} = \mathrm{M}_{\psi} 2\varepsilon_{xy}. \tag{18.80}$$

By substituting this into Equation (18.54) one obtains

$$p_{ja} = \tau = \mathrm{M}_{\psi} 2\left(\frac{\tau}{2\mathrm{G}}\right). \tag{18.81}$$

This yields

$$\mathrm{M}_{\psi_{\alpha\beta}} = \mathrm{M}_{\psi_{\beta\gamma}} = \mathrm{M}_{\psi_{\gamma\alpha}} = \mathrm{G}. \tag{18.82}$$

c. By considering the pure volumetric strain as shown in Figure 17.35, after employing Hooke's law one obtains

$$\varepsilon_x = \varepsilon_y = \varepsilon_z = \frac{\sigma}{\mathrm{E}}(1-2\nu). \tag{18.83}$$

In the case of infinitesimal strains, the volumetric strain is simply

$$\varepsilon_v = \varepsilon_x + \varepsilon_y + \varepsilon_z = \frac{\sigma}{\mathrm{E}} 3(1-2\nu). \tag{18.84}$$

When substituted into equation (18.69), this yields

$$\begin{aligned} m_v &= \mathrm{M}_v e_v = \mathrm{M}_v \varepsilon_v = \mathrm{M}_v \frac{\sigma}{\mathrm{E}} 3(1-2\nu), \\ m_\alpha &= \mathrm{M}_\alpha e_\alpha = \mathrm{M}_\alpha \varepsilon_x = \mathrm{M}_\alpha \frac{\sigma}{\mathrm{E}}(1-2\nu) \end{aligned} \tag{18.85}$$

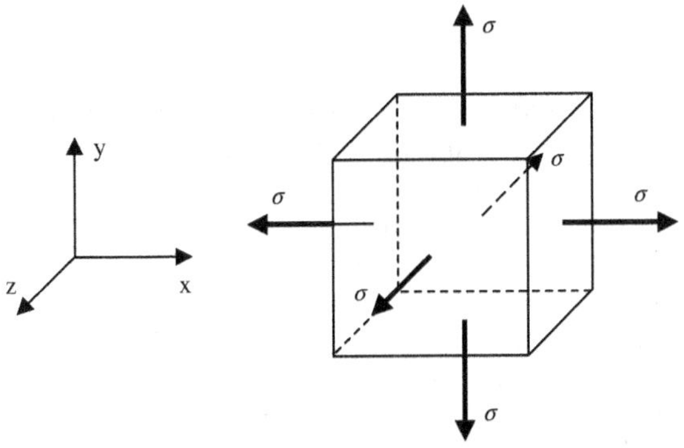

Figure 18.11 A pure volumetric strain

and

$$m_v + m_\alpha = \frac{\sigma}{\mathrm{E}}(1-2\nu)\,(3\mathrm{M}_v + \mathrm{M}_\alpha). \tag{18.86}$$

In the case of infinitesimal strains, it is necessary that $p_{ia} = \sigma$. When substituted into Equation (18.54) this yields

$$m_v + m_\alpha = \sigma. \tag{18.87}$$

After substituting for $m_v + m_\alpha$ from Equation (18.85) one obtains

$$\frac{\sigma}{\mathrm{E}}(1-2\nu)\,(3\mathrm{M}_v + \mathrm{M}_\alpha) = \sigma. \tag{18.88}$$

After substitution of M_α from Equation (18.75) this yields

$$\left(3\mathrm{M}_v + \frac{\mathrm{E}}{(1+\nu)}\right) = \frac{\mathrm{E}}{(1-2\nu)} \tag{18.89}$$

or

$$\mathrm{M}_v = \frac{\nu\mathrm{E}}{(1+\nu)(1-2\nu)}. \tag{18.90}$$

In summary, the material constants are as follows:

$$\begin{aligned}
\mathrm{M}_\alpha = \mathrm{M}_\beta = \mathrm{M}_\gamma &= \frac{\mathrm{E}}{(1+\nu)} \\
\mathrm{M}_{\psi_{\alpha\beta}} = \mathrm{M}_{\psi_{\beta\gamma}} = \mathrm{M}_{\psi_{\gamma\alpha}} = \mathrm{G} &= \frac{\mathrm{E}}{2(1+\nu)} \\
\mathrm{M}_v &= \frac{\nu\mathrm{E}}{(1+\nu)(1-2\nu)}.
\end{aligned} \tag{18.91}$$

18.6 Summary

In this chapter a generalized approach for the systematic development of constitutive laws for anisotropic materials under large strains has been presented. The approach presented first calculates the generalized internal forces on the Munjiza generalized material element. These forces are deformation-invariant and as such, are stored within the material package.

In the case of hypo-deformability, this means that at each step an increment in the generalized internal forces is calculated and simply added to the generalized internal forces already accumulated from the previous steps. As the already accumulated generalized internal forces do not change with further rotation and/or further stretching of the generalized material element, this means that they are deformation-invariant (deformation-objective). Consequently, adding the generalized internal forces due to deformation at different steps is straightforward and it involves no "stress correction" nor the need to use corrections such as the say Jaumann rate. This is a very important difference between the often used co-rotational formulations and the full multiplicative decomposition based theoretically exact formulation presented here.

Some examples of "simple" 3D constitutive laws have also been explained. These include an anisotropic constitutive law based on the generalized 3D material element. In addition, an elastic constitutive law based on the cube shaped material element for isotropic solids has been presented as a logical extension (from small strains to large strains) of Hooke's law.

Finally, one is shown how to calculate the Cauchy stress tensor matrix. This matrix describes the internal forces on the deformed solid's surfaces (the Cauchy material element). As such, it is calculated on demand. It is never stored within the material package. This is because it is not deformation invariant (objective). The generalized internal forces are stored within the material package instead; however, the generalized internal forces are never passed outside of the material package.

Further Reading

[1] Anand, L. (1985) Constitutive equations for hot-working of metals. *Int. J. Plasticity*, **1**(3): 213–31.

[2] Argyris, J. H. and Kleiber, M. (1977) Incremental formulation in nonlinear mechanics and large strain elasto-plasticity – natural approach. Part 1. *Comput. Method Appl. M.*, **11**(2): 215–47.

[3] Argyris, J. H. and St. Doltsinis, J. (1979) On the large strain inelastic analysis in natural formulation. Part I: Quasistatic problems. *Comput. Method Appl. M.*, **20**(2): 213–51.

[4] Argyris, J. H. and St. Doltsinis, J. (1980) On the large strain inelastic analysis in natural formulation. Part II. Dynamic problems. *Comput. Method Appl. M.*, **21**(1): 91–126.

[5] Bathe, K. J. (1982) *Finite Element Procedures Engineering Analysis*. New Jersey, Prentice-Hall.

[6] Bathe, K. J. (1996) *Finite Element Procedures*. New Jersey, Prentice-Hall.

[7] Bonet, J. and Burton, A. J. (1998) A simple orthotropic, transversely isotropic hyperelastic constitutive equation for large strain computations. *Comput Method Appl M.*, **162**(1–4): 151–64.

[8] Bonet, J. and Wood, R. D. (1997) *Nonlinear Continuum Mechanics for Finite Element Analysis*. Cambridge, Cambridge University Press.

[9] de Souza Neto, E. A. and Peric, D. (1996) A computational framework for a class of fully coupled models for elastoplastic damage at finite strains with reference to the linearization aspects. *Comput. Method Appl. M.*, **130**(1–2): 179–93.

[10] de Souza Neto, E. A., Pires, F. M. A. and Owen, D. R. J. (2005) F-bar-based linear triangles and tetrahedra for finite strain analysis of nearly incompressible solids. Part I: formulation and benchmarking. *International Journal for Numerical Methods in Engineering*, **62**(3): 353–83.

[11] de Souza Neto, E. A. and Feijoo, R. A. (2008) On the equivalence between spatial and material volume averaging of stress in large strain multi-scale solid constitutive models. *Mech. Mater.*, **40**(10): 803–11.

[12] de Souza Neto, E. A., Peric, D., Huang, G. C. and Owen, D. R. J. (1995) Remarks on the stability of enhanced strain elements in finite elasticity and elastoplasticity. *Communications in Numerical Methods in Engineering*, **11**(11): 951–61.

[13] Green, A. E. and Naghdi, P. M. (1971) Some remarks on elastic-plastic deformation at finite strain. *International Journal of Engineering Science*, **9**(12): 1219–29.

[14] Gurtin, M. E. (1981) *An Introduction to Continuum Mechanics*. New York, Academic Press.

[15] Hill, R. (1979) *Aspects of Invariance in Solid Mechanics*, ed. Y. Chia-Shun, Elsevier, pp. 1–75.

[16] Hughes, T.J.R. (2000) *The Finite Element Method: Linear Static and Dynamic Finite Element Analysis*. New York, Dover Publications.

[17] Malvern, L. E. (1969) *Introduction to the Mechanics of a Continuous Medium*. New Jersey, Prentice Hall.

[18] Munjiza, A. (2004) *The Combined Finite-Discrete Element Method*. Chichester, John Wiley & Sons, Ltd.

[19] Peric, D. (1992) On consistent stress rates in solid mechanics – computational implications. *International Journal for Numerical Methods in Engineering*, **33**(4): 799–817.

[20] Peric, D. and Owen, D. R. J. (1998) Finite-element applications to the nonlinear mechanics of solids. *Reports on Progress in Physics*, **61**(11): 1495–1574.

[21] Peric, D., Owen, D. R. J. and Honnor, M. E. (1992) A model for finite strain elastoplasticity based on logarithmic strains – computational issues. *Comput. Method Appl. M.*, **94**(1): 35–61.

[22] Simo, J. C. (1985) On the computational significance of the intermediate configuration and hyperelastic stress relations in finite deformation elastoplasticity. *Mech. Mater.*, **4**(3–4): 439–51.
[23] Simo, J. C. (1988) A framework for finite strain elastoplasticity based on maximum plastic dissipation and the multiplicative decomposition. 1. Continuum formulation. *Comput. Method Appl. M.*, **66**(2): 199–219.
[24] Simo, J. C. (1988) A framework for finite strain elastoplasticity based on maximum plastic dissipation and the multiplicative decomposition. 2. Computational aspects. *Comput. Method Appl. M.*, **68**(1): 1–31.
[25] Simo, J. C. and Ortiz, M. (1985) A unified approach to finite deformation elastoplastic analysis based on the use of hyperelastic constitutive equations. *Comput. Method Appl. M.*, **49**(2): 221–45.
[26] Simo, J. C. and Ju, J. W. (1989) Finite deformation damage-elastoplasticity: a non conventional framework. *International Journal of Computational Mechanics*, **5**: 375–400.
[27] Simo, J. C. and Hughes, T. J. R. (1998) *Computational Inelasticity*. New York, Springer.
[28] Simo, J. C., Armero, F. and Taylor, R. L. (1993) Improved versions of assumed enhanced strain tri-linear elements for 3D finite deformation problems. *Comput. Method Appl. M.*, **110**(3–4): 359–86.
[29] Zienkiewicz, O. C. (1971) *The Finite Element Method in Engineering Science*. New York, McGraw-Hill.
[30] Zienkiewicz, O. C. and Taylor, R. L. (2005) *The Finite Element Method Set*. Oxford, Elsevier Science.

Part Four

The Finite Element Method in 2D

19

2D Finite Element: Deformation Kinematics Using the Homogeneous Deformation Triangle

19.1 The Finite Element Mesh

The key idea of the finite element method is the discretization of a complex solid into finite sub-domains, Figure 19.1. With discretization, instead of dealing with complex irregular solid domain geometry, one deals with the relatively simple individual triangular sub-domains. Each of the triangles is uniquely defined by three material points P_1, P_2 and P_3, with straight edges in between, Figure 19.2. Because the neighboring finite elements share these material points, these material points are called nodes. As the solid deforms, these material points move in space, Figure 19.3.

19.2 The Homogeneous Deformation Finite Element

The geometry of the three-noded 2D finite element is uniquely defined by the coordinates of its nodes, which can be; Figure 19.4:

a. Initial coordinates

$$(\bar{x}_1 \ \ \bar{y}_1) \ ; \ (\bar{x}_2 \ \ \bar{y}_2) \ ; \ (\bar{x}_3 \ \ \bar{y}_3). \tag{19.1}$$

b. Previous coordinates

$$(\hat{x}_1 \ \ \hat{y}_1) \ ; \ (\hat{x}_2 \ \ \hat{y}_2) \ ; \ (\hat{x}_3 \ \ \hat{y}_3). \tag{19.2}$$

c. Current coordinates

$$(\tilde{x}_1 \ \ \tilde{y}_1) \ ; \ (\tilde{x}_2 \ \ \tilde{y}_2) \ ; \ (\tilde{x}_3 \ \ \tilde{y}_3). \tag{19.3}$$

Large Strain Finite Element Method: A Practical Course, First Edition. Antonio Munjiza, Esteban Rougier and Earl E. Knight.

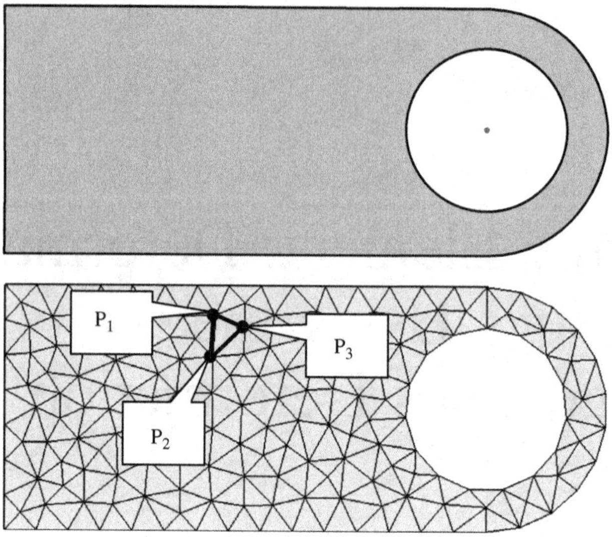

Figure 19.1 Discretization of an irregular solid domain into simple triangular subdomains (3-noded triangular finite elements)

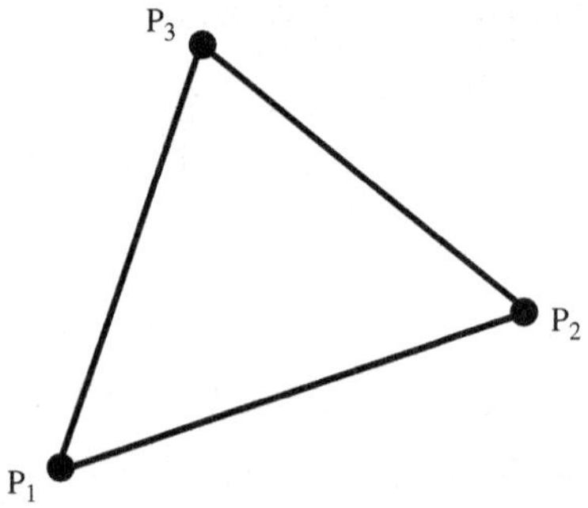

Figure 19.2 Three material points (nodes) defining the geometry of the three-noded triangular finite element

Material Points Inside the Finite Element. In order to define the material point's position inside the finite element (triangle), it is convenient to introduce a local non-Cartesian coordinate system

$$(\xi \;\; \eta) \tag{19.4}$$

in such a way that, Figure 19.5

$$\begin{aligned} \bar{P}_1 &= (\xi_1, \;\; \eta_1) = (0 \;\; 0) \\ \bar{P}_2 &= (\xi_2, \;\; \eta_2) = (1 \;\; 0) \\ \bar{P}_3 &= (\xi_3, \;\; \eta_3) = (0 \;\; 1). \end{aligned} \tag{19.5}$$

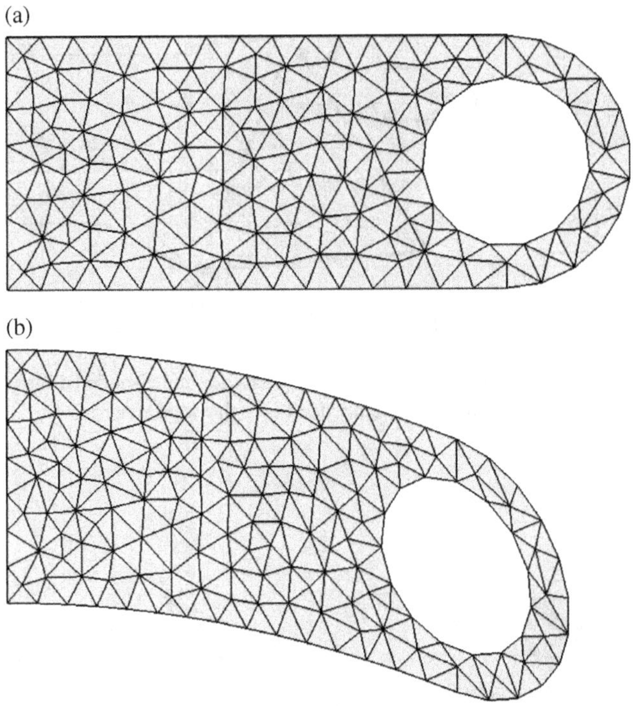

Figure 19.3 Initial (a) and current (b) positions of nodes (material points) and the mesh

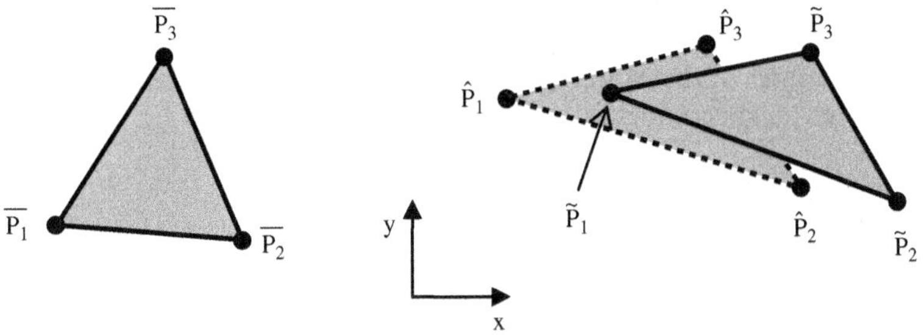

Figure 19.4 Initial, previous and current positions of a single finite element

The coordinate lines $\xi = \text{const}$ and $\eta = \text{const}$ are fixed to the material points of the solid and they move with these material points. As described previously, this coordinate system is called the solid-embedded coordinate system.

The Shape Functions. Using the solid-embedded coordinate system the shape functions are defined for the finite element as follows (Figure 19.6).

$$\begin{aligned} N_1 &= 1 - \xi - \eta \\ N_2 &= \xi \\ N_3 &= \eta \end{aligned} \tag{19.6}$$

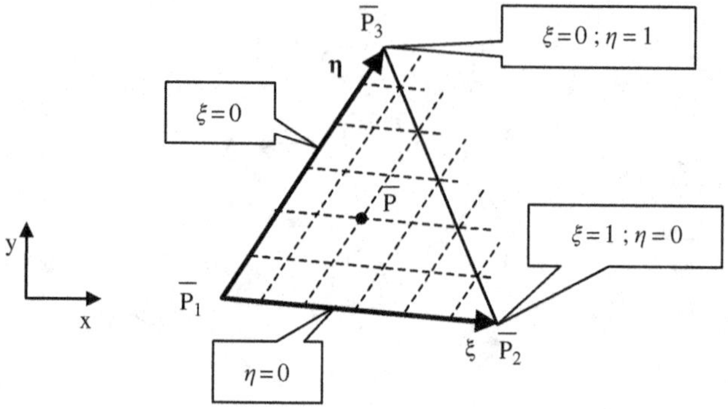

Figure 19.5 The local non-Cartesian coordinate system for the 3-noded triangular finite element. The dashed lines show the coordinate lines embedded into the solid material

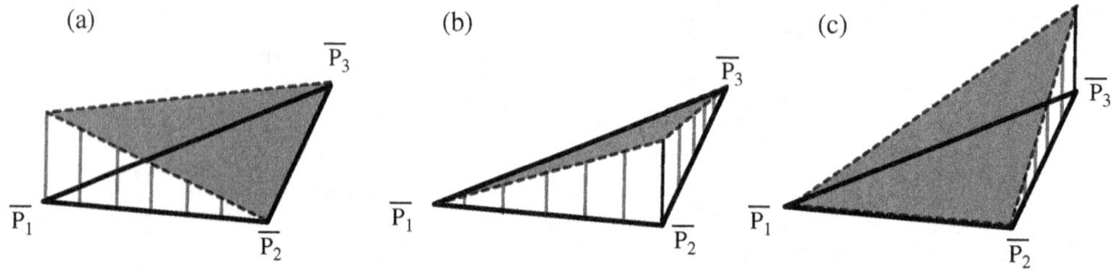

Figure 19.6 Graphical representation of the shape functions for the three-noded triangle finite element: (a) N_1; (b) N_2; (c) N_3

Global Coordinates of the Material Points. Any material point P inside the finite element is described by its local coordinates

$$P=(\xi \ \ \eta). \tag{19.7}$$

The same material point is defined by its initial global coordinates

$$\overline{P}=(\overline{x} \ \ \overline{y}). \tag{19.8}$$

The global coordinates are conveniently approximated from the local coordinates using the shape functions

$$\begin{aligned}\overline{x} &= \overline{x}_1 N_1 + \overline{x}_2 N_2 + \overline{x}_3 N_3\\ \overline{y} &= \overline{y}_1 N_1 + \overline{y}_2 N_2 + \overline{y}_3 N_3,\end{aligned} \tag{19.9}$$

where $(\overline{x}_1, \overline{y}_1)$, $(\overline{x}_2, \overline{y}_2)$ and $(\overline{x}_3, \overline{y}_3)$ are the initial global coordinates of the triangle's nodes P_1, P_2 and P_3.

The previous global coordinates of the same material point are given by

$$\hat{P}=(\hat{x} \ \ \hat{y}) \tag{19.10}$$

and are approximated from the previous global nodal coordinates using the same shape functions

$$\begin{aligned} \hat{x} &= \hat{x}_1 N_1 + \hat{x}_2 N_2 + \hat{x}_3 N_3 \\ \hat{y} &= \hat{y}_1 N_1 + \hat{y}_2 N_2 + \hat{y}_3 N_3 . \end{aligned} \tag{19.11}$$

In a similar manner, the current coordinates of the same material point

$$\tilde{P} = (\tilde{x} \;\; \tilde{y}) \tag{19.12}$$

are approximated from the current global nodal coordinates as follows

$$\begin{aligned} \tilde{x} &= \tilde{x}_1 N_1 + \tilde{x}_2 N_2 + \tilde{x}_3 N_3 \\ \tilde{y} &= \tilde{y}_1 N_1 + \tilde{y}_2 N_2 + \tilde{y}_3 N_3 . \end{aligned} \tag{19.13}$$

The Material-Embedded Infinitesimal Base. At any material point

$$P = (\bar{\xi} \;\; \bar{\eta}) = (\hat{\xi} \;\; \hat{\eta}) = (\tilde{\xi} \;\; \tilde{\eta}) = (\xi \;\; \eta) \tag{19.14}$$

inside the triangle, the material axes for an anisotropic material are defined by the initial material vectors

$$[\bar{\boldsymbol{\alpha}} \;\; \bar{\boldsymbol{\beta}}](\text{im}), \tag{19.15}$$

i.e., by the material-embedded infinitesimal vector base. These vectors define the homogenized material element. The homogenized material element is obtained from the representative volume by employing the assumption of continuum. The representative volume is always of a finite dimension, while the homogenized material element is infinitesimally small, Figure 19.7.

Note that the dimensions of this homogenized material element are α im and β im, where im stands for infinitesimally small unit of length (infi-meter). For each triangle finite element the initial position of base vectors $\bar{\boldsymbol{\alpha}}$ and $\bar{\boldsymbol{\beta}}$ is supplied as an input to the finite element analysis.

The Solid-Embedded Infinitesimal Vector Base. The global vector base $[\mathbf{i} \;\; \mathbf{j}]$ is made of vectors **i** and **j** that coincide with the axes of the global Cartesian coordinate system. In a similar manner, an infinitesimal vector base $[\boldsymbol{\xi} \;\; \boldsymbol{\eta}]$ im is introduced such that the vectors $\boldsymbol{\xi}$ and $\boldsymbol{\eta}$ are defined as tangents on the coordinate lines ξ and η of the solid-embedded coordinate system (ξ, η).

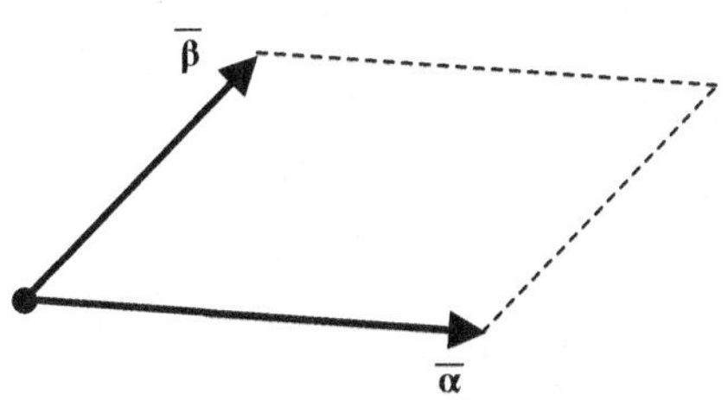

Figure 19.7 Initial shape and orientation of the homogenized infinitesimal material element defined by the initial material vectors $\bar{\boldsymbol{\alpha}}$ and $\bar{\boldsymbol{\beta}}$. This material element is called generalized material element

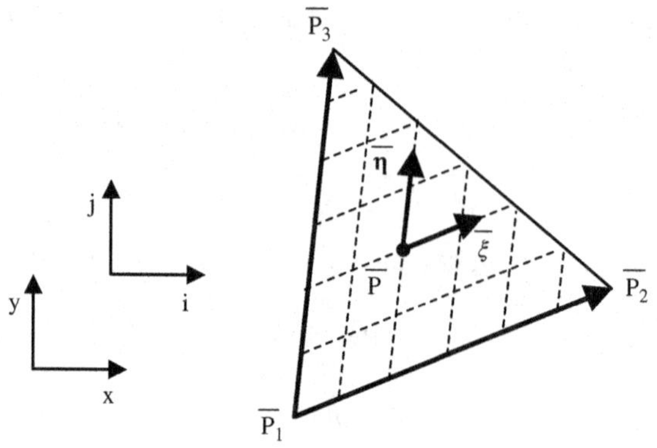

Figure 19.8 Initial position of the solid-embedded infinitesimal vector base [$\boldsymbol{\xi}$ $\boldsymbol{\eta}$] im

From the above definition, it follows that the initial position of the solid-embedded vector base is given by

$$\bar{\boldsymbol{\xi}} = \frac{\partial \bar{x}}{\partial \xi}\mathbf{i} + \frac{\partial \bar{y}}{\partial \xi}\mathbf{j}$$
$$\bar{\boldsymbol{\eta}} = \frac{\partial \bar{x}}{\partial \eta}\mathbf{i} + \frac{\partial \bar{y}}{\partial \eta}\mathbf{j}, \tag{19.16}$$

as shown in Figure 19.8.

As the initial coordinates of the triangular finite element's material points are given by

$$\bar{x} = \bar{x}_1 N_1 + \bar{x}_2 N_2 + \bar{x}_3 N_3$$
$$\bar{y} = \bar{y}_1 N_1 + \bar{y}_2 N_2 + \bar{y}_3 N_3, \tag{19.17}$$

it follows that

$$\begin{aligned}\frac{\partial \bar{x}}{\partial \xi} &= \bar{x}_1 \frac{\partial N_1}{\partial \xi} + \bar{x}_2 \frac{\partial N_2}{\partial \xi} + \bar{x}_3 \frac{\partial N_3}{\partial \xi}\\ &= \bar{x}_1 \frac{\partial (1-\bar{\xi}-\bar{\eta})}{\partial \xi} + \bar{x}_2 \frac{\partial \bar{\xi}}{\partial \xi} + \bar{x}_3 \frac{\partial \bar{\eta}}{\partial \xi}\\ &= -\bar{x}_1 + \bar{x}_2\\ &= \bar{x}_2 - \bar{x}_1\end{aligned} \tag{19.18}$$

and also

$$\frac{\partial \bar{x}}{\partial \eta} = \bar{x}_3 - \bar{x}_1, \tag{19.19}$$

$$\frac{\partial \bar{y}}{\partial \xi} = \bar{y}_2 - \bar{y}_1, \tag{19.20}$$

$$\frac{\partial \bar{y}}{\partial \eta} = \bar{y}_3 - \bar{y}_1. \tag{19.21}$$

When substituted into Equation (19.16) these yield

$$\begin{bmatrix} \bar{\boldsymbol{\xi}} & \bar{\boldsymbol{\eta}} \end{bmatrix} = \begin{bmatrix} \bar{x}_2 - \bar{x}_1 & \bar{x}_3 - \bar{x}_1 \\ \bar{y}_2 - \bar{y}_1 & \bar{y}_3 - \bar{y}_1 \end{bmatrix}. \tag{19.22}$$

Embedded Vector Bases. As both the solid-embedded and the material-embedded vector bases are fixed to the material points, it follows that one can be obtained from the other, Figure 19.9.

The material-embedded vector base can be expressed in terms of the solid-embedded vector base as follows

$$\begin{aligned} \bar{\boldsymbol{\alpha}} &= \bar{\alpha}_{\bar{\xi}} \bar{\boldsymbol{\xi}} + \bar{\alpha}_{\bar{\eta}} \bar{\boldsymbol{\eta}} \\ &= \begin{bmatrix} \bar{\alpha}_{\bar{\xi}} \\ \bar{\alpha}_{\bar{\eta}} \end{bmatrix}, \end{aligned} \tag{19.23}$$

$$\begin{aligned} \bar{\boldsymbol{\beta}} &= \bar{\beta}_{\bar{\xi}} \bar{\boldsymbol{\xi}} + \bar{\beta}_{\bar{\eta}} \bar{\boldsymbol{\eta}} \\ &= \begin{bmatrix} \bar{\beta}_{\bar{\xi}} \\ \bar{\beta}_{\bar{\eta}} \end{bmatrix}, \end{aligned} \tag{19.24}$$

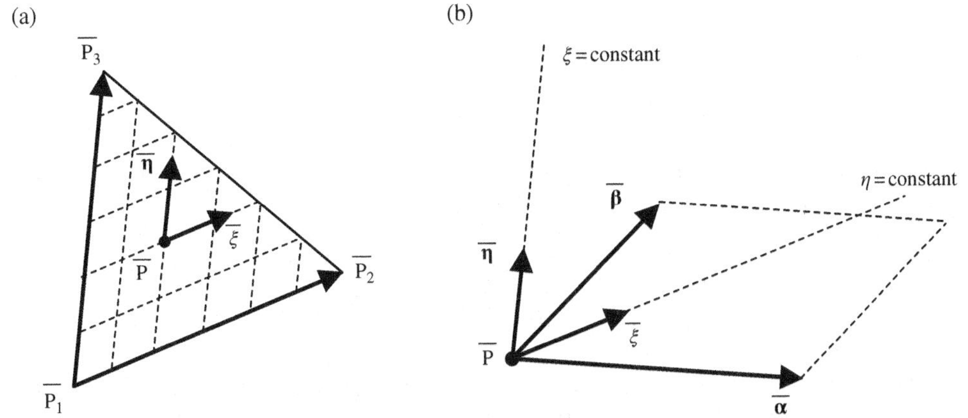

Figure 19.9 (a) Solid-embedded infinitesimal base $\begin{bmatrix} \bar{\boldsymbol{\xi}} & \bar{\boldsymbol{\eta}} \end{bmatrix}$ obtained as tangential vectors to coordinate lines $\boldsymbol{\xi}$ and $\boldsymbol{\eta}$; (b) vectors of the material-embedded base $\begin{bmatrix} \bar{\boldsymbol{\alpha}} & \bar{\boldsymbol{\beta}} \end{bmatrix}$

or

$$\begin{bmatrix}\bar{\boldsymbol{\alpha}} & \bar{\boldsymbol{\beta}}\end{bmatrix} = \begin{bmatrix}\bar{\alpha}_{\bar{\xi}} & \bar{\beta}_{\bar{\xi}} \\ \bar{\alpha}_{\bar{\eta}} & \bar{\beta}_{\bar{\eta}}\end{bmatrix}. \tag{19.25}$$

This means that in the initial configuration, the generalized material element is defined by

$$\begin{aligned}\begin{bmatrix}\bar{\boldsymbol{\alpha}} & \bar{\boldsymbol{\beta}}\end{bmatrix} &= \begin{bmatrix}(\bar{\alpha}_{\bar{\xi}}\bar{\boldsymbol{\xi}} + \bar{\alpha}_{\bar{\eta}}\bar{\boldsymbol{\eta}}) & (\bar{\beta}_{\bar{\xi}}\bar{\boldsymbol{\xi}} + \bar{\beta}_{\bar{\eta}}\bar{\boldsymbol{\eta}})\end{bmatrix} \\ &= \begin{bmatrix}\bar{\boldsymbol{\xi}} & \bar{\boldsymbol{\eta}}\end{bmatrix}\begin{bmatrix}\bar{\alpha}_{\bar{\xi}} & \bar{\beta}_{\bar{\xi}} \\ \bar{\alpha}_{\bar{\eta}} & \bar{\beta}_{\bar{\eta}}\end{bmatrix}.\end{aligned} \tag{19.26}$$

In the previous configuration the same infinitesimal volume is defined by

$$\begin{aligned}\begin{bmatrix}\hat{\boldsymbol{\alpha}} & \hat{\boldsymbol{\beta}}\end{bmatrix} &= \begin{bmatrix}\left(\hat{\alpha}_{\hat{\xi}}\hat{\boldsymbol{\xi}} + \hat{\alpha}_{\hat{\eta}}\hat{\boldsymbol{\eta}}\right) & \left(\hat{\beta}_{\hat{\xi}}\hat{\boldsymbol{\xi}} + \hat{\beta}_{\hat{\eta}}\hat{\boldsymbol{\eta}}\right)\end{bmatrix} \\ &= \begin{bmatrix}\hat{\boldsymbol{\xi}} & \hat{\boldsymbol{\eta}}\end{bmatrix}\begin{bmatrix}\hat{\alpha}_{\hat{\xi}} & \hat{\beta}_{\hat{\xi}} \\ \hat{\alpha}_{\hat{\eta}} & \hat{\beta}_{\hat{\eta}}\end{bmatrix}.\end{aligned} \tag{19.27}$$

In the current configuration the same infinitesimal material element is given by

$$\begin{aligned}\begin{bmatrix}\tilde{\boldsymbol{\alpha}} & \tilde{\boldsymbol{\beta}}\end{bmatrix} &= \begin{bmatrix}\left(\tilde{\alpha}_{\tilde{\xi}}\tilde{\boldsymbol{\xi}} + \tilde{\alpha}_{\tilde{\eta}}\tilde{\boldsymbol{\eta}}\right) & \left(\tilde{\beta}_{\tilde{\xi}}\tilde{\boldsymbol{\xi}} + \tilde{\beta}_{\tilde{\eta}}\tilde{\boldsymbol{\eta}}\right)\end{bmatrix} \\ &= \begin{bmatrix}\tilde{\boldsymbol{\xi}} & \tilde{\boldsymbol{\eta}}\end{bmatrix}\begin{bmatrix}\tilde{\alpha}_{\tilde{\xi}} & \tilde{\beta}_{\tilde{\xi}} \\ \tilde{\alpha}_{\tilde{\eta}} & \tilde{\beta}_{\tilde{\eta}}\end{bmatrix}.\end{aligned} \tag{19.28}$$

As both infinitesimal bases $\begin{bmatrix}\bar{\boldsymbol{\alpha}} & \bar{\boldsymbol{\beta}}\end{bmatrix}$ and $\begin{bmatrix}\bar{\boldsymbol{\xi}} & \bar{\boldsymbol{\eta}}\end{bmatrix}$ are fixed to the solid's material points, the matrix shown in Equation (19.25) does not change with the solid's deformation, i.e. it stays constant, which means

$$\begin{bmatrix}\bar{\alpha}_{\bar{\xi}} & \bar{\beta}_{\bar{\xi}} \\ \bar{\alpha}_{\bar{\eta}} & \bar{\beta}_{\bar{\eta}}\end{bmatrix} = \begin{bmatrix}\tilde{\alpha}_{\tilde{\xi}} & \tilde{\beta}_{\tilde{\xi}} \\ \tilde{\alpha}_{\tilde{\eta}} & \tilde{\beta}_{\tilde{\eta}}\end{bmatrix} = \begin{bmatrix}\hat{\alpha}_{\hat{\xi}} & \hat{\beta}_{\hat{\xi}} \\ \hat{\alpha}_{\hat{\eta}} & \hat{\beta}_{\hat{\eta}}\end{bmatrix} = \begin{bmatrix}\alpha_{\xi} & \beta_{\xi} \\ \alpha_{\eta} & \beta_{\eta}\end{bmatrix}. \tag{19.29}$$

By substituting Equation (19.22) into Equation (19.26) one obtains the matrix of the base $\begin{bmatrix}\bar{\boldsymbol{\alpha}} & \bar{\boldsymbol{\beta}}\end{bmatrix}$ expressed using the $\begin{bmatrix}\mathbf{i} & \mathbf{j}\end{bmatrix}$ global base.

$$\begin{bmatrix}\bar{\boldsymbol{\alpha}} & \bar{\boldsymbol{\beta}}\end{bmatrix} = \begin{bmatrix}\bar{x}_2 - \bar{x}_1 & \bar{x}_3 - \bar{x}_1 \\ \bar{y}_2 - \bar{y}_1 & \bar{y}_3 - \bar{y}_1\end{bmatrix}\begin{bmatrix}\alpha_{\xi} & \beta_{\xi} \\ \alpha_{\eta} & \beta_{\eta}\end{bmatrix}. \tag{19.30}$$

As explained before, since $\begin{bmatrix}\bar{\boldsymbol{\alpha}} & \bar{\boldsymbol{\beta}}\end{bmatrix}$ define the initial generalized material element, for each finite element they are given as input to the finite element analysis and may, for example, reflect the orientation of geological layers in geosciences problems or the orientation of carbon fibers in composite material problems.

Thus, at the beginning of the finite element analysis, the matrix linking the material-embedded vector base and the solid-embedded vector base is obtained

$$\begin{bmatrix} \alpha_\xi & \beta_\xi \\ \alpha_\eta & \beta_\eta \end{bmatrix} = \begin{bmatrix} \bar{x}_2-\bar{x}_1 & \bar{x}_3-\bar{x}_1 \\ \bar{y}_2-\bar{y}_1 & \bar{y}_3-\bar{y}_1 \end{bmatrix}^{-1} \begin{bmatrix} \bar{\alpha}_i & \bar{\beta}_i \\ \bar{\alpha}_j & \bar{\beta}_j \end{bmatrix}, \tag{19.31}$$

where, as explained before

$$\begin{bmatrix} \bar{x}_2-\bar{x}_1 & \bar{x}_3-\bar{x}_1 \\ \bar{y}_2-\bar{y}_1 & \bar{y}_3-\bar{y}_1 \end{bmatrix} \tag{19.32}$$

is known from the input geometry (initial nodal coordinates) of the problem being solved.

The previous position of the material-embedded vector base is obtained from Equation (19.27) and is given by

$$[\hat{\boldsymbol{\alpha}} \;\; \hat{\boldsymbol{\beta}}] = [\mathbf{i} \;\; \mathbf{j}] \begin{bmatrix} \hat{\alpha}_i & \hat{\beta}_i \\ \hat{\alpha}_j & \hat{\beta}_j \end{bmatrix} = [\hat{\boldsymbol{\xi}} \;\; \hat{\boldsymbol{\eta}}] \begin{bmatrix} \alpha_\xi & \beta_\xi \\ \alpha_\eta & \beta_\eta \end{bmatrix}, \tag{19.33}$$

where

$$[\hat{\boldsymbol{\xi}} \;\; \hat{\boldsymbol{\eta}}] = \begin{bmatrix} \hat{x}_2-\hat{x}_1 & \hat{x}_3-\hat{x}_1 \\ \hat{y}_2-\hat{y}_1 & \hat{y}_3-\hat{y}_1 \end{bmatrix} \tag{19.34}$$

are the previous nodal coordinates of the finite element.

The current position of the material-embedded vector base is obtained from Equation (19.28)

$$[\tilde{\boldsymbol{\alpha}} \;\; \tilde{\boldsymbol{\beta}}] = [\mathbf{i} \;\; \mathbf{j}] \begin{bmatrix} \tilde{\alpha}_i & \tilde{\beta}_i \\ \tilde{\alpha}_j & \tilde{\beta}_j \end{bmatrix} = [\tilde{\boldsymbol{\xi}} \;\; \tilde{\boldsymbol{\eta}}] \begin{bmatrix} \alpha_\xi & \beta_\xi \\ \alpha_\eta & \beta_\eta \end{bmatrix}, \tag{19.35}$$

where

$$[\tilde{\boldsymbol{\xi}} \;\; \tilde{\boldsymbol{\eta}}] = \begin{bmatrix} \tilde{x}_2-\tilde{x}_1 & \tilde{x}_3-\tilde{x}_1 \\ \tilde{y}_2-\tilde{y}_1 & \tilde{y}_3-\tilde{y}_1 \end{bmatrix}. \tag{19.36}$$

As it can be seen from Equations (19.32), (19.34) and (19.36), the material-embedded vector base does not depend on the local coordinates ξ and η, i.e. it is constant over the domain of the finite element. This implies that the generalized material element is of constant shape over the finite element. Stretches, strains and stresses are therefore also constant over the domain of the finite element. This finite element is therefore called the constant strain finite element or more correctly (given the context of the geometric nonlinearity) the homogeneous deformation finite element, or simply the homogeneous deformation triangle.

19.3 Summary

In this chapter it has been shown how the solid domain of any shape is discretized into triangles. Each triangle has a set of material axes attached to its material points. These come from the input of the problem to be solved and they also define the generalized material element.

As the triangle moves, the coordinates of its nodes uniquely define the position of any material point within it and therefore the position of the solid-embedded vector base, $[\boldsymbol{\xi}\ \boldsymbol{\eta}]$. From the position of these solid-embedded base vectors the geometry of the previous and current generalized material element are calculated using Equations (19.33) and (19.35). These are then passed to the material package for calculation of the Munjiza and Cauchy stress matrices, see Chapter 17.

The Cauchy stress matrix is passed back to the finite element simulation and from it the nodal forces are calculated.

Further Reading

[1] Munjiza, A. (2004) *The Combined Finite-Discrete Element Method.* Chichester, John Wiley & Sons, Ltd.

[2] Oden, J. T. (1972) *Finite elements of nonlinear continua.* New York, McGraw-Hill.

[3] Hughes, T.J.R. (2000) *The Finite Element Method: Linear Static and Dynamic Finite Element Analysis.* New York, Dover Publications.

[4] Zienkiewicz, O. C. (1971) *The Finite Element Method in Engineering Science.* New York, McGraw-Hill.

[5] Zienkiewicz, O. C., Rojek, J., Taylor, R. L. and Pastor, M. (1998) Triangles and tetrahera in explicit dynamic codes for solids. *International Journal for Numerical Methods in Engineering*, **43**(3): 565–83.

[6] Zienkiewicz, O. C. and Taylor, R. L. (2005) *The Finite Element Method Set.* Oxford, Elsevier Science.

20

2D Finite Element: Deformation Kinematics Using Iso-Parametric Finite Elements

20.1 The Finite Element Library

In Chapter 19 a homogeneous deformation three-noded triangular finite element was introduced. In theory, any shape of finite element can be employed with any number of nodes. Some of the possible elements are shown in Figure 20.1, Figure 20.2 and Figure 20.3.

Within a finite element package, finite elements such as those shown above are conveniently packaged into a finite element library. The user chooses which elements to use when creating the finite element mesh (meshing).

20.2 The Shape Functions

For a node i of any given finite element, a single shape function

$$N_i = N_i(\xi, \eta) \tag{20.1}$$

is defined in such a way that

$$\begin{gathered} N_i(\xi_j, \eta_j) = 0 \quad \text{for all } j \neq i \\ \text{and} \\ N_i(\xi_j, \eta_j) = 1 \quad \text{for every } j = i, \end{gathered} \tag{20.2}$$

i.e., a given shape function N_i has a value of 1 at node i and zero at all other nodes. In addition, the C^0 continuity conditions are enforced on the shape functions:

Large Strain Finite Element Method: A Practical Course, First Edition. Antonio Munjiza, Esteban Rougier and Earl E. Knight.

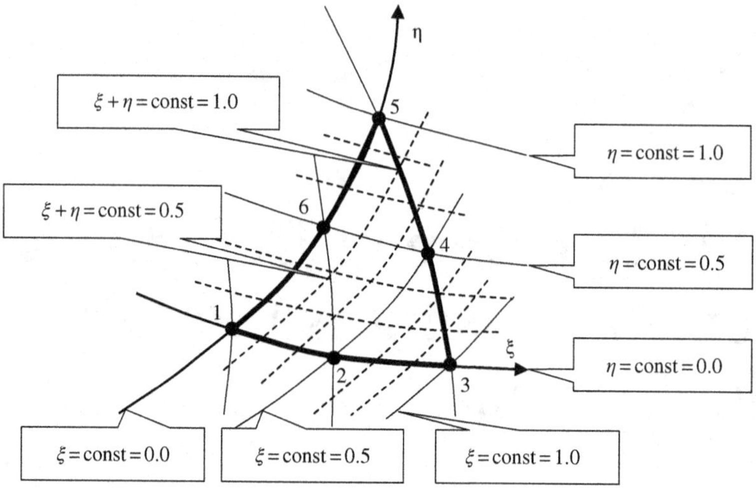

Figure 20.1 The six-noded triangle finite element with the local solid-embedded curved coordinate system. The coordinate lines are fixed to the material points

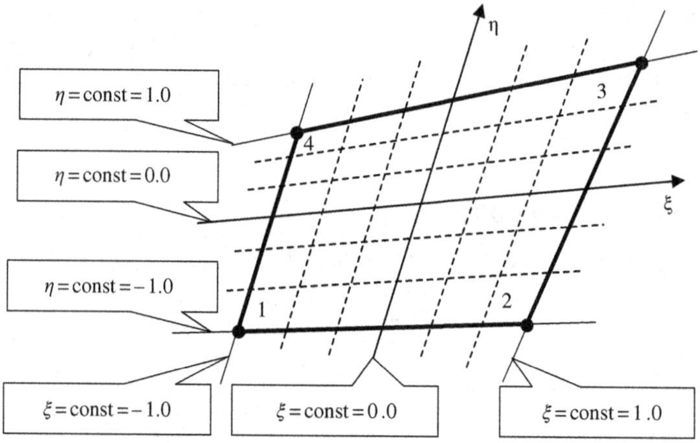

Figure 20.2 The four-noded quadrilateral finite element with straight coordinate lines fixed to the material points defining the solid-embedded coordinate system

The shape function for any node is equal to zero at all finite element edges except for the edges the particular node belongs to; also any shape function along a given edge depends only on the nodes on that edge.

The shape functions for the three elements shown in Figure 20.1, Figure 20.2, and Figure 20.3 are as follows:

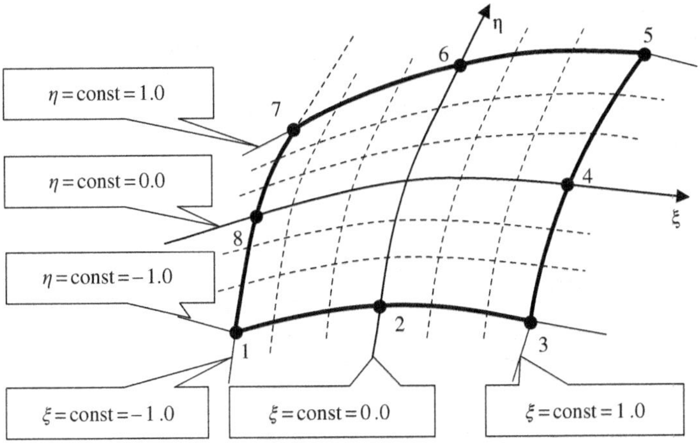

Figure 20.3 The eight-noded quadrilateral finite element with the solid-embedded coordinate system being defined by the curved coordinate lines that are fixed to the material points

a. Six-Noded Triangle:

$$\begin{aligned} & N_1 = 2(1-\xi-\eta)\left(\frac{1}{2}-\xi-\eta\right) && N_4 = 4\xi\eta \\ & N_2 = 4(1-\xi-\eta)\xi && N_5 = 2\eta\left(\eta-\frac{1}{2}\right) \\ & N_3 = 2\xi\left(\xi-\frac{1}{2}\right) && N_6 = 4(1-\xi-\eta)\eta \end{aligned} \tag{20.3}$$

b. Four-Noded Quadrilateral:

$$\begin{aligned} & N_1 = \frac{1}{4}(1-\xi)(1-\eta) && N_3 = \frac{1}{4}(1+\xi)(1+\eta) \\ & N_2 = \frac{1}{4}(1+\xi)(1-\eta) && N_4 = \frac{1}{4}(1-\xi)(1+\eta) \end{aligned} \tag{20.4}$$

c. Eight-Noded Quadrilateral:

$$\begin{aligned} & N_1 = \frac{1}{4}(1-\xi)(1-\eta)(-\xi-\eta-1) && N_5 = \frac{1}{4}(1+\xi)(1+\eta)(\xi+\eta-1) \\ & N_2 = \frac{1}{2}(1-\xi^2)(1-\eta) && N_6 = \frac{1}{2}(1-\xi^2)(1+\eta) \\ & N_3 = \frac{1}{4}(1+\xi)(1-\eta)(\xi-\eta-1) && N_7 = \frac{1}{4}(1-\xi)(1+\eta)(-\xi+\eta-1) \\ & N_4 = \frac{1}{2}(1+\xi)(1-\eta^2) && N_8 = \frac{1}{2}(1-\xi)(1-\eta^2) \end{aligned} \tag{20.5}$$

The above shape functions are derived in an "ad hoc" way; for this reason, they are called serendipity shape functions and the corresponding finite elements are called serendipity finite elements.

There are other ways to define shape functions in a systematic way, such as Lagrange polynomials. These yield Lagrange shape functions and consequently, Lagrange finite elements. It is also possible to use special numerically derived functions [1], [4].

20.3 Nodal Positions

The position of the finite element is defined by the position of all the nodes. These are conveniently stored in the finite element package as follows:

a. Initial nodal positions

$$
\begin{aligned}
\bar{P}_1 &= (\bar{x}_1, \bar{y}_1) \\
\bar{P}_2 &= (\bar{x}_2, \bar{y}_2) \\
&\ldots
\end{aligned}
\tag{20.6}
$$

b. Previous nodal positions

$$
\begin{aligned}
\hat{P}_1 &= (\hat{x}_1, \hat{y}_1) \\
\hat{P}_2 &= (\hat{x}_2, \hat{y}_2) \\
&\ldots
\end{aligned}
\tag{20.7}
$$

c. Current nodal positions

$$
\begin{aligned}
\tilde{P}_1 &= (\tilde{x}_1, \tilde{y}_1) \\
\tilde{P}_2 &= (\tilde{x}_2, \tilde{y}_2) \\
&\ldots,
\end{aligned}
\tag{20.8}
$$

where x and y define the global coordinates using the Cartesian coordinate system, while "–" stand for "initial," "^" stands for "previous" and "~" stands for "current."

The initial nodal positions come from the input to the finite element method. These define the initial geometry of the finite element mesh and consequently the initial position of the solid.

As the solid deforms, the nodal coordinates change. In the finite element method this deformation usually occurs in steps (anything from a few steps to a few billion steps). At each step the solid deforms further relative to the previous step.

The current step is of particular importance because it defines the latest deformation state of the solid body. The step immediately before the current is also important because it defines the latest (current) change in the solid's deformation; if one knows the size of the time step, the rate of deformation can then be obtained from the current and the previous positions.

20.4 Positions of Material Points inside a Single Finite Element

By knowing the nodal positions described above, one knows the initial, the current and the previous coordinates of the nodes only; i.e., one does not know anything about the points inside the finite element. The position of any material point P within a single finite element is approximated using the local coordinates

$$\begin{aligned}\bar{\mathrm{P}} &= (\bar{\xi}, \bar{\eta})\\ \hat{\mathrm{P}} &= (\hat{\xi}, \hat{\eta})\\ \tilde{\mathrm{P}} &= (\tilde{\xi}, \tilde{\eta}),\end{aligned} \tag{20.9}$$

where $\bar{\mathrm{P}}$, $\hat{\mathrm{P}}$ and $\tilde{\mathrm{P}}$ respectively represent the initial, previous and current positions of a given material point.

As the solid-embedded coordinate lines are fixed to the material points, the material point's local coordinates do not change as the finite element deforms (moves in space), i.e.,

$$(\bar{\xi}, \bar{\eta}) = (\hat{\xi}, \hat{\eta}) = (\tilde{\xi}, \tilde{\eta}) = (\xi, \eta). \tag{20.10}$$

In contrast, the material points' global (x, y) coordinates keep changing as the finite element moves, i.e.,

$$\begin{aligned}\bar{\mathrm{P}} &= (\bar{x}, \bar{y})\\ \hat{\mathrm{P}} &= (\hat{x}, \hat{y})\\ \tilde{\mathrm{P}} &= (\tilde{x}, \tilde{y}),\end{aligned} \tag{20.11}$$

where $(\bar{x}, \bar{y})$, $(\hat{x}, \hat{y})$ and $(\tilde{x}, \tilde{y})$ represent the initial, previous and current global coordinates respectively. In general

$$(\bar{x}, \bar{y}) \neq (\hat{x}, \hat{y}) \neq (\tilde{x}, \tilde{y}). \tag{20.12}$$

The global (x, y) coordinates are approximated from the local (ξ, η) coordinates using the shape functions:

a. Initial global coordinates of material points (ξ, η)

$$\begin{aligned}\bar{\mathrm{x}} &= \bar{x}_1 \mathrm{N}_1(\xi,\eta) + \bar{x}_2 \mathrm{N}_2(\xi,\eta) + \bar{x}_3 \mathrm{N}_3(\xi,\eta) + \ldots + \bar{x}_n \mathrm{N}_n(\xi,\eta)\\ \bar{\mathrm{y}} &= \bar{y}_1 \mathrm{N}_1(\xi,\eta) + \bar{y}_2 \mathrm{N}_2(\xi,\eta) + \bar{y}_3 \mathrm{N}_3(\xi,\eta) + \ldots + \bar{y}_n \mathrm{N}_n(\xi,\eta),\end{aligned} \tag{20.13}$$

where $\bar{x}_i$ and $\bar{y}_i$ are the initial nodal coordinates and N_i are the shape functions.

b. Previous global coordinates of material points (ξ, η)

$$\begin{aligned}\hat{\mathrm{x}} &= \hat{x}_1 \mathrm{N}_1(\xi,\eta) + \hat{x}_2 \mathrm{N}_2(\xi,\eta) + \hat{x}_3 \mathrm{N}_3(\xi,\eta) + \ldots + \hat{x}_n \mathrm{N}_n(\xi,\eta)\\ \hat{\mathrm{y}} &= \hat{y}_1 \mathrm{N}_1(\xi,\eta) + \hat{y}_2 \mathrm{N}_2(\xi,\eta) + \hat{y}_3 \mathrm{N}_3(\xi,\eta) + \ldots + \hat{y}_n \mathrm{N}_n(\xi,\eta),\end{aligned} \tag{20.14}$$

where $\hat{x}_i$ and $\hat{y}_i$ are the previous nodal coordinates.

c. Current global coordinates of material points (ξ, η)

$$\tilde{x}=\tilde{x}_1 N_1(\xi,\eta)+\tilde{x}_2 N_2(\xi,\eta)+\tilde{x}_3 N_3(\xi,\eta)+\ldots+\tilde{x}_n N_n(\xi,\eta)$$
$$\tilde{y}=\tilde{y}_1 N_1(\xi,\eta)+\tilde{y}_2 N_2(\xi,\eta)+\tilde{y}_3 N_3(\xi,\eta)+\ldots+\tilde{y}_n N_n(\xi,\eta), \tag{20.15}$$

where $\tilde{x}_i$ and $\tilde{y}_i$ are the current nodal coordinates.

20.5 The Solid-Embedded Vector Base

The solid-embedded coordinate lines $\xi=\text{const}$ and $\eta=\text{const}$ are local to each finite element and in general are neither straight lines nor orthogonal to each other. At each material point, the coordinate lines intersect each other at different angles.

For a given point P, it is possible to define vectors $\boldsymbol{\xi}$ and $\boldsymbol{\eta}$ such that vector $\boldsymbol{\xi}$ is tangential to the corresponding ξ line, and vector $\boldsymbol{\eta}$ is tangential to the corresponding η line, Figure 20.4.

These two vectors form the solid-embedded vector base

$$[\boldsymbol{\xi}\ \boldsymbol{\eta}](\text{im}), \tag{20.16}$$

where the base vectors $\boldsymbol{\xi}$ and $\boldsymbol{\eta}$ are of the unit length im. These base vectors are generally different for every material point

$$\text{P}=(\xi,\eta). \tag{20.17}$$

By definition vector $\boldsymbol{\xi}$ is given by

$$\begin{aligned}\boldsymbol{\xi}&=\frac{\partial x}{\partial \xi}\mathbf{i}+\frac{\partial y}{\partial \xi}\mathbf{j}\\ &=[\mathbf{i}\ \ \mathbf{j}]\begin{bmatrix}\dfrac{\partial x}{\partial \xi}\\ \dfrac{\partial y}{\partial \xi}\end{bmatrix}\end{aligned} \tag{20.18}$$

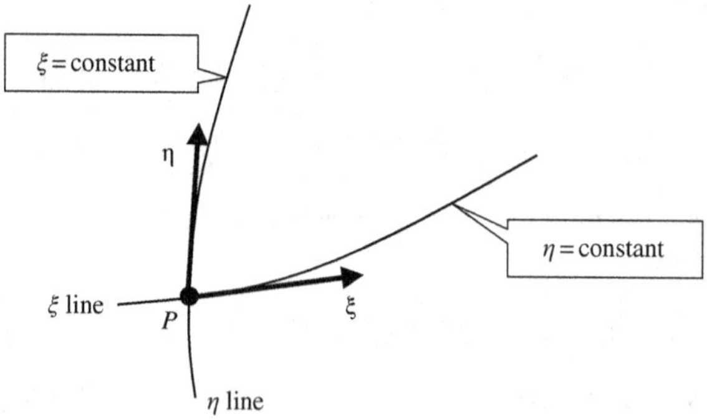

Figure 20.4 Solid-embedded infinitesimal vector base $[\boldsymbol{\xi}\ \ \boldsymbol{\eta}]$ with vectors $\boldsymbol{\xi}$ and $\boldsymbol{\eta}$ being tangential to the curved coordinate lines ξ and η respectively

and vector $\boldsymbol{\eta}$ is given by

$$\boldsymbol{\eta} = \frac{\partial x}{\partial \eta}\mathbf{i} + \frac{\partial y}{\partial \eta}\mathbf{j}$$
$$= [\mathbf{i}\ \ \mathbf{j}]\begin{bmatrix} \frac{\partial x}{\partial \eta} \\ \frac{\partial y}{\partial \eta} \end{bmatrix}. \tag{20.19}$$

The solid-embedded vector base is therefore given by

$$[\boldsymbol{\xi}\ \ \boldsymbol{\eta}] = [\mathbf{i}\ \ \mathbf{j}]\begin{bmatrix} \frac{\partial x}{\partial \xi} & \frac{\partial x}{\partial \eta} \\ \frac{\partial y}{\partial \xi} & \frac{\partial y}{\partial \eta} \end{bmatrix}$$
$$= \begin{bmatrix} \frac{\partial x}{\partial \xi} & \frac{\partial x}{\partial \eta} \\ \frac{\partial y}{\partial \xi} & \frac{\partial y}{\partial \eta} \end{bmatrix}. \tag{20.20}$$

This vector base is local to each finite element's material point. Its base vectors are fixed to the material points in the infinitesimal vicinity of a given material point P. As such, these base vectors deform together with the material in the infinitesimal vicinity of the solid body's material point P.

In engineering terms, one can think of them as infinitesimally short lines plotted on the surface of a latex sheet. As the sheet stretches (deforms) the lines change in length, orientation and also the angle between them changes, as shown in Figure 20.5.

By definition, the initial position of the solid-embedded base is given by

$$[\bar{\boldsymbol{\xi}}\ \ \bar{\boldsymbol{\eta}}] = \begin{bmatrix} \bar{\xi}_i & \bar{\eta}_i \\ \bar{\xi}_j & \bar{\eta}_i \end{bmatrix} = \begin{bmatrix} \frac{\partial \bar{x}}{\partial \xi} & \frac{\partial \bar{x}}{\partial \eta} \\ \frac{\partial \bar{y}}{\partial \xi} & \frac{\partial \bar{y}}{\partial \eta} \end{bmatrix}. \tag{20.21}$$

In a similar manner, the previous position of the solid-embedded base is

$$[\hat{\boldsymbol{\xi}}\ \ \hat{\boldsymbol{\eta}}] = \begin{bmatrix} \hat{\xi}_i & \hat{\eta}_i \\ \hat{\xi}_j & \hat{\eta}_i \end{bmatrix} = \begin{bmatrix} \frac{\partial \hat{x}}{\partial \xi} & \frac{\partial \hat{x}}{\partial \eta} \\ \frac{\partial \hat{y}}{\partial \xi} & \frac{\partial \hat{y}}{\partial \eta} \end{bmatrix}. \tag{20.22}$$

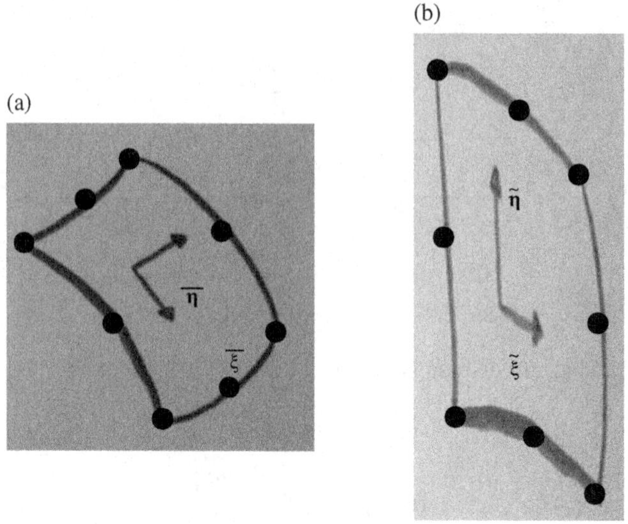

Figure 20.5 Vector base $[\boldsymbol{\xi} \;\; \boldsymbol{\eta}]$ plotted on a latex sheet: (a) initial position; (b) current position. It is worth noting that the length and orientation of vectors $\boldsymbol{\xi}$ and $\boldsymbol{\eta}$ changes as the latex sheet is deformed

The current position of the solid-embedded base is given by

$$[\tilde{\boldsymbol{\xi}} \;\; \tilde{\boldsymbol{\eta}}] = \begin{bmatrix} \tilde{\xi}_i & \tilde{\eta}_i \\ \tilde{\xi}_j & \tilde{\eta}_i \end{bmatrix} = \begin{bmatrix} \dfrac{\partial \tilde{x}}{\partial \xi} & \dfrac{\partial \tilde{x}}{\partial \eta} \\ \dfrac{\partial \tilde{y}}{\partial \xi} & \dfrac{\partial \tilde{y}}{\partial \eta} \end{bmatrix}. \tag{20.23}$$

20.6 The Material-Embedded Vector Base

Depending on the problem being solved, the finite element's material is defined by the infinitesimal material element of parallelogram shape conveniently called the generalized material element. The generalized material element is uniquely defined by two infinitesimally short vectors $\boldsymbol{\alpha}$ and $\boldsymbol{\beta}$, which are fixed to the solid body's material points.

The initial position of these two vectors is provided as an input to the finite element analysis (in terms of global orthonormal base $[\mathbf{i} \;\; \mathbf{j}]$):

$$\begin{aligned} [\bar{\boldsymbol{\alpha}} \;\; \bar{\boldsymbol{\beta}}] &= [\mathbf{i} \;\; \mathbf{j}] \begin{bmatrix} \bar{\alpha}_i & \bar{\beta}_i \\ \bar{\alpha}_j & \bar{\beta}_j \end{bmatrix} \\ &= \begin{bmatrix} \bar{\alpha}_i & \bar{\beta}_i \\ \bar{\alpha}_j & \bar{\beta}_j \end{bmatrix}. \end{aligned} \tag{20.24}$$

As at every material point P of the solid there exists the local solid-embedded vector base

$$\begin{bmatrix} \bar{\boldsymbol{\xi}} & \bar{\boldsymbol{\eta}} \end{bmatrix} = \begin{bmatrix} \bar{\xi}_i & \bar{\eta}_i \\ \bar{\xi}_j & \bar{\eta}_j \end{bmatrix} \tag{20.25}$$

and it is possible to express vectors $\bar{\boldsymbol{\alpha}}$ and $\bar{\boldsymbol{\beta}}$ using the base $\begin{bmatrix} \bar{\boldsymbol{\xi}} & \bar{\boldsymbol{\eta}} \end{bmatrix}$.

$$\begin{bmatrix} \bar{\boldsymbol{\alpha}} & \bar{\boldsymbol{\beta}} \end{bmatrix} = \begin{bmatrix} \bar{\boldsymbol{\xi}} & \bar{\boldsymbol{\eta}} \end{bmatrix} \begin{bmatrix} \alpha_\xi & \beta_\xi \\ \alpha_\eta & \beta_\eta \end{bmatrix}, \tag{20.26}$$

or simply

$$\begin{bmatrix} \bar{\boldsymbol{\alpha}} & \bar{\boldsymbol{\beta}} \end{bmatrix} = \begin{bmatrix} \alpha_\xi & \beta_\xi \\ \alpha_\eta & \beta_\eta \end{bmatrix}, \tag{20.27}$$

where (as customary throughout this book) the base $\begin{bmatrix} \bar{\boldsymbol{\xi}} & \bar{\boldsymbol{\eta}} \end{bmatrix}$ is assumed from the context.

By combining Equations (20.24) and (20.26), one obtains

$$\begin{bmatrix} \bar{\alpha}_i & \bar{\beta}_i \\ \bar{\alpha}_j & \bar{\beta}_j \end{bmatrix} = \begin{bmatrix} \bar{\xi}_i & \bar{\eta}_i \\ \bar{\xi}_j & \bar{\eta}_j \end{bmatrix} \begin{bmatrix} \alpha_\xi & \beta_\xi \\ \alpha_\eta & \beta_\eta \end{bmatrix}, \tag{20.28}$$

which yields

$$\begin{bmatrix} \alpha_\xi & \beta_\xi \\ \alpha_\eta & \beta_\eta \end{bmatrix} = \begin{bmatrix} \bar{\xi}_i & \bar{\eta}_i \\ \bar{\xi}_j & \bar{\eta}_j \end{bmatrix}^{-1} \begin{bmatrix} \bar{\alpha}_i & \bar{\beta}_i \\ \bar{\alpha}_j & \bar{\beta}_j \end{bmatrix}, \tag{20.29}$$

where (from Equation (20.21))

$$\begin{bmatrix} \bar{\xi}_i & \bar{\eta}_i \\ \bar{\xi}_j & \bar{\eta}_j \end{bmatrix} = \begin{bmatrix} \dfrac{\partial \bar{\mathrm{x}}}{\partial \xi} & \dfrac{\partial \bar{\mathrm{x}}}{\partial \eta} \\ \dfrac{\partial \bar{\mathrm{y}}}{\partial \xi} & \dfrac{\partial \bar{\mathrm{y}}}{\partial \eta} \end{bmatrix} \tag{20.30}$$

and, as explained above,

$$\begin{bmatrix} \bar{\alpha}_i & \bar{\beta}_i \\ \bar{\alpha}_j & \bar{\beta}_j \end{bmatrix} \tag{20.31}$$

are provided as input to the finite element analysis.

As the material-embedded base $\begin{bmatrix} \bar{\boldsymbol{\alpha}} & \bar{\boldsymbol{\beta}} \end{bmatrix}$ and the solid-embedded base $[\bar{\xi} \;\; \bar{\eta}]$ are both fixed to the material points of the solid body, it follows that the relationship between them does not change with the deformation. From this, it follows that the previous and current material axes are simply

$$\begin{bmatrix} \hat{\boldsymbol{\alpha}} & \hat{\boldsymbol{\beta}} \end{bmatrix} = \begin{bmatrix} \hat{\alpha}_i & \hat{\beta}_i \\ \hat{\alpha}_j & \hat{\beta}_j \end{bmatrix} = \begin{bmatrix} \hat{\xi}_i & \hat{\eta}_i \\ \hat{\xi}_j & \hat{\eta}_j \end{bmatrix} \begin{bmatrix} \alpha_\xi & \beta_\xi \\ \alpha_\eta & \beta_\eta \end{bmatrix}, \tag{20.32}$$

$$\begin{bmatrix}\tilde{\boldsymbol{\alpha}} & \tilde{\boldsymbol{\beta}}\end{bmatrix} = \begin{bmatrix}\tilde{\alpha}_i & \tilde{\beta}_i \\ \tilde{\alpha}_j & \tilde{\beta}_j\end{bmatrix} = \begin{bmatrix}\tilde{\xi}_i & \tilde{\eta}_i \\ \tilde{\xi}_j & \tilde{\eta}_j\end{bmatrix}\begin{bmatrix}\alpha_\xi & \beta_\xi \\ \alpha_\eta & \beta_\eta\end{bmatrix}, \tag{20.33}$$

where

$$\begin{bmatrix}\hat{\xi}_i & \hat{\eta}_i \\ \hat{\xi}_j & \hat{\eta}_j\end{bmatrix} = \begin{bmatrix}\dfrac{\partial \hat{\mathrm{x}}}{\partial \xi} & \dfrac{\partial \hat{\mathrm{x}}}{\partial \eta} \\ \dfrac{\partial \hat{\mathrm{y}}}{\partial \xi} & \dfrac{\partial \hat{\mathrm{y}}}{\partial \eta}\end{bmatrix}, \tag{20.34}$$

$$\begin{bmatrix}\tilde{\xi}_i & \tilde{\eta}_i \\ \tilde{\xi}_j & \tilde{\eta}_j\end{bmatrix} = \begin{bmatrix}\dfrac{\partial \tilde{\mathrm{x}}}{\partial \xi} & \dfrac{\partial \tilde{\mathrm{x}}}{\partial \eta} \\ \dfrac{\partial \tilde{\mathrm{y}}}{\partial \xi} & \dfrac{\partial \tilde{\mathrm{y}}}{\partial \eta}\end{bmatrix}. \tag{20.35}$$

As the material point's global coordinates are calculated from the local coordinates using the shape functions, Equations (20.13), (20.14) and (20.15), it follows that

$$\begin{aligned}
\frac{\partial \bar{\mathrm{x}}}{\partial \xi} &= \bar{x}_1\frac{\partial \mathrm{N}_1}{\partial \xi} + \bar{x}_2\frac{\partial \mathrm{N}_2}{\partial \xi} + \bar{x}_3\frac{\partial \mathrm{N}_3}{\partial \xi} + \ldots + \bar{x}_n\frac{\partial \mathrm{N}_n}{\partial \xi} \\
\frac{\partial \bar{\mathrm{x}}}{\partial \eta} &= \bar{x}_1\frac{\partial \mathrm{N}_1}{\partial \eta} + \bar{x}_2\frac{\partial \mathrm{N}_2}{\partial \eta} + \bar{x}_3\frac{\partial \mathrm{N}_3}{\partial \eta} + \ldots + \bar{x}_n\frac{\partial \mathrm{N}_n}{\partial \eta} \\
\frac{\partial \bar{\mathrm{y}}}{\partial \xi} &= \bar{y}_1\frac{\partial \mathrm{N}_1}{\partial \xi} + \bar{y}_2\frac{\partial \mathrm{N}_2}{\partial \xi} + \bar{y}_3\frac{\partial \mathrm{N}_3}{\partial \xi} + \ldots + \bar{y}_n\frac{\partial \mathrm{N}_n}{\partial \xi} \\
\frac{\partial \bar{\mathrm{y}}}{\partial \eta} &= \bar{y}_1\frac{\partial \mathrm{N}_1}{\partial \eta} + \bar{y}_2\frac{\partial \mathrm{N}_2}{\partial \eta} + \bar{y}_3\frac{\partial \mathrm{N}_3}{\partial \eta} + \ldots + \bar{y}_n\frac{\partial \mathrm{N}_n}{\partial \eta}.
\end{aligned} \tag{20.36}$$

which can be written in matrix form

$$\begin{bmatrix}\dfrac{\partial \bar{\mathrm{x}}}{\partial \xi} & \dfrac{\partial \bar{\mathrm{x}}}{\partial \eta} \\ \dfrac{\partial \bar{\mathrm{y}}}{\partial \xi} & \dfrac{\partial \bar{\mathrm{y}}}{\partial \eta}\end{bmatrix} = \begin{bmatrix}\bar{x}_1 & \bar{x}_2 & \bar{x}_3 & \ldots & \bar{x}_n \\ \bar{y}_1 & \bar{y}_2 & \bar{y}_3 & \ldots & \bar{y}_n\end{bmatrix}\begin{bmatrix}\dfrac{\partial \mathrm{N}_1}{\partial \xi} & \dfrac{\partial \mathrm{N}_1}{\partial \eta} \\ \dfrac{\partial \mathrm{N}_2}{\partial \xi} & \dfrac{\partial \mathrm{N}_2}{\partial \eta} \\ \dfrac{\partial \mathrm{N}_3}{\partial \xi} & \dfrac{\partial \mathrm{N}_3}{\partial \eta} \\ \vdots & \vdots \\ \dfrac{\partial \mathrm{N}_n}{\partial \xi} & \dfrac{\partial \mathrm{N}_n}{\partial \eta}\end{bmatrix}. \tag{20.37}$$

In a similar manner

$$\begin{bmatrix} \frac{\partial \hat{x}}{\partial \xi} & \frac{\partial \hat{x}}{\partial \eta} \\ \frac{\partial \hat{y}}{\partial \xi} & \frac{\partial \hat{y}}{\partial \eta} \end{bmatrix} = \begin{bmatrix} \hat{x}_1 & \hat{x}_2 & \hat{x}_3 & \dots & \hat{x}_n \\ \hat{y}_1 & \hat{y}_2 & \hat{y}_3 & \dots & \hat{y}_n \end{bmatrix} \begin{bmatrix} \frac{\partial N_1}{\partial \xi} & \frac{\partial N_1}{\partial \eta} \\ \frac{\partial N_2}{\partial \xi} & \frac{\partial N_2}{\partial \eta} \\ \frac{\partial N_3}{\partial \xi} & \frac{\partial N_3}{\partial \eta} \\ \vdots & \vdots \\ \frac{\partial N_n}{\partial \xi} & \frac{\partial N_n}{\partial \eta} \end{bmatrix}. \tag{20.38}$$

and

$$\begin{bmatrix} \frac{\partial \tilde{x}}{\partial \xi} & \frac{\partial \tilde{x}}{\partial \eta} \\ \frac{\partial \tilde{y}}{\partial \xi} & \frac{\partial \tilde{y}}{\partial \eta} \end{bmatrix} = \begin{bmatrix} \tilde{x}_1 & \tilde{x}_2 & \tilde{x}_3 & \dots & \tilde{x}_n \\ \tilde{y}_1 & \tilde{y}_2 & \tilde{y}_3 & \dots & \tilde{y}_n \end{bmatrix} \begin{bmatrix} \frac{\partial N_1}{\partial \xi} & \frac{\partial N_1}{\partial \eta} \\ \frac{\partial N_2}{\partial \xi} & \frac{\partial N_2}{\partial \eta} \\ \frac{\partial N_3}{\partial \xi} & \frac{\partial N_3}{\partial \eta} \\ \vdots & \vdots \\ \frac{\partial N_n}{\partial \xi} & \frac{\partial N_n}{\partial \eta} \end{bmatrix}. \tag{20.39}$$

20.7 Some Examples of 2D Finite Elements

The Four-Noded Quadrilateral. For the four-noded quadrilateral (see Equation (20.4)), the shape functions are given by

$$\begin{aligned} N_1 &= \frac{1}{4}(1-\xi)(1-\eta) \\ N_2 &= \frac{1}{4}(1+\xi)(1-\eta) \\ N_3 &= \frac{1}{4}(1+\xi)(1+\eta) \\ N_4 &= \frac{1}{4}(1-\xi)(1+\eta), \end{aligned} \tag{20.40}$$

which yields

$$\begin{bmatrix} \dfrac{\partial N_1}{\partial \xi} & \dfrac{\partial N_1}{\partial \eta} \\ \dfrac{\partial N_2}{\partial \xi} & \dfrac{\partial N_2}{\partial \eta} \\ \dfrac{\partial N_3}{\partial \xi} & \dfrac{\partial N_3}{\partial \eta} \\ \dfrac{\partial N_4}{\partial \xi} & \dfrac{\partial N_4}{\partial \eta} \end{bmatrix} = \frac{1}{4}\begin{bmatrix} -1+\eta & -1+\xi \\ 1-\eta & -1-\xi \\ 1+\eta & 1+\xi \\ -1-\eta & 1-\xi \end{bmatrix}. \tag{20.41}$$

When substituted into Equation (20.37) this results in

$$\begin{bmatrix} \bar{\xi}_i & \bar{\eta}_i \\ \bar{\xi}_j & \bar{\eta}_j \end{bmatrix} = \begin{bmatrix} \dfrac{\partial \bar{x}}{\partial \xi} & \dfrac{\partial \bar{x}}{\partial \eta} \\ \dfrac{\partial \bar{y}}{\partial \xi} & \dfrac{\partial \bar{y}}{\partial \eta} \end{bmatrix} = \frac{1}{4}\begin{bmatrix} \bar{x}_1 & \bar{x}_2 & \bar{x}_3 & \bar{x}_4 \\ \bar{y}_1 & \bar{y}_2 & \bar{y}_3 & \bar{y}_4 \end{bmatrix}\begin{bmatrix} -1+\eta & -1+\xi \\ 1-\eta & -1-\xi \\ 1+\eta & 1+\xi \\ -1-\eta & 1-\xi \end{bmatrix}. \tag{20.42}$$

Also, by substituting Equation (20.41) into Equation (20.38) one obtains

$$\begin{bmatrix} \hat{\xi}_i & \hat{\eta}_i \\ \hat{\xi}_j & \hat{\eta}_j \end{bmatrix} = \begin{bmatrix} \dfrac{\partial \hat{x}}{\partial \xi} & \dfrac{\partial \hat{x}}{\partial \eta} \\ \dfrac{\partial \hat{y}}{\partial \xi} & \dfrac{\partial \hat{y}}{\partial \eta} \end{bmatrix} = \frac{1}{4}\begin{bmatrix} \hat{x}_1 & \hat{x}_2 & \hat{x}_3 & \hat{x}_4 \\ \hat{y}_1 & \hat{y}_2 & \hat{y}_3 & \hat{y}_4 \end{bmatrix}\begin{bmatrix} -1+\eta & -1+\xi \\ 1-\eta & -1-\xi \\ 1+\eta & 1+\xi \\ -1-\eta & 1-\xi \end{bmatrix}. \tag{20.43}$$

In a similar manner, substitution of Equation (20.41) into Equation (20.39) yields

$$\begin{bmatrix} \tilde{\xi}_i & \tilde{\eta}_i \\ \tilde{\xi}_j & \tilde{\eta}_j \end{bmatrix} = \begin{bmatrix} \dfrac{\partial \tilde{x}}{\partial \xi} & \dfrac{\partial \tilde{x}}{\partial \eta} \\ \dfrac{\partial \tilde{y}}{\partial \xi} & \dfrac{\partial \tilde{y}}{\partial \eta} \end{bmatrix} = \frac{1}{4}\begin{bmatrix} \tilde{x}_1 & \tilde{x}_2 & \tilde{x}_3 & \tilde{x}_4 \\ \tilde{y}_1 & \tilde{y}_2 & \tilde{y}_3 & \tilde{y}_4 \end{bmatrix}\begin{bmatrix} -1+\eta & -1+\xi \\ 1-\eta & -1-\xi \\ 1+\eta & 1+\xi \\ -1-\eta & 1-\xi \end{bmatrix}. \tag{20.44}$$

The Six-Noded Triangle. For the six-noded triangle the shape functions are given by (see Section 20.2):

$$\begin{aligned}
N_1 &= 2(1-\xi-\eta)\left(\frac{1}{2}-\xi-\eta\right)\\
N_2 &= 4(1-\xi-\eta)\xi\\
N_3 &= 2\xi\left(\xi-\frac{1}{2}\right)\\
N_4 &= 4\xi\eta\\
N_5 &= 2\eta\left(\eta-\frac{1}{2}\right)\\
N_6 &= 4(1-\xi-\eta)\eta.
\end{aligned} \tag{20.45}$$

The derivatives of these are as follows

$$\begin{bmatrix} \frac{\partial N_1}{\partial \xi} & \frac{\partial N_1}{\partial \eta} \\ \frac{\partial N_2}{\partial \xi} & \frac{\partial N_2}{\partial \eta} \\ \frac{\partial N_3}{\partial \xi} & \frac{\partial N_3}{\partial \eta} \\ \frac{\partial N_4}{\partial \xi} & \frac{\partial N_4}{\partial \eta} \\ \frac{\partial N_5}{\partial \xi} & \frac{\partial N_5}{\partial \eta} \\ \frac{\partial N_6}{\partial \xi} & \frac{\partial N_6}{\partial \eta} \end{bmatrix} = \begin{bmatrix} -3+4\xi+4\eta & -3+4\xi+4\eta \\ 4-8\xi-4\eta & -4\xi \\ 4\xi-1 & 0 \\ 4\eta & 4\xi \\ 0 & 4\eta-1 \\ -4\eta & 4-4\xi-8\eta \end{bmatrix}, \tag{20.46}$$

which, when substituted into Equations (20.37), (20.38) and (20.39) these yield:

a. Initial position of the solid-embedded vector base

$$\begin{bmatrix} \bar{\xi}_i & \bar{\eta}_i \\ \bar{\xi}_j & \bar{\eta}_j \end{bmatrix} = \begin{bmatrix} \frac{\partial \bar{x}}{\partial \xi} & \frac{\partial \bar{x}}{\partial \eta} \\ \frac{\partial \bar{y}}{\partial \xi} & \frac{\partial \bar{y}}{\partial \eta} \end{bmatrix} = \begin{bmatrix} \bar{x}_1 & \bar{x}_2 & \bar{x}_3 & \bar{x}_4 & \bar{x}_6 & \bar{x}_6 \\ \bar{y}_1 & \bar{y}_2 & \bar{y}_3 & \bar{y}_4 & \bar{y}_5 & \bar{y}_6 \end{bmatrix} \begin{bmatrix} -3+4\xi+4\eta & -3+4\xi+4\eta \\ 4-8\xi-4\eta & -4\xi \\ 4\xi-1 & 0 \\ 4\eta & 4\xi \\ 0 & 4\eta-1 \\ -4\eta & 4-4\xi-8\eta \end{bmatrix}. \tag{20.47}$$

b. Previous position of the solid-embedded vector base

$$\begin{bmatrix} \hat{\xi}_i & \hat{\eta}_i \\ \hat{\xi}_j & \hat{\eta}_j \end{bmatrix} = \begin{bmatrix} \frac{\partial \hat{x}}{\partial \xi} & \frac{\partial \hat{x}}{\partial \eta} \\ \frac{\partial \hat{y}}{\partial \xi} & \frac{\partial \hat{y}}{\partial \eta} \end{bmatrix} = \begin{bmatrix} \hat{x}_1 & \hat{x}_2 & \hat{x}_3 & \hat{x}_4 & \hat{x}_6 & \hat{x}_6 \\ \hat{y}_1 & \hat{y}_2 & \hat{y}_3 & \hat{y}_4 & \hat{y}_5 & \hat{y}_6 \end{bmatrix} \begin{bmatrix} -3+4\xi+4\eta & -3+4\xi+4\eta \\ 4-8\xi-4\eta & -4\xi \\ 4\xi-1 & 0 \\ 4\eta & 4\xi \\ 0 & 4\eta-1 \\ -4\eta & 4-4\xi-8\eta \end{bmatrix}. \tag{20.48}$$

c. Current position of the solid-embedded vector base

$$\begin{bmatrix} \tilde{\xi}_i & \tilde{\eta}_i \\ \tilde{\xi}_j & \tilde{\eta}_j \end{bmatrix} = \begin{bmatrix} \frac{\partial \tilde{x}}{\partial \xi} & \frac{\partial \tilde{x}}{\partial \eta} \\ \frac{\partial \tilde{y}}{\partial \xi} & \frac{\partial \tilde{y}}{\partial \eta} \end{bmatrix} = \begin{bmatrix} \tilde{x}_1 & \tilde{x}_2 & \tilde{x}_3 & \tilde{x}_4 & \tilde{x}_6 & \tilde{x}_6 \\ \tilde{y}_1 & \tilde{y}_2 & \tilde{y}_3 & \tilde{y}_4 & \tilde{y}_5 & \tilde{y}_6 \end{bmatrix} \begin{bmatrix} -3+4\xi+4\eta & -3+4\xi+4\eta \\ 4-8\xi-4\eta & -4\xi \\ 4\xi-1 & 0 \\ 4\eta & 4\xi \\ 0 & 4\eta-1 \\ -4\eta & 4-4\xi-8\eta \end{bmatrix}. \tag{20.49}$$

As explained before, the current, the previous and the initial positions of the material-embedded vector base $[\boldsymbol{\alpha} \;\; \boldsymbol{\beta}]$ are obtained from the solid-embedded vector base. In this manner, the deformation kinematics are resolved. These deformation kinematics are obtained from the initial, the previous and the current positions of the solid-embedded vector base $[\boldsymbol{\xi} \;\; \boldsymbol{\eta}]$ and are described by the initial, the previous and the current positions of the material-embedded vector base $[\boldsymbol{\alpha} \;\; \boldsymbol{\beta}]$. This base describes the generalized solid element's deformation and is used directly in constitutive law calculations.

20.8 Summary

In theory, it is possible to introduce finite elements of any shape and any number of nodes using the above described procedure. In all cases, the final result is the determination of the deformation kinematics at material points inside the domain of the finite element.

In this chapter the kinematics of deformation for finite elements of general shapes has been defined using the shape functions and the nodal coordinates.

In all cases, for any given material point $P = (\xi, \eta)$, the initial position of the material-embedded vector base $[\boldsymbol{\alpha} \ \boldsymbol{\beta}]$ is given. The current and the previous positions of the same vectors are obtained from the current and the previous nodal coordinates of the finite element. This approach has been demonstrated for a number of 2D finite elements.

Further Reading

[1] Bonet, J. and Wood, R. D. (1997) *Nonlinear Continuum Mechanics for Finite Element Analysis*. Cambridge, Cambridge University Press.

[2] Hughes, T.J.R. (2000) *The Finite Element Method: Linear Static and Dynamic Finite Element Analysis*. New York, Dover Publications.

[3] Munjiza, A. (1992) Discrete elements in transient dynamics of fractured media, PhD Thesis, Civ. Eng. Dept., Swansea.

[4] Munjiza, A. (1992) Numerically derived shape functions in finite element analysis, MSc Thesis, University of Split.

[5] Oden, J. T. (1972) *Finite Elements of Nonlinear Continua*. New York, McGraw-Hill.

[6] Rvachev, V. L., G. P. Manko and A. N. Shevchenko (1986) The R-function approach and software for the analysis of physical and mechanical fields. In J. P. Crestin and J. F. McWaters (eds), *Software for Discrete Manufacturing*. Paris, North-Holland.

[7] Zienkiewicz, O. C. (1971) *The Finite Element Method in Engineering Science*. New York, McGraw-Hill.

[8] Zienkiewicz, O. C., Rojek, J., Taylor, R. L. and Pastor, M. (1998) Triangles and tetrahera in explicit dynamic codes for solids. *International Journal for Numerical Methods in Engineering*, **43**(3), 565–83.

[9] Zienkiewicz, O. C. and Taylor, R. L. (2005) *The Finite Element Method Set*. Oxford, Elsevier Science.

21

Integration of Nodal Forces over Volume of 2D Finite Elements

From the initial, previous, and current positions of the generalized material element, using the constitutive law, the Munjiza stress tensor matrix is calculated as described in Chapter 17. With this matrix in hand, the Cauchy stress tensor matrix is calculated and passed back to the finite element package.

The next step in the finite element analysis is the equivalent nodal forces calculations. Due to the finite element deformation, internal forces are produced. In order to simplify the finite element calculation, these internal forces are replaced by the equivalent nodal forces, Figure 21.1. The equivalent nodal forces are calculated by considering equilibrium. Both static and dynamic equilibrium can be considered. The only difference between the static and dynamic equilibrium is the presence of inertia forces in the dynamic analysis, which are simply calculated as mass multiplied by the acceleration.

A number of different approaches can be employed to calculate the equivalent nodal forces corresponding to the internal forces. One of these approaches is the principle of virtual work.

21.1 The Principle of Virtual Work in the 2D Finite Element Method

It is worth remembering that the Cauchy stress defines the internal forces on the surfaces of the Cauchy material element. These surfaces coincide with the global orthonormal base

$$[\mathbf{i}\ \ \mathbf{j}]. \tag{21.1}$$

The Cauchy material element is taken from the current position (not initial, not previous, or any other position). As such, large displacements are taken into account exactly. This is achieved by writing equilibrium equations for the deformed solid.

Large Strain Finite Element Method: A Practical Course, First Edition. Antonio Munjiza, Esteban Rougier and Earl E. Knight.

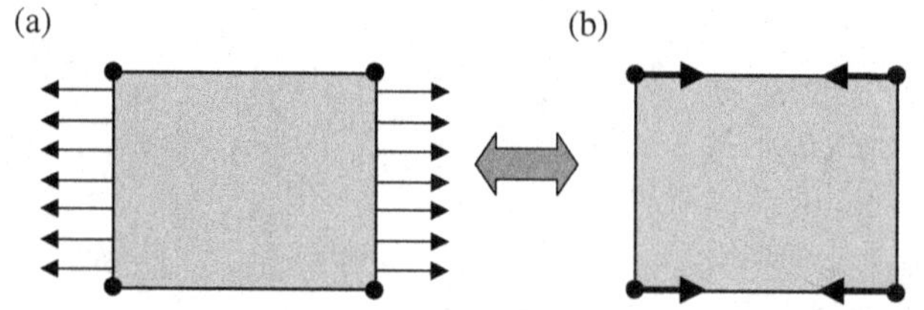

Figure 21.1 (a) The internal forces; (b) the equivalent nodal forces with which the finite element pulls its nodes

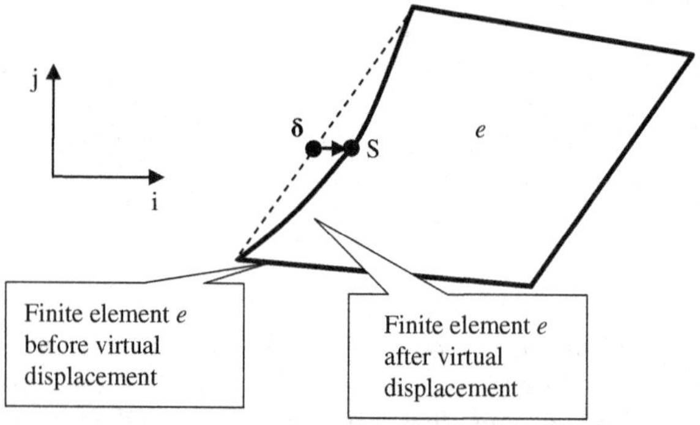

Figure 21.2 Virtual displacement of node S on the finite element e. The virtual displacement is applied in the global direction **i**

Once the current position is known, the question becomes:

What are the forces the finite element pulls its nodes with?

It is obvious that the conventional method of the "sum of horizontal" and, the "sum of vertical" forces is hard to employ here. The most convenient way for solving the above problem is via the principle of virtual work.

In order to know the nodal forces **f** on node S, one needs to produce a virtual displacement $\boldsymbol{\delta}$ of node S as shown in Figure 21.2.

The virtual displacements of any material point $(\xi,\ \eta)$ inside the finite element are approximated from the nodal virtual displacements using the shape functions

$$\begin{aligned}\boldsymbol{\delta}_i &= \boldsymbol{\delta}_{i1}\mathrm{N}_1 + \boldsymbol{\delta}_{i2}\mathrm{N}_2 + \boldsymbol{\delta}_{i3}\mathrm{N}_3 + \cdots + \boldsymbol{\delta}_{in}\mathrm{N}_n \\ \boldsymbol{\delta}_j &= \boldsymbol{\delta}_{j1}\mathrm{N}_1 + \boldsymbol{\delta}_{j2}\mathrm{N}_2 + \boldsymbol{\delta}_{j3}\mathrm{N}_3 + \cdots + \boldsymbol{\delta}_{jn}\mathrm{N}_n.\end{aligned} \tag{21.2}$$

Virtual Displacement in the i Direction. It is convenient to provide virtual nodal displacements in turn in order to probe the equilibrium. For example, by providing a virtual displacement for node 1

$$\boldsymbol{\delta}_1 = \begin{bmatrix} u_1 \\ v_1 \end{bmatrix} = \begin{bmatrix} 1 \\ 0 \end{bmatrix} \tag{21.3}$$

and substituting it into Equation (21.2), one obtains the corresponding virtual displacements of all the material points of the finite element

$$\begin{aligned}\delta_i &= u = u_1 \mathrm{N}_1 \\ \delta_j &= v = 0.\end{aligned} \tag{21.4}$$

From Equation (21.4), it follows that

$$\begin{aligned}\frac{\partial u}{\partial \tilde{x}} &= u_1 \frac{\partial \mathrm{N}_1}{\partial \tilde{x}} = \frac{\partial \mathrm{N}_1}{\partial \tilde{x}} \\ \frac{\partial u}{\partial \tilde{y}} &= u_1 \frac{\partial \mathrm{N}_1}{\partial \tilde{y}} = \frac{\partial \mathrm{N}_1}{\partial \tilde{y}}.\end{aligned} \tag{21.5}$$

By substituting from Equation (21.5) into Equation (15.41) one obtains

$$dW_i = \left(\begin{bmatrix} c_{ii} & c_{ij} \end{bmatrix} \begin{bmatrix} \dfrac{\partial \mathrm{N}_1}{\partial \tilde{x}} \\ \dfrac{\partial \mathrm{N}_1}{\partial \tilde{y}} \end{bmatrix} + \begin{bmatrix} \dfrac{\partial c_{ii}}{\partial \tilde{x}} & \dfrac{\partial c_{ij}}{\partial \tilde{y}} \end{bmatrix} \begin{bmatrix} \mathrm{N}_1 \\ \mathrm{N}_1 \end{bmatrix} \right) d\tilde{x} d\tilde{y}. \tag{21.6}$$

The total virtual work for the whole finite element is obtained by integration over the current volume (area) $\tilde{V}_e$ of the finite element,

$$W_i = \iint_{\tilde{V}_e} \left(\begin{bmatrix} c_{ii} & c_{ij} \end{bmatrix} \begin{bmatrix} \dfrac{\partial \mathrm{N}_1}{\partial \tilde{x}} \\ \dfrac{\partial \mathrm{N}_1}{\partial \tilde{y}} \end{bmatrix} + \begin{bmatrix} \dfrac{\partial c_{ii}}{\partial \tilde{x}} & \dfrac{\partial c_{ij}}{\partial \tilde{y}} \end{bmatrix} \begin{bmatrix} \mathrm{N}_1 \\ \mathrm{N}_1 \end{bmatrix} \right) d\tilde{x} d\tilde{y}. \tag{21.7}$$

According to the principle of virtual work, to obtain equilibrium it is necessary that the total virtual work be zero, thus

$$f_{i1} \cdot 1 + \iint_{\tilde{V}_e} \left(\begin{bmatrix} c_{ii} & c_{ij} \end{bmatrix} \begin{bmatrix} \dfrac{\partial \mathrm{N}_1}{\partial \tilde{x}} \\ \dfrac{\partial \mathrm{N}_1}{\partial \tilde{y}} \end{bmatrix} + \begin{bmatrix} \dfrac{\partial c_{ii}}{\partial \tilde{x}} & \dfrac{\partial c_{ij}}{\partial \tilde{y}} \end{bmatrix} \begin{bmatrix} \mathrm{N}_1 \\ \mathrm{N}_1 \end{bmatrix} \right) d\tilde{x} d\tilde{y} = 0, \tag{21.8}$$

which yields

$$f_{i1} = -\iint_{\tilde{V}_e} \left(\begin{bmatrix} c_{ii} & c_{ij} \end{bmatrix} \begin{bmatrix} \dfrac{\partial \mathrm{N}_1}{\partial \tilde{x}} \\ \dfrac{\partial \mathrm{N}_1}{\partial \tilde{y}} \end{bmatrix} + \begin{bmatrix} \dfrac{\partial c_{ii}}{\partial \tilde{x}} & \dfrac{\partial c_{ij}}{\partial \tilde{y}} \end{bmatrix} \begin{bmatrix} \mathrm{N}_1 \\ \mathrm{N}_1 \end{bmatrix} \right) d\tilde{x} d\tilde{y}, \tag{21.9}$$

where f_{i1} represents the force by which the finite element pulls its node 1 in the **i** direction.

Virtual Displacement in the j Direction. By providing only a virtual displacement $\boldsymbol{\delta}_1$ at node 1, such that

$$\boldsymbol{\delta}_1 = \begin{bmatrix} u_1 \\ v_1 \end{bmatrix} = \begin{bmatrix} 0 \\ 1 \end{bmatrix}, \tag{21.10}$$

one obtains the virtual displacements of material points inside the finite element being given by

$$\begin{aligned} u &= 0 \\ v &= v_1 \mathrm{N}_1. \end{aligned} \tag{21.11}$$

From Equation (21.11) it follows that

$$\begin{aligned} \frac{\partial v}{\partial \tilde{x}} &= v_1 \frac{\partial \mathrm{N}_1}{\partial \tilde{x}} = \frac{\partial \mathrm{N}_1}{\partial \tilde{x}} \\ \frac{\partial v}{\partial \tilde{y}} &= v_1 \frac{\partial \mathrm{N}_1}{\partial \tilde{y}} = \frac{\partial \mathrm{N}_1}{\partial \tilde{y}}, \end{aligned} \tag{21.12}$$

which when substituted into the equation for the virtual work in the **j** direction, Equation (15.43), yields

$$dW_j = \left(\begin{bmatrix} c_{ji} & c_{jj} \end{bmatrix} \begin{bmatrix} \dfrac{\partial \mathrm{N}_1}{\partial \tilde{x}} \\[2ex] \dfrac{\partial \mathrm{N}_1}{\partial \tilde{y}} \end{bmatrix} + \begin{bmatrix} \dfrac{\partial c_{ji}}{\partial \tilde{x}} & \dfrac{\partial c_{jj}}{\partial \tilde{y}} \end{bmatrix} \begin{bmatrix} \mathrm{N}_1 \\ \mathrm{N}_1 \end{bmatrix} \right) d\tilde{x} d\tilde{y}. \tag{21.13}$$

The total virtual work for the whole finite element is obtained by integration over the volume (area) V_e of the finite element

$$W_j = \iint_{\tilde{V}_e} \left(\begin{bmatrix} c_{ji} & c_{jj} \end{bmatrix} \begin{bmatrix} \dfrac{\partial \mathrm{N}_1}{\partial \tilde{x}} \\[2ex] \dfrac{\partial \mathrm{N}_1}{\partial \tilde{y}} \end{bmatrix} + \begin{bmatrix} \dfrac{\partial c_{ji}}{\partial \tilde{x}} & \dfrac{\partial c_{jj}}{\partial \tilde{y}} \end{bmatrix} \begin{bmatrix} \mathrm{N}_1 \\ \mathrm{N}_1 \end{bmatrix} \right) d\tilde{x} d\tilde{y}. \tag{21.14}$$

This virtual work should be counterbalanced by the virtual work of the equivalent nodal force f_{j1}, thus

$$f_{j1} \cdot 1 + \iint_{\tilde{V}_e} \left(\begin{bmatrix} c_{ji} & c_{jj} \end{bmatrix} \begin{bmatrix} \dfrac{\partial \mathrm{N}_1}{\partial \tilde{x}} \\[2ex] \dfrac{\partial \mathrm{N}_1}{\partial \tilde{y}} \end{bmatrix} + \begin{bmatrix} \dfrac{\partial c_{ji}}{\partial \tilde{x}} & \dfrac{\partial c_{jj}}{\partial \tilde{y}} \end{bmatrix} \begin{bmatrix} \mathrm{N}_1 \\ \mathrm{N}_1 \end{bmatrix} \right) d\tilde{x} d\tilde{y} = 0. \tag{21.15}$$

From Equation (21.15), it follows that the equivalent nodal force in the global **j** direction is given by

$$f_{j1} = -\iint\limits_{\tilde{V}_e} \left(\begin{bmatrix} c_{ji} & c_{jj} \end{bmatrix} \begin{bmatrix} \dfrac{\partial \mathrm{N}_1}{\partial \tilde{x}} \\ \dfrac{\partial \mathrm{N}_1}{\partial \tilde{y}} \end{bmatrix} + \begin{bmatrix} \dfrac{\partial c_{ji}}{\partial \tilde{x}} & \dfrac{\partial c_{jj}}{\partial \tilde{y}} \end{bmatrix} \begin{bmatrix} \mathrm{N}_1 \\ \mathrm{N}_1 \end{bmatrix} \right) d\tilde{x} d\tilde{y}. \tag{21.16}$$

By combining Equations (21.9) and (21.16), one obtains the equivalent nodal force for node 1 of the finite element

$$\mathbf{f}_1 = \begin{bmatrix} f_{i1} \\ f_{j1} \end{bmatrix} = -\iint\limits_{\tilde{V}_e} \left(\begin{bmatrix} c_{ii} & c_{ij} \\ c_{ji} & c_{jj} \end{bmatrix} \begin{bmatrix} \dfrac{\partial \mathrm{N}_1}{\partial \tilde{x}} \\ \dfrac{\partial \mathrm{N}_1}{\partial \tilde{y}} \end{bmatrix} + \mathrm{N}_1 \begin{bmatrix} c_{ii} & c_{ij} \\ c_{ji} & c_{jj} \end{bmatrix} \begin{bmatrix} \dfrac{\partial}{\partial \tilde{x}} \\ \dfrac{\partial}{\partial \tilde{y}} \end{bmatrix} \right) d\tilde{x} d\tilde{y}, \tag{21.17}$$

where the above "multiplication" is defined as

$$\begin{bmatrix} c_{ii} & c_{ij} \\ c_{ji} & c_{jj} \end{bmatrix} \begin{bmatrix} \dfrac{\partial}{\partial \tilde{x}} \\ \dfrac{\partial}{\partial \tilde{y}} \end{bmatrix} = \begin{bmatrix} \dfrac{\partial c_{ii}}{\partial \tilde{x}} + \dfrac{\partial c_{ij}}{\partial \tilde{y}} \\ \dfrac{\partial c_{ji}}{\partial \tilde{x}} + \dfrac{\partial c_{jj}}{\partial \tilde{y}} \end{bmatrix}. \tag{21.18}$$

Equivalent Nodal Forces. By repeating the same procedure for other nodes of the finite element, the equivalent nodal forces for all the nodes are obtained as follows

$$\begin{aligned}
\mathbf{f}_1 &= \begin{bmatrix} f_{i1} \\ f_{j1} \end{bmatrix} = -\iint\limits_{\tilde{V}_e} \left(\begin{bmatrix} c_{ii} & c_{ij} \\ c_{ji} & c_{jj} \end{bmatrix} \begin{bmatrix} \dfrac{\partial \mathrm{N}_1}{\partial \tilde{x}} \\ \dfrac{\partial \mathrm{N}_1}{\partial \tilde{y}} \end{bmatrix} + \mathrm{N}_1 \begin{bmatrix} c_{ii} & c_{ij} \\ c_{ji} & c_{jj} \end{bmatrix} \begin{bmatrix} \dfrac{\partial}{\partial \tilde{x}} \\ \dfrac{\partial}{\partial \tilde{y}} \end{bmatrix} \right) d\tilde{x} d\tilde{y} \\
\mathbf{f}_2 &= \begin{bmatrix} f_{i2} \\ f_{j2} \end{bmatrix} = -\iint\limits_{\tilde{V}_e} \left(\begin{bmatrix} c_{ii} & c_{ij} \\ c_{ji} & c_{jj} \end{bmatrix} \begin{bmatrix} \dfrac{\partial \mathrm{N}_2}{\partial \tilde{x}} \\ \dfrac{\partial \mathrm{N}_2}{\partial \tilde{y}} \end{bmatrix} + \mathrm{N}_2 \begin{bmatrix} c_{ii} & c_{ij} \\ c_{ji} & c_{jj} \end{bmatrix} \begin{bmatrix} \dfrac{\partial}{\partial \tilde{x}} \\ \dfrac{\partial}{\partial \tilde{y}} \end{bmatrix} \right) d\tilde{x} d\tilde{y} \\
\mathbf{f}_3 &= \begin{bmatrix} f_{i3} \\ f_{j3} \end{bmatrix} = -\iint\limits_{\tilde{V}_e} \left(\begin{bmatrix} c_{ii} & c_{ij} \\ c_{ji} & c_{jj} \end{bmatrix} \begin{bmatrix} \dfrac{\partial \mathrm{N}_3}{\partial \tilde{x}} \\ \dfrac{\partial \mathrm{N}_3}{\partial \tilde{y}} \end{bmatrix} + \mathrm{N}_3 \begin{bmatrix} c_{ii} & c_{ij} \\ c_{ji} & c_{jj} \end{bmatrix} \begin{bmatrix} \dfrac{\partial}{\partial \tilde{x}} \\ \dfrac{\partial}{\partial \tilde{y}} \end{bmatrix} \right) d\tilde{x} d\tilde{y} \\
&\vdots \\
\mathbf{f}_n &= \begin{bmatrix} f_{in} \\ f_{jn} \end{bmatrix} = -\iint\limits_{\tilde{V}_e} \left(\begin{bmatrix} c_{ii} & c_{ij} \\ c_{ji} & c_{jj} \end{bmatrix} \begin{bmatrix} \dfrac{\partial \mathrm{N}_n}{\partial \tilde{x}} \\ \dfrac{\partial \mathrm{N}_n}{\partial \tilde{y}} \end{bmatrix} + \mathrm{N}_n \begin{bmatrix} c_{ii} & c_{ij} \\ c_{ji} & c_{jj} \end{bmatrix} \begin{bmatrix} \dfrac{\partial}{\partial \tilde{x}} \\ \dfrac{\partial}{\partial \tilde{y}} \end{bmatrix} \right) d\tilde{x} d\tilde{y}.
\end{aligned} \tag{21.19}$$

All of the above equations can conveniently be combined into one equation using matrix notation

$$
\begin{aligned}
\begin{bmatrix} \mathbf{f}_1 & \mathbf{f}_2 & \mathbf{f}_3 & \cdots & \mathbf{f}_n \end{bmatrix} &= \begin{bmatrix} f_{i1} & f_{i2} & f_{i3} & \cdots & f_{in} \\ f_{j1} & f_{j2} & f_{j3} & \cdots & f_{jn} \end{bmatrix} \\
&= -\iint_{\tilde{V}_e} \begin{bmatrix} c_{ii} & c_{ij} \\ c_{ji} & c_{jj} \end{bmatrix} \begin{bmatrix} \frac{\partial \mathrm{N}_1}{\partial \tilde{x}} & \frac{\partial \mathrm{N}_2}{\partial \tilde{x}} & \frac{\partial \mathrm{N}_3}{\partial \tilde{x}} & \cdots & \frac{\partial \mathrm{N}_n}{\partial \tilde{x}} \\ \frac{\partial \mathrm{N}_1}{\partial \tilde{y}} & \frac{\partial \mathrm{N}_2}{\partial \tilde{y}} & \frac{\partial \mathrm{N}_3}{\partial \tilde{y}} & \cdots & \frac{\partial \mathrm{N}_n}{\partial \tilde{y}} \end{bmatrix} d\tilde{x}d\tilde{y} \\
&\quad -\iint_{\tilde{V}_e} \begin{bmatrix} \frac{\partial c_{ii}}{\partial \tilde{x}} & \frac{\partial c_{ij}}{\partial \tilde{y}} \\ \frac{\partial c_{ji}}{\partial \tilde{x}} & \frac{\partial c_{jj}}{\partial \tilde{y}} \end{bmatrix} \begin{bmatrix} \mathrm{N}_1 & \mathrm{N}_2 & \mathrm{N}_3 & \cdots & \mathrm{N}_n \\ \mathrm{N}_1 & \mathrm{N}_2 & \mathrm{N}_3 & \cdots & \mathrm{N}_n \end{bmatrix} d\tilde{x}d\tilde{y}.
\end{aligned}
\tag{21.20}
$$

One should note that the above equations are very often introduced in the literature without the proper derivation for many state that the virtual work is a product of the real stress and virtual strains; as it can be seen from Equation (21.20), this holds true only for very simple elements, e.g., the homogeneous deformation triangle. In contrast, the general consideration of the virtual work introduced above is valid for any finite element regardless of the number of nodes.

21.2 Nodal Forces for the Homogeneous Deformation Triangle

The homogeneous deformation triangle has three nodes, as shown in Figure 21.3. Integration over the homogeneous deformation triangle is usually done using just one integration point. In this case Equation (21.20) yields

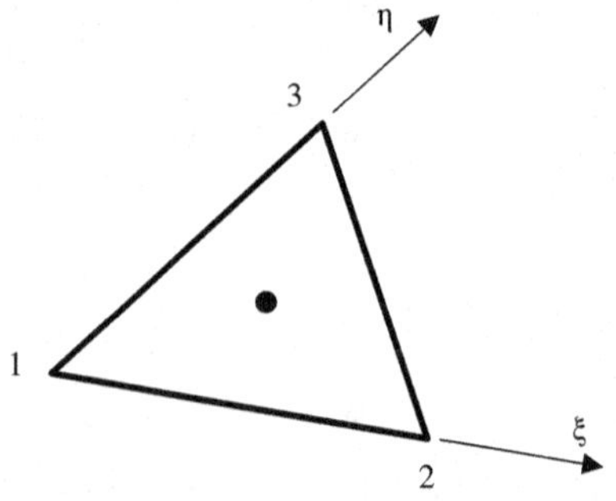

Figure 21.3 Gauss point for the homogeneous deformation triangle

$$
\begin{aligned}
[\mathbf{f}_1\ \ \mathbf{f}_2\ \ \mathbf{f}_3] &= \begin{bmatrix} f_{i1} & f_{i2} & f_{i3} \\ f_{j1} & f_{j2} & f_{j3} \end{bmatrix} \\
&= -\iint_{\tilde{V}_e} \begin{bmatrix} c_{ii} & c_{ij} \\ c_{ji} & c_{jj} \end{bmatrix} \begin{bmatrix} \frac{\partial N_1}{\partial \tilde{x}} & \frac{\partial N_2}{\partial \tilde{x}} & \frac{\partial N_3}{\partial \tilde{x}} \\ \frac{\partial N_1}{\partial \tilde{y}} & \frac{\partial N_2}{\partial \tilde{y}} & \frac{\partial N_3}{\partial \tilde{y}} \end{bmatrix} d\tilde{x}d\tilde{y} \\
&\quad -\iint_{\tilde{V}_e} \begin{bmatrix} \frac{\partial c_{ii}}{\partial \tilde{x}} & \frac{\partial c_{ij}}{\partial \tilde{y}} \\ \frac{\partial c_{ji}}{\partial \tilde{x}} & \frac{\partial c_{jj}}{\partial \tilde{y}} \end{bmatrix} \begin{bmatrix} N_1 & N_2 & N_3 \\ N_1 & N_2 & N_3 \end{bmatrix} d\tilde{x}d\tilde{y}.
\end{aligned}
\tag{21.21}
$$

As the stress over the finite element is constant the Cauchy stress tensor matrix is also constant and the second term in the above equation is identical to zero, thus

$$
\begin{aligned}
[\mathbf{f}_1\ \ \mathbf{f}_2\ \ \mathbf{f}_3] &= \begin{bmatrix} f_{i1} & f_{i2} & f_{i3} \\ f_{j1} & f_{j2} & f_{j3} \end{bmatrix} \\
&= -\iint_{\tilde{V}_e} \begin{bmatrix} c_{ii} & c_{ij} \\ c_{ji} & c_{jj} \end{bmatrix} \begin{bmatrix} \frac{\partial N_1}{\partial \tilde{x}} & \frac{\partial N_2}{\partial \tilde{x}} & \frac{\partial N_3}{\partial \tilde{x}} \\ \frac{\partial N_1}{\partial \tilde{y}} & \frac{\partial N_2}{\partial \tilde{y}} & \frac{\partial N_3}{\partial \tilde{y}} \end{bmatrix} d\tilde{x}d\tilde{y}.
\end{aligned}
\tag{21.22}
$$

From the local derivatives of the shape functions

$$
\begin{bmatrix} \frac{\partial N_1}{\partial \xi} \\ \frac{\partial N_1}{\partial \eta} \end{bmatrix} = \begin{bmatrix} \frac{\partial x}{\partial \xi} & \frac{\partial y}{\partial \xi} \\ \frac{\partial x}{\partial \eta} & \frac{\partial y}{\partial \eta} \end{bmatrix} \begin{bmatrix} \frac{\partial N_1}{\partial \tilde{x}} \\ \frac{\partial N_1}{\partial \tilde{y}} \end{bmatrix},
\tag{21.23}
$$

one easily obtains the global derivatives of the shape functions as follows

$$
\begin{bmatrix} \frac{\partial N_1}{\partial \tilde{x}} \\ \frac{\partial N_1}{\partial \tilde{y}} \end{bmatrix} = \begin{bmatrix} \frac{\partial x}{\partial \xi} & \frac{\partial y}{\partial \xi} \\ \frac{\partial x}{\partial \eta} & \frac{\partial y}{\partial \eta} \end{bmatrix}^{-1} \begin{bmatrix} \frac{\partial N_1}{\partial \xi} \\ \frac{\partial N_1}{\partial \eta} \end{bmatrix} = \begin{bmatrix} \frac{\partial x}{\partial \xi} & \frac{\partial x}{\partial \eta} \\ \frac{\partial y}{\partial \xi} & \frac{\partial y}{\partial \eta} \end{bmatrix}^{-T} \begin{bmatrix} \frac{\partial N_1}{\partial \xi} \\ \frac{\partial N_1}{\partial \eta} \end{bmatrix}.
\tag{21.24}
$$

And for the homogeneous deformation triangle, the shape functions are given by

$$
\begin{aligned}
N_1 &= 1-\xi-\eta \\
N_2 &= \xi \\
N_3 &= \eta,
\end{aligned}
\tag{21.25}
$$

it follows that

$$\begin{bmatrix} \dfrac{\partial N_1}{\partial \xi} \\ \dfrac{\partial N_1}{\partial \eta} \end{bmatrix} = \begin{bmatrix} -1 \\ -1 \end{bmatrix}. \tag{21.26}$$

As the current coordinates are given by

$$\begin{aligned} \tilde{x} &= \tilde{x}_1 N_1 + \tilde{x}_2 N_2 + \tilde{x}_3 N_3 \\ \tilde{y} &= \tilde{y}_1 N_1 + \tilde{y}_2 N_2 + \tilde{y}_3 N_3, \end{aligned} \tag{21.27}$$

it follows that

$$\begin{bmatrix} \dfrac{\partial x}{\partial \xi} & \dfrac{\partial x}{\partial \eta} \\ \dfrac{\partial y}{\partial \xi} & \dfrac{\partial y}{\partial \eta} \end{bmatrix}^T = \begin{bmatrix} \dfrac{\partial \tilde{x}}{\partial \xi} & \dfrac{\partial \tilde{y}}{\partial \xi} \\ \dfrac{\partial \tilde{x}}{\partial \eta} & \dfrac{\partial \tilde{y}}{\partial \eta} \end{bmatrix} = \begin{bmatrix} \tilde{x}_2 - \tilde{x}_1 & \tilde{y}_2 - \tilde{y}_1 \\ \tilde{x}_3 - \tilde{x}_1 & \tilde{y}_3 - \tilde{y}_1 \end{bmatrix}. \tag{21.28}$$

This yields

$$\begin{bmatrix} \dfrac{\partial x}{\partial \xi} & \dfrac{\partial x}{\partial \eta} \\ \dfrac{\partial y}{\partial \xi} & \dfrac{\partial y}{\partial \eta} \end{bmatrix}^{-T} = \frac{1}{\det \begin{bmatrix} \tilde{x}_2 - \tilde{x}_1 & \tilde{y}_2 - \tilde{y}_1 \\ \tilde{x}_3 - \tilde{x}_1 & \tilde{y}_3 - \tilde{y}_1 \end{bmatrix}} \begin{bmatrix} \tilde{y}_3 - \tilde{y}_1 & \tilde{y}_1 - \tilde{y}_2 \\ \tilde{x}_1 - \tilde{x}_3 & \tilde{x}_2 - \tilde{x}_1 \end{bmatrix}. \tag{21.29}$$

By substituting into Equation (21.24) one obtains

$$\begin{bmatrix} \dfrac{\partial N_1}{\partial \tilde{x}} \\ \dfrac{\partial N_1}{\partial \tilde{y}} \end{bmatrix} = \frac{1}{\det \begin{bmatrix} \tilde{x}_2 - \tilde{x}_1 & \tilde{y}_2 - \tilde{y}_1 \\ \tilde{x}_3 - \tilde{x}_1 & \tilde{y}_3 - \tilde{y}_1 \end{bmatrix}} \begin{bmatrix} \tilde{y}_3 - \tilde{y}_1 & \tilde{y}_1 - \tilde{y}_2 \\ \tilde{x}_1 - \tilde{x}_3 & \tilde{x}_2 - \tilde{x}_1 \end{bmatrix} \begin{bmatrix} -1 \\ -1 \end{bmatrix}. \tag{21.30}$$

By repeating the above process for N_2 and N_3 and substituting into Equation (21.22) the equivalent nodal forces are obtained

$$\begin{aligned} [\mathbf{f}_1 \ \ \mathbf{f}_2 \ \ \mathbf{f}_3] &= \begin{bmatrix} f_{i1} & f_{i2} & f_{i3} \\ f_{j1} & f_{j2} & f_{j3} \end{bmatrix} \\ &= -\iint_{\tilde{V}_e} \begin{bmatrix} c_{ii} & c_{ij} \\ c_{ji} & c_{jj} \end{bmatrix} \left(\begin{bmatrix} \dfrac{\partial x}{\partial \xi} & \dfrac{\partial x}{\partial \eta} \\ \dfrac{\partial y}{\partial \xi} & \dfrac{\partial y}{\partial \eta} \end{bmatrix}^{-T} \begin{bmatrix} \dfrac{\partial N_1}{\partial \xi} & \dfrac{\partial N_2}{\partial \xi} & \dfrac{\partial N_3}{\partial \xi} \\ \dfrac{\partial N_1}{\partial \eta} & \dfrac{\partial N_2}{\partial \eta} & \dfrac{\partial N_3}{\partial \eta} \end{bmatrix} \right) d\tilde{V}_e, \end{aligned} \tag{21.31}$$

where $\tilde{V}_e$ is the current volume of the triangle. The above integration is done using one Gauss point ($\xi = 1/3$, $\eta = 1/3$), which yields

$$\begin{aligned}[\mathbf{f}_1 \ \mathbf{f}_2 \ \mathbf{f}_3] &= \begin{bmatrix} f_{i1} & f_{i2} & f_{i3} \\ f_{j1} & f_{j2} & f_{j3} \end{bmatrix} \\ &= -\begin{bmatrix} c_{ii} & c_{ij} \\ c_{ji} & c_{jj} \end{bmatrix} \left(\begin{bmatrix} \frac{\partial x}{\partial \xi} & \frac{\partial x}{\partial \eta} \\ \frac{\partial y}{\partial \xi} & \frac{\partial y}{\partial \eta} \end{bmatrix}^{-T} \begin{bmatrix} \frac{\partial N_1}{\partial \xi} & \frac{\partial N_2}{\partial \xi} & \frac{\partial N_3}{\partial \xi} \\ \frac{\partial N_1}{\partial \eta} & \frac{\partial N_2}{\partial \eta} & \frac{\partial N_3}{\partial \eta} \end{bmatrix} \right) \tilde{V}_e, \end{aligned} \tag{21.32}$$

where

$$\begin{aligned} \tilde{V}_e &= \frac{1}{2} \det \begin{bmatrix} \tilde{x}_2 - \tilde{x}_1 & \tilde{y}_2 - \tilde{y}_1 \\ \tilde{x}_3 - \tilde{x}_1 & \tilde{y}_3 - \tilde{y}_1 \end{bmatrix} \\ &= \frac{1}{2}((\tilde{x}_2 - \tilde{x}_1)(\tilde{y}_3 - \tilde{y}_1) - (\tilde{x}_3 - \tilde{x}_1)(\tilde{y}_2 - \tilde{y}_1)). \end{aligned} \tag{21.33}$$

By substituting

$$\begin{aligned} \frac{\partial N_1}{\partial \xi} &= \frac{\partial N_1}{\partial \eta} = -1 \\ \frac{\partial N_2}{\partial \xi} &= \frac{\partial N_3}{\partial \eta} = 1 \\ \frac{\partial N_2}{\partial \eta} &= \frac{\partial N_3}{\partial \xi} = 0, \end{aligned} \tag{21.34}$$

one obtains

$$[\mathbf{f}_1 \ \mathbf{f}_2 \ \mathbf{f}_3] = -\begin{bmatrix} c_{ii} & c_{ij} \\ c_{ji} & c_{jj} \end{bmatrix} \begin{bmatrix} \tilde{y}_3 - \tilde{y}_1 & \tilde{y}_1 - \tilde{y}_2 \\ \tilde{x}_1 - \tilde{x}_3 & \tilde{x}_2 - \tilde{x}_1 \end{bmatrix} \begin{bmatrix} -1 & 1 & 0 \\ -1 & 0 & 1 \end{bmatrix} \frac{\tilde{V}_e}{\det \begin{bmatrix} \tilde{x}_2 - \tilde{x}_1 & \tilde{y}_2 - \tilde{y}_1 \\ \tilde{x}_3 - \tilde{x}_1 & \tilde{y}_3 - \tilde{y}_1 \end{bmatrix}}. \tag{21.35}$$

After substitution of Equation (21.33) into Equation (21.35) one obtains

$$[\mathbf{f}_1 \ \mathbf{f}_2 \ \mathbf{f}_3] = \frac{1}{2} \begin{bmatrix} c_{ii} & c_{ij} \\ c_{ji} & c_{jj} \end{bmatrix} \left(\begin{bmatrix} \tilde{y}_3 - \tilde{y}_1 & \tilde{y}_1 - \tilde{y}_2 \\ \tilde{x}_1 - \tilde{x}_3 & \tilde{x}_2 - \tilde{x}_1 \end{bmatrix} \begin{bmatrix} 1 & -1 & 0 \\ 1 & 0 & -1 \end{bmatrix} \right). \tag{21.36}$$

21.3 Nodal Forces for the Six-Noded Triangle

For the six-noded triangle, the nodal forces obtained using the principle of virtual work are as follows:

$$\begin{aligned}[\mathbf{f}_1\ \mathbf{f}_2\ \mathbf{f}_3\ \mathbf{f}_4\ \mathbf{f}_5\ \mathbf{f}_6] &= \begin{bmatrix} f_{i1} & f_{i2} & f_{i3} & f_{i4} & f_{i5} & f_{i6} \\ f_{j1} & f_{j2} & f_{j3} & f_{j4} & f_{j5} & f_{j6} \end{bmatrix} \\ &= -\iint_{\tilde{V}_e} \begin{bmatrix} c_{ii} & c_{ij} \\ c_{ji} & c_{jj} \end{bmatrix} \begin{bmatrix} \frac{\partial N_1}{\partial \tilde{x}} & \frac{\partial N_2}{\partial \tilde{x}} & \frac{\partial N_3}{\partial \tilde{x}} & \frac{\partial N_4}{\partial \tilde{x}} & \frac{\partial N_5}{\partial \tilde{x}} & \frac{\partial N_6}{\partial \tilde{x}} \\ \frac{\partial N_1}{\partial \tilde{y}} & \frac{\partial N_2}{\partial \tilde{y}} & \frac{\partial N_3}{\partial \tilde{y}} & \frac{\partial N_4}{\partial \tilde{y}} & \frac{\partial N_5}{\partial \tilde{y}} & \frac{\partial N_6}{\partial \tilde{y}} \end{bmatrix} d\tilde{x}d\tilde{y} \\ &\quad -\iint_{\tilde{V}_e} \begin{bmatrix} \frac{\partial c_{ii}}{\partial \tilde{x}} & \frac{\partial c_{ij}}{\partial \tilde{y}} \\ \frac{\partial c_{ji}}{\partial \tilde{x}} & \frac{\partial c_{jj}}{\partial \tilde{y}} \end{bmatrix} \begin{bmatrix} N_1 & N_2 & N_3 & N_4 & N_5 & N_6 \\ N_1 & N_2 & N_3 & N_4 & N_5 & N_6 \end{bmatrix} d\tilde{x}d\tilde{y}. \end{aligned} \tag{21.37}$$

$$\begin{aligned}[\mathbf{f}_1\ \mathbf{f}_2\ \mathbf{f}_3\ \mathbf{f}_4\ \mathbf{f}_5\ \mathbf{f}_6] &= \begin{bmatrix} f_{i1} & f_{i2} & f_{i3} & f_{i4} & f_{i5} & f_{i6} \\ f_{j1} & f_{j2} & f_{j3} & f_{j4} & f_{j5} & f_{j6} \end{bmatrix} \\ &= -\iint_{\tilde{V}_e} \begin{bmatrix} c_{ii} & c_{ij} \\ c_{ji} & c_{jj} \end{bmatrix} \begin{bmatrix} \frac{\partial \tilde{x}}{\partial \xi} & \frac{\partial \tilde{x}}{\partial \eta} \\ \frac{\partial \tilde{y}}{\partial \xi} & \frac{\partial \tilde{y}}{\partial \eta} \end{bmatrix}^{-T} \begin{bmatrix} \frac{\partial N_1}{\partial \xi} & \frac{\partial N_2}{\partial \xi} & \frac{\partial N_3}{\partial \xi} & \frac{\partial N_4}{\partial \xi} & \frac{\partial N_5}{\partial \xi} & \frac{\partial N_6}{\partial \xi} \\ \frac{\partial N_1}{\partial \eta} & \frac{\partial N_2}{\partial \eta} & \frac{\partial N_3}{\partial \eta} & \frac{\partial N_4}{\partial \eta} & \frac{\partial N_5}{\partial \eta} & \frac{\partial N_6}{\partial \eta} \end{bmatrix} d\tilde{x}d\tilde{y} \\ &\quad -\iint_{\tilde{V}_e} \begin{bmatrix} \frac{\partial c_{ii}}{\partial \tilde{x}} & \frac{\partial c_{ij}}{\partial \tilde{y}} \\ \frac{\partial c_{ji}}{\partial \tilde{x}} & \frac{\partial c_{jj}}{\partial \tilde{y}} \end{bmatrix} \begin{bmatrix} N_1 & N_2 & N_3 & N_4 & N_5 & N_6 \\ N_1 & N_2 & N_3 & N_4 & N_5 & N_6 \end{bmatrix} d\tilde{x}d\tilde{y}. \end{aligned} \tag{21.38}$$

By calculating these at Gauss points and applying Gaussian numerical integration, the nodal forces are obtained.

In order to facilitate the numerical integration the infinitesimal 2D volume (i.e., area)

$$d\tilde{V}_e = d\tilde{x}d\tilde{y}. \tag{21.39}$$

is calculated by considering the curved axes ξ and η at any material point $P = (\xi,\ \eta)$. The infinitesimal vectors tangential to the curved axes $\xi = \text{const}$ and $\eta = \text{const}$ are obtained as

$$[\tilde{\boldsymbol{\xi}}d\xi\ \ \tilde{\boldsymbol{\eta}}d\eta] = \begin{bmatrix} \frac{\partial \tilde{x}}{\partial \xi} & \frac{\partial \tilde{x}}{\partial \eta} \\ \frac{\partial \tilde{y}}{\partial \xi} & \frac{\partial \tilde{y}}{\partial \eta} \end{bmatrix} \begin{bmatrix} d\xi \\ d\eta \end{bmatrix}. \tag{21.40}$$

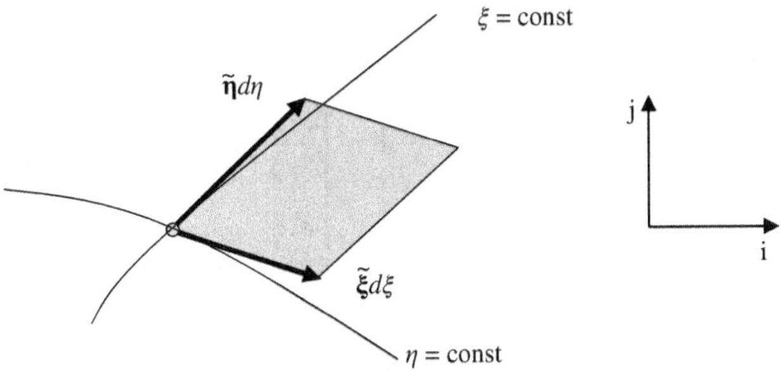

Figure 21.4 Infinitesimal volume

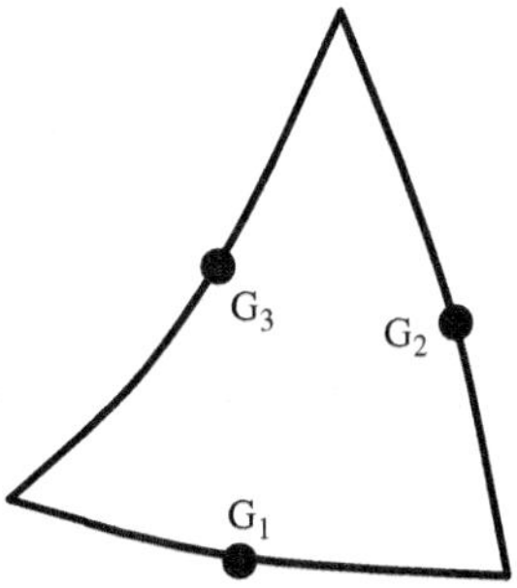

Figure 21.5 Gauss integration points for the 6-noded triangle finite element

The corresponding infinitesimal volume is shown in Figure 21.4 and it is given as the cross product

$$d\tilde{V}_e = (\tilde{\boldsymbol{\xi}}d\xi) \times (\tilde{\boldsymbol{\eta}}d\eta) = (\tilde{\boldsymbol{\xi}} \times \tilde{\boldsymbol{\eta}})d\xi d\eta$$
$$= \det \begin{bmatrix} \frac{\partial \tilde{x}}{\partial \xi} & \frac{\partial \tilde{x}}{\partial \eta} \\ \frac{\partial \tilde{y}}{\partial \xi} & \frac{\partial \tilde{y}}{\partial \eta} \end{bmatrix} d\xi d\eta. \tag{21.41}$$

The integration is performed using the solid-embedded coordinate system in conjunction with the Gauss integration points as shown in Figure 21.5.

21.4 Nodal Forces for the Four-Noded Quadrilateral

From the Cauchy stress, using the principle of virtual work, the equivalent nodal forces are calculated

$$[\mathbf{f}_1\ \mathbf{f}_2\ \mathbf{f}_2\ \mathbf{f}_4] = \begin{bmatrix} f_{i1} & f_{i2} & f_{i3} & f_{i4} \\ f_{j1} & f_{j2} & f_{j3} & f_{j4} \end{bmatrix}$$
$$= -\iint_{\tilde{V}_e} \begin{bmatrix} c_{ii} & c_{ij} \\ c_{ji} & c_{jj} \end{bmatrix} \begin{bmatrix} \frac{\partial \tilde{x}}{\partial \xi} & \frac{\partial \tilde{x}}{\partial \eta} \\ \frac{\partial \tilde{y}}{\partial \xi} & \frac{\partial \tilde{y}}{\partial \eta} \end{bmatrix}^{-T} \begin{bmatrix} \frac{\partial N_1}{\partial \xi} & \frac{\partial N_2}{\partial \xi} & \frac{\partial N_3}{\partial \xi} & \frac{\partial N_4}{\partial \xi} \\ \frac{\partial N_1}{\partial \eta} & \frac{\partial N_2}{\partial \eta} & \frac{\partial N_3}{\partial \eta} & \frac{\partial N_4}{\partial \eta} \end{bmatrix} d\tilde{x}d\tilde{y} \tag{21.42}$$
$$-\iint_{\tilde{V}_e} \begin{bmatrix} \frac{\partial c_{ii}}{\partial \tilde{x}} & \frac{\partial c_{ij}}{\partial \tilde{y}} \\ \frac{\partial c_{ji}}{\partial \tilde{x}} & \frac{\partial c_{jj}}{\partial \tilde{y}} \end{bmatrix} \begin{bmatrix} N_1 & N_2 & N_3 & N_4 \\ N_1 & N_2 & N_3 & N_4 \end{bmatrix} d\tilde{x}d\tilde{y}.$$

The infinitesimal volume $d\tilde{x}d\tilde{y}$ is obtained (see Figure 21.4) as

$$d\tilde{V}_e = d\tilde{x}d\tilde{y} = \det \begin{bmatrix} \frac{\partial \tilde{x}}{\partial \xi} & \frac{\partial \tilde{x}}{\partial \eta} \\ \frac{\partial \tilde{y}}{\partial \xi} & \frac{\partial \tilde{y}}{\partial \eta} \end{bmatrix} d\xi d\eta. \tag{21.43}$$

By substituting Equation (21.43) into (21.42) one obtains

$$[\mathbf{f}_1\ \mathbf{f}_2\ \mathbf{f}_2\ \mathbf{f}_4] = \begin{bmatrix} f_{i1} & f_{i2} & f_{i3} & f_{i4} \\ f_{j1} & f_{j2} & f_{j3} & f_{j4} \end{bmatrix}$$
$$= -\iint_{\tilde{V}_e} \begin{bmatrix} c_{ii} & c_{ij} \\ c_{ji} & c_{jj} \end{bmatrix} \begin{bmatrix} \frac{\partial \tilde{x}}{\partial \xi} & \frac{\partial \tilde{x}}{\partial \eta} \\ \frac{\partial \tilde{y}}{\partial \xi} & \frac{\partial \tilde{y}}{\partial \eta} \end{bmatrix}^{-T} \begin{bmatrix} \frac{\partial N_1}{\partial \xi} & \frac{\partial N_2}{\partial \xi} & \frac{\partial N_3}{\partial \xi} & \frac{\partial N_4}{\partial \xi} \\ \frac{\partial N_1}{\partial \eta} & \frac{\partial N_2}{\partial \eta} & \frac{\partial N_3}{\partial \eta} & \frac{\partial N_4}{\partial \eta} \end{bmatrix} \det \begin{bmatrix} \frac{\partial \tilde{x}}{\partial \xi} & \frac{\partial \tilde{x}}{\partial \eta} \\ \frac{\partial \tilde{y}}{\partial \xi} & \frac{\partial \tilde{y}}{\partial \eta} \end{bmatrix} d\xi d\eta$$
$$-\iint_{\tilde{V}_e} \begin{bmatrix} \frac{\partial c_{ii}}{\partial \tilde{x}} & \frac{\partial c_{ij}}{\partial \tilde{y}} \\ \frac{\partial c_{ji}}{\partial \tilde{x}} & \frac{\partial c_{jj}}{\partial \tilde{y}} \end{bmatrix} \begin{bmatrix} N_1 & N_2 & N_3 & N_4 \\ N_1 & N_2 & N_3 & N_4 \end{bmatrix} \det \begin{bmatrix} \frac{\partial \tilde{x}}{\partial \xi} & \frac{\partial \tilde{x}}{\partial \eta} \\ \frac{\partial \tilde{y}}{\partial \xi} & \frac{\partial \tilde{y}}{\partial \eta} \end{bmatrix} d\xi d\eta. \tag{21.44}$$

The above integral is calculated using four Gauss integration points, Figure 21.6. The numerical integration yields

$$[\mathbf{f}_1\ \mathbf{f}_2\ \mathbf{f}_2\ \mathbf{f}_4] =$$
$$= -\sum_{G=1}^{4} \begin{bmatrix} c_{ii} & c_{ij} \\ c_{ji} & c_{jj} \end{bmatrix} \begin{bmatrix} \frac{\partial \tilde{x}}{\partial \xi} & \frac{\partial \tilde{x}}{\partial \eta} \\ \frac{\partial \tilde{y}}{\partial \xi} & \frac{\partial \tilde{y}}{\partial \eta} \end{bmatrix}^{-T} \begin{bmatrix} \frac{\partial N_1}{\partial \xi} & \frac{\partial N_2}{\partial \xi} & \frac{\partial N_3}{\partial \xi} & \frac{\partial N_4}{\partial \xi} \\ \frac{\partial N_1}{\partial \eta} & \frac{\partial N_2}{\partial \eta} & \frac{\partial N_3}{\partial \eta} & \frac{\partial N_4}{\partial \eta} \end{bmatrix} \det \begin{bmatrix} \frac{\partial \tilde{x}}{\partial \xi} & \frac{\partial \tilde{x}}{\partial \eta} \\ \frac{\partial \tilde{y}}{\partial \xi} & \frac{\partial \tilde{y}}{\partial \eta} \end{bmatrix} \tag{21.45}$$
$$-\sum_{G=1}^{4} \begin{bmatrix} \frac{\partial c_{ii}}{\partial \tilde{x}} & \frac{\partial c_{ij}}{\partial \tilde{y}} \\ \frac{\partial c_{ji}}{\partial \tilde{x}} & \frac{\partial c_{jj}}{\partial \tilde{y}} \end{bmatrix} \begin{bmatrix} N_1 & N_2 & N_3 & N_4 \\ N_1 & N_2 & N_3 & N_4 \end{bmatrix} \det \begin{bmatrix} \frac{\partial \tilde{x}}{\partial \xi} & \frac{\partial \tilde{x}}{\partial \eta} \\ \frac{\partial \tilde{y}}{\partial \xi} & \frac{\partial \tilde{y}}{\partial \eta} \end{bmatrix},$$

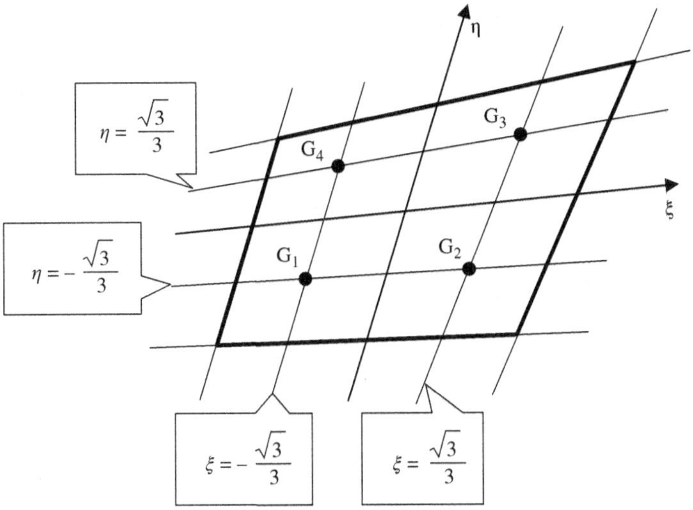

Figure 21.6 Gauss integration points for the 4-noded quadrilateral finite element

where the weight for each Gauss point is equal to one.

21.5 Summary

In this chapter, starting from the Cauchy stress tensor matrix obtained from the material package, the forces by which the finite element pulls its nodes are obtained. This is done by employing the principle of virtual work in conjunction with Gaussian numerical integration over the volume of the finite element.

Further Reading

[1] Bonet, J. and Wood, R. D. (1997) *Nonlinear Continuum Mechanics for Finite Element Analysis.* Cambridge, Cambridge University Press.
[2] Hughes, T.J.R. (2000) *The Finite Element Method: Linear Static and Dynamic Finite Element Analysis.* New York, Dover Publications.
[3] Oden, J. T. (1972) *Finite Elements of Nonlinear Continua.* New York, McGraw-Hill.
[4] Zienkiewicz, O. C. (1971) *The Finite Element Method in Engineering Science.* New York, McGraw-Hill.
[5] Zienkiewicz, O. C., Rojek, J., Taylor, R. L. and Pastor, M. (1998) Triangles and tetrahera in explicit dynamic codes for solids. *International Journal for Numerical Methods in Engineering*, **43**(3), 565–83.
[6] Zienkiewicz, O. C. and Taylor, R. L. (2005) *The Finite Element Method Set.* Oxford, Elsevier Science.

22

Reduced and Selective Integration of Nodal Forces over Volume of 2D Finite Elements

22.1 Volumetric Locking

Within the finite element method, full integration of the equilibrium equations often leads to problems associated with locking. Locking is a situation wherein two or more physical realities compete for the available degrees of freedom within each finite element.

The most publicized locking phenomenon is the volumetric locking. For example, consider the problem shown in Figure 22.1.

The problem shown consist of two homogeneous deformation triangular finite elements and four nodes, out of which three are fixed and one is free to move (node 3).

Node 3 is loaded with force P, as shown. Now consider a material that is relatively soft in shear and has a very high bulk modulus. In other words, it is relatively easy to shear the material, but very hard to change its volume. A representative material would be rubber.

Should node 3 move in the vertical direction, the volume of the finite element 1 changes. Thus, relatively speaking the vertical displacement of node 3 is locked by pressure induced through the volume change of element 1.

In a similar manner, should node 3 move horizontally, the volume of element 2 changes, giving rise to a high volumetric stress that locks the horizontal displacement of node 3. As a consequence, both degrees of freedom are consumed by the change of volume (volumetric strain) in the material and there is no degree of freedom left to describe the shearing of the material.

As a result, the displacements obtained by the finite element analysis are wrong (much smaller than the real world displacements) by an order of magnitude. Moreover, the obtained displacements do not get better by employing smaller finite elements. In other words, there is no convergence.

Should one attempt to solve the same problem with a much finer mesh, the result is the same: node 1 is locked by volumetric stress in elements 1 and 2 and in turn, node 2 is locked by volumetric stress in elements 3 and 4; etc., Figure 22.2.

Large Strain Finite Element Method: A Practical Course, First Edition. Antonio Munjiza, Esteban Rougier and Earl E. Knight.

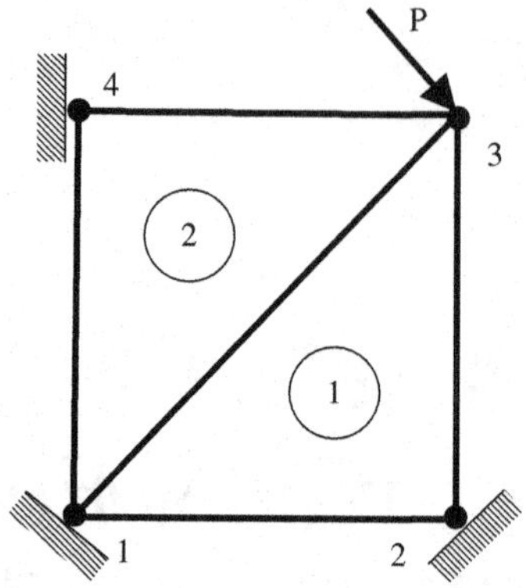

Figure 22.1 A simple example of volumetric locking. Numbers inside the circles denote the finite elements

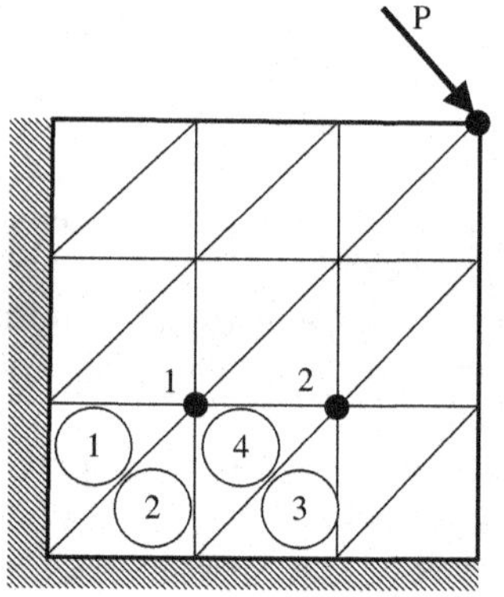

Figure 22.2 A simple example of volumetric locking with a finer mesh

In order to circumvent volumetric locking in the finite element method, two approaches are often adopted:

a. Reduced integration
b. Selective integration

22.2 Reduced Integration

Reduced integration is often used with four-noded quadrilateral finite elements. For full integration of a four-noded quadrilateral element four Gauss integration points are used (see Chapter 21). However, for the reduced integration case only one Gauss integration point is used, Figure 22.3.

Utilizing this approach one finds that the integrals for virtual work are under-integrated; this fact leads to another problem, which is the existence of unconstrained degrees of freedom.

For example, the problem shown in Figure 22.4 has two degrees of freedom and the integration done through a single Gauss point constraints effectively only one degree of freedom, which yields the zero energy modes: the solid deforms and no internal forces resisting the deformation

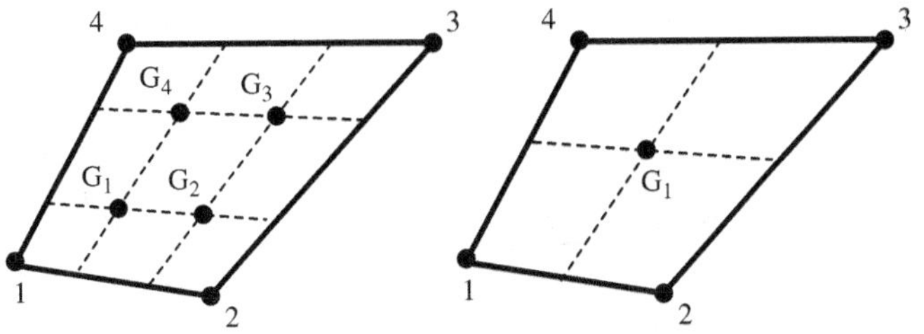

Figure 22.3 Full and reduce integration formula for the 4-noded quadrilateral finite element

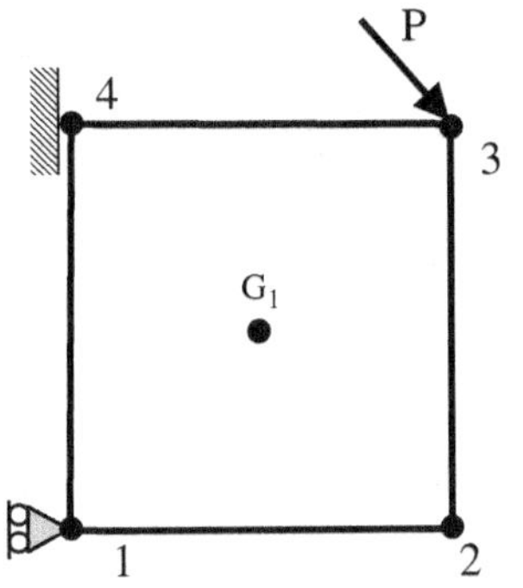

Figure 22.4 Reduced integration leads to unconstrained degrees of freedom

are produced. For four-noded quadrilateral finite elements these modes are defined as the classic hour-glassing modes because of the particular shape of the zero energy modes [1], [3].

22.3 Selective Integration

Full integration may lead to locking, while reduced integration may lead to zero energy modes. In order to fix both of the above problems, selective integration is employed. The idea is relatively simple:

Calculate different constitutive components at different selectively chosen Gauss points

The easiest way to explain the above idea is to use a six-noded composite triangle finite element, as shown in Figure 22.5.

This element is composed of four initially identical three-noded triangles. However, a single initial infinitesimal volume element is defined for the composite triangle as a whole

$$\begin{bmatrix} \bar{\boldsymbol{\alpha}} & \bar{\boldsymbol{\beta}} \end{bmatrix} = \begin{bmatrix} \bar{\alpha}_i & \bar{\beta}_i \\ \bar{\alpha}_j & \bar{\beta}_j \end{bmatrix}. \tag{22.1}$$

Now, for the composite triangle as a whole, the volume change is calculated and the volumetric stretch s_v is obtained.

For each of the three-noded triangles, separate local ξ and η axes are introduced and the position of the material base is calculated as follows:

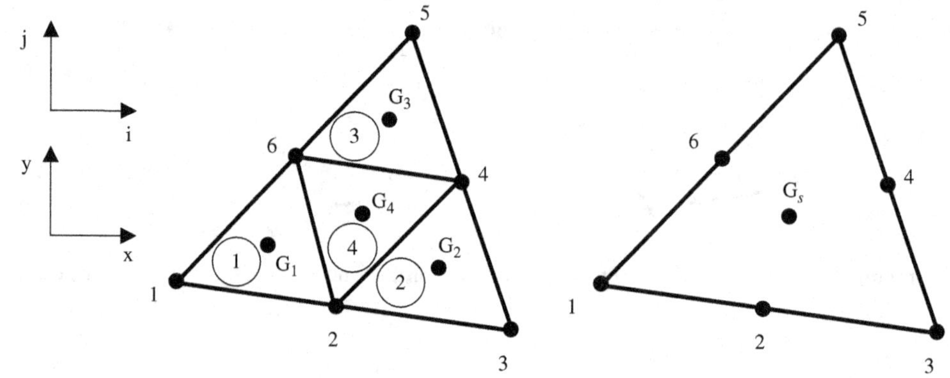

Figure 22.5 The six-noded composite triangle finite element with integration points based on selective integration – volumetric stretch is considered only at point G_s (which is the same as G_4) all other stretches are considered at all Gauss points (G_1, G_2, G_3, G_4)

a. For element 1:

$$\begin{bmatrix} \dfrac{\partial \bar{x}}{\partial \xi} & \dfrac{\partial \bar{x}}{\partial \eta} \\[2ex] \dfrac{\partial \bar{y}}{\partial \xi} & \dfrac{\partial \bar{y}}{\partial \eta} \end{bmatrix} = \begin{bmatrix} \bar{x}_2-\bar{x}_1 & \bar{x}_6-\bar{x}_1 \\ \bar{y}_2-\bar{y}_1 & \bar{y}_6-\bar{y}_1 \end{bmatrix}. \tag{22.2}$$

The current infinitesimal solid element is defined by

$$\begin{aligned} \begin{bmatrix} \tilde{\boldsymbol{\alpha}} & \tilde{\boldsymbol{\beta}} \end{bmatrix} &= \begin{bmatrix} \tilde{\alpha}_i & \tilde{\beta}_i \\ \tilde{\alpha}_j & \tilde{\beta}_j \end{bmatrix} \\ &= \begin{bmatrix} \tilde{x}_2-\tilde{x}_1 & \tilde{x}_6-\tilde{x}_1 \\ \tilde{y}_2-\tilde{y}_1 & \tilde{y}_6-\tilde{y}_1 \end{bmatrix} \left(\begin{bmatrix} \bar{x}_2-\bar{x}_1 & \bar{x}_6-\bar{x}_1 \\ \bar{y}_2-\bar{y}_1 & \bar{y}_6-\bar{y}_1 \end{bmatrix}^{-1} \begin{bmatrix} \bar{\alpha}_i & \bar{\beta}_i \\ \bar{\alpha}_j & \bar{\beta}_j \end{bmatrix} \right) \end{aligned} \tag{22.3}$$

The previous infinitesimal solid element is defined by

$$\begin{aligned} \begin{bmatrix} \hat{\boldsymbol{\alpha}} & \hat{\boldsymbol{\beta}} \end{bmatrix} &= \begin{bmatrix} \hat{\alpha}_i & \hat{\beta}_i \\ \hat{\alpha}_j & \hat{\beta}_j \end{bmatrix} \\ &= \begin{bmatrix} \hat{x}_2-\hat{x}_1 & \hat{x}_6-\hat{x}_1 \\ \hat{y}_2-\hat{y}_1 & \hat{y}_6-\hat{y}_1 \end{bmatrix} \left(\begin{bmatrix} \bar{x}_2-\bar{x}_1 & \bar{x}_6-\bar{x}_1 \\ \bar{y}_2-\bar{y}_1 & \bar{y}_6-\bar{y}_1 \end{bmatrix}^{-1} \begin{bmatrix} \bar{\alpha}_i & \bar{\beta}_i \\ \bar{\alpha}_j & \bar{\beta}_j \end{bmatrix} \right). \end{aligned} \tag{22.4}$$

b. For element 2: the current and previous infinitesimal solid elements are respectively

$$\begin{aligned} \begin{bmatrix} \tilde{\boldsymbol{\alpha}} & \tilde{\boldsymbol{\beta}} \end{bmatrix} &= \begin{bmatrix} \tilde{\alpha}_i & \tilde{\beta}_i \\ \tilde{\alpha}_j & \tilde{\beta}_j \end{bmatrix} \\ &= \begin{bmatrix} \tilde{x}_3-\tilde{x}_2 & \tilde{x}_4-\tilde{x}_2 \\ \tilde{y}_3-\tilde{y}_2 & \tilde{y}_4-\tilde{y}_2 \end{bmatrix} \left(\begin{bmatrix} \bar{x}_3-\bar{x}_2 & \bar{x}_4-\bar{x}_2 \\ \bar{y}_3-\bar{y}_2 & \bar{y}_4-\bar{y}_2 \end{bmatrix}^{-1} \begin{bmatrix} \bar{\alpha}_i & \bar{\beta}_i \\ \bar{\alpha}_j & \bar{\beta}_j \end{bmatrix} \right). \end{aligned} \tag{22.5}$$

$$
\begin{aligned}
[\hat{\boldsymbol{\alpha}}\ \hat{\boldsymbol{\beta}}] &= \begin{bmatrix} \hat{\alpha}_i & \hat{\beta}_i \\ \hat{\alpha}_j & \hat{\beta}_j \end{bmatrix} \\
&= \begin{bmatrix} \hat{x}_3-\hat{x}_2 & \hat{x}_4-\hat{x}_2 \\ \hat{y}_3-\hat{y}_2 & \hat{y}_4-\hat{y}_2 \end{bmatrix} \left(\begin{bmatrix} \bar{x}_3-\bar{x}_2 & \bar{x}_4-\bar{x}_2 \\ \bar{y}_3-\bar{y}_2 & \bar{y}_4-\bar{y}_2 \end{bmatrix}^{-1} \begin{bmatrix} \bar{\alpha}_i & \bar{\beta}_i \\ \bar{\alpha}_j & \bar{\beta}_j \end{bmatrix} \right).
\end{aligned} \tag{22.6}
$$

c. For element 3, the current infinitesimal solid element is defined by

$$
\begin{aligned}
[\tilde{\boldsymbol{\alpha}}\ \tilde{\boldsymbol{\beta}}] &= \begin{bmatrix} \tilde{\alpha}_i & \tilde{\beta}_i \\ \tilde{\alpha}_j & \tilde{\beta}_j \end{bmatrix} \\
&= \begin{bmatrix} \tilde{x}_5-\tilde{x}_4 & \tilde{x}_6-\tilde{x}_4 \\ \tilde{y}_5-\tilde{y}_4 & \tilde{y}_6-\tilde{y}_4 \end{bmatrix} \left(\begin{bmatrix} \bar{x}_5-\bar{x}_4 & \bar{x}_6-\bar{x}_4 \\ \bar{y}_5-\bar{y}_4 & \bar{y}_6-\bar{y}_4 \end{bmatrix}^{-1} \begin{bmatrix} \bar{\alpha}_i & \bar{\beta}_i \\ \bar{\alpha}_j & \bar{\beta}_j \end{bmatrix} \right)
\end{aligned} \tag{22.7}
$$

and the previous infinitesimal solid element is given by

$$
\begin{aligned}
[\hat{\boldsymbol{\alpha}}\ \hat{\boldsymbol{\beta}}] &= \begin{bmatrix} \hat{\alpha}_i & \hat{\beta}_i \\ \hat{\alpha}_j & \hat{\beta}_j \end{bmatrix} \\
&= \begin{bmatrix} \hat{x}_5-\hat{x}_4 & \hat{x}_6-\hat{x}_4 \\ \hat{y}_5-\hat{y}_4 & \hat{y}_6-\hat{y}_4 \end{bmatrix} \left(\begin{bmatrix} \bar{x}_5-\bar{x}_4 & \bar{x}_6-\bar{x}_4 \\ \bar{y}_5-\bar{y}_4 & \bar{y}_6-\bar{y}_4 \end{bmatrix}^{-1} \begin{bmatrix} \bar{\alpha}_i & \bar{\beta}_i \\ \bar{\alpha}_j & \bar{\beta}_j \end{bmatrix} \right).
\end{aligned} \tag{22.8}
$$

d. For element 4, the current and previous infinitesimal solid elements are given by

$$
\begin{aligned}
[\tilde{\boldsymbol{\alpha}}\ \tilde{\boldsymbol{\beta}}] &= \begin{bmatrix} \tilde{\alpha}_i & \tilde{\beta}_i \\ \tilde{\alpha}_j & \tilde{\beta}_j \end{bmatrix} \\
&= \begin{bmatrix} \tilde{x}_4-\tilde{x}_2 & \tilde{x}_6-\tilde{x}_2 \\ \tilde{y}_4-\tilde{y}_2 & \tilde{y}_6-\tilde{y}_2 \end{bmatrix} \left(\begin{bmatrix} \bar{x}_4-\bar{x}_2 & \bar{x}_6-\bar{x}_2 \\ \bar{y}_4-\bar{y}_2 & \bar{y}_6-\bar{y}_2 \end{bmatrix}^{-1} \begin{bmatrix} \bar{\alpha}_i & \bar{\beta}_i \\ \bar{\alpha}_j & \bar{\beta}_j \end{bmatrix} \right)
\end{aligned} \tag{22.9}
$$

$$
\begin{aligned}
[\hat{\boldsymbol{\alpha}}\ \hat{\boldsymbol{\beta}}] &= \begin{bmatrix} \hat{\alpha}_i & \hat{\beta}_i \\ \hat{\alpha}_j & \hat{\beta}_j \end{bmatrix} \\
&= \begin{bmatrix} \hat{x}_4-\hat{x}_2 & \hat{x}_6-\hat{x}_2 \\ \hat{y}_4-\hat{y}_2 & \hat{y}_6-\hat{y}_2 \end{bmatrix} \left(\begin{bmatrix} \bar{x}_4-\bar{x}_2 & \bar{x}_6-\bar{x}_2 \\ \bar{y}_4-\bar{y}_2 & \bar{y}_6-\bar{y}_2 \end{bmatrix}^{-1} \begin{bmatrix} \bar{\alpha}_i & \bar{\beta}_i \\ \bar{\alpha}_j & \bar{\beta}_j \end{bmatrix} \right).
\end{aligned} \tag{22.10}
$$

The obtained four sets of base vectors define four different generalized material elements. These are passed to the material package together with the volumetric stretch s_v (which is therefore the same for all four elements) resulting in a single Gauss point for volumetric change and four Gauss points for all the other stretches.

Many other combinations of composite elements are possible, for example, a quadrilateral consisting of either two or four triangles, Figure 22.6. In both cases the volumetric stretch is

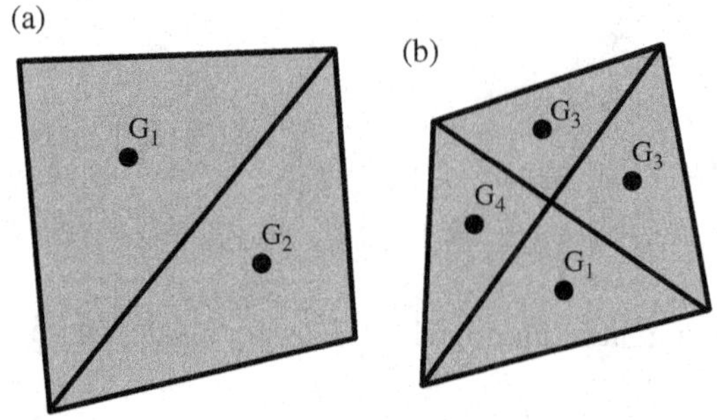

Figure 22.6 Some examples of composite finite elements: (a) quadrilateral finite element composed of two triangles; (b) quadrilateral finite element composed of four triangles

calculated from the volume change of the whole finite element, while all other stretches are calculated for the separate homogeneous deformation triangles. Composite elements such as these are very convenient for contact problems for they reach the accuracy of the higher order elements while preserving the simplicity of the lowest order elements.

22.4 Shear Locking

The Shear Locking Phenomenon. Take a four-noded quadrilateral finite element in pure bending, as shown in Figure 22.7. The normal strain at the Gauss point G_1 is given by

$$\varepsilon_{G_1} = \frac{\Delta l}{l}\frac{h_g}{h} = \frac{\sqrt{3}}{3}\frac{\Delta l}{l} = \frac{\sqrt{3}}{3}\varepsilon, \tag{22.11}$$

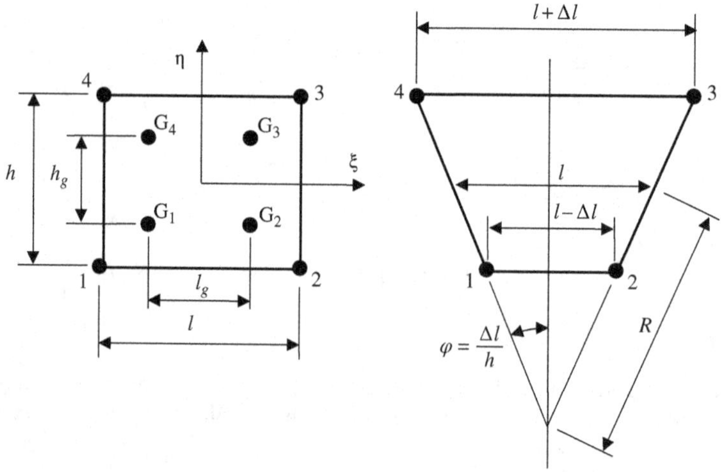

Figure 22.7 Quadrilateral finite element in pure bending

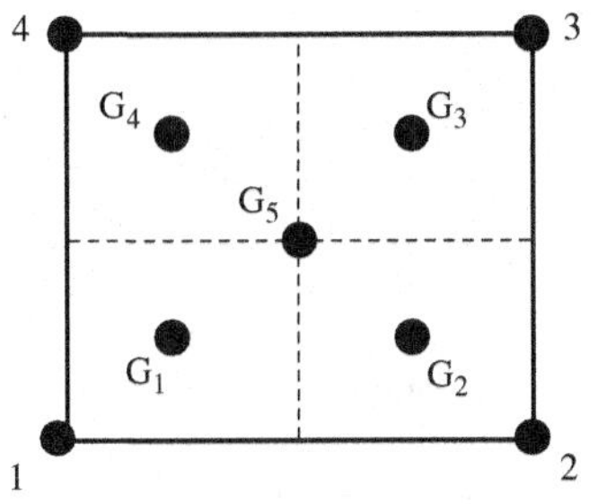

Figure 22.8 Selective integration for the four-noded quadrilateral

where ε is the normal strain in the top fiber of the element. The engineering shear strain at the Gauss point G_1 is given by

$$\gamma_{G_1} = \frac{\Delta l}{h}\frac{h_g}{h} = \frac{\sqrt{3}}{3}\frac{\Delta l}{h}. \tag{22.12}$$

This can be expressed in terms of the normal strain at the top fiber of the element, as follows

$$\begin{aligned}\gamma_{G_1} &= \frac{\sqrt{3}}{3}\frac{\Delta l}{l}\frac{l}{h} \\ &= \frac{\sqrt{3}}{3}\varepsilon\frac{l}{h} \\ &= \varepsilon_{G_1}\frac{l}{h}.\end{aligned} \tag{22.13}$$

From Equation (22.13) it is evident that γ_{G_1} is not objective, but it is a function of the initial geometry of the element. It is worth noting that when solving a real physical problem strains are an objective reality created from the problem itself and should not be a function of the finite element mesh (grid).

Of course, stresses and strains are always a function of the finite element mesh, however there is a convergence to the theoretical results as the grid size becomes infinitely small.

With the shear stress defined by Equation (22.13) there is no convergence and shear stress is simply a function of the aspect ratio l/h. As the aspect ratio increases, so does the shear stress.

In order to avoid shear locking, one can introduce reduced integration, which leads to the problem of zero energy modes (as explained with volumetric locking).

A better solution is selective integration, Figure 22.8. The approach is relatively simple: the shear (angle) stretch of the generalized material element is calculated at point G_5 and is assumed to be constant over the whole finite element. The volumetric stretch is calculated by considering the volume change of the whole element and it is assumed to be constant over the finite element. All the other stretches are calculated at Gauss points G_1, G_2, G_3 and G_4. At these points, these stretches are combined with the constant shear and volumetric stretches obtained before. The Cauchy stress tensor matrix is calculated as usual using the constitutive law (the material package). The nodal forces are obtained using numerical integration of virtual work over the Gauss points G_1, G_2, G_3 and G_4.

22.5 Summary

With both pressure and shear locking it is possible to employ selective integration using stresses. When using stresses, it is necessary to calculate different stress components at different material points. This may be a problem for some constitutive law formulations. It may also lead to systematic errors due to different stress components being combined in a manner that in effect bypasses the constitutive law. Constitutive laws are, in general, implemented separate from the finite element packages and cannot be expected to take into account the fine details of the selective integration. For this reason, selective integration in terms of stresses should be avoided and as such is not considered in this book.

In this book a general approach is considered instead. It employs the selective integration in terms of stretches, as described. Stretches are calculated at specific points (not necessarily integration points). These points are called stretch sampling points; for example, volumetric stretch and shear stretch may be sampled at the center of the quadrilateral finite element and simply copied into the Gauss points, where it is combined with the rest of the stretch components in order to obtain the full description of the stretching of the generalized material element. These are then passed to the material package for stress calculation at the Gauss points. This approach has been demonstrated in this chapter using composite finite elements.

It is possible to extend this principle even further and to completely separate stretch sampling points, stress calculation points, and integration points. This is often done in conjunction with highly nonlinear material laws.

Further Reading

[1] Armero, F. (2000) On the locking and stability of finite elements in finite deformation plane strain problems. *Comput Struct.*, **75**(3): 261–90.

[2] Belytschko, T., J.S.-J. Ong, W.K. Liu and J.M. Kennedy (1984) Hourglass control in linear and nonlinear problems. *Computer Methods in Applied Mechanics and Engineering*, **43**: 251–76.

[3] Belytschko, T. and Bindeman, L. P. (1991) Assumed strain stabilization of the 4-node quadrilateral with 1-point quadrature for nonlinear problems. *Comput. Method Appl. M.*, **88**(3): 311–40.

[4] Belytschko, T. and Bindeman, L. P. (1993) Assumed strain stabilization of the 8 node hexahedral element. *Comput. Method Appl. M.*, **105**(2): 225–60.

[5] Flanagan, D.P. and T. Belytschko (1981) A uniform strain hexahedron and quadrilateral with orthogonal hourglass control, *International Journal for Numerical Methods in Engineering*, **17**: 679–706.

[6] Flanagan, D. P. and Belytschko, T. (1984) Eigenvalues and stable time steps for the uniform strain hexahedron and quadrilateral. *Journal of Applied Mechanics-Transactions of the Asme*, **51**(1): 35–40.

[7] Oden, J. T. (1972) *Finite Elements of Nonlinear Continua*. New York, McGraw-Hill.

[8] Zienkiewicz, O. C. (1971) *The Finite Element Method in Engineering Science*. New York, McGraw-Hill.

[9] Zienkiewicz, O. C., Rojek, J., Taylor, R. L. and Pastor, M. (1998) Triangles and tetrahera in explicit dynamic codes for solids. *International Journal for Numerical Methods in Engineering*, **43**(3): 565–83.

[10] Zienkiewicz, O. C. and Taylor, R. L. (2005) *The Finite Element Method Set*. Oxford, Elsevier Science.

23

3D Deformation Kinematics Using the Homogeneous Deformation Tetrahedron Finite Element

23.1 Introduction

Similar to the 2D finite element method, finite strain-based deformability for 3D solids is usually implemented in 3D through either an explicit dynamic formulation or through an explicit iterative static formulation. In both cases, no stiffness matrix is calculated. Calculating a consistent tangential stiffness matrix in a multiplicative decomposition-based 3D problem would be quite difficult, or even impossible, especially when the problem combines solid deformability along with contact among the different solid parts.

For the sake of simplicity of the formulation, and for CPU efficiency, explicit and iterative formulations are nearly exclusively deployed. This is also the most used case in the context of modern multi-core computer hardware, where parallelization of finite element implementations is much more efficient when no stiffness matrix is present.

Explicit and iterative solvers are also better suited for problems that have a large spectral radius. Spectral radius is the ratio between the largest and the smallest frequency in the system. Gaussian elimination-based solvers and/or matrix inversions tend to accumulate relatively large errors due to the rounding errors associated with large spectral radii. In contrast, explicit and iterative solvers tend to smooth these errors, especially in the highest frequency region.

Massive scale and grand scale nonlinear computations are usually based on the solutions where no stiffness matrix is calculated and preferably where no system of equations is solved. One important reason for this is the fact that nonlinear simulations usually have nonunique solutions, most of which are not physically correct. Explicit approaches such as dynamic relaxation naturally lead to the correct physical solution. In contrast, solvers such as Newton-Raphson or other related types of solvers may lead to nonphysical solutions.

Large Strain Finite Element Method: A Practical Course, First Edition. Antonio Munjiza, Esteban Rougier and Earl E. Knight.

In the rest of this chapter the 3D finite element method based on the multiplicative decomposition is described. Different finite elements are introduced and for all the cases the deformation kinematics are obtained.

23.2 The Homogeneous Deformation Four-Noded Tetrahedron Finite Element

Geometry. The simplest finite element in 3D is the four-noded homogeneous deformation tetrahedron whose geometry is uniquely defined by 4 nodes as shown in Figure 23.1. In order to describe the deformation of the tetrahedron, two coordinate systems are introduced (Figure 23.2):

a. The global Cartesian coordinate system, (x, y, z). This is a coordinate system that is fixed in space.
b. The local solid-embedded coordinate system (ξ, η, ζ). This is a system with coordinate surfaces fixed to the material points of the solid. As the solid translates, rotates and stretches (i.e., deforms), it carries the coordinate surfaces with it. Thus, the coordinate surfaces and coordinate lines deform (translate, rotate and stretch) together with the solid.

Shape Functions. At this point, it is convenient to introduce the shape functions for the tetrahedron finite element

$$\mathbf{N} = \begin{bmatrix} N_1 \\ N_2 \\ N_3 \\ N_4 \end{bmatrix} = \begin{bmatrix} 1-\xi-\eta-\zeta \\ \xi \\ \eta \\ \zeta \end{bmatrix}. \tag{23.1}$$

The graphical representation of these shape functions is shown in Figure 23.3. The shape function N_1 is equal to 1 at node 1 and equal to zero at all other nodes. In a similar manner shape functions N_2, N_3 and N_4 are equal to 1 at nodes 2, 3 and 4 respectively and zero at all other nodes.

Derivations of the Shape Functions. Very often one needs the following derivation of the shape functions

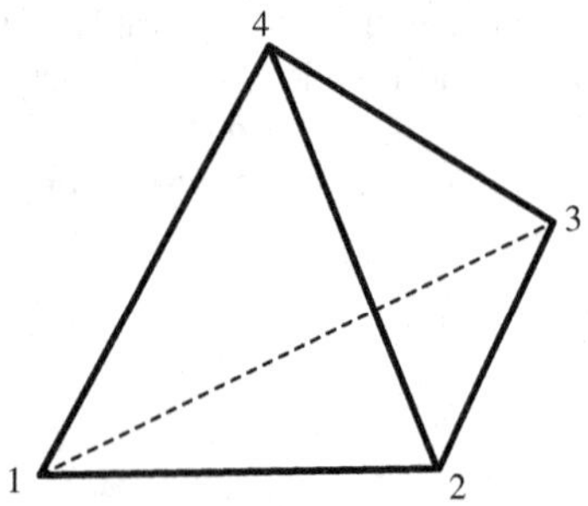

Figure 23.1 The geometry of the 4-noded tetrahedron finite element

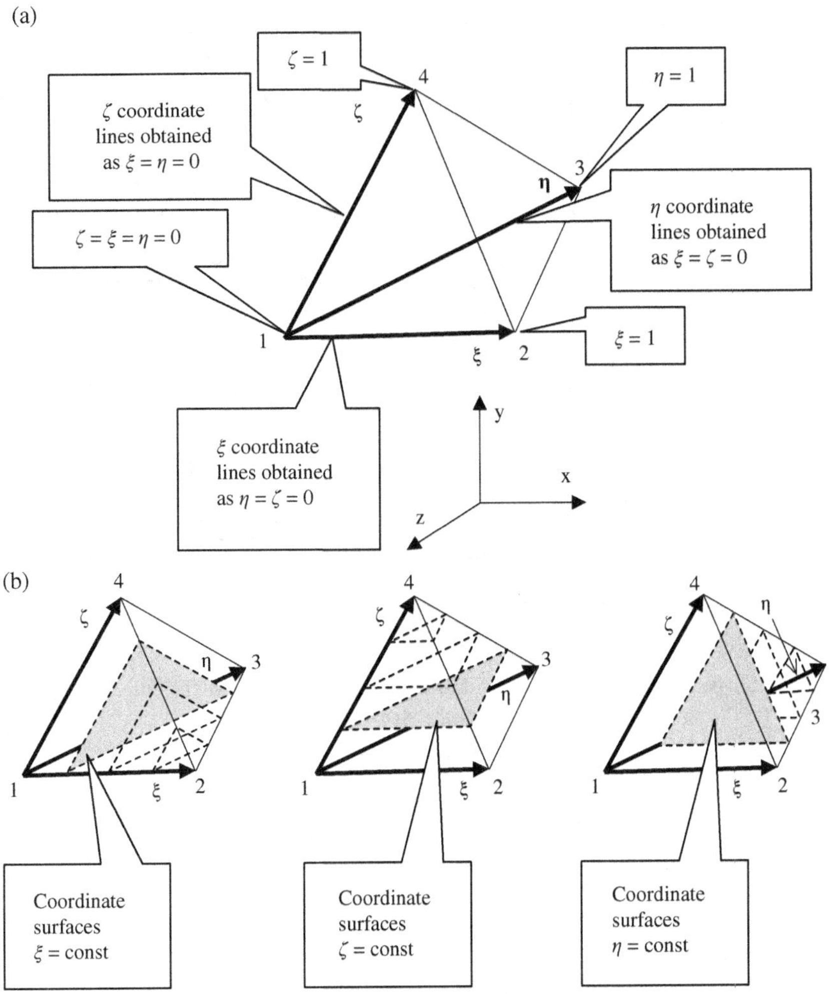

Figure 23.2 The solid-embedded coordinate system: (a) the coordinate lines; (b) the coordinate surfaces

$$\begin{bmatrix} \frac{\partial N_1}{\partial \xi} \\ \frac{\partial N_1}{\partial \eta} \\ \frac{\partial N_1}{\partial \zeta} \end{bmatrix} = \begin{bmatrix} -1 \\ -1 \\ -1 \end{bmatrix}; \quad \begin{bmatrix} \frac{\partial N_2}{\partial \xi} \\ \frac{\partial N_2}{\partial \eta} \\ \frac{\partial N_2}{\partial \zeta} \end{bmatrix} = \begin{bmatrix} 1 \\ 0 \\ 0 \end{bmatrix}; \quad \begin{bmatrix} \frac{\partial N_3}{\partial \xi} \\ \frac{\partial N_3}{\partial \eta} \\ \frac{\partial N_3}{\partial \zeta} \end{bmatrix} = \begin{bmatrix} 0 \\ 1 \\ 0 \end{bmatrix}; \quad \begin{bmatrix} \frac{\partial N_4}{\partial \xi} \\ \frac{\partial N_4}{\partial \eta} \\ \frac{\partial N_4}{\partial \zeta} \end{bmatrix} = \begin{bmatrix} 0 \\ 0 \\ 1 \end{bmatrix}. \tag{23.2}$$

Global Coordinates of Material Points. For a given material point

$$P = (\xi, \eta, \zeta), \tag{23.3}$$

located inside the tetrahedron, the initial, the previous and the current global coordinates are approximated from the global nodal coordinates using the shape functions

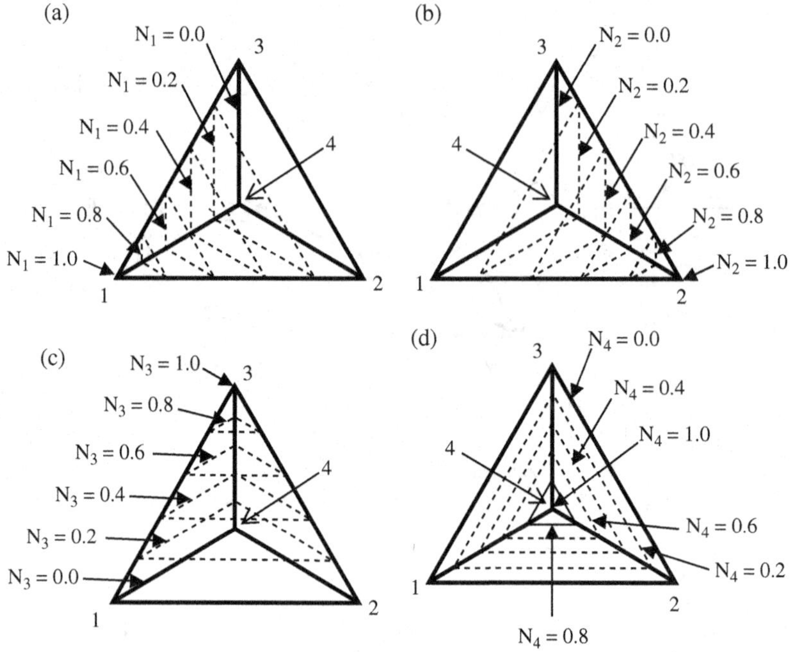

Figure 23.3 Shape functions for the tetrahedron finite element: (a) the shape function N_1; (b) the shape function N_2; (c) the shape function N_3 and (d) the shape function N_4

$$
\begin{aligned}
\bar{x} &= \bar{x}(\xi,\eta,\zeta) \\
&= \bar{x}_1 N_1(\xi,\eta,\zeta) + \bar{x}_2 N_2(\xi,\eta,\zeta) + \bar{x}_3 N_3(\xi,\eta,\zeta) + \bar{x}_4 N_4(\xi,\eta,\zeta) \\
\bar{y} &= \bar{y}(\xi,\eta,\zeta) \\
&= \bar{y}_1 N_1(\xi,\eta,\zeta) + \bar{y}_2 N_2(\xi,\eta,\zeta) + \bar{y}_3 N_3(\xi,\eta,\zeta) + \bar{y}_4 N_4(\xi,\eta,\zeta) \\
\bar{z} &= \bar{z}(\xi,\eta,\zeta) \\
&= \bar{z}_1 N_1(\xi,\eta,\zeta) + \bar{z}_2 N_2(\xi,\eta,\zeta) + \bar{z}_3 N_3(\xi,\eta,\zeta) + \bar{z}_4 N_4(\xi,\eta,\zeta)
\end{aligned}
\tag{23.4}
$$

$$
\begin{aligned}
\hat{x} &= \hat{x}(\xi,\eta,\zeta) \\
&= \hat{x}_1 N_1(\xi,\eta,\zeta) + \hat{x}_2 N_2(\xi,\eta,\zeta) + \hat{x}_3 N_3(\xi,\eta,\zeta) + \hat{x}_4 N_4(\xi,\eta,\zeta) \\
\hat{y} &= \hat{y}(\xi,\eta,\zeta) \\
&= \hat{y}_1 N_1(\xi,\eta,\zeta) + \hat{y}_2 N_2(\xi,\eta,\zeta) + \hat{y}_3 N_3(\xi,\eta,\zeta) + \hat{y}_4 N_4(\xi,\eta,\zeta) \\
\hat{z} &= \hat{z}(\xi,\eta,\zeta) \\
&= \hat{z}_1 N_1(\xi,\eta,\zeta) + \hat{z}_2 N_2(\xi,\eta,\zeta) + \hat{z}_3 N_3(\xi,\eta,\zeta) + \hat{z}_4 N_4(\xi,\eta,\zeta)
\end{aligned}
\tag{23.5}
$$

$$
\begin{aligned}
\tilde{x} &= \tilde{x}(\xi,\eta,\zeta) \\
&= \tilde{x}_1 N_1(\xi,\eta,\zeta) + \tilde{x}_2 N_2(\xi,\eta,\zeta) + \tilde{x}_3 N_3(\xi,\eta,\zeta) + \tilde{x}_4 N_4(\xi,\eta,\zeta) \\
\tilde{y} &= \tilde{y}(\xi,\eta,\zeta) \\
&= \tilde{y}_1 N_1(\xi,\eta,\zeta) + \tilde{y}_2 N_2(\xi,\eta,\zeta) + \tilde{y}_3 N_3(\xi,\eta,\zeta) + \tilde{y}_4 N_4(\xi,\eta,\zeta) \\
\tilde{z} &= \tilde{z}(\xi,\eta,\zeta) \\
&= \tilde{z}_1 N_1(\xi,\eta,\zeta) + \tilde{z}_2 N_2(\xi,\eta,\zeta) + \tilde{z}_3 N_3(\xi,\eta,\zeta) + \tilde{z}_4 N_4(\xi,\eta,\zeta),
\end{aligned}
\tag{23.6}
$$

where it is worth being reminded that

$$\begin{aligned}\xi &= \bar{\xi} = \hat{\xi} = \tilde{\xi}\\ \eta &= \bar{\eta} = \hat{\eta} = \tilde{\eta}\\ \zeta &= \bar{\zeta} = \hat{\zeta} = \tilde{\zeta}.\end{aligned} \tag{23.7}$$

This condition occurs because ξ, η and ζ are embedded with the material of the solid. In the above equations "–" stands for initial position, "^" stands for previous position and "~" stands for current position. Equations (23.4), (23.5) and (23.6) are then conveniently written in matrix form:

$$\begin{aligned}\begin{bmatrix}\bar{x}\\ \bar{y}\\ \bar{z}\end{bmatrix} &= \begin{bmatrix}\bar{x}_1 & \bar{x}_2 & \bar{x}_3 & \bar{x}_4\\ \bar{y}_1 & \bar{y}_2 & \bar{y}_3 & \bar{y}_4\\ \bar{z}_1 & \bar{z}_2 & \bar{z}_3 & \bar{z}_4\end{bmatrix}\begin{bmatrix}N_1(\xi,\eta,\zeta)\\ N_2(\xi,\eta,\zeta)\\ N_3(\xi,\eta,\zeta)\\ N_4(\xi,\eta,\zeta)\end{bmatrix}\\ &= \begin{bmatrix}\bar{x}_1 & \bar{x}_2 & \bar{x}_3 & \bar{x}_4\\ \bar{y}_1 & \bar{y}_2 & \bar{y}_3 & \bar{y}_4\\ \bar{z}_1 & \bar{z}_2 & \bar{z}_3 & \bar{z}_4\end{bmatrix}\begin{bmatrix}1-\xi-\eta-\zeta\\ \xi\\ \eta\\ \zeta\end{bmatrix}\end{aligned} \tag{23.8}$$

$$\begin{aligned}\begin{bmatrix}\hat{x}\\ \hat{y}\\ \hat{z}\end{bmatrix} &= \begin{bmatrix}\hat{x}_1 & \hat{x}_2 & \hat{x}_3 & \hat{x}_4\\ \hat{y}_1 & \hat{y}_2 & \hat{y}_3 & \hat{y}_4\\ \hat{z}_1 & \hat{z}_2 & \hat{z}_3 & \hat{z}_4\end{bmatrix}\begin{bmatrix}N_1(\xi,\eta,\zeta)\\ N_2(\xi,\eta,\zeta)\\ N_3(\xi,\eta,\zeta)\\ N_4(\xi,\eta,\zeta)\end{bmatrix}\\ &= \begin{bmatrix}\hat{x}_1 & \hat{x}_2 & \hat{x}_3 & \hat{x}_4\\ \hat{y}_1 & \hat{y}_2 & \hat{y}_3 & \hat{y}_4\\ \hat{z}_1 & \hat{z}_2 & \hat{z}_3 & \hat{z}_4\end{bmatrix}\begin{bmatrix}1-\xi-\eta-\zeta\\ \xi\\ \eta\\ \zeta\end{bmatrix}\end{aligned} \tag{23.9}$$

$$\begin{aligned}\begin{bmatrix}\tilde{x}\\ \tilde{y}\\ \tilde{z}\end{bmatrix} &= \begin{bmatrix}\tilde{x}_1 & \tilde{x}_2 & \tilde{x}_3 & \tilde{x}_4\\ \tilde{y}_1 & \tilde{y}_2 & \tilde{y}_3 & \tilde{y}_4\\ \tilde{z}_1 & \tilde{z}_2 & \tilde{z}_3 & \tilde{z}_4\end{bmatrix}\begin{bmatrix}N_1(\xi,\eta,\zeta)\\ N_2(\xi,\eta,\zeta)\\ N_3(\xi,\eta,\zeta)\\ N_4(\xi,\eta,\zeta)\end{bmatrix}\\ &= \begin{bmatrix}\tilde{x}_1 & \tilde{x}_2 & \tilde{x}_3 & \tilde{x}_4\\ \tilde{y}_1 & \tilde{y}_2 & \tilde{y}_3 & \tilde{y}_4\\ \tilde{z}_1 & \tilde{z}_2 & \tilde{z}_3 & \tilde{z}_4\end{bmatrix}\begin{bmatrix}1-\xi-\eta-\zeta\\ \xi\\ \eta\\ \zeta\end{bmatrix}.\end{aligned} \tag{23.10}$$

Derivation of a Scalar Field. Very often a derivation of a scalar field

$$u = \mathrm{u}(x, y, z) \tag{23.11}$$

over the finite element needs to be calculated. This is done by considering that

$$\begin{bmatrix} \frac{\partial u}{\partial \xi} \\ \frac{\partial u}{\partial \eta} \\ \frac{\partial u}{\partial \zeta} \end{bmatrix} = \begin{bmatrix} \frac{\partial u}{\partial x} & \frac{\partial u}{\partial y} & \frac{\partial u}{\partial z} \end{bmatrix} \begin{bmatrix} \frac{\partial x}{\partial \xi} & \frac{\partial x}{\partial \eta} & \frac{\partial x}{\partial \zeta} \\ \frac{\partial y}{\partial \xi} & \frac{\partial y}{\partial \eta} & \frac{\partial y}{\partial \zeta} \\ \frac{\partial z}{\partial \xi} & \frac{\partial z}{\partial \eta} & \frac{\partial z}{\partial \zeta} \end{bmatrix} = \begin{bmatrix} \frac{\partial x}{\partial \xi} & \frac{\partial x}{\partial \eta} & \frac{\partial x}{\partial \zeta} \\ \frac{\partial y}{\partial \xi} & \frac{\partial y}{\partial \eta} & \frac{\partial y}{\partial \zeta} \\ \frac{\partial z}{\partial \xi} & \frac{\partial z}{\partial \eta} & \frac{\partial z}{\partial \zeta} \end{bmatrix}^T \begin{bmatrix} \frac{\partial u}{\partial x} \\ \frac{\partial u}{\partial y} \\ \frac{\partial u}{\partial z} \end{bmatrix}, \tag{23.12}$$

from which one obtains

$$\begin{bmatrix} \frac{\partial u}{\partial x} \\ \frac{\partial u}{\partial y} \\ \frac{\partial u}{\partial z} \end{bmatrix} = \begin{bmatrix} \frac{\partial x}{\partial \xi} & \frac{\partial x}{\partial \eta} & \frac{\partial x}{\partial \zeta} \\ \frac{\partial y}{\partial \xi} & \frac{\partial y}{\partial \eta} & \frac{\partial y}{\partial \zeta} \\ \frac{\partial z}{\partial \xi} & \frac{\partial z}{\partial \eta} & \frac{\partial z}{\partial \zeta} \end{bmatrix}^{-T} \begin{bmatrix} \frac{\partial u}{\partial \xi} \\ \frac{\partial u}{\partial \eta} \\ \frac{\partial u}{\partial \zeta} \end{bmatrix}. \tag{23.13}$$

Solid-embedded Vector Base. When using the solid-embedded coordinate system, it is also convenient to introduce the infinitesimal solid-embedded vector base $[\boldsymbol{\xi} \;\; \boldsymbol{\eta} \;\; \boldsymbol{\zeta}]$, Figure 23.4.

For a given material point P the solid embedded local vectors $\boldsymbol{\xi}$, $\boldsymbol{\eta}$ and $\boldsymbol{\zeta}$ are by definition given by

$$\begin{aligned} \boldsymbol{\xi} &= \xi_i \mathbf{i} + \xi_j \mathbf{j} + \xi_k \mathbf{k} \\ &= \left(\frac{\partial x}{\partial \xi}\mathbf{i} + \frac{\partial y}{\partial \xi}\mathbf{j} + \frac{\partial z}{\partial \xi}\mathbf{k} \right) (\mathrm{im}), \end{aligned} \tag{23.14}$$

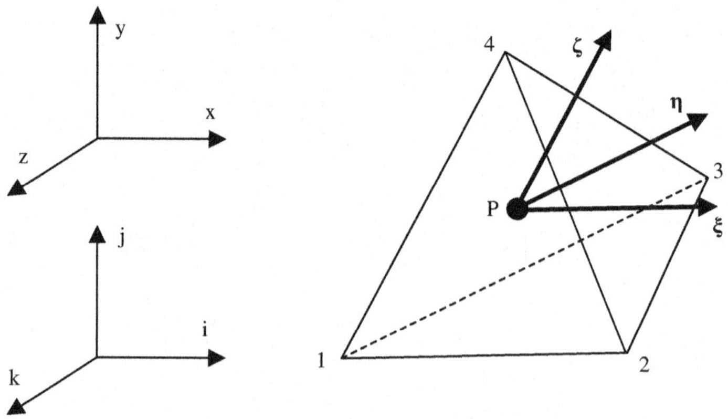

Figure 23.4 The infinitesimal solid-embedded vector base fixed to the material points in the infinitesimal vicinity of a given material point P

where im stands for infi-meter, i.e., an infinitesimally small length unit, thus making the above vectors infinitesimally small. The above equation can be written in matrix form

$$\boldsymbol{\xi} = \begin{bmatrix} \mathbf{i} & \mathbf{j} & \mathbf{k} \end{bmatrix} \begin{bmatrix} \frac{\partial x}{\partial \xi} \\ \frac{\partial y}{\partial \xi} \\ \frac{\partial z}{\partial \xi} \end{bmatrix}, \tag{23.15}$$

or simply (by implicit assumption of the global base)

$$\boldsymbol{\xi} = \begin{bmatrix} \frac{\partial x}{\partial \xi} \\ \frac{\partial y}{\partial \xi} \\ \frac{\partial z}{\partial \xi} \end{bmatrix}, \tag{23.16}$$

where the equal sign means that: the vector $\boldsymbol{\xi}$ is represented by the matrix

$$\begin{bmatrix} \frac{\partial x}{\partial \xi} \\ \frac{\partial y}{\partial \xi} \\ \frac{\partial z}{\partial \xi} \end{bmatrix}, \tag{23.17}$$

obtained using the global base $\begin{bmatrix} \mathbf{i} & \mathbf{j} & \mathbf{k} \end{bmatrix}$, i.e., in Equation (23.16) the base is implied by the context.

In a similar manner

$$\boldsymbol{\eta} = \begin{bmatrix} \frac{\partial x}{\partial \eta} \\ \frac{\partial y}{\partial \eta} \\ \frac{\partial z}{\partial \eta} \end{bmatrix} \quad \text{and} \quad \boldsymbol{\zeta} = \begin{bmatrix} \frac{\partial x}{\partial \zeta} \\ \frac{\partial y}{\partial \zeta} \\ \frac{\partial z}{\partial \zeta} \end{bmatrix}. \tag{23.18}$$

By substituting the initial global coordinates $\bar{x}$, $\bar{y}$ and $\bar{z}$ from Equation (23.4), and by performing the required differentiation using Equation (23.2) one obtains the initial position of the solid-embedded vector base

$$[\bar{\xi}\ \bar{\eta}\ \bar{\zeta}]=\begin{bmatrix}\frac{\partial\bar{x}}{\partial\xi} & \frac{\partial\bar{x}}{\partial\eta} & \frac{\partial\bar{x}}{\partial\zeta}\\ \frac{\partial\bar{y}}{\partial\xi} & \frac{\partial\bar{y}}{\partial\eta} & \frac{\partial\bar{y}}{\partial\zeta}\\ \frac{\partial\bar{z}}{\partial\xi} & \frac{\partial\bar{z}}{\partial\eta} & \frac{\partial\bar{z}}{\partial\zeta}\end{bmatrix} = \begin{bmatrix}\bar{x}_2-\bar{x}_1 & \bar{x}_3-\bar{x}_1 & \bar{x}_4-\bar{x}_1\\ \bar{y}_2-\bar{y}_1 & \bar{y}_3-\bar{y}_1 & \bar{y}_4-\bar{y}_1\\ \bar{z}_2-\bar{z}_1 & \bar{z}_3-\bar{z}_1 & \bar{z}_4-\bar{z}_1\end{bmatrix}(\text{im}). \tag{23.19}$$

In a similar manner, by substituting the previous global coordinates from Equation (23.5), the previous position of the solid-embedded vector base is obtained

$$[\hat{\xi}\ \hat{\eta}\ \hat{\zeta}]=\begin{bmatrix}\frac{\partial\hat{x}}{\partial\xi} & \frac{\partial\hat{x}}{\partial\eta} & \frac{\partial\hat{x}}{\partial\zeta}\\ \frac{\partial\hat{y}}{\partial\xi} & \frac{\partial\hat{y}}{\partial\eta} & \frac{\partial\hat{y}}{\partial\zeta}\\ \frac{\partial\hat{z}}{\partial\xi} & \frac{\partial\hat{z}}{\partial\eta} & \frac{\partial\hat{z}}{\partial\zeta}\end{bmatrix} = \begin{bmatrix}\hat{x}_2-\hat{x}_1 & \hat{x}_3-\hat{x}_1 & \hat{x}_4-\hat{x}_1\\ \hat{y}_2-\hat{y}_1 & \hat{y}_3-\hat{y}_1 & \hat{y}_4-\hat{y}_1\\ \hat{z}_2-\hat{z}_1 & \hat{z}_3-\hat{z}_1 & \hat{z}_4-\hat{z}_1\end{bmatrix}(\text{im}). \tag{23.20}$$

The current position of the solid-embedded vector base is obtained by substituting the current global coordinates from Equation (23.6), which after differentiation yields

$$[\tilde{\xi}\ \tilde{\eta}\ \tilde{\zeta}]=\begin{bmatrix}\frac{\partial\tilde{x}}{\partial\xi} & \frac{\partial\tilde{x}}{\partial\eta} & \frac{\partial\tilde{x}}{\partial\zeta}\\ \frac{\partial\tilde{y}}{\partial\xi} & \frac{\partial\tilde{y}}{\partial\eta} & \frac{\partial\tilde{y}}{\partial\zeta}\\ \frac{\partial\tilde{z}}{\partial\xi} & \frac{\partial\tilde{z}}{\partial\eta} & \frac{\partial\tilde{z}}{\partial\zeta}\end{bmatrix} = \begin{bmatrix}\tilde{x}_2-\tilde{x}_1 & \tilde{x}_3-\tilde{x}_1 & \tilde{x}_4-\tilde{x}_1\\ \tilde{y}_2-\tilde{y}_1 & \tilde{y}_3-\tilde{y}_1 & \tilde{y}_4-\tilde{y}_1\\ \tilde{z}_2-\tilde{z}_1 & \tilde{z}_3-\tilde{z}_1 & \tilde{z}_4-\tilde{z}_1\end{bmatrix}(\text{im}). \tag{23.21}$$

The physical meaning of these three bases can be explained by "plotting" solid-embedded coordinate lines ξ, η, ζ through a tetrahedron made of rubber. At any point P, these

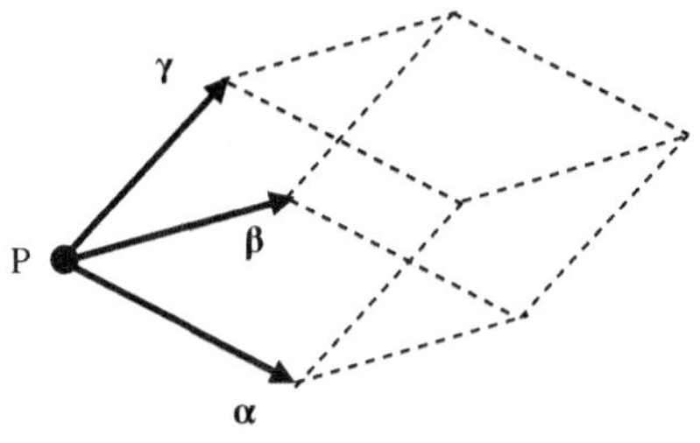

Figure 23.5 Generalized infinitesimal material element and material-embedded vector base

three lines define an infinitesimal local solid element, the edges of which are defined by the vectors

$$[\boldsymbol{\xi} \ \ \boldsymbol{\eta} \ \ \boldsymbol{\zeta}]. \tag{23.22}$$

As the rubber is deformed these three vectors deform as well (for they are fixed to the material points in the infinitesimal vicinity of point P).

Generalized Infinitesimal Material Element. When dealing with anisotropic materials, the material itself is defined by the generalized infinitesimal material element, Figure 23.5, defined by the infinitesimal vector base

$$[\boldsymbol{\alpha} \ \ \boldsymbol{\beta} \ \ \boldsymbol{\gamma}]. \tag{23.23}$$

This vector base is fixed to the solid's material points in the infinitesimal vicinity of a given material point P. For this reason, it is called the material-embedded vector base.

The material-embedded vectors $\boldsymbol{\alpha}$, $\boldsymbol{\beta}$ and $\boldsymbol{\gamma}$ are material characteristics of the solid; they follow the specific directions that define the anisotropy of the material, such as geologic layering, direction of cold rolling, direction of fiber reinforcement, etc.

The material-embedded vector base versus the solid-embedded vector base. As the solid deforms, the solid-embedded vectors $[\boldsymbol{\xi} \ \ \boldsymbol{\eta} \ \ \boldsymbol{\zeta}]$ are obtained from the nodal coordinates of the tetrahedron, equations (23.19), (23.20) and (23.21).

Any change of the shape of the generalized infinitesimal material element is obtained by expressing material-embedded vectors

$$[\boldsymbol{\alpha} \ \ \boldsymbol{\beta} \ \ \boldsymbol{\gamma}] \tag{23.24}$$

in terms of vectors

$$[\boldsymbol{\xi} \ \ \boldsymbol{\eta} \ \ \boldsymbol{\zeta}], \tag{23.25}$$

which for the base vector $\boldsymbol{\alpha}$ yields

$$\begin{aligned} \boldsymbol{\alpha} &= \alpha_\xi \boldsymbol{\xi} + \alpha_\eta \boldsymbol{\eta} + \alpha_\zeta \boldsymbol{\zeta} \\ &= [\boldsymbol{\xi} \ \ \boldsymbol{\eta} \ \ \boldsymbol{\zeta}] \begin{bmatrix} \alpha_\xi \\ \alpha_\eta \\ \alpha_\zeta \end{bmatrix}, \end{aligned} \tag{23.26}$$

or simply

$$\boldsymbol{\alpha} = \begin{bmatrix} \xi_i & \eta_i & \zeta_i \\ \xi_j & \eta_j & \zeta_j \\ \xi_k & \eta_k & \zeta_k \end{bmatrix} \begin{bmatrix} \alpha_\xi \\ \alpha_\eta \\ \alpha_\zeta \end{bmatrix}. \tag{23.27}$$

In a similar manner, for the base vector $\boldsymbol{\beta}$, one obtains

$$\boldsymbol{\beta} = \begin{bmatrix} \xi_i & \eta_i & \zeta_i \\ \xi_j & \eta_j & \zeta_j \\ \xi_k & \eta_k & \zeta_k \end{bmatrix} \begin{bmatrix} \beta_\xi \\ \beta_\eta \\ \beta_\zeta \end{bmatrix} \tag{23.28}$$

and for the base vector $\boldsymbol{\gamma}$

$$\boldsymbol{\gamma} = \begin{bmatrix} \xi_i & \eta_i & \zeta_i \\ \xi_j & \eta_j & \zeta_j \\ \xi_k & \eta_k & \zeta_k \end{bmatrix} \begin{bmatrix} \gamma_\xi \\ \gamma_\eta \\ \gamma_\zeta \end{bmatrix}. \tag{23.29}$$

By combining Equations (23.27), (23.28) and (23.29) one obtains the material-embedded vector base in terms of the solid-embedded vector base

$$[\boldsymbol{\alpha}\ \boldsymbol{\beta}\ \boldsymbol{\gamma}] = \begin{bmatrix} \alpha_i & \beta_i & \gamma_i \\ \alpha_j & \beta_j & \gamma_j \\ \alpha_k & \beta_k & \gamma_k \end{bmatrix} = \begin{bmatrix} \xi_i & \eta_i & \zeta_i \\ \xi_j & \eta_j & \zeta_j \\ \xi_k & \eta_k & \zeta_k \end{bmatrix} \begin{bmatrix} \alpha_\xi & \beta_\xi & \gamma_\xi \\ \alpha_\eta & \beta_\eta & \gamma_\eta \\ \alpha_\zeta & \beta_\zeta & \gamma_\zeta \end{bmatrix}. \tag{23.30}$$

As both

$$[\boldsymbol{\alpha}\ \boldsymbol{\beta}\ \boldsymbol{\gamma}] \tag{23.31}$$

and

$$[\boldsymbol{\xi}\ \boldsymbol{\eta}\ \boldsymbol{\zeta}] \tag{23.32}$$

are fixed to the material points in the infinitesimal vicinity of a given material point P, it follows that the initial, the previous and the current positions of vectors $\boldsymbol{\alpha}$, $\boldsymbol{\beta}$ and $\boldsymbol{\gamma}$ are all given by Equation (23.30), where the initial position of the material-embedded vector base is

$$[\bar{\boldsymbol{\alpha}}\ \bar{\boldsymbol{\beta}}\ \bar{\boldsymbol{\gamma}}] = \begin{bmatrix} \bar{\alpha}_i & \bar{\beta}_i & \bar{\gamma}_i \\ \bar{\alpha}_j & \bar{\beta}_j & \bar{\gamma}_j \\ \bar{\alpha}_k & \bar{\beta}_k & \bar{\gamma}_k \end{bmatrix} = \begin{bmatrix} \bar{\xi}_i & \bar{\eta}_i & \bar{\zeta}_i \\ \bar{\xi}_j & \bar{\eta}_j & \bar{\zeta}_j \\ \bar{\xi}_k & \bar{\eta}_k & \bar{\zeta}_k \end{bmatrix} \begin{bmatrix} \alpha_\xi & \beta_\xi & \gamma_\xi \\ \alpha_\eta & \beta_\eta & \gamma_\eta \\ \alpha_\zeta & \beta_\zeta & \gamma_\zeta \end{bmatrix}, \tag{23.33}$$

while the previous position of the material-embedded vector base is

$$\begin{bmatrix}\hat{\boldsymbol{\alpha}} & \hat{\boldsymbol{\beta}} & \hat{\boldsymbol{\gamma}}\end{bmatrix} = \begin{bmatrix} \hat{\alpha}_i & \hat{\beta}_i & \hat{\gamma}_i \\ \hat{\alpha}_j & \hat{\beta}_j & \hat{\gamma}_j \\ \hat{\alpha}_k & \hat{\beta}_k & \hat{\gamma}_k \end{bmatrix} = \begin{bmatrix} \hat{\xi}_i & \hat{\eta}_i & \hat{\zeta}_i \\ \hat{\xi}_j & \hat{\eta}_j & \hat{\zeta}_j \\ \hat{\xi}_k & \hat{\eta}_k & \hat{\zeta}_k \end{bmatrix} \begin{bmatrix} \alpha_\xi & \beta_\xi & \gamma_\xi \\ \alpha_\eta & \beta_\eta & \gamma_\eta \\ \alpha_\zeta & \beta_\zeta & \gamma_\zeta \end{bmatrix} \tag{23.34}$$

and the current position of the material-embedded vector base is given by

$$\begin{bmatrix}\tilde{\boldsymbol{\alpha}} & \tilde{\boldsymbol{\beta}} & \tilde{\boldsymbol{\gamma}}\end{bmatrix} = \begin{bmatrix} \tilde{\alpha}_i & \tilde{\beta}_i & \tilde{\gamma}_i \\ \tilde{\alpha}_j & \tilde{\beta}_j & \tilde{\gamma}_j \\ \tilde{\alpha}_k & \tilde{\beta}_k & \tilde{\gamma}_k \end{bmatrix} = \begin{bmatrix} \tilde{\xi}_i & \tilde{\eta}_i & \tilde{\zeta}_i \\ \tilde{\xi}_j & \tilde{\eta}_j & \tilde{\zeta}_j \\ \tilde{\xi}_k & \tilde{\eta}_k & \tilde{\zeta}_k \end{bmatrix} \begin{bmatrix} \alpha_\xi & \beta_\xi & \gamma_\xi \\ \alpha_\eta & \beta_\eta & \gamma_\eta \\ \alpha_\zeta & \beta_\zeta & \gamma_\zeta \end{bmatrix}. \tag{23.35}$$

As the matrix

$$\begin{bmatrix} \bar{\alpha}_i & \bar{\beta}_i & \bar{\gamma}_i \\ \bar{\alpha}_j & \bar{\beta}_j & \bar{\gamma}_j \\ \bar{\alpha}_k & \bar{\beta}_k & \bar{\gamma}_k \end{bmatrix} \tag{23.36}$$

defines the material properties of the solid itself, it must be supplied as an input to the finite element analysis. This is done by supplying a single matrix from Equation (23.36) for each finite element, thus allowing the material properties to change over the domain. At the start of the finite element analysis, Equation (23.33) is used to calculate the relationship between bases $[\boldsymbol{\xi} \;\; \boldsymbol{\eta} \;\; \boldsymbol{\zeta}]$ and $[\boldsymbol{\alpha} \;\; \boldsymbol{\beta} \;\; \boldsymbol{\gamma}]$, thus yielding

$$\begin{bmatrix} \alpha_\xi & \beta_\xi & \gamma_\xi \\ \alpha_\eta & \beta_\eta & \gamma_\eta \\ \alpha_\zeta & \beta_\zeta & \gamma_\zeta \end{bmatrix} = \begin{bmatrix} \bar{\xi}_i & \bar{\eta}_i & \bar{\zeta}_i \\ \bar{\xi}_j & \bar{\eta}_j & \bar{\zeta}_j \\ \bar{\xi}_k & \bar{\eta}_k & \bar{\zeta}_k \end{bmatrix}^{-1} \begin{bmatrix} \bar{\alpha}_i & \bar{\beta}_i & \bar{\gamma}_i \\ \bar{\alpha}_j & \bar{\beta}_j & \bar{\gamma}_j \\ \bar{\alpha}_k & \bar{\beta}_k & \bar{\gamma}_k \end{bmatrix}. \tag{23.37}$$

Constitutive Law. By passing bases $\begin{bmatrix}\bar{\boldsymbol{\alpha}} & \bar{\boldsymbol{\beta}} & \bar{\boldsymbol{\gamma}}\end{bmatrix}$, $\begin{bmatrix}\hat{\boldsymbol{\alpha}} & \hat{\boldsymbol{\beta}} & \hat{\boldsymbol{\gamma}}\end{bmatrix}$ and $\begin{bmatrix}\tilde{\boldsymbol{\alpha}} & \tilde{\boldsymbol{\beta}} & \tilde{\boldsymbol{\gamma}}\end{bmatrix}$ to the material package, the Munjiza stress tensor matrix is calculated and stored within the material package. Within the material package, the Cauchy stress matrix is calculated on demand as described in Chapter 18. The Cauchy matrix is passed back to the finite element package for calculation of the nodal forces.

23.3 Summary

In this chapter, the kinematics of deformation of solid contained within a four-noded tetrahedron finite element has been explained. The current, the previous and the initial positions of the solid-embedded vector base $[\boldsymbol{\xi} \;\; \boldsymbol{\eta} \;\; \boldsymbol{\zeta}]$ are obtained from the current, the previous and the initial nodal coordinates respectively. From these three bases, the previous and the current positions of the material-embedded vector base $[\boldsymbol{\alpha} \;\; \boldsymbol{\beta} \;\; \boldsymbol{\gamma}]$ are obtained. The initial position of the material-embedded vector base $\begin{bmatrix}\bar{\boldsymbol{\alpha}} & \bar{\boldsymbol{\beta}} & \bar{\boldsymbol{\gamma}}\end{bmatrix}$ is given in the input to the finite element analysis. These nine vectors uniquely define the initial, the previous and the current position of the generalized material element. As such, they uniquely define the deformation kinematics of the generalized material element and are used to calculate the internal forces (stresses).

This is done by passing the vectors to the material package. The material package calculates and stores the stress using a deformation invariant matrix or in our case, the Munjiza stress tensor matrix. However, the material package returns only the Cauchy stress tensor matrix to the finite element package, which is then used to calculate the nodal forces.

Further Reading

[1] Argyris, J. H., Fried, I. and Scharpf, D. W. (1968) Tet 20 and Tea 8 elements for matrix displacement method. *Aeronaut J.*, **72**(691): 618–23.

[2] Bonet, J. and Burton, A. J. (1998) A simple average nodal pressure tetrahedral element for incompressible and nearly incompressible dynamic explicit applications. *Communications in Numerical Methods in Engineering*, **14**(5): 437–49.

[3] Bonet, J., Marriott, H. and Hassan, O. (2001) An averaged nodal deformation gradient linear tetrahedral element for large strain explicit dynamic applications. *Communications in Numerical Methods in Engineering*, **17**(8): 551–61.

[4] Bonet, J. and Wood, R. D. (1997) *Nonlinear Continuum Mechanics for Finite Element Analysis*. Cambridge, Cambridge University Press.

[5] Hughes, T.J.R. (2000) *The Finite Element Method: Linear Static and Dynamic Finite Element Analysis*. New York, Dover Publications.

[6] Liu, W. K., Guo, Y., Tang, S. and Belytschko. T. (1998) A multiple-quadrature eight-node hexahedral finite element for large deformation elastoplastic analysis. *Comput. Method Appl. M.*, **154**(1–2): 69–132.

[7] Oden, J. T. (1972) *Finite Elements of Nonlinear Continua*. New York, McGraw-Hill.

[8] Peric, D. and Owen, D. R. J. (1998) Finite-element applications to the nonlinear mechanics of solids. *Reports on Progress in Physics*, **61**(11): 1495–1574.

[9] Zienkiewicz, O. C. (1971) *The Finite Element Method in Engineering Science*. New York, McGraw-Hill.

[10] Zienkiewicz, O. C., Rojek, J., Taylor, R. L. and Pastor, M. (1998) Triangles and tetrahera in explicit dynamic codes for solids. *International Journal for Numerical Methods in Engineering*, **43**(3), 565–83.

[11] Zienkiewicz, O. C. and Taylor, R. L. (2005) *The Finite Element Method Set*. Oxford, Elsevier Science.

24

3D Deformation Kinematics Using Iso-Parametric Finite Elements

24.1 The Finite Element Library

In Chapter 23 the deformation kinematics for a four-noded tetrahedron finite element was introduced. There are a number of other finite elements, all of which are derived using similar procedures. Some of the other possible elements are shown in Figure 24.1 and Figure 24.2.

24.2 The Shape Functions

For any 3D finite element geometry, it is convenient to define the shape functions. For a node i of any given finite element, a single shape function

$$\mathrm{N}_i = \mathrm{N}_i(\xi, \eta) \tag{24.1}$$

is defined in such a way that

$$\begin{gathered} \mathrm{N}_i\left(\xi_j, \eta_j\right) = 0 \quad \text{for } j \neq i \\ \text{and} \\ \mathrm{N}_i\left(\xi_j, \eta_j\right) = 1 \quad \text{for } j = i. \end{gathered} \tag{24.2}$$

The shape functions for the finite elements shown in Figure 24.1 and Figure 24.2 are as follows:

Large Strain Finite Element Method: A Practical Course, First Edition. Antonio Munjiza, Esteban Rougier and Earl E. Knight.

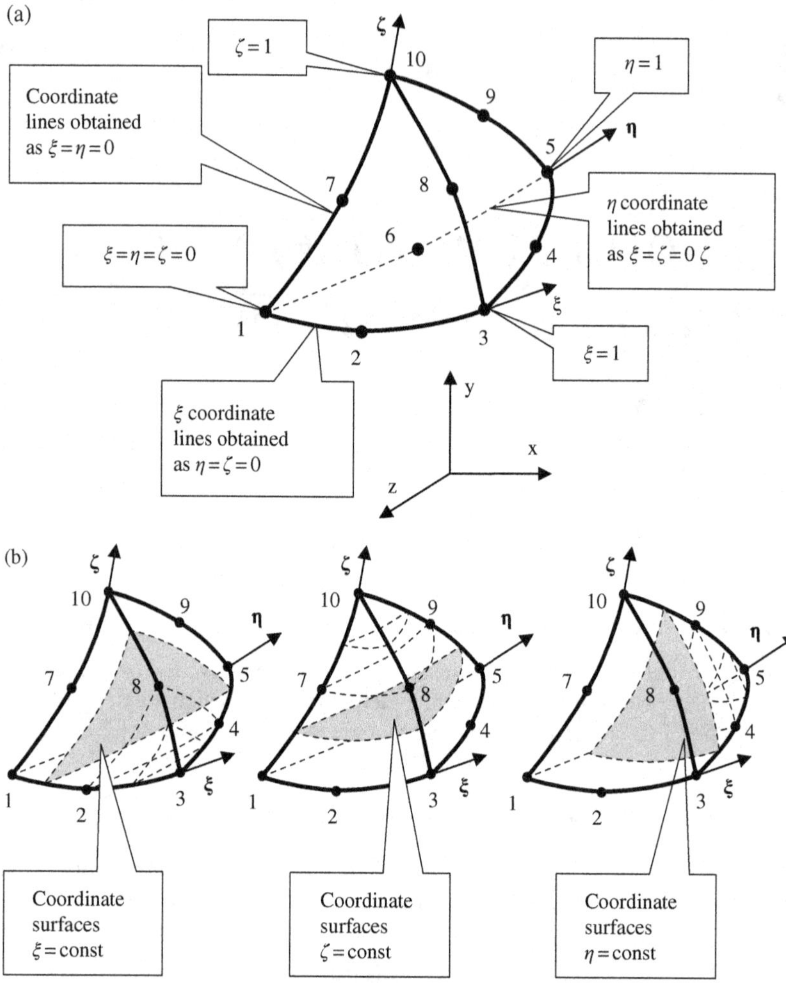

Figure 24.1 The ten-noded tetrahedron finite element: (a) the solid-embedded coordinate lines; (b) the solid-embedded coordinate surfaces shown using dashed lines

a. Ten-Noded Solid:

$$\begin{aligned}
&N_1 = 2(1-\xi-\eta-\zeta)\left(\frac{1}{2}-\xi-\eta-\zeta\right) && N_6 = 4\eta(1-\xi-\eta-\zeta)\\
&N_2 = 4\xi(1-\xi-\eta-\zeta) && N_7 = 4\zeta(1-\xi-\eta-\zeta)\\
&N_3 = 2\xi\left(\xi-\frac{1}{2}\right) && N_8 = 4\xi\zeta\\
&N_4 = 4\xi\eta && N_9 = 4\eta\zeta\\
&N_5 = 2\eta\left(\eta-\frac{1}{2}\right) && N_{10} = 2\zeta\left(\zeta-\frac{1}{2}\right)
\end{aligned} \tag{24.3}$$

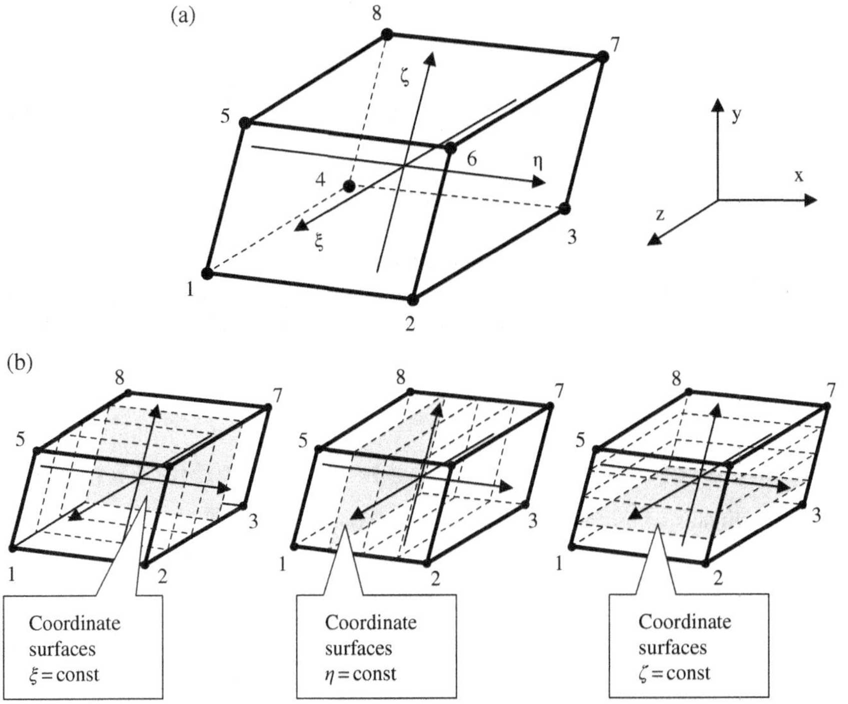

Figure 24.2 The eight-noded solid finite element. (a) The solid-embedded coordinate lines; (b) the solid-embedded coordinate surfaces shown using dashed lines

b. Eight-Noded Solid:

$$
\begin{aligned}
&N_1=\frac{1}{8}(1-\xi)(1-\eta)(1-\zeta) \quad N_5=\frac{1}{8}(1-\xi)(1-\eta)(1+\zeta)\\
&N_2=\frac{1}{8}(1+\xi)(1-\eta)(1-\zeta) \quad N_6=\frac{1}{8}(1+\xi)(1-\eta)(1+\zeta)\\
&N_3=\frac{1}{8}(1+\xi)(1+\eta)(1-\zeta) \quad N_7=\frac{1}{8}(1+\xi)(1+\eta)(1+\zeta)\\
&N_4=\frac{1}{8}(1-\xi)(1+\eta)(1-\zeta) \quad N_8=\frac{1}{8}(1-\xi)(1+\eta)(1+\zeta).
\end{aligned}
\tag{24.4}
$$

These shape functions have been defined in an "ad-hoc" way. Thus, these elements form what is called the serendipity family of finite elements. There are other types of shape functions such as Lagrange shape functions, which are used with the Lagrange family of finite elements.

24.3 Nodal Positions

The finite element's node positions are conveniently stored in the finite element package as follows

$$\begin{array}{lll} \bar{P}_1 = (\bar{x}_1, \bar{y}_1, \bar{z}_1) & \hat{P}_1 = (\hat{x}_1, \hat{y}_1, \hat{z}_1) & \tilde{P}_1 = (\tilde{x}_1, \tilde{y}_1, \tilde{z}_1) \\ \bar{P}_2 = (\bar{x}_2, \bar{y}_2, \bar{z}_2) & \hat{P}_2 = (\hat{x}_2, \hat{y}_2, \hat{z}_2) & \tilde{P}_2 = (\tilde{x}_2, \tilde{y}_2, \tilde{z}_2) \\ \vdots & \vdots & \vdots \\ \bar{P}_n = (\bar{x}_n, \bar{y}_n, \bar{z}_n) & \hat{P}_n = (\hat{x}_n, \hat{y}_n, \hat{z}_n) & \tilde{P}_n = (\tilde{x}_n, \tilde{y}_n, \tilde{z}_n), \end{array} \tag{24.5}$$

where x, y and z define the global coordinates using the Cartesian coordinate system, and "–" stands for initial position, "ˆ" stands for previous position and "~" stands for current position.

The initial nodal positions define the initial geometry of the finite element mesh and are supplied as input to the finite element analysis.

24.4 Positions of Material Points inside a Single Finite Element

The local coordinates of any material point P are given by

$$\bar{P} = (\bar{\xi}, \bar{\eta}, \bar{\zeta}) \; ; \hat{P} = (\hat{\xi}, \hat{\eta}, \hat{\zeta}) \; ; \tilde{P} = (\tilde{\xi}, \tilde{\eta}, \tilde{\zeta}), \tag{24.6}$$

where

$$(\bar{\xi}, \bar{\eta}, \bar{\zeta}) = (\hat{\xi}, \hat{\eta}, \hat{\zeta}) = (\tilde{\xi}, \tilde{\eta}, \tilde{\zeta}) = (\xi, \eta, \zeta) \tag{24.7}$$

and $\bar{P}$, $\hat{P}$ and $\tilde{P}$ represent the initial, the previous and the current positions of point P respectively.

The global (x, y, z) coordinates of the same material point keep changing as the solid deforms and are given by

$$\bar{P} = (\bar{x}, \bar{y}, \bar{z}) \; ; \hat{P} = (\hat{x}, \hat{y}, \hat{z}) \; ; \tilde{P} = (\tilde{x}, \tilde{y}, \tilde{z}), \tag{24.8}$$

where $(\bar{x}, \bar{y}, \bar{z})$, $(\hat{x}, \hat{y}, \hat{z})$ and $(\tilde{x}, \tilde{y}, \tilde{z})$ represent the initial, the previous and the current global coordinates respectively. In general

$$(\bar{x}, \bar{y}, \bar{z}) \neq (\hat{x}, \hat{y}, \hat{z}) \neq (\tilde{x}, \tilde{y}, \tilde{z}). \tag{24.9}$$

These coordinates are approximated from the local (ξ, η, ζ) coordinates using the shape functions:

$$\begin{aligned} \bar{x} &= \bar{\mathrm{x}}(\xi,\eta,\zeta) \\ &= \bar{x}_1 N_1(\xi,\eta,\zeta) + \bar{x}_2 N_2(\xi,\eta,\zeta) + \bar{x}_3 N_3(\xi,\eta,\zeta) + \cdots + \bar{x}_n N_n(\xi,\eta,\zeta) \\ \bar{y} &= \bar{\mathrm{y}}(\xi,\eta,\zeta) \\ &= \bar{y}_1 N_1(\xi,\eta,\zeta) + \bar{y}_2 N_2(\xi,\eta,\zeta) + \bar{y}_3 N_3(\xi,\eta,\zeta) + \cdots + \bar{y}_n N_n(\xi,\eta,\zeta) \\ \bar{z} &= \bar{\mathrm{z}}(\xi,\eta,\zeta) \\ &= \bar{z}_1 N_1(\xi,\eta,\zeta) + \bar{z}_2 N_2(\xi,\eta,\zeta) + \bar{z}_3 N_3(\xi,\eta,\zeta) + \cdots + \bar{z}_n N_n(\xi,\eta,\zeta) \end{aligned} \tag{24.10}$$

$$\begin{aligned}\hat{x} &= \hat{\mathrm{x}}(\xi,\eta,\zeta)\\ &= \hat{x}_1\mathrm{N}_1(\xi,\eta,\zeta)+\hat{x}_2\mathrm{N}_2(\xi,\eta,\zeta)+\hat{x}_3\mathrm{N}_3(\xi,\eta,\zeta)+\cdots+\hat{x}_n\mathrm{N}_n(\xi,\eta,\zeta)\\ \hat{y} &= \hat{\mathrm{y}}(\xi,\eta,\zeta)\\ &= \hat{y}_1\mathrm{N}_1(\xi,\eta,\zeta)+\hat{y}_2\mathrm{N}_2(\xi,\eta,\zeta)+\hat{y}_3\mathrm{N}_3(\xi,\eta,\zeta)+\cdots+\hat{y}_n\mathrm{N}_n(\xi,\eta,\zeta)\\ \hat{z} &= \hat{\mathrm{z}}(\xi,\eta,\zeta)\\ &= \hat{z}_1\mathrm{N}_1(\xi,\eta,\zeta)+\hat{z}_2\mathrm{N}_2(\xi,\eta,\zeta)+\hat{z}_3\mathrm{N}_3(\xi,\eta,\zeta)+\cdots+\hat{z}_n\mathrm{N}_n(\xi,\eta,\zeta)\end{aligned} \tag{24.11}$$

$$\begin{aligned}\tilde{x} &= \tilde{\mathrm{x}}(\xi,\eta,\zeta)\\ &= \tilde{x}_1\mathrm{N}_1(\xi,\eta,\zeta)+\tilde{x}_2\mathrm{N}_2(\xi,\eta,\zeta)+\tilde{x}_3\mathrm{N}_3(\xi,\eta,\zeta)+\cdots+\tilde{x}_n\mathrm{N}_n(\xi,\eta,\zeta)\\ \tilde{y} &= \tilde{\mathrm{y}}(\xi,\eta,\zeta)\\ &= \tilde{y}_1\mathrm{N}_1(\xi,\eta,\zeta)+\tilde{y}_2\mathrm{N}_2(\xi,\eta,\zeta)+\tilde{y}_3\mathrm{N}_3(\xi,\eta,\zeta)+\cdots+\tilde{y}_n\mathrm{N}_n(\xi,\eta,\zeta)\\ \tilde{z} &= \tilde{\mathrm{z}}(\xi,\eta,\zeta)\\ &= \tilde{z}_1\mathrm{N}_1(\xi,\eta,\zeta)+\tilde{z}_2\mathrm{N}_2(\xi,\eta,\zeta)+\tilde{z}_3\mathrm{N}_3(\xi,\eta,\zeta)+\cdots+\tilde{z}_n\mathrm{N}_n(\xi,\eta,\zeta),\end{aligned} \tag{24.12}$$

where $(\bar{x}_i,\bar{y}_i,\bar{z}_i)$, $(\hat{x}_i,\hat{y}_i,\hat{z}_i)$ and $(\tilde{x}_i,\tilde{y}_i,\tilde{z}_i)$ are the initial, previous and current nodal coordinates respectively.

24.5 The Solid-Embedded Infinitesimal Vector Base

The solid-embedded coordinate lines ($\xi=$const, $\eta=$const), ($\eta=$const, $\zeta=$const) and ($\xi=$const, $\zeta=$const) are local to each finite element and are in general neither straight lines nor orthogonal to each other. At each material point, the coordinate lines intersect each other at different angles.

For a given point P, it is possible to define vectors $\boldsymbol{\xi}$, $\boldsymbol{\eta}$ and $\boldsymbol{\xi}$ such that the vector $\boldsymbol{\xi}$ is tangential to the corresponding ξ line, and vector $\boldsymbol{\eta}$ is tangential to the corresponding η line, and vector $\boldsymbol{\zeta}$ is tangential to the corresponding ζ line, Figure 24.3.

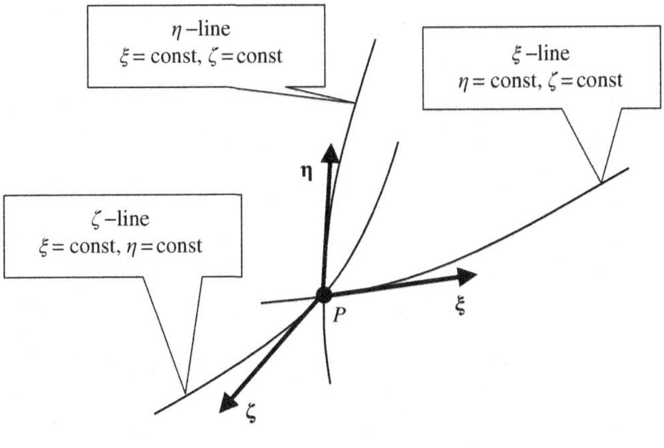

Figure 24.3 The solid-embedded vector base

These three vectors form the solid-embedded vector base

$$[\boldsymbol{\xi} \;\; \boldsymbol{\eta} \;\; \boldsymbol{\zeta}\,](\mathrm{im}). \tag{24.13}$$

The base vectors $\boldsymbol{\xi}$, $\boldsymbol{\eta}$ and $\boldsymbol{\zeta}$ are infinitesimally short, which is indicated by the infinitesimally short unit for length called the infi-meter (im).

By definition, the solid-embedded vector base is given by

$$[\boldsymbol{\xi} \;\; \boldsymbol{\eta} \;\; \boldsymbol{\zeta}] = [\mathbf{i} \;\; \mathbf{j} \;\; \mathbf{k}] \begin{bmatrix} \frac{\partial \mathrm{x}}{\partial \xi} & \frac{\partial \mathrm{x}}{\partial \eta} & \frac{\partial \mathrm{x}}{\partial \zeta} \\ \frac{\partial \mathrm{y}}{\partial \xi} & \frac{\partial \mathrm{y}}{\partial \eta} & \frac{\partial \mathrm{y}}{\partial \zeta} \\ \frac{\partial \mathrm{z}}{\partial \xi} & \frac{\partial \mathrm{z}}{\partial \eta} & \frac{\partial \mathrm{z}}{\partial \zeta} \end{bmatrix}. \tag{24.14}$$

This can be written as

$$[\boldsymbol{\xi} \;\; \boldsymbol{\eta} \;\; \boldsymbol{\zeta}] = \begin{bmatrix} \frac{\partial \mathrm{x}}{\partial \xi} & \frac{\partial \mathrm{x}}{\partial \eta} & \frac{\partial \mathrm{x}}{\partial \zeta} \\ \frac{\partial \mathrm{y}}{\partial \xi} & \frac{\partial \mathrm{y}}{\partial \eta} & \frac{\partial \mathrm{y}}{\partial \zeta} \\ \frac{\partial \mathrm{z}}{\partial \xi} & \frac{\partial \mathrm{z}}{\partial \eta} & \frac{\partial \mathrm{z}}{\partial \zeta} \end{bmatrix}, \tag{24.15}$$

where the global $[\mathbf{i} \;\; \mathbf{j} \;\; \mathbf{k}]$ base is implicitly implied from the context.

In engineering terms, one can think of these vectors as infinitesimally short lines plotted on the solid; which will deform with the solid.

The initial, the current and the previous positions of the solid-embedded vector base are respectively given by

$$[\bar{\boldsymbol{\xi}} \;\; \bar{\boldsymbol{\eta}} \;\; \bar{\boldsymbol{\zeta}}] = \begin{bmatrix} \bar{\xi}_i & \bar{\eta}_i & \bar{\zeta}_i \\ \bar{\xi}_j & \bar{\eta}_i & \bar{\zeta}_j \\ \bar{\xi}_k & \bar{\eta}_k & \bar{\zeta}_k \end{bmatrix} = \begin{bmatrix} \frac{\partial \bar{\mathrm{x}}}{\partial \xi} & \frac{\partial \bar{\mathrm{x}}}{\partial \eta} & \frac{\partial \bar{\mathrm{x}}}{\partial \zeta} \\ \frac{\partial \bar{\mathrm{y}}}{\partial \xi} & \frac{\partial \bar{\mathrm{y}}}{\partial \eta} & \frac{\partial \bar{\mathrm{y}}}{\partial \zeta} \\ \frac{\partial \bar{\mathrm{z}}}{\partial \xi} & \frac{\partial \bar{\mathrm{z}}}{\partial \eta} & \frac{\partial \bar{\mathrm{z}}}{\partial \zeta} \end{bmatrix}, \tag{24.16}$$

$$[\hat{\boldsymbol{\xi}} \;\; \hat{\boldsymbol{\eta}} \;\; \hat{\boldsymbol{\zeta}}] = \begin{bmatrix} \hat{\xi}_i & \hat{\eta}_i & \hat{\zeta}_i \\ \hat{\xi}_j & \hat{\eta}_i & \hat{\zeta}_j \\ \hat{\xi}_k & \hat{\eta}_k & \hat{\zeta}_k \end{bmatrix} = \begin{bmatrix} \frac{\partial \hat{\mathrm{x}}}{\partial \xi} & \frac{\partial \hat{\mathrm{x}}}{\partial \eta} & \frac{\partial \hat{\mathrm{x}}}{\partial \zeta} \\ \frac{\partial \hat{\mathrm{y}}}{\partial \xi} & \frac{\partial \hat{\mathrm{y}}}{\partial \eta} & \frac{\partial \hat{\mathrm{y}}}{\partial \zeta} \\ \frac{\partial \hat{\mathrm{z}}}{\partial \xi} & \frac{\partial \hat{\mathrm{z}}}{\partial \eta} & \frac{\partial \hat{\mathrm{z}}}{\partial \zeta} \end{bmatrix}, \tag{24.17}$$

$$
[\tilde{\boldsymbol{\xi}} \;\; \tilde{\boldsymbol{\eta}} \;\; \tilde{\boldsymbol{\zeta}}] = \begin{bmatrix} \tilde{\xi}_i & \tilde{\eta}_i & \tilde{\zeta}_i \\ \tilde{\xi}_j & \tilde{\eta}_i & \tilde{\zeta}_j \\ \tilde{\xi}_k & \tilde{\eta}_k & \tilde{\zeta}_k \end{bmatrix} = \begin{bmatrix} \dfrac{\partial \tilde{x}}{\partial \xi} & \dfrac{\partial \tilde{x}}{\partial \eta} & \dfrac{\partial \tilde{x}}{\partial \zeta} \\ \dfrac{\partial \tilde{y}}{\partial \xi} & \dfrac{\partial \tilde{y}}{\partial \eta} & \dfrac{\partial \tilde{y}}{\partial \zeta} \\ \dfrac{\partial \tilde{z}}{\partial \xi} & \dfrac{\partial \tilde{z}}{\partial \eta} & \dfrac{\partial \tilde{z}}{\partial \zeta} \end{bmatrix}. \tag{24.18}
$$

As the global coordinates of the material points are calculated from the local coordinates using the shape functions, it follows that

$$
\begin{aligned}
\frac{\partial \bar{x}}{\partial \xi} &= \bar{x}_1 \frac{\partial N_1}{\partial \xi} + \bar{x}_2 \frac{\partial N_2}{\partial \xi} + \bar{x}_3 \frac{\partial N_3}{\partial \xi} + \cdots + \bar{x}_n \frac{\partial N_n}{\partial \xi} \\
\frac{\partial \bar{x}}{\partial \eta} &= \bar{x}_1 \frac{\partial N_1}{\partial \eta} + \bar{x}_2 \frac{\partial N_2}{\partial \eta} + \bar{x}_3 \frac{\partial N_3}{\partial \eta} + \cdots + \bar{x}_n \frac{\partial N_n}{\partial \eta} \\
\frac{\partial \bar{x}}{\partial \zeta} &= \bar{x}_1 \frac{\partial N_1}{\partial \zeta} + \bar{x}_2 \frac{\partial N_2}{\partial \zeta} + \bar{x}_3 \frac{\partial N_3}{\partial \zeta} + \cdots + \bar{x}_n \frac{\partial N_n}{\partial \zeta} \\
\frac{\partial \bar{y}}{\partial \xi} &= \bar{y}_1 \frac{\partial N_1}{\partial \xi} + \bar{y}_2 \frac{\partial N_2}{\partial \xi} + \bar{y}_3 \frac{\partial N_3}{\partial \xi} + \cdots + \bar{y}_n \frac{\partial N_n}{\partial \xi} \\
\frac{\partial \bar{y}}{\partial \eta} &= \bar{y}_1 \frac{\partial N_1}{\partial \eta} + \bar{y}_2 \frac{\partial N_2}{\partial \eta} + \bar{y}_3 \frac{\partial N_3}{\partial \eta} + \cdots + \bar{y}_n \frac{\partial N_n}{\partial \eta} \\
\frac{\partial \bar{y}}{\partial \zeta} &= \bar{y}_1 \frac{\partial N_1}{\partial \zeta} + \bar{y}_2 \frac{\partial N_2}{\partial \zeta} + \bar{y}_3 \frac{\partial N_3}{\partial \zeta} + \cdots + \bar{y}_n \frac{\partial N_n}{\partial \zeta} \\
\frac{\partial \bar{z}}{\partial \xi} &= \bar{z}_1 \frac{\partial N_1}{\partial \xi} + \bar{z}_2 \frac{\partial N_2}{\partial \xi} + \bar{z}_3 \frac{\partial N_3}{\partial \xi} + \cdots + \bar{z}_n \frac{\partial N_n}{\partial \xi} \\
\frac{\partial \bar{z}}{\partial \eta} &= \bar{z}_1 \frac{\partial N_1}{\partial \eta} + \bar{z}_2 \frac{\partial N_2}{\partial \eta} + \bar{z}_3 \frac{\partial N_3}{\partial \eta} + \cdots + \bar{z}_n \frac{\partial N_n}{\partial \eta} \\
\frac{\partial \bar{z}}{\partial \zeta} &= \bar{z}_1 \frac{\partial N_1}{\partial \zeta} + \bar{z}_2 \frac{\partial N_2}{\partial \zeta} + \bar{z}_3 \frac{\partial N_3}{\partial \zeta} + \cdots + \bar{z}_n \frac{\partial N_n}{\partial \zeta},
\end{aligned} \tag{24.19}
$$

which can be written in a matrix form

$$
\begin{bmatrix} \dfrac{\partial \bar{x}}{\partial \xi} & \dfrac{\partial \bar{x}}{\partial \eta} & \dfrac{\partial \bar{x}}{\partial \zeta} \\ \dfrac{\partial \bar{y}}{\partial \xi} & \dfrac{\partial \bar{y}}{\partial \eta} & \dfrac{\partial \bar{y}}{\partial \zeta} \\ \dfrac{\partial \bar{z}}{\partial \xi} & \dfrac{\partial \bar{z}}{\partial \eta} & \dfrac{\partial \bar{z}}{\partial \zeta} \end{bmatrix} = \begin{bmatrix} \bar{x}_1 & \bar{x}_2 & \bar{x}_3 & \ldots & \bar{x}_n \\ \bar{y}_1 & \bar{y}_2 & \bar{y}_3 & \ldots & \bar{y}_n \\ \bar{z}_1 & \bar{z}_2 & \bar{z}_3 & \ldots & \bar{z}_n \end{bmatrix} \begin{bmatrix} \dfrac{\partial N_1}{\partial \xi} & \dfrac{\partial N_1}{\partial \eta} & \dfrac{\partial N_1}{\partial \zeta} \\ \dfrac{\partial N_2}{\partial \xi} & \dfrac{\partial N_2}{\partial \eta} & \dfrac{\partial N_2}{\partial \zeta} \\ \dfrac{\partial N_3}{\partial \xi} & \dfrac{\partial N_3}{\partial \eta} & \dfrac{\partial N_3}{\partial \zeta} \\ \vdots & \vdots & \vdots \\ \dfrac{\partial N_n}{\partial \xi} & \dfrac{\partial N_n}{\partial \eta} & \dfrac{\partial N_n}{\partial \zeta} \end{bmatrix}. \tag{24.20}
$$

In a similar manner,

$$\begin{bmatrix} \frac{\partial \hat{x}}{\partial \xi} & \frac{\partial \hat{x}}{\partial \eta} & \frac{\partial \hat{x}}{\partial \zeta} \\ \frac{\partial \hat{y}}{\partial \xi} & \frac{\partial \hat{y}}{\partial \eta} & \frac{\partial \hat{y}}{\partial \zeta} \\ \frac{\partial \hat{z}}{\partial \xi} & \frac{\partial \hat{z}}{\partial \eta} & \frac{\partial \hat{z}}{\partial \zeta} \end{bmatrix} = \begin{bmatrix} \hat{x}_1 & \hat{x}_2 & \hat{x}_3 & \dots & \hat{x}_n \\ \hat{y}_1 & \hat{y}_2 & \hat{y}_3 & \dots & \hat{y}_n \\ \hat{z}_1 & \hat{z}_2 & \hat{z}_3 & \dots & \hat{z}_n \end{bmatrix} \begin{bmatrix} \frac{\partial N_1}{\partial \xi} & \frac{\partial N_1}{\partial \eta} & \frac{\partial N_1}{\partial \zeta} \\ \frac{\partial N_2}{\partial \xi} & \frac{\partial N_2}{\partial \eta} & \frac{\partial N_2}{\partial \zeta} \\ \frac{\partial N_3}{\partial \xi} & \frac{\partial N_3}{\partial \eta} & \frac{\partial N_3}{\partial \zeta} \\ \vdots & \vdots & \vdots \\ \frac{\partial N_n}{\partial \xi} & \frac{\partial N_n}{\partial \eta} & \frac{\partial N_n}{\partial \zeta} \end{bmatrix} \tag{24.21}$$

$$\begin{bmatrix} \frac{\partial \tilde{x}}{\partial \xi} & \frac{\partial \tilde{x}}{\partial \eta} & \frac{\partial \tilde{x}}{\partial \zeta} \\ \frac{\partial \tilde{y}}{\partial \xi} & \frac{\partial \tilde{y}}{\partial \eta} & \frac{\partial \tilde{y}}{\partial \zeta} \\ \frac{\partial \tilde{z}}{\partial \xi} & \frac{\partial \tilde{z}}{\partial \eta} & \frac{\partial \tilde{z}}{\partial \zeta} \end{bmatrix} = \begin{bmatrix} \tilde{x}_1 & \tilde{x}_2 & \tilde{x}_3 & \dots & \tilde{x}_n \\ \tilde{y}_1 & \tilde{y}_2 & \tilde{y}_3 & \dots & \tilde{y}_n \\ \tilde{z}_1 & \tilde{z}_2 & \tilde{z}_3 & \dots & \tilde{z}_n \end{bmatrix} \begin{bmatrix} \frac{\partial N_1}{\partial \xi} & \frac{\partial N_1}{\partial \eta} & \frac{\partial N_1}{\partial \zeta} \\ \frac{\partial N_2}{\partial \xi} & \frac{\partial N_2}{\partial \eta} & \frac{\partial N_2}{\partial \zeta} \\ \frac{\partial N_3}{\partial \xi} & \frac{\partial N_3}{\partial \eta} & \frac{\partial N_3}{\partial \zeta} \\ \vdots & \vdots & \vdots \\ \frac{\partial N_n}{\partial \xi} & \frac{\partial N_n}{\partial \eta} & \frac{\partial N_n}{\partial \zeta} \end{bmatrix}. \tag{24.22}$$

24.6 The Material-Embedded Infinitesimal Vector Base

Depending on the problem being solved, a finite element's material is defined by the infinitesimal material element of parallelepiped shape conveniently called the generalized material element. This generalized material element is uniquely defined by three infinitesimally short vectors $\bar{\boldsymbol{\alpha}}$, $\bar{\boldsymbol{\beta}}$, $\bar{\boldsymbol{\gamma}}$. These vectors are fixed to the material points of the solid body.

The initial position of these vectors is provided as an input to the finite element analysis (in terms of global orthonormal base $[\mathbf{i}\ \ \mathbf{j}\ \ \mathbf{k}]$):

$$\begin{aligned} \begin{bmatrix} \bar{\boldsymbol{\alpha}} & \bar{\boldsymbol{\beta}} & \bar{\boldsymbol{\gamma}} \end{bmatrix} &= \begin{bmatrix} \mathbf{i} & \mathbf{j} & \mathbf{k} \end{bmatrix} \begin{bmatrix} \bar{\alpha}_i & \bar{\beta}_i & \bar{\gamma}_i \\ \bar{\alpha}_j & \bar{\beta}_j & \bar{\gamma}_j \\ \bar{\alpha}_k & \bar{\beta}_k & \bar{\gamma}_k \end{bmatrix} \\ &= \begin{bmatrix} \bar{\alpha}_i & \bar{\beta}_i & \bar{\gamma}_i \\ \bar{\alpha}_j & \bar{\beta}_j & \bar{\gamma}_j \\ \bar{\alpha}_k & \bar{\beta}_k & \bar{\gamma}_k \end{bmatrix} \end{aligned} \tag{24.23}$$

By considering the initial position of the solid-embedded base (Equation (24.16)), it is possible to express vectors $\bar{\boldsymbol{\alpha}}$, $\bar{\boldsymbol{\beta}}$ and $\bar{\boldsymbol{\gamma}}$ using the base $\begin{bmatrix} \bar{\boldsymbol{\alpha}} & \bar{\boldsymbol{\beta}} & \bar{\boldsymbol{\gamma}} \end{bmatrix}$

$$\begin{bmatrix} \bar{\alpha}_i & \bar{\beta}_i & \bar{\gamma}_i \\ \bar{\alpha}_j & \bar{\beta}_j & \bar{\gamma}_j \\ \bar{\alpha}_k & \bar{\beta}_k & \bar{\gamma}_k \end{bmatrix} = \begin{bmatrix} \bar{\xi}_i & \bar{\eta}_i & \bar{\zeta}_i \\ \bar{\xi}_j & \bar{\eta}_j & \bar{\zeta}_j \\ \bar{\xi}_k & \bar{\eta}_k & \bar{\zeta}_k \end{bmatrix} \begin{bmatrix} \alpha_\xi & \beta_\xi & \gamma_\xi \\ \alpha_\eta & \beta_\eta & \gamma_\eta \\ \alpha_\zeta & \beta_\zeta & \gamma_\zeta \end{bmatrix}, \tag{24.24}$$

which yields

$$\begin{bmatrix} \alpha_\xi & \beta_\xi & \gamma_\xi \\ \alpha_\eta & \beta_\eta & \gamma_\eta \\ \alpha_\zeta & \beta_\zeta & \gamma_\zeta \end{bmatrix} = \begin{bmatrix} \bar{\xi}_i & \bar{\eta}_i & \bar{\zeta}_i \\ \bar{\xi}_j & \bar{\eta}_j & \bar{\zeta}_j \\ \bar{\xi}_k & \bar{\eta}_k & \bar{\zeta}_k \end{bmatrix}^{-1} \begin{bmatrix} \bar{\alpha}_i & \bar{\beta}_i & \bar{\gamma}_i \\ \bar{\alpha}_j & \bar{\beta}_j & \bar{\gamma}_j \\ \bar{\alpha}_k & \bar{\beta}_k & \bar{\gamma}_k \end{bmatrix}. \tag{24.25}$$

The previous and the current position of the material-embedded base are then respectively

$$\begin{bmatrix} \hat{\alpha}_i & \hat{\beta}_i & \hat{\gamma}_i \\ \hat{\alpha}_j & \hat{\beta}_j & \hat{\gamma}_j \\ \hat{\alpha}_k & \hat{\beta}_k & \hat{\gamma}_k \end{bmatrix} = \begin{bmatrix} \hat{\xi}_i & \hat{\eta}_i & \hat{\zeta}_i \\ \hat{\xi}_j & \hat{\eta}_j & \hat{\zeta}_j \\ \hat{\xi}_k & \hat{\eta}_k & \hat{\zeta}_k \end{bmatrix} \begin{bmatrix} \alpha_\xi & \beta_\xi & \gamma_\xi \\ \alpha_\eta & \beta_\eta & \gamma_\eta \\ \alpha_\zeta & \beta_\zeta & \gamma_\zeta \end{bmatrix}, \tag{24.26}$$

$$\begin{bmatrix} \tilde{\alpha}_i & \tilde{\beta}_i & \tilde{\gamma}_i \\ \tilde{\alpha}_j & \tilde{\beta}_j & \tilde{\gamma}_j \\ \tilde{\alpha}_k & \tilde{\beta}_k & \tilde{\gamma}_k \end{bmatrix} = \begin{bmatrix} \tilde{\xi}_i & \tilde{\eta}_i & \tilde{\zeta}_i \\ \tilde{\xi}_j & \tilde{\eta}_j & \tilde{\zeta}_j \\ \tilde{\xi}_k & \tilde{\eta}_k & \tilde{\zeta}_k \end{bmatrix} \begin{bmatrix} \alpha_\xi & \beta_\xi & \gamma_\xi \\ \alpha_\eta & \beta_\eta & \gamma_\eta \\ \alpha_\zeta & \beta_\zeta & \gamma_\zeta \end{bmatrix}. \tag{24.27}$$

24.7 Examples of Deformation Kinematics

Using the procedure developed in the previous sections, it is possible to derive deformation kinematics for any iso-parametric finite element.

The Ten-Noded Tetrahedron. For the ten-noded tetrahedron the shape functions are given by Equation (24.3). From these, the derivatives of the shape functions are obtained as follows

$$\begin{bmatrix} \frac{\partial N_1}{\partial \xi} & \frac{\partial N_1}{\partial \eta} & \frac{\partial N_1}{\partial \zeta} \\ \frac{\partial N_2}{\partial \xi} & \frac{\partial N_2}{\partial \eta} & \frac{\partial N_2}{\partial \zeta} \\ \frac{\partial N_3}{\partial \xi} & \frac{\partial N_3}{\partial \eta} & \frac{\partial N_3}{\partial \zeta} \\ \frac{\partial N_4}{\partial \xi} & \frac{\partial N_4}{\partial \eta} & \frac{\partial N_4}{\partial \zeta} \\ \frac{\partial N_5}{\partial \xi} & \frac{\partial N_5}{\partial \eta} & \frac{\partial N_5}{\partial \zeta} \\ \frac{\partial N_6}{\partial \xi} & \frac{\partial N_6}{\partial \eta} & \frac{\partial N_6}{\partial \zeta} \\ \frac{\partial N_7}{\partial \xi} & \frac{\partial N_7}{\partial \eta} & \frac{\partial N_7}{\partial \zeta} \\ \frac{\partial N_8}{\partial \xi} & \frac{\partial N_8}{\partial \eta} & \frac{\partial N_8}{\partial \zeta} \\ \frac{\partial N_9}{\partial \xi} & \frac{\partial N_9}{\partial \eta} & \frac{\partial N_9}{\partial \zeta} \\ \frac{\partial N_{10}}{\partial \xi} & \frac{\partial N_{10}}{\partial \eta} & \frac{\partial N_{10}}{\partial \zeta} \end{bmatrix} = \begin{bmatrix} -3+4(\xi+\eta+\zeta) & -3+4(\xi+\eta+\zeta) & -3+4(\xi+\eta+\zeta) \\ -4(2\xi+\eta+\zeta) & -4\xi & -4\xi \\ 4\xi-1 & 0 & 0 \\ 4\eta & 4\xi & 0 \\ 0 & 4\eta-1 & 0 \\ -4\eta & -4(\xi+2\eta+\zeta) & -4\eta \\ -4\zeta & -4\zeta & -4(\xi+\eta+2\zeta) \\ 4\zeta & 0 & 4\xi \\ 0 & 4\zeta & 4\eta \\ 0 & 0 & 4\zeta-1 \end{bmatrix} \tag{24.28}$$

When substituted into Equations (24.20), (24.21) and (24.22), these yield:

a. the initial position of the solid-embedded vector base

$$[\bar{\boldsymbol{\xi}}\ \bar{\boldsymbol{\eta}}\ \bar{\boldsymbol{\zeta}}] = [\mathbf{i}\ \mathbf{j}\ \mathbf{k}]\begin{bmatrix} \frac{\partial \bar{x}}{\partial \xi} & \frac{\partial \bar{x}}{\partial \eta} & \frac{\partial \bar{x}}{\partial \zeta} \\ \frac{\partial \bar{y}}{\partial \xi} & \frac{\partial \bar{y}}{\partial \eta} & \frac{\partial \bar{y}}{\partial \zeta} \\ \frac{\partial \bar{z}}{\partial \xi} & \frac{\partial \bar{z}}{\partial \eta} & \frac{\partial \bar{z}}{\partial \zeta} \end{bmatrix} =$$

$$\begin{bmatrix} \bar{x}_1 & \bar{x}_2 & \bar{x}_3 & \bar{x}_4 & \bar{x}_5 & \bar{x}_6 & \bar{x}_7 & \bar{x}_8 & \bar{x}_9 & \bar{x}_{10} \\ \bar{y}_1 & \bar{y}_2 & \bar{y}_3 & \bar{y}_4 & \bar{y}_5 & \bar{y}_6 & \bar{y}_7 & \bar{y}_8 & \bar{y}_9 & \bar{y}_{10} \\ \bar{z}_1 & \bar{z}_2 & \bar{z}_3 & \bar{z}_4 & \bar{z}_5 & \bar{z}_6 & \bar{z}_7 & \bar{z}_8 & \bar{z}_9 & \bar{z}_{10} \end{bmatrix} \begin{bmatrix} -3+4(\xi+\eta+\zeta) & -3+4(\xi+\eta+\zeta) & -3+4(\xi+\eta+\zeta) \\ -4(2\xi+\eta+\zeta) & -4\xi & -4\xi \\ 4\xi-1 & 0 & 0 \\ 4\eta & 4\xi & 0 \\ 0 & 4\eta-1 & 0 \\ -4\eta & -4(\xi+2\eta+\zeta) & -4\eta \\ -4\zeta & -4\zeta & -4(\xi+\eta+2\zeta) \\ 4\zeta & 0 & 4\xi \\ 0 & 4\zeta & 4\eta \\ 0 & 0 & 4\zeta-1 \end{bmatrix} \tag{24.29}$$

b. the previous position of the solid-embedded vector base

$$[\hat{\boldsymbol{\xi}}\ \hat{\boldsymbol{\eta}}\ \hat{\boldsymbol{\zeta}}] = [\mathbf{i}\ \mathbf{j}\ \mathbf{k}]\begin{bmatrix} \frac{\partial \hat{x}}{\partial \xi} & \frac{\partial \hat{x}}{\partial \eta} & \frac{\partial \hat{x}}{\partial \zeta} \\ \frac{\partial \hat{y}}{\partial \xi} & \frac{\partial \hat{y}}{\partial \eta} & \frac{\partial \hat{y}}{\partial \zeta} \\ \frac{\partial \hat{z}}{\partial \xi} & \frac{\partial \hat{z}}{\partial \eta} & \frac{\partial \hat{z}}{\partial \zeta} \end{bmatrix} =$$

$$\begin{bmatrix} \hat{x}_1 & \hat{x}_2 & \hat{x}_3 & \hat{x}_4 & \hat{x}_5 & \hat{x}_6 & \hat{x}_7 & \hat{x}_8 & \hat{x}_9 & \hat{x}_{10} \\ \hat{y}_1 & \hat{y}_2 & \hat{y}_3 & \hat{y}_4 & \hat{y}_5 & \hat{y}_6 & \hat{y}_7 & \hat{y}_8 & \hat{y}_9 & \hat{y}_{10} \\ \hat{z}_1 & \hat{z}_2 & \hat{z}_3 & \hat{z}_4 & \hat{z}_5 & \hat{z}_6 & \hat{z}_7 & \hat{z}_8 & \hat{z}_9 & \hat{z}_{10} \end{bmatrix} \begin{bmatrix} -3+4(\xi+\eta+\zeta) & -3+4(\xi+\eta+\zeta) & -3+4(\xi+\eta+\zeta) \\ -4(2\xi+\eta+\zeta) & -4\xi & -4\xi \\ 4\xi-1 & 0 & 0 \\ 4\eta & 4\xi & 0 \\ 0 & 4\eta-1 & 0 \\ -4\eta & -4(\xi+2\eta+\zeta) & -4\eta \\ -4\zeta & -4\zeta & -4(\xi+\eta+2\zeta) \\ 4\zeta & 0 & 4\xi \\ 0 & 4\zeta & 4\eta \\ 0 & 0 & 4\zeta-1 \end{bmatrix} \tag{24.30}$$

c. the current position of the solid-embedded vector base

$$
\begin{bmatrix}\tilde{\boldsymbol{\xi}} & \tilde{\boldsymbol{\eta}} & \tilde{\boldsymbol{\zeta}}\end{bmatrix} = \begin{bmatrix}\mathbf{i} & \mathbf{j} & \mathbf{k}\end{bmatrix}\begin{bmatrix}\frac{\partial \tilde{x}}{\partial \xi} & \frac{\partial \tilde{x}}{\partial \eta} & \frac{\partial \tilde{x}}{\partial \zeta}\\ \frac{\partial \tilde{y}}{\partial \xi} & \frac{\partial \tilde{y}}{\partial \eta} & \frac{\partial \tilde{y}}{\partial \zeta}\\ \frac{\partial \tilde{z}}{\partial \xi} & \frac{\partial \tilde{z}}{\partial \eta} & \frac{\partial \tilde{z}}{\partial \zeta}\end{bmatrix} =
$$

$$
\begin{bmatrix}\tilde{x}_1 & \tilde{x}_2 & \tilde{x}_3 & \tilde{x}_4 & \tilde{x}_5 & \tilde{x}_6 & \tilde{x}_7 & \tilde{x}_8 & \tilde{x}_9 & \tilde{x}_{10}\\ \tilde{y}_1 & \tilde{y}_2 & \tilde{y}_3 & \tilde{y}_4 & \tilde{y}_5 & \tilde{y}_6 & \tilde{y}_7 & \tilde{y}_8 & \tilde{y}_9 & \tilde{y}_{10}\\ \tilde{z}_1 & \tilde{z}_2 & \tilde{z}_3 & \tilde{z}_4 & \tilde{z}_5 & \tilde{z}_6 & \tilde{z}_7 & \tilde{z}_8 & \tilde{z}_9 & \tilde{z}_{10}\end{bmatrix}
\begin{bmatrix}-3+4(\xi+\eta+\zeta) & -3+4(\xi+\eta+\zeta) & -3+4(\xi+\eta+\zeta)\\ -4(2\xi+\eta+\zeta) & -4\xi & -4\xi\\ 4\xi-1 & 0 & 0\\ 4\eta & 4\xi & 0\\ 0 & 4\eta-1 & 0\\ -4\eta & -4(\xi+2\eta+\zeta) & -4\eta\\ -4\zeta & -4\zeta & -4(\xi+\eta+2\zeta)\\ 4\zeta & 0 & 4\xi\\ 0 & 4\zeta & 4\eta\\ 0 & 0 & 4\zeta-1\end{bmatrix} \tag{24.31}
$$

As can be seen, all of these base vectors are obtained directly from the initial, the current and the previous nodal coordinates. From these six vectors the current and the previous positions of the material-embedded vector base are obtained using Equations (24.24), (24.26) and (24.27).

As the initial position of the material vector base is given in the input to the finite element analysis, it follows that $\begin{bmatrix}\bar{\boldsymbol{\alpha}} & \bar{\boldsymbol{\beta}} & \bar{\boldsymbol{\gamma}}\end{bmatrix}$, $\begin{bmatrix}\hat{\boldsymbol{\alpha}} & \hat{\boldsymbol{\beta}} & \hat{\boldsymbol{\gamma}}\end{bmatrix}$ and $\begin{bmatrix}\tilde{\boldsymbol{\alpha}} & \tilde{\boldsymbol{\beta}} & \tilde{\boldsymbol{\gamma}}\end{bmatrix}$ are known for any material point $P=(\xi, \eta, \zeta)$ inside the finite element. As already explained, these nine vectors define the deformation kinematics of the generalized material element.

The Eight-Noded Solid. The shape functions for the eight-noded solid are given by Equation (24.4). From these, the derivatives of the shape functions are obtained

$$\begin{bmatrix} \frac{\partial N_1}{\partial \xi} & \frac{\partial N_1}{\partial \eta} & \frac{\partial N_1}{\partial \zeta} \\ \frac{\partial N_2}{\partial \xi} & \frac{\partial N_2}{\partial \eta} & \frac{\partial N_2}{\partial \zeta} \\ \frac{\partial N_3}{\partial \xi} & \frac{\partial N_3}{\partial \eta} & \frac{\partial N_3}{\partial \zeta} \\ \frac{\partial N_4}{\partial \xi} & \frac{\partial N_4}{\partial \eta} & \frac{\partial N_4}{\partial \zeta} \\ \frac{\partial N_5}{\partial \xi} & \frac{\partial N_5}{\partial \eta} & \frac{\partial N_5}{\partial \zeta} \\ \frac{\partial N_6}{\partial \xi} & \frac{\partial N_6}{\partial \eta} & \frac{\partial N_6}{\partial \zeta} \\ \frac{\partial N_7}{\partial \xi} & \frac{\partial N_7}{\partial \eta} & \frac{\partial N_7}{\partial \zeta} \\ \frac{\partial N_8}{\partial \xi} & \frac{\partial N_8}{\partial \eta} & \frac{\partial N_8}{\partial \zeta} \end{bmatrix} = \frac{1}{8} \begin{bmatrix} -(1-\eta)(1-\zeta) & -(1-\xi)(1-\zeta) & -(1-\xi)(1-\eta) \\ +(1-\eta)(1-\zeta) & +(1+\xi)(1-\zeta) & -(1+\xi)(1-\eta) \\ +(1+\eta)(1-\zeta) & +(1+\xi)(1-\zeta) & -(1+\xi)(1+\eta) \\ -(1+\eta)(1-\zeta) & +(1-\xi)(1-\zeta) & -(1-\xi)(1+\eta) \\ -(1-\eta)(1+\zeta) & -(1-\xi)(1+\zeta) & +(1-\xi)(1-\eta) \\ +(1-\eta)(1+\zeta) & -(1+\xi)(1+\zeta) & +(1+\xi)(1-\eta) \\ +(1+\eta)(1+\zeta) & +(1+\xi)(1+\zeta) & +(1+\xi)(1+\eta) \\ -(1+\eta)(1+\zeta) & +(1-\xi)(1+\zeta) & +(1-\xi)(1+\eta) \end{bmatrix} \tag{24.32}$$

When substituted into Equations (24.20), (24.21) and (24.22), these yield:

a. the initial position of the solid-embedded vector base

$$[\bar{\boldsymbol{\xi}} \;\; \bar{\boldsymbol{\eta}} \;\; \bar{\boldsymbol{\zeta}}] = [\mathbf{i} \;\; \mathbf{j} \;\; \mathbf{k}] \begin{bmatrix} \frac{\partial \bar{x}}{\partial \xi} & \frac{\partial \bar{x}}{\partial \eta} & \frac{\partial \bar{x}}{\partial \zeta} \\ \frac{\partial \bar{y}}{\partial \xi} & \frac{\partial \bar{y}}{\partial \eta} & \frac{\partial \bar{y}}{\partial \zeta} \\ \frac{\partial \bar{z}}{\partial \xi} & \frac{\partial \bar{z}}{\partial \eta} & \frac{\partial \bar{z}}{\partial \zeta} \end{bmatrix} =$$

$$\begin{bmatrix} \bar{x}_1 & \bar{x}_2 & \bar{x}_3 & \bar{x}_4 & \bar{x}_5 & \bar{x}_6 & \bar{x}_7 & \bar{x}_8 \\ \bar{y}_1 & \bar{y}_2 & \bar{y}_3 & \bar{y}_4 & \bar{y}_5 & \bar{y}_6 & \bar{y}_7 & \bar{y}_8 \\ \bar{z}_1 & \bar{z}_2 & \bar{z}_3 & \bar{z}_4 & \bar{z}_5 & \bar{z}_6 & \bar{z}_7 & \bar{z}_8 \end{bmatrix} \frac{1}{8} \begin{bmatrix} -(1-\eta)(1-\zeta) & -(1-\xi)(1-\zeta) & -(1-\xi)(1-\eta) \\ +(1-\eta)(1-\zeta) & +(1+\xi)(1-\zeta) & -(1+\xi)(1-\eta) \\ +(1+\eta)(1-\zeta) & +(1+\xi)(1-\zeta) & -(1+\xi)(1+\eta) \\ -(1+\eta)(1-\zeta) & +(1-\xi)(1-\zeta) & -(1-\xi)(1+\eta) \\ -(1-\eta)(1+\zeta) & -(1-\xi)(1+\zeta) & +(1-\xi)(1-\eta) \\ +(1-\eta)(1+\zeta) & -(1+\xi)(1+\zeta) & +(1+\xi)(1-\eta) \\ +(1+\eta)(1+\zeta) & +(1+\xi)(1+\zeta) & +(1+\xi)(1+\eta) \\ -(1+\eta)(1+\zeta) & +(1-\xi)(1+\zeta) & +(1-\xi)(1+\eta) \end{bmatrix} \tag{24.33}$$

b. the previous position of the solid-embedded vector base

$$[\hat{\boldsymbol{\xi}}\ \hat{\boldsymbol{\eta}}\ \hat{\boldsymbol{\zeta}}] = [\mathbf{i}\ \mathbf{j}\ \mathbf{k}] \begin{bmatrix} \frac{\partial \hat{x}}{\partial \xi} & \frac{\partial \hat{x}}{\partial \eta} & \frac{\partial \hat{x}}{\partial \zeta} \\ \frac{\partial \hat{y}}{\partial \xi} & \frac{\partial \hat{y}}{\partial \eta} & \frac{\partial \hat{y}}{\partial \zeta} \\ \frac{\partial \hat{z}}{\partial \xi} & \frac{\partial \hat{z}}{\partial \eta} & \frac{\partial \hat{z}}{\partial \zeta} \end{bmatrix} =$$

$$\begin{bmatrix} \hat{x}_1 & \hat{x}_2 & \hat{x}_3 & \hat{x}_4 & \hat{x}_5 & \hat{x}_6 & \hat{x}_7 & \hat{x}_8 \\ \hat{y}_1 & \hat{y}_2 & \hat{y}_3 & \hat{y}_4 & \hat{y}_5 & \hat{y}_6 & \hat{y}_7 & \hat{y}_8 \\ \hat{z}_1 & \hat{z}_2 & \hat{z}_3 & \hat{z}_4 & \hat{z}_5 & \hat{z}_6 & \hat{z}_7 & \hat{z}_8 \end{bmatrix} \frac{1}{8} \begin{bmatrix} -(1-\eta)(1-\zeta) & -(1-\xi)(1-\zeta) & -(1-\xi)(1-\eta) \\ +(1-\eta)(1-\zeta) & +(1+\xi)(1-\zeta) & -(1+\xi)(1-\eta) \\ +(1+\eta)(1-\zeta) & +(1+\xi)(1-\zeta) & -(1+\xi)(1+\eta) \\ -(1+\eta)(1-\zeta) & +(1-\xi)(1-\zeta) & -(1-\xi)(1+\eta) \\ -(1-\eta)(1+\zeta) & -(1-\xi)(1+\zeta) & +(1-\xi)(1-\eta) \\ +(1-\eta)(1+\zeta) & -(1+\xi)(1+\zeta) & +(1+\xi)(1-\eta) \\ +(1+\eta)(1+\zeta) & +(1+\xi)(1+\zeta) & +(1+\xi)(1+\eta) \\ -(1+\eta)(1+\zeta) & +(1-\xi)(1+\zeta) & +(1-\xi)(1+\eta) \end{bmatrix} \tag{24.34}$$

c. the current position of the solid-embedded vector base

$$[\tilde{\boldsymbol{\xi}}\ \tilde{\boldsymbol{\eta}}\ \tilde{\boldsymbol{\zeta}}] = [\mathbf{i}\ \mathbf{j}\ \mathbf{k}] \begin{bmatrix} \frac{\partial \tilde{x}}{\partial \xi} & \frac{\partial \tilde{x}}{\partial \eta} & \frac{\partial \tilde{x}}{\partial \zeta} \\ \frac{\partial \tilde{y}}{\partial \xi} & \frac{\partial \tilde{y}}{\partial \eta} & \frac{\partial \tilde{y}}{\partial \zeta} \\ \frac{\partial \tilde{z}}{\partial \xi} & \frac{\partial \tilde{z}}{\partial \eta} & \frac{\partial \tilde{z}}{\partial \zeta} \end{bmatrix} =$$

$$\begin{bmatrix} \tilde{x}_1 & \tilde{x}_2 & \tilde{x}_3 & \tilde{x}_4 & \tilde{x}_5 & \tilde{x}_6 & \tilde{x}_7 & \tilde{x}_8 \\ \tilde{y}_1 & \tilde{y}_2 & \tilde{y}_3 & \tilde{y}_4 & \tilde{y}_5 & \tilde{y}_6 & \tilde{y}_7 & \tilde{y}_8 \\ \tilde{z}_1 & \tilde{z}_2 & \tilde{z}_3 & \tilde{z}_4 & \tilde{z}_5 & \tilde{z}_6 & \tilde{z}_7 & \tilde{z}_8 \end{bmatrix} \frac{1}{8} \begin{bmatrix} -(1-\eta)(1-\zeta) & -(1-\xi)(1-\zeta) & -(1-\xi)(1-\eta) \\ +(1-\eta)(1-\zeta) & +(1+\xi)(1-\zeta) & -(1+\xi)(1-\eta) \\ +(1+\eta)(1-\zeta) & +(1+\xi)(1-\zeta) & -(1+\xi)(1+\eta) \\ -(1+\eta)(1-\zeta) & +(1-\xi)(1-\zeta) & -(1-\xi)(1+\eta) \\ -(1-\eta)(1+\zeta) & -(1-\xi)(1+\zeta) & +(1-\xi)(1-\eta) \\ +(1-\eta)(1+\zeta) & -(1+\xi)(1+\zeta) & +(1+\xi)(1-\eta) \\ +(1+\eta)(1+\zeta) & +(1+\xi)(1+\zeta) & +(1+\xi)(1+\eta) \\ -(1+\eta)(1+\zeta) & +(1-\xi)(1+\zeta) & +(1-\xi)(1+\eta) \end{bmatrix} \tag{24.35}$$

By substituting Equations (24.33), (24.34) and (24.35) into equations (24.20), (24.21) and (24.22) the previous and the current positions of the material-embedded vector base, $[\hat{\boldsymbol{\alpha}}\ \hat{\boldsymbol{\beta}}\ \hat{\boldsymbol{\gamma}}]$ and $[\tilde{\boldsymbol{\alpha}}\ \tilde{\boldsymbol{\beta}}\ \tilde{\boldsymbol{\gamma}}]$ are obtained, while the initial position $[\bar{\boldsymbol{\alpha}}\ \bar{\boldsymbol{\beta}}\ \bar{\boldsymbol{\gamma}}]$ is supplied as an input to the analysis.

When these are substituted into Equation (24.27) the material-embedded vector base is obtained for any material point of the finite element. From these vectors, the generalized material element is known and by employing the constitutive law stress can be calculated for any material point. These nine vectors therefore uniquely define the deformation kinematics of the generalized material element for any given point $P = (\xi, \eta, \zeta)$ within the domain of the finite element.

24.8 Summary

Deformation kinematics of a 3D solid within a single 3D finite element has been obtained in terms of the initial, the previous and the current nodal coordinates. In this chapter, the general procedure for describing these kinematics for a n-noded 3D finite element has been explained.

This has been further elucidated by applying the general procedures to the particular examples of a ten-noded tetrahedron finite element and an eight-noded 3D solid finite element.

The obtained deformation kinematics are suitable to directly pass to the material package. The material package returns the Cauchy stress tensor matrix (internal forces on the Cauchy material element's surfaces). The Cauchy stress tensor matrix is then used to calculate the nodal forces by which the finite element pulls its nodes.

Further Reading

[1] Bonet, J. and Burton, A. J. (1998) A simple average nodal pressure tetrahedral element for incompressible and nearly incompressible dynamic explicit applications. *Communications in Numerical Methods in Engineering*, **14**(5): 437–49.

[2] Bonet, J., Marriott, H. and Hassan, O. (2001) An averaged nodal deformation gradient linear tetrahedral element for large strain explicit dynamic applications. *Communications in Numerical Methods in Engineering*, **17**(8): 551–61.

[3] Bonet, J. and Wood, R. D. (1997) *Nonlinear Continuum Mechanics for Finite Element Analysis*. Cambridge, Cambridge University Press.

[4] Hughes, T.J.R. (2000) *The Finite Element Method: Linear Static and Dynamic Finite Element Analysis*. New York, Dover Publications.

[5] Liu, W. K., Guo, Y., Tang, S. and Belytschko. T. (1998) A multiple-quadrature eight-node hexahedral finite element for large deformation elastoplastic analysis. *Comput. Method Appl. M.*, **154**(1–2): 69–132.

[6] Oden, J. T. (1972) *Finite Elements of Nonlinear Continua*, New York, McGraw-Hill.

[7] Peric, D. and Owen, D. R. J. (1998) Finite-element applications to the nonlinear mechanics of solids. *Reports on Progress in Physics*, **61**(11): 1495–1574.

[8] Zienkiewicz, O. C. (1971) *The Finite Element Method in Engineering Science*. New York, McGraw-Hill.

[9] Zienkiewicz, O. C., Rojek, J., Taylor, R. L. and Pastor, M. (1998) Triangles and tetrahera in explicit dynamic codes for solids. *International Journal for Numerical Methods in Engineering*, **43**(3): 565–83.

[10] Zienkiewicz, O. C. and Taylor, R. L. (2005) *The Finite Element Method Set*. Oxford, Elsevier Science.

Part Five

The Finite Element Method in 3D

25

Integration of Nodal Forces over Volume of 3D Finite Elements

25.1 Nodal Forces Using Virtual Work

One way to calculate the nodal forces due to the internal forces is by using the principle of virtual work. For example, the nodal force on node 1 in the **i** direction is obtained by supplying a virtual displacement to node 1 equal to 1 im in the **i** direction

$$u_1 = 1\ \text{im}. \tag{25.1}$$

The virtual displacements of the material points inside the finite element are approximated using the shape functions, which yields

$$\begin{aligned} \mathrm{u}_1(\xi, \eta, \zeta) &= \mathrm{N}_1(\xi, \eta, \zeta) \\ &= 1 - \xi - \eta - \zeta. \end{aligned} \tag{25.2}$$

This way, a given material point

$$\mathrm{P} = (\xi, \eta, \zeta) \tag{25.3}$$

is moved in the **i** direction by an infinitesimally small amount equal to $1 - \xi - \eta - \zeta$ im (infi-meters).

Large Strain Finite Element Method: A Practical Course, First Edition. Antonio Munjiza, Esteban Rougier and Earl E. Knight.

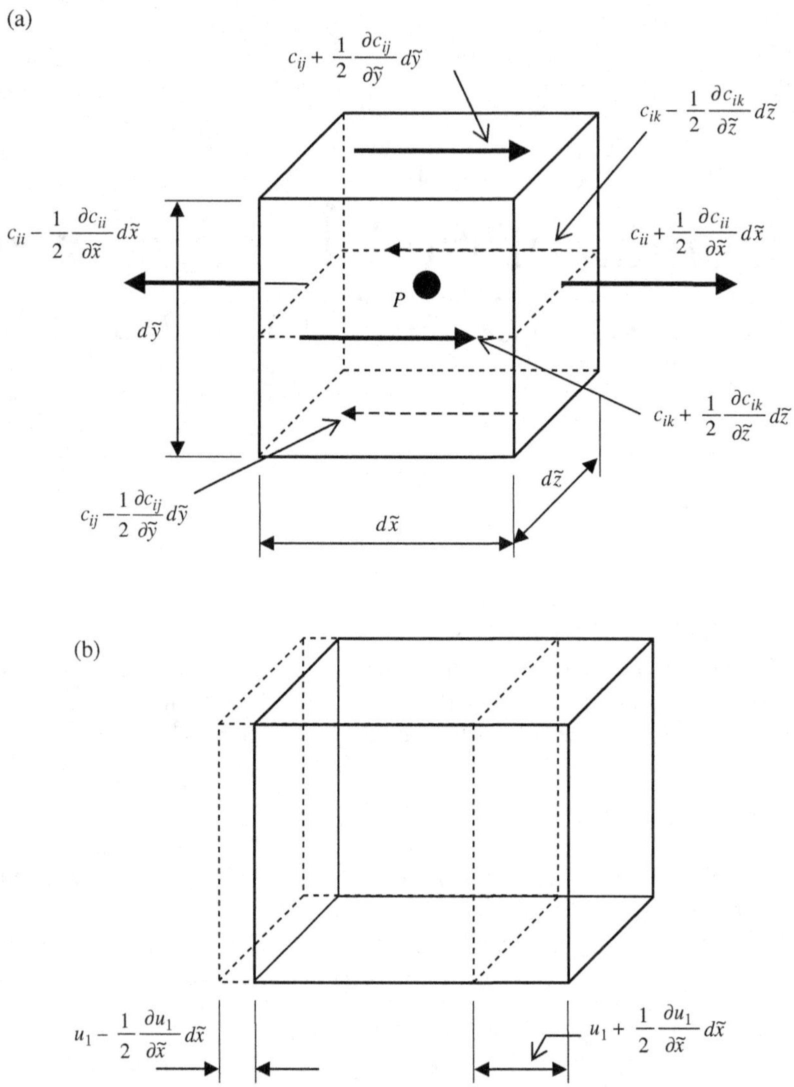

Figure 25.1 Virtual work: (a) The Cauchy infinitesimal material element with components of Cauchy stress tensor matrix representing internal forces in the **i** direction; (b) virtual displacements in the **i** direction of the Cauchy infinitesimal material element due to the virtual displacement of node 1 in the **i** direction, $u_1(\xi, \eta, \zeta)$

By considering all the internal forces in the **i** direction shown in Figure 25.1a and the corresponding virtual displacements shown in Figure 25.1b, the virtual work of the internal forces is obtained by multiplying the virtual displacement with the corresponding internal forces and integrating over the volume of the finite element. The total virtual work should be equal to zero, i.e.,

$$
\begin{aligned}
f_{i1}\cdot 1 + \Bigg(& -\iiint\limits_{\tilde{V}_e}\left(c_{ii}-\frac{1}{2}\frac{\partial c_{ii}}{\partial \tilde{x}}d\tilde{x}\right)\left(\mathrm{u}_1-\frac{1}{2}\frac{\partial \mathrm{u}_1}{\partial \tilde{x}}d\tilde{x}\right)d\tilde{y}d\tilde{z} \\
& +\iiint\limits_{\tilde{V}_e}\left(c_{ii}+\frac{1}{2}\frac{\partial c_{ii}}{\partial \tilde{x}}d\tilde{x}\right)\left(\mathrm{u}_1+\frac{1}{2}\frac{\partial \mathrm{u}_1}{\partial \tilde{x}}d\tilde{x}\right)d\tilde{y}d\tilde{z} \\
& -\iiint\limits_{\tilde{V}_e}\left(c_{ij}-\frac{1}{2}\frac{\partial c_{ij}}{\partial \tilde{y}}d\tilde{y}\right)\left(\mathrm{u}_1-\frac{1}{2}\frac{\partial \mathrm{u}_1}{\partial \tilde{y}}d\tilde{y}\right)d\tilde{x}d\tilde{z} \\
& +\iiint\limits_{\tilde{V}_e}\left(c_{ij}+\frac{1}{2}\frac{\partial c_{ij}}{\partial \tilde{y}}d\tilde{y}\right)\left(\mathrm{u}_1+\frac{1}{2}\frac{\partial \mathrm{u}_1}{\partial \tilde{y}}d\tilde{y}\right)d\tilde{x}d\tilde{z} \\
& -\iiint\limits_{\tilde{V}_e}\left(c_{ik}-\frac{1}{2}\frac{\partial c_{ik}}{\partial \tilde{z}}d\tilde{z}\right)\left(\mathrm{u}_1-\frac{1}{2}\frac{\partial \mathrm{u}_1}{\partial \tilde{z}}d\tilde{z}\right)d\tilde{x}d\tilde{y} \\
& +\iiint\limits_{\tilde{V}_e}\left(c_{ik}+\frac{1}{2}\frac{\partial c_{ik}}{\partial \tilde{z}}d\tilde{z}\right)\left(\mathrm{u}_1+\frac{1}{2}\frac{\partial \mathrm{u}_1}{\partial \tilde{z}}d\tilde{z}\right)d\tilde{x}d\tilde{y}\Bigg)=0.
\end{aligned}
\tag{25.4}
$$

After discarding the high order terms in Equation (25.4), one obtains

$$
\begin{aligned}
f_{i1} = & -\iiint\limits_{\tilde{V}_e}\left(\frac{\partial c_{ii}}{\partial \tilde{x}}\mathrm{u}_1 d\tilde{x}d\tilde{y}d\tilde{z}+c_{ii}\frac{\partial \mathrm{u}_1}{\partial \tilde{x}}d\tilde{x}d\tilde{y}d\tilde{z}\right) \\
& -\iiint\limits_{\tilde{V}_e}\left(\frac{\partial c_{ij}}{\partial \tilde{y}}\mathrm{u}_1 d\tilde{x}d\tilde{y}d\tilde{z}+c_{ij}\frac{\partial \mathrm{u}_1}{\partial \tilde{y}}d\tilde{x}d\tilde{y}d\tilde{z}\right) \\
& -\iiint\limits_{\tilde{V}_e}\left(\frac{\partial c_{ik}}{\partial \tilde{z}}\mathrm{u}_1 d\tilde{x}d\tilde{y}d\tilde{z}+c_{ik}\frac{\partial \mathrm{u}_1}{\partial \tilde{z}}d\tilde{x}d\tilde{y}d\tilde{z}\right).
\end{aligned}
\tag{25.5}
$$

The above equation can be written in a matrix form

$$
f_{i1} = -\iiint\limits_{\tilde{V}_e}\left\{\begin{bmatrix} c_{ii} & c_{ij} & c_{ik}\end{bmatrix}\begin{bmatrix}\dfrac{\partial \mathrm{u}_1}{\partial \tilde{x}} \\[2ex] \dfrac{\partial \mathrm{u}_1}{\partial \tilde{y}} \\[2ex] \dfrac{\partial \mathrm{u}_1}{\partial \tilde{z}}\end{bmatrix}+\left(\frac{\partial c_{ii}}{\partial \tilde{x}}+\frac{\partial c_{ij}}{\partial \tilde{y}}+\frac{\partial c_{ik}}{\partial \tilde{z}}\right)\mathrm{u}_1\right\}d\tilde{x}d\tilde{y}d\tilde{z}. \tag{25.6}
$$

25.2 Four-Noded Tetrahedron Finite Element

As the Cauchy stress tensor matrix is constant over the tetrahedron finite element, it follows that

$$\frac{\partial c_{ii}}{\partial \tilde{x}} = \frac{\partial c_{ij}}{\partial \tilde{y}} = \frac{\partial c_{ik}}{\partial \tilde{z}} = 0, \tag{25.7}$$

which, when substituted into Equation (25.6), yields

$$f_{i1} = -\iiint_{\tilde{V}_e} \left\{ \begin{bmatrix} c_{ii} & c_{ij} & c_{ik} \end{bmatrix} \begin{bmatrix} \frac{\partial u_1}{\partial \tilde{x}} \\ \frac{\partial u_1}{\partial \tilde{y}} \\ \frac{\partial u_1}{\partial \tilde{z}} \end{bmatrix} \right\} d\tilde{x}d\tilde{y}d\tilde{z}. \tag{25.8}$$

By substituting the derivatives from Equation (23.13) into Equation (25.8) one obtains

$$f_{i1} = -\iiint_{\tilde{V}_e} \left\{ \begin{bmatrix} c_{ii} & c_{ij} & c_{ik} \end{bmatrix} \begin{bmatrix} \frac{\partial \tilde{x}}{\partial \xi} & \frac{\partial \tilde{x}}{\partial \eta} & \frac{\partial \tilde{x}}{\partial \zeta} \\ \frac{\partial \tilde{y}}{\partial \xi} & \frac{\partial \tilde{y}}{\partial \eta} & \frac{\partial \tilde{y}}{\partial \zeta} \\ \frac{\partial \tilde{z}}{\partial \xi} & \frac{\partial \tilde{z}}{\partial \eta} & \frac{\partial \tilde{z}}{\partial \zeta} \end{bmatrix}^{-T} \begin{bmatrix} \frac{\partial u_1}{\partial \xi} \\ \frac{\partial u_1}{\partial \eta} \\ \frac{\partial u_1}{\partial \zeta} \end{bmatrix} \right\} d\tilde{x}d\tilde{y}d\tilde{z}. \tag{25.9}$$

From Equation (25.2) it follows that

$$\begin{bmatrix} \frac{\partial u_1}{\partial \xi} \\ \frac{\partial u_1}{\partial \eta} \\ \frac{\partial u_1}{\partial \zeta} \end{bmatrix} = \begin{bmatrix} \frac{\partial N_1}{\partial \xi} \\ \frac{\partial N_1}{\partial \eta} \\ \frac{\partial N_1}{\partial \zeta} \end{bmatrix}. \tag{25.10}$$

When substituted into Equation (25.9) this yields

$$f_{i1} = -\iiint_{\tilde{V}_e} \left\{ \begin{bmatrix} c_{ii} & c_{ij} & c_{ik} \end{bmatrix} \begin{bmatrix} \frac{\partial \tilde{x}}{\partial \xi} & \frac{\partial \tilde{x}}{\partial \eta} & \frac{\partial \tilde{x}}{\partial \zeta} \\ \frac{\partial \tilde{y}}{\partial \xi} & \frac{\partial \tilde{y}}{\partial \eta} & \frac{\partial \tilde{y}}{\partial \zeta} \\ \frac{\partial \tilde{z}}{\partial \xi} & \frac{\partial \tilde{z}}{\partial \eta} & \frac{\partial \tilde{z}}{\partial \zeta} \end{bmatrix}^{-T} \begin{bmatrix} \frac{\partial N_1}{\partial \xi} \\ \frac{\partial N_1}{\partial \eta} \\ \frac{\partial N_1}{\partial \zeta} \end{bmatrix} \right\} d\tilde{x}d\tilde{y}d\tilde{z}. \tag{25.11}$$

By substituting x̃, ỹ and z̃ one obtains

$$f_{i1} = -\iiint_{\tilde{V}_e} \begin{bmatrix} c_{ii} & c_{ij} & c_{ik} \end{bmatrix} \begin{bmatrix} \tilde{x}_2-\tilde{x}_1 & \tilde{x}_3-\tilde{x}_1 & \tilde{x}_4-\tilde{x}_1 \\ \tilde{y}_2-\tilde{y}_1 & \tilde{y}_3-\tilde{y}_1 & \tilde{y}_4-\tilde{y}_1 \\ \tilde{z}_2-\tilde{z}_1 & \tilde{z}_3-\tilde{z}_1 & \tilde{z}_4-\tilde{z}_1 \end{bmatrix}^{-T} \begin{bmatrix} \dfrac{\partial N_1}{\partial \xi} \\ \dfrac{\partial N_1}{\partial \eta} \\ \dfrac{\partial N_1}{\partial \zeta} \end{bmatrix} d\tilde{x}d\tilde{y}d\tilde{z}, \tag{25.12}$$

where

$$\begin{bmatrix} \dfrac{\partial N_1}{\partial \xi} \\ \dfrac{\partial N_1}{\partial \eta} \\ \dfrac{\partial N_1}{\partial \zeta} \end{bmatrix} = \begin{bmatrix} -1 \\ -1 \\ -1 \end{bmatrix}. \tag{25.13}$$

The above integration can be done either analytically or by using the Gaussian integration. One Gauss point is used

$$G = (\xi_G, \eta_G, \zeta_G) = \left(\frac{1}{3}, \frac{1}{3}, \frac{1}{3}\right). \tag{25.14}$$

The result of the integration is as follows:

$$f_{i1} = \begin{bmatrix} c_{ii} & c_{ij} & c_{ik} \end{bmatrix} \begin{bmatrix} \tilde{x}_2-\tilde{x}_1 & \tilde{x}_3-\tilde{x}_1 & \tilde{x}_4-\tilde{x}_1 \\ \tilde{y}_2-\tilde{y}_1 & \tilde{y}_3-\tilde{y}_1 & \tilde{y}_4-\tilde{y}_1 \\ \tilde{z}_2-\tilde{z}_1 & \tilde{z}_3-\tilde{z}_1 & \tilde{z}_4-\tilde{z}_1 \end{bmatrix}^{-T} \begin{bmatrix} 1 \\ 1 \\ 1 \end{bmatrix} \tilde{v}, \tag{25.15}$$

where $\tilde{v}$ is the current volume of the tetrahedron.

The same procedure is repeated for the virtual displacement of nodes 2, 3 and 4, all in the **i** direction, thus producing the equivalent nodal forces f_{i1}, f_{i2}, f_{i3} and f_{i4}:

$$\begin{aligned} f_{i2} &= -\begin{bmatrix} c_{ii} & c_{ij} & c_{ik} \end{bmatrix} \begin{bmatrix} \tilde{x}_2-\tilde{x}_1 & \tilde{x}_3-\tilde{x}_1 & \tilde{x}_4-\tilde{x}_1 \\ \tilde{y}_2-\tilde{y}_1 & \tilde{y}_3-\tilde{y}_1 & \tilde{y}_4-\tilde{y}_1 \\ \tilde{z}_2-\tilde{z}_1 & \tilde{z}_3-\tilde{z}_1 & \tilde{z}_4-\tilde{z}_1 \end{bmatrix}^{-T} \begin{bmatrix} \dfrac{\partial N_2}{\partial \xi} \\ \dfrac{\partial N_2}{\partial \eta} \\ \dfrac{\partial N_2}{\partial \zeta} \end{bmatrix} \tilde{v} \\ &= \begin{bmatrix} c_{ii} & c_{ij} & c_{ik} \end{bmatrix} \begin{bmatrix} \tilde{x}_2-\tilde{x}_1 & \tilde{x}_3-\tilde{x}_1 & \tilde{x}_4-\tilde{x}_1 \\ \tilde{y}_2-\tilde{y}_1 & \tilde{y}_3-\tilde{y}_1 & \tilde{y}_4-\tilde{y}_1 \\ \tilde{z}_2-\tilde{z}_1 & \tilde{z}_3-\tilde{z}_1 & \tilde{z}_4-\tilde{z}_1 \end{bmatrix}^{-T} \begin{bmatrix} -1 \\ 0 \\ 0 \end{bmatrix} \tilde{v} \end{aligned} \tag{25.16}$$

$$f_{i3} = -\begin{bmatrix} c_{ii} & c_{ij} & c_{ik} \end{bmatrix} \begin{bmatrix} \tilde{x}_2-\tilde{x}_1 & \tilde{x}_3-\tilde{x}_1 & \tilde{x}_4-\tilde{x}_1 \\ \tilde{y}_2-\tilde{y}_1 & \tilde{y}_3-\tilde{y}_1 & \tilde{y}_4-\tilde{y}_1 \\ \tilde{z}_2-\tilde{z}_1 & \tilde{z}_3-\tilde{z}_1 & \tilde{z}_4-\tilde{z}_1 \end{bmatrix}^{-T} \begin{bmatrix} \frac{\partial N_3}{\partial \xi} \\ \frac{\partial N_3}{\partial \eta} \\ \frac{\partial N_3}{\partial \zeta} \end{bmatrix} \tilde{v}$$

$$= \begin{bmatrix} c_{ii} & c_{ij} & c_{ik} \end{bmatrix} \begin{bmatrix} \tilde{x}_2-\tilde{x}_1 & \tilde{x}_3-\tilde{x}_1 & \tilde{x}_4-\tilde{x}_1 \\ \tilde{y}_2-\tilde{y}_1 & \tilde{y}_3-\tilde{y}_1 & \tilde{y}_4-\tilde{y}_1 \\ \tilde{z}_2-\tilde{z}_1 & \tilde{z}_3-\tilde{z}_1 & \tilde{z}_4-\tilde{z}_1 \end{bmatrix}^{-T} \begin{bmatrix} 0 \\ -1 \\ 0 \end{bmatrix} \tilde{v} \tag{25.17}$$

$$f_{i4} = -\begin{bmatrix} c_{ii} & c_{ij} & c_{ik} \end{bmatrix} \begin{bmatrix} \tilde{x}_2-\tilde{x}_1 & \tilde{x}_3-\tilde{x}_1 & \tilde{x}_4-\tilde{x}_1 \\ \tilde{y}_2-\tilde{y}_1 & \tilde{y}_3-\tilde{y}_1 & \tilde{y}_4-\tilde{y}_1 \\ \tilde{z}_2-\tilde{z}_1 & \tilde{z}_3-\tilde{z}_1 & \tilde{z}_4-\tilde{z}_1 \end{bmatrix}^{-T} \begin{bmatrix} \frac{\partial N_4}{\partial \xi} \\ \frac{\partial N_4}{\partial \eta} \\ \frac{\partial N_4}{\partial \zeta} \end{bmatrix} \tilde{v}$$

$$= \begin{bmatrix} c_{ii} & c_{ij} & c_{ik} \end{bmatrix} \begin{bmatrix} \tilde{x}_2-\tilde{x}_1 & \tilde{x}_3-\tilde{x}_1 & \tilde{x}_4-\tilde{x}_1 \\ \tilde{y}_2-\tilde{y}_1 & \tilde{y}_3-\tilde{y}_1 & \tilde{y}_4-\tilde{y}_1 \\ \tilde{z}_2-\tilde{z}_1 & \tilde{z}_3-\tilde{z}_1 & \tilde{z}_4-\tilde{z}_1 \end{bmatrix}^{-T} \begin{bmatrix} 0 \\ 0 \\ -1 \end{bmatrix} \tilde{v}. \tag{25.18}$$

Equations (25.15), (25.16), (25.17) and (25.18) can be expressed using matrices as follows:

$$\begin{bmatrix} f_{i1} & f_{i2} & f_{i3} & f_{i4} \end{bmatrix}$$
$$= -\begin{bmatrix} c_{ii} & c_{ij} & c_{ik} \end{bmatrix} \begin{bmatrix} \tilde{x}_2-\tilde{x}_1 & \tilde{x}_3-\tilde{x}_1 & \tilde{x}_4-\tilde{x}_1 \\ \tilde{y}_2-\tilde{y}_1 & \tilde{y}_3-\tilde{y}_1 & \tilde{y}_4-\tilde{y}_1 \\ \tilde{z}_2-\tilde{z}_1 & \tilde{z}_3-\tilde{z}_1 & \tilde{z}_4-\tilde{z}_1 \end{bmatrix}^{-T} \begin{bmatrix} \frac{\partial N_1}{\partial \xi} & \frac{\partial N_2}{\partial \xi} & \frac{\partial N_3}{\partial \xi} & \frac{\partial N_4}{\partial \xi} \\ \frac{\partial N_1}{\partial \eta} & \frac{\partial N_2}{\partial \eta} & \frac{\partial N_3}{\partial \eta} & \frac{\partial N_4}{\partial \eta} \\ \frac{\partial N_1}{\partial \zeta} & \frac{\partial N_2}{\partial \zeta} & \frac{\partial N_3}{\partial \zeta} & \frac{\partial N_4}{\partial \zeta} \end{bmatrix} \tilde{v} \tag{25.19}$$

By an analogous repetition of the above derivation process for the **j** and **k** global directions, the **j** and **k** components of the equivalent nodal forces are obtained

$$\begin{bmatrix} f_{j1} & f_{j2} & f_{j3} & f_{j4} \end{bmatrix}$$
$$= -\begin{bmatrix} c_{ji} & c_{jj} & c_{jk} \end{bmatrix} \begin{bmatrix} \tilde{x}_2-\tilde{x}_1 & \tilde{x}_3-\tilde{x}_1 & \tilde{x}_4-\tilde{x}_1 \\ \tilde{y}_2-\tilde{y}_1 & \tilde{y}_3-\tilde{y}_1 & \tilde{y}_4-\tilde{y}_1 \\ \tilde{z}_2-\tilde{z}_1 & \tilde{z}_3-\tilde{z}_1 & \tilde{z}_4-\tilde{z}_1 \end{bmatrix}^{-T} \begin{bmatrix} \frac{\partial N_1}{\partial \xi} & \frac{\partial N_2}{\partial \xi} & \frac{\partial N_3}{\partial \xi} & \frac{\partial N_4}{\partial \xi} \\ \frac{\partial N_1}{\partial \eta} & \frac{\partial N_2}{\partial \eta} & \frac{\partial N_3}{\partial \eta} & \frac{\partial N_4}{\partial \eta} \\ \frac{\partial N_1}{\partial \zeta} & \frac{\partial N_2}{\partial \zeta} & \frac{\partial N_3}{\partial \zeta} & \frac{\partial N_4}{\partial \zeta} \end{bmatrix} \tilde{v} \tag{25.20}$$

$$\begin{bmatrix} f_{k1} & f_{k2} & f_{k3} & f_{k4} \end{bmatrix} = -\begin{bmatrix} c_{ki} & c_{kj} & c_{kk} \end{bmatrix} \begin{bmatrix} \tilde{x}_2-\tilde{x}_1 & \tilde{x}_3-\tilde{x}_1 & \tilde{x}_4-\tilde{x}_1 \\ \tilde{y}_2-\tilde{y}_1 & \tilde{y}_3-\tilde{y}_1 & \tilde{y}_4-\tilde{y}_1 \\ \tilde{z}_2-\tilde{z}_1 & \tilde{z}_3-\tilde{z}_1 & \tilde{z}_4-\tilde{z}_1 \end{bmatrix}^{-T} \begin{bmatrix} \frac{\partial N_1}{\partial \xi} & \frac{\partial N_2}{\partial \xi} & \frac{\partial N_3}{\partial \xi} & \frac{\partial N_4}{\partial \xi} \\ \frac{\partial N_1}{\partial \eta} & \frac{\partial N_2}{\partial \eta} & \frac{\partial N_3}{\partial \eta} & \frac{\partial N_4}{\partial \eta} \\ \frac{\partial N_1}{\partial \zeta} & \frac{\partial N_2}{\partial \zeta} & \frac{\partial N_3}{\partial \zeta} & \frac{\partial N_4}{\partial \zeta} \end{bmatrix} \tilde{v} \tag{25.21}$$

The equations can be combined into one matrix equation giving equivalent internal nodal forces in terms of the Cauchy stress tensor matrix

$$\begin{bmatrix} f_{i1} & f_{i2} & f_{i3} & f_{i4} \\ f_{j1} & f_{j2} & f_{j3} & f_{j4} \\ f_{k1} & f_{k2} & f_{k3} & f_{k4} \end{bmatrix} = -\begin{bmatrix} c_{ii} & c_{ij} & c_{ik} \\ c_{ji} & c_{jj} & c_{jk} \\ c_{ki} & c_{kj} & c_{kk} \end{bmatrix} \begin{bmatrix} \tilde{x}_2-\tilde{x}_1 & \tilde{x}_3-\tilde{x}_1 & \tilde{x}_4-\tilde{x}_1 \\ \tilde{y}_2-\tilde{y}_1 & \tilde{y}_3-\tilde{y}_1 & \tilde{y}_4-\tilde{y}_1 \\ \tilde{z}_2-\tilde{z}_1 & \tilde{z}_3-\tilde{z}_1 & \tilde{z}_4-\tilde{z}_1 \end{bmatrix}^{-T} \begin{bmatrix} \frac{\partial N_1}{\partial \xi} & \frac{\partial N_2}{\partial \xi} & \frac{\partial N_3}{\partial \xi} & \frac{\partial N_4}{\partial \xi} \\ \frac{\partial N_1}{\partial \eta} & \frac{\partial N_2}{\partial \eta} & \frac{\partial N_3}{\partial \eta} & \frac{\partial N_4}{\partial \eta} \\ \frac{\partial N_1}{\partial \zeta} & \frac{\partial N_2}{\partial \zeta} & \frac{\partial N_3}{\partial \zeta} & \frac{\partial N_4}{\partial \zeta} \end{bmatrix} \tilde{v} \tag{25.22}$$

or

$$\begin{bmatrix} \mathbf{f}_1 & \mathbf{f}_2 & \mathbf{f}_3 & \mathbf{f}_4 \end{bmatrix} = \begin{bmatrix} f_{i1} & f_{i2} & f_{i3} & f_{i4} \\ f_{j1} & f_{j2} & f_{j3} & f_{j4} \\ f_{k1} & f_{k2} & f_{k3} & f_{k4} \end{bmatrix} = \tilde{v} \begin{bmatrix} c_{ii} & c_{ij} & c_{ik} \\ c_{ji} & c_{jj} & c_{jk} \\ c_{ki} & c_{kj} & c_{kk} \end{bmatrix} \begin{bmatrix} \tilde{x}_2-\tilde{x}_1 & \tilde{x}_3-\tilde{x}_1 & \tilde{x}_4-\tilde{x}_1 \\ \tilde{y}_2-\tilde{y}_1 & \tilde{y}_3-\tilde{y}_1 & \tilde{y}_4-\tilde{y}_1 \\ \tilde{z}_2-\tilde{z}_1 & \tilde{z}_3-\tilde{z}_1 & \tilde{z}_4-\tilde{z}_1 \end{bmatrix}^{-T} \begin{bmatrix} 1 & -1 & 0 & 0 \\ 1 & 0 & -1 & 0 \\ 1 & 0 & 0 & -1 \end{bmatrix} \tag{25.23}$$

25.3 Reduce Integration for Eight-Noded 3D Solid

Reduced integration is often used with the eight-noded 3D solid finite element. Normally for the full integration case eight Gauss integration points are used with this type of element, while for the reduced integration case only one Gauss integration point is used, Figure 25.2.

Reduced integration is often used with this element in order to avoid the counter-productive numerical artifact known as shear locking.

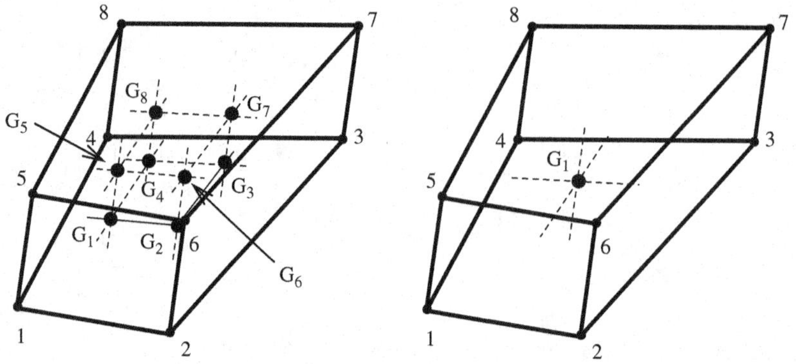

Figure 25.2 Full (eight Gauss points) and reduced (one Gauss point) integration formula for the eight-noded hexahedron finite element

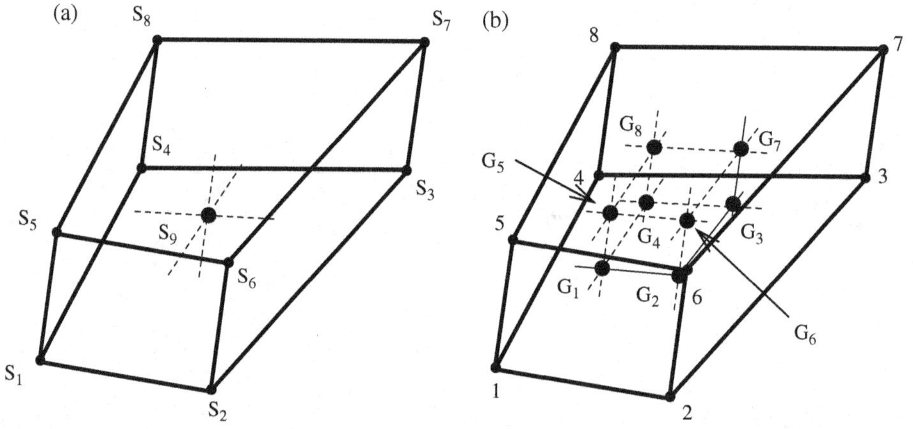

Figure 25.3 Selective integration for the eight-noded hexahedron finite element: (a) stretch sampling points; (b) Gauss integration points

25.4 Selective Stretch Sampling-Based Integration for the Eight-Noded Solid Finite Element

Reduced integration helps to solve the problem of shear and volumetric locking. However, due to the virtual work equations being under-integrated, it does lead to unconstrained degrees of freedom, also known as the zero energy modes. In order to address both locking and zero energy modes, the reduced integration method is replaced by selective integration. The concepts of selective and reduced integration were initially introduced in the context of small strain elasticity, where it really meant one simply changes the number of Gauss points.

In the context of the finite strain elasticity, this can be misleading and it is better to think in terms of a selective stretch sampling. In this manner, the integration formula need not change from full to the selective integration; the stretch sampling points change instead.

For example, in Figure 25.3 eight Gauss integration points are used. These are combined with nine stretch sampling points. Volumetric stretch is sampled at point S_9, and is assumed constant over the finite element. All other stretches are sampled at points S_1, S_2, S_3, S_4, S_5, S_6, S_7, S_8 and are combined with the constant shear and volumetric stretches from point S_9. As such, they are passed to the material package and the stress at points $S_1, S_2, \ldots, S_8$ is obtained. This stress is then interpolated over the domain of the finite element using the shape functions

$$\begin{aligned}\mathbf{C}(\xi,\eta,\zeta) = {} & \mathbf{C}_1\mathrm{N}_1(\xi,\eta,\zeta) + \mathbf{C}_2\mathrm{N}_2(\xi,\eta,\zeta) \\ & + \mathbf{C}_3\mathrm{N}_3(\xi,\eta,\zeta) + \mathbf{C}_4\mathrm{N}_4(\xi,\eta,\zeta) \\ & + \mathbf{C}_5\mathrm{N}_5(\xi,\eta,\zeta) + \mathbf{C}_6\mathrm{N}_6(\xi,\eta,\zeta) \\ & + \mathbf{C}_7\mathrm{N}_7(\xi,\eta,\zeta) + \mathbf{C}_8\mathrm{N}_8(\xi,\eta,\zeta).\end{aligned} \tag{25.24}$$

From Equation (25.24) the stress at the Gauss points is calculated and numerical integration of the virtual work is performed.

25.5 Summary

In this chapter a general procedure for obtaining nodal forces (the forces by which the finite element pulls its nodes) for 3D finite elements is described. The procedure uses: (a) stretch sampling points; (b) stress sampling points; (c) stress interpolation at Gauss points; and (d) numerical integration of virtual work over Gauss points.

Further Reading

[1] Bonet, J. and Burton, A. J. (1998) A simple average nodal pressure tetrahedral element for incompressible and nearly incompressible dynamic explicit applications. *Communications in Numerical Methods in Engineering*, **14**(5): 437–49.

[2] Bonet, J., Marriott, H. and Hassan, O. (2001) An averaged nodal deformation gradient linear tetrahedral element for large strain explicit dynamic applications. *Communications in Numerical Methods in Engineering*, **17**(8): 551–61.

[3] Bonet, J. and Wood, R. D. (1997) *Nonlinear Continuum Mechanics for Finite Element Analysis*. Cambridge, Cambridge University Press.

[4] Hughes, T.J.R. (2000) *The Finite Element Method: Linear Static and Dynamic Finite Element Analysis*. New York, Dover Publications.

[5] Liu, W. K., Guo, Y., Tang, S. and Belytschko. T. (1998) A multiple-quadrature eight-node hexahedral finite element for large deformation elastoplastic analysis. *Comput. Method Appl. M.*, **154**(1–2): 69–132.

[6] Munjiza, A. (2004) *The Combined Finite-Discrete Element Method*. Chichester, John Wiley & Sons, Ltd.

[7] Oden, J. T. (1972) *Finite Elements of Nonlinear Continua*. New York, McGraw-Hill.

[8] Peric, D. and Owen, D. R. J. (1998) Finite-element applications to the nonlinear mechanics of solids. *Reports on Progress in Physics*, **61**(11): 1495–1574.

[9] Zienkiewicz, O. C. (1971) *The Finite Element Method in Engineering Science*. New York, McGraw-Hill.

[10] Zienkiewicz, O. C., Rojek, J., Taylor, R. L. and Pastor, M. (1998) Triangles and tetrahera in explicit dynamic codes for solids. *International Journal for Numerical Methods in Engineering*, **43**(3): 565–83.

[11] Zienkiewicz, O. C. and Taylor, R. L. (2005) *The Finite Element Method Set*. Oxford, Elsevier Science.

26

Integration of Nodal Forces over Boundaries of Finite Elements

26.1 Stress at Element Boundaries

For highly nonlinear material laws even lower order displacement fields can produce higher order stress fields. For these cases it is not logical that the nodal forces on one boundary of the finite element are influenced by a stress state close to a completely different boundary, Figure 26.1.

To address this, it is convenient to obtain the nodal forces by considering stresses only at the boundaries of the finite element, Figure 26.2. Stress on the boundary between two adjacent nodes produces stress on those nodes only. Thus, in Figure 26.2 the nodal force on node 1 is produced by integrating the stress over boundaries 1–2 and 1–6.

This approach is especially useful with higher order and composite elements. One such composite elements is the six-noded-composite-triangle, as shown in Figure 26.3. This triangle is composed of four initially identical three-noded triangles.

The composite triangle produces better accuracy than the three-noded triangle for the same number of degrees of freedom. In many cases the accuracy is as good as the six-noded triangle because the mid-edge nodes are placed exactly at the mid-points between the corner nodes.

A big advantage of composite elements is the fact that the edges remain two-noded and straight, thus significantly simplifying the numerical solution of the other physical processes involved in the simulation, such as contact between the elements. In addition, the finite element formulation and implementation is simpler. Moreover, there is plenty of flexibility in how the equivalent nodal forces are obtained and full, reduced and selective integration are all relatively straight forward to implement.

Large Strain Finite Element Method: A Practical Course, First Edition. Antonio Munjiza, Esteban Rougier and Earl E. Knight.

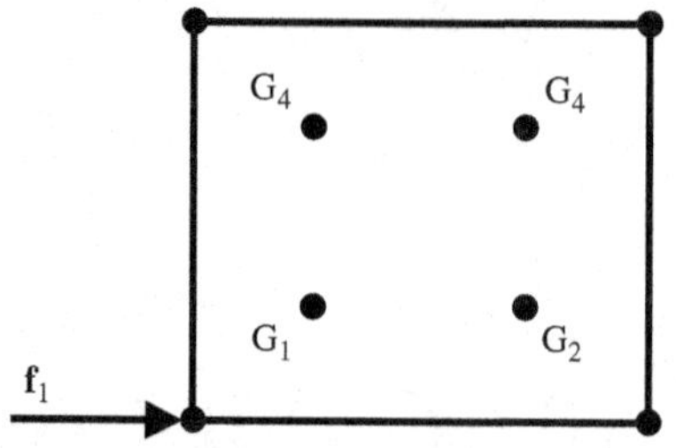

Figure 26.1 Nodal force at node 1 is influenced by the stress state at all Gauss points

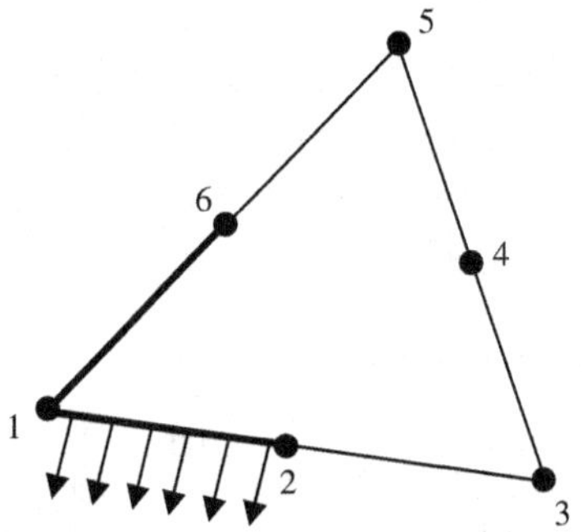

Figure 26.2 Stress between nodes 1 and 2 produces nodal forces on nodes 1 and 2 only

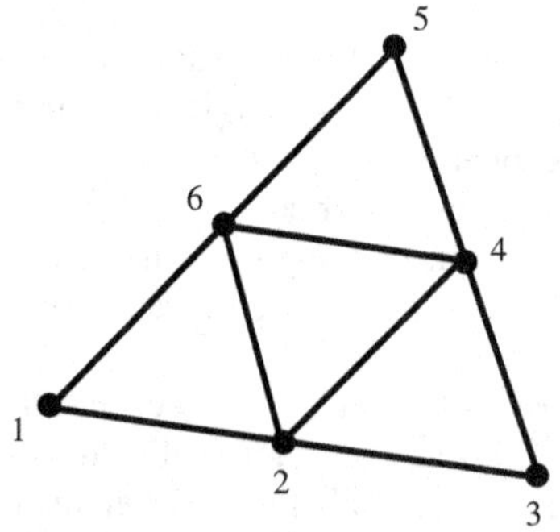

Figure 26.3 Six-noded composite triangle

26.2 Integration of the Equivalent Nodal Forces over the Triangle Finite Element

The three-noded triangle in 2D is usually integrated using one Gauss integration point, Figure 26.4.

The Volume of the Finite Element. The volume of the finite element in the current position is given by

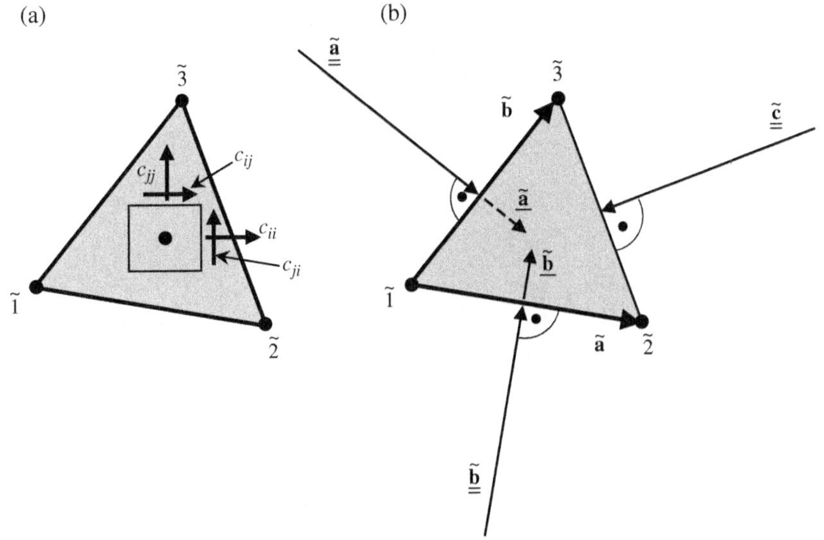

Figure 26.4 (a) One Gauss point integration using virtual work principle; (b) direct integration of nodal forces over the surfaces (edges) of the finite element

$$\tilde v = \frac{1}{2}\tilde{\mathbf a}\times\tilde{\mathbf b} = \frac{1}{2}\det\begin{bmatrix}\tilde a_i & \tilde b_i\\ \tilde a_j & \tilde b_j\end{bmatrix} = \frac{1}{2}\left(\tilde a_i\tilde b_j - \tilde a_j\tilde b_i\right), \tag{26.1}$$

where

$$\begin{bmatrix}\tilde{\mathbf a} & \tilde{\mathbf b}\end{bmatrix} = \begin{bmatrix}\tilde a_i & \tilde b_i\\ \tilde a_j & \tilde b_j\end{bmatrix} = \begin{bmatrix}\tilde x_2-\tilde x_1 & \tilde x_3-\tilde x_1\\ \tilde y_2-\tilde y_1 & \tilde y_3-\tilde y_1\end{bmatrix} \tag{26.2}$$

are the current edges of the finite element. These edges form a vector base, the dual base, which by definition, is given by

$$\begin{bmatrix}\underline{\tilde{\mathbf a}} & \underline{\tilde{\mathbf b}}\end{bmatrix} = \begin{bmatrix}\underline{\tilde a}_i & \underline{\tilde b}_i\\ \underline{\tilde a}_j & \underline{\tilde b}_j\end{bmatrix} = \left(\begin{bmatrix}\tilde a_i & \tilde b_i\\ \tilde a_j & \tilde b_j\end{bmatrix}^{-1}\right)^{\mathrm T} = \begin{bmatrix}\tilde a_i & \tilde b_i\\ \tilde a_j & \tilde b_j\end{bmatrix}^{-\mathrm T}. \tag{26.3}$$

Surface Vectors. The volume of the triangle is given by

$$\tilde v = \frac{1}{2}\underline{\underline{\tilde{\mathbf a}}}\bullet\tilde{\mathbf a} = \frac{1}{2}\underline{\underline{\tilde{\mathbf b}}}\bullet\tilde{\mathbf b}, \tag{26.4}$$

where $\underline{\underline{\tilde{\mathbf a}}}$ and $\underline{\underline{\tilde{\mathbf b}}}$ are the surface vectors shown in Figure 26.4. Since the surface vectors are parallel to the corresponding dual base vectors, it follows that

$$\tilde{v} = \frac{1}{2}\underline{\underline{\tilde{\mathbf{a}}}} \bullet \tilde{\mathbf{a}} = \frac{1}{2}\left(\underline{\tilde{\mathbf{a}}} \bullet \tilde{\mathbf{a}}\right)\tilde{v}$$

$$\tilde{v} = \frac{1}{2}\underline{\underline{\tilde{\mathbf{b}}}} \bullet \tilde{\mathbf{b}} = \frac{1}{2}\left(\underline{\tilde{\mathbf{b}}} \bullet \tilde{\mathbf{b}}\right)\tilde{v}, \tag{26.5}$$

which yields

$$\left[\underline{\underline{\tilde{\mathbf{a}}}} \;\; \underline{\underline{\tilde{\mathbf{b}}}}\right] = \tilde{v}\left[\underline{\tilde{\mathbf{a}}} \;\; \underline{\tilde{\mathbf{b}}}\right] = \begin{bmatrix} \tilde{y}_3 - \tilde{y}_1 & \tilde{y}_1 - \tilde{y}_2 \\ \tilde{x}_1 - \tilde{x}_3 & \tilde{x}_2 - \tilde{x}_1 \end{bmatrix} \tag{26.6}$$

and

$$\underline{\underline{\tilde{\mathbf{c}}}} = -\left(\underline{\underline{\tilde{\mathbf{a}}}} + \underline{\underline{\tilde{\mathbf{b}}}}\right). \tag{26.7}$$

The Equivalent Nodal Forces. The equivalent nodal forces on each of the nodes of the triangle are obtained by considering the internal forces across each of the surfaces of the triangle and splitting them between the two nodes that define the particular surface. Thus, the force on nodes 1, 2 and 3 are given by

$$\mathbf{f}_1 = \frac{1}{2}\underline{\underline{\mathbf{f}}}_a + \frac{1}{2}\underline{\underline{\mathbf{f}}}_b \tag{26.8}$$

$$\mathbf{f}_2 = \frac{1}{2}\underline{\underline{\mathbf{f}}}_b + \frac{1}{2}\underline{\underline{\mathbf{f}}}_c = \frac{1}{2}\underline{\underline{\mathbf{f}}}_b - \frac{1}{2}\left(\underline{\underline{\mathbf{f}}}_a + \underline{\underline{\mathbf{f}}}_b\right) = -\frac{1}{2}\underline{\underline{\mathbf{f}}}_a \tag{26.9}$$

$$\mathbf{f}_3 = \frac{1}{2}\underline{\underline{\mathbf{f}}}_a + \frac{1}{2}\underline{\underline{\mathbf{f}}}_c = \frac{1}{2}\underline{\underline{\mathbf{f}}}_a - \frac{1}{2}\left(\underline{\underline{\mathbf{f}}}_a + \underline{\underline{\mathbf{f}}}_b\right) = -\frac{1}{2}\underline{\underline{\mathbf{f}}}_b. \tag{26.10}$$

The resultant internal forces through surfaces $\underline{\underline{\tilde{\mathbf{a}}}}$ and $\underline{\underline{\tilde{\mathbf{b}}}}$ are obtained from the Cauchy stress tensor matrix

$$\left[\underline{\underline{\mathbf{f}}}_a \;\; \underline{\underline{\mathbf{f}}}_b\right] = \begin{bmatrix} c_{ii} & c_{ij} \\ c_{ji} & c_{jj} \end{bmatrix}\left[\underline{\underline{\tilde{\mathbf{a}}}} \;\; \underline{\underline{\tilde{\mathbf{b}}}}\right] = \begin{bmatrix} c_{ii} & c_{ij} \\ c_{ji} & c_{jj} \end{bmatrix}\begin{bmatrix} \tilde{y}_3 - \tilde{y}_1 & \tilde{y}_1 - \tilde{y}_2 \\ \tilde{x}_1 - \tilde{x}_3 & \tilde{x}_2 - \tilde{x}_1 \end{bmatrix}, \tag{26.11}$$

which, when substituted into equation (26.8), (26.9), and (26.10) yields

$$\begin{aligned} \left[\mathbf{f}_1 \;\; \mathbf{f}_2 \;\; \mathbf{f}_3\right] &= \begin{bmatrix} f_{i1} & f_{i2} & f_{i3} \\ f_{j1} & f_{j2} & f_{j3} \end{bmatrix} \\ &= \begin{bmatrix} c_{ii} & c_{ij} \\ c_{ji} & c_{jj} \end{bmatrix}\begin{bmatrix} \tilde{y}_3 - \tilde{y}_1 & \tilde{y}_1 - \tilde{y}_2 \\ \tilde{x}_1 - \tilde{x}_3 & \tilde{x}_2 - \tilde{x}_1 \end{bmatrix}\begin{bmatrix} 1/2 & -1/2 & 0 \\ 1/2 & 0 & -1/2 \end{bmatrix}, \end{aligned} \tag{26.12}$$

which yields

$$\begin{aligned}[\mathbf{f}_1\ \mathbf{f}_2\ \mathbf{f}_3] &= \begin{bmatrix} f_{i1} & f_{i2} & f_{i3} \\ f_{j1} & f_{j2} & f_{j3} \end{bmatrix} \\ &= \frac{1}{2}\begin{bmatrix} c_{ii} & c_{ij} \\ c_{ji} & c_{jj} \end{bmatrix}\begin{bmatrix} \tilde{y}_3-\tilde{y}_1 & \tilde{y}_1-\tilde{y}_2 \\ \tilde{x}_1-\tilde{x}_3 & \tilde{x}_2-\tilde{x}_1 \end{bmatrix}\begin{bmatrix} 1 & -1 & 0 \\ 1 & 0 & -1 \end{bmatrix}.\end{aligned} \tag{26.13}$$

The obtained forces represent the forces by which the finite element pulls its nodes. The final result is the same as the result obtained using the method of virtual work, see Equation (21.36).

26.3 Integration over the Boundary of the Composite Triangle

The composite triangle has six surfaces, as shown in Figure 26.5. The resultant internal forces for each surface are obtained by integrating the components of the Cauchy stress tensor matrix over that surface and splitting the resultant force to the two nodes that define the surface; for example, for surface ${}_{5\_6}\underline{\underline{\tilde{\mathbf{a}}}}$, the force is split between nodes 5 and 6. As the Cauchy stress over the surface ${}_{5\_6}\underline{\underline{\tilde{\mathbf{a}}}}$ is constant, this means that the force is

$${}_{5\_6}\mathbf{f} = {}_{5\_6}\mathbf{C}_{5\_6}\underline{\underline{\tilde{\mathbf{a}}}} = \begin{bmatrix} {}_{5\_6}c_{ii} & {}_{5\_6}c_{ij} \\ {}_{5\_6}c_{ji} & {}_{5\_6}c_{jj} \end{bmatrix}\begin{bmatrix} \tilde{y}_5 - \tilde{y}_6 \\ \tilde{x}_6 - \tilde{x}_5 \end{bmatrix}. \tag{26.14}$$

The stress tensor is calculated at the center of the sub-triangle $\tilde{4}-\tilde{5}-\tilde{6}$ (stress sampling point S_4). This is done by considering the stretches, which can be chosen selectively. For instance, the volumetric stretch can be calculated for the composite triangle as a whole and combined with the other stretches calculated at the stress sampling point S_4, Figure 26.5.

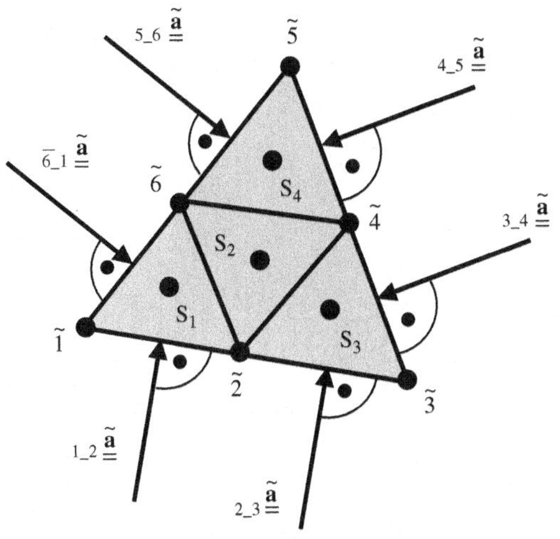

Figure 26.5 The surfaces of the composite triangle together with stress sampling points

26.4 Integration over the Boundary of the Six-Noded Triangle

In the case of the six-noded triangle, Figure 26.6, a number of options are available to integrate the internal forces over its boundaries.

One option is to integrate the stress over the boundary numerically. Take, for example the surface defined by nodes $\tilde{5}$ and $\tilde{6}$. The stress over this surface is not constant. This means that, in order to properly represent the stress distribution through the equivalent nodal forces, one needs two Gauss integration points, as shown in Figure 26.7.

First the local coordinate system for the surface is defined such that it passes through nodes $\tilde{5}$ and $\tilde{6}$, and that it follows the deformed surface. Now, two Gauss points G_1 and G_2 are introduced, such that

$$\xi_1 = \frac{1}{\sqrt{3}}; \ \xi_2 = -\frac{1}{\sqrt{3}} \tag{26.15}$$

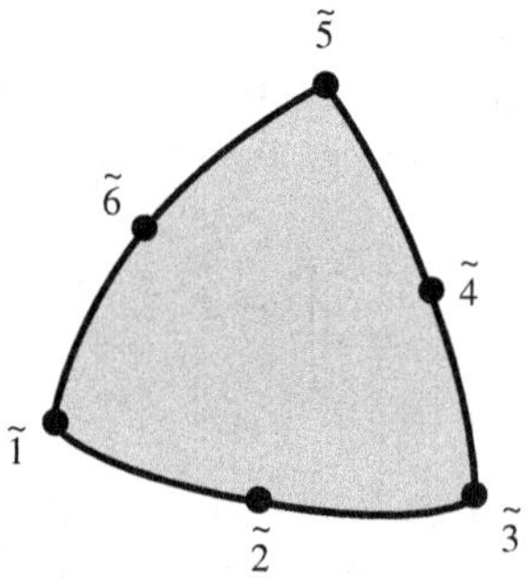

Figure 26.6 Stress integration over the boundary of the six-noded triangle

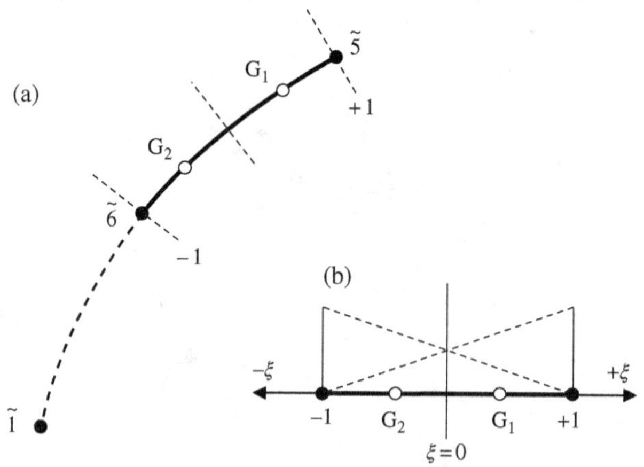

Figure 26.7 Integration of internal forces over the surface $\tilde{5} - \tilde{6}$: (a) the Gauss points; (b) the virtual displacement

The equivalent nodal force is integrated using these two Gauss points and then distributed to the nodes $\tilde{5}$ and $\tilde{6}$.

The force distribution is best done by supplying a virtual displacement of the boundary only, Figure 26.7.

Stress sampling is usually not done at the Gauss points, but at the nodes of the finite element. Thus, a different stress state may be obtained at each of the nodes.

In order to obtain these nodal stresses, a selective sampling of the stretches is employed enabling the full, reduced and selective integration concepts to be implemented. The stress at the Gauss points is obtained by interpolating the stress over the nodes using the shape functions

$$\begin{aligned}\mathbf{C}(\xi,\eta)=&\,\mathbf{C}_1\mathrm{N}_1(\xi,\eta)+\mathbf{C}_2\mathrm{N}_2(\xi,\eta)+\mathbf{C}_3\mathrm{N}_3(\xi,\eta)\\&+\mathbf{C}_4\mathrm{N}_4(\xi,\eta)+\mathbf{C}_5\mathrm{N}_5(\xi,\eta)+\mathbf{C}_6\mathrm{N}_6(\xi,\eta).\end{aligned} \tag{26.16}$$

26.5 Integration of the Equivalent Internal Nodal Forces over the Tetrahedron Boundaries

The surface 2–3–4 of the tetrahedron finite element shown in Figure 26.8 is defined by the vector ${}_{2\_3\_4}\underline{\underline{\tilde{\mathbf{n}}}}$, where "~" indicates that the surface ${}_{2\_3\_4}\mathbf{n}$ is considered in its current (not previous, not initial) position.

The components of the Cauchy stress tensor matrix define the internal forces on the surfaces of the Cauchy material element. The Cauchy material element is a cube made of material points in their current positions. By definition, it follows that the resultant internal force on surface 2–3–4 is obtained by multiplying the Cauchy stress tensor matrix by the matrix of the surface vector,

$${}_{2\_3\_4}\mathbf{f}=\begin{bmatrix}{}_{2\_3\_4}f\\{}_{2\_3\_4}f\\{}_{2\_3\_4}f\end{bmatrix}=\begin{bmatrix}c_{ii}&c_{ij}&c_{ik}\\c_{ji}&c_{jj}&c_{jk}\\c_{ki}&c_{kj}&c_{kk}\end{bmatrix}\begin{bmatrix}{}_{2\_3\_4}n_i\\{}_{2\_3\_4}n_j\\{}_{2\_3\_4}n_k\end{bmatrix}. \tag{26.17}$$

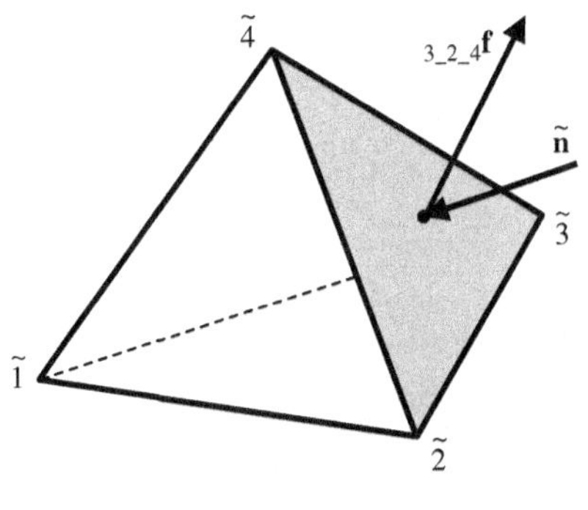

Figure 26.8 Integration of nodal forces over the boundary of the tetrahedron finite element

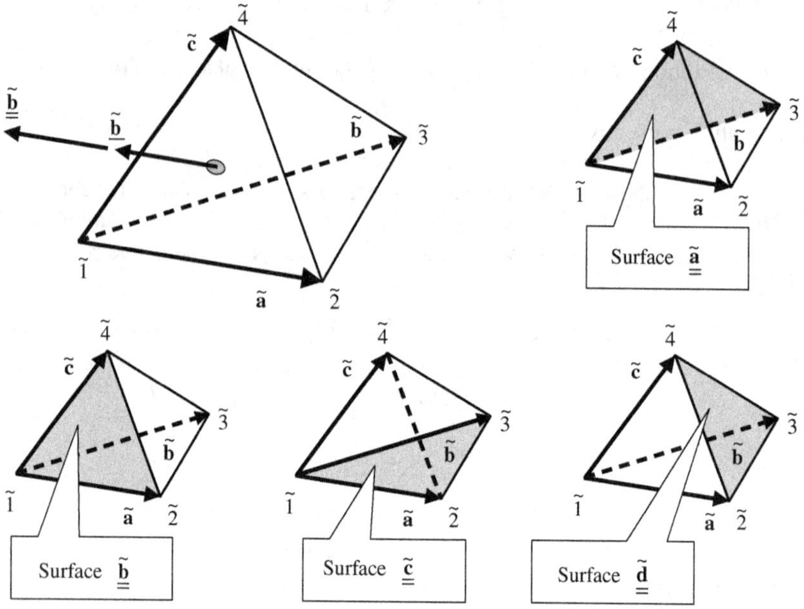

Figure 26.9 The edges and surfaces of the tetrahedron

Since the stress is constant over the surface, the resultant force is equally distributed among nodes 2, 3 and 4, thus producing the equivalent nodal forces

$$
\begin{aligned}
{}_2\mathbf{f} &= \tfrac{1}{3}\left({}_{2\_3\_4}\mathbf{f}\right) \\
{}_3\mathbf{f} &= \tfrac{1}{3}\left({}_{2\_3\_4}\mathbf{f}\right). \\
{}_4\mathbf{f} &= \tfrac{1}{3}\left({}_{2\_3\_4}\mathbf{f}\right)
\end{aligned}
\tag{26.18}
$$

The surfaces of the tetrahedron. The surfaces of the tetrahedron are obtained by considering the edges of the tetrahedron as shown in Figure 26.9. These edges form a vector base

$$
\begin{bmatrix} \tilde{\mathbf{a}} & \tilde{\mathbf{b}} & \tilde{\mathbf{c}} \end{bmatrix} = \begin{bmatrix} \tilde{x}_2 - \tilde{x}_1 & \tilde{x}_3 - \tilde{x}_1 & \tilde{x}_4 - \tilde{x}_1 \\ \tilde{y}_2 - \tilde{y}_1 & \tilde{y}_3 - \tilde{y}_1 & \tilde{y}_4 - \tilde{y}_1 \\ \tilde{z}_2 - \tilde{z}_1 & \tilde{z}_3 - \tilde{z}_1 & \tilde{z}_4 - \tilde{z}_1 \end{bmatrix}. \tag{26.19}
$$

The dual base is given by

$$
\begin{bmatrix} \underline{\tilde{\mathbf{a}}} & \underline{\tilde{\mathbf{b}}} & \underline{\tilde{\mathbf{c}}} \end{bmatrix} = \begin{bmatrix} \underline{\tilde{a}}_i & \underline{\tilde{b}}_i & \underline{\tilde{c}}_i \\ \underline{\tilde{a}}_j & \underline{\tilde{b}}_j & \underline{\tilde{c}}_j \\ \underline{\tilde{a}}_k & \underline{\tilde{b}}_k & \underline{\tilde{c}}_k \end{bmatrix} = \begin{bmatrix} \tilde{x}_2 - \tilde{x}_1 & \tilde{x}_3 - \tilde{x}_1 & \tilde{x}_4 - \tilde{x}_1 \\ \tilde{y}_2 - \tilde{y}_1 & \tilde{y}_3 - \tilde{y}_1 & \tilde{y}_4 - \tilde{y}_1 \\ \tilde{z}_2 - \tilde{z}_1 & \tilde{z}_3 - \tilde{z}_1 & \tilde{z}_4 - \tilde{z}_1 \end{bmatrix}^{-T}. \tag{26.20}
$$

The corresponding surfaces (Figure 26.9) are obtained from the dual base as follows

$$\begin{bmatrix} \underline{\underline{\tilde{\mathbf{a}}}} & \underline{\underline{\tilde{\mathbf{b}}}} & \underline{\underline{\tilde{\mathbf{c}}}} \end{bmatrix} = 3\tilde{v}\begin{bmatrix} \underline{\tilde{\mathbf{a}}} & \underline{\tilde{\mathbf{b}}} & \underline{\tilde{\mathbf{c}}} \end{bmatrix} = 3\tilde{v}\begin{bmatrix} \tilde{x}_2-\tilde{x}_1 & \tilde{x}_3-\tilde{x}_1 & \tilde{x}_4-\tilde{x}_1 \\ \tilde{y}_2-\tilde{y}_1 & \tilde{y}_3-\tilde{y}_1 & \tilde{y}_4-\tilde{y}_1 \\ \tilde{z}_2-\tilde{z}_1 & \tilde{z}_3-\tilde{z}_1 & \tilde{z}_4-\tilde{z}_1 \end{bmatrix}^{-T} \tag{26.21}$$

where the current volume of the tetrahedron, $\tilde{v}$, is given by

$$\begin{aligned} \tilde{v} &= \frac{1}{6}\det\begin{bmatrix} \tilde{x}_2-\tilde{x}_1 & \tilde{x}_3-\tilde{x}_1 & \tilde{x}_4-\tilde{x}_1 \\ \tilde{y}_2-\tilde{y}_1 & \tilde{y}_3-\tilde{y}_1 & \tilde{y}_4-\tilde{y}_1 \\ \tilde{z}_2-\tilde{z}_1 & \tilde{z}_3-\tilde{z}_1 & \tilde{z}_4-\tilde{z}_1 \end{bmatrix} \\ &= \frac{1}{6}\begin{vmatrix} \tilde{x}_2-\tilde{x}_1 & \tilde{x}_3-\tilde{x}_1 & \tilde{x}_4-\tilde{x}_1 \\ \tilde{y}_2-\tilde{y}_1 & \tilde{y}_3-\tilde{y}_1 & \tilde{y}_4-\tilde{y}_1 \\ \tilde{z}_2-\tilde{z}_1 & \tilde{z}_3-\tilde{z}_1 & \tilde{z}_4-\tilde{z}_1 \end{vmatrix}. \end{aligned} \tag{26.22}$$

Equation (26.21) is obtained by considering that

$$\tilde{v} = \frac{1}{3}\underline{\underline{\tilde{\mathbf{a}}}} \bullet \tilde{\mathbf{a}} = \frac{1}{3}\underline{\underline{\tilde{\mathbf{b}}}} \bullet \tilde{\mathbf{b}} = \frac{1}{3}\underline{\underline{\tilde{\mathbf{c}}}} \bullet \tilde{\mathbf{c}}, \tag{26.23}$$

while

$$\underline{\tilde{\mathbf{a}}} \bullet \tilde{\mathbf{a}} = \underline{\tilde{\mathbf{b}}} \bullet \tilde{\mathbf{b}} = \underline{\tilde{\mathbf{c}}} \bullet \tilde{\mathbf{c}} = 1 \tag{26.24}$$

and vectors $\underline{\underline{\tilde{\mathbf{a}}}}$ and $\underline{\tilde{\mathbf{a}}}$; $\underline{\underline{\tilde{\mathbf{b}}}}$ and $\underline{\tilde{\mathbf{b}}}$; $\underline{\underline{\tilde{\mathbf{c}}}}$ and $\underline{\tilde{\mathbf{c}}}$ are parallel to each other, Figure 26.9.

From Equation (26.23), it follows that

$$\underline{\underline{\tilde{\mathbf{a}}}} \bullet \tilde{\mathbf{a}} = 3\tilde{v} \quad \text{and} \quad 3\left(\underline{\tilde{\mathbf{a}}} \bullet \tilde{\mathbf{a}}\right)\tilde{v} = 3\tilde{v}. \tag{26.25}$$

This yields

$$\underline{\underline{\tilde{\mathbf{a}}}} \bullet \tilde{\mathbf{a}} = 3\left(\underline{\tilde{\mathbf{a}}} \bullet \tilde{\mathbf{a}}\right)\tilde{v} \tag{26.26}$$

and

$$\underline{\underline{\tilde{\mathbf{a}}}} = 3\tilde{v}\,\underline{\tilde{\mathbf{a}}}. \tag{26.27}$$

The surface $\underline{\underline{\tilde{\mathbf{d}}}}$ shown in Figure 26.9 is simply

$$\underline{\underline{\tilde{\mathbf{d}}}} = -\left(\underline{\underline{\tilde{\mathbf{a}}}} + \underline{\underline{\tilde{\mathbf{b}}}} + \underline{\underline{\tilde{\mathbf{c}}}}\right). \tag{26.28}$$

The equivalent internal force on each of the nodes is obtained by summing the contributions from the three surfaces that meet at that node. This yields the equivalent nodal forces

$$
\begin{aligned}
[\mathbf{f}_1\ \mathbf{f}_2\ \mathbf{f}_3\ \mathbf{f}_4] &= [\mathbf{i}\ \mathbf{j}\ \mathbf{k}]\begin{bmatrix} f_{i1} & f_{i2} & f_{i3} & f_{i4} \\ f_{j1} & f_{j2} & f_{j3} & f_{j4} \\ f_{k1} & f_{k2} & f_{k3} & f_{k4} \end{bmatrix} \\
&= \begin{bmatrix} f_{i1} & f_{i2} & f_{i3} & f_{i4} \\ f_{j1} & f_{j2} & f_{j3} & f_{j4} \\ f_{k1} & f_{k2} & f_{k3} & f_{k4} \end{bmatrix} \\
&= \begin{bmatrix} c_{ii} & c_{ij} & c_{ik} \\ c_{ji} & c_{jj} & c_{jk} \\ c_{ki} & c_{kj} & c_{kk} \end{bmatrix}\left(3\tilde{v}\begin{bmatrix} \tilde{x}_2-\tilde{x}_1 & \tilde{x}_3-\tilde{x}_1 & \tilde{x}_4-\tilde{x}_1 \\ \tilde{y}_2-\tilde{y}_1 & \tilde{y}_3-\tilde{y}_1 & \tilde{y}_4-\tilde{y}_1 \\ \tilde{z}_2-\tilde{z}_1 & \tilde{z}_3-\tilde{z}_1 & \tilde{z}_4-\tilde{z}_1 \end{bmatrix}^{-T}\begin{bmatrix} 1/3 & -1/3 & 0 & 0 \\ 1/3 & 0 & -1/3 & 0 \\ 1/3 & 0 & 0 & -1/3 \end{bmatrix}\right) \\
&= \begin{bmatrix} c_{ii} & c_{ij} & c_{ik} \\ c_{ji} & c_{jj} & c_{jk} \\ c_{ki} & c_{kj} & c_{kk} \end{bmatrix}\left(\tilde{v}\begin{bmatrix} \tilde{x}_2-\tilde{x}_1 & \tilde{x}_3-\tilde{x}_1 & \tilde{x}_4-\tilde{x}_1 \\ \tilde{y}_2-\tilde{y}_1 & \tilde{y}_3-\tilde{y}_1 & \tilde{y}_4-\tilde{y}_1 \\ \tilde{z}_2-\tilde{z}_1 & \tilde{z}_3-\tilde{z}_1 & \tilde{z}_4-\tilde{z}_1 \end{bmatrix}^{-T}\begin{bmatrix} 1 & -1 & 0 & 0 \\ 1 & 0 & -1 & 0 \\ 1 & 0 & 0 & -1 \end{bmatrix}\right).
\end{aligned}
\tag{26.29}
$$

The obtained forces represent the forces by which the tetrahedron pulls its four nodes. The forces are the same as those obtained using the principle of virtual work, see Equation (25.23).

26.6 Summary

Apart from using the principle of virtual work to calculate the equivalent nodal forces, it is possible to integrate the nodal forces over the boundaries of the finite element.

The integration over the boundary of the finite element is possible both in 2D and in 3D problems. In a sense, it generalizes the process of integration of equivalent nodal forces. The stretch sampling points are separated from the stress calculation points. In this manner, different locking mechanisms (shear, volumetric, etc.) are eliminated by using a suitable set of stretch sampling points. By combining different stretch components at the stress calculation points, the Cauchy stress matrix at each of the stress calculation points is obtained. By interpolation of this stress over the domain of the finite element, the Cauchy stress matrix at the surfaces of the finite elements is calculated. By integration over the surface, the equivalent nodal forces at each of the nodes making a particular surface are obtained.

Selective sampling of stretches for stress calculations comes naturally when either volumetric or shear locking problems need to be resolved.

For problems with high material nonlinearity, stress calculation points are usually the nodes of the finite element. Consequently, the stress distribution over the finite element is conveniently described using the shape functions. This produces a nonuniform stress distribution over the surfaces of the finite elements. Integration of this stress is usually done using Gaussian integration over the boundary, while the distribution of forces to nodes is done by taking into account the nonuniform stress distribution over the surface.

Further Reading

[1] Bonet, J. and Burton, A. J. (1998) A simple average nodal pressure tetrahedral element for incompressible and nearly incompressible dynamic explicit applications. *Communications in Numerical Methods in Engineering*, **14**(5): 437–49.
[2] Bonet, J., Marriott, H. and Hassan, O. (2001) An averaged nodal deformation gradient linear tetrahedral element for large strain explicit dynamic applications. *Communications in Numerical Methods in Engineering*, **17**(8): 551–61.
[3] Bonet, J. and Wood, R. D. (1997) *Nonlinear Continuum Mechanics for Finite Element Analysis*. Cambridge, Cambridge University Press.
[4] Hughes, T.J.R. (2000) *The Finite Element Method: Linear Static and Dynamic Finite Element Analysis*. New York, Dover Publications.
[5] Liu, W. K., Guo, Y., Tang, S. and Belytschko. T. (1998) A multiple-quadrature eight-node hexahedral finite element for large deformation elastoplastic analysis. *Comput. Method Appl. M.*, **154**(1–2): 69–132.
[6] Munjiza, A. (2004) *The Combined Finite-Discrete Element Method*. Chichester, John Wiley & Sons, Ltd.
[7] Oden, J. T. (1972) *Finite Elements of Nonlinear Continua*. New York, McGraw-Hill.
[8] Peric, D. and Owen, D. R. J. (1998) Finite-element applications to the nonlinear mechanics of solids. *Reports on Progress in Physics*, **61**(11): 1495–1574.
[9] Zienkiewicz, O. C. (1971) *The Finite Element Method in Engineering Science*. New York, McGraw-Hill.
[10] Zienkiewicz, O. C., Rojek, J., Taylor, R. L. and Pastor, M. (1998) Triangles and tetrahera in explicit dynamic codes for solids. *International Journal for Numerical Methods in Engineering*, **43**(3), 565–83.
[11] Zienkiewicz, O. C. and Taylor, R. L. (2005) *The Finite Element Method Set*. Oxford, Elsevier Science.

Part Six

The Finite Element Method in 2.5D

27

Deformation in 2.5D Using Membrane Finite Elements

27.1 Solids in 2.5D

Large displacements are especially important when dealing with 2.5D problems. These include plates, membranes and shells. A typical 2.5D membrane is shown in Figure 27.1.

The problem shown is a tent, which is nothing more than a piece of cloth that offers no resistance in bending. The only internal forces it can produce are in the plane of the cloth. These forces are therefore called membrane internal forces. As the wind blows, the tent changes its shape in order to produce an equilibrium of internal and external forces. In other words, equilibrium is possible only through large displacements producing a load-dependent final shape.

A special case of another membrane is a simple rope. As the rope is loaded (think of a washing line), it changes its shape and as it does equilibrium is produced, Figure 27.2.

Membranes are widely used as lightweight load bearing structures in architecture (London dome), mechanical engineering (fuel pumps), aerospace engineering (parachutes, helium balloons and hot air balloons), naval architecture (sails), medical engineering (cell membranes), electronics, chemical engineering, etc.

As explained above, membranes always involve large displacements. Large displacements simply mean that the equilibrium equations have to be written considering the geometry of the deformed membrane (the current geometry) and not the initial load-free geometry.

Many membranes are made of glass, steel or similar types of materials. In normal conditions, such materials cannot undergo large strains, i.e. they cannot stretch by say 10–50%. Consequently, this is the case wherein one encounters large displacements, but small strains.

Membranes are also made of materials such as rubber, which can stretch to the extent that the original length is doubled or tripled. In these cases, the small strain formulations produce incorrect results and the only way to properly describe the problem is to use the finite

Large Strain Finite Element Method: A Practical Course, First Edition. Antonio Munjiza, Esteban Rougier and Earl E. Knight.

Figure 27.1 A membrane (tent) subject to wind loads

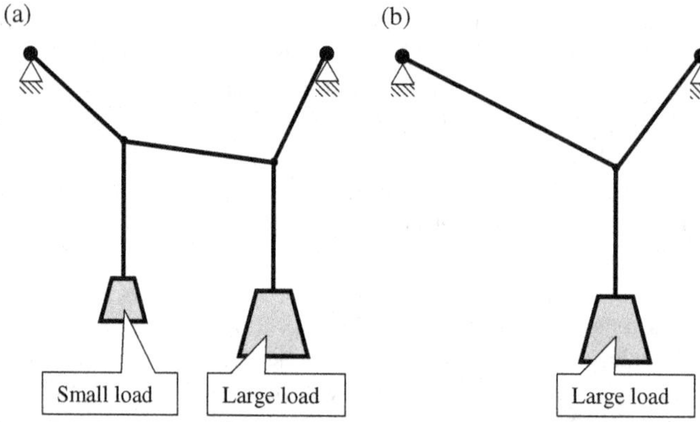

Figure 27.2 (a) A piece of string under two-point loads, one heavy (large) and another light (small); (b) a piece of string under a heavy one-point load

(large) strain formulation. This is therefore the case of both large strains and large displacements.

In the past, the additive decomposition within the co-rotational framework has been applied to solve these types of problems. The co-rotational formulation is a convenient way of reusing the small-strain-small-displacement formulation to approximate large-strain-large-displacement physics. However, the co-rotational formulation is only an approximation of both large displacements and large strains.

In contrast, the consistent large-strain-large-displacement formulation described in this book is based on multiplicative decomposition of deformation. As such, it is the exact formulation for both large displacements and large strains. In practical implementations it can be as simple as its small-strain-small-displacement counterpart. In other words, it is actually simpler to implement than the co-rotational formulation.

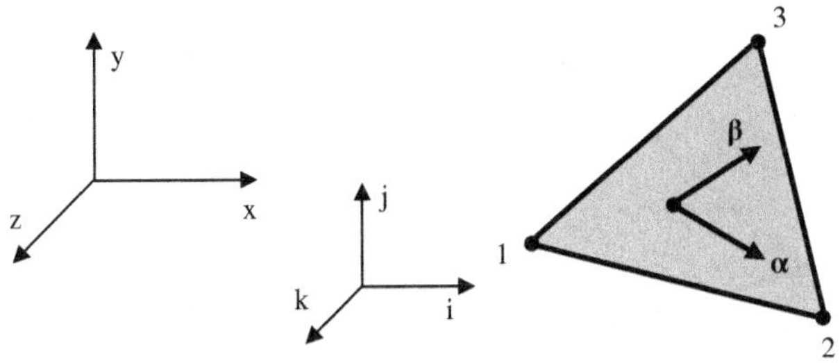

Figure 27.3 The geometry of the three-noded membrane finite element with the material-embedded vector base $[\boldsymbol{\alpha} \ \boldsymbol{\beta}]$

27.2 The Homogeneous Deformation Three-Noded Triangular Membrane Finite Element

One of the simplest membrane finite elements comprising the exact multiplicative decomposition-based formulation for both large-strains and large-displacements is the three-noded triangular membrane finite element.

The Geometry of the Element. The geometry of the finite element is described by three nodes, Figure 27.3.

Material-Embedded Material Element. For each finite element the directional characterization of the material within the finite element (anisotropic material) is defined by the 2.5D generalized material element and it is defined as input to the finite element simulation via the following two vectors

$$\boldsymbol{\alpha} = \begin{bmatrix} \alpha_i \\ \alpha_j \\ \alpha_k \end{bmatrix} \quad \boldsymbol{\beta} = \begin{bmatrix} \beta_i \\ \beta_j \\ \beta_k \end{bmatrix}. \tag{27.1}$$

These two vectors are conveniently called material-embedded vectors and they form the material-embedded base, or simply the material base

$$[\boldsymbol{\alpha} \ \boldsymbol{\beta}] = \begin{bmatrix} \alpha_i & \beta_i \\ \alpha_j & \beta_j \\ \alpha_k & \beta_k \end{bmatrix} \text{(im)}. \tag{27.2}$$

Note that vectors $\boldsymbol{\alpha}$ and $\boldsymbol{\beta}$ are infinitesimally small, thus an "im" ("infi-meter") unit for length is used in Equation (27.2).

Solid-embedded Coordinate System. It is convenient to describe the position of any material point P within the finite element by its local coordinates ξ and η

$$\text{P} = (\xi, \eta), \tag{27.3}$$

where the coordinate lines are embedded in the solid and move (deform) together with the material points of the solid, Figure 27.4. This coordinate system is therefore called the solid-embedded coordinate system.

Using the local coordinates ξ and η, the shape functions for the finite element are introduced as follows

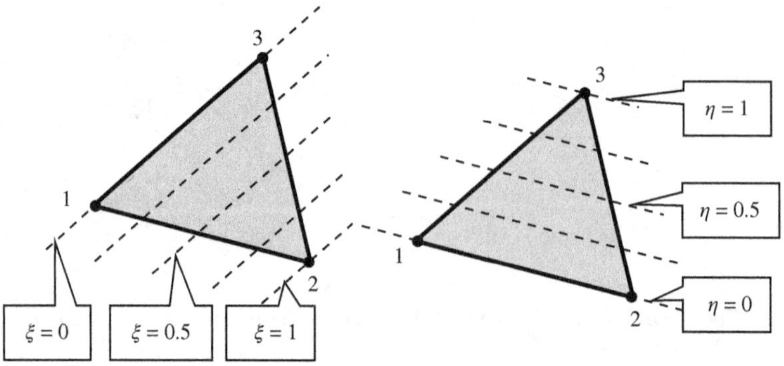

Figure 27.4 The solid-embedded local coordinate system – dashed lines show the solid-embedded coordinate lines

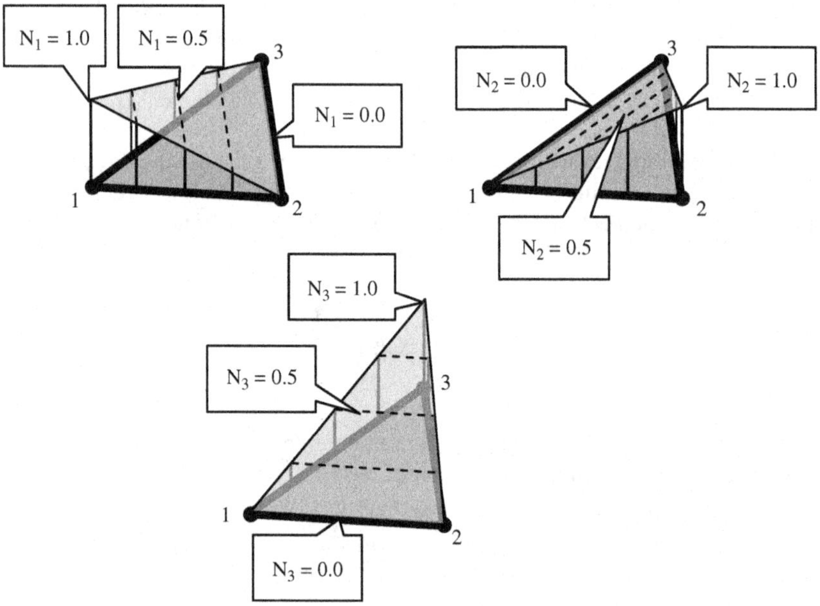

Figure 27.5 The shape functions

$$
\begin{aligned}
N_1(\xi, \eta) &= 1 - \xi - \eta \\
N_2(\xi, \eta) &= \xi \\
N_3(\xi, \eta) &= \eta,
\end{aligned}
\tag{27.4}
$$

as shown in Figure 27.5.

Global Coordinates of Material Points. As the solid deforms, its material points move in space; nevertheless, the local coordinates (ξ, η) for any given material point P stay constant.

The following positions of the finite element are of particular interest:

- Initial position
- Previous position
- Current position .

The initial position of the finite element is given as an input to the finite element analysis. This is done by supplying the initial coordinates of the three nodes of the triangle.

The finite element analysis is done in an explicit manner using either time steps (for dynamic problems) or load steps (for static problems). In this manner, the position of the finite element corresponding to the current step is called the current position, while the position of the finite element from the step just before the current step is called the previous position.

The difference in nodal positions between the current and the previous position is usually relatively small in comparison to the difference between the current position and the initial position, which can be very large, Figure 27.6.

The global coordinates of the nodes of the finite element are updated each time step and as such are stored in the finite element analysis database. The initial coordinates of the nodes are as follows

$$\begin{aligned} \bar{1} &= (\bar{x}_1 \;\; \bar{y}_1 \;\; \bar{z}_1) \\ \bar{2} &= (\bar{x}_2 \;\; \bar{y}_2 \;\; \bar{z}_2) \\ \bar{3} &= (\bar{x}_3 \;\; \bar{y}_3 \;\; \bar{z}_3). \end{aligned} \tag{27.5}$$

These can be described in matrix form

$$\bar{\mathbf{X}} = \begin{bmatrix} \bar{x}_1 & \bar{y}_1 & \bar{z}_1 \\ \bar{x}_2 & \bar{y}_2 & \bar{z}_2 \\ \bar{x}_3 & \bar{y}_3 & \bar{z}_3 \end{bmatrix}. \tag{27.6}$$

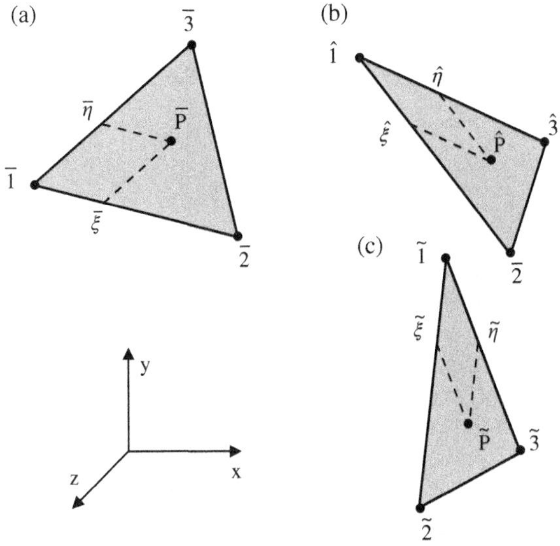

Figure 27.6 The initial (a), previous (b) and current (c) position of the membrane finite element. It is worth noting that for any material point P: $\bar{\xi} = \hat{\xi} = \tilde{\xi} = \xi$ and $\bar{\eta} = \hat{\eta} = \tilde{\eta} = \eta$, i.e. the local coordinates do not change with deformation

In a similar manner, the previous and current nodal coordinates are given by

$$\hat{\mathbf{X}} = \begin{bmatrix} \hat{x}_1 & \hat{y}_1 & \hat{z}_1 \\ \hat{x}_2 & \hat{y}_2 & \hat{z}_2 \\ \hat{x}_3 & \hat{y}_3 & \hat{z}_3 \end{bmatrix} \tag{27.7}$$

and

$$\tilde{\mathbf{X}} = \begin{bmatrix} \tilde{x}_1 & \tilde{y}_1 & \tilde{z}_1 \\ \tilde{x}_2 & \tilde{y}_2 & \tilde{z}_2 \\ \tilde{x}_3 & \tilde{y}_3 & \tilde{z}_3 \end{bmatrix}. \tag{27.8}$$

The global initial coordinates of any point P

$$\bar{\mathrm{P}} = (\bar{x}, \bar{y}, \bar{z}), \tag{27.9}$$

within the membrane finite element are interpolated from the initial nodal coordinates

$$\begin{aligned} \bar{x} &= \bar{x}_1 \mathrm{N}_1 + \bar{x}_2 \mathrm{N}_2 + \bar{x}_3 \mathrm{N}_3 \\ \bar{y} &= \bar{y}_1 \mathrm{N}_1 + \bar{y}_2 \mathrm{N}_2 + \bar{y}_3 \mathrm{N}_3 \\ \bar{z} &= \bar{z}_1 \mathrm{N}_1 + \bar{z}_2 \mathrm{N}_2 + \bar{z}_3 \mathrm{N}_3, \end{aligned} \tag{27.10}$$

where N_1, N_2 and N_3 are the shape functions.

In a similar way, the previous global coordinates of any point P are given by

$$\hat{\mathrm{P}} = (\hat{x}, \hat{y}, \hat{z}) \tag{27.11}$$

and are interpolated from the previous nodal coordinates using the shape functions as follows

$$\begin{aligned} \hat{x} &= \hat{x}_1 \mathrm{N}_1 + \hat{x}_2 \mathrm{N}_2 + \hat{x}_3 \mathrm{N}_3 \\ \hat{y} &= \hat{y}_1 \mathrm{N}_1 + \hat{y}_2 \mathrm{N}_2 + \hat{y}_3 \mathrm{N}_3 \\ \hat{z} &= \hat{z}_1 \mathrm{N}_1 + \hat{z}_2 \mathrm{N}_2 + \hat{z}_3 \mathrm{N}_3. \end{aligned} \tag{27.12}$$

The current global coordinates of any point P are given by

$$\tilde{\mathrm{P}} = (\tilde{x}, \tilde{y}, \tilde{z}) \tag{27.13}$$

and are interpolated from the current nodal coordinates using the shape functions

$$\begin{aligned} \tilde{x} &= \tilde{x}_1 \mathrm{N}_1 + \tilde{x}_2 \mathrm{N}_2 + \tilde{x}_3 \mathrm{N}_3 \\ \tilde{y} &= \tilde{y}_1 \mathrm{N}_1 + \tilde{y}_2 \mathrm{N}_2 + \tilde{y}_3 \mathrm{N}_3 \\ \tilde{z} &= \tilde{z}_1 \mathrm{N}_1 + \tilde{z}_2 \mathrm{N}_2 + \tilde{z}_3 \mathrm{N}_3. \end{aligned} \tag{27.14}$$

Equations (27.10), (27.12) and (27.14) can now be conveniently written in matrix form

$$\begin{bmatrix}\bar{x}\\ \bar{y}\\ \bar{z}\end{bmatrix}=\begin{bmatrix}\bar{x}_1 & \bar{x}_2 & \bar{x}_3\\ \bar{y}_1 & \bar{y}_2 & \bar{y}_3\\ \bar{z}_1 & \bar{z}_2 & \bar{z}_3\end{bmatrix}\begin{bmatrix}N_1(\xi,\eta)\\ N_2(\xi,\eta)\\ N_3(\xi,\eta)\end{bmatrix}=\begin{bmatrix}\bar{x}_1 & \bar{x}_2 & \bar{x}_3\\ \bar{y}_1 & \bar{y}_2 & \bar{y}_3\\ \bar{z}_1 & \bar{z}_2 & \bar{z}_3\end{bmatrix}\begin{bmatrix}1-\xi-\eta\\ \xi\\ \eta\end{bmatrix} \tag{27.15}$$

$$\begin{bmatrix}\hat{x}\\ \hat{y}\\ \hat{z}\end{bmatrix}=\begin{bmatrix}\hat{x}_1 & \hat{x}_2 & \hat{x}_3\\ \hat{y}_1 & \hat{y}_2 & \hat{y}_3\\ \hat{z}_1 & \hat{z}_2 & \hat{z}_3\end{bmatrix}\begin{bmatrix}N_1(\xi,\eta)\\ N_2(\xi,\eta)\\ N_3(\xi,\eta)\end{bmatrix}=\begin{bmatrix}\hat{x}_1 & \hat{x}_2 & \hat{x}_3\\ \hat{y}_1 & \hat{y}_2 & \hat{y}_3\\ \hat{z}_1 & \hat{z}_2 & \hat{z}_3\end{bmatrix}\begin{bmatrix}1-\xi-\eta\\ \xi\\ \eta\end{bmatrix} \tag{27.16}$$

$$\begin{bmatrix}\tilde{x}\\ \tilde{y}\\ \tilde{z}\end{bmatrix}=\begin{bmatrix}\tilde{x}_1 & \tilde{x}_2 & \tilde{x}_3\\ \tilde{y}_1 & \tilde{y}_2 & \tilde{y}_3\\ \tilde{z}_1 & \tilde{z}_2 & \tilde{z}_3\end{bmatrix}\begin{bmatrix}N_1(\xi,\eta)\\ N_2(\xi,\eta)\\ N_3(\xi,\eta)\end{bmatrix}=\begin{bmatrix}\tilde{x}_1 & \tilde{x}_2 & \tilde{x}_3\\ \tilde{y}_1 & \tilde{y}_2 & \tilde{y}_3\\ \tilde{z}_1 & \tilde{z}_2 & \tilde{z}_3\end{bmatrix}\begin{bmatrix}1-\xi-\eta\\ \xi\\ \eta\end{bmatrix}. \tag{27.17}$$

The Solid-Embedded Vector Base. Using the solid-embedded local coordinate system $[\xi \;\; \eta]$ it is convenient to define the local vector base $[\boldsymbol{\xi} \;\; \boldsymbol{\eta}]$, which is embedded in the solid, i.e., fixed to the solid's material points. The base vectors of this base are defined as follows

$$\boldsymbol{\xi}=\left(\frac{\partial x}{\partial \xi}\mathbf{i}+\frac{\partial y}{\partial \xi}\mathbf{j}+\frac{\partial z}{\partial \xi}\mathbf{k}\right)\text{ (im)}, \tag{27.18}$$

$$\boldsymbol{\eta}=\left(\frac{\partial x}{\partial \eta}\mathbf{i}+\frac{\partial y}{\partial \eta}\mathbf{j}+\frac{\partial z}{\partial \eta}\mathbf{k}\right)\text{ (im)}. \tag{27.19}$$

In actual fact, these vectors follow the tangents on the solid-embedded coordinate lines shown in Figure 27.4. It is worth noting that these vectors are infinitesimally small, which is achieved by using the infinitesimally small unit for length.

The initial position of these vectors is defined by the initial global coordinates of the material points to which these vectors are attached

$$\begin{aligned}\bar{\boldsymbol{\xi}}&=\frac{\partial \bar{x}}{\partial \xi}\mathbf{i}+\frac{\partial \bar{y}}{\partial \xi}\mathbf{j}+\frac{\partial \bar{z}}{\partial \xi}\mathbf{k}=\bar{\xi}_i\mathbf{i}+\bar{\xi}_j\mathbf{j}+\bar{\xi}_k\mathbf{k}\\ \bar{\boldsymbol{\eta}}&=\frac{\partial \bar{x}}{\partial \eta}\mathbf{i}+\frac{\partial \bar{y}}{\partial \eta}\mathbf{j}+\frac{\partial \bar{z}}{\partial \eta}\mathbf{k}=\bar{\eta}_i\mathbf{i}+\bar{\eta}_j\mathbf{j}+\bar{\eta}_k\mathbf{k}.\end{aligned} \tag{27.20}$$

In a similar manner, the previous position of these vectors is defined by the previous position of the material points to which vectors $\boldsymbol{\xi}$ and $\boldsymbol{\eta}$ are fixed

$$\hat{\boldsymbol{\xi}} = \frac{\partial \hat{x}}{\partial \xi}\mathbf{i} + \frac{\partial \hat{y}}{\partial \xi}\mathbf{j} + \frac{\partial \hat{z}}{\partial \xi}\mathbf{k} = \hat{\xi}_i\mathbf{i} + \hat{\xi}_j\mathbf{j} + \hat{\xi}_k\mathbf{k}$$
$$\hat{\boldsymbol{\eta}} = \frac{\partial \hat{x}}{\partial \eta}\mathbf{i} + \frac{\partial \hat{y}}{\partial \eta}\mathbf{j} + \frac{\partial \hat{z}}{\partial \eta}\mathbf{k} = \hat{\eta}_i\mathbf{i} + \hat{\eta}_j\mathbf{j} + \hat{\eta}_k\mathbf{k}. \tag{27.21}$$

Consequently, the current position of these vectors is defined by the current position of the same material points

$$\tilde{\boldsymbol{\xi}} = \frac{\partial \tilde{x}}{\partial \xi}\mathbf{i} + \frac{\partial \tilde{y}}{\partial \xi}\mathbf{j} + \frac{\partial \tilde{z}}{\partial \xi}\mathbf{k} = \tilde{\xi}_i\mathbf{i} + \tilde{\xi}_j\mathbf{j} + \tilde{\xi}_k\mathbf{k}$$
$$\tilde{\boldsymbol{\eta}} = \frac{\partial \tilde{x}}{\partial \eta}\mathbf{i} + \frac{\partial \tilde{y}}{\partial \eta}\mathbf{j} + \frac{\partial \tilde{z}}{\partial \eta}\mathbf{k} = \tilde{\eta}_i\mathbf{i} + \tilde{\eta}_j\mathbf{j} + \tilde{\eta}_k\mathbf{k}. \tag{27.22}$$

Equations (27.20), (27.21) and (27.22) are conveniently written in matrix form as follows:

$$\begin{bmatrix}\bar{\boldsymbol{\xi}} & \bar{\boldsymbol{\eta}}\end{bmatrix} = \begin{bmatrix}\mathbf{i} & \mathbf{j} & \mathbf{k}\end{bmatrix}\begin{bmatrix} \frac{\partial \bar{x}}{\partial \xi} & \frac{\partial \bar{x}}{\partial \eta} \\ \frac{\partial \bar{y}}{\partial \xi} & \frac{\partial \bar{y}}{\partial \eta} \\ \frac{\partial \bar{z}}{\partial \xi} & \frac{\partial \bar{z}}{\partial \eta} \end{bmatrix} \tag{27.23}$$

$$\begin{bmatrix}\hat{\boldsymbol{\xi}} & \hat{\boldsymbol{\eta}}\end{bmatrix} = \begin{bmatrix}\mathbf{i} & \mathbf{j} & \mathbf{k}\end{bmatrix}\begin{bmatrix} \frac{\partial \hat{x}}{\partial \xi} & \frac{\partial \hat{x}}{\partial \eta} \\ \frac{\partial \hat{y}}{\partial \xi} & \frac{\partial \hat{y}}{\partial \eta} \\ \frac{\partial \hat{z}}{\partial \xi} & \frac{\partial \hat{z}}{\partial \eta} \end{bmatrix} \tag{27.24}$$

$$\begin{bmatrix}\tilde{\boldsymbol{\xi}} & \tilde{\boldsymbol{\eta}}\end{bmatrix} = \begin{bmatrix}\mathbf{i} & \mathbf{j} & \mathbf{k}\end{bmatrix}\begin{bmatrix} \frac{\partial \tilde{x}}{\partial \xi} & \frac{\partial \tilde{x}}{\partial \eta} \\ \frac{\partial \tilde{y}}{\partial \xi} & \frac{\partial \tilde{y}}{\partial \eta} \\ \frac{\partial \tilde{z}}{\partial \xi} & \frac{\partial \tilde{z}}{\partial \eta} \end{bmatrix}. \tag{27.25}$$

After substituting $(\bar{x},\bar{y},\bar{z})$, $(\bar{x},\bar{y},\bar{z})$ and $(\bar{x},\bar{y},\bar{z})$ from Equations (27.15), (27.16) and (27.17) respectively, one obtains:

a. Initial position of the solid-embedded vector base $[\boldsymbol{\xi} \ \boldsymbol{\eta}]$

$$[\bar{\boldsymbol{\xi}} \ \bar{\boldsymbol{\eta}}] = [\mathbf{i} \ \mathbf{j} \ \mathbf{k}] \begin{bmatrix} \bar{\xi}_i & \bar{\eta}_i \\ \bar{\xi}_j & \bar{\eta}_j \\ \bar{\xi}_k & \bar{\eta}_k \end{bmatrix} = \begin{bmatrix} \bar{\xi}_i & \bar{\eta}_i \\ \bar{\xi}_j & \bar{\eta}_j \\ \bar{\xi}_k & \bar{\eta}_k \end{bmatrix} = \begin{bmatrix} \bar{x}_1 & \bar{x}_2 & \bar{x}_3 \\ \bar{y}_1 & \bar{y}_2 & \bar{y}_3 \\ \bar{z}_1 & \bar{z}_2 & \bar{z}_3 \end{bmatrix} \begin{bmatrix} \frac{\partial N_1}{\partial \xi} & \frac{\partial N_1}{\partial \eta} \\ \frac{\partial N_2}{\partial \xi} & \frac{\partial N_2}{\partial \eta} \\ \frac{\partial N_3}{\partial \xi} & \frac{\partial N_3}{\partial \eta} \end{bmatrix}, \tag{27.26}$$

where the global base $[\mathbf{i} \ \mathbf{j} \ \mathbf{k}]$ is omitted from the second part of the equation, which means that it is implicit from the context.

b. Previous position of the solid-embedded vector base $[\boldsymbol{\xi} \ \boldsymbol{\eta}]$

$$[\hat{\boldsymbol{\xi}} \ \hat{\boldsymbol{\eta}}] = [\mathbf{i} \ \mathbf{j} \ \mathbf{k}] \begin{bmatrix} \hat{\xi}_i & \hat{\eta}_i \\ \hat{\xi}_j & \hat{\eta}_j \\ \hat{\xi}_k & \hat{\eta}_k \end{bmatrix} = \begin{bmatrix} \hat{\xi}_i & \hat{\eta}_i \\ \hat{\xi}_j & \hat{\eta}_j \\ \hat{\xi}_k & \hat{\eta}_k \end{bmatrix} = \begin{bmatrix} \hat{x}_1 & \hat{x}_2 & \hat{x}_3 \\ \hat{y}_1 & \hat{y}_2 & \hat{y}_3 \\ \hat{z}_1 & \hat{z}_2 & \hat{z}_3 \end{bmatrix} \begin{bmatrix} \frac{\partial N_1}{\partial \xi} & \frac{\partial N_1}{\partial \eta} \\ \frac{\partial N_2}{\partial \xi} & \frac{\partial N_2}{\partial \eta} \\ \frac{\partial N_3}{\partial \xi} & \frac{\partial N_3}{\partial \eta} \end{bmatrix}, \tag{27.27}$$

where the omitted global base $[\mathbf{i} \ \mathbf{j} \ \mathbf{k}]$ is implicitly assumed from the context.

c. Current position of the solid-embedded vector base $[\boldsymbol{\xi} \ \boldsymbol{\eta}]$

$$[\tilde{\boldsymbol{\xi}} \ \tilde{\boldsymbol{\eta}}] = [\mathbf{i} \ \mathbf{j} \ \mathbf{k}] \begin{bmatrix} \tilde{\xi}_i & \tilde{\eta}_i \\ \tilde{\xi}_j & \tilde{\eta}_j \\ \tilde{\xi}_k & \tilde{\eta}_k \end{bmatrix} = \begin{bmatrix} \tilde{\xi}_i & \tilde{\eta}_i \\ \tilde{\xi}_j & \tilde{\eta}_j \\ \tilde{\xi}_k & \tilde{\eta}_k \end{bmatrix} = \begin{bmatrix} \tilde{x}_1 & \tilde{x}_2 & \tilde{x}_3 \\ \tilde{y}_1 & \tilde{y}_2 & \tilde{y}_3 \\ \tilde{z}_1 & \tilde{z}_2 & \tilde{z}_3 \end{bmatrix} \begin{bmatrix} \frac{\partial N_1}{\partial \xi} & \frac{\partial N_1}{\partial \eta} \\ \frac{\partial N_2}{\partial \xi} & \frac{\partial N_2}{\partial \eta} \\ \frac{\partial N_3}{\partial \xi} & \frac{\partial N_3}{\partial \eta} \end{bmatrix} \tag{27.28}$$

with the global base being implied by the context.

By substitution of N_1, N_2 and N_3 from Equation (27.4) one obtains

$$[\bar{\boldsymbol{\xi}}\ \bar{\boldsymbol{\eta}}] = \begin{bmatrix} \bar{\xi}_i & \bar{\eta}_i \\ \bar{\xi}_j & \bar{\eta}_j \\ \bar{\xi}_k & \bar{\eta}_k \end{bmatrix} = \begin{bmatrix} \bar{x}_1 & \bar{x}_2 & \bar{x}_3 \\ \bar{y}_1 & \bar{y}_2 & \bar{y}_3 \\ \bar{z}_1 & \bar{z}_2 & \bar{z}_3 \end{bmatrix} \begin{bmatrix} \frac{\partial(1-\xi-\eta)}{\partial\xi} & \frac{\partial(1-\xi-\eta)}{\partial\eta} \\ \frac{\partial\xi}{\partial\xi} & \frac{\partial\xi}{\partial\eta} \\ \frac{\partial\eta}{\partial\xi} & \frac{\partial\eta}{\partial\eta} \end{bmatrix} \tag{27.29}$$

$$[\hat{\boldsymbol{\xi}}\ \hat{\boldsymbol{\eta}}] = \begin{bmatrix} \hat{\xi}_i & \hat{\eta}_i \\ \hat{\xi}_j & \hat{\eta}_j \\ \hat{\xi}_k & \hat{\eta}_k \end{bmatrix} = \begin{bmatrix} \hat{x}_1 & \hat{x}_2 & \hat{x}_3 \\ \hat{y}_1 & \hat{y}_2 & \hat{y}_3 \\ \hat{z}_1 & \hat{z}_2 & \hat{z}_3 \end{bmatrix} \begin{bmatrix} \frac{\partial(1-\xi-\eta)}{\partial\xi} & \frac{\partial(1-\xi-\eta)}{\partial\eta} \\ \frac{\partial\xi}{\partial\xi} & \frac{\partial\xi}{\partial\eta} \\ \frac{\partial\eta}{\partial\xi} & \frac{\partial\eta}{\partial\eta} \end{bmatrix} \tag{27.30}$$

$$[\tilde{\boldsymbol{\xi}}\ \tilde{\boldsymbol{\eta}}] = \begin{bmatrix} \tilde{\xi}_i & \tilde{\eta}_i \\ \tilde{\xi}_j & \tilde{\eta}_j \\ \tilde{\xi}_k & \tilde{\eta}_k \end{bmatrix} = \begin{bmatrix} \tilde{x}_1 & \tilde{x}_2 & \tilde{x}_3 \\ \tilde{y}_1 & \tilde{y}_2 & \tilde{y}_3 \\ \tilde{z}_1 & \tilde{z}_2 & \tilde{z}_3 \end{bmatrix} \begin{bmatrix} \frac{\partial(1-\xi-\eta)}{\partial\xi} & \frac{\partial(1-\xi-\eta)}{\partial\eta} \\ \frac{\partial\xi}{\partial\xi} & \frac{\partial\xi}{\partial\eta} \\ \frac{\partial\eta}{\partial\xi} & \frac{\partial\eta}{\partial\eta} \end{bmatrix} \tag{27.31}$$

After the differentiation, the following equations for the initial, previous, and current positions of the solid-embedded vector base are obtained

$$[\bar{\boldsymbol{\xi}}\ \bar{\boldsymbol{\eta}}] = \begin{bmatrix} \bar{x}_1 & \bar{x}_2 & \bar{x}_3 \\ \bar{y}_1 & \bar{y}_2 & \bar{y}_3 \\ \bar{z}_1 & \bar{z}_2 & \bar{z}_3 \end{bmatrix} \begin{bmatrix} -1 & -1 \\ 1 & 0 \\ 0 & 1 \end{bmatrix} \tag{27.32}$$

$$[\hat{\boldsymbol{\xi}}\ \hat{\boldsymbol{\eta}}] = \begin{bmatrix} \hat{x}_1 & \hat{x}_2 & \hat{x}_3 \\ \hat{y}_1 & \hat{y}_2 & \hat{y}_3 \\ \hat{z}_1 & \hat{z}_2 & \hat{z}_3 \end{bmatrix} \begin{bmatrix} -1 & -1 \\ 1 & 0 \\ 0 & 1 \end{bmatrix} \tag{27.33}$$

$$[\tilde{\boldsymbol{\xi}}\ \tilde{\boldsymbol{\eta}}] = \begin{bmatrix} \tilde{x}_1 & \tilde{x}_2 & \tilde{x}_3 \\ \tilde{y}_1 & \tilde{y}_2 & \tilde{y}_3 \\ \tilde{z}_1 & \tilde{z}_2 & \tilde{z}_3 \end{bmatrix} \begin{bmatrix} -1 & -1 \\ 1 & 0 \\ 0 & 1 \end{bmatrix}, \tag{27.34}$$

which finally yields:

a. The initial position of the solid-embedded vector base

$$\begin{bmatrix} \bar{\boldsymbol{\xi}} & \bar{\boldsymbol{\eta}} \end{bmatrix} = \begin{bmatrix} \bar{x}_2 - \bar{x}_1 & \bar{x}_3 - \bar{x}_1 \\ \bar{y}_2 - \bar{y}_1 & \bar{y}_3 - \bar{y}_1 \\ \bar{z}_2 - \bar{z}_1 & \bar{z}_3 - \bar{z}_1 \end{bmatrix}. \tag{27.35}$$

b. The previous position of the solid-embedded vector base

$$\begin{bmatrix} \hat{\boldsymbol{\xi}} & \hat{\boldsymbol{\eta}} \end{bmatrix} = \begin{bmatrix} \hat{x}_2 - \hat{x}_1 & \hat{x}_3 - \hat{x}_1 \\ \hat{y}_2 - \hat{y}_1 & \hat{y}_3 - \hat{y}_1 \\ \hat{z}_2 - \hat{z}_1 & \hat{z}_3 - \hat{z}_1 \end{bmatrix}. \tag{27.36}$$

c. The current position of the solid-embedded vector base

$$\begin{bmatrix} \tilde{\boldsymbol{\xi}} & \tilde{\boldsymbol{\eta}} \end{bmatrix} = \begin{bmatrix} \tilde{x}_2 - \tilde{x}_1 & \tilde{x}_3 - \tilde{x}_1 \\ \tilde{y}_2 - \tilde{y}_1 & \tilde{y}_3 - \tilde{y}_1 \\ \tilde{z}_2 - \tilde{z}_1 & \tilde{z}_3 - \tilde{z}_1 \end{bmatrix}. \tag{27.37}$$

The Material-Embedded Vector Base as a Function of the Solid-Embedded Vector Base. Since the base vectors of the solid-embedded base and the base vectors of the material-embedded base are all fixed to the material points of the solid, both sets of vectors deform (move and stretch) together with the solid. As a consequence, in the infinitesimally close vicinity of any material point P, there exists a deformation-independent (invariant) relationship between vectors $[\boldsymbol{\alpha} \;\; \boldsymbol{\beta}]$ and $[\boldsymbol{\xi} \;\; \boldsymbol{\eta}]$, i.e.,

$$\begin{bmatrix} \bar{\boldsymbol{\alpha}} & \bar{\boldsymbol{\beta}} \end{bmatrix} = \begin{bmatrix} \bar{\boldsymbol{\xi}} & \bar{\boldsymbol{\eta}} \end{bmatrix} \begin{bmatrix} \alpha_\xi & \beta_\xi \\ \alpha_\eta & \beta_\eta \end{bmatrix}, \tag{27.38}$$

$$\begin{bmatrix} \hat{\boldsymbol{\alpha}} & \hat{\boldsymbol{\beta}} \end{bmatrix} = \begin{bmatrix} \hat{\boldsymbol{\xi}} & \hat{\boldsymbol{\eta}} \end{bmatrix} \begin{bmatrix} \alpha_\xi & \beta_\xi \\ \alpha_\eta & \beta_\eta \end{bmatrix}, \tag{27.39}$$

$$\begin{bmatrix} \tilde{\boldsymbol{\alpha}} & \tilde{\boldsymbol{\beta}} \end{bmatrix} = \begin{bmatrix} \tilde{\boldsymbol{\xi}} & \tilde{\boldsymbol{\eta}} \end{bmatrix} \begin{bmatrix} \alpha_\xi & \beta_\xi \\ \alpha_\eta & \beta_\eta \end{bmatrix}. \tag{27.40}$$

As explained before, for each membrane finite element the initial position of the material-embedded base vectors $[\boldsymbol{\alpha} \;\; \boldsymbol{\beta}]$ is given as an input to the finite element analysis. This is done by supplying the following matrix

$$\begin{bmatrix} \bar{\boldsymbol{\alpha}} & \bar{\boldsymbol{\beta}} \end{bmatrix} = \begin{bmatrix} \mathbf{i} & \mathbf{j} & \mathbf{k} \end{bmatrix} \begin{bmatrix} \bar{\alpha}_i & \bar{\beta}_i \\ \bar{\alpha}_j & \bar{\beta}_j \\ \bar{\alpha}_k & \bar{\beta}_k \end{bmatrix} = \begin{bmatrix} \bar{\alpha}_i & \bar{\beta}_i \\ \bar{\alpha}_j & \bar{\beta}_j \\ \bar{\alpha}_k & \bar{\beta}_k \end{bmatrix}. \tag{27.41}$$

In contrast, the initial position of the solid-embedded base vectors $[\boldsymbol{\xi}\ \boldsymbol{\eta}]$ is obtained from the initial nodal coordinates, which are given as an input to the finite element analysis. As explained above, the relationship between the initial position of the solid-embedded base and the initial coordinates is given by

$$\begin{aligned}[\boldsymbol{\xi}\ \boldsymbol{\eta}] &= [\mathbf{i}\ \mathbf{j}\ \mathbf{k}]\begin{bmatrix}\bar{\xi}_i & \bar{\eta}_i\\ \bar{\xi}_j & \bar{\eta}_j\\ \bar{\xi}_k & \bar{\eta}_k\end{bmatrix}\\ &= \begin{bmatrix}\bar{\xi}_i & \bar{\eta}_i\\ \bar{\xi}_j & \bar{\eta}_j\\ \bar{\xi}_k & \bar{\eta}_k\end{bmatrix} = \begin{bmatrix}\bar{x}_2-\bar{x}_1 & \bar{x}_3-\bar{x}_1\\ \bar{y}_2-\bar{y}_1 & \bar{y}_3-\bar{y}_1\\ \bar{z}_2-\bar{z}_1 & \bar{z}_3-\bar{z}_1\end{bmatrix}.\end{aligned} \tag{27.42}$$

By combining Equations (27.38), (27.41), and (27.42) one obtains:

$$\begin{aligned}\begin{bmatrix}\bar{\alpha}_i & \bar{\beta}_i\\ \bar{\alpha}_j & \bar{\beta}_j\\ \bar{\alpha}_k & \bar{\beta}_k\end{bmatrix} &= \begin{bmatrix}\bar{\xi}_i & \bar{\eta}_i\\ \bar{\xi}_j & \bar{\eta}_j\\ \bar{\xi}_k & \bar{\eta}_k\end{bmatrix}\begin{bmatrix}\alpha_\xi & \beta_\xi\\ \alpha_\eta & \beta_\eta\end{bmatrix}\\ &= \begin{bmatrix}\bar{x}_2-\bar{x}_1 & \bar{x}_3-\bar{x}_1\\ \bar{y}_2-\bar{y}_1 & \bar{y}_3-\bar{y}_1\\ \bar{z}_2-\bar{z}_1 & \bar{z}_3-\bar{z}_1\end{bmatrix}\begin{bmatrix}\alpha_\xi & \beta_\xi\\ \alpha_\eta & \beta_\eta\end{bmatrix}.\end{aligned} \tag{27.43}$$

By crossing out either **i**, **j** or **k** component in turn, one obtains:

$$\begin{bmatrix}\bar{\alpha}_j & \bar{\beta}_j\\ \bar{\alpha}_k & \bar{\beta}_k\end{bmatrix} = \begin{bmatrix}\bar{y}_2-\bar{y}_1 & \bar{y}_3-\bar{y}_1\\ \bar{z}_2-\bar{z}_1 & \bar{z}_3-\bar{z}_1\end{bmatrix}\begin{bmatrix}\alpha_\xi & \beta_\xi\\ \alpha_\eta & \beta_\eta\end{bmatrix}, \tag{27.44}$$

$$\begin{bmatrix}\bar{\alpha}_i & \bar{\beta}_i\\ \bar{\alpha}_k & \bar{\beta}_k\end{bmatrix} = \begin{bmatrix}\bar{x}_2-\bar{x}_1 & \bar{x}_3-\bar{x}_1\\ \bar{z}_2-\bar{z}_1 & \bar{z}_3-\bar{z}_1\end{bmatrix}\begin{bmatrix}\alpha_\xi & \beta_\xi\\ \alpha_\eta & \beta_\eta\end{bmatrix}, \tag{27.45}$$

$$\begin{bmatrix}\bar{\alpha}_i & \bar{\beta}_i\\ \bar{\alpha}_j & \bar{\beta}_j\end{bmatrix} = \begin{bmatrix}\bar{x}_2-\bar{x}_1 & \bar{x}_3-\bar{x}_1\\ \bar{y}_2-\bar{y}_1 & \bar{y}_3-\bar{y}_1\end{bmatrix}\begin{bmatrix}\alpha_\xi & \beta_\xi\\ \alpha_\eta & \beta_\eta\end{bmatrix}. \tag{27.46}$$

At this stage the following determinants are calculated:

$$\bar{d}_i = \begin{vmatrix}\bar{y}_2-\bar{y}_1 & \bar{y}_3-\bar{y}_1\\ \bar{z}_2-\bar{z}_1 & \bar{z}_3-\bar{z}_1\end{vmatrix} \tag{27.47}$$

$$\bar{d}_j = \begin{vmatrix}\bar{x}_2-\bar{x}_1 & \bar{x}_3-\bar{x}_1\\ \bar{z}_2-\bar{z}_1 & \bar{z}_3-\bar{z}_1\end{vmatrix} \tag{27.48}$$

$$\bar{d}_k = \begin{vmatrix}\bar{x}_2-\bar{x}_1 & \bar{x}_3-\bar{x}_1\\ \bar{y}_2-\bar{y}_1 & \bar{y}_3-\bar{y}_1\end{vmatrix}. \tag{27.49}$$

The equation with the maximum absolute value of the determinant is considered and the

$$\begin{bmatrix} \alpha_\xi & \beta_\xi \\ \alpha_\eta & \beta_\eta \end{bmatrix} \tag{27.50}$$

matrix is obtained from that equation, i.e. from either:

$$\begin{bmatrix} \alpha_\xi & \beta_\xi \\ \alpha_\eta & \beta_\eta \end{bmatrix} = \begin{bmatrix} \bar{y}_2-\bar{y}_1 & \bar{y}_3-\bar{y}_1 \\ \bar{z}_2-\bar{z}_1 & \bar{z}_3-\bar{z}_1 \end{bmatrix}^{-1} \begin{bmatrix} \bar{\alpha}_j & \bar{\beta}_j \\ \bar{\alpha}_k & \bar{\beta}_k \end{bmatrix} \tag{27.51}$$

or

$$\begin{bmatrix} \alpha_\xi & \beta_\xi \\ \alpha_\eta & \beta_\eta \end{bmatrix} = \begin{bmatrix} \bar{x}_2-\bar{x}_1 & \bar{x}_3-\bar{x}_1 \\ \bar{z}_2-\bar{z}_1 & \bar{z}_3-\bar{z}_1 \end{bmatrix}^{-1} \begin{bmatrix} \bar{\alpha}_i & \bar{\beta}_i \\ \bar{\alpha}_k & \bar{\beta}_k \end{bmatrix} \tag{27.52}$$

or

$$\begin{bmatrix} \alpha_\xi & \beta_\xi \\ \alpha_\eta & \beta_\eta \end{bmatrix} = \begin{bmatrix} \bar{x}_2-\bar{x}_1 & \bar{x}_3-\bar{x}_1 \\ \bar{y}_2-\bar{y}_1 & \bar{y}_3-\bar{y}_1 \end{bmatrix}^{-1} \begin{bmatrix} \bar{\alpha}_i & \bar{\beta}_i \\ \bar{\alpha}_j & \bar{\beta}_j \end{bmatrix}, \tag{27.53}$$

depending on the determinant.

The Current and Previous Material-Embedded Vector Bases. The previous material-embedded vector base is obtained from the previous solid-embedded base

$$\begin{aligned} [\hat{\boldsymbol{\alpha}} \;\; \hat{\boldsymbol{\beta}}] &= \begin{bmatrix} \hat{\alpha}_i & \hat{\beta}_i \\ \hat{\alpha}_j & \hat{\beta}_j \\ \hat{\alpha}_k & \hat{\beta}_k \end{bmatrix} = \begin{bmatrix} \hat{\xi}_i & \hat{\eta}_i \\ \hat{\xi}_j & \hat{\eta}_j \\ \hat{\xi}_k & \hat{\eta}_k \end{bmatrix} \begin{bmatrix} \alpha_\xi & \beta_\xi \\ \alpha_\eta & \beta_\eta \end{bmatrix} \\ &= \begin{bmatrix} \hat{x}_2-\hat{x}_1 & \hat{x}_3-\hat{x}_1 \\ \hat{y}_2-\hat{y}_1 & \hat{y}_3-\hat{y}_1 \\ \hat{z}_2-\hat{z}_1 & \hat{z}_3-\hat{z}_1 \end{bmatrix} \begin{bmatrix} \alpha_\xi & \beta_\xi \\ \alpha_\eta & \beta_\eta \end{bmatrix}. \end{aligned} \tag{27.54}$$

In a similar manner, the current material-embedded base is obtained using the current solid-embedded base

$$\begin{aligned} [\tilde{\boldsymbol{\alpha}} \;\; \tilde{\boldsymbol{\beta}}] &= \begin{bmatrix} \tilde{\alpha}_i & \tilde{\beta}_i \\ \tilde{\alpha}_j & \tilde{\beta}_j \\ \tilde{\alpha}_k & \tilde{\beta}_k \end{bmatrix} = \begin{bmatrix} \tilde{\xi}_i & \tilde{\eta}_i \\ \tilde{\xi}_j & \tilde{\eta}_j \\ \tilde{\xi}_k & \tilde{\eta}_k \end{bmatrix} \begin{bmatrix} \alpha_\xi & \beta_\xi \\ \alpha_\eta & \beta_\eta \end{bmatrix} \\ &= \begin{bmatrix} \tilde{x}_2-\tilde{x}_1 & \tilde{x}_3-\tilde{x}_1 \\ \tilde{y}_2-\tilde{y}_1 & \tilde{y}_3-\tilde{y}_1 \\ \tilde{z}_2-\tilde{z}_1 & \tilde{z}_3-\tilde{z}_1 \end{bmatrix} \begin{bmatrix} \alpha_\xi & \beta_\xi \\ \alpha_\eta & \beta_\eta \end{bmatrix}. \end{aligned} \tag{27.55}$$

Calculation of the Cauchy Stress Tensor Matrix. The previous and current material-embedded vector bases are obtained from the previous and current coordinates of the three-noded membrane finite element while the initial material-embedded base is supplied as an input to the finite element analysis. As a result, the shapes of the initial, the current, and the previous generalized infinitesimal material elements are known. From them, the Munjiza stress tensor matrix is calculated using the constitutive law for plane stress (material package).

The Cauchy stress tensor matrix is then calculated from the Munjiza stress tensor matrix, thus yielding

$$\mathbf{C}=\begin{bmatrix} c_{ii} & c_{ij} & c_{ik} \\ c_{ji} & c_{jj} & c_{jk} \\ c_{ki} & c_{kj} & c_{kk} \end{bmatrix}, \tag{27.56}$$

where c_{ij} signifies the global **i** component of the internal force on the surface defined by the global **j** base vector. It is worth being reminded that components of the Cauchy stress tensor matrix define internal forces on surfaces of the solid in its current position (the Cauchy material element).

Nodal Forces Using the Principle of Virtual Work. Equivalent nodal forces corresponding to the stress tensor obtained above can be calculated in several different ways. One approach is to determine the work of internal forces on virtual displacements.

The virtual displacements of a three-noded membrane finite element are supplied as a succession of infinitesimally small nodal displacements. For example, the nodal displacement applied to node 1 in the direction of the global base vector **i** is given by

$$\delta_{i1}=1\ \mathrm{im}. \tag{27.57}$$

The virtual displacements of the points inside the triangle are interpolated using the shape functions

$$\begin{aligned}\delta_{i1}&=1\cdot \mathrm{N}_1(\xi,\eta)\\ &=1\cdot(1-\xi-\eta)\\ &=(1-\xi-\eta).\end{aligned} \tag{27.58}$$

This is repeated for all the other virtual displacements, thus yielding the equivalent nodal forces due to the internal forces within the finite element, see Equation 21.20

$$\begin{aligned}\left[\mathbf{f}_1\ \ \mathbf{f}_2\ \ \mathbf{f}_3\right]=\begin{bmatrix} f_{i1} & f_{i2} & f_{i3} \\ f_{j1} & f_{j2} & f_{j3} \\ f_{k1} & f_{k2} & f_{k3} \end{bmatrix}&=-\iiint\limits_{\tilde{V}_e}\left(\begin{bmatrix} c_{ii} & c_{ij} & c_{ik} \\ c_{ji} & c_{jj} & c_{jk} \\ c_{ki} & c_{kj} & c_{kk} \end{bmatrix}\begin{bmatrix} \frac{\partial \mathrm{N}_1}{\partial \tilde{x}} & \frac{\partial \mathrm{N}_2}{\partial \tilde{x}} & \frac{\partial \mathrm{N}_3}{\partial \tilde{x}} \\ \frac{\partial \mathrm{N}_1}{\partial \tilde{y}} & \frac{\partial \mathrm{N}_2}{\partial \tilde{y}} & \frac{\partial \mathrm{N}_3}{\partial \tilde{y}} \\ \frac{\partial \mathrm{N}_1}{\partial \tilde{z}} & \frac{\partial \mathrm{N}_2}{\partial \tilde{z}} & \frac{\partial \mathrm{N}_3}{\partial \tilde{z}} \end{bmatrix}\right)d\tilde{x}d\tilde{y}d\tilde{z}\\ &-\iiint\limits_{\tilde{V}_e}\left(\begin{bmatrix} \frac{\partial c_{ii}}{\partial \tilde{x}} & \frac{\partial c_{ij}}{\partial \tilde{y}} & \frac{\partial c_{ik}}{\partial \tilde{z}} \\ \frac{\partial c_{ji}}{\partial \tilde{x}} & \frac{\partial c_{jj}}{\partial \tilde{y}} & \frac{\partial c_{jk}}{\partial \tilde{z}} \\ \frac{\partial c_{ki}}{\partial \tilde{x}} & \frac{\partial c_{kj}}{\partial \tilde{y}} & \frac{\partial c_{kk}}{\partial \tilde{z}} \end{bmatrix}\begin{bmatrix} \mathrm{N}_1 & \mathrm{N}_2 & \mathrm{N}_3 \\ \mathrm{N}_1 & \mathrm{N}_2 & \mathrm{N}_3 \\ \mathrm{N}_1 & \mathrm{N}_2 & \mathrm{N}_3 \end{bmatrix}\right)d\tilde{x}d\tilde{y}d\tilde{z}.\end{aligned} \tag{27.59}$$

One should note that the integration is done over the current volume of the finite element (thus $\tilde{V}_e$, and $d\tilde{x}$, $d\tilde{y}$, $d\tilde{z}$). In addition, the virtual displacements are applied to the material point's current position, thus $\partial \mathrm{N}_1/\partial \tilde{x}$, not $\partial \mathrm{N}_1/\partial \hat{x}$ and not $\partial \mathrm{N}_1/\partial \bar{x}$.

Since the Cauchy stress over the three-noded membrane finite element is constant, it follows that

$$\begin{bmatrix} f_{i1} & f_{i2} & f_{i3} \\ f_{j1} & f_{j2} & f_{j3} \\ f_{k1} & f_{k2} & f_{k3} \end{bmatrix} = -\iiint_{V_e} \left(\begin{bmatrix} c_{ii} & c_{ij} & c_{ik} \\ c_{ji} & c_{jj} & c_{jk} \\ c_{ki} & c_{kj} & c_{kk} \end{bmatrix} \begin{bmatrix} \frac{\partial \mathrm{N}_1}{\partial \tilde{x}} & \frac{\partial \mathrm{N}_2}{\partial \tilde{x}} & \frac{\partial \mathrm{N}_3}{\partial \tilde{x}} \\ \frac{\partial \mathrm{N}_1}{\partial \tilde{y}} & \frac{\partial \mathrm{N}_2}{\partial \tilde{y}} & \frac{\partial \mathrm{N}_3}{\partial \tilde{y}} \\ \frac{\partial \mathrm{N}_1}{\partial \tilde{z}} & \frac{\partial \mathrm{N}_2}{\partial \tilde{z}} & \frac{\partial \mathrm{N}_3}{\partial \tilde{z}} \end{bmatrix} \right) d\tilde{x}d\tilde{y}d\tilde{z}. \tag{27.60}$$

In order to calculate the above integral, one needs to calculate the global derivatives of the shape functions, i.e.,

$$\begin{bmatrix} \frac{\partial \mathrm{N}_1}{\partial \tilde{x}} & \frac{\partial \mathrm{N}_2}{\partial \tilde{x}} & \frac{\partial \mathrm{N}_3}{\partial \tilde{x}} \\ \frac{\partial \mathrm{N}_1}{\partial \tilde{y}} & \frac{\partial \mathrm{N}_2}{\partial \tilde{y}} & \frac{\partial \mathrm{N}_3}{\partial \tilde{y}} \\ \frac{\partial \mathrm{N}_1}{\partial \tilde{z}} & \frac{\partial \mathrm{N}_2}{\partial \tilde{z}} & \frac{\partial \mathrm{N}_3}{\partial \tilde{z}} \end{bmatrix}. \tag{27.61}$$

This is done by starting with the shape function's local derivatives, which are readily available. For example,

$$\begin{bmatrix} \frac{\partial \mathrm{N}_1}{\partial \xi} & \frac{\partial \mathrm{N}_2}{\partial \xi} & \frac{\partial \mathrm{N}_3}{\partial \xi} \\ \frac{\partial \mathrm{N}_1}{\partial \eta} & \frac{\partial \mathrm{N}_2}{\partial \eta} & \frac{\partial \mathrm{N}_3}{\partial \eta} \end{bmatrix} = \begin{bmatrix} -1 & 1 & 0 \\ -1 & 0 & 1 \end{bmatrix}. \tag{27.62}$$

By employing differentiation rules, one obtains

$$\begin{bmatrix} \frac{\partial \mathrm{N}_1}{\partial \xi} \\ \frac{\partial \mathrm{N}_1}{\partial \eta} \end{bmatrix} = \begin{bmatrix} \frac{\partial \mathrm{N}_1}{\partial \tilde{x}}\frac{\partial \tilde{x}}{\partial \xi} + \frac{\partial \mathrm{N}_1}{\partial \tilde{y}}\frac{\partial \tilde{y}}{\partial \xi} + \frac{\partial \mathrm{N}_1}{\partial \tilde{z}}\frac{\partial \tilde{z}}{\partial \xi} \\ \frac{\partial \mathrm{N}_1}{\partial \tilde{x}}\frac{\partial \tilde{x}}{\partial \eta} + \frac{\partial \mathrm{N}_1}{\partial \tilde{y}}\frac{\partial \tilde{y}}{\partial \eta} + \frac{\partial \mathrm{N}_1}{\partial \tilde{z}}\frac{\partial \tilde{z}}{\partial \eta} \end{bmatrix}, \tag{27.63}$$

which can be written in matrix form

$$\begin{bmatrix} \dfrac{\partial N_1}{\partial \xi} \\ \dfrac{\partial N_1}{\partial \eta} \end{bmatrix} = \begin{bmatrix} \dfrac{\partial \tilde{x}}{\partial \xi} & \dfrac{\partial \tilde{y}}{\partial \xi} & \dfrac{\partial \tilde{z}}{\partial \xi} \\ \dfrac{\partial \tilde{x}}{\partial \eta} & \dfrac{\partial \tilde{y}}{\partial \eta} & \dfrac{\partial \tilde{z}}{\partial \eta} \end{bmatrix} \begin{bmatrix} \dfrac{\partial N_1}{\partial \tilde{x}} \\ \dfrac{\partial N_1}{\partial \tilde{y}} \\ \dfrac{\partial N_1}{\partial \tilde{z}} \end{bmatrix}. \tag{27.64}$$

Since the current global coordinates are given by

$$\begin{bmatrix} \tilde{x} \\ \tilde{y} \\ \tilde{z} \end{bmatrix} = \begin{bmatrix} \tilde{x}_1 N_1 + \tilde{x}_2 N_2 + \tilde{x}_3 N_3 \\ \tilde{y}_1 N_1 + \tilde{y}_2 N_2 + \tilde{y}_3 N_3 \\ \tilde{z}_1 N_1 + \tilde{z}_2 N_2 + \tilde{z}_3 N_3 \end{bmatrix}, \tag{27.65}$$

it follows that

$$\begin{bmatrix} \dfrac{\partial \tilde{x}}{\partial \xi} & \dfrac{\partial \tilde{y}}{\partial \xi} & \dfrac{\partial \tilde{z}}{\partial \xi} \\ \dfrac{\partial \tilde{x}}{\partial \eta} & \dfrac{\partial \tilde{y}}{\partial \eta} & \dfrac{\partial \tilde{z}}{\partial \eta} \end{bmatrix} = \begin{bmatrix} \tilde{x}_2 - \tilde{x}_1 & \tilde{y}_2 - \tilde{y}_1 & \tilde{z}_2 - \tilde{z}_1 \\ \tilde{x}_3 - \tilde{x}_1 & \tilde{y}_3 - \tilde{y}_1 & \tilde{z}_3 - \tilde{z}_1 \end{bmatrix}. \tag{27.66}$$

When this is substituted into Equation (27.63), it yields

$$\begin{bmatrix} \dfrac{\partial N_1}{\partial \xi} \\ \dfrac{\partial N_1}{\partial \eta} \end{bmatrix} = \begin{bmatrix} \tilde{x}_2 - \tilde{x}_1 & \tilde{y}_2 - \tilde{y}_1 & \tilde{z}_2 - \tilde{z}_1 \\ \tilde{x}_3 - \tilde{x}_1 & \tilde{y}_3 - \tilde{y}_1 & \tilde{z}_3 - \tilde{z}_1 \end{bmatrix} \begin{bmatrix} \dfrac{\partial N_1}{\partial \tilde{x}} \\ \dfrac{\partial N_1}{\partial \tilde{y}} \\ \dfrac{\partial N_1}{\partial \tilde{z}} \end{bmatrix}. \tag{27.67}$$

In order to resolve the above equation, a temporary local vector

$$\tilde{\boldsymbol{\zeta}} = \tilde{\boldsymbol{\xi}} \times \tilde{\boldsymbol{\eta}} = \begin{vmatrix} \mathbf{i} & \mathbf{j} & \mathbf{k} \\ \tilde{x}_2 - \tilde{x}_1 & \tilde{y}_2 - \tilde{y}_1 & \tilde{z}_2 - \tilde{z}_1 \\ \tilde{x}_3 - \tilde{x}_1 & \tilde{y}_3 - \tilde{y}_1 & \tilde{z}_3 - \tilde{z}_1 \end{vmatrix} = \begin{bmatrix} \tilde{\zeta}_i \\ \tilde{\zeta}_j \\ \tilde{\zeta}_k \end{bmatrix} \tag{27.68}$$

is introduced and

$$\frac{\partial N_1}{\partial \zeta} = 0 \tag{27.69}$$

is assumed. By substituting into Equation (27.67), one obtains

$$\begin{bmatrix} \dfrac{\partial N_1}{\partial \xi} \\ \dfrac{\partial N_1}{\partial \eta} \\ 0 \end{bmatrix} = \begin{bmatrix} \tilde{x}_2-\tilde{x}_1 & \tilde{y}_2-\tilde{y}_1 & \tilde{z}_2-\tilde{z}_1 \\ \tilde{x}_3-\tilde{x}_1 & \tilde{y}_3-\tilde{y}_1 & \tilde{z}_3-\tilde{z}_1 \\ \tilde{\zeta}_i & \tilde{\zeta}_j & \tilde{\zeta}_k \end{bmatrix} \begin{bmatrix} \dfrac{\partial N_1}{\partial \tilde{x}} \\ \dfrac{\partial N_1}{\partial \tilde{y}} \\ \dfrac{\partial N_1}{\partial \tilde{z}} \end{bmatrix}, \tag{27.70}$$

which yields

$$\begin{bmatrix} \dfrac{\partial N_1}{\partial \tilde{x}} \\ \dfrac{\partial N_1}{\partial \tilde{y}} \\ \dfrac{\partial N_1}{\partial \tilde{z}} \end{bmatrix} = \begin{bmatrix} \tilde{x}_2-\tilde{x}_1 & \tilde{y}_2-\tilde{y}_1 & \tilde{z}_2-\tilde{z}_1 \\ \tilde{x}_3-\tilde{x}_1 & \tilde{y}_3-\tilde{y}_1 & \tilde{z}_3-\tilde{z}_1 \\ \tilde{\zeta}_i & \tilde{\zeta}_j & \tilde{\zeta}_k \end{bmatrix}^{-1} \begin{bmatrix} \dfrac{\partial N_1}{\partial \xi} \\ \dfrac{\partial N_1}{\partial \eta} \\ 0 \end{bmatrix}. \tag{27.71}$$

After differentiation one obtains

$$\begin{bmatrix} \dfrac{\partial N_1}{\partial \tilde{x}} \\ \dfrac{\partial N_1}{\partial \tilde{y}} \\ \dfrac{\partial N_1}{\partial \tilde{z}} \end{bmatrix} = \begin{bmatrix} \tilde{x}_2-\tilde{x}_1 & \tilde{y}_2-\tilde{y}_1 & \tilde{z}_2-\tilde{z}_1 \\ \tilde{x}_3-\tilde{x}_1 & \tilde{y}_3-\tilde{y}_1 & \tilde{z}_3-\tilde{z}_1 \\ \tilde{\zeta}_i & \tilde{\zeta}_j & \tilde{\zeta}_k \end{bmatrix}^{-1} \begin{bmatrix} -1 \\ -1 \\ 0 \end{bmatrix}. \tag{27.72}$$

By repeating the above explained derivation process for shape functions N_2 and N_3 one obtains

$$\begin{aligned} \begin{bmatrix} \dfrac{\partial N_2}{\partial \tilde{x}} \\ \dfrac{\partial N_2}{\partial \tilde{y}} \\ \dfrac{\partial N_2}{\partial \tilde{z}} \end{bmatrix} &= \begin{bmatrix} \tilde{x}_2-\tilde{x}_1 & \tilde{y}_2-\tilde{y}_1 & \tilde{z}_2-\tilde{z}_1 \\ \tilde{x}_3-\tilde{x}_1 & \tilde{y}_3-\tilde{y}_1 & \tilde{z}_3-\tilde{z}_1 \\ \tilde{\zeta}_i & \tilde{\zeta}_j & \tilde{\zeta}_k \end{bmatrix}^{-1} \begin{bmatrix} \dfrac{\partial N_2}{\partial \xi} \\ \dfrac{\partial N_2}{\partial \eta} \\ 0 \end{bmatrix} \\ &= \begin{bmatrix} \tilde{x}_2-\tilde{x}_1 & \tilde{y}_2-\tilde{y}_1 & \tilde{z}_2-\tilde{z}_1 \\ \tilde{x}_3-\tilde{x}_1 & \tilde{y}_3-\tilde{y}_1 & \tilde{z}_3-\tilde{z}_1 \\ \tilde{\zeta}_i & \tilde{\zeta}_j & \tilde{\zeta}_k \end{bmatrix}^{-1} \begin{bmatrix} 1 \\ 0 \\ 0 \end{bmatrix} \end{aligned} \tag{27.73}$$

and

$$\begin{bmatrix} \dfrac{\partial N_3}{\partial \tilde{x}} \\ \dfrac{\partial N_3}{\partial \tilde{y}} \\ \dfrac{\partial N_3}{\partial \tilde{z}} \end{bmatrix} = \begin{bmatrix} \tilde{x}_2-\tilde{x}_1 & \tilde{y}_2-\tilde{y}_1 & \tilde{z}_2-\tilde{z}_1 \\ \tilde{x}_3-\tilde{x}_1 & \tilde{y}_3-\tilde{y}_1 & \tilde{z}_3-\tilde{z}_1 \\ \tilde{\zeta}_i & \tilde{\zeta}_j & \tilde{\zeta}_k \end{bmatrix}^{-1} \begin{bmatrix} \dfrac{\partial N_3}{\partial \xi} \\ \dfrac{\partial N_3}{\partial \eta} \\ 0 \end{bmatrix} \tag{27.74}$$

$$= \begin{bmatrix} \tilde{x}_2-\tilde{x}_1 & \tilde{y}_2-\tilde{y}_1 & \tilde{z}_2-\tilde{z}_1 \\ \tilde{x}_3-\tilde{x}_1 & \tilde{y}_3-\tilde{y}_1 & \tilde{z}_3-\tilde{z}_1 \\ \tilde{\zeta}_i & \tilde{\zeta}_j & \tilde{\zeta}_k \end{bmatrix}^{-1} \begin{bmatrix} 0 \\ 1 \\ 0 \end{bmatrix}.$$

Equations (27.72), (27.73), and (27.74) can be written in matrix form

$$\begin{bmatrix} \dfrac{\partial N_1}{\partial \tilde{x}} & \dfrac{\partial N_2}{\partial \tilde{x}} & \dfrac{\partial N_3}{\partial \tilde{x}} \\ \dfrac{\partial N_1}{\partial \tilde{y}} & \dfrac{\partial N_2}{\partial \tilde{y}} & \dfrac{\partial N_3}{\partial \tilde{y}} \\ \dfrac{\partial N_1}{\partial \tilde{z}} & \dfrac{\partial N_2}{\partial \tilde{z}} & \dfrac{\partial N_3}{\partial \tilde{z}} \end{bmatrix} = \begin{bmatrix} \tilde{x}_2-\tilde{x}_1 & \tilde{y}_2-\tilde{y}_1 & \tilde{z}_2-\tilde{z}_1 \\ \tilde{x}_3-\tilde{x}_1 & \tilde{y}_3-\tilde{y}_1 & \tilde{z}_3-\tilde{z}_1 \\ \tilde{\zeta}_i & \tilde{\zeta}_j & \tilde{\zeta}_k \end{bmatrix}^{-1} \begin{bmatrix} -1 & 1 & 0 \\ -1 & 0 & 1 \\ 0 & 0 & 0 \end{bmatrix} \tag{27.75}$$

$$= \begin{bmatrix} \tilde{x}_2-\tilde{x}_1 & \tilde{x}_3-\tilde{x}_1 & \tilde{\zeta}_i \\ \tilde{y}_2-\tilde{y}_1 & \tilde{y}_3-\tilde{y}_1 & \tilde{\zeta}_j \\ \tilde{z}_2-\tilde{z}_1 & \tilde{z}_3-\tilde{z}_1 & \tilde{\zeta}_k \end{bmatrix}^{-T} \begin{bmatrix} -1 & 1 & 0 \\ -1 & 0 & 1 \\ 0 & 0 & 0 \end{bmatrix}$$

After substitution of Equation (27.75) into Equation (27.60), the equivalent nodal forces are obtained

$$\begin{bmatrix} f_{i1} & f_{i2} & f_{i3} \\ f_{j1} & f_{j2} & f_{j3} \\ f_{k1} & f_{k2} & f_{k3} \end{bmatrix} = -\iiint_{V_e} \left(\begin{bmatrix} c_{ii} & c_{ij} & c_{ik} \\ c_{ji} & c_{jj} & c_{jk} \\ c_{ki} & c_{kj} & c_{kk} \end{bmatrix} \begin{bmatrix} \tilde{x}_2-\tilde{x}_1 & \tilde{x}_3-\tilde{x}_1 & \tilde{\zeta}_i \\ \tilde{y}_2-\tilde{y}_1 & \tilde{y}_3-\tilde{y}_1 & \tilde{\zeta}_j \\ \tilde{z}_2-\tilde{z}_1 & \tilde{z}_3-\tilde{z}_1 & \tilde{\zeta}_k \end{bmatrix}^{-T} \begin{bmatrix} -1 & 1 & 0 \\ -1 & 0 & 1 \\ 0 & 0 & 0 \end{bmatrix} \right) d\tilde{x}d\tilde{y}d\tilde{z}. \tag{27.76}$$

Since all the terms being integrated are constant over the finite element, the above integration can be obtained analytically (alternatively, one Gauss integration point can be used).

$$\begin{bmatrix} f_{i1} & f_{i2} & f_{i3} \\ f_{j1} & f_{j2} & f_{j3} \\ f_{k1} & f_{k2} & f_{k3} \end{bmatrix} = \begin{bmatrix} c_{ii} & c_{ij} & c_{ik} \\ c_{ji} & c_{jj} & c_{jk} \\ c_{ki} & c_{kj} & c_{kk} \end{bmatrix} \begin{bmatrix} \tilde{x}_2-\tilde{x}_1 & \tilde{x}_3-\tilde{x}_1 & \tilde{\zeta}_i \\ \tilde{y}_2-\tilde{y}_1 & \tilde{y}_3-\tilde{y}_1 & \tilde{\zeta}_j \\ \tilde{z}_2-\tilde{z}_1 & \tilde{z}_3-\tilde{z}_1 & \tilde{\zeta}_k \end{bmatrix}^{-T} \begin{bmatrix} 1 & -1 & 0 \\ 1 & 0 & -1 \\ 0 & 0 & 0 \end{bmatrix} \tilde{A}\bar{t}. \quad (27.77)$$

where $\tilde{A}$ and $\bar{t}$ are respectively the current area and the initial thickness of the triangular finite element. It is worth noting that the initial thickness has been used. This is due to the fact that the plane stress constitutive law has been employed. In other words, the 2D constitutive law from the material package takes into account the change of thickness of say, a rubber membrane, and corrects the returned Cauchy stress tensor matrix accordingly.

Integration of Nodal Forces Over the Edges. Another approach for integrating nodal forces is the integration of nodal forces over the edges (boundary) of the triangular membrane finite element, Figure 27.7.

In Figure 27.7, vectors $\tilde{\boldsymbol{\xi}}$, $\tilde{\boldsymbol{\eta}}$ and $\tilde{\boldsymbol{\zeta}}$ are given by

$$\begin{bmatrix} \tilde{\boldsymbol{\xi}} & \tilde{\boldsymbol{\eta}} & \tilde{\boldsymbol{\zeta}} \end{bmatrix} = \begin{bmatrix} \tilde{x}_2-\tilde{x}_1 & \tilde{x}_3-\tilde{x}_1 & \tilde{\zeta}_i \\ \tilde{y}_2-\tilde{y}_1 & \tilde{y}_3-\tilde{y}_1 & \tilde{\zeta}_j \\ \tilde{z}_2-\tilde{z}_1 & \tilde{z}_3-\tilde{z}_1 & \tilde{\zeta}_k \end{bmatrix}, \quad (27.78)$$

where

$$\tilde{\boldsymbol{\zeta}} = \frac{\tilde{\boldsymbol{\xi}} \times \tilde{\boldsymbol{\eta}}}{|\tilde{\boldsymbol{\xi}} \times \tilde{\boldsymbol{\eta}}|}. \quad (27.79)$$

The dual base to the base shown in Equation (27.78) is

$$\begin{bmatrix} \underline{\tilde{\boldsymbol{\xi}}} \\ \underline{\tilde{\boldsymbol{\eta}}} \\ \underline{\tilde{\boldsymbol{\zeta}}} \end{bmatrix} = \begin{bmatrix} \tilde{x}_2-\tilde{x}_1 & \tilde{x}_3-\tilde{x}_1 & \tilde{\zeta}_i \\ \tilde{y}_2-\tilde{y}_1 & \tilde{y}_3-\tilde{y}_1 & \tilde{\zeta}_j \\ \tilde{z}_2-\tilde{z}_1 & \tilde{z}_3-\tilde{z}_1 & \tilde{\zeta}_k \end{bmatrix}^{-1}. \quad (27.80)$$

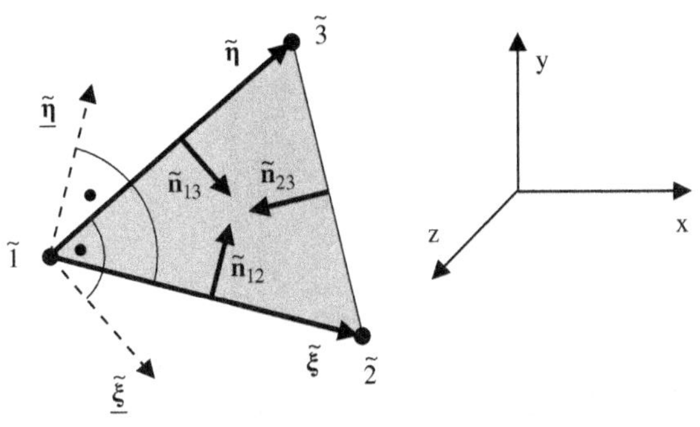

Figure 27.7 The equivalent nodal forces obtained by integration over the boundary of the finite element

The duality of bases

$$[\tilde{\boldsymbol{\xi}} \;\; \tilde{\boldsymbol{\eta}} \;\; \tilde{\boldsymbol{\zeta}}] \quad \text{and} \quad \begin{bmatrix} \underline{\tilde{\boldsymbol{\xi}}} \\ \underline{\tilde{\boldsymbol{\eta}}} \\ \underline{\tilde{\boldsymbol{\zeta}}} \end{bmatrix} \tag{27.81}$$

means that the dot product

$$\underline{\tilde{\boldsymbol{\xi}}} \bullet \tilde{\boldsymbol{\xi}} = 1. \tag{27.82}$$

This implies that surface $\tilde{\mathbf{n}}_{13}$ is given by

$$\tilde{\mathbf{n}}_{13} = \underline{\tilde{\boldsymbol{\xi}}} 2\tilde{A}\bar{t}, \tag{27.83}$$

which is because

$$\tilde{\mathbf{n}}_{13} \bullet \tilde{\boldsymbol{\xi}} = 2\tilde{A}\bar{t}, \tag{27.84}$$

thus

$$\underline{\tilde{\boldsymbol{\xi}}} 2\tilde{A}\bar{t} \bullet \tilde{\boldsymbol{\xi}} = 2\tilde{A}\bar{t} \Rightarrow 2\tilde{A}\bar{t} = 2\tilde{A}\bar{t}. \tag{27.85}$$

In a similar way, since the dot product

$$\underline{\tilde{\boldsymbol{\eta}}} \bullet \tilde{\boldsymbol{\eta}} = 1. \tag{27.86}$$

It follows that surface $\tilde{\mathbf{n}}_{12}$ is simply

$$\tilde{\mathbf{n}}_{12} = \underline{\tilde{\boldsymbol{\eta}}} 2\tilde{A}\bar{t}, \tag{27.87}$$

where $\tilde{A}$ is the current area of the triangle and $\bar{t}$ is the initial thickness of the triangle.

The surface $\tilde{\mathbf{n}}_{23}$ is given by

$$\begin{aligned} \tilde{\mathbf{n}}_{23} &= -(\tilde{\mathbf{n}}_{12} + \tilde{\mathbf{n}}_{13}) \\ &= -\underline{\tilde{\boldsymbol{\eta}}} 2\tilde{A}\bar{t} - \underline{\tilde{\boldsymbol{\xi}}} 2\tilde{A}\bar{t}. \end{aligned} \tag{27.88}$$

The force on node 1 is obtained by integrating the internal forces over boundaries $\tilde{\mathbf{n}}_{12}$ and $\tilde{\mathbf{n}}_{13}$, which yields

$$\mathbf{f}_1 = \frac{1}{2}\mathbf{C}(\tilde{\mathbf{n}}_{12} + \tilde{\mathbf{n}}_{13}), \tag{27.89}$$

or, after the substitution

$$\mathbf{f}_1 = \begin{bmatrix} f_{i1} \\ f_{j1} \\ f_{k1} \end{bmatrix} = \tilde{A}\bar{t}\left(\begin{bmatrix} c_{ii} & c_{ij} & c_{ik} \\ c_{ji} & c_{jj} & c_{jk} \\ c_{ki} & c_{kj} & c_{kk} \end{bmatrix} \begin{bmatrix} \underline{\tilde{\xi}_i} \\ \underline{\tilde{\xi}_j} \\ \underline{\tilde{\xi}_k} \end{bmatrix} + \begin{bmatrix} c_{ii} & c_{ij} & c_{ik} \\ c_{ji} & c_{jj} & c_{jk} \\ c_{ki} & c_{kj} & c_{kk} \end{bmatrix} \begin{bmatrix} \underline{\tilde{\eta}_i} \\ \underline{\tilde{\eta}_j} \\ \underline{\tilde{\eta}_k} \end{bmatrix} \right) \tag{27.90}$$

In a similar manner

$$\mathbf{f}_2 = \begin{bmatrix} f_{i2} \\ f_{j2} \\ f_{k2} \end{bmatrix} = -\tilde{A}\bar{t} \begin{bmatrix} c_{ii} & c_{ij} & c_{ik} \\ c_{ji} & c_{jj} & c_{jk} \\ c_{ki} & c_{kj} & c_{kk} \end{bmatrix} \begin{bmatrix} \underline{\tilde{\xi}_i} \\ \underline{\tilde{\xi}_j} \\ \underline{\tilde{\xi}_k} \end{bmatrix} \tag{27.91}$$

$$\mathbf{f}_3 = \begin{bmatrix} f_{i3} \\ f_{j3} \\ f_{k3} \end{bmatrix} = -\tilde{A}\bar{t} \begin{bmatrix} c_{ii} & c_{ij} & c_{ik} \\ c_{ji} & c_{jj} & c_{jk} \\ c_{ki} & c_{kj} & c_{kk} \end{bmatrix} \begin{bmatrix} \underline{\tilde{\eta}_i} \\ \underline{\tilde{\eta}_j} \\ \underline{\tilde{\eta}_k} \end{bmatrix}. \tag{27.92}$$

These forces can be written in matrix form

$$\begin{aligned}
[\mathbf{f}_1 \ \ \mathbf{f}_2 \ \ \mathbf{f}_3] &= \begin{bmatrix} f_{i1} & f_{i2} & f_{i3} \\ f_{j1} & f_{j2} & f_{j3} \\ f_{k1} & f_{k2} & f_{k3} \end{bmatrix} \\
&= \begin{bmatrix} c_{ii} & c_{ij} & c_{ik} \\ c_{ji} & c_{jj} & c_{jk} \\ c_{ki} & c_{kj} & c_{kk} \end{bmatrix} \begin{bmatrix} \left(\underline{\tilde{\xi}_i} + \underline{\tilde{\eta}_i}\right) & -\underline{\tilde{\xi}_i} & -\underline{\tilde{\eta}_i} \\ \left(\underline{\tilde{\xi}_j} + \underline{\tilde{\eta}_j}\right) & -\underline{\tilde{\xi}_j} & -\underline{\tilde{\eta}_j} \\ \left(\underline{\tilde{\xi}_k} + \underline{\tilde{\eta}_k}\right) & -\underline{\tilde{\xi}_k} & -\underline{\tilde{\eta}_k} \end{bmatrix} \tilde{A}t \\
&= \begin{bmatrix} c_{ii} & c_{ij} & c_{ik} \\ c_{ji} & c_{jj} & c_{jk} \\ c_{ki} & c_{kj} & c_{kk} \end{bmatrix} \begin{bmatrix} \tilde{\xi}_i & \tilde{\eta}_i & \tilde{\zeta}_i \\ \tilde{\xi}_j & \tilde{\eta}_j & \tilde{\zeta}_j \\ \tilde{\xi}_k & \tilde{\eta}_k & \tilde{\zeta}_k \end{bmatrix} \begin{bmatrix} 1 & -1 & 0 \\ 1 & 0 & -1 \\ 0 & 0 & 0 \end{bmatrix} \tilde{A}t \\
&= \tilde{A}t \begin{bmatrix} c_{ii} & c_{ij} & c_{ik} \\ c_{ji} & c_{jj} & c_{jk} \\ c_{ki} & c_{kj} & c_{kk} \end{bmatrix} \begin{bmatrix} \tilde{x}_2 - \tilde{x}_1 & \tilde{x}_3 - \tilde{x}_1 & \tilde{\zeta}_i \\ \tilde{y}_2 - \tilde{y}_1 & \tilde{y}_3 - \tilde{y}_1 & \tilde{\zeta}_j \\ \tilde{z}_2 - \tilde{z}_1 & \tilde{z}_3 - \tilde{z}_1 & \tilde{\zeta}_k \end{bmatrix}^{-T} \begin{bmatrix} 1 & -1 & 0 \\ 1 & 0 & -1 \\ 0 & 0 & 0 \end{bmatrix}
\end{aligned} \tag{27.93}$$

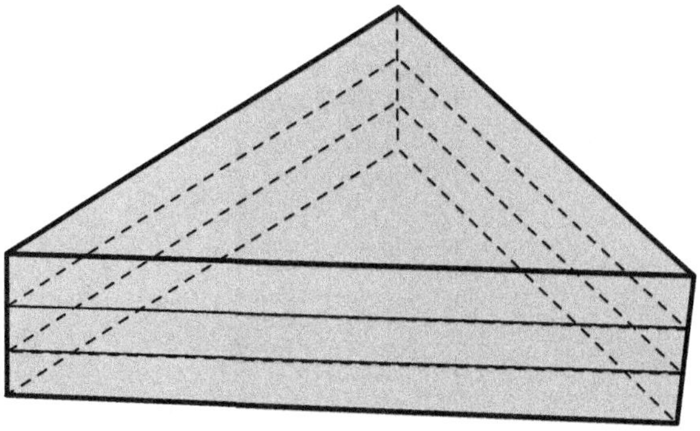

Figure 27.8 Integration of nodal forces over the multi-layered membrane finite element

The obtained nodal forces are the same as those calculated using the principle of virtual work, see Equation (27.77).

Multi-Layered Membranes. Multi-layered membranes are integrated using either the virtual work volumetric integration formula or the boundary integration formula. Each layer membrane layer is integrated separately and the forces are then added together, Figure 27.8.

It is worth noting that the material-embedded vector base can be different from layer to layer. Also, different constitutive laws can be applied to different layers.

27.3 Summary

In this chapter, the large displacement large deformation multiplicative decomposition formulation for 2.5D membranes using the three noded homogeneous deformation finite element has been introduced. First the geometry of the triangle was described using the nodal coordinates and the shape functions. This was followed by the calculation of the deformation kinematics using the solid-embedded vector base and the material-embedded vector base.

Finally the nodal forces were calculated from the Cauchy stress tensor matrix. The formulation presented also covers multi-layered anisotropic membranes. Unlike the alternative co-rotational formulation, this formulation represents both large displacements and large strains exactly.

Examples shown in Chapter 1 of this book were obtained using this formulation.

Further Reading

[1] Bonet, J. and Wood, R. D. (1997) *Nonlinear Continuum Mechanics for Finite Element Analysis*. Cambridge, Cambridge University Press.

[2] de Souza Neto, E. A., Peric, D. and Owen, D. R. J. (1995) Finite elasticity in spatial description – linearization aspects with 3-D membrane applications. *International Journal for Numerical Methods in Engineering*, **38**(20): 3365–81.

[3] Hughes, T.J.R. (2000) *The Finite Element Method: Linear Static and Dynamic Finite Element Analysis*. New York, Dover Publications.

[4] Hughes, T. J. R. and Liu, W. K. (1981) Nonlinear finite element analysis of shells: Part I. Three-dimensional shells. *Comput. Method Appl. M.*, **26**(3): 331–62.

[5] Hughes, T. J. R. and Liu, W. K. (1981) Nonlinear finite element analysis of shells: Part II. Two-dimensional shells. *Comput. Method Appl. M.*, **27**(2): 167–81.

[6] Oden, J. T. (1972) *Finite Elements of Nonlinear Continua*. New York: McGraw-Hill.

[7] Parisch, H. (1986) Efficient non-linear finite element shell formulation involving large strains. *Eng. Computation*, **3**(2): 121–8.

[8] Zienkiewicz, O. C. (1971) *The Finite Element Method in Engineering Science*. New York, McGraw-Hill.

[9] Zienkiewicz, O. C., Rojek, J., Taylor, R. L. and Pastor, M. (1998) Triangles and tetrahera in explicit dynamic codes for solids. *International Journal for Numerical Methods in Engineering*, **43**(3), 565–83.

[10] Zienkiewicz, O. C. and Taylor, R. L. (2005) *The Finite Element Method Set*. Oxford, Elsevier Science.

28

Deformation in 2.5D Using Shell Finite Elements

28.1 Introduction

Unlike membranes, shells can support both in-plane stresses (membrane stresses) and bending stresses.

One way to obtain the anisotropic material formulation for shells is to introduce a vector base $[\boldsymbol{\alpha} \;\; \boldsymbol{\beta}]$ comprising base vectors $\boldsymbol{\alpha}$ and $\boldsymbol{\beta}$ both of which are fixed to the solid's material points. As such, they deform with the material (rotate, stretch, etc.).

Both of these vectors are defined as input to the finite element analysis and, in essence, they define the preferred directions that characterize the given finite element's material. Both vectors are assumed constant over a given finite element, but can vary from element to element.

The length (magnitude) of these vectors is measured in "im" (infinitesimally small length unit, i.e., "infimeter") making the vectors infinitesimally small. As such, they define an infinitesimally small volume of material, which is of parallelepiped shape and is called the generalized material element.

It has been customary to use rotational degrees of freedom with shell elements. In recent years, it has been demonstrated that shell elements can be introduced using only translational degrees of freedom at the finite element nodes. This is especially convenient when dealing with anisotropic materials and/or multi-layered shells.

In Figure 28.1 a multi-layered shell is shown with each layer being made of anisotropic material.

For each layer, a material-embedded vector base is defined

$$\begin{bmatrix} \boldsymbol{\alpha} \\ \boldsymbol{\beta} \\ \boldsymbol{\gamma} \end{bmatrix} \text{ (im).} \tag{28.1}$$

Large Strain Finite Element Method: A Practical Course, First Edition. Antonio Munjiza, Esteban Rougier and Earl E. Knight.

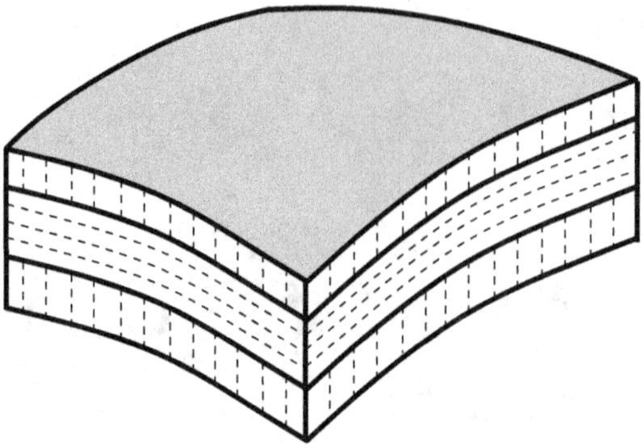

Figure 28.1 A multi-layered anisotropic shell. Dotted lines indicate the preferred material directions with a material embedded vector base defining the generalized material element

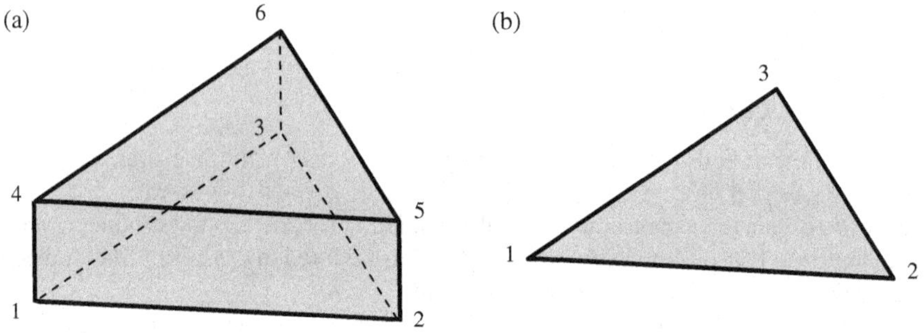

Figure 28.2 (a) Six-noded shell element versus; (b) three-noded shell element

28.2 The Six-Noded Triangular Shell Finite Element

As the shell deforms, the material points of the shell move (displace) in space. In order to represent the deformation of each multi-layered anisotropic shell subject to large displacements and large strains, (for example a shell made of rubber, or a shell made of plastically deforming folding metal), it is convenient to introduce a six-noded shell element, instead of the customary three-noded shell element, Figure 28.2.

In the case of the three-noded shell element, the displacements of the individual nodes in the global x, y and z directions are combined with the node's rotation about the global x, y and z axes, thus in total, 18 degrees of freedom are considered.

In contrast, in the case of the six-noded shell elements, there is no need to consider nodal rotations; thus at each node only the three displacements in the x, y and z directions are considered. This also produces 18 degrees of freedom.

In other words, the number of degrees of freedom for the six-noded shell element is the same as the number of degrees of freedom for the three-noded shell element. Nevertheless, the large-strain-large-displacement formulation happens to be much simpler for a six-noded shell element. For this reason, only the six-noded shell element is considered in this chapter.

28.3 The Solid-Embedded Coordinate System

In order to simplify the deformation formulation for the six-noded shell element, it is convenient to introduce the solid-embedded coordinate system

$$(\xi, \eta, \zeta), \tag{28.2}$$

such that the surfaces of constant ζ follow the shell layers, while the surface of constant ξ and constant η go across the shell layers, Figure 28.3.

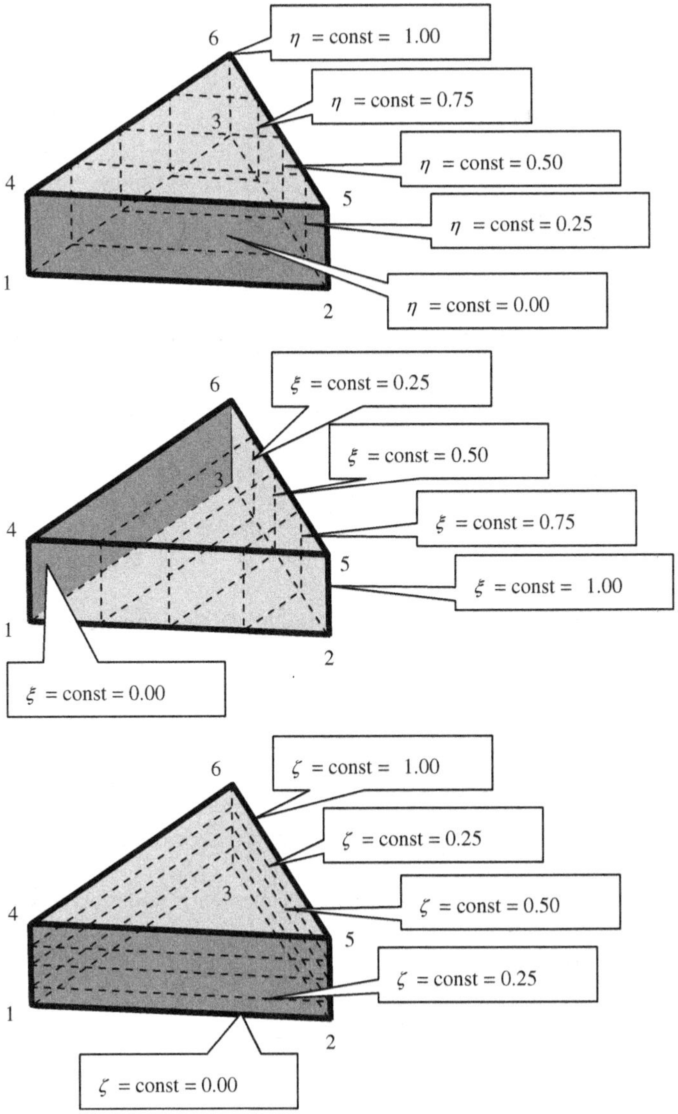

Figure 28.3 The solid-embedded local coordinate system, with solid-embedded coordinate surfaces being shown

28.4 Nodal Coordinates

The position of the finite element at any instance is defined by the position of its six-nodes

$$\begin{aligned}1 &= (x_1, y_1, z_1)\\ 2 &= (x_2, y_2, z_2)\\ 3 &= (x_3, y_3, z_3)\\ 4 &= (x_4, y_4, z_4)\\ 5 &= (x_5, y_5, z_5)\\ 6 &= (x_6, y_6, z_6).\end{aligned} \tag{28.3}$$

In a matrix form this can be written as follows,

$$\begin{bmatrix}1 & 2 & 3 & 4 & 5 & 6\end{bmatrix} = \begin{bmatrix} x_1 & x_2 & x_3 & x_4 & x_5 & x_6 \\ y_1 & y_2 & y_3 & y_4 & y_5 & y_6 \\ z_1 & z_2 & z_3 & z_4 & z_5 & z_6 \end{bmatrix}. \tag{28.4}$$

Of particular interest are the following sets of nodal coordinates:

a. The initial nodal coordinates

$$\begin{bmatrix}\bar{1} & \bar{2} & \bar{3} & \bar{4} & \bar{5} & \bar{6}\end{bmatrix} = \begin{bmatrix} \bar{x}_1 & \bar{x}_2 & \bar{x}_3 & \bar{x}_4 & \bar{x}_5 & \bar{x}_6 \\ \bar{y}_1 & \bar{y}_2 & \bar{y}_3 & \bar{y}_4 & \bar{y}_5 & \bar{y}_6 \\ \bar{z}_1 & \bar{z}_2 & \bar{z}_3 & \bar{z}_4 & \bar{z}_5 & \bar{z}_6 \end{bmatrix}. \tag{28.5}$$

These coordinates define the finite element's initial position and are supplied as an input to the finite element analysis.

b. The previous nodal coordinates

$$\begin{bmatrix}\hat{1} & \hat{2} & \hat{3} & \hat{4} & \hat{5} & \hat{6}\end{bmatrix} = \begin{bmatrix} \hat{x}_1 & \hat{x}_2 & \hat{x}_3 & \hat{x}_4 & \hat{x}_5 & \hat{x}_6 \\ \hat{y}_1 & \hat{y}_2 & \hat{y}_3 & \hat{y}_4 & \hat{y}_5 & \hat{y}_6 \\ \hat{z}_1 & \hat{z}_2 & \hat{z}_3 & \hat{z}_4 & \hat{z}_5 & \hat{z}_6 \end{bmatrix}. \tag{28.6}$$

These coordinates define the position of the finite element's nodes at the step just before the current step – the previous step.

c. The current nodal coordinates

$$\begin{bmatrix}\tilde{1} & \tilde{2} & \tilde{3} & \tilde{4} & \tilde{5} & \tilde{6}\end{bmatrix} = \begin{bmatrix} \tilde{x}_1 & \tilde{x}_2 & \tilde{x}_3 & \tilde{x}_4 & \tilde{x}_5 & \tilde{x}_6 \\ \tilde{y}_1 & \tilde{y}_2 & \tilde{y}_3 & \tilde{y}_4 & \tilde{y}_5 & \tilde{y}_6 \\ \tilde{z}_1 & \tilde{z}_2 & \tilde{z}_3 & \tilde{z}_4 & \tilde{z}_5 & \tilde{z}_6 \end{bmatrix}. \tag{28.7}$$

These coordinates define the position of the six nodes at the current step.

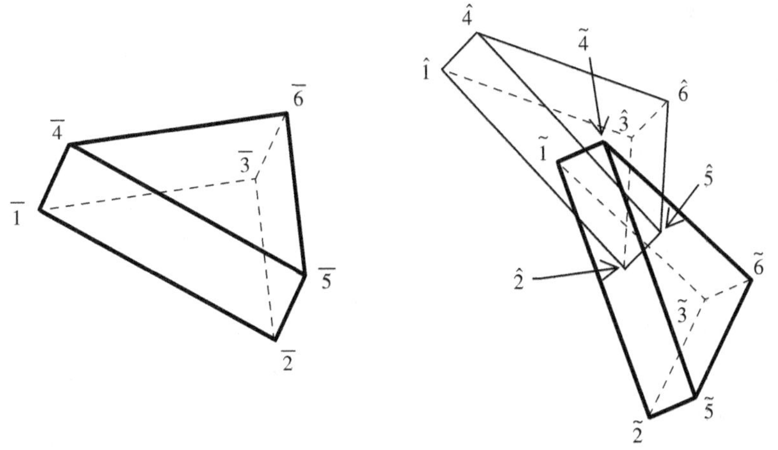

Figure 28.4 The initial, the previous and the current position of the finite element

The initial, the previous and the current coordinates are shown in Figure 28.4, it is evident that the difference between the initial and the previous coordinates can be very large, while the difference between the current and the previous coordinate is usually relatively small depending on the step size employed.

28.5 The Coordinates of the Finite Element's Material Points

The finite element's nodal coordinates are known at each step of the finite element analysis and are obtained using either explicit or iterative solver, Chapter 3. However, the coordinates of the material points inside the domain of the finite element are not known and have to be interpolated (approximated) from the known nodal coordinates. This is done using the solid-embedded coordinate system (ξ, η, ζ) and the shape functions

$$[\mathrm{N}_1(\xi,\eta) \;\; \mathrm{N}_2(\xi,\eta) \;\; \mathrm{N}_3(\xi,\eta) \;\; \mathrm{N}_4(\xi,\eta) \;\; \mathrm{N}_5(\xi,\eta) \;\; \mathrm{N}_6(\xi,\eta)]. \tag{28.8}$$

Using the shape functions, the global coordinates of a given material point

$$\mathrm{P} = (\xi, \eta, \zeta) \tag{28.9}$$

are approximated as follows

$$\begin{bmatrix} x \\ y \\ z \end{bmatrix} = \begin{bmatrix} x_1 & x_2 & x_3 & x_4 & x_5 & x_6 \\ y_1 & y_2 & y_3 & y_4 & y_5 & y_6 \\ z_1 & z_2 & z_3 & z_4 & z_5 & z_6 \end{bmatrix} \begin{bmatrix} \mathrm{N}_1 \\ \mathrm{N}_2 \\ \mathrm{N}_3 \\ \mathrm{N}_4 \\ \mathrm{N}_5 \\ \mathrm{N}_6 \end{bmatrix}. \tag{28.10}$$

These produce:

a. Initial coordinates

$$\begin{bmatrix} \bar{x} \\ \bar{y} \\ \bar{z} \end{bmatrix} = \begin{bmatrix} \bar{x}_1 & \bar{x}_2 & \bar{x}_3 & \bar{x}_4 & \bar{x}_5 & \bar{x}_6 \\ \bar{y}_1 & \bar{y}_2 & \bar{y}_3 & \bar{y}_4 & \bar{y}_5 & \bar{y}_6 \\ \bar{z}_1 & \bar{z}_2 & \bar{z}_3 & \bar{z}_4 & \bar{z}_5 & \bar{z}_6 \end{bmatrix} \begin{bmatrix} N_1 \\ N_2 \\ N_3 \\ N_4 \\ N_5 \\ N_6 \end{bmatrix}. \tag{28.11}$$

b. Previous coordinates

$$\begin{bmatrix} \hat{x} \\ \hat{y} \\ \hat{z} \end{bmatrix} = \begin{bmatrix} \hat{x}_1 & \hat{x}_2 & \hat{x}_3 & \hat{x}_4 & \hat{x}_5 & \hat{x}_6 \\ \hat{y}_1 & \hat{y}_2 & \hat{y}_3 & \hat{y}_4 & \hat{y}_5 & \hat{y}_6 \\ \hat{z}_1 & \hat{z}_2 & \hat{z}_3 & \hat{z}_4 & \hat{z}_5 & \hat{z}_6 \end{bmatrix} \begin{bmatrix} N_1 \\ N_2 \\ N_3 \\ N_4 \\ N_5 \\ N_6 \end{bmatrix}. \tag{28.12}$$

c. Current coordinates

$$\begin{bmatrix} \tilde{x} \\ \tilde{y} \\ \tilde{z} \end{bmatrix} = \begin{bmatrix} \tilde{x}_1 & \tilde{x}_2 & \tilde{x}_3 & \tilde{x}_4 & \tilde{x}_5 & \tilde{x}_6 \\ \tilde{y}_1 & \tilde{y}_2 & \tilde{y}_3 & \tilde{y}_4 & \tilde{y}_5 & \tilde{y}_6 \\ \tilde{z}_1 & \tilde{z}_2 & \tilde{z}_3 & \tilde{z}_4 & \tilde{z}_5 & \tilde{z}_6 \end{bmatrix} \begin{bmatrix} N_1 \\ N_2 \\ N_3 \\ N_4 \\ N_5 \\ N_6 \end{bmatrix}. \tag{28.13}$$

In this manner, the initial, the previous and the current coordinates of all material points inside the finite element are known at every step of the finite element analysis.

28.6 The Solid-Embedded Infinitesimal Vector Base

Using the solid-embedded coordinate system (ξ, η, ζ), for any material point

$$P = (\xi, \eta, \zeta) \tag{28.14}$$

it is possible to define a solid-embedded vector base

$$[\boldsymbol{\xi} \;\; \boldsymbol{\eta} \;\; \boldsymbol{\zeta}], \tag{28.15}$$

such that

$$\boldsymbol{\xi} = \left(\frac{\partial x}{\partial \xi}\mathbf{i} + \frac{\partial y}{\partial \xi}\mathbf{j} + \frac{\partial z}{\partial \xi}\mathbf{k}\right)(\text{im})$$
$$\boldsymbol{\eta} = \left(\frac{\partial x}{\partial \eta}\mathbf{i} + \frac{\partial y}{\partial \eta}\mathbf{j} + \frac{\partial z}{\partial \eta}\mathbf{k}\right)(\text{im}) \quad (28.16)$$
$$\boldsymbol{\zeta} = \left(\frac{\partial x}{\partial \zeta}\mathbf{i} + \frac{\partial y}{\partial \zeta}\mathbf{j} + \frac{\partial z}{\partial \zeta}\mathbf{k}\right)(\text{im}).$$

where "im" indicates that these base vectors are infinitesimally short.

These vectors represent tangents on the lines defined by

$$\left.\begin{matrix}\eta = \text{const}\\ \zeta = \text{const}\end{matrix}\right\} \text{for vector } \boldsymbol{\xi}, \quad (28.17)$$

$$\left.\begin{matrix}\xi = \text{const}\\ \zeta = \text{const}\end{matrix}\right\} \text{for vector } \boldsymbol{\eta}, \quad (28.18)$$

$$\left.\begin{matrix}\xi = \text{const}\\ \eta = \text{const}\end{matrix}\right\} \text{for vector } \boldsymbol{\zeta}. \quad (28.19)$$

As the coordinate lines defined by Equations (28.17), (28.18) and (28.19) are embedded in the solid, i.e., they are fixed to the solid's material points, it follows that the infinitesimal base vectors $[\boldsymbol{\xi}\ \ \boldsymbol{\eta}\ \ \boldsymbol{\zeta}]$ are also fixed to point P; for this reason this vector base is called the solid-embedded vector base.

By substituting nodal coordinates from Equation (28.10) into (28.16) one obtains

$$[\boldsymbol{\xi}\ \ \boldsymbol{\eta}\ \ \boldsymbol{\zeta}] = \begin{bmatrix}\xi_i & \eta_i & \zeta_i\\ \xi_j & \eta_j & \zeta_j\\ \xi_k & \eta_k & \zeta_k\end{bmatrix}$$
$$= \begin{bmatrix}x_1 & x_2 & x_3 & x_4 & x_5 & x_6\\ y_1 & y_2 & y_3 & y_4 & y_5 & y_6\\ z_1 & z_2 & z_3 & z_4 & z_5 & z_6\end{bmatrix} \begin{bmatrix}\frac{\partial N_1}{\partial \xi} & \frac{\partial N_1}{\partial \eta} & \frac{\partial N_1}{\partial \zeta}\\ \frac{\partial N_2}{\partial \xi} & \frac{\partial N_2}{\partial \eta} & \frac{\partial N_2}{\partial \zeta}\\ \frac{\partial N_3}{\partial \xi} & \frac{\partial N_3}{\partial \eta} & \frac{\partial N_3}{\partial \zeta}\\ \frac{\partial N_4}{\partial \xi} & \frac{\partial N_4}{\partial \eta} & \frac{\partial N_4}{\partial \zeta}\\ \frac{\partial N_5}{\partial \xi} & \frac{\partial N_5}{\partial \eta} & \frac{\partial N_5}{\partial \zeta}\\ \frac{\partial N_6}{\partial \xi} & \frac{\partial N_6}{\partial \eta} & \frac{\partial N_6}{\partial \zeta}\end{bmatrix} \quad (28.20)$$

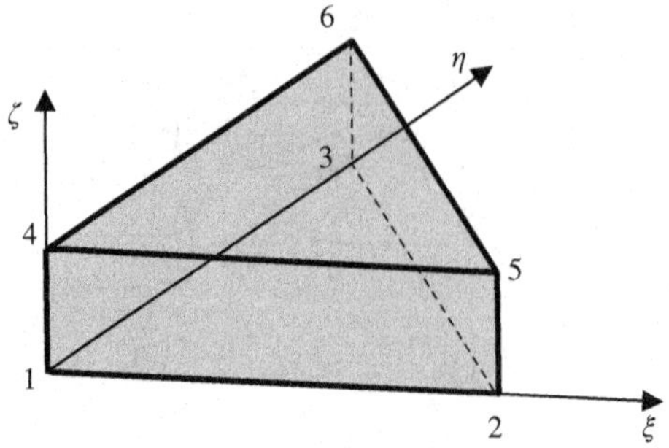

Figure 28.5 Shape functions

After differentiation this yields

$$\begin{bmatrix} \xi_i & \eta_i & \zeta_i \\ \xi_j & \eta_j & \zeta_j \\ \xi_k & \eta_k & \zeta_k \end{bmatrix} = \begin{bmatrix} x_1 & x_2 & x_3 & x_4 & x_5 & x_6 \\ y_1 & y_2 & y_3 & y_4 & y_5 & y_6 \\ z_1 & z_2 & z_3 & z_4 & z_5 & z_6 \end{bmatrix} \begin{bmatrix} (\zeta-1) & (\xi-1) & (\xi+\eta-1) \\ (1-\zeta) & 0 & -\xi \\ 0 & (1-\zeta) & -\eta \\ -\zeta & -\zeta & (1-\xi-\eta) \\ \zeta & 0 & \xi \\ 0 & \zeta & \eta \end{bmatrix} \tag{28.21}$$

where the following shape functions are used (see Figure 28.5):

$$\begin{aligned} N_1 &= (1-\zeta)(1-\xi-\eta) \\ N_2 &= (1-\zeta)\xi \\ N_3 &= (1-\zeta)\eta \\ N_4 &= \zeta(1-\xi-\eta) \\ N_5 &= \zeta\xi \\ N_6 &= \zeta\eta. \end{aligned} \tag{28.22}$$

This means that the initial, the previous and the current position of the solid-embedded vector base are as follows:

a. the initial position

$$\begin{bmatrix} \bar{\xi}_i & \bar{\eta}_i & \bar{\zeta}_i \\ \bar{\xi}_j & \bar{\eta}_j & \bar{\zeta}_j \\ \bar{\xi}_k & \bar{\eta}_k & \bar{\zeta}_k \end{bmatrix} = \begin{bmatrix} \bar{x}_1 & \bar{x}_2 & \bar{x}_3 & \bar{x}_4 & \bar{x}_5 & \bar{x}_6 \\ \bar{y}_1 & \bar{y}_2 & \bar{y}_3 & \bar{y}_4 & \bar{y}_5 & \bar{y}_6 \\ \bar{z}_1 & \bar{z}_2 & \bar{z}_3 & \bar{z}_4 & \bar{z}_5 & \bar{z}_6 \end{bmatrix} \begin{bmatrix} -(1-\zeta) & -(1-\zeta) & -(1-\xi-\eta) \\ (1-\zeta) & 0 & -\xi \\ 0 & (1-\zeta) & -\eta \\ -\zeta & -\zeta & (1-\xi-\eta) \\ \zeta & 0 & \xi \\ 0 & \zeta & \eta \end{bmatrix} \tag{28.23}$$

b. the previous position

$$\begin{bmatrix} \hat{\xi}_i & \hat{\eta}_i & \hat{\zeta}_i \\ \hat{\xi}_j & \hat{\eta}_j & \hat{\zeta}_j \\ \hat{\xi}_k & \hat{\eta}_k & \hat{\zeta}_k \end{bmatrix} = \begin{bmatrix} \hat{x}_1 & \hat{x}_2 & \hat{x}_3 & \hat{x}_4 & \hat{x}_5 & \hat{x}_6 \\ \hat{y}_1 & \hat{y}_2 & \hat{y}_3 & \hat{y}_4 & \hat{y}_5 & \hat{y}_6 \\ \hat{z}_1 & \hat{z}_2 & \hat{z}_3 & \hat{z}_4 & \hat{z}_5 & \hat{z}_6 \end{bmatrix} \begin{bmatrix} -(1-\zeta) & -(1-\zeta) & -(1-\xi-\eta) \\ (1-\zeta) & 0 & -\xi \\ 0 & (1-\zeta) & -\eta \\ -\zeta & -\zeta & (1-\xi-\eta) \\ \zeta & 0 & \xi \\ 0 & \zeta & \eta \end{bmatrix} \tag{28.24}$$

c. the current position

$$\begin{bmatrix} \tilde{\xi}_i & \tilde{\eta}_i & \tilde{\zeta}_i \\ \tilde{\xi}_j & \tilde{\eta}_j & \tilde{\zeta}_j \\ \tilde{\xi}_k & \tilde{\eta}_k & \tilde{\zeta}_k \end{bmatrix} = \begin{bmatrix} \tilde{x}_1 & \tilde{x}_2 & \tilde{x}_3 & \tilde{x}_4 & \tilde{x}_5 & \tilde{x}_6 \\ \tilde{y}_1 & \tilde{y}_2 & \tilde{y}_3 & \tilde{y}_4 & \tilde{y}_5 & \tilde{y}_6 \\ \tilde{z}_1 & \tilde{z}_2 & \tilde{z}_3 & \tilde{z}_4 & \tilde{z}_5 & \tilde{z}_6 \end{bmatrix} \begin{bmatrix} -(1-\zeta) & -(1-\zeta) & -(1-\xi-\eta) \\ (1-\zeta) & 0 & -\xi \\ 0 & (1-\zeta) & -\eta \\ -\zeta & -\zeta & (1-\xi-\eta) \\ \zeta & 0 & \xi \\ 0 & \zeta & \eta \end{bmatrix} \tag{28.25}$$

28.7 The Solid-Embedded Vector Base versus the Material-Embedded Vector Base

The base vectors of the solid-embedded vector base are fixed to the material points and so are the vectors of the material-embedded vector base. From this, it follows that there is a uniquely defined relationship between these two bases such that

$$\begin{bmatrix} \alpha_i & \beta_i & \gamma_i \\ \alpha_j & \beta_j & \gamma_j \\ \alpha_k & \beta_k & \gamma_k \end{bmatrix} = \begin{bmatrix} \xi_i & \eta_i & \zeta_i \\ \xi_j & \eta_j & \zeta_j \\ \xi_k & \eta_k & \zeta_k \end{bmatrix} \begin{bmatrix} \alpha_\xi & \beta_\xi & \gamma_\xi \\ \alpha_\eta & \beta_\eta & \gamma_\eta \\ \alpha_\zeta & \beta_\zeta & \gamma_\zeta \end{bmatrix}. \tag{28.26}$$

By substituting from Equation (28.21) one obtains

$$\begin{bmatrix} \alpha_i & \beta_i & \gamma_i \\ \alpha_j & \beta_j & \gamma_j \\ \alpha_k & \beta_k & \gamma_k \end{bmatrix} = \begin{bmatrix} x_1 & x_2 & x_3 & x_4 & x_5 & x_6 \\ y_1 & y_2 & y_3 & y_4 & y_5 & y_6 \\ z_1 & z_2 & z_3 & z_4 & z_5 & z_6 \end{bmatrix} \begin{bmatrix} -(1-\zeta) & -(1-\zeta) & -(1-\xi-\eta) \\ (1-\zeta) & 0 & -\xi \\ 0 & (1-\zeta) & -\eta \\ -\zeta & -\zeta & (1-\xi-\eta) \\ \zeta & 0 & \xi \\ 0 & \zeta & \eta \end{bmatrix} \begin{bmatrix} \alpha_\xi & \beta_\xi & \gamma_\xi \\ \alpha_\eta & \beta_\eta & \gamma_\eta \\ \alpha_\zeta & \beta_\zeta & \gamma_\zeta \end{bmatrix} \tag{28.27}$$

This means that the initial position of the material-embedded vector base is given by

$$\begin{bmatrix} \bar{\alpha}_i & \bar{\beta}_i & \bar{\gamma}_i \\ \bar{\alpha}_j & \bar{\beta}_j & \bar{\gamma}_j \\ \bar{\alpha}_k & \bar{\beta}_k & \bar{\gamma}_k \end{bmatrix} = \begin{bmatrix} \bar{x}_1 & \bar{x}_2 & \bar{x}_3 & \bar{x}_4 & \bar{x}_5 & \bar{x}_6 \\ \bar{y}_1 & \bar{y}_2 & \bar{y}_3 & \bar{y}_4 & \bar{y}_5 & \bar{y}_6 \\ \bar{z}_1 & \bar{z}_2 & \bar{z}_3 & \bar{z}_4 & \bar{z}_5 & \bar{z}_6 \end{bmatrix} \begin{bmatrix} -(1-\zeta) & -(1-\zeta) & -(1-\xi-\eta) \\ (1-\zeta) & 0 & -\xi \\ 0 & (1-\zeta) & -\eta \\ -\zeta & -\zeta & (1-\xi-\eta) \\ \zeta & 0 & \xi \\ 0 & \zeta & \eta \end{bmatrix} \begin{bmatrix} \bar{\alpha}_\xi & \bar{\beta}_\xi & \bar{\gamma}_\xi \\ \bar{\alpha}_\eta & \bar{\beta}_\eta & \bar{\gamma}_\eta \\ \bar{\alpha}_\zeta & \bar{\beta}_\zeta & \bar{\gamma}_\zeta \end{bmatrix} \tag{28.28}$$

As the initial position of the material-embedded vector base is given as an input to the finite element analysis, it is used to calculate

$$\begin{bmatrix} \bar{\alpha}_\xi & \bar{\beta}_\xi & \bar{\gamma}_\xi \\ \bar{\alpha}_\eta & \bar{\beta}_\eta & \bar{\gamma}_\eta \\ \bar{\alpha}_\zeta & \bar{\beta}_\zeta & \bar{\gamma}_\zeta \end{bmatrix} = \left(\begin{bmatrix} \bar{x}_1 & \bar{x}_2 & \bar{x}_3 & \bar{x}_4 & \bar{x}_5 & \bar{x}_6 \\ \bar{y}_1 & \bar{y}_2 & \bar{y}_3 & \bar{y}_4 & \bar{y}_5 & \bar{y}_6 \\ \bar{z}_1 & \bar{z}_2 & \bar{z}_3 & \bar{z}_4 & \bar{z}_5 & \bar{z}_6 \end{bmatrix} \begin{bmatrix} -(1-\zeta) & -(1-\zeta) & -(1-\xi-\eta) \\ (1-\zeta) & 0 & -\xi \\ 0 & (1-\zeta) & -\eta \\ -\zeta & -\zeta & (1-\xi-\eta) \\ \zeta & 0 & \xi \\ 0 & \zeta & \eta \end{bmatrix} \right)^{-1} \begin{bmatrix} \bar{\alpha}_i & \bar{\beta}_i & \bar{\gamma}_i \\ \bar{\alpha}_j & \bar{\beta}_j & \bar{\gamma}_j \\ \bar{\alpha}_k & \bar{\beta}_k & \bar{\gamma}_k \end{bmatrix} \tag{28.29}$$

The matrix obtained from Equation (28.29) is used to calculate the previous position of the material-embedded vector base

$$\begin{bmatrix} \hat{\alpha}_\xi & \hat{\beta}_\xi & \hat{\gamma}_\xi \\ \hat{\alpha}_\eta & \hat{\beta}_\eta & \hat{\gamma}_\eta \\ \hat{\alpha}_\zeta & \hat{\beta}_\zeta & \hat{\gamma}_\zeta \end{bmatrix} = \left(\begin{bmatrix} \hat{x}_1 & \hat{x}_2 & \hat{x}_3 & \hat{x}_4 & \hat{x}_5 & \hat{x}_6 \\ \hat{y}_1 & \hat{y}_2 & \hat{y}_3 & \hat{y}_4 & \hat{y}_5 & \hat{y}_6 \\ \hat{z}_1 & \hat{z}_2 & \hat{z}_3 & \hat{z}_4 & \hat{z}_5 & \hat{z}_6 \end{bmatrix} \begin{bmatrix} -(1-\zeta) & -(1-\zeta) & -(1-\xi-\eta) \\ (1-\zeta) & 0 & -\xi \\ 0 & (1-\zeta) & -\eta \\ -\zeta & -\zeta & (1-\xi-\eta) \\ \zeta & 0 & \xi \\ 0 & \zeta & \eta \end{bmatrix} \right)^{-1} \begin{bmatrix} \hat{\alpha}_i & \hat{\beta}_i & \hat{\gamma}_i \\ \hat{\alpha}_j & \hat{\beta}_j & \hat{\gamma}_j \\ \hat{\alpha}_k & \hat{\beta}_k & \hat{\gamma}_k \end{bmatrix} \tag{28.30}$$

The same matrix is also used to calculate the current position of the material-embedded vector base

$$\begin{bmatrix} \tilde{\alpha}_\xi & \tilde{\beta}_\xi & \tilde{\gamma}_\xi \\ \tilde{\alpha}_\eta & \tilde{\beta}_\eta & \tilde{\gamma}_\eta \\ \tilde{\alpha}_\zeta & \tilde{\beta}_\zeta & \tilde{\gamma}_\zeta \end{bmatrix} = \left(\begin{bmatrix} \hat{x}_1 & \hat{x}_2 & \hat{x}_3 & \hat{x}_4 & \hat{x}_5 & \hat{x}_6 \\ \hat{y}_1 & \hat{y}_2 & \hat{y}_3 & \hat{y}_4 & \hat{y}_5 & \hat{y}_6 \\ \hat{z}_1 & \hat{z}_2 & \hat{z}_3 & \hat{z}_4 & \hat{z}_5 & \hat{z}_6 \end{bmatrix} \begin{bmatrix} -(1-\zeta) & -(1-\zeta) & -(1-\xi-\eta) \\ (1-\zeta) & 0 & -\xi \\ 0 & (1-\zeta) & -\eta \\ -\zeta & -\zeta & (1-\xi-\eta) \\ \zeta & 0 & \xi \\ 0 & \zeta & \eta \end{bmatrix} \right)^{-1} \begin{bmatrix} \tilde{\alpha}_i & \tilde{\beta}_i & \tilde{\gamma}_i \\ \tilde{\alpha}_j & \tilde{\beta}_j & \tilde{\gamma}_j \\ \tilde{\alpha}_k & \tilde{\beta}_k & \tilde{\gamma}_k \end{bmatrix} \tag{28.31}$$

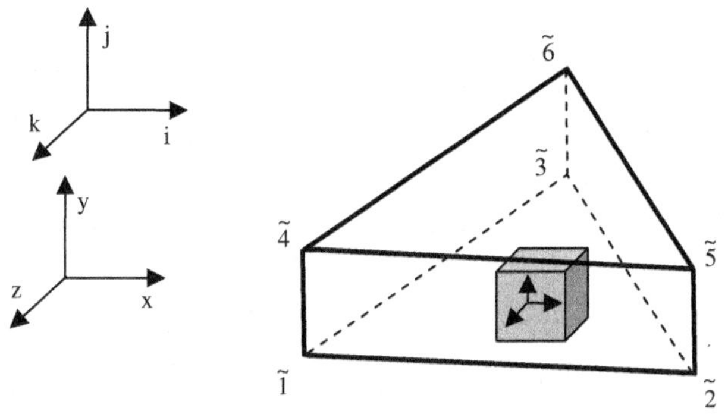

Figure 28.6 The Cauchy material element and the Cauchy stress tensor matrix. Note that the Cauchy material element is a cube in the current position

In this manner, the initial, the previous and the current position of the material-embedded vector base are all known at every material point of the shell finite element. These three vectors define the kinematics of deformation.

28.8 The Constitutive Law

The three infinitesimal base vectors $[\boldsymbol{\alpha} \;\; \boldsymbol{\beta} \;\; \boldsymbol{\gamma}]$ define the generalized material element. Using the material package (i.e. the constitutive law) from the initial, the current and the previous shape of the generalized material element, the Munjiza stress tensor matrix is calculated.

Within the material package the Cauchy stress tensor matrix is calculated from the Munjiza stress tensor matrix. The Cauchy stress tensor matrix is then returned to the finite element analysis and is used to evaluate the current nodal forces.

It is worth being reminded that the components of the Cauchy stress tensor matrix define the internal forces in the global $[\mathbf{i} \;\; \mathbf{j} \;\; \mathbf{k}]$ directions on surfaces of the Cauchy material element, i.e., the cube-shaped element in global $[\mathbf{i} \;\; \mathbf{j} \;\; \mathbf{k}]$ directions (see Figure 28.6).

Unlike the generalized material element, which always consists of the same material points, the Cauchy material element comprises different material points, as the deformation progresses. However, it is always cube shaped.

28.9 Selective Stretch Sampling Based Integration of the Equivalent Nodal Forces

Shear Locking and Stretch Sampling Points. One problem that can occur with the six-noded shell element is the phenomenon of shear-locking. This is illustrated in Figure 28.7. Theoretically, in pure bending there should be no shear stress. However, due to the description of the finite element's deformation (the position of the material points with the finite element) being only approximated (not exactly represented) through the shape functions, the shear stress is different from zero everywhere, except at the line A-B as shown in Figure 28.7.

This spurious shear stress is disproportionally large in comparison with the normal (bending) stress – as such, it has a dominant influence on the nodal forces, thus producing erroneous results. This phenomenon is called shear locking, for the shear stress "locks" all the degrees of freedom, making the finite element very stiff.

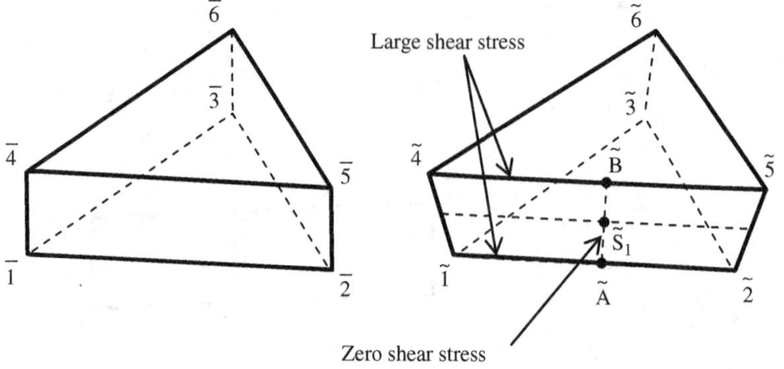

Figure 28.7 Shear locking in pure bending: (a) the initial state; (b) the current state of a shell element in pure bending

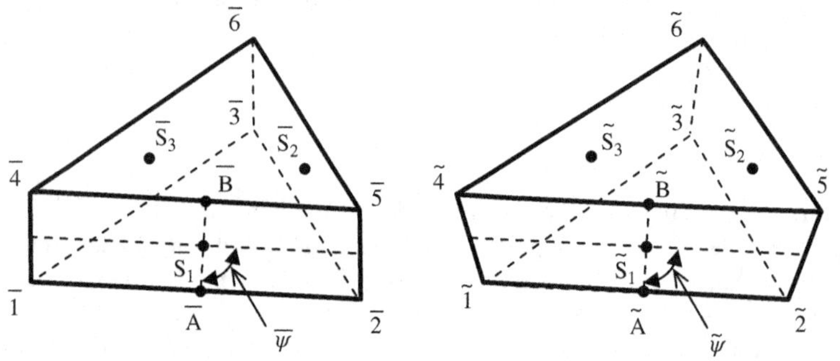

Figure 28.8 Initial and current angle ψ at stretch sampling point S_1

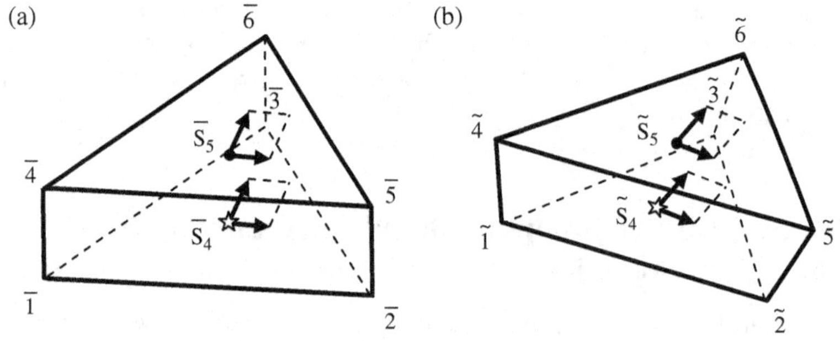

Figure 28.9 Stretch sampling points for membrane deformation: (a) initial and (b) current position

In order to counteract shear locking, the selective sampling of stretches is employed. This is done through the calculation of shear stretch (angle stretch) along the line A–B shown in Figure 28.7. The simplest option is to calculate the shear stretch at point S_1, and assume it to be constant over the surface $\tilde{1}-\tilde{2}-\tilde{5}-\tilde{4}$, Figure 28.8.

The membrane stretches are best sampled at the center (point S_4 and S_5) of the top and bottom triangles, Figure 28.9.

In addition, stretching perpendicular to the shell layers is calculated from the stretching in the ζ direction with stretch sampling points S_6, S_7, S_8, Figure 28.10.

Stress Sampling Points. For a shell with a single layer, the above stretch sampling points are combined with six stress sampling points, Figure 28.11. These points, in essence, describe the stress distribution over the shell element's domain.

For a multi-layered shell, the stress sampling points are supplied for each shell layer, Figure 28.12.

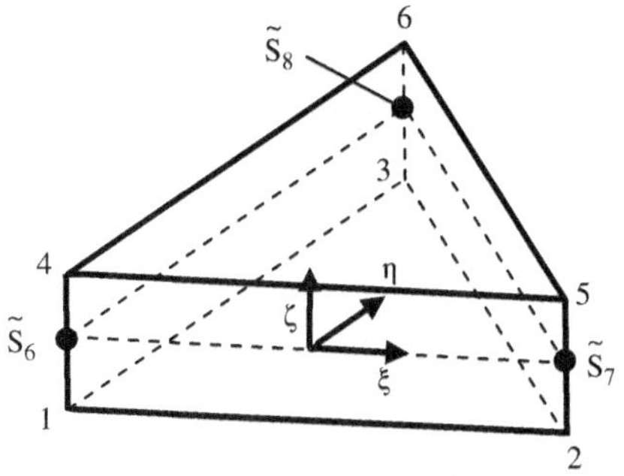

Figure 28.10 Stretch sampling points for stretching in ζ direction

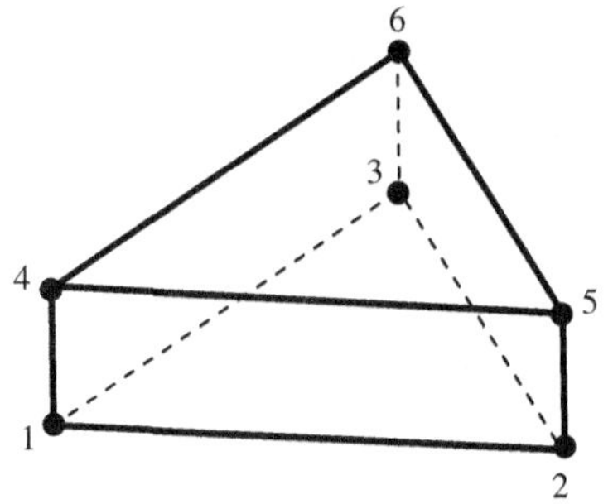

Figure 28.11 Stress sampling point for homogeneous shell

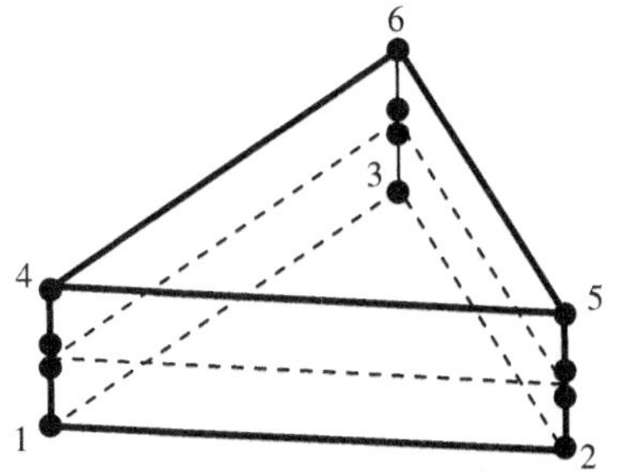

Figure 28.12 Stress sampling points for a multilayered shell – six stress sampling points are supplied for each layer

Integration of Nodal Forces Using Virtual Work. One way to integrate the nodal forces over the finite element's volume is to use the principle of virtual work (work of real forces on virtual displacements).

First, the virtual displacements at the finite element's nodes are supplied in turn. For example, virtual displacement at node 1 in the **i** direction

$$u_1 = \mathrm{u}_1(\xi, \eta, \zeta) = u_1 \mathrm{N}_1(\xi, \eta, \zeta), \tag{28.32}$$

where u_1 is the virtual displacement of node 1 and N_1 is the shape function for node 1.

The equivalent nodal forces due to the internal forces are obtained from the principle of virtual work. In the case of the virtual displacement given by Equation (28.32) one obtains the equivalent nodal force in the **i** direction on node 1

$$f_{i1}\mathrm{u}_1 = -\iiint_{\tilde{V}_e} \left\{ \begin{bmatrix} c_{ii} & c_{ij} & c_{ik} \end{bmatrix} \begin{bmatrix} \dfrac{\partial \mathrm{u}_1}{\partial \tilde{x}} \\ \dfrac{\partial \mathrm{u}_1}{\partial \tilde{y}} \\ \dfrac{\partial \mathrm{u}_1}{\partial \tilde{z}} \end{bmatrix} + \left(\frac{\partial c_{ii}}{\partial \tilde{x}} + \frac{\partial c_{ij}}{\partial \tilde{y}} + \frac{\partial c_{ik}}{\partial \tilde{z}} \right) \mathrm{u}_1 \right\} d\tilde{x} d\tilde{y} d\tilde{z} \tag{28.33}$$

The integration is done using Gaussian integration. Six Gauss integration points are selected, Figure 28.13.

The solid-embedded coordinates of the Gauss points are as follows

$$\begin{aligned} \mathrm{G}_1 &= \left(\frac{1}{2}, 0, -\frac{\sqrt{3}}{3} \right), \quad \mathrm{G}_2 = \left(\frac{1}{2}, 0, +\frac{\sqrt{3}}{3} \right) \\ \mathrm{G}_3 &= \left(\frac{1}{2}, \frac{1}{2}, -\frac{\sqrt{3}}{3} \right), \quad \mathrm{G}_4 = \left(\frac{1}{2}, \frac{1}{2}, +\frac{\sqrt{3}}{3} \right) \\ \mathrm{G}_5 &= \left(0, \frac{1}{2}, -\frac{\sqrt{3}}{3} \right), \quad \mathrm{G}_6 = \left(0, \frac{1}{2}, +\frac{\sqrt{3}}{3} \right). \end{aligned} \tag{28.34}$$

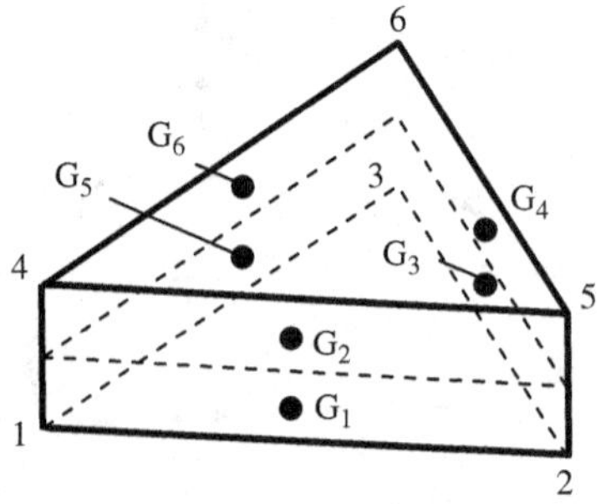

Figure 28.13 Gauss integration points for the six-noded shell element

After substituting these into Equation (25.6), one obtains the integral by summing over the Gauss points

$$f_{i1} = -\frac{\tilde{V}_e}{6}\sum_{j=1}^{6}\begin{bmatrix} c_{ii} & c_{ij} & c_{ik} \end{bmatrix}\begin{bmatrix} \frac{\partial \mathrm{u}_1}{\partial \tilde{x}} \\ \frac{\partial \mathrm{u}_1}{\partial \tilde{y}} \\ \frac{\partial \mathrm{u}_1}{\partial \tilde{z}} \end{bmatrix} - \frac{\tilde{V}_e}{6}\sum_{j=1}^{6}\left(\frac{\partial c_{ii}}{\partial \tilde{x}} + \frac{\partial c_{ij}}{\partial \tilde{y}} + \frac{\partial c_{ik}}{\partial \tilde{z}}\right)\mathrm{u}_1, \tag{28.35}$$

where $\tilde{V}_e$ is the finite element's current volume. The second half of Equation (28.35) represents the change of stress over the finite element and is identical to zero (for the homogeneous deformation finite element) which yields the equivalent nodal force

$$f_{i1} = -\frac{\tilde{V}_e}{6}\sum_{j=1}^{6}\begin{bmatrix} c_{ii} & c_{ij} & c_{ik} \end{bmatrix}\begin{bmatrix} \frac{\partial \mathrm{u}_1}{\partial \tilde{x}} \\ \frac{\partial \mathrm{u}_1}{\partial \tilde{y}} \\ \frac{\partial \mathrm{u}_1}{\partial \tilde{z}} \end{bmatrix}. \tag{28.36}$$

By substituting

$$\begin{aligned} \begin{bmatrix} \frac{\partial \mathrm{u}_1}{\partial \tilde{x}} \\ \frac{\partial \mathrm{u}_1}{\partial \tilde{y}} \\ \frac{\partial \mathrm{u}_1}{\partial \tilde{z}} \end{bmatrix} &= \begin{bmatrix} \frac{\partial \tilde{\mathrm{x}}}{\partial \xi} & \frac{\partial \tilde{\mathrm{x}}}{\partial \eta} & \frac{\partial \tilde{\mathrm{x}}}{\partial \zeta} \\ \frac{\partial \tilde{\mathrm{y}}}{\partial \xi} & \frac{\partial \tilde{\mathrm{y}}}{\partial \eta} & \frac{\partial \tilde{\mathrm{y}}}{\partial \zeta} \\ \frac{\partial \tilde{\mathrm{z}}}{\partial \xi} & \frac{\partial \tilde{\mathrm{z}}}{\partial \eta} & \frac{\partial \tilde{\mathrm{z}}}{\partial \zeta} \end{bmatrix}^{-\mathrm{T}} \begin{bmatrix} \frac{\partial \mathrm{u}_1}{\partial \xi} \\ \frac{\partial \mathrm{u}_1}{\partial \eta} \\ \frac{\partial \mathrm{u}_1}{\partial \zeta} \end{bmatrix} \\ &= \begin{bmatrix} \frac{\partial \tilde{\mathrm{x}}}{\partial \xi} & \frac{\partial \tilde{\mathrm{x}}}{\partial \eta} & \frac{\partial \tilde{\mathrm{x}}}{\partial \zeta} \\ \frac{\partial \tilde{\mathrm{y}}}{\partial \xi} & \frac{\partial \tilde{\mathrm{y}}}{\partial \eta} & \frac{\partial \tilde{\mathrm{y}}}{\partial \zeta} \\ \frac{\partial \tilde{\mathrm{z}}}{\partial \xi} & \frac{\partial \tilde{\mathrm{z}}}{\partial \eta} & \frac{\partial \tilde{\mathrm{z}}}{\partial \zeta} \end{bmatrix}^{-\mathrm{T}} \begin{bmatrix} \frac{\partial \mathrm{N}_1}{\partial \xi} \\ \frac{\partial \mathrm{N}_1}{\partial \eta} \\ \frac{\partial \mathrm{N}_1}{\partial \zeta} \end{bmatrix}, \end{aligned} \tag{28.37}$$

one obtains

$$f_{i1} = -\frac{\tilde{V}_e}{6}\sum_{j=1}^{6}\left[c_{ii}c_{ij}c_{ik}\right]\begin{bmatrix}\frac{\partial\tilde{x}}{\partial\xi} & \frac{\partial\tilde{x}}{\partial\eta} & \frac{\partial\tilde{x}}{\partial\zeta}\\ \frac{\partial\tilde{y}}{\partial\xi} & \frac{\partial\tilde{y}}{\partial\eta} & \frac{\partial\tilde{y}}{\partial\zeta}\\ \frac{\partial\tilde{z}}{\partial\xi} & \frac{\partial\tilde{z}}{\partial\eta} & \frac{\partial\tilde{z}}{\partial\zeta}\end{bmatrix}^{-T}\begin{bmatrix}\frac{\partial N_1}{\partial\xi}\\ \frac{\partial N_1}{\partial\eta}\\ \frac{\partial N_1}{\partial\zeta}\end{bmatrix}, \tag{28.38}$$

where

$$\begin{bmatrix}\frac{\partial\tilde{x}}{\partial\xi} & \frac{\partial\tilde{x}}{\partial\eta} & \frac{\partial\tilde{x}}{\partial\zeta}\\ \frac{\partial\tilde{y}}{\partial\xi} & \frac{\partial\tilde{y}}{\partial\eta} & \frac{\partial\tilde{y}}{\partial\zeta}\\ \frac{\partial\tilde{z}}{\partial\xi} & \frac{\partial\tilde{z}}{\partial\eta} & \frac{\partial\tilde{z}}{\partial\zeta}\end{bmatrix}^{-T} \tag{28.39}$$

is obtained from the fact that

$$\begin{bmatrix}\tilde{x}\\ \tilde{y}\\ \tilde{z}\end{bmatrix} = \begin{bmatrix}\tilde{x}_1 & \tilde{x}_2 & \tilde{x}_3 & \tilde{x}_4 & \tilde{x}_5 & \tilde{x}_6\\ \tilde{y}_1 & \tilde{y}_2 & \tilde{y}_3 & \tilde{y}_4 & \tilde{y}_5 & \tilde{y}_6\\ \tilde{z}_1 & \tilde{z}_2 & \tilde{z}_3 & \tilde{z}_4 & \tilde{z}_5 & \tilde{z}_6\end{bmatrix}\begin{bmatrix}N_1\\ N_2\\ N_3\\ N_4\\ N_5\\ N_6\end{bmatrix}. \tag{28.40}$$

The other equivalent forces are by analogy

$$f_{j1} = -\frac{\tilde{V}_e}{6}\sum_{j=1}^{6}\left[c_{ji}\quad c_{jj}\quad c_{jk}\right]\begin{bmatrix}\frac{\partial\tilde{x}}{\partial\xi} & \frac{\partial\tilde{x}}{\partial\eta} & \frac{\partial\tilde{x}}{\partial\zeta}\\ \frac{\partial\tilde{y}}{\partial\xi} & \frac{\partial\tilde{y}}{\partial\eta} & \frac{\partial\tilde{y}}{\partial\zeta}\\ \frac{\partial\tilde{z}}{\partial\xi} & \frac{\partial\tilde{z}}{\partial\eta} & \frac{\partial\tilde{z}}{\partial\zeta}\end{bmatrix}^{-T}\begin{bmatrix}\frac{\partial N_1}{\partial\xi}\\ \frac{\partial N_1}{\partial\eta}\\ \frac{\partial N_1}{\partial\zeta}\end{bmatrix}, \tag{28.41}$$

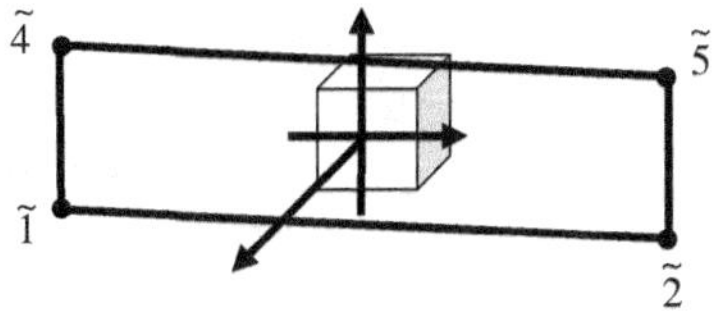

Figure 28.14 Components of the Cauchy stress matrix over the boundary $\tilde{1}-\tilde{2}-\tilde{5}-\tilde{4}$

$$f_{k1} = -\frac{\tilde{V}_e}{6}\sum_{j=1}^{6}\begin{bmatrix} c_{ki} & c_{kj} & c_{kk}\end{bmatrix}\begin{bmatrix} \frac{\partial \tilde{x}}{\partial \xi} & \frac{\partial \tilde{x}}{\partial \eta} & \frac{\partial \tilde{x}}{\partial \zeta} \\ \frac{\partial \tilde{y}}{\partial \xi} & \frac{\partial \tilde{y}}{\partial \eta} & \frac{\partial \tilde{y}}{\partial \zeta} \\ \frac{\partial \tilde{z}}{\partial \xi} & \frac{\partial \tilde{z}}{\partial \eta} & \frac{\partial \tilde{z}}{\partial \zeta}\end{bmatrix}^{-T}\begin{bmatrix}\frac{\partial N_1}{\partial \xi} \\ \frac{\partial N_1}{\partial \eta} \\ \frac{\partial N_1}{\partial \zeta}\end{bmatrix}. \tag{28.42}$$

In matrix form this yields

$$\begin{bmatrix} f_{i1} & f_{i2} & f_{i3} & f_{i4} & f_{i5} & f_{i6} \\ f_{j1} & f_{j2} & f_{j3} & f_{j4} & f_{j5} & f_{j6} \\ f_{k1} & f_{k2} & f_{k3} & f_{k4} & f_{k5} & f_{k6}\end{bmatrix} =$$

$$\frac{-\tilde{V}_e}{6}\begin{bmatrix} c_{ii} & c_{ij} & c_{ik} \\ c_{ji} & c_{jj} & c_{jk} \\ c_{ki} & c_{kj} & c_{kk}\end{bmatrix}\begin{bmatrix} \frac{\partial \tilde{x}}{\partial \xi} & \frac{\partial \tilde{x}}{\partial \eta} & \frac{\partial \tilde{x}}{\partial \zeta} \\ \frac{\partial \tilde{y}}{\partial \xi} & \frac{\partial \tilde{y}}{\partial \eta} & \frac{\partial \tilde{y}}{\partial \zeta} \\ \frac{\partial \tilde{z}}{\partial \xi} & \frac{\partial \tilde{z}}{\partial \eta} & \frac{\partial \tilde{z}}{\partial \zeta}\end{bmatrix}^{-T}\begin{bmatrix} \frac{\partial N_1}{\partial \xi} & \frac{\partial N_2}{\partial \xi} & \frac{\partial N_3}{\partial \xi} & \frac{\partial N_4}{\partial \xi} & \frac{\partial N_5}{\partial \xi} & \frac{\partial N_6}{\partial \xi} \\ \frac{\partial N_1}{\partial \eta} & \frac{\partial N_2}{\partial \eta} & \frac{\partial N_3}{\partial \eta} & \frac{\partial N_4}{\partial \eta} & \frac{\partial N_5}{\partial \eta} & \frac{\partial N_6}{\partial \eta} \\ \frac{\partial N_1}{\partial \zeta} & \frac{\partial N_2}{\partial \zeta} & \frac{\partial N_3}{\partial \zeta} & \frac{\partial N_4}{\partial \zeta} & \frac{\partial N_5}{\partial \zeta} & \frac{\partial N_6}{\partial \zeta}\end{bmatrix} \tag{28.43}$$

Integration of Nodal Forces Over Boundary. An alternative approach to the virtual work formulation is to integrate the nodal forces over the boundary of the finite element, Figure 28.14.

In the Figure 28.14, the components of the Cauchy stress tensor matrix are integrated over the surface $\tilde{1}-\tilde{2}-\tilde{5}-\tilde{4}$ and the obtained forces are distributed to the nodes of $\tilde{1}$, $\tilde{2}$, $\tilde{5}$, and $\tilde{4}$.

28.10 Multi-Layered Shell as an Assembly of Single Layer Shells

Very often one is interested in delamination or fracture of the multi-layered shells, Figure 28.15. In this case, it is convenient to model each layer separately and "glue" them together through a joint element.

The individual layers are "glued" together using either a penalty approach or more elaborate joint model that represents the deformation physics between the individual layers.

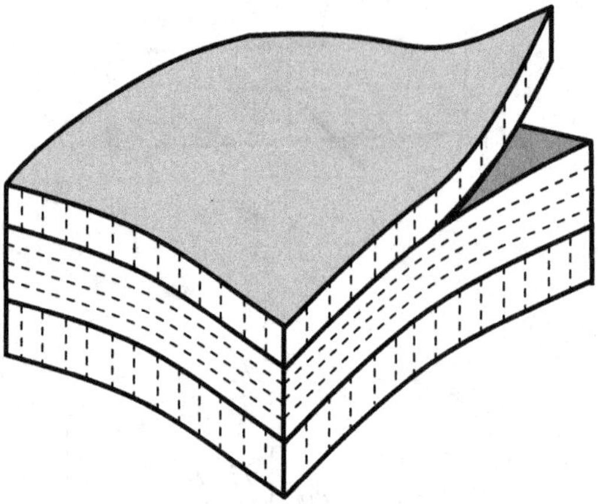

Figure 28.15 Delamination of multi-layered shells

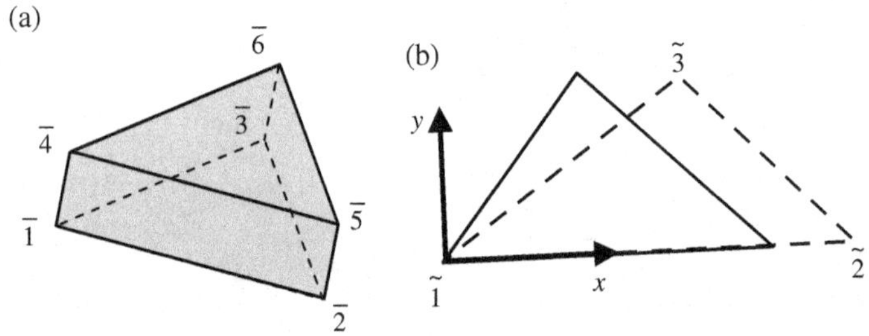

Figure 28.16 The lower surface of the shell element: (a) the initial position, (b) the lower surface of the shell

28.11 Improving the CPU Performance of the Shell Element

Membrane Stretches of the Lower Surface. For the actual implementation of the shell element described in Section 28.9, it is convenient to change the global coordinate system in such a way that the origin coincides with node $\tilde{1}$, the axis x goes through edge $\tilde{1}-\tilde{2}$, axis z is orthogonal to the plane of the lower surface of the shell and axis y is orthogonal to axis z and axis x, Figure 28.16.

By using this approach, after deducting translation and rotation of the coordinate system x, y, z itself (which do not contribute to stretches), one can write the initial, the previous and the current coordinates of the nodes of the lower surface as follows:

$$\begin{bmatrix} \bar{x}_1 & \bar{x}_2 & \bar{x}_3 \\ \bar{y}_1 & \bar{y}_2 & \bar{y}_3 \\ \bar{z}_1 & \bar{z}_2 & \bar{z}_3 \end{bmatrix} = \begin{bmatrix} 0 & \bar{x}_2 & \bar{x}_3 \\ 0 & 0 & \bar{y}_3 \\ 0 & 0 & 0 \end{bmatrix}, \tag{28.44}$$

$$\begin{bmatrix} \hat{x}_1 & \hat{x}_2 & \hat{x}_3 \\ \hat{y}_1 & \hat{y}_2 & \hat{y}_3 \\ \hat{z}_1 & \hat{z}_2 & \hat{z}_3 \end{bmatrix} = \begin{bmatrix} 0 & \hat{x}_2 & \hat{x}_3 \\ 0 & 0 & \hat{y}_3 \\ 0 & 0 & 0 \end{bmatrix}, \tag{28.45}$$

$$\begin{bmatrix} \tilde{x}_1 & \tilde{x}_2 & \tilde{x}_3 \\ \tilde{y}_1 & \tilde{y}_2 & \tilde{y}_3 \\ \tilde{z}_1 & \tilde{z}_2 & \tilde{z}_3 \end{bmatrix} = \begin{bmatrix} 0 & \tilde{x}_2 & \tilde{x}_3 \\ 0 & 0 & \tilde{y}_3 \\ 0 & 0 & 0 \end{bmatrix}. \tag{28.46}$$

Note that the nodal coordinates have in essence become 2D. As such, the z dimension can be eliminated which yields

$$\begin{bmatrix} \bar{x}_1 & \bar{x}_2 & \bar{x}_3 \\ \bar{y}_1 & \bar{y}_2 & \bar{y}_3 \end{bmatrix} = \begin{bmatrix} 0 & \bar{x}_2 & \bar{x}_3 \\ 0 & 0 & \bar{y}_3 \end{bmatrix}, \tag{28.47}$$

$$\begin{bmatrix} \hat{x}_1 & \hat{x}_2 & \hat{x}_3 \\ \hat{y}_1 & \hat{y}_2 & \hat{y}_3 \end{bmatrix} = \begin{bmatrix} 0 & \hat{x}_2 & \hat{x}_3 \\ 0 & 0 & \hat{y}_3 \end{bmatrix}, \tag{28.48}$$

$$\begin{bmatrix} \tilde{x}_1 & \tilde{x}_2 & \tilde{x}_3 \\ \tilde{y}_1 & \tilde{y}_2 & \tilde{y}_3 \end{bmatrix} = \begin{bmatrix} 0 & \tilde{x}_2 & \tilde{x}_3 \\ 0 & 0 & \tilde{y}_3 \end{bmatrix}. \tag{28.49}$$

Any material point inside the lower triangle is then described by its initial coordinates

$$\begin{aligned} \bar{x} &= \mathrm{N}_1\bar{x}_1 + \mathrm{N}_2\bar{x}_2 + \mathrm{N}_3\bar{x}_3 \\ \bar{y} &= \mathrm{N}_1\bar{y}_1 + \mathrm{N}_2\bar{y}_2 + \mathrm{N}_3\bar{y}_3. \end{aligned} \tag{28.50}$$

As

$$\bar{x}_1 = \bar{y}_1 = 0, \tag{28.51}$$

this yields

$$\begin{aligned} \bar{x} &= \mathrm{N}_2\bar{x}_2 + \mathrm{N}_3\bar{x}_3 \\ \bar{y} &= \mathrm{N}_2\bar{y}_2 + \mathrm{N}_3\bar{y}_3. \end{aligned} \tag{28.52}$$

In a similar manner any material point inside the triangle is defined by its previous coordinate

$$\begin{aligned} \hat{x} &= \mathrm{N}_2\hat{x}_2 + \mathrm{N}_3\hat{x}_3 \\ \hat{y} &= \mathrm{N}_2\hat{y}_2 + \mathrm{N}_3\hat{y}_3 \end{aligned} \tag{28.53}$$

and its current coordinate

$$\begin{aligned} \tilde{x} &= N_2\tilde{x}_2 + N_3\tilde{x}_3 \\ \tilde{y} &= N_2\tilde{y}_2 + N_3\tilde{y}_3, \end{aligned} \tag{28.54}$$

where the shape functions are given by

$$\begin{aligned} \mathrm{N}_1 &= 1 - \xi - \eta \\ \mathrm{N}_2 &= \xi \\ \mathrm{N}_3 &= \eta. \end{aligned} \tag{28.55}$$

The solid-embedded vector base is by definition given by

$$\begin{bmatrix} \bar{\xi}_i & \bar{\eta}_i \\ \bar{\xi}_j & \bar{\eta}_j \end{bmatrix} = \begin{bmatrix} \dfrac{\partial \bar{x}}{\partial \xi} & \dfrac{\partial \bar{x}}{\partial \eta} \\ \dfrac{\partial \bar{y}}{\partial \xi} & \dfrac{\partial \bar{y}}{\partial \eta} \end{bmatrix} = \begin{bmatrix} \bar{x}_2 & \bar{x}_3 \\ 0 & \bar{y}_3 \end{bmatrix}, \tag{28.56}$$

$$\begin{bmatrix} \hat{\xi}_i & \hat{\eta}_i \\ \hat{\xi}_j & \hat{\eta}_j \end{bmatrix} = \begin{bmatrix} \dfrac{\partial \hat{x}}{\partial \xi} & \dfrac{\partial \hat{x}}{\partial \eta} \\ \dfrac{\partial \hat{y}}{\partial \xi} & \dfrac{\partial \hat{y}}{\partial \eta} \end{bmatrix} = \begin{bmatrix} \hat{x}_2 & \hat{x}_3 \\ 0 & \hat{y}_3 \end{bmatrix}, \tag{28.57}$$

$$\begin{bmatrix} \tilde{\xi}_i & \tilde{\eta}_i \\ \tilde{\xi}_j & \tilde{\eta}_j \end{bmatrix} = \begin{bmatrix} \dfrac{\partial \tilde{x}}{\partial \xi} & \dfrac{\partial \tilde{x}}{\partial \eta} \\ \dfrac{\partial \tilde{y}}{\partial \xi} & \dfrac{\partial \tilde{y}}{\partial \eta} \end{bmatrix} = \begin{bmatrix} \tilde{x}_2 & \tilde{x}_3 \\ 0 & \tilde{y}_3 \end{bmatrix}. \tag{28.58}$$

From this, the material-embedded vectors $\boldsymbol{\alpha}$ and $\boldsymbol{\beta}$ are obtained

$$\begin{bmatrix} \bar{\alpha}_i & \bar{\beta}_i \\ \bar{\alpha}_j & \bar{\beta}_j \end{bmatrix}; \begin{bmatrix} \hat{\alpha}_i & \hat{\beta}_i \\ \hat{\alpha}_j & \hat{\beta}_j \end{bmatrix}; \begin{bmatrix} \tilde{\alpha}_i & \tilde{\beta}_i \\ \tilde{\alpha}_j & \tilde{\beta}_j \end{bmatrix}. \tag{28.59}$$

These six vectors define the initial, the previous and the current positions of the generalized anisotropic material element in the infinitesimal vicinity of the shell's lower surface, Figure 28.17.

Membrane Stretches of the Upper Surface. In order to analyze the membrane stretches of the upper surface, the global coordinates system is moved to node $\tilde{4}$, Figure 28.18.

The initial, the current and the previous coordinates of nodes 4, 5 and 6 are given by

$$\begin{bmatrix} \bar{x}_4 & \bar{x}_5 & \bar{x}_6 \\ \bar{y}_4 & \bar{y}_5 & \bar{y}_6 \end{bmatrix} = \begin{bmatrix} 0 & \bar{x}_5 & \bar{x}_6 \\ 0 & 0 & \bar{y}_6 \end{bmatrix}, \tag{28.60}$$

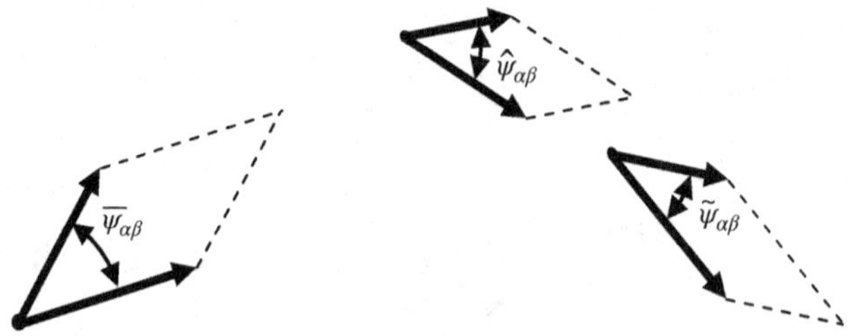

Figure 28.17 Membrane stretches of the shell's lower surface. These stretches are constant over the shell's lower surface

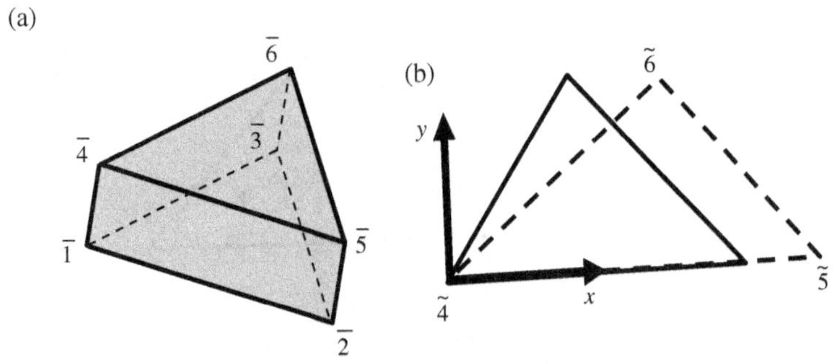

Figure 28.18 The upper surface of the shell element: (a) the initial position, (b) the upper surface of the shell

$$\begin{bmatrix} \hat{x}_4 & \hat{x}_5 & \hat{x}_6 \\ \hat{y}_4 & \hat{y}_5 & \hat{y}_6 \end{bmatrix} = \begin{bmatrix} 0 & \hat{x}_5 & \hat{x}_6 \\ 0 & 0 & \hat{y}_6 \end{bmatrix}, \tag{28.61}$$

$$\begin{bmatrix} \tilde{x}_4 & \tilde{x}_5 & \tilde{x}_6 \\ \tilde{y}_4 & \tilde{y}_5 & \tilde{y}_6 \end{bmatrix} = \begin{bmatrix} 0 & \tilde{x}_5 & \tilde{x}_6 \\ 0 & 0 & \tilde{y}_6 \end{bmatrix}. \tag{28.62}$$

The solid-embedded infinitesimal vector base $[\boldsymbol{\xi} \;\; \boldsymbol{\eta}]$ is therefore

$$\begin{bmatrix} \bar{\xi}_i & \bar{\eta}_i \\ \bar{\xi}_j & \bar{\eta}_j \end{bmatrix} = \begin{bmatrix} \dfrac{\partial \bar{x}}{\partial \xi} & \dfrac{\partial \bar{x}}{\partial \eta} \\ \dfrac{\partial \bar{y}}{\partial \xi} & \dfrac{\partial \bar{y}}{\partial \eta} \end{bmatrix} = \begin{bmatrix} \bar{x}_5 & \bar{x}_6 \\ 0 & \bar{y}_6 \end{bmatrix}, \tag{28.63}$$

$$\begin{bmatrix} \hat{\xi}_i & \hat{\eta}_i \\ \hat{\xi}_j & \hat{\eta}_j \end{bmatrix} = \begin{bmatrix} \dfrac{\partial \hat{x}}{\partial \xi} & \dfrac{\partial \hat{x}}{\partial \eta} \\ \dfrac{\partial \hat{y}}{\partial \xi} & \dfrac{\partial \hat{y}}{\partial \eta} \end{bmatrix} = \begin{bmatrix} \hat{x}_5 & \hat{x}_6 \\ 0 & \hat{y}_6 \end{bmatrix}, \tag{28.64}$$

$$\begin{bmatrix} \tilde{\xi}_i & \tilde{\eta}_i \\ \tilde{\xi}_j & \tilde{\eta}_j \end{bmatrix} = \begin{bmatrix} \dfrac{\partial \tilde{x}}{\partial \xi} & \dfrac{\partial \tilde{x}}{\partial \eta} \\ \dfrac{\partial \tilde{y}}{\partial \xi} & \dfrac{\partial \tilde{y}}{\partial \eta} \end{bmatrix} = \begin{bmatrix} \tilde{x}_5 & \tilde{x}_6 \\ 0 & \tilde{y}_6 \end{bmatrix}. \tag{28.65}$$

From these, the material-embedded vector base $[\boldsymbol{\alpha} \;\; \boldsymbol{\beta}]$ is obtained

$$\begin{bmatrix} \bar{\alpha}_i & \bar{\beta}_i \\ \bar{\alpha}_j & \bar{\beta}_j \end{bmatrix}; \begin{bmatrix} \hat{\alpha}_i & \hat{\beta}_i \\ \hat{\alpha}_j & \hat{\beta}_j \end{bmatrix}; \begin{bmatrix} \tilde{\alpha}_i & \tilde{\beta}_i \\ \tilde{\alpha}_j & \tilde{\beta}_j \end{bmatrix}. \tag{28.66}$$

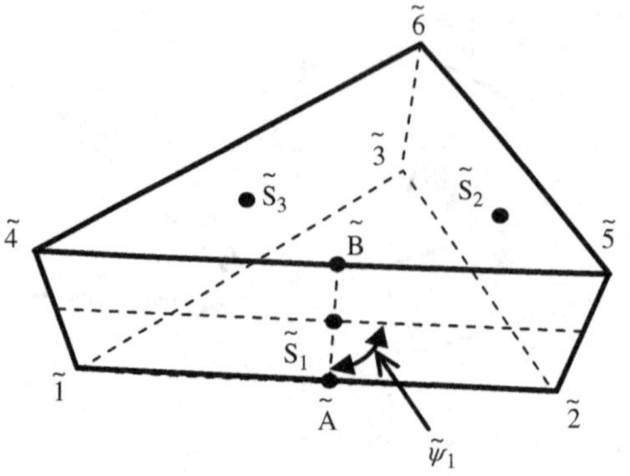

Figure 28.19 Shear stretches

This vector base describes the generalized anisotropic material element's kinematics of deformation. As such, it yields membrane stretches in the infinitesimal vicinity of the top surface of the shell. These stretches are constant over the surface of the shell.

Shear Stretches. In order to avoid shear locking, the shear stretches are calculated at points S_1, S_2 and S_3 as shown in Figure 28.19.

Shear stretch at point S_1 is calculated from the initial, the current and the previous position of vectors

$$\begin{bmatrix} \frac{(x_2+x_5)}{2}-\frac{(x_1+x_4)}{2} \\ \frac{(y_2+y_5)}{2}-\frac{(y_1+y_4)}{2} \\ \frac{(z_2+z_5)}{2}-\frac{(z_1+z_4)}{2} \end{bmatrix} \tag{28.67}$$

and

$$\begin{bmatrix} \frac{(x_4+x_5)}{2}-\frac{(x_1+x_2)}{2} \\ \frac{(y_4+y_5)}{2}-\frac{(y_1+y_2)}{2} \\ \frac{(z_4+z_5)}{2}-\frac{(z_1+z_2)}{2} \end{bmatrix}, \tag{28.68}$$

which produce the in-plane angle ψ_1, see Figure 28.19.

In a similar manner, the in-plane angle ψ_2 is calculated by considering the initial, the current and the previous positions of vectors

$$\begin{bmatrix} \frac{(x_3+x_6)}{2}-\frac{(x_2+x_5)}{2} \\ \frac{(y_3+y_6)}{2}-\frac{(y_2+y_5)}{2} \\ \frac{(z_3+z_6)}{2}-\frac{(z_2+z_5)}{2} \end{bmatrix} \tag{28.69}$$

and

$$\begin{bmatrix} \frac{(x_5+x_6)}{2}-\frac{(x_2+x_3)}{2} \\ \frac{(y_5+y_6)}{2}-\frac{(y_2+y_3)}{2} \\ \frac{(z_5+z_6)}{2}-\frac{(z_2+z_3)}{2} \end{bmatrix}. \tag{28.70}$$

Finally, the in-plane angle ψ_3 is calculated by considering the initial, the current and the previous positions of vectors

$$\begin{bmatrix} \frac{(x_1+x_4)}{2}-\frac{(x_3+x_6)}{2} \\ \frac{(y_1+y_4)}{2}-\frac{(y_3+y_6)}{2} \\ \frac{(z_1+z_4)}{2}-\frac{(z_3+z_6)}{2} \end{bmatrix} \tag{28.71}$$

and

$$\begin{bmatrix} \frac{(x_4+x_6)}{2}-\frac{(x_1+x_3)}{2} \\ \frac{(y_4+y_6)}{2}-\frac{(y_1+y_3)}{2} \\ \frac{(z_4+z_6)}{2}-\frac{(z_1+z_3)}{2} \end{bmatrix}. \tag{28.72}$$

The stretches in ζ direction, perpendicular to the surface of the shell are obtained from the initial, the previous and the current positions of vectors

$$\begin{bmatrix} \bar{x}_4-\bar{x}_1 \\ \bar{y}_4-\bar{y}_1 \\ \bar{z}_4-\bar{z}_1 \end{bmatrix};\begin{bmatrix} \hat{x}_4-\hat{x}_1 \\ \hat{y}_4-\hat{y}_1 \\ \hat{z}_4-\hat{z}_1 \end{bmatrix};\begin{bmatrix} \tilde{x}_4-\tilde{x}_1 \\ \tilde{y}_4-\tilde{y}_1 \\ \tilde{z}_4-\tilde{z}_1 \end{bmatrix}, \tag{28.73}$$

$$\begin{bmatrix} \bar{x}_5-\bar{x}_2 \\ \bar{y}_5-\bar{y}_2 \\ \bar{z}_5-\bar{z}_2 \end{bmatrix};\begin{bmatrix} \hat{x}_5-\hat{x}_2 \\ \hat{y}_5-\hat{y}_2 \\ \hat{z}_5-\hat{z}_2 \end{bmatrix};\begin{bmatrix} \tilde{x}_5-\tilde{x}_2 \\ \tilde{y}_5-\tilde{y}_2 \\ \tilde{z}_5-\tilde{z}_2 \end{bmatrix}, \tag{28.74}$$

$$\begin{bmatrix} \bar{x}_6-\bar{x}_3 \\ \bar{y}_6-\bar{y}_3 \\ \bar{z}_6-\bar{z}_3 \end{bmatrix};\begin{bmatrix} \hat{x}_6-\hat{x}_3 \\ \hat{y}_6-\hat{y}_3 \\ \hat{z}_6-\hat{z}_3 \end{bmatrix};\begin{bmatrix} \tilde{x}_6-\tilde{x}_3 \\ \tilde{y}_6-\tilde{y}_3 \\ \tilde{z}_6-\tilde{z}_3 \end{bmatrix}. \tag{28.75}$$

From the above stretches, the stress tensor at any point of the solid domain can be obtained using the material law as described in Chapter 18. The material law returns the Cauchy stress tensor matrix, which is used in the calculation of the equivalent nodal forces as described in previous sections of this chapter.

28.12 Summary

In this chapter a multiplicative decomposition based large strain large displacement anisotropic material based formulation for multi-layered shells has been presented. This was accomplished through a detailed description of a six-noded shell element.

Equivalent nodal forces are calculated using either the virtual work formulation or using the integration of internal forces over boundaries of the finite element. In both cases, selective sampling of stretches is employed to avoid the shear locking phenomena. Shell examples shown in Chapter 1 of this book were obtained using this finite element.

Further Reading

[1] Oden, J. T. (1972) *Finite Elements of Nonlinear Continua*. New York: McGraw-Hill.

[2] Belytschko, T., Lin, J. I. and Tsay, C. S. (1984) Explicit algorithms for the nonlinear dynamics of shells. *Comput. Method Appl. M.*, **42**(2): 225–51.

[3] Belytschko, T., Wong, B. L. and Stolarski, H. (1989) Assumed strain stabilization procedure for the 9-node Lagrange shell element. *International Journal for Numerical Methods in Engineering*, **28**(2): 385–414.

[4] Belytschko, T., Wong, B. L. and Chiang, H. Y. (1992) Advances in one-point quadrature shell elements. *Comput. Method Appl. M.*, **96**(1): 93–107.

[5] Bonet, J. and Wood, R. D. (1997) *Nonlinear Continuum Mechanics for Finite Element Analysis*. Cambridge, Cambridge University Press.

[6] de Souza Neto, E. A., Peric, D. and Owen, D. R. J. (1995) Finite elasticity in spatial description – linearization aspects with 3-D membrane applications. *International Journal for Numerical Methods in Engineering*, **38**(20): 3365–81.

[7] Hughes, T.J.R. (2000) *The Finite Element Method: Linear Static and Dynamic Finite Element Analysis*. New York, Dover Publications.

[8] Hughes, T. J. R. and Liu, W. K. (1981) Nonlinear finite element analysis of shells: Part I. Three-dimensional shells. *Comput. Method Appl. M.*, **26**(3): 331–62.

[9] Hughes, T. J. R. and Liu, W. K. (1981) Nonlinear finite element analysis of shells: Part II. Two-dimensional shells. *Comput. Method Appl. M.*, **27**(2): 167–81.

[10] Parisch, H. (1986) Efficient non-linear finite element shell formulation involving large strains. *Eng. Computation*, **3**(2): 121–8.

[11] Zienkiewicz, O. C. (1971) *The Finite Element Method in Engineering Science*. New York, McGraw-Hill.

[12] Zienkiewicz, O. C., Rojek, J., Taylor, R. L. and Pastor, M. (1998) Triangles and tetrahera in explicit dynamic codes for solids. *International Journal for Numerical Methods in Engineering*, **43**(3), 565–83.

[13] Zienkiewicz, O. C. and Taylor, R. L. (2005) *The Finite Element Method Set*. Oxford, Elsevier Science.

Index

acceleration, 37, 38, 42, 43, 85, 90, 94, 343
a-conjugate, 51
addition of matrices, 16
additive decomposition, 203, 259, 418
angle stretch, 269, 270, 289, 290, 297–9, 301, 307, 450
anisotropic, 8, 10, 13, 203, 218, 220–222, 238, 246, 255, 259, 260, 263–5, 267, 272, 280, 293, 299, 307, 312, 313, 321, 375, 419, 438–40, 458, 460, 462
anisotropic constitutive law, 313
anisotropic elastic material, 307
anisotropic material element, 220, 264, 265, 458, 460
anisotropic material nonlinearity, 10
anisotropic materials, 10, 13, 203, 218, 220–222, 246, 255, 259, 260, 263–5, 272, 312, 321, 375, 419, 439, 458, 460
a-orthogonal, 49, 53, 55, 57
a-orthonormal column matrices, 32
associativity of the summation, 87
assumption of continuum, 209, 214, 261, 263, 264, 321
axial stretches, 297
base vectors, 91–3, 95, 96, 98, 99, 102, 111–14, 117, 122, 123, 130, 136, 141, 144, 147, 152, 170, 179, 182, 191, 192, 228, 231, 243, 244, 255, 260, 321, 326, 332, 333, 361, 375, 376, 384, 389, 405, 423, 427, 428, 430, 439, 445, 447, 449
bending moment, 77, 78

cartesian coordinate system, 44, 227, 241, 242, 318, 320, 321, 330, 368, 382
Cauchy material element, 223, 225–7, 231, 232, 234, 238, 249–55, 302, 303, 306, 313, 343, 409, 430, 449
Cauchy stress, 167, 168, 173, 186, 188, 203, 225–7, 231, 235–8, 250, 252, 255, 258, 266, 267, 274, 278–80, 299, 302, 313, 326, 343, 349, 353, 355, 363, 377, 378, 392, 396, 399, 406, 407, 409, 412, 429–31, 435, 438, 449, 455, 462
central difference, 35, 37, 38, 40–42, 62
change of angle, 257
change of volume, 258, 357
commutativity, 89, 90
components of stress tensor matrix, 185

Large Strain Finite Element Method: A Practical Course, First Edition. Antonio Munjiza, Esteban Rougier and Earl E. Knight.

components of the Cauchy stress, 226, 227, 299, 407, 409, 449, 455
composite elements, 361, 362, 403
composite triangle, 359, 360, 403, 404, 407
computational mechanics, 14, 168, 174, 209, 215, 256, 259, 294, 314
conditional stability, 41
conjugate directions method, 13, 50, 51, 53, 60, 63
conjugate gradient method, 62, 63
constant strain finite element, 325
constitutive law, 13, 213, 231, 258–60, 264–7, 279, 292, 293, 295, 296, 299, 312, 313, 340, 343, 363, 364, 377, 392, 435, 438, 449
contact, 14, 204, 362, 367, 403
continuity conditions, 327
continuum mechanics, 174, 209, 214, 238, 256, 259, 261, 294, 313, 341, 355, 378, 392, 401, 413, 438, 462
convergence, 50, 357, 363
co-rotational formulation, 203, 293, 302, 313, 418, 438
crack, 14, 157, 175, 176, 209, 214
cross method, 48
current configuration, 204, 265, 324
current position, 204, 209, 214, 225, 227, 229–32, 235, 236, 238, 244, 246, 249, 250, 255, 259, 266–8, 275, 288, 292, 296, 319, 325, 331, 334, 340, 343, 344, 371, 374, 376–8, 382, 387, 389, 391, 392, 404, 409, 421, 424–7, 430, 431, 443, 446–50, 458, 461
current volume, 268, 275, 288, 296, 345, 351, 397, 411, 431, 453

deformation-dependent, 275
deformation independent, 228
deformation-independent matrices, 267
deformation invariant, 293, 312, 313, 378
deformation kinematics, 13, 249, 250, 255, 260, 264, 267, 279, 288, 292, 295, 319, 321, 323, 325, 329, 331, 333, 335, 337, 339–41, 367, 369, 371, 373, 375, 377–9, 381, 383, 385, 387, 389, 391, 392, 438
deformation-objective, 312
deformed solid element, 223, 302
derivation of a scalar field, 372
derivation of the shape functions, 368
determinant, 22–4, 28, 428, 429
diagonal matrix, 21, 60
differential, 35, 63, 73, 78, 86, 108, 115, 127, 137, 139, 149, 150, 152, 154, 174, 189, 200, 209, 214
  calculus, 108, 115, 127, 154, 174, 189, 200, 209
  equations, 35, 63, 73, 78, 86, 209
differentiating, 137
differentiation, 373, 374, 426, 431, 433, 446
direction of the force, 178
direction of the surface 168, 178
discretization, 37, 317, 318
displacement vector, 104, 130, 131
divergence, 47
dot product, 30, 44, 51, 52, 55, 92, 93, 436
dual base vectors, 98, 99, 170, 405
dual bases, 949, 105, 111, 112, 118, 123, 127, 136, 170, 405, 410, 411, 435
duality condition, 99, 119, 123
dynamic finite element analysis, 35, 256, 294, 313, 326, 341, 355, 392, 401, 413, 438, 462
dynamic relaxation, 13, 37, 43, 367

edge stretches, 269, 272, 277, 288, 289, 297, 300, 307
eigenmatrix, 25, 26
eigenvalues, 24–6, 39, 40, 61, 364
eight-noded quadrilateral finite element, 329
eight-noded solid finite element, 381
elastic constants, 280, 284, 287, 288, 302, 308, 309
engineering definition of stress, 174
engineering strain, 6, 209–12
equilibrium, 3–6, 36, 43, 75, 76, 78, 231, 232, 235, 238, 250, 255, 256, 273, 343–5, 357, 417
equilibrium equations, 343, 357, 417
equilibrium of forces, 3, 4
equilibrium of moments, 273
equivalent nodal forces, 13, 343, 344, 346, 347, 350, 353, 397, 398, 403, 404, 406, 408–10, 412, 430, 434, 435, 449, 452, 453, 462
Euler-Almansi strain, 210–212
exact solution, 6, 45, 47, 52
explicit, 13, 35, 37, 39, 41, 43, 45, 47, 49, 51, 53, 55, 57, 59, 61, 63, 203, 204, 326, 341, 355, 364, 367, 378, 392, 401, 413, 421, 438, 443, 462

explicit dynamic, 367, 378, 392, 401, 413
explicit integration, 37
explicit iterative static formulation, 367
external damping, 42

finite element library, 327, 379
finite strain deformability theory, 203
first derivative, 140, 151
first order tensor, 13, 129, 131–5, 137–41, 143, 145, 147, 149–51, 153, 154, 158, 159, 177
first order tensor to an n-dimensional space, 147
first Piola-Kirchhoff, 168, 169, 172, 181, 186–8, 258, 275–80, 305, 306
force base, 179, 182, 196, 198
force components, 106, 162, 250
four gauss points integration, 71
four-noded quadrilateral finite elements, 328, 358, 359, 362
fourth dimension, 117, 191

Gaussian elimination, 26, 28, 29, 62, 367
Gaussian integration, 65
Gaussian numerical integration, 67, 352, 355
Gauss integration point, 353–5, 358, 399, 400, 404, 408, 434, 452
Gauss point, 68, 70–72, 348, 352, 355, 358–64, 397, 400, 401, 404, 405, 408, 409, 452, 453
Gauss-seidel method, 46–8, 53
general base, 12, 98, 100, 102–4, 106, 108, 111, 114, 117, 120, 124, 136, 140, 142, 151, 167, 179, 182, 184
general base for forces and surfaces, 184
general force base, 182, 196
generalized material element, 222–5, 231, 238, 246, 247, 249, 255, 257–60, 264, 265, 267–9, 271, 273–7, 279–81, 288–90, 293, 295–300, 302, 303, 307, 308, 312, 321, 324–6, 334, 343, 361, 363, 364, 386, 389, 392, 419, 439, 440, 449
generalized material-embedded base vectors, 247–9, 260
generalized stress matrix components, 275
generalized stretch, 280, 292
general virtual displacement, 235, 254
global base, 98–100, 109, 111, 113, 161, 164, 166, 167, 177–82, 192, 196, 324, 373, 425, 430
global base vector, 111, 430
global coordinates, 229, 232, 244, 320, 330, 331, 336, 369, 373, 374, 382, 385, 420–423, 432, 443, 458
global derivatives, 349, 431
global orthonormal base, 98, 101, 111, 134, 141, 168, 170, 176, 186, 187, 191, 196, 219, 226, 299, 302, 334, 343, 386
global orthonormal vector base, 144
global vector base, 165, 260, 321
Green-Lagrange strain, 210–212
Green-Naghdi rate, 259, 267

heterogeneous solid, 217
homogeneous and anisotropic solids, 218, 238
homogeneous and isotropic solid, 219
homogeneous deformation triangle, 317
homogeneous deformation finite element, 317, 438, 453
homogeneous internal forces, 175, 176
homogenization, 260, 262–4, 272, 273
homogenized material element, 321
Hooke' s law, 213, 281, 282, 284–7, 309–11, 313
hyper-elastic material, 13, 258, 259, 264, 265, 267, 280, 295
hyper-lines, 48, 49
hyper-point, 44, 45
hyper-space, 44
hyper-surfaces, 45
hypo-elastic formulation, 13, 258, 259, 264, 266, 288, 299

identity matrix, 21
implicit equilibrium formulation, 4
implicit formulation, 42
inertia forces, 36, 43, 250, 256, 343
infi-meter, 77, 207, 208, 221, 243, 259, 373, 393, 439
infinitesimal base, 321, 323, 324, 445, 449
infinitesimally small, 5–7, 75, 77, 139, 150, 176, 206–8, 221, 222, 224, 227, 228, 232, 233, 236, 243, 246, 249, 251, 252, 259, 284, 321, 373, 393, 419, 423, 430, 439
infinitesimally small vector, 150
infinitesimal vicinity, 7, 176, 210, 231, 238, 243, 247, 255, 333, 372, 375, 376, 458, 460

initial geometry, 3, 330, 382
initial position, 204, 206, 209, 223, 224, 227, 236, 245, 248, 249, 260, 266, 267, 269, 295, 297, 321, 322, 330, 333, 334, 339–41, 371, 373, 376, 377, 382, 386, 388–90, 392, 421, 423, 425, 427, 428, 442, 446, 448, 456, 459
inner product, 30, 110
instability, 5, 6, 60–62
integrating over the volume, 394
integrating the stress over boundaries, 403
integration scheme, 37, 38, 40–42, 62
internal damping, 42, 43
internal forces, 3–5, 7, 13, 35, 36, 38, 59, 75, 77, 155–8, 160, 162, 164, 168, 173, 175–7, 184, 185, 187, 188, 225–7, 231–4, 236–8, 249–55, 258, 261, 267, 271–4, 276, 277, 279, 292, 293, 299–302, 304–8, 312, 313, 343, 344, 358, 378, 392–4, 406–9, 412, 417, 430, 436, 449, 452, 462
inverse transpose matrix, 100
iso-parametric finite elements, 329, 331, 333, 335, 337–9, 341
isotropic, 154, 217–19, 238, 256, 259, 263, 280, 281, 284, 285, 292–4, 307–9, 313
isotropic plane strain, 293
isotropic plane stress, 292
isotropic solid, 217–20, 238, 280, 299, 307, 308, 313

Jaumann rate, 203, 259, 267, 275

kinematic hardening, 259
kinematics of deformation, 203, 205, 207, 209, 211, 213, 214, 217, 219, 221, 223, 225, 227, 229, 231, 233, 235, 237, 239, 241, 243, 245, 247, 249, 251, 253, 255, 267, 299, 340, 377, 449

Lagrange polynomials, 330
Lagrange shape functions, 330, 381
large displacement formulation, 6, 10
large displacements, 6, 8–10, 12, 13, 203, 343, 417, 418, 438, 440
  formulation, 6, 10
  and large strains, 440
  small strain, 8
large strains, 6, 8, 10, 12, 203, 213, 214, 255, 312, 313, 417, 418, 440, 462
large strains large displacement shells, 10
linear mapping, 131–4, 141, 143, 144, 147, 153, 158, 163, 174, 176, 177, 181, 184, 192, 196, 225, 258
linear mapping of surfaces to forces, 158, 174
load, 3, 5, 43, 50–53, 155, 156, 175, 213, 417, 418, 421
local coordinates, 229, 244, 320, 325, 331, 336, 382, 385, 419–21
local derivatives, 349, 431
logarithmic strains, 210–212, 214, 280, 294, 307, 308, 313
lower triangular matrix, 21

mapping, 130–134, 141, 143, 144, 147, 153, 158, 163, 164, 174, 176, 177, 181, 184, 185, 192, 196, 225, 258
material axes, 218, 219, 259, 321, 326, 335
material element, 208, 214, 218, 221–7, 231, 232, 234, 238, 246, 247, 249–51, 253–5, 257–61, 264, 265, 267–9, 271, 273–7, 280, 281, 288–90, 293, 295–300, 302, 303, 306–8, 312, 313, 321, 324–6, 334, 343, 363, 364, 375, 378, 386, 389, 392, 394, 409, 419, 439, 440, 449, 458, 460
material-embedded infinitesimal base, 321
material-embedded vector base, 228, 231, 238, 247, 249, 323, 325, 375–7, 389, 392, 427, 429, 438, 439, 447, 448, 459
material model, 13, 14, 264, 267, 280, 301, 302, 307
material package, 13, 264–7, 268, 274, 278, 292, 293, 295, 296, 299, 302, 312, 313, 326, 355, 361, 363, 364, 377, 378, 392, 401, 430, 435, 449
material point, 7, 75, 76, 78, 168, 203–5, 217, 219, 221, 224–9, 231, 232, 235, 243, 244, 246, 247, 249–51, 255, 260, 264, 317–24, 326, 328, 329, 331–6, 340, 341, 344–6, 352, 364, 368, 369, 372, 375, 376, 382, 383, 385, 386, 389, 392, 393, 419–21, 423, 424, 427, 439, 440, 443, 444, 447, 449, 457
material properties, 218, 219, 377
matrices, 15–19, 21, 23, 25–7, 29–33, 50, 51, 53, 60, 62, 83, 94, 110, 114, 115, 121, 132–4, 141, 142, 153, 154, 167, 168, 170, 179, 186, 230, 258, 264, 267, 292, 326, 398

matrix algebra, 13, 16, 32, 108, 115, 127
matrix inversions, 367
matrix multiplication, 19, 108
mechanics of discontinua, 209
mesh, 317, 319, 327, 330, 357, 358, 363, 382
method of residuals, 75
microstructure, 209, 210, 261, 262
multiplication of matrices by a scalar, 18
multiplicative decomposition, 7, 8, 13, 203, 204, 214, 215, 239, 256, 259, 266, 267, 288–90, 292, 294, 300, 301, 313, 314, 367, 368, 418, 438, 462
Munjiza generalized material element, 222, 255, 312

n-dimensional space, 44, 117, 119, 121, 123, 125, 127, 147, 152, 200
nd space, 45, 49, 50, 127, 152, 154
Newton-Cotes integration, 65–7
nodal forces, 13, 343–5, 347–53, 355, 357, 359, 361, 363, 377, 378, 393, 395, 397–9, 401, 403–13, 430, 434, 435, 437, 438, 449, 452, 455, 462
non-cartesian coordinate system, 318, 320
non-homogeneous, 247
nonlinear algebraic equations, 13, 58
nonlinear elasticity, 213
nonlinearity, 6, 8, 10, 203, 235, 325, 412
nonlinear material laws, 364, 403
nonlinear materials, 10, 302
non-orthonormal bases, 101, 111, 168
numerical integration, 65, 67–9, 71–3, 352, 354, 355, 401
numerical integration of virtual work, 401

objective, 280, 293, 302, 313, 363
one Gauss point integration, 70, 405
one Gauss point numerical integration, 68
orthonormal base, 91–3, 97, 98, 101, 104, 106, 109, 111, 117, 120–122, 134, 139, 141, 158, 168, 170, 176, 186, 187, 195, 196, 219, 226, 299, 302, 343, 386
orthonormal column matrices, 32
orthonormal vector base, 127, 144, 155

parallelepiped, 246, 386
parallelization, 367
parallelogram-shaped infinitesimal material element, 222
physical meaning, 94, 102, 159, 160, 163, 173, 185, 222, 374
physical reality, 89, 131–3, 168, 185
plane strain, 284, 285, 287, 288, 293, 364
plane strain isotropic material, 284
plane stress, 281, 283, 284, 288, 292, 430, 435
plane stress formulation, 283
plastically, 213, 440
Poisson's ratio, 288, 309
polygon, 89–91, 117, 118
polygon of vectors, 89, 90, 117, 118
positive definite matrix, 26, 50, 53, 63
potential energy method, 75
preconditioner, 61
preconditioning, 60, 61
previous position, 238, 245, 248, 260, 266, 288, 292, 325, 330, 333, 340, 341, 371, 374, 377, 382, 384, 388, 389, 391, 421, 424, 425, 427, 447, 448, 460, 461
principle of virtual work, 75, 78, 232, 250, 343–5, 352, 353, 355, 393, 412, 430, 438, 452
pure shear, 284, 287, 310

quadratic form, 24, 26, 62

rate of deformation, 330
rate of stretching, 292
reduced integration, 358, 359, 363, 399, 400
reference configuration, 204
representative sample, 262–4
representative volume, 261, 263, 272, 273, 321
residual, 53–7, 59–61, 75
resultant internal forces, 177, 188, 226, 258, 274, 299, 301, 302, 305, 406, 407, 409
rotated, 222, 257
rotation, 7, 8, 77, 222–4, 257, 280, 293, 302, 312, 440, 456

sampling points, 37, 364, 400, 401, 407, 412, 449–51
scalar field, 45, 137–40, 149–52, 154, 372
scalar function, 81, 82, 85
scalar graphs, 82, 85
scalar variable, 81–3, 85, 86, 204
scouting, 45
second order formulation, 6

second order tensors, 8, 13, 155, 157–61, 163, 165, 167, 169, 171, 173–5, 177, 179, 181, 183, 185–7, 189, 191, 193, 195, 197, 199, 200
second Piola-Kirchhoff stress tensor matrix, 169, 172, 188, 258
selective integration, 13, 357–61, 363, 364, 400, 403, 409
serendipity, 330, 381
serendipity shape functions, 330
shape function, 319, 320, 327, 328, 330, 331, 336, 337, 339–41, 344, 349, 368–70, 379, 381, 385, 387, 389, 393, 401, 412, 419, 420, 422, 431, 433, 438, 443, 446, 449, 452, 457
shear forces, 168
shear locking, 362–4, 399, 412, 449, 450, 460, 462
shear strain, 271, 363
shear stress, 155, 363, 449, 450
shear stretches, 460
shearing, 258, 271, 290, 297, 357
simply supported beam, 76–8
singular square matrix, 28
six-noded shell element, 440, 441, 449, 452, 462
six-noded triangle finite element, 328
slope, 62, 129–34, 137, 139, 140, 149, 151, 154
slope of the tangential line, 140, 151
slope of the tangential plane, 139, 140, 149
slope tensor, 129, 132–4, 154
small strain formulations, 271, 417
small strains, 6, 8–10, 203, 213, 271, 281–4, 286, 287, 309, 310, 313
solid-embedded coordinate lines, 374, 380, 381, 383
solid-embedded coordinate surfaces, 380, 381, 441
solid-embedded coordinate system, 227, 241, 242, 319, 328, 329, 353, 368, 369, 372, 419, 441, 444
solid-embedded infinitesimal vector base, 322, 332, 383
solid-embedded vector base, 228, 231, 243, 245, 247, 248, 322, 323, 325, 326, 332, 333, 339, 340, 372–7, 383, 384, 388–91, 425–7, 438, 445, 447, 458
spectral radius, 25, 40, 41, 367
speed, 50, 67, 81, 82, 85
square matrices, 16, 30
square matrix, 16, 21–3, 24, 26, 28, 30–32, 98, 99
square-shaped material element, 221, 222, 225
square-shaped solid body, 235, 236
stability, 5, 6, 38, 41, 256, 294, 313, 364
static equilibrium, 43
static problems, 43, 204, 256, 293, 421
static solutions, 37
stiffness matrix, 35, 36, 50, 204, 367
strain increments, 301
strain measures, 8, 209–12, 301, 307, 308
strain rate, 214, 292
stress distribution, 408, 412, 451
stress increments, 291
stress integration, 13, 408
stress sampling, 401, 407, 409, 451
stress-strain relationship, 213
stress tensor, 13, 155, 158–61, 163, 164, 167–9, 172–8, 180, 181, 184–9, 226, 227, 231, 238, 249, 250, 252, 255, 258, 265, 267, 272, 278, 279, 292, 293, 302, 305, 306, 313, 343, 349, 355, 363, 377, 378, 392, 394, 396, 399, 407, 409, 429, 430, 435, 438, 449, 455, 462
stress transformation, 302
stress updates, 275
stretch 7, 8, 13, 205, 206, 208–12, 214, 223, 224, 257, 259, 268–72, 288–90, 292, 293, 296–302, 359–61, 363, 364, 368, 400, 401, 407, 412, 417, 427, 439, 449–51, 460
stretching, 204, 206, 207, 209, 214, 223, 225, 255, 257, 258, 263, 266, 268–72, 288–92, 297–300, 312, 364, 451
stretching modes, 291, 298–300
structural elements, 42, 249, 261
subtraction of matrices, 17
surface base vectors, 192
surface's base, 179–81, 186, 187, 192, 195, 198
symmetric matrix, 21, 26, 62
system of equations, 44–6, 62, 63, 367

tangential stiffness matrix, 204, 367
tangential vectors, 323
ten-noded tetrahedron, 380, 392
tensorial calculus, 13
tensor matrix, 159, 160, 163, 164, 167–9, 172, 173, 178, 181, 185–8, 194, 198,

226, 227, 231, 238, 250, 252, 255, 258, 265, 267, 272, 278, 292, 302, 305, 306, 313, 343, 349, 355, 363, 377, 378, 392, 394, 396, 399, 406, 407, 409, 429, 430, 435, 438, 449, 455, 462
tensor transformation, 133, 185
tetrahedron finite element, 367–71, 373, 375, 377, 379, 380, 392, 396, 409
theoretically exact, 6, 7, 13, 313
three points Newton-Cotes integration, 67
time integration, 38, 40–42, 62, 63
total virtual work, 235, 237, 238, 254, 345, 346, 394
transformation, 95, 133, 185, 302
translating, 225
translation, 7, 8, 203, 222, 257, 456
transposition of a matrix, 18
triangular finite element, 318, 320, 322, 327, 357, 435
Truesdell rate, 259, 267
two Gauss points numerical integration, 68
two points Newton-Cotes integration, 66

under-integrated, 358, 400
unified constitutive approach, 257, 259, 261, 263, 265, 267, 269, 271, 273, 275, 277, 279, 281, 283, 285, 287, 289, 291–3, 295, 297, 299, 301, 303, 305, 307, 309, 311, 313
uniformly distributed load, 155
unit length, 32, 207, 208, 332
unit length column matrix, 32
unit prefixes, 83, 84
unit set, 84, 85
unit-length, 206, 207
upper triangular matrix, 21

vector base, 93–7, 108, 111, 113, 114, 127, 131–4, 142, 144, 147, 153–5, 161, 164, 165, 167, 168, 172, 174, 178, 186, 218, 219, 221, 228, 230, 231, 238, 243, 245–9, 260, 321–3, 325, 326, 332–4, 339–41, 372–7, 383, 384, 386, 388–92, 405, 410, 419, 423, 425–7, 429, 438–40, 444–9, 458–60
vector of the dual base, 112, 123
vector transformation, 95
vector variable equations, 91
vertical displacement, 357
virtual displacement, 75–7, 78, 232–4, 236–8, 250–254, 344–6, 393, 394, 397, 409, 430, 452
virtual work, 75–8, 231–4, 236–8, 249–54, 256, 343–6, 348, 353, 355, 358, 363, 393, 394, 400, 401, 405, 407, 412, 430, 438, 452, 455, 462
virtual work of the internal forces, 232, 234, 236, 237, 394
viscosity, 14, 267, 292, 293
volume change, 269, 274, 286, 287, 357, 359, 362, 363
volumetric integration, 438
volumetric locking, 357, 358, 363, 400
volumetric strain, 281, 283, 286, 287, 309, 311
volumetric stretch, 7, 268, 269, 271, 288, 290, 296, 297, 301, 307, 360, 361, 363, 364, 400, 401, 407

Young's modulus, 288, 309

zero energy modes, 358, 359, 363, 400
zero matrix, 21

Printed in the United States
By Bookmasters